Haim Abramovich

Advanced Aerospace Materials

Also of interest

Hybrid Composite Materials and Manufacturing.
Fibers, Nano-Fillers and Integrated Additive Processes
Edited by: Vipin Kumar, 2025
ISBN 978-3-11-101934-5, e-ISBN (PDF) 978-3-11-101954-3

Polymer Matrix Composite Materials.
Structural and Functional Applications
Debdatta Ratna and Bikash Chandra Chakraborty, 2024
ISBN 978-3-11-078148-9, e-ISBN 978-3-11-078157-1

Self-Reinforced Polymer Composites.
The Science, Engineering and Technology
Padmanabhan Krishnan and Sharan Chandran M, 2022
ISBN 978-3-11-064729-7, e-ISBN 978-3-11-064733-4

Intelligent Materials and Structures.
2nd Edition
Haim Abramovich, 2021
ISBN 978-3-11-072669-5, e-ISBN 978-3-11-072670-1

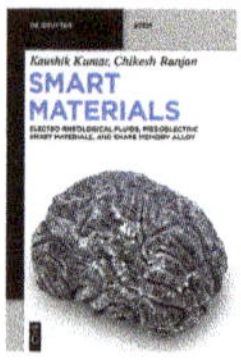

Smart Materials.
Electro-Rheological Fluids, Piezoelectric Smart Materials, and Shape Memory Alloys
Kaushik Kumar and Chikesh Ranjan, 2025
ISBN 978-3-11-137901-2, e-ISBN 978-3-11-137962-3

Haim Abramovich

Advanced Aerospace Materials

Aluminum-Based Structures, Composite Structures

3rd, Completely Revised and Extended Edition

DE GRUYTER

Author
Prof. Dr. Haim Abramovich
Aerospace Structural Laboratory
Technion-Israel Institute of Technology
Technion City
32000 Haifa
Israel
abramovich.haim@gmail.com

ISBN 978-3-11-162076-3
e-ISBN (PDF) 978-3-11-162110-4
e-ISBN (EPUB) 978-3-11-162126-5

Library of Congress Control Number: 2025943174

Bibliographic information published by the Deutsche Nationalbibliothek
The Deutsche Nationalbibliothek lists this publication in the Deutsche Nationalbibliografie; detailed bibliographic data are available on the Internet at http://dnb.dnb.de.

Cover image: claudio.arnese/iStock/Getty Images Plus
Typesetting: Integra Software Services Pvt. Ltd.

www.degruyterbrill.com
Questions about General Product Safety Regulation:
productsafety@degruyterbrill.com

Preface

The first edition of the book Advanced Aerospace Materials was published in 2019, and it was aimed at dissemination of advanced engineering ideas using isotropic and composite materials. The book was written to provide students and scholars a good understanding of new emerging technologies using composite materials, mainly in the aerospace sector. It can serve as an introductory book for graduate students wishing to study about composite made structures, enabling them to acquire the necessary physical and mathematical tools to understand the various aspects of composite based structures as compared to isotropic ones. The book was based on the various studies and investigations performed during the years by the author on metal and composite structures.

The second edition of the book was published in 2023 and was aimed at updating the content of the first edition based on additional research performed by the author on smart structures and innovative research published in the literature during the past years.

The second edition of the book contained ten chapters from the first edition namely: an extended introductory on thin-walled structures, characterizing aerospace and aeronautical structures. The chapter stressed the transition from aluminum-based structures to laminated composite ones, to save weight. An introductory to elasticity topics and equations of motion is also presented to enable the understanding of structural analysis of aerospace-aeronautical structures. Chapter 2 is devoted to laminated composite materials and presents the classical lamination theory and the first-order shear deformation theory. This chapter aims at providing mathematical and physical insight of composite materials. To complement this topic, higher-order theories are also displayed. Chapter 3 is devoted to some design formulas to be used by engineers on solving various problems of thin-walled structures. Fatigue issues are presented in Chapter 4 of the book. This chapter presents introductory topics in this well-documented subject of engineering. The crack propagation subject is described and reviewed in Chapter 5 of the book, thus complementing the topics presented in Chapter 4, enabling the reader to understand and use the various equations to predict the life of a single structural component. Chapter 6 displays various issues for the buckling for columns and plates, using isotropic and laminated composite materials. A new section 6.5 dealing with buckling of shell structures was added to the second edition. A complementary chapter, Chapter 7, presents the vibrational aspects for the structures presented in the previous chapter, with an additional section 7.5 been added to deal with vibration of shells. Another important topic in the analysis of thin-walled structure, the dynamic buckling of structural components due to pulse loading, is described and presented in detail in Chapter 8, highlighting both analytical and experimental aspects of the subject. Chapter 9 initiates the reader to the optimization of thin-walled structures issues, while Chapter 10 deals with structural health monitoring topic. Five new chapters were added to the second addition of the book: Chapter 11

https://doi.org/10.1515/9783111621104-202

describing one of the most applied nondestructive experimental methods to determine buckling of thin walled structures, the VCT (Vibration Correlation technique) method; Chapter 12 is devoted to another important issue, namely the morphing of flying vehicles, with a dedicated example being discussed in this chapter; Chapter 13 displays the topic of large deflections for columns, beams and plates; Chapter 14 presents the shearography and acoustic emission non-destructive testing (NDT) methods for metal and laminated composite structures; and the last chapter, Chapter 15, deals with a useful method which monitors the natural frequency of a structure to yield its material properties.

The present third edition of the book updated the chapters of the second edition, with a special focus on Chapter 6 in which a new section 6.6 dealing with miscellaneous stability problems was added. Moreover, four new chapters were added to the book to reflect other important engineering issues: Chapter 16 presenting a various important topic of buckling and vibration of arches, trusses and frames, Chapter 17 containing important issues like Rayleigh approach, Southwell's plot, and new advanced excitation devices; Chapter 18 treating the important issue of the nonlinear experimental and numerical behavior of multiwall rectangular isotropic plates under transverse pressure; and finally; Chapter 19 which adds a thorough investigation on the behavior of long, thin isotropic rectangular plates under normal pressure with interesting conclusions.

I wish that this third edition book containing 19 chapters summarizing the state-of-the-art in structural analysis, together with other books and the huge numbers of references existing in the literature, would enable readers to become familiar with the treatment of thin-walled structures, and how to investigate and calculate their structural behavior.

I would like to thank all my former graduate students for their dedicated research in the field of composite-based structures, morphing, large-deflected structures and NDT methods, Mr Joshua Harris, Mr Osher Shapira, Mr Amit Geva, Dr Gilad Hakim, Mr Yair Elbaz and Mr Yaniv Seri for their contributions to the present book.

I want to thank, my wife, Dorit, and my children, Chen, Oz, Shir and Or, for their support, understanding, love and devotion. Their continuous support throughout the writing period enabled the publishing of the present book.

Nesher, December 2025
Haim Abramovich

Contents

1 Introduction

1.1 Introduction

This chapter aims at presenting the reader with aerospace structures, its basic structural parts and the transfer from aluminum to laminated composite materials, as a main way of saving weight. To be able to analyze these aerospace structures, which are thin-walled structures, basic elasticity equations will be derived and presented. The following chapters will present additional topics and equations regarding the behavior of thin-walled structures and how to analyze it.

1.2 Aerospace structures

Prior to defining the structure itself, one must define the word *aerospace*. As it is defined, aerospace combines the science, engineering and business of flying vehicles in the Earth's atmosphere, leading to the word *aeronautics*, while flying in the surrounding space is called *astronautics*. The vehicles used for aerospace traveling are called *aerospace structures*.

A machine or vehicle such as an airplane, helicopter, glider or any autonomous device capable of flying in the Earth's atmosphere is denoted as *aircraft*. Its main parts are schematically presented in Fig. 1.1.

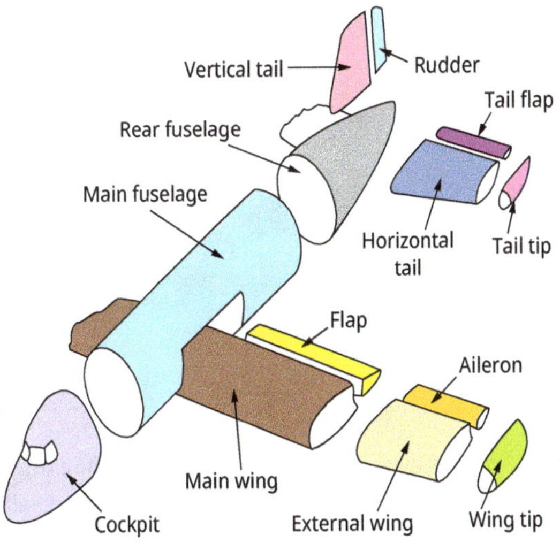

Fig. 1.1: Main parts of a typical aircraft (adapted from www.clker.com/clipart-parts-of-an-airplane.html).

https://doi.org/10.1515/9783111621104-001

The vehicle capable of flying in space, like a satellite, space station or any other machine, would be named as *spacecraft*.

Figure 1.2 presents one of the famous old aircraft, the Lockheed Vega, a six-passenger high-wing monoplane by Lockheed Corporation, which was used to break flight records. The famous Amelia Earhart used this aircraft to fly the Atlantic. From a structural point of view, one should note the wooden monologue fuselage and the plywood-covered wings [1]. To increase the performance of the aircraft, the wood was replaced by aluminum, which is the basic material to manufacture the structural parts of the aircraft. A typical metal skin aircraft fuselage assembly is shown in Fig. 1.3, depicting the frames and their clips and the longitudinal riveted stringers. The wings would be made of an assembly of spars giving the aerodynamic shape to the wing, ribs connecting the spars and all covered by skin panels. A space aircraft has similar structural design, like an aircraft (see Fig. 1.4), if it has to return to the Earth. If the space device is designed to spend its life in space, like satellites, solar panels and/or space antennas (see Figs. 1.5 and 1.6), then its shape and the adjacent structure would be according to the loads expected to be applied on it during its space mission.

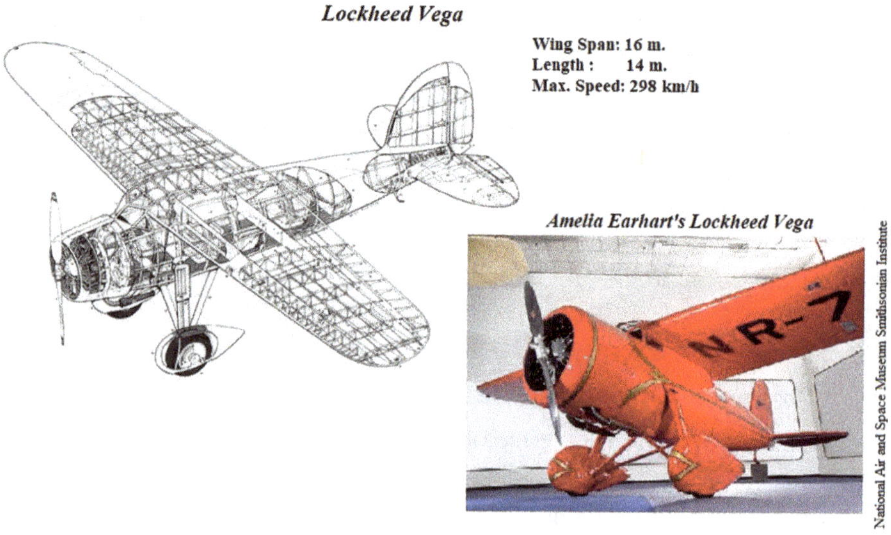

Fig. 1.2: Lockheed Vega aircraft.

The aerospace structures aim at providing operational demands and safety within a minimum weight. These structures comprise thin load skins, frames, stiffeners and spars, all made of high strength and stiffness materials to comply with the minimum weight criterion. To be able to design and calculate those structures, three basic structures, namely beams (or column for buckling and rods for tension), plates and shells, have to be analyzed and their behavior under various types of loads be understood (Fig. 1.7). As shown,

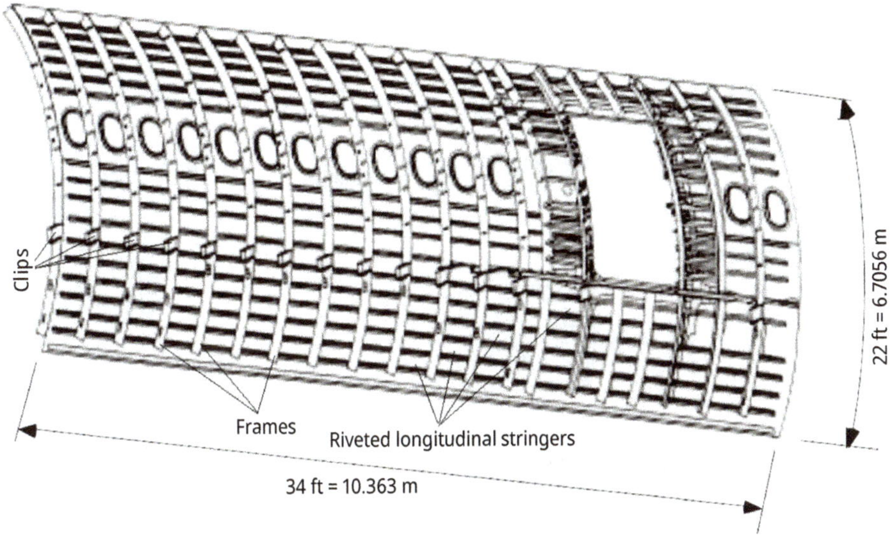

Fig. 1.3: Typical metal skin aircraft fuselage assembly (from NASA CR4730, Ref. [2]).

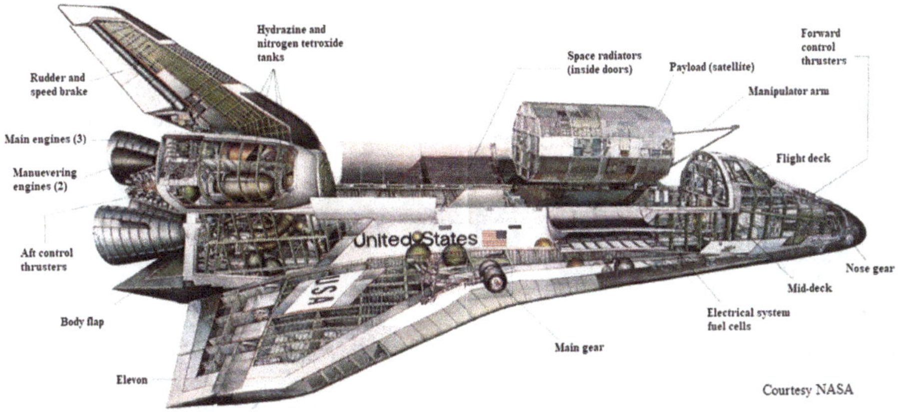

Fig. 1.4: A spacecraft – the NASA Shuttle.

a beam is a 1D structure (h, $b \ll L$), a thin plate is considered a 2D structure (a, $b \gg t$), while a cylindrical shell forms a 3D thin-walled structure.

Many books and articles have been written on aerospace structures regarding how to analyze and calculate their response to static and dynamic loads. Typical references and their contents can be found in [2–6].

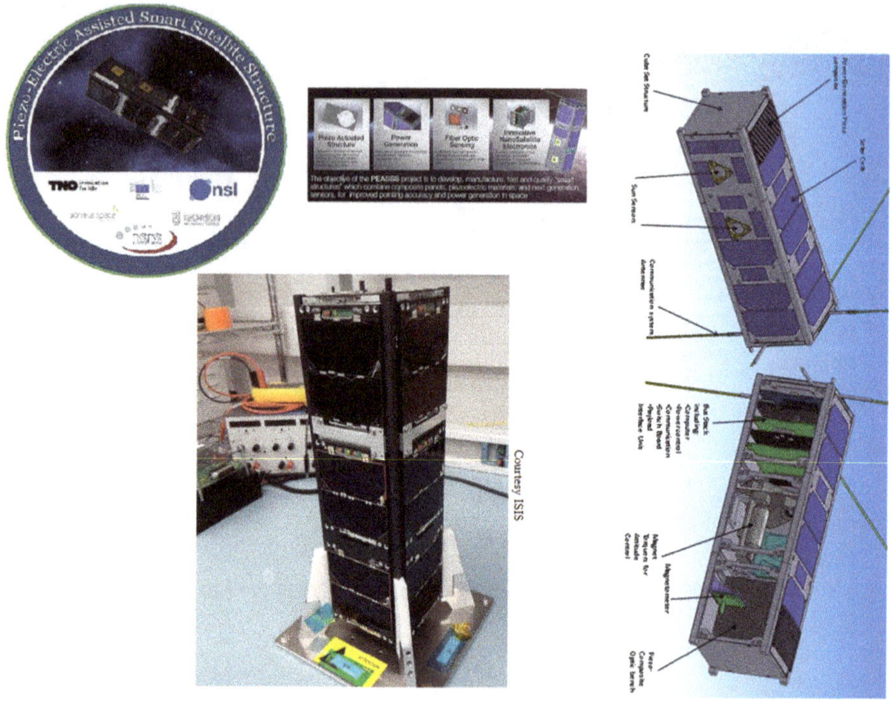

Fig. 1.5: The PEASSS (piezoelectric-*a*ssisted *s*mart *s*atellite *s*tructure) nanosatellite launched on February 15, 2017, by an Indian PSLV rocket together with other 103 nanosatellites (adapted from [7]).

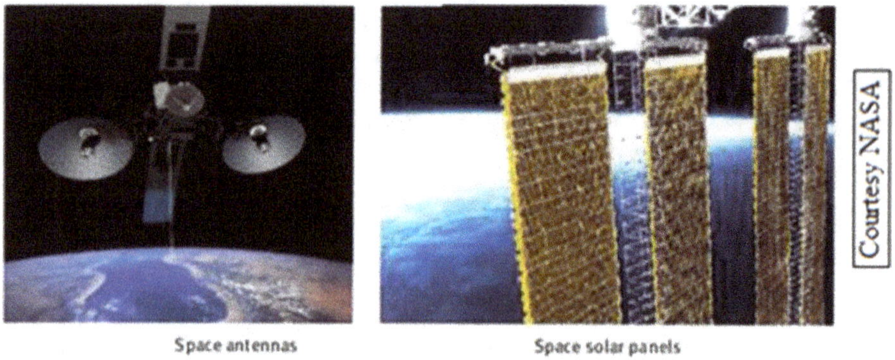

Fig. 1.6: Space antennas and solar panels.

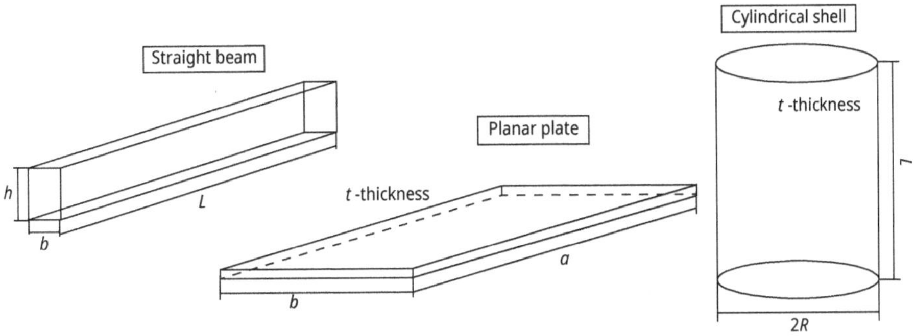

Fig. 1.7: Schematic drawings of the three basic structures: a straight beam, a planar plate and a cylindrical shell.

1.3 Aerospace structures – transition to composite materials

The constant strive to improve efficiency and increase performance of the aircraft in parallel with the need to reduce their development and operating costs is the motto of the aircraft industry [8]. Using composite materials for primary aircraft structures and thus reducing the aircraft weight may be the answer to improve aircraft efficiency and performance.

Intensive investigation has been carried out on the introduction of composite materials in commercial and military aircrafts [9–12]. Figures 1.8 and 1.9 present the realization of a fuselage panel, originally manufactured from aluminum (see Fig. 1.3), using either laminated graphite epoxy stringers (Fig. 1.8) or laminated graphite epoxy sandwich faces (Fig. 1.9).

One should note that the composite structure depicted in Fig. 1.9 does not have any stringers. The fibers have been oriented to support the loads normally carried by stringers, as in Figs. 1.3 and 1.8.

The reduction in the panel weight in comparison with the aluminum baseline model is presented in Fig. 1.10 from Polland et al. [9]. One can observe that using skin-stringer or sandwich structures would reduce the total weight by 24% or 13%, respectively.

It is interesting to note that in the 1950s composite materials were first used on commercial aircraft, and 2% of the Boeing 707 was made of fiberglass (see Fig. 1.11). Airbus introduced composite materials in its aircraft in the 1980s, using 5% composites on the A310-300.

This trend continued, and by the turn of the century, the advancements made in composite manufacturing allowed both Boeing and Airbus to significantly increase the use of composites. Boeing jumped from 12% on the Boeing 777 to 50% on the Boeing 787, better known as the "Dreamliner" (see Tab. 1.1), while Airbus moved from 10% on the A340 to 25% on the A380 (see Tab. 1.2) and finally to 53% on the A350XWB (Fig. 1.12).

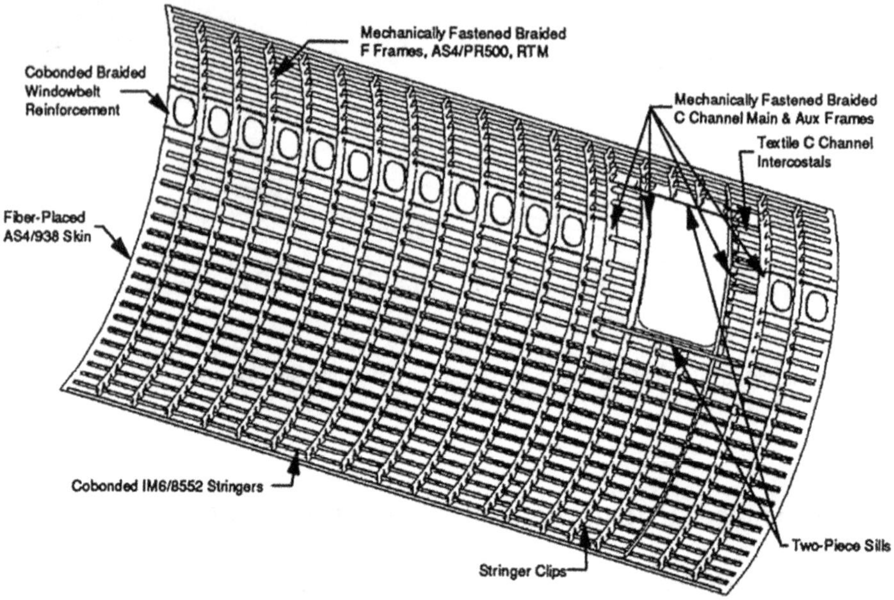

Fig. 1.8: Typical skin-stringer-frame design composite concept (from NASA CR4735, Ref. [12]).

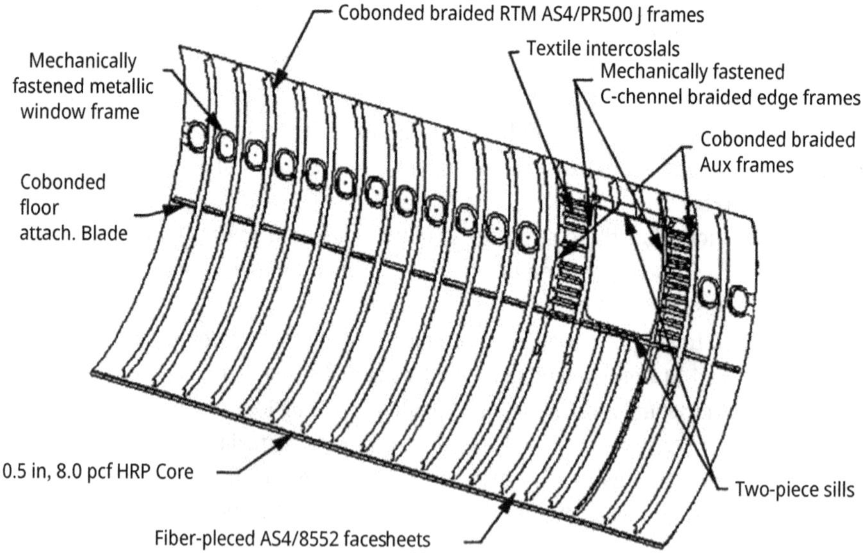

Fig. 1.9: Typical sandwich-frame side design composite concept (from NASA CR4735 [12]).

With its 787 "Dreamliner," Boeing became the first airliner to launch a full-size commercial aircraft using composite materials for the almost full fuselage, upper and lower wing skin, radome, wing flaps, elevators, ailerons, vertical fin and horizontal stabilizer. One should note that the use of composites in the Boeing 787 is 80% by volume and 50% by weight.

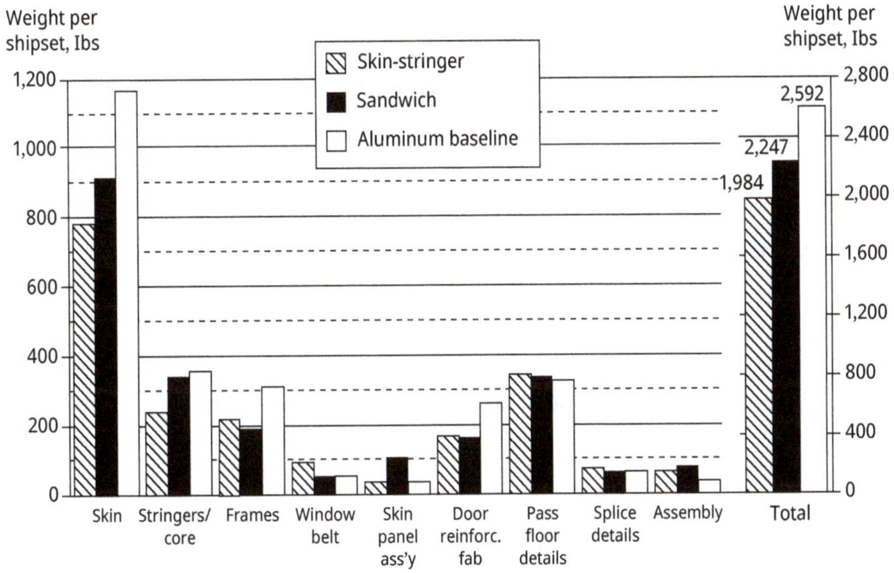

Fig. 1.10: Weight comparison of side panel designs (from NASA CR4730 [9]).

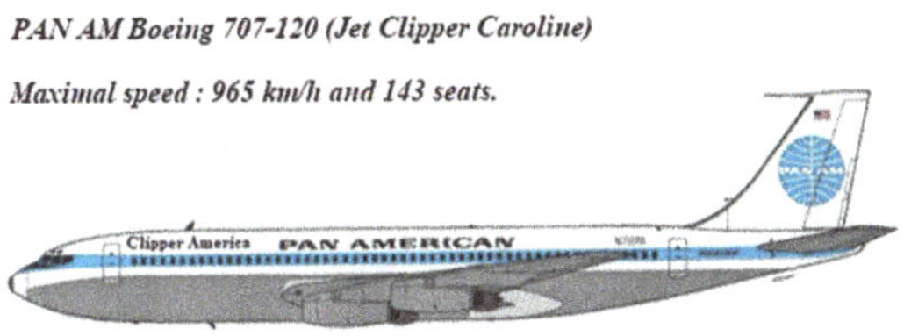

Fig. 1.11: Boeing 707–120 from PAN AMERICAN airlines (adapted from Pan Am: An Airline and Its Aircraft, by R.E.G. Davies, 1987).

Tab. 1.1: Boeing 787 Dreamliner composite components.
Boeing 787 Dreamliner structure – typical data
787 body only: materials used are fiberglass, aluminum, carbon-based composite, sandwich (carbon type) and metals (steel, aluminum and titanium)
787 whole plane: materials used (by weight) are composites (50%), aluminum (20%), titanium (15%), steel (10%) and others (5%)
Distribution of composites per various parts of the 787 plane

Part names	***Materials***
Nose landing gear doors	*Graphite*
Body main landing gear doors	*Graphite*
Environment control system ducts	*Kevlar*
Wing to body fairings	*Graphite/Kevlar/fiberglass + graphite/Kevlar + non-woven Kevlar mat*
Trunnion fairings and wing landing gear	*Graphite/Kevlar*
Brakes	*Structural carbon*
Cowl components	*Graphite*
Spoilers	*Graphite*
Wing leading edge lower panels	*Kevlar/fiberglass*
Fixed trailing edge panels	*Graphite/Kevlar + non-woven Kevlar mat*
Auxiliary power inlet	*Graphite*
Elevators	*Graphite*
Fixed trailing edge upper panels	*Graphite/fiberglass*
Fixed trailing edge lower panels	*Graphite/Kevlar + non-woven Kevlar mat*
Rudder	*Graphite*
Top fairings	*Fiberglass*
Aft flaps – outboard	*Graphite*
Aft flaps – inboard	*Graphite/fiberglass*
Flap support fairings – foreword segment	*Graphite/Kevlar + non-woven Kevlar mat*
Flap support fairings – aft segment	*Graphite/fiberglass*
Ailerons	*Graphite*
Engine strut fairings	*Kevlar/fiberglass*

Tab. 1.2: Airbus A380 composite components.
Airbus A380 structure – typical data
Distribution of composites per various parts of the A380 plane

Part names	*Materials*
Radome	*Fiberglass*
Vertical stabilizer	*Aramid fiber*
Belly fairing skins	*Carbon fiber + fiberglass*
Pylon fairings	*Carbon fiber + fiberglass*
Central torsion box	*Carbon fiber*
Nose landing gear doors	*Carbon fiber*
Main and center landing gear doors	*Carbon fiber*
Main landing gear leg-fairing door	*Carbon fiber*
Nacelle cowlings	*Carbon fiber*
Spoilers	*Carbon fiber*
Training edge upper and lower panels	*Carbon fiber*
Shroud box	*Carbon fiber*
Overwing panel	*Carbon fiber*
Pressure bulkhead	*Carbon fiber*
Keel beam	*Carbon fiber*
Tail cone	*Carbon fiber*
Horizontal stabilizer outer boxes	*Carbon fiber*
Outer flap	*Carbon fiber*
Flap – track fairings	*Carbon fiber*
Ailerons	*Carbon fiber*
Outer wing	*Carbon fiber*
Fixed leading-edge upper panels	*Carbon fiber*
Fixed leading-edge lower panels	*Carbon fiber*

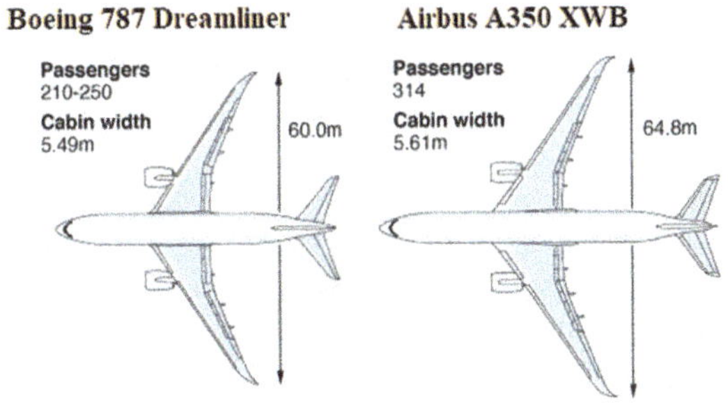

Fig. 1.12: Boeing 787 versus Airbus A350 XWB (adapted from www.boeing.com and airbus.com, respectively).

Chapter 2 of this book is devoted to composite materials and their uses. Additional information regarding composite materials can be found in [13, 14].

1.4 Basic topics in elasticity

1.4.1 Stresses, strains and rigid body rotations

To be able to calculate the structural behavior of basic and complicate components of aerospace structures, the basic concepts of elasticity will next be presented. More in-depth notions can be found in [15].

In a 3D structure, the components of the mechanical stress (defined as force per unit area) are presented in Fig. 1.13 for a cubic point in a structural continuum (for a Cartesian coordinate system).

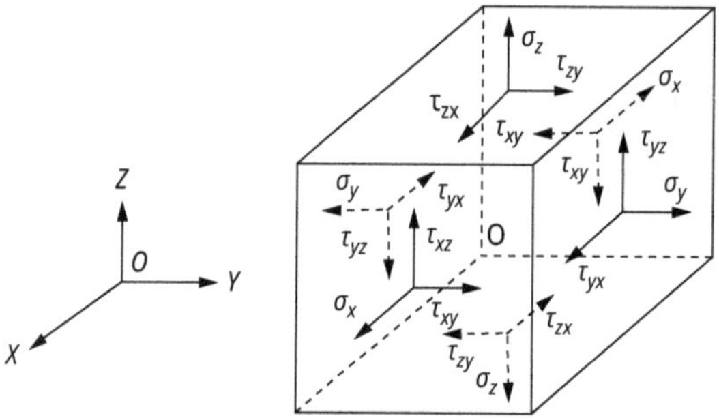

Fig. 1.13: Mechanical stresses in a Cartesian coordinate system.

The expressions with only one index, σ_x, σ_y, σ_z, are normal stresses acting in the directions x, y and z, respectively. Positive values will mean tension stress, while a negative value will stand for compression stress. The other expressions τ_{xz}, τ_{zx}, τ_{yz}, τ_{zy}, τ_{xy}, τ_{yx} are shear stresses, where the first index denotes the normal to the surface on which the shear stress acts, while the second index denotes the direction of the stress. One should note that due to equilibrium, the following identities hold for the shear stresses:

$$\tau_{xz} = \tau_{zx}, \quad \tau_{yz} = \tau_{zy}, \quad \tau_{xy} = \tau_{yx}. \tag{1.1}$$

The stress tensor will have the following form:

$$\begin{bmatrix} \sigma_x & \tau_{xy} & \tau_{xz} \\ \tau_{yx} & \sigma_y & \tau_{yz} \\ \tau_{zx} & \tau_{zy} & \sigma_z \end{bmatrix} = \begin{bmatrix} \sigma_x & \tau_{xy} & \tau_{xz} \\ \tau_{xy} & \sigma_y & \tau_{yz} \\ \tau_{xz} & \tau_{yz} & \sigma_z \end{bmatrix} \tag{1.2}$$

The stresses in cylindrical coordinate system (r, θ, z) (see Fig. 1.14) are presented in Fig. 1.15. The normal stresses are σ_r, σ_θ, σ_z. All the other stresses τ_{rz}, τ_{zr}, $\tau_{\theta z}$, $\tau_{z\theta}$, $\tau_{r\theta}$, $\tau_{\theta r}$ would stand for shear stresses:

$$\tau_{r\theta} = \tau_{\theta r}, \quad \tau_{rz} = \tau_{zr}, \quad \tau_{\theta z} = \tau_{z\theta}. \tag{1.3}$$

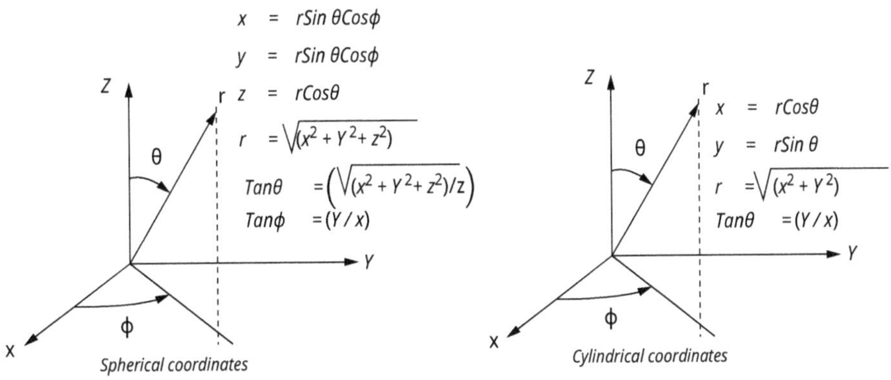

Fig. 1.14: Spherical and cylindrical coordinate systems.

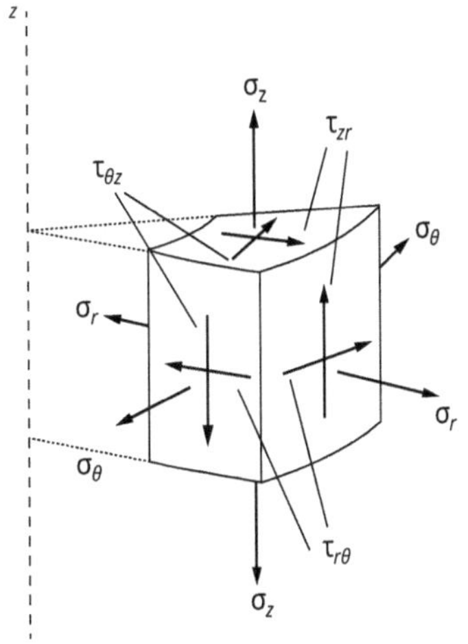

Fig. 1.15: Mechanical stresses in cylindrical coordinate system.

The stress tensor will have the following form:

$$\begin{bmatrix} \sigma_r & \tau_{r\theta} & \tau_{rz} \\ \tau_{\theta r} & \sigma_\theta & \tau_{\theta z} \\ \tau_{zr} & \tau_{z\theta} & \sigma_z \end{bmatrix} = \begin{bmatrix} \sigma_r & \tau_{r\theta} & \tau_{rz} \\ \tau_{r\theta} & \sigma_\theta & \tau_{\theta z} \\ \tau_{rz} & \tau_{\theta z} & \sigma_z \end{bmatrix} \tag{1.4}$$

One should note that a polar coordinate system (r, θ) is a 2D planar system, with only three stresses σ_r, σ_θ, $\tau_{r\theta}$.

Writing the stresses in spherical coordinates (see Fig. 1.14) yields

$$\begin{bmatrix} \sigma_r & \tau_{r\theta} & \tau_{r\phi} \\ \tau_{\theta r} & \sigma_\theta & \tau_{\theta\phi} \\ \tau_{\phi r} & \tau_{\phi\theta} & \sigma_\phi \end{bmatrix} = \begin{bmatrix} \sigma_r & \tau_{r\theta} & \tau_{r\phi} \\ \tau_{r\theta} & \sigma_\theta & \tau_{\theta\phi} \\ \tau_{r\phi} & \tau_{\theta\phi} & \sigma_\phi \end{bmatrix} \tag{1.5}$$

The associated strains for a Cartesian coordinate system are defined as

$$\begin{gathered} \varepsilon_x = \frac{\partial u}{\partial x}, \quad \varepsilon_y = \frac{\partial v}{\partial y}, \quad \varepsilon_z = \frac{\partial w}{\partial z} \\ \gamma_{xy} = \frac{\partial u}{\partial y} + \frac{\partial v}{\partial x}, \quad \gamma_{yz} = \frac{\partial v}{\partial z} + \frac{\partial w}{\partial y}, \quad \gamma_{xz} = \frac{\partial u}{\partial z} + \frac{\partial w}{\partial x} \end{gathered} \tag{1.6}$$

where u, v and w are the displacements in the x, y and z directions, respectively; ε_x, ε_y and ε_z are the tension/compression strains in the x, y and z directions, respectively, while the shear strains are represented by $\gamma_{xy} = \gamma_{yx}$, $\gamma_{yz} = \gamma_{zy}$ and $\gamma_{xz} = \gamma_{zx}$. In matrix notation, the strains are written as

$$\begin{bmatrix} \varepsilon_x & \varepsilon_{xy} & \varepsilon_{xz} \\ \varepsilon_{xy} & \varepsilon_y & \varepsilon_{yz} \\ \varepsilon_{xz} & \varepsilon_{yz} & \varepsilon_z \end{bmatrix} = \begin{bmatrix} \varepsilon_x & 0.5\gamma_{xy} & 0.5\gamma_{xz} \\ 0.5\gamma_{xy} & \varepsilon_y & 0.5\gamma_{yz} \\ 0.5\gamma_{xz} & 0.5\gamma_{yz} & \varepsilon_z \end{bmatrix} \tag{1.7}$$

For a cylindrical coordinate system, having the displacements u_r, u_θ and u_z in the r, θ, and z directions, respectively, the strains are defined as follows:

$$\begin{gathered} \varepsilon_r = \frac{\partial u_r}{\partial r}, \quad \varepsilon_\theta = \frac{u_r}{r} + \frac{1}{r}\frac{\partial u_\theta}{\partial \theta}, \quad \varepsilon_z = \frac{\partial u_z}{\partial z} \\ \gamma_{r\theta} = \gamma_{\theta r} = \frac{\partial u_\theta}{\partial r} + \frac{1}{r}\frac{\partial u_r}{\partial \theta} - \frac{u_\theta}{r}, \quad \gamma_{\theta z} = \gamma_{z\theta} = \frac{1}{r}\frac{\partial u_z}{\partial \theta} + \frac{\partial u_\theta}{\partial z}, \quad \gamma_{zr} = \gamma_{rz} = \frac{\partial u_r}{\partial z} + \frac{\partial u_z}{\partial r} \end{gathered} \tag{1.8}$$

The matrix notation would be

$$\begin{bmatrix} \varepsilon_r & \varepsilon_{r\theta} & \varepsilon_{rz} \\ \varepsilon_{r\theta} & \varepsilon_\theta & \varepsilon_{\theta z} \\ \varepsilon_{rz} & \varepsilon_{\theta z} & \varepsilon_z \end{bmatrix} = \begin{bmatrix} \varepsilon_r & 0.5\gamma_{r\theta} & 0.5\gamma_{rz} \\ 0.5\gamma_{r\theta} & \varepsilon_\theta & 0.5\gamma_{\theta z} \\ 0.5\gamma_{rz} & 0.5\gamma_{\theta z} & \varepsilon_z \end{bmatrix} \tag{1.9}$$

where ε_r, ε_θ and ε_z are the tension/compression strains in the r, θ and z directions, respectively, while the shear strains are represented by $\gamma_{r\theta} = \gamma_{\theta r}$, $\gamma_{\theta z} = \gamma_{z\theta}$ and $\gamma_{zr} = \gamma_{rz}$.

For a spherical coordinate system, having the displacements u_r, u_θ and u_ϕ in the r, θ and ϕ directions, respectively, the strains are defined as follows:

$$\varepsilon_r = \frac{\partial u_r}{\partial r}; \; \varepsilon_\theta = \frac{u_r}{r} + \frac{1}{r}\frac{\partial u_\theta}{\partial \theta}; \; \varepsilon_\phi = \frac{1}{r\,\sin\theta}\frac{\partial u_\phi}{\partial \phi} + \frac{u_r}{r} + \frac{u_\theta \cos\theta}{r}$$

$$\gamma_{r\theta} = \gamma_{\theta r} = r\frac{\partial}{\partial r}\left(\frac{u_\theta}{r}\right) + \frac{1}{r}\frac{\partial u_r}{\partial \theta} = \frac{\partial u_\theta}{\partial r} - \frac{u_\theta}{r} + \frac{1}{r}\frac{\partial u_r}{\partial \theta}$$

$$\gamma_{\theta\phi} = \gamma_{\phi\theta} = \frac{\sin\theta}{r}\frac{\partial}{\partial \theta}\left(\frac{u_\phi}{\sin\theta}\right) + \frac{1}{r\sin\theta}\frac{\partial u_\theta}{\partial \phi} = \frac{1}{r}\frac{\partial u_\phi}{\partial \theta} - \frac{u_\phi}{r}\cot\theta + \frac{1}{r\sin\theta}\frac{\partial u_\theta}{\partial \phi} \tag{1.10}$$

$$\gamma_{r\phi} = \gamma_{\phi r} = \frac{1}{r\sin\theta}\frac{\partial u_r}{\partial \phi} + r\frac{\partial}{\partial r}\left(\frac{u_\phi}{r}\right) = \frac{1}{r\sin\theta}\frac{\partial u_r}{\partial \phi} + \frac{\partial u_\phi}{\partial r} - \frac{u_\phi}{r}$$

where ε_r, ε_θ and ε_ϕ are the tension/compression strains in the r, θ and ϕ directions, respectively, while the shear strains are represented by $\gamma_{r\theta} = \gamma_{\theta r}$, $\gamma_{\theta\phi} = \gamma_{\phi\theta}$ and $\gamma_{\phi r} = \gamma_{r\phi}$. The matrix of the strains in spherical coordinates has the following form:

$$\begin{bmatrix} \varepsilon_r & \varepsilon_{r\theta} & \varepsilon_{r\phi} \\ \varepsilon_{r\theta} & \varepsilon_\theta & \varepsilon_{\theta\phi} \\ \varepsilon_{r\phi} & \varepsilon_{\theta\phi} & \varepsilon_\phi \end{bmatrix} = \begin{bmatrix} \varepsilon_r & 0.5\gamma_{r\theta} & 0.5\gamma_{r\phi} \\ 0.5\gamma_{r\theta} & \varepsilon_y & 0.5\gamma_{\theta\phi} \\ 0.5\gamma_{r\phi} & 0.5\gamma_{\theta\phi} & \varepsilon_z \end{bmatrix} \tag{1.11}$$

One should note that during deforming, an element on a given body would change its shape, translate and rotate. So, besides strains, one can write the rigid body rotations as a function of the assumed displacements. For a Cartesian coordinate system, these rotations are defined as (see also [15])

$$\omega_x = \frac{1}{2}\left(\frac{\partial w}{\partial y} - \frac{\partial v}{\partial z}\right), \quad \omega_y = \frac{1}{2}\left(\frac{\partial u}{\partial z} - \frac{\partial w}{\partial x}\right), \quad \omega_z = \frac{1}{2}\left(\frac{\partial v}{\partial x} - \frac{\partial u}{\partial y}\right) \tag{1.12}$$

and

$$\omega_{zy} = -\omega_{yz} \equiv \omega_x, \quad \omega_{xz} = -\omega_{zx} \equiv \omega_y, \quad \omega_{yx} = -\omega_{xy} \equiv \omega_z$$

The definitions for the rotations in a cylindrical coordinate system will have the following expressions:

$$\omega_r = \frac{1}{2}\left(\frac{1}{r}\frac{\partial u_z}{\partial \theta} - \frac{\partial u_\theta}{\partial z}\right), \quad \omega_\theta = \frac{1}{2}\left(\frac{\partial u_r}{\partial z} - \frac{\partial u_z}{\partial r}\right), \quad \omega_z = \frac{1}{2r}\left(\frac{\partial}{\partial r}(r \cdot u_\theta) - \frac{\partial u_r}{\partial \theta}\right) \tag{1.13}$$

and

$$\omega_{z\theta} = -\omega_{\theta z} \equiv \omega_r, \quad \omega_{rz} = -\omega_{zr} \equiv \omega_\theta, \quad \omega_{\theta r} = -\omega_{r\theta} \equiv \omega_z$$

while for a spherical coordinate system, these rotations will be given by

$$\omega_r = \frac{1}{2r^2 \sin\theta}\left[\frac{\partial}{\partial\theta}(r \cdot u_\phi \cdot \sin\theta) - \frac{\partial}{\partial\phi}(r \cdot u_\theta)\right]$$
$$\omega_\theta = \frac{1}{2r\sin\theta}\left[\frac{\partial u_r}{\partial\phi} - \frac{\partial}{\partial r}(r \cdot u_\phi \cdot \sin\theta)\right], \omega_\phi = \frac{1}{2r}\left[\frac{\partial}{\partial r}(r \cdot u_\theta) - \frac{\partial u_r}{\partial\theta}\right] \tag{1.14}$$

and

$$\omega_{\phi\theta} = -\,\omega_{\theta\phi} \equiv \omega_r, \quad \omega_{r\phi} = -\omega_{\phi r} \equiv \omega_\theta, \quad \omega_{\theta r} = -\omega_{r\theta} \equiv \omega_\phi$$

Assuming Hook's law for isotropic materials, one can write the strain–stress relations in a Cartesian coordinate system, yielding the following expressions:

$$\varepsilon_x = \frac{1}{E}\left[\sigma_x - \nu\left(\sigma_y + \sigma_z\right)\right], \quad \varepsilon_y = \frac{1}{E}\left[\sigma_y - \nu\left(\sigma_x + \sigma_z\right)\right]$$
$$\varepsilon_z = \frac{1}{E}\left[\sigma_z - \nu\left(\sigma_x + \sigma_y\right)\right] \tag{1.15}$$
$$\gamma_{xy} = \frac{\tau_{xy}}{G}, \quad \gamma_{yz} = \frac{\tau_{yz}}{G}, \quad \gamma_{xz} = \frac{\tau_{xz}}{G}$$

where E, $G = \frac{E}{2(1+\nu)}$ and ν are the Young's modulus, the shear modulus and the Poisson's ratio, respectively. For a cylindrical coordinate system, the strain–stress relations can be written as

$$\varepsilon_r = \frac{1}{E}[\sigma_r - \nu(\sigma_\theta + \sigma_z)], \quad \varepsilon_\theta = \frac{1}{E}[\sigma_\theta - \nu(\sigma_r + \sigma_z)],$$
$$\varepsilon_z = \frac{1}{E}[\sigma_z - \nu(\sigma_r + \sigma_\theta)], \tag{1.16}$$
$$\gamma_{r\theta} = \frac{\tau_{r\theta}}{G}, \quad \gamma_{\theta z} = \frac{\tau_{\theta z}}{G}, \quad \gamma_{rz} = \frac{\tau_{rz}}{G}$$

In a similar way, the strain–stress relations for a spherical coordinate system can be shown to have the following expressions:

$$\varepsilon_r = \frac{1}{E}\left[\sigma_r - \nu\left(\sigma_\theta + \sigma_\phi\right)\right], \quad \varepsilon_\theta = \frac{1}{E}\left[\sigma_\theta - \nu\left(\sigma_r + \sigma_\phi\right)\right]$$
$$\varepsilon_\phi = \frac{1}{E}\left[\sigma_\phi - \nu\left(\sigma_r + \sigma_\theta\right)\right] \tag{1.17}$$
$$\gamma_{r\theta} = \frac{\tau_{r\theta}}{G}, \quad \gamma_{\theta\phi} = \frac{\tau_{\theta\phi}}{G}, \quad \gamma_{r\phi} = \frac{\tau_{r\phi}}{G}$$

The stress–strain relations, namely the stresses being expressed as a function of the strains, for Cartesian, cylindrical and spherical coordinate systems are presented in the following equations:

$$\sigma_x = \frac{E}{(\nu+1)(1-2\nu)}\left[(1-\nu)\varepsilon_x + \nu\left(\varepsilon_y + \varepsilon_z\right)\right]$$
$$\sigma_y = \frac{E}{(\nu+1)(1-2\nu)}\left[(1-\nu)\varepsilon_y + \nu(\varepsilon_x + \varepsilon_z)\right] \tag{1.18}$$
$$\sigma_z = \frac{E}{(\nu+1)(1-2\nu)}\left[(1-\nu)\varepsilon_z + \nu\left(\varepsilon_x + \varepsilon_y\right)\right]$$
$$\tau_{xy} = G\gamma_{xy}, \quad \tau_{yz} = G\gamma_{yz}, \quad \tau_{xz} = G\gamma_{xz}$$

$$\sigma_r = \frac{E}{(\nu+1)(1-2\nu)}\left[(1-\nu)\varepsilon_r + \nu(\varepsilon_\theta + \varepsilon_z)\right]$$
$$\sigma_\theta = \frac{E}{(\nu+1)(1-2\nu)}\left[(1-\nu)\varepsilon_\theta + \nu(\varepsilon_r + \varepsilon_z)\right] \tag{1.19}$$
$$\sigma_z = \frac{E}{(\nu+1)(1-2\nu)}\left[(1-\nu)\varepsilon_z + \nu(\varepsilon_r + \varepsilon_\theta)\right]$$
$$\tau_{r\theta} = G\gamma_{r\theta}, \quad \tau_{\theta z} = G\gamma_{\theta z}, \quad \tau_{rz} = G\gamma_{rz}$$

$$\sigma_r = \frac{E}{(\nu+1)(1-2\nu)}\left[(1-\nu)\varepsilon_r + \nu\left(\varepsilon_\theta + \varepsilon_\phi\right)\right]$$
$$\sigma_\theta = \frac{E}{(\nu+1)(1-2\nu)}\left[(1-\nu)\varepsilon_\theta + \nu\left(\varepsilon_r + \varepsilon_\phi\right)\right] \tag{1.20}$$
$$\sigma_\phi = \frac{E}{(\nu+1)(1-2\nu)}\left[(1-\nu)\varepsilon_\phi + \nu(\varepsilon_r + \varepsilon_\theta)\right]$$
$$\tau_{r\theta} = G\gamma_{r\theta}, \quad \tau_{\theta\phi} = G\gamma_{\theta\phi}, \quad \tau_{r\phi} = G\gamma_{r\phi}$$

1.4.2 Equilibrium and compatibility equations in elasticity

For a Cartesian coordinate system, the equilibrium equations have the following form:

$$\frac{\partial \sigma_x}{\partial x} + \frac{\partial \tau_{xy}}{\partial y} + \frac{\partial \tau_{xz}}{\partial z} + \bar{X} = 0$$
$$\frac{\partial \tau_{xy}}{\partial x} + \frac{\partial \sigma_y}{\partial y} + \frac{\partial \tau_{yz}}{\partial z} + \bar{Y} = 0 \tag{1.21}$$
$$\frac{\partial \tau_{xz}}{\partial x} + \frac{\partial \tau_{yz}}{\partial y} + \frac{\partial \sigma_z}{\partial z} + \bar{Z} = 0$$

while $\bar{X}$, $\bar{Y}$ and $\bar{Z}$ being body forces in the x, y and z directions, respectively.

Similarly, for a cylindrical coordinate system, the three equilibrium equations would read

$$\begin{aligned}
&\frac{\partial\sigma_r}{\partial r}+\frac{1}{r}\frac{\partial\tau_{r\theta}}{\partial\theta}+\frac{\partial\tau_{rz}}{\partial z}+\frac{\sigma_r-\sigma_\theta}{r}+\bar{R}=0\\
&\frac{\partial\tau_{r\theta}}{\partial r}+\frac{1}{r}\frac{\partial\sigma_\theta}{\partial\theta}+\frac{\partial\tau_{\theta z}}{\partial z}+\frac{2\tau_{r\theta}}{r}+\bar{\Theta}=0\\
&\frac{\partial\tau_{rz}}{\partial r}+\frac{1}{r}\frac{\partial\tau_{\theta z}}{\partial\theta}+\frac{\partial\sigma_z}{\partial z}+\frac{\tau_{rz}}{r}+\bar{Z}=0
\end{aligned} \tag{1.22}$$

with $\bar{R}$, $\bar{\Theta}$ and $\bar{Z}$ being body forces in the r, θ and z directions, respectively.

Finally, the equilibrium equations in a spherical coordinate system can be written as

$$\begin{aligned}
&\frac{\partial\sigma_r}{\partial r}+\frac{1}{r}\frac{\partial\tau_{r\theta}}{\partial\theta}+\frac{1}{r\cdot\sin\theta}\frac{\partial\tau_{r\phi}}{\partial\phi}+\frac{1}{r}\left(2\sigma_r-\sigma_\theta-\sigma_\phi+\tau_{r\theta}\cot\theta\right)+\bar{R}=0\\
&\frac{\partial\tau_{r\theta}}{\partial r}+\frac{1}{r}\frac{\partial\sigma_\theta}{\partial\theta}+\frac{1}{r\cdot\sin\theta}\frac{\partial\tau_{\theta\phi}}{\partial\phi}+\frac{1}{r}\left[\left(\sigma_\theta-\sigma_\phi\right)\cot\theta+3\tau_{r\theta}\right]+\bar{\Theta}=0\\
&\frac{\partial\tau_{r\phi}}{\partial r}+\frac{1}{r}\frac{\partial\tau_{\theta\phi}}{\partial\theta}+\frac{1}{r\cdot\sin\theta}\frac{\partial\sigma_\phi}{\partial\phi}+\frac{1}{r}\left(3\tau_{r\phi}+2\tau_{\theta\phi}\right)+\bar{\Phi}=0
\end{aligned} \tag{1.23}$$

with $\bar{R}$, $\bar{\Theta}$ and $\bar{\Phi}$ being body forces in the r, θ and ϕ directions, respectively.

With the strains and stresses being interconnected through the Hook's law, the question remains how to get the displacements if the expressions for the strains are known. To ensure a unique solution for the displacements, once the strains are given, the compatibility equations should be satisfied. These equations, for a Cartesian coordinate system, are given in a compact form as

$$\varepsilon_{ij,\,kl}+\varepsilon_{kl,\,ij}-\varepsilon_{ik,\,jl}-\varepsilon_{jl,\,ik}=0$$

where

$$i,j,k,l=x,y,z \tag{1.24}$$

The compact form of eq. (1.24) transforms into the following six equations, which have to be fulfilled to ensure a unique solution for the displacement field, for a prescribed strain field:

$$
\begin{aligned}
&\frac{\partial^2 \varepsilon_x}{\partial y^2} + \frac{\partial^2 \varepsilon_y}{\partial x^2} = 2\frac{\partial^2 \varepsilon_{xy}}{\partial x \partial y} \\
&\frac{\partial^2 \varepsilon_x}{\partial z^2} + \frac{\partial^2 \varepsilon_z}{\partial x^2} = 2\frac{\partial^2 \varepsilon_{xz}}{\partial x \partial z} \\
&\frac{\partial^2 \varepsilon_y}{\partial z^2} + \frac{\partial^2 \varepsilon_z}{\partial y^2} = 2\frac{\partial^2 \varepsilon_{yz}}{\partial y \partial z} \\
&\frac{\partial^2 \varepsilon_x}{\partial y \partial z} + \frac{\partial^2 \varepsilon_{yz}}{\partial x^2} = \frac{\partial^2 \varepsilon_{xz}}{\partial x \partial y} + \frac{\partial^2 \varepsilon_{xy}}{\partial x \partial z} \\
&\frac{\partial^2 \varepsilon_y}{\partial x \partial z} + \frac{\partial^2 \varepsilon_{xz}}{\partial y^2} = \frac{\partial^2 \varepsilon_{yz}}{\partial x \partial y} + \frac{\partial^2 \varepsilon_{xy}}{\partial y \partial z} \\
&\frac{\partial^2 \varepsilon_z}{\partial x \partial y} + \frac{\partial^2 \varepsilon_{xy}}{\partial z^2} = \frac{\partial^2 \varepsilon_{xz}}{\partial y \partial z} + \frac{\partial^2 \varepsilon_{yz}}{\partial x \partial z}
\end{aligned}
\tag{1.25}
$$

where

$$
\begin{aligned}
&\varepsilon_{xy} \equiv 0.5\gamma_{xy}, \quad \varepsilon_{xz} \equiv 0.5\gamma_{xz}, \quad \varepsilon_{yz} \equiv 0.5\gamma_{yz} \\
&\varepsilon_{yx} \equiv 0.5\gamma_{yx}, \quad \varepsilon_{zx} \equiv 0.5\gamma_{zx}, \quad \varepsilon_{zy} \equiv 0.5\gamma_{zy}
\end{aligned}
\tag{1.26}
$$

Equations (1.25) can be written in terms of stresses and are usually referred to as the compatibility equations in terms of stress, or the Beltrami–Michell (see [15, 16]) compatibility equations. Before presenting these compatibility equations, it is instructive to present the equilibrium equations presented in eq. (1.21), using the assumed Cartesian displacements. These are called the equilibrium equations in terms of displacement or the Navier's equations. Their form is given as

$$
\begin{aligned}
&(\lambda + G)\frac{\partial \varepsilon}{\partial x} + G\nabla^2 u + F_x = 0 \\
&(\lambda + G)\frac{\partial \varepsilon}{\partial y} + G\nabla^2 v + F_y = 0 \\
&(\lambda + G)\frac{\partial \varepsilon}{\partial z} + G\nabla^2 w + F_z = 0
\end{aligned}
\tag{1.27}
$$

where the Laplace operator, ∇^2, is defined (for a Cartesian coordinate system[1]) as

$$
\nabla^2 \equiv \frac{\partial^2}{\partial x^2} + \frac{\partial^2}{\partial y^2} + \frac{\partial^2}{\partial z^2}
\tag{1.28}
$$

and

1 For a cylindrical coordinate system $\nabla^2 \equiv \frac{\partial^2}{\partial r^2} + \frac{1}{r}\frac{\partial}{\partial r} + \frac{1}{r^2}\frac{\partial^2}{\partial \theta^2} + \frac{\partial^2}{\partial z^2}$, while for a spherical one, the expression is $\nabla^2 \equiv \frac{\partial^2}{\partial r^2} + \frac{2}{r}\frac{\partial}{\partial r} + \left(\frac{1}{r\sin(\phi)}\right)^2 \frac{\partial}{\partial \theta^2} + \frac{1}{r^2}\frac{\partial^2}{\partial \phi^2} + \frac{1}{r^2 \tan(\phi)}\frac{\partial}{\partial \phi}$.

$$\varepsilon \equiv \varepsilon_x + \varepsilon_y + \varepsilon_z = \frac{\partial u}{\partial x} + \frac{\partial v}{\partial y} + \frac{\partial w}{\partial z}$$

and the body forces F_x, F_y and F_z are defined according to the body forces defined in eq. (1.21), yielding

$$F_x \equiv \bar{X}, \quad F_y \equiv \bar{Y}, \quad F_z \equiv \bar{Z} \tag{1.29}$$

Finally, the constants λ and G, also called the Lame's constants, are defined as

$$G \equiv \frac{E}{2(1+\nu)}, \quad \lambda = \frac{\nu E}{(1+\nu)(1-2\nu)}. \tag{1.30}$$

where ν and E are the Poisson's ratio and the Young's modulus, respectively.

The Navier's equations in a cylindrical coordinate system can be presented as

$$\begin{aligned}
&(\lambda+2G)\frac{\partial H_\varepsilon}{\partial r} - 2G\left(\frac{1}{r}\frac{\partial \rho_z}{\partial \theta} - \frac{\partial \rho_\theta}{\partial z}\right) + F_r = 0\\
&(\lambda+2G)\frac{\partial H_\varepsilon}{r\partial \theta} - 2G\left(\frac{\partial \rho_r}{\partial z} - \frac{\partial \rho_z}{\partial r}\right) + F_\theta = 0\\
&(\lambda+2G)\frac{\partial H_\varepsilon}{\partial z} - 2G\left(\frac{\partial (r\rho_\theta)}{\partial r} - \frac{\partial \rho_r}{\partial \theta}\right) + F_z = 0
\end{aligned} \tag{1.31}$$

where, according to eq. (1.22), the body forces are defined as

$$F_r \equiv \bar{R}, \quad F_\theta \equiv \bar{\Theta}, \quad F_z \equiv \bar{Z} \tag{1.32}$$

and

$$\begin{aligned}
H_\varepsilon &\equiv \frac{1}{r}\frac{(ru_r)}{\partial r} + \frac{1}{r}\frac{\partial u_\theta}{\partial \theta} + \frac{\partial u_z}{\partial z}\\
\rho_r &\equiv \frac{1}{2}\left[\frac{1}{r}\frac{\partial u_z}{\partial \theta} - \frac{\partial u_\theta}{\partial z}\right]\\
\rho_\theta &\equiv \frac{1}{2}\left[\frac{\partial u_r}{\partial z} - \frac{\partial u_z}{\partial r}\right]\\
\rho_z &\equiv \frac{1}{2}\left[\frac{\partial (ru_\theta)}{\partial r} - \frac{\partial u_r}{\partial \theta}\right]
\end{aligned} \tag{1.33}$$

while for a spherical coordinate system, the relevant expressions are

$$
\begin{aligned}
&(\lambda+2G)\frac{\partial \bar{H}_\varepsilon}{\partial r}-\frac{2G}{r\sin\theta}\left[\frac{\partial\left(\bar{\rho}_\phi\sin\theta\right)}{\partial\theta}-\frac{\partial\bar{\rho}_\theta}{\partial\phi}\right]+\bar{F}_r=0\\
&(\lambda+2G)\frac{\partial \bar{H}_\varepsilon}{r\partial\theta}-\frac{2G}{r}\left[\frac{1}{\sin\theta}\frac{\partial(\bar{\rho}_r)}{\partial\phi}-\frac{\partial\left(r\bar{\rho}_\phi\right)}{\partial r}\right]+\bar{F}_\theta=0\\
&\frac{(\lambda+2G)}{r\sin\theta}\frac{\partial \bar{H}_\varepsilon}{\partial\phi}-\frac{2G}{r}\left[\frac{\partial(r\bar{\rho}_\theta)}{\partial r}-\frac{\partial(r\bar{\rho}_r)}{\partial\theta}\right]+\bar{F}_\phi=0
\end{aligned}
\tag{1.34}
$$

where, according to eq. (1.23), the body forces are defined as

$$\bar{F}_r\equiv\bar{R},\qquad \bar{F}_\theta\equiv\bar{\Theta},\qquad \bar{F}_z\equiv\bar{\Phi}\tag{1.35}$$

and

$$
\begin{aligned}
\bar{H}_\varepsilon&\equiv\frac{1}{r^2}\frac{\partial(r^2u_r)}{\partial r}+\frac{1}{r\sin\theta}\frac{\partial(u_\theta\sin\theta)}{\partial\theta}+\frac{1}{r\sin\theta}\frac{\partial u_\phi}{\partial\phi}\\
\bar{\rho}_r&\equiv\frac{1}{2r\sin\theta}\left[\frac{\partial\left(u_\phi\sin\theta\right)}{\partial\theta}-\frac{\partial u_\theta}{\partial\phi}\right]\\
\bar{\rho}_\theta&\equiv\frac{1}{2}\left[\frac{1}{r\sin\theta}\frac{\partial u_r}{\partial\phi}-\frac{1}{r}\frac{\partial(ru_\phi)}{\partial r}\right]\\
\bar{\rho}_z&\equiv\frac{1}{2r}\left[\frac{\partial(ru_\theta)}{\partial r}-\frac{\partial u_r}{\partial\theta}\right]
\end{aligned}
\tag{1.36}
$$

The Beltrami–Michell compatibility equations are given in a compact form[2] as

$$\sigma_{ij,kk}+\frac{1}{1+\nu}\sigma_{kk,uj}=-\frac{\nu}{1-\nu}F_{k,k}\delta_{ij}-F_{i,j}-F_{j,i}\tag{1.37}$$

2 δ_{ij} is the Kronecker delta, defined as $\delta_{ij}=0$ if $i\neq j$ and $\delta_{ij}=1$ if $i=j$.

Expanding eq. (1.32) leads to the following six equations (for a Cartesian system):

$$\begin{aligned}
&\nabla^2\sigma_x + \frac{1}{1+\nu}\frac{\partial^2\Lambda}{\partial x^2} = -\frac{1}{1-\nu}\left[\frac{\partial F_x}{\partial x}+\frac{\partial F_y}{\partial y}+\frac{\partial F_z}{\partial z}\right]-2\frac{\partial F_x}{\partial x}\\
&\nabla^2\sigma_y + \frac{1}{1+\nu}\frac{\partial^2\Lambda}{\partial y^2} = -\frac{1}{1-\nu}\left[\frac{\partial F_x}{\partial x}+\frac{\partial F_y}{\partial y}+\frac{\partial F_z}{\partial z}\right]-2\frac{\partial F_y}{\partial y}\\
&\nabla^2\sigma_z + \frac{1}{1+\nu}\frac{\partial^2\Lambda}{\partial z^2} = -\frac{1}{1-\nu}\left[\frac{\partial F_x}{\partial x}+\frac{\partial F_y}{\partial y}+\frac{\partial F_z}{\partial z}\right]-2\frac{\partial F_z}{\partial z}\\
&\nabla^2\tau_{yz} + \frac{1}{1+\nu}\frac{\partial^2\Lambda}{\partial y\partial z} = -\left(\frac{\partial F_z}{\partial y}+\frac{\partial F_y}{\partial z}\right)\\
&\nabla^2\tau_{zx} + \frac{1}{1+\nu}\frac{\partial^2\Lambda}{\partial z\partial x} = -\left(\frac{\partial F_x}{\partial z}+\frac{\partial F_z}{\partial x}\right)\\
&\nabla^2\tau_{xy} + \frac{1}{1+\nu}\frac{\partial^2\Lambda}{\partial x\partial y} = -\left(\frac{\partial F_y}{\partial x}+\frac{\partial F_x}{\partial y}\right)
\end{aligned}\tag{1.38}$$

where $\Lambda \equiv \sigma_x + \sigma_y + \sigma_z$.

For a cylindrical coordinate system, the strain compatibility equations have the following form (see derivation in [17]):

$$\begin{aligned}
&\frac{1}{r}\frac{\partial^2\varepsilon_r}{\partial\theta^2}+\frac{\partial}{\partial r}\left\{r\frac{\partial\varepsilon_\theta}{\partial r}-(\varepsilon_r-\varepsilon_\theta)\right\}=\frac{\partial}{\partial\theta}\left\{\frac{\partial\gamma_{r\theta}}{\partial r}+\frac{\gamma_{r\theta}}{r}\right\}\\
&\frac{1}{r^2}\frac{\partial^2\varepsilon_z}{\partial\theta^2}+\frac{\partial^2\varepsilon_\theta}{\partial z^2}+\frac{1}{r}\frac{\partial\varepsilon_z}{\partial r}=\frac{1}{r}\frac{\partial}{\partial z}\left\{\frac{\partial\gamma_{\theta z}}{\partial\theta}+\gamma_{rz}\right\}\\
&\frac{\partial^2\varepsilon_z}{\partial r^2}+\frac{\partial^2\varepsilon_r}{\partial z^2}=\frac{\partial^2\gamma_{rz}}{\partial r\partial z}\\
&\frac{2}{r}\frac{\partial^2\varepsilon_r}{\partial\theta\partial z}=\frac{1}{r}\frac{\partial}{\partial r}\left\{r\frac{\partial\gamma_{r\theta}}{\partial z}-\gamma_{\theta z}\right\}+\frac{\partial}{\partial r}\left\{\frac{1}{r}\frac{\partial\gamma_{rz}}{\partial\theta}-\frac{\partial\gamma_{\theta z}}{\partial r}\right\}+\frac{1}{r}\frac{\partial\gamma_{r\theta}}{\partial z}+\frac{\gamma_{\theta z}}{r^2}\\
&2\frac{\partial}{\partial z}\left\{\frac{\partial\varepsilon_\theta}{\partial r}-\left(\frac{\varepsilon_r-\varepsilon_\theta}{r}\right)\right\}=\frac{1}{r}\frac{\partial}{\partial\theta}\left\{\frac{\partial\gamma_{\theta z}}{\partial r}-\frac{1}{r}\frac{\partial\gamma_{rz}}{\partial\theta}+\frac{\partial\gamma_{r\theta}}{\partial z}\right\}+\frac{1}{r^2}\frac{\partial\gamma_{\theta z}}{\partial\theta}\\
&\frac{2}{r}\frac{\partial}{\partial\theta}\left\{\frac{\partial\varepsilon_z}{\partial r}-\frac{\varepsilon_z}{r}\right\}=\frac{\partial}{\partial z}\left\{\frac{\partial\gamma_{\theta z}}{\partial r}+\frac{1}{r}\frac{\partial\gamma_{rz}}{\partial\theta}-\frac{\partial\gamma_{r\theta}}{\partial z}-\frac{\gamma_{\theta z}}{r}\right\}
\end{aligned}\tag{1.39}$$

where the various strain components were defined in eq. (1.8).

The associated Beltrami–Michell compatibility equations given for a cylindrical coordinate system are given as

$$\frac{1}{r}\frac{\partial}{\partial r}\left(r\frac{\partial\sigma_r}{\partial r}\right)+\frac{1}{r^2}\frac{\partial\sigma_r}{\partial\theta^2}+\frac{\partial^2\sigma_r}{\partial z^2}+\frac{1}{1+\nu}\frac{\partial^2\Omega}{\partial r^2}=-\frac{1}{1-\nu}\frac{1}{r}\left[\frac{\partial(rF_r)}{\partial r}+\frac{\partial F_\theta}{\partial\theta}+\frac{r\partial F_z}{\partial z}\right]-2\frac{\partial F_r}{\partial r}$$

$$\frac{1}{r}\frac{\partial}{\partial r}\left(r\frac{\partial\sigma_\theta}{\partial r}\right)+\frac{1}{r^2}\frac{\partial\sigma_\theta}{\partial\theta^2}+\frac{\partial^2\sigma_\theta}{\partial z^2}+\frac{1}{1+\nu}\frac{\partial^2\Omega}{\partial\theta^2}=-\frac{1}{1-\nu}\frac{1}{r}\left[\frac{\partial(rF_r)}{\partial r}+\frac{\partial F_\theta}{\partial\theta}+\frac{r\partial F_z}{\partial z}\right]-2\frac{\partial F_\theta}{\partial\theta}$$

$$\frac{1}{r}\frac{\partial}{\partial r}\left(r\frac{\partial\sigma_z}{\partial r}\right)+\frac{1}{r^2}\frac{\partial\sigma_z}{\partial\theta^2}+\frac{\partial^2\sigma_z}{\partial z^2}+\frac{1}{1+\nu}\frac{\partial^2\Omega}{\partial z^2}=-\frac{1}{1-\nu}\frac{1}{r}\left[\frac{\partial(rF_r)}{\partial r}+\frac{\partial F_\theta}{\partial\theta}+\frac{r\partial F_z}{\partial z}\right]-2\frac{\partial F_z}{\partial z}$$

$$\frac{1}{r}\frac{\partial}{\partial r}\left(r\frac{\partial\sigma_{\theta z}}{\partial r}\right)+\frac{1}{r^2}\frac{\partial\sigma_{\theta z}}{\partial\theta^2}+\frac{\partial^2\sigma_{\theta z}}{\partial z^2}+\frac{1}{1+\nu}\frac{\partial^2\Omega}{\partial\theta\partial z}=-\left(\frac{\partial F_\theta}{\partial z}+\frac{\partial F_z}{\partial\theta}\right)$$

$$\frac{1}{r}\frac{\partial}{\partial r}\left(r\frac{\partial\sigma_{rz}}{\partial r}\right)+\frac{1}{r^2}\frac{\partial\sigma_{rz}}{\partial\theta^2}+\frac{\partial^2\sigma_{rz}}{\partial z^2}+\frac{1}{1+\nu}\frac{\partial^2\Omega}{\partial r\partial z}=-\left(\frac{\partial F_z}{\partial r}+\frac{\partial F_r}{\partial z}\right)$$

$$\frac{1}{r}\frac{\partial}{\partial r}\left(r\frac{\partial\sigma_{r\theta}}{\partial r}\right)+\frac{1}{r^2}\frac{\partial\sigma_{r\theta}}{\partial\theta^2}+\frac{\partial^2\sigma_{r\theta}}{\partial z^2}+\frac{1}{1+\nu}\frac{\partial^2\Omega}{\partial r\partial\theta}=-\left(\frac{\partial F_r}{\partial\theta}+\frac{\partial F_\theta}{\partial r}\right)$$

(1.40)

where $\Omega\equiv\sigma_r+\sigma_\theta+\sigma_z$.

The strain compatibility equations in a spherical coordinate system, defined in eq. (1.10), have the following form (for clarity, the expressions for the various strains are not presented using the three assumed displacements u_r, u_θ and u_ϕ in the directions r, θ, ϕ, respectively):

$$\frac{\partial^2\varepsilon_r}{\partial\theta^2}+\frac{\partial^2\varepsilon_\theta}{\partial r^2}-2\frac{\partial^2\varepsilon_{r\theta}}{\partial r\partial\theta}=0$$

$$\frac{\partial^2\varepsilon_\theta}{\partial\phi^2}+\frac{\partial^2\varepsilon_\phi}{\partial\theta^2}-2\frac{\partial^2\varepsilon_{\theta\phi}}{\partial\theta\partial\phi}=0$$

$$\frac{\partial^2\varepsilon_\phi}{\partial r^2}+\frac{\partial^2\varepsilon_r}{\partial\phi^2}-2\frac{\partial^2\varepsilon_{\phi r}}{\partial\phi\partial r}=0$$

$$\frac{\partial^2\varepsilon_r}{\partial\theta\partial\phi}+\frac{\partial^2\varepsilon_{\theta\phi}}{\partial r^2}-\frac{\partial^2\varepsilon_{r\theta}}{\partial r\partial\phi}-\frac{\partial^2\varepsilon_{r\phi}}{\partial r\partial\theta}=0 \quad (1.41)$$

$$\frac{\partial^2\varepsilon_\theta}{\partial\phi\partial r}+\frac{\partial^2\varepsilon_{\phi r}}{\partial\theta^2}-\frac{\partial^2\varepsilon_{\theta\phi}}{\partial\theta\partial r}-\frac{\partial^2\varepsilon_{\theta r}}{\partial\theta\partial\phi}=0$$

$$\frac{\partial^2\varepsilon_\phi}{\partial r\partial\theta}+\frac{\partial^2\varepsilon_{r\theta}}{\partial\phi^2}-\frac{\partial^2\varepsilon_{\phi r}}{\partial\phi\partial\theta}-\frac{\partial^2\varepsilon_{\phi\theta}}{\partial\phi\partial r}=0$$

where

$$\varepsilon_{r\theta}\equiv0.5\gamma_{r\theta},\quad \varepsilon_{r\phi}\equiv0.5\gamma_{r\phi},\quad \varepsilon_{\theta\phi}\equiv0.5\gamma_{\theta\phi}$$
$$\varepsilon_{\theta r}\equiv0.5\gamma_{\theta r},\quad \varepsilon_{\phi r}\equiv0.5\gamma_{\phi r},\quad \varepsilon_{\phi\theta}\equiv0.5\gamma_{\phi\theta} \quad (1.42)$$

Finally, the associated Beltrami–Michell compatibility equations given for a spherical coordinate system are given as

$$\frac{1}{r^2}\frac{\partial}{\partial r}\left(r^2\frac{\partial\sigma_r}{\partial r}\right)+\frac{1}{r^2\sin\theta}\frac{\partial}{\partial\theta}\left(\sin\theta\frac{\partial\sigma_r}{\partial\theta}\right)+\frac{1}{r^2\sin^2\theta}\frac{\partial^2\sigma_r}{\partial\phi^2}+\frac{1}{1+\nu}\frac{\partial^2\Psi}{\partial r^2}=$$
$$-\frac{1}{1-\nu}\frac{1}{r}\left[\frac{1}{r}\frac{\partial(r^2F_r)}{\partial r}+\frac{1}{\sin\theta}\frac{\partial}{\partial\theta}(\sin\theta\cdot F_\theta)+\frac{1}{\sin\theta}\frac{\partial F_\phi}{\partial\phi}\right]-2\frac{\partial F_r}{\partial r}$$
$$\frac{1}{r^2}\frac{\partial}{\partial r}\left(r^2\frac{\partial\sigma_\theta}{\partial r}\right)+\frac{1}{r^2\sin\theta}\frac{\partial}{\partial\theta}\left(\sin\theta\frac{\partial\sigma_\theta}{\partial\theta}\right)+\frac{1}{r^2\sin^2\theta}\frac{\partial^2\sigma_\theta}{\partial\phi^2}+\frac{1}{1+\nu}\frac{\partial^2\Psi}{\partial\theta^2}=$$
$$-\frac{1}{1-\nu}\frac{1}{r}\left[\frac{1}{r}\frac{\partial(r^2F_r)}{\partial r}+\frac{1}{\sin\theta}\frac{\partial}{\partial\theta}(\sin\theta\cdot F_\theta)+\frac{1}{\sin\theta}\frac{\partial F_\phi}{\partial\phi}\right]-2\frac{\partial F_\theta}{\partial\theta}$$
$$\frac{1}{r^2}\frac{\partial}{\partial r}\left(r^2\frac{\partial\sigma_\phi}{\partial r}\right)+\frac{1}{r^2\sin\theta}\frac{\partial}{\partial\theta}\left(\sin\theta\frac{\partial\sigma_\phi}{\partial\theta}\right)+\frac{1}{r^2\sin^2\theta}\frac{\partial^2\sigma_\phi}{\partial\phi^2}+\frac{1}{1+\nu}\frac{\partial^2\Psi}{\partial\phi^2}=$$
$$-\frac{1}{1-\nu}\frac{1}{r}\left[\frac{1}{r}\frac{\partial(r^2F_r)}{\partial r}+\frac{1}{\sin\theta}\frac{\partial}{\partial\theta}(\sin\theta\cdot F_\theta)+\frac{1}{\sin\theta}\frac{\partial F_\phi}{\partial\phi}\right]-2\frac{\partial F_\phi}{\partial\phi}$$
$$\frac{1}{r^2}\frac{\partial}{\partial r}\left(r^2\frac{\partial\sigma_{\theta\phi}}{\partial r}\right)+\frac{1}{r^2\sin\theta}\frac{\partial}{\partial\theta}\left(\sin\theta\frac{\partial\sigma_{\theta\phi}}{\partial\theta}\right)+\frac{1}{r^2\sin^2\theta}\frac{\partial^2\sigma_{\theta\phi}}{\partial\phi^2}+\frac{1}{1+\nu}\frac{\partial^2\Psi}{\partial\theta\partial\phi}=-\left(\frac{\partial F_\theta}{\partial\phi}+\frac{\partial F_\phi}{\partial\theta}\right)$$
$$\frac{1}{r^2}\frac{\partial}{\partial r}\left(r^2\frac{\partial\sigma_{r\phi}}{\partial r}\right)+\frac{1}{r^2\sin\theta}\frac{\partial}{\partial\theta}\left(\sin\theta\frac{\partial\sigma_{r\phi}}{\partial\theta}\right)+\frac{1}{r^2\sin^2\theta}\frac{\partial^2\sigma_{r\phi}}{\partial\phi^2}+\frac{1}{1+\nu}\frac{\partial^2\Psi}{\partial\theta\partial\phi}=-\left(\frac{\partial F_r}{\partial\phi}+\frac{\partial F_\phi}{\partial r}\right)$$
$$\frac{1}{r^2}\frac{\partial}{\partial r}\left(r^2\frac{\partial\sigma_{r\theta}}{\partial r}\right)+\frac{1}{r^2\sin\theta}\frac{\partial}{\partial\theta}\left(\sin\theta\frac{\partial\sigma_{r\theta}}{\partial\theta}\right)+\frac{1}{r^2\sin^2\theta}\frac{\partial^2\sigma_{r\theta}}{\partial\phi^2}+\frac{1}{1+\nu}\frac{\partial^2\Psi}{\partial r\partial\theta}=-\left(\frac{\partial F_r}{\partial\theta}+\frac{\partial F_\theta}{\partial r}\right) \tag{1.43}$$

where $\Psi\equiv\sigma_r+\sigma_\theta+\sigma_\phi$

1.4.3 Plane stress and plane strain (2D representations)

Often 3D elasticity problems can be reduced to only 2D representations, thus enabling an easier solution. These cases are either called *plane stress* and/or *plane strain* problems.

1.4.3.1 Plane stress problems

Three-dimensional cases for which one dimension is much smaller than the other two, like in the case of thin flat plates, can be approximated using the plane stress assumption. For these cases, the plate thickness would be in the z direction, with loads being applied perpendicular to it. Then, under these assumptions, all stress components having a subscript z are assumed to be zero. The remaining stress components are then assumed to be functions of only x and y coordinates. This will transform the 3D equations, presented in eq. (1.15) (for a Cartesian coordinate system) into the following simplified 2D form:

$$\begin{aligned}
&\varepsilon_x = \frac{1}{E}\left[\sigma_x - \nu\sigma_y\right], \quad \varepsilon_y = \frac{1}{E}\left[\sigma_y - \nu\sigma_x\right] \\
&\varepsilon_z = -\frac{1}{E}\left[\nu\left(\sigma_x + \sigma_y\right)\right] \\
&\gamma_{xy} = \frac{\tau_{xy}}{G}, \quad \gamma_{yz} = 0, \quad \gamma_{xz} = 0
\end{aligned} \tag{1.44}$$

Accordingly, for a cylindrical coordinate system, we will get the following equations (based on eq. (1.16), under the assumption that there is no z dependence), which are also known as *polar* representation:

$$\begin{aligned}
&\varepsilon_r = \frac{1}{E}\left[\sigma_r - \nu\sigma_\theta\right], \quad \varepsilon_\theta = \frac{1}{E}\left[\sigma_\theta - \nu\sigma_r\right] \\
&\varepsilon_z = -\frac{1}{E}\left[\nu\left(\sigma_r + \sigma_\theta\right)\right] \\
&\gamma_{r\theta} = \frac{\tau_{r\theta}}{G}, \quad \gamma_{\theta z} = 0, \quad \gamma_{rz} = 0
\end{aligned} \tag{1.45}$$

The inverse equations, namely stresses as a function of strains, are next presented:[3]

$$\begin{aligned}
&\sigma_x = \frac{E}{(1-\nu^2)}\left[\varepsilon_x + \nu\varepsilon_y\right] \\
&\sigma_y = \frac{E}{(1-\nu^2)}\left[\varepsilon_y + \nu\varepsilon_x\right] \\
&\sigma_z = 0 = \left[(1-\nu)\varepsilon_z + \nu\left(\varepsilon_x + \varepsilon_y\right)\right] \\
&\tau_{xy} = G\gamma_{xy}, \quad \tau_{yz} = 0, \quad \tau_{xz} = 0
\end{aligned} \tag{1.46}$$

$$\begin{aligned}
&\sigma_r = \frac{E}{(1-\nu^2)}\left[\varepsilon_r + \nu\varepsilon_\theta\right] \\
&\sigma_\theta = \frac{E}{(1-\nu^2)}\left[\varepsilon_\theta + \nu\varepsilon_r\right] \\
&\sigma_z = 0 = \left[(1-\nu)\varepsilon_z + \nu(\varepsilon_r + \varepsilon_\theta)\right] \\
&\tau_{r\theta} = G\gamma_{r\theta}, \quad \tau_{\theta z} = 0, \quad \tau_{rz} = 0
\end{aligned} \tag{1.47}$$

The strain compatibility equations, as presented in eq. (1.25), will have the following compact form in the plane stress case for a Cartesian system:

3 One should note that the strain in the z direction $\varepsilon_z \neq 0$!

$$\frac{\partial^2 \varepsilon_x}{\partial y^2} + \frac{\partial^2 \varepsilon_y}{\partial x^2} = 2\frac{\partial^2 \varepsilon_{xy}}{\partial x \partial y}$$
$$\frac{\partial^2 \varepsilon_z}{\partial x^2} = 0$$
$$\frac{\partial^2 \varepsilon_z}{\partial y^2} = 0 \tag{1.48}$$
$$\frac{\partial^2 \varepsilon_z}{\partial x \partial y} = 0$$

Note that the last three equations in eq. (1.48) might cause some difficulties for plane stress problems. For a cylindrical coordinate system, the strain compatibility equations would be (see eq. (1.39) for the 3D case)

$$\frac{1}{r}\frac{\partial^2 \varepsilon_r}{\partial \theta^2} + \frac{\partial}{\partial r}\left\{ r\frac{\partial \varepsilon_\theta}{\partial r} - (\varepsilon_r - \varepsilon_\theta) \right\} = \frac{\partial}{\partial \theta}\left\{ \frac{\partial \gamma_{r\theta}}{\partial r} + \frac{\gamma_{r\theta}}{r} \right\}$$
$$\frac{1}{r}\frac{\partial^2 \varepsilon_z}{\partial \theta^2} + \frac{\partial \varepsilon_z}{\partial r} = 0$$
$$\frac{\partial^2 \varepsilon_z}{\partial r^2} = 0 \tag{1.49}$$
$$\frac{\partial}{\partial \theta}\left\{ \frac{\partial \varepsilon_z}{\partial r} - \frac{\varepsilon_z}{r} \right\} = 0$$

1.4.3.2 Plane strain problems

Three-dimensional cases for which one dimension is much larger than the other two, as in the case of long cylindrical bodies or water dam-type structures, can be approximated using the plane strain assumption. For these cases, the larger dimension would be in the z direction, and for any x–y plane (perpendicular to z), it is assumed that the loads are independent of z. Then, under these assumptions, all strain components having a subscript z are assumed to be zero. The remaining strain components are then assumed to be functions of only x and y coordinates.[4] This will transform the 3D equations presented in eq. (1.15) (for a Cartesian coordinate system) into the following simplified 2D form:

4 Note that in this case $\sigma_z \neq 0$!

$$\begin{aligned}
&\varepsilon_x = \frac{1}{E}\left[\sigma_x - \nu\left(\sigma_y + \sigma_z\right)\right], \quad \varepsilon_y = \frac{1}{E}\left[\sigma_y - \nu(\sigma_z + \sigma_x)\right] \\
&\varepsilon_z = 0 = \frac{1}{E}\left[\sigma_z - \nu\left(\sigma_x + \sigma_y\right)\right] \\
&\gamma_{xy} = \frac{\tau_{xy}}{G}, \quad \gamma_{yz} = 0, \quad \gamma_{xz} = 0
\end{aligned} \tag{1.50}$$

Accordingly, for a cylindrical coordinate system, we will obtain

$$\begin{aligned}
&\varepsilon_r = \frac{1}{E}[\sigma_r - \nu(\sigma_\theta + \sigma_z)], \quad \varepsilon_\theta = \frac{1}{E}[\sigma_\theta - \nu(\sigma_z + \sigma_r)] \\
&\varepsilon_z = 0 = -\frac{1}{E}[\sigma_z - \nu(\sigma_r + \sigma_\theta)] \\
&\gamma_{r\theta} = \frac{\tau_{r\theta}}{G}, \quad \gamma_{\theta z} = 0, \quad \gamma_{rz} = 0
\end{aligned} \tag{1.51}$$

The inverse equations, namely stresses as a function of strains, are next presented:

$$\begin{aligned}
&\sigma_x = \frac{E}{(1+\nu)(1-2\nu)}\left[(1-\nu)\varepsilon_x + \nu\varepsilon_y\right] \\
&\sigma_y = \frac{E}{(1+\nu)(1-2\nu)}\left[(1-\nu)\varepsilon_y + \nu\varepsilon_x\right] \\
&\sigma_z = \frac{E}{(1+\nu)(1-2\nu)}\left[\nu\left(\varepsilon_x + \varepsilon_y\right)\right] \\
&\tau_{xy} = G\gamma_{xy}, \quad \tau_{yz} = 0, \quad \tau_{xz} = 0
\end{aligned} \tag{1.52}$$

$$\begin{aligned}
&\sigma_r = \frac{E}{(1+\nu)(1-2\nu)}[(1-\nu)\varepsilon_r + \nu\varepsilon_\theta] \\
&\sigma_\theta = \frac{E}{(1+\nu)(1-2\nu)}[(1-\nu)\varepsilon_\theta + \nu\varepsilon_r] \\
&\sigma_z = \frac{E}{(1+\nu)(1-2\nu)}[\nu(\varepsilon_r + \varepsilon_\theta)] \\
&\tau_{r\theta} = G\gamma_{r\theta}, \quad \tau_{\theta z} = 0, \quad \tau_{rz} = 0
\end{aligned} \tag{1.53}$$

The strain compatibility equations, as presented in eq. (1.25), will have the following compact form in the plane strain case for a Cartesian system:

$$
\begin{aligned}
&\frac{\partial^2 \varepsilon_x}{\partial y^2} + \frac{\partial^2 \varepsilon_y}{\partial x^2} = 2\frac{\partial^2 \varepsilon_{xy}}{\partial x \partial y}\\
&\frac{\partial^2 \varepsilon_z}{\partial x^2} = 0\\
&\frac{\partial^2 \varepsilon_z}{\partial y^2} = 0\\
&\frac{\partial^2 \varepsilon_z}{\partial x \partial y} = 0
\end{aligned}
\tag{1.54}
$$

Note that the last three equations in eq. (1.54) might complicate the solutions for plane strain problems (the same as the plane stress case, presented above). For a cylindrical coordinate system, the strain compatibility equations would be (see eq. (1.39) for the 3D case)

$$
\begin{aligned}
&\frac{1}{r}\frac{\partial^2 \varepsilon_r}{\partial \theta^2} + \frac{\partial}{\partial r}\left\{ r\frac{\partial \varepsilon_\theta}{\partial r} - (\varepsilon_r - \varepsilon_\theta) \right\} = \frac{\partial}{\partial \theta}\left\{ \frac{\partial \gamma_{r\theta}}{\partial r} + \frac{\gamma_{r\theta}}{r} \right\}\\
&\frac{1}{r}\frac{\partial^2 \varepsilon_z}{\partial \theta^2} + \frac{\partial \varepsilon_z}{\partial r} = 0\\
&\frac{\partial^2 \varepsilon_z}{\partial r^2} = 0\\
&\frac{\partial}{\partial \theta}\left\{ \frac{\partial \varepsilon_z}{\partial r} - \frac{\varepsilon_z}{r} \right\} = 0
\end{aligned}
\tag{1.55}
$$

1.4.4 The Airy function $\phi(x, y)$

One of the most efficient ways to solve 2D problems in elasticity is to introduce a new variable, known as the Airy stress function, $\phi(x, y)$, suggested by sir George Airy in 1862.[5] According to his idea, the stresses are defined as a function of the new variable, leading to a new differential equation, which can be shown to be solvable in a much easier way compared to the solution of the equations of equilibrium, the Navier's equations (see eq. (1.27)). This leads to the following expressions for the three planar stresses (for a Cartesian coordinate system):

5 Sir George Biddeli Airy (1801–1892), English astronomer and mathematician (https://en.wikipedia.org/wiki/George_Biddell_Airy).

$$\sigma_x(x,y) \equiv \frac{\partial^2\phi(x,y)}{\partial y^2}$$
$$\sigma_y(x,y) \equiv \frac{\partial^2\phi(x,y)}{\partial x^2} \tag{1.56}$$
$$\tau_{xy}(x,y) \equiv -\frac{\partial^2\phi(x,y)}{\partial x \partial y}$$

Having the stresses expressed by the Airy function, $\phi(x, y)$, we can substitute eq. (1.56) into eq. (1.54), yielding the compatibility equation for a state of plane strain, with no body forces, where the unknown is the Airy function:

$$\frac{\partial^4\phi(x,y)}{\partial x^4} + 2\frac{\partial^4\phi(x,y)}{\partial x^2\partial y^2} + \frac{\partial^4\phi(x,y)}{\partial y^4} \equiv \nabla^4\phi(x,y) = 0 \tag{1.57}$$

Equation (1.57) is also known as the biharmonic equation or $\nabla^4\phi(x, y) = 0$.

For a cylindrical coordinate system, the biharmonic equation has the following form:

$$\begin{gathered}\nabla^4\phi(r,\theta) \equiv \nabla^2\left[\nabla^2\phi(r,\theta)\right] = \\ \left(\frac{\partial^2}{\partial r^2} + \frac{1}{r}\frac{\partial}{\partial r} + \frac{1}{r^2}\frac{\partial^2}{\partial \theta^2}\right)\left(\frac{\partial^2\phi(r,\theta)}{\partial r^2} + \frac{1}{r}\frac{\partial\phi(r,\theta)}{\partial r} + \frac{1}{r^2}\frac{\partial^2\phi(r,\theta)}{\partial \theta^2}\right) = 0\end{gathered} \tag{1.58}$$

where the stress components are defined as

$$\sigma_r(r,\theta) \equiv \frac{1}{r}\frac{\partial\phi(r,\theta)}{\partial r} + \frac{1}{r^2}\frac{\partial^2\phi(r,\theta)}{\partial \theta^2}$$
$$\sigma_\theta(r,\theta) \equiv \frac{\partial^2\phi(r,\theta)}{\partial r^2} \tag{1.59}$$
$$\tau_{r\theta}(r,\theta) \equiv -\frac{1}{r}\frac{\partial^2\phi(r,\theta)}{\partial r\partial \theta} + \frac{1}{r^2}\frac{\partial\phi(r,\theta)}{\partial \theta}$$

Often, there are problems in which the applied loads have an axisymmetric distribution, namely there is no dependence on the θ variable, and all derivatives with respect to it would vanish. This results in the following stress field: $\sigma_r(r)$, $\sigma_\theta(r)$ and $\sigma_{r\theta}(r) = 0$. For the axisymmetric stress distribution, the biharmonic function (eq. (1.58)) reduces to

$$\begin{gathered}\nabla^4\phi(r) \equiv \left(\frac{\partial^2}{\partial r^2} + \frac{1}{r}\frac{\partial}{\partial r}\right)\left(\frac{\partial^2\phi(r)}{\partial r^2} + \frac{1}{r}\frac{\partial\phi(r)}{\partial r}\right) = \\ \frac{d^4\phi(r)}{dr^4} + \frac{2}{r}\frac{d^3\phi(r)}{dr^3} - \frac{1}{r^2}\frac{d^2\phi(r)}{dr^2} + \frac{1}{r^3}\frac{d\phi(r)}{dr} = 0\end{gathered} \tag{1.60}$$

The solution for eq. (1.60) has the following form:

$$\phi(r) = A_1 + A_2 \log r + A_3 r^2 + A_4 r^2 \log r \tag{1.61}$$

and, according to eq. (1.59), the expressions for the stresses are

$$\begin{aligned}
\sigma_r(r) &\equiv \frac{1}{r}\frac{d\phi(r)}{dr} = \frac{A_2}{r^2} + 2A_3 + A_4(2\log r + 1) \\
\sigma_\theta(r) &\equiv \frac{d^2\phi(r,\theta)}{dr^2} = -\frac{A_2}{r^2} + 2A_3 + A_4(2\log r + 3) \\
\tau_{r\theta}(r,\theta) &\equiv 0
\end{aligned} \tag{1.62}$$

where A_1, A_2, A_3 and A_4 are constants to be determined from the boundary conditions of the given problem.

For the case of plane stress, it can be shown that for the case of no body forces with symmetrically distributed applied loads, the Airy stress function $\phi(x, y)$ has the following form:

$$\phi(x,y) \equiv \psi(x,y) - \frac{vz^2}{2(1+v)r^2}\nabla^2\psi(x,y) \tag{1.63}$$

where $\psi(x,y)$ is the solution of the following biharmonic function $\nabla^2\psi(x,y)$. Assuming the condition for the plane strain, namely, z is a very small quantity, leads to $\phi(x,y) \approx \psi(x,y)$, which makes also the plane stress cases being solved by the biharmonic equation presented above.

In conclusion, the use of the Airy stress function representation reduces the problem of solving the stresses for an elastic body to that of finding a solution for the biharmonic partial differential equation, $\nabla^4\phi(x,y) = 0$, whose derivatives would satisfy certain boundary conditions, according to the posed problem.

1.4.5 Thermal field

All the previous equations were derived assuming the temperature of the given structure is constant, without any changes. However, for an unconstrained body, changing the uniform temperature, either by heating or by cooling, would cause the body to expand or contract, leading to normal strains. Preventing the expansion or contraction would give rise to thermal stresses inside the structure.

To take this effect into consideration, one must redefine the stress–strain relationships, superimposing the thermal strains onto the mechanical one. Assuming the change in temperature is given spatially by $T(x, y)$, the corresponding change in the length would be written as

$$\Delta L = \alpha L T(x,y) \tag{1.64}$$

where α is the linear thermal expansion coefficient. The associated thermal strain is then defined as

$$\varepsilon_t = \frac{\Delta L}{L} = \alpha T(x,y) \tag{1.65}$$

Adding thermal strains to the mechanical ones (see, e.g., eq. (1.44)) leads the plane stress case to

$$\varepsilon_x = \frac{1}{E}\left[\sigma_x - \nu\sigma_y\right] + \alpha T, \quad \varepsilon_y = \frac{1}{E}\left[\sigma_y - \nu\sigma_x\right] + \alpha T, \quad \gamma_{xy} = \frac{\tau_{xy}}{G} \tag{1.66}$$

Expressing stresses as a function of strains leads to

$$\begin{aligned} \sigma_x &= \frac{E}{(1-\nu^2)}\left[\varepsilon_x + \nu\varepsilon_y\right] - \frac{E\alpha T}{(1-\nu)} \\ \sigma_y &= \frac{E}{(1-\nu^2)}\left[\varepsilon_y + \nu\varepsilon_x\right] - \frac{E\alpha T}{(1-\nu)} \\ \tau_{xy} &= G\gamma_{xy} \end{aligned} \tag{1.67}$$

Similar expressions can also be obtained for the plane strain case.

The biharmonic equation in the presence of a temperature field, $T(x,y)$, can be shown to have the following form:

$$\nabla^4\phi(x,y) + \alpha E\nabla^2 T(x,y) = 0 \tag{1.68}$$

This equation is true for both the plane stress and plane strains, provided the body forces can be assumed to be negligible.

References

[1] Francillon, R.J. Lockheed Aircraft since, Naval Institute Press, Annapolis, Maryland, 1913, 1987, ISBN 0-85177-835-6.

[2] Niu, M. C.-Y. Airframe Structural Design –Practical Design Information and Data on Aircraft Structures, CONMILIT Press Ltd., ©, 1988, 607.

[3] Bruhn, E.F. Analysis and Design of Flight Vehicle Structures, Jacobs Publication ©, 1973, 650.

[4] Megson, T.H.G. Aircraft Structures for Engineering Students, 4th Edition, Butterworth-Heinemann (an imprint of Elsevier Ltd.,) ©, 2007, 804.

[5] Gran, B. Bruhn Errata- a Companion to Analysis of Flight Vehicle Structures, GRAN Corporation, 2nd Edition, 2008, 309.

[6] Weisshaar, T.A. Aerospace Structures- an Introduction to Fundamental Problems, Purdue University, USA, 28th of July, 2011, 197.

[7] Rockberger, D. and Abramovich, H. Piezoelectric Assisted Smart Satellite Structure (PEASSS) – An Innovative Low Cost Nano-Satellite, SPIE's Annual International Symposium on Smart Structures and Materials, Conference 9057: Active and Passive Smart Structures and Integrated Systems VIII, 9–13 March 2014, San Diego, CA, USA.

[8] Kalanchiam, M., and Chinnasamy, M. Advantages of Composite Materials in Aircraft Structures, International Journal of Aerospace and Mechanical Engineering, 6(11), 2012, 2428–2432.

[9] Polland, D.R., Finn, S.R., Griess, K.H., Hafenrichter, J. L., Hanson, C.T., Ilsewicz, L.B., Metschan, S.L., Scholz, D.B., and Smith, P.J. Global Cost Weight Evaluation of Fuselage Panel Design Concepts, NASA, CR4730, April 1997, 319.

[10] Walker, T.H., Minguet, P.J., Flynn, B.W., Carbery, D.J., Swanson, G.D., and Ilsewicz, L.B Advanced Technology Composite Fuselage-Structural Performance, NASA CR4732, April 1997, 101.

[11] Flynn, B.W., Bodine, J.B., Dopker, B., Finn, S.R., Griess, K.H., Hanson, C.T., Harris, C.G., Nelson, K.M., Walker, T.H., Kennedy, T.C., and Nahan, M.F Advanced Technology Composite Fuselage-Repair and Damage Assessment Supporting Maintenance, NASA CR4733, April 1997, 154.

[12] Willden, K.S., Harris, C.G., Flynn, B.W., Gessel, M.G., Scholz, D.B., Stawski, S., and Winson, V. Advanced Technology Composite Fuselage-Manufacturing, NASA CR4735, April 1997, 189.

[13] Birch, H. Aerospace Materials-Changing Planes, Chemistry World, Oct, 2013, 60–63.

[14] Abramovich, H. Stability and Vibrations of Thin Walled Composite Structures, ©, Elsevier Ltd, Woodhead Publishing Limited, The Officers' Mess Business Centre, Royston Road, Duxford, CB22 4QH, United Kingdom; 50 Hampshire Street, 5th Floor, Cambridge, MA 02139, United States; The Boulevard, Langford Lane, Kidlington, OX5 1GB, United Kingdom, 2017, 778.

[15] Timoshenko, S.P., and Goodier, J.N. Theory of Elasticity, 3rd Ed, International Student Edition, McGraw-Hill Book Company, Kögakusha Company, Ltd, Tokyo, Japan, ©, 1970, McGraw-Hill Inc., 567.

[16] Chou, P.C., and Pagano, N.J. Elasticity Tensor, Dyadic and Engineering Approaches, Dover Publications, Inc., New York, 1992, 290.

[17] Carlucci, D., Payne, N., and Mehmedagic, I. Small Strain Compatibility Conditions of an Elastic Solid in Cylindrical Coordinates, Technical Report ARDSM-TR-12001, U.S. Army Armament Research, Development and Engineering Center, Munitions Engineering Technology Center, Picatinny Arsenal, New Jersey, USA, April 2013, 14.

2 Composite materials

2.1 Introduction

2.1.1 General introduction

One of the common definitions for a composite material, usually made of two constituents, one being the fiber (the reinforcement) and the other the glue (the matrix), states that a combination of the two materials would result in better properties than those of the individual components when they are used alone. The main advantages of composite materials over other existing materials, like metals or plastics, are their high strength and stiffness, combined with low density, allowing for a weight reduction in the finished part. The various types of composites are usually referred to in the literature, as a block diagram, as depicted in Fig. 2.1. In this chapter when we are talking about the composite material, we restrict ourselves to only continuous fibers (reinforcements) being embedded into a carrying matrix that is formed from an adequate glue.

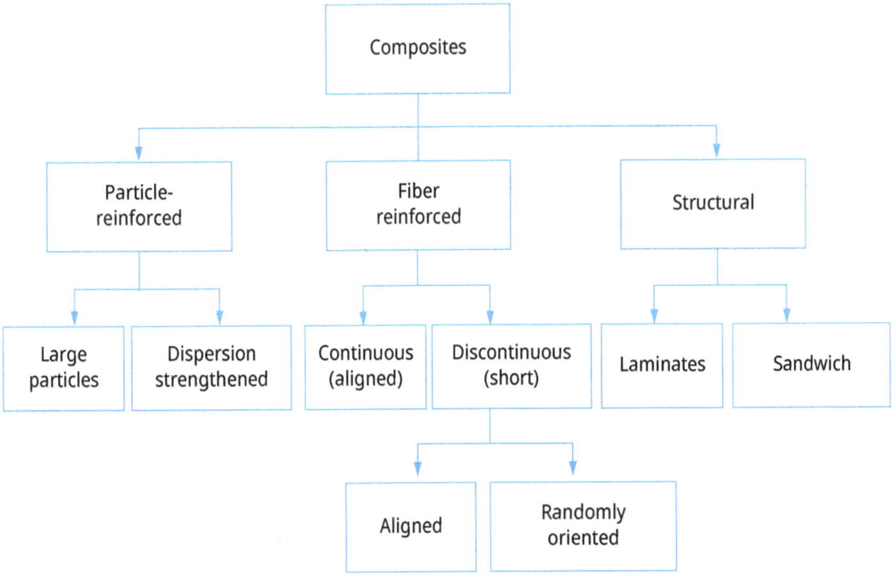

Fig. 2.1: Typical composite materials.

Examples of such continuous reinforcements include unidirectional, woven cloth and helical winding (see Fig. 2.2). Continuous fiber composites are often made into laminates by stacking single sheets of continuous fibers in different orientations to obtain the desired strength and stiffness properties with fiber volumes as high as 60–70%.

https://doi.org/10.1515/9783111621104-002

Fibers produce high-strength composites because of their small diameter; they contain far fewer defects (normally surface defects) compared to the material produced in bulk. On top of it, due to their small diameter the fibers are flexible and suitable for complicated manufacturing processes, such as small radii or weaving. Materials like glass, graphite, carbon or aramid are used to produce fibers (see typical properties in Tab. 2.1). Today's usage of composite materials is mainly driven by the aerospace sector, with large percentage of the modern airplane structures, like Boeing 787 or Airbus A380, being manufactured from carbon, glass and aramid fibers (see Fig. 2.3). The main material for the matrix is a polymer, which has low strength and stiffness. The main functions of the matrix are to keep the fibers in the proper orientation and spacing and providing protection to the fiber from abrasion and the environment. In polymer matrix composites, the good and strong bond between the matrix and the reinforcement allows the matrix to transmit the outside loads from the matrix to the fibers through shear loading at the interface. Two types of polymer matrices are available: thermosets and thermoplastics. A thermoset starts as a low-viscosity resin that reacts and cures during processing, forming a solid. A thermoplastic is a high-viscosity resin that is processed by heating it above its melting temperature. Because a thermoset resin sets up and cures during processing, it cannot be reprocessed by reheating. A thermoplastic can be reheated above its melting temperature for additional processing.

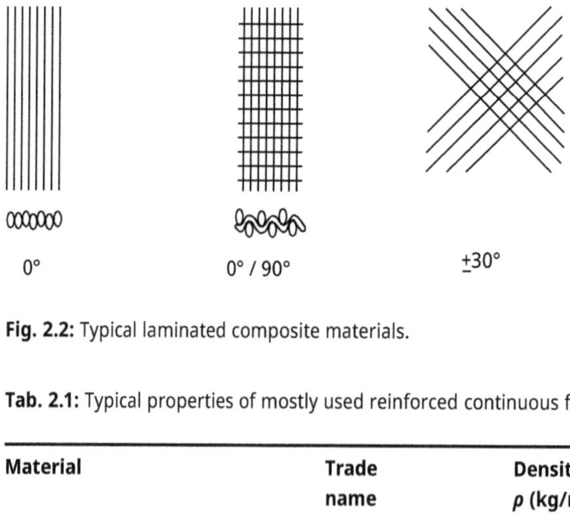

Fig. 2.2: Typical laminated composite materials.

Tab. 2.1: Typical properties of mostly used reinforced continuous fibers (from [1] and [2]).

Material	Trade name	Density, ρ (kg/m^2)	Typical fiber diameter (μm)	Young's modulus, E (GPa)	Tensile strength (GPa)
α-Al_2O_3 (aluminum oxide)	FP (US)	3,960	20	385	1.8
$Al_2O_3 + SiO_2 + B_2O_3$ (mullite)	Nextel480 (USA)	3,050	11	224	2.3
$Al_2O_3 + SiO_2$ (alumina–silica)	Altex (Japan)	3,300	10–15	210	2.0
Boron (CVD* on tungsten)	VMC (Japan)	2,600	140	410	4.0

Tab. 2.1 (continued)

Material	Trade name	Density, ρ (kg/m^2)	Typical fiber diameter (μm)	Young's modulus, E (GPa)	Tensile strength (GPa)
Carbon (PAN** precursor)	T300 (Japan)	1,800	7	230	3.5
Carbon (PAN** precursor)	T800 (Japan)	1,800	5.5	295	5.6
Carbon (pitch*** precursor)	Thorne IP755 (USA)	2,060	10	517	2.1
SiC (+O) (silicon carbide)	Nicalon (Japan)	2,600	15	190	2.5–3.3
SiC (low O) (silicon carbide)	Hi-Nicalon (Japan)	2,740	14	270	2.8
SiC (+O + Ti) (silicon carbide)	Tyranno (Japan)	2,400	9	200	2.8
SiC (monofilament) (silicon carbide)	Sigma	3,100	100	400	3.5
E-glass (silica)		2,500	10	70	1.5–2.0
E-glass (silica)		2,500	10	70	1.5–2.0
Quartz(silica)		2,200	3–15	80	3.5
Aromatic polyamide	Kevlar 49 (USA)	1,500	12	130	3.6
Polyethylene (UHMW)†	Spectra 100 (USA)	970	38	175	3.0
High carbon steel	E.g., piano wire	7,800	250	210	2.8
Aluminum	Electrical wire	2,680	1670	75	0.27
Titanium	Wire	4,700	250	115	0.434

*CVD, chemical vapor deposition.
**PAN, polyacrylonitrile. About 90% of the carbon fibers produced are made from PAN.
***Pitch is a viscoelastic material that is composed of aromatic hydrocarbons. Pitch is produced via the distillation of carbon-based materials, such as plants, crude oil and coal.
†UHMW, ultra-high-molecular-weight polyethylene (or polyethene, the most common plastic produced in the world) is a subset of the thermoplastic polyethylene.

Boeing 787 Dreamliner structure-typical data

787 body only – materials used: fiberglass, aluminum, carbon based composite, sandwich (carbon type), metals (steel, aluminum and titanium).

787 whole plane – materials used (by weight): composites (50%), aluminum (20%), titanium (15%), steel (10%), other (5%).

(a)

Airbus A380 structure- typical data

A380 whole plane – materials used (by weight): composites (22%), aluminum (61%), titanium & steel (10%), other (7%).

(b)

Fig. 2.3: Usage of composite materials in aerospace structures: (a) Boeing 787 and (b) Airbus A380.

2.2 Unidirectional composites

Unidirectional composites are usually composed of two constituents, the fiber and the matrix (which is the glue holding together the two components). Based on the rule of mixtures, one can calculate the properties of the unidirectional layer based on the properties of the fibers and the matrix and their volume fracture. The assumption to be made when applying the rule of mixtures is that the two constituents are bonded together and they behave like a single body. The longitudinal modulus (or the major modulus), E_{11}, of the layer can be written as

$$E_{11} = E_f V_f + E_m V_m \tag{2.1}$$

where E_f and E_m are the longitudinal moduli for the fibers and the matrix, respectively, and V_f and V_m are their volume fractions.[1]

The major Poisson's coefficient, υ_{12}, is given by

$$\upsilon_{12} = \upsilon_f V_f + \upsilon_m V_m \tag{2.2}$$

where υ_f and υ_m are the longitudinal moduli for the fibers and the matrix, respectively.

One should note that the minor Poisson's coefficient, υ_{21}, will be calculated to be

$$\frac{\upsilon_{12}}{E_{11}} = \frac{\upsilon_{21}}{E_{22}} \Rightarrow \upsilon_{21} = \upsilon_{12}\frac{E_{22}}{E_{11}} \tag{2.3}$$

The transverse modulus (or the minor modulus), E_{22}, of the layer is given as

$$\frac{1}{E_{22}} = \frac{V_f}{E_f} + \frac{V_m}{E_m} \Rightarrow E_{22} = \frac{E_m}{V_f \frac{E_m}{E_f} + V_m} = \frac{E_m}{V_f \frac{E_m}{E_f} + (1 - V_f)} \tag{2.4}$$

The shear modulus of the layer, G_{12}, is given as

$$\frac{1}{G_{12}} = \frac{V_f}{G_f} + \frac{V_m}{G_m} \Rightarrow G_{12} = \frac{G_m}{V_f \frac{G_m}{G_f} + V_m} = \frac{G_m}{V_f \frac{G_m}{G_f} + (1 - V_f)} \tag{2.5}$$

where G_f and G_m are the shear moduli for the fibers and the matrix, respectively.

To be able to assess the differences between the properties of the fiber and to compare to those of the matrix, the reader is referred to Tab. 2.2 (from [1]).

As described in [1], the simple micromechanics model used in the rule of mixtures predicts well the values of the four variables, E_{11}, E_{22}, G_{12}, and υ_{12}, when compared to experimental values, as mentioned in Tab. 2.3.

1 Note that $V_f + V_m = 1$.

Tab. 2.2: Typical properties of T300 carbon fibers and 914 epoxy resin.

Property	T300 carbon fibers	914 epoxy resin matrix
Young's modulus, E (GPa)	220	3.3
Shear modulus, G (GPa)	25	1.2
Poisson's ratio, υ	0.15	0.37

Tab. 2.3: Predictions of unidirectional composite properties by simple micromechanic models (adapted from [1]).

Equation	Relationship	Predicted values (moduli in GPa)	Experimental values (moduli in GPa)
2.1	$E_{11} = E_f V_f + E_m(1 - V_f)$	124.7	125.0
2.4	$\frac{1}{E_{22}} = \frac{V_f}{E_f} + \frac{(1-V_f)}{E_m}$	7.4	9.1
2.5	$\frac{1}{G_{12}} = \frac{V_f}{G_f} + \frac{(1-V_f)}{G_m}$	2.6	5.0
2.2	$\upsilon_{12} = \upsilon_f V_f + \upsilon_m(1 - V_f)$	0.25	0.34

2.3 Properties of a single ply

A ply has two major dimensions, and one, the thickness, is very small as compared to the two major ones. Therefore, the 3D presentation of an orthotropic material will be simplified to a 2D presentation (plane stress) by assuming that $\sigma_{33} = 0$ [1, 2]. This leads to a reduced compliance matrix for the ply, in the form:

$$\begin{Bmatrix} \varepsilon_{11} \\ \varepsilon_{22} \\ \gamma_{12} \end{Bmatrix} = \begin{bmatrix} \frac{1}{E_1} & -\frac{\nu_{21}}{E_2} & 0 \\ -\frac{\nu_{12}}{E_1} & \frac{1}{E_2} & 0 \\ 0 & 0 & \frac{1}{G_{12}} \end{bmatrix} \begin{Bmatrix} \sigma_{11} \\ \sigma_{22} \\ \sigma_{12} \end{Bmatrix} \tag{2.6}$$

a third equation, for the strain in the thickness direction, ε_{33}, which is seldom used, has the form

$$\varepsilon_{33} = -\frac{\gamma_{13}}{E_1}\sigma_{11} - \frac{\gamma_{23}}{E_2}\sigma_{22} \tag{2.7}$$

and the remaining two equations for the shear strains are written as

$$\begin{Bmatrix} \gamma_{23} \\ \gamma_{13} \end{Bmatrix} = \begin{bmatrix} \frac{1}{G_{23}} & 0 \\ 0 & \frac{1}{G_{13}} \end{bmatrix} \begin{Bmatrix} \sigma_{23} \\ \sigma_{13} \end{Bmatrix} \tag{2.8}$$

Calculation of the stresses as a function of strains would yield (by the use of eqs. (2.6) and (2.8)):

$$\begin{Bmatrix} \sigma_{11} \\ \sigma_{22} \\ \sigma_{12} \end{Bmatrix} = \begin{bmatrix} Q_{11} & Q_{12} & 0 \\ Q_{21} & Q_{22} & 0 \\ 0 & 0 & Q_{66} \end{bmatrix} \begin{Bmatrix} \varepsilon_{11} \\ \varepsilon_{22} \\ \gamma_{12} \end{Bmatrix} = \begin{bmatrix} \frac{E_1}{(1-\upsilon_{12}\upsilon_{21})} & \frac{\upsilon_{21}E_1}{(1-\upsilon_{12}\upsilon_{21})} & 0 \\ \frac{\upsilon_{12}E_2}{(1-\upsilon_{12}\upsilon_{21})} & \frac{E_2}{(1-\upsilon_{12}\upsilon_{21})} & 0 \\ 0 & 0 & G_{12} \end{bmatrix} \begin{Bmatrix} \varepsilon_{11} \\ \varepsilon_{22} \\ \gamma_{12} \end{Bmatrix} \tag{2.9}$$

with $Q_{12} = Q_{21}$ and $\nu_{12} \neq \nu_{21}$

$$\begin{Bmatrix} \sigma_{23} \\ \sigma_{13} \end{Bmatrix} = \begin{bmatrix} Q_{23} & 0 \\ 0 & Q_{13} \end{bmatrix} \begin{Bmatrix} \gamma_{23} \\ \gamma_{13} \end{Bmatrix} = \begin{bmatrix} G_{23} & 0 \\ 0 & G_{13} \end{bmatrix} \begin{Bmatrix} \gamma_{23} \\ \gamma_{13} \end{Bmatrix} \tag{2.10}$$

2.4 Transformation of stresses and strains

Consider the two coordinate systems described in Fig. 2.4. The one with indexes 1 and 2 describes the ply orthotropic coordinate system, while the other one (x, y) is an arbitrary one, rotated at a given angle θ relative to the 1, 2 system. The transformation of the stresses and the strains from the 1, 2 coordinate system to the x, y coordinate system is done by multiplication of both the stresses and the strains at the ply level by the transformation matrix **T** as given by[2]

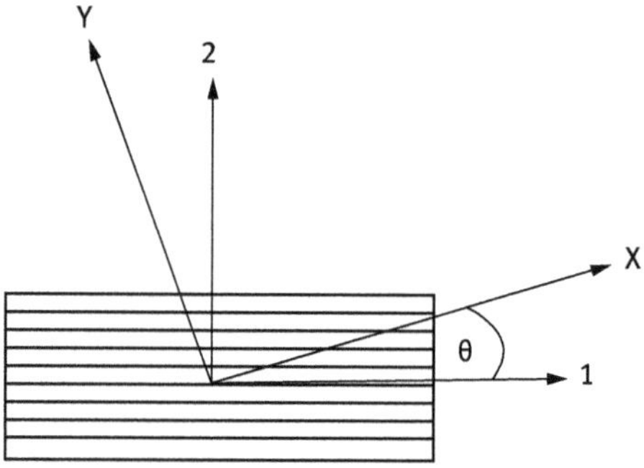

Fig. 2.4: Two coordinate systems: 1, 2 the ply orthotropic axes; *X*, *Y* arbitrary axes.

2 See, for example, Ref. [1]: Primer on Composite Materials: Analysis by J. E. Ashton and J.C. Halpin, TECHNOMIC Publishing Co., Inc., 750 Summer St., Stamford, Conn. 06901, USA, 1969.

$$\begin{Bmatrix} \sigma_1 \\ \sigma_2 \\ \tau_{12} \end{Bmatrix}^k = [T] \begin{Bmatrix} \sigma_x \\ \sigma_y \\ \tau_{xy} \end{Bmatrix}^k \tag{2.11}$$

$$\begin{Bmatrix} \varepsilon_1 \\ \varepsilon_2 \\ \frac{\gamma_{12}}{2} \end{Bmatrix}^k = [T] \begin{Bmatrix} \varepsilon_x \\ \varepsilon_y \\ \frac{\gamma_{xy}}{2} \end{Bmatrix}^k \tag{2.12}$$

where k is the number of the ply, for which the transformation of strains and stresses is performed.[3] The transformation matrix T is given by

$$[T] = \begin{bmatrix} c^2 & s^2 & 2cs \\ s^2 & c^2 & -2cs \\ -cs & cs & c^2 - s^2 \end{bmatrix} \quad \text{where} \quad \begin{matrix} c \equiv \cos\theta \\ s \equiv \sin\theta \end{matrix} \tag{2.13}$$

To obtain the inverse of the matrix T, one needs simply to insert $-\theta$ instead of θ in eq. (2.13) to yield

$$[T]^{-1} = [T(-\theta)] = \begin{bmatrix} c^2 & s^2 & -2cs \\ s^2 & c^2 & 2cs \\ cs & -cs & c^2 - s^2 \end{bmatrix} \tag{2.14}$$

The ply (or lamina) strain–stress relationships transformed to the laminate references axis (x, y) is written as

$$\begin{Bmatrix} \sigma_1 \\ \sigma_2 \\ \tau_{12} \end{Bmatrix}^k = [T]^{-1}[Q]^k[T] \begin{Bmatrix} \varepsilon_x \\ \varepsilon_y \\ \gamma_{xy} \end{Bmatrix}^k \tag{2.15}$$

where

$$[Q]^k = \begin{bmatrix} Q_{11} & Q_{12} & 0 \\ Q_{12} & Q_{22} & 0 \\ 0 & 0 & 2Q_{66} \end{bmatrix}^k$$

and the expressions for Q_{11}, Q_{12}, Q_{22} and Q_{66} are given in eq. (2.9). Performing the matrix multiplication in eq. (2.15) yields

3 Note that $\sigma_{11} \equiv \sigma_1$; $\sigma_{22} \equiv \sigma_2$; $\varepsilon_{11} \equiv \varepsilon_1$; $\varepsilon_{22} \equiv \varepsilon_2$.

$$\begin{Bmatrix} \sigma_1 \\ \sigma_2 \\ \tau_{12} \end{Bmatrix}^k = [\bar{Q}]^k \begin{Bmatrix} \varepsilon_x \\ \varepsilon_y \\ \gamma_{xy} \end{Bmatrix}^k \tag{2.16}$$

where

$$[\bar{Q}]^k = \begin{Bmatrix} \overline{Q_{11}} & \overline{Q_{12}} & \overline{Q_{16}} \\ \overline{Q_{12}} & \overline{Q_{22}} & \overline{Q_{26}} \\ \overline{Q_{16}} & \overline{Q_{26}} & \overline{Q_{66}} \end{Bmatrix}^k$$

where

$$\begin{aligned}
\overline{Q_{11}} &= Q_{11}\cos^4\theta + 2(Q_{12} + 2Q_{66})\sin^2\theta\cos^2\theta + Q_{22}\sin^4\theta \\
\overline{Q_{12}} &= (Q_{11} + Q_{22} - 4Q_{66})\sin^2\theta\cos^2\theta + Q_{12}\left(\sin^4\theta + \cos^4\theta\right) \\
\overline{Q_{22}} &= Q_{11}\sin^4\theta + 2(Q_{12} + 2Q_{66})\sin^2\theta\cos^2\theta + Q_{22}\cos^4\theta \\
\overline{Q_{16}} &= (Q_{11} - Q_{12} - 2Q_{66})\sin\theta\cos^3\theta + (Q_{12} - Q_{22} + 2Q_{66})\sin^3\theta\cos\theta \\
\overline{Q_{26}} &= (Q_{11} - Q_{12} - 2Q_{66})\sin^3\theta\cos\theta + (Q_{12} - Q_{22} + 2Q_{66})\sin\theta\cos^3\theta \\
\overline{Q_{66}} &= (Q_{11} + Q_{22} - 2Q_{12} - 2Q_{66})\sin^2\theta\cos^2\theta + Q_{66}\left(\sin^4\theta + \cos^4\theta\right)
\end{aligned} \tag{2.17}$$

Another useful way of presenting the various terms of the matrix $[\bar{Q}]^k$ is the invariant procedure suggested by Tsai & Pagano, described in detail in [2]:

$$\begin{aligned}
\overline{Q_{11}} &= U_1 + U_2\cos(2\theta) + U_3\cos(4\theta) \\
\overline{Q_{12}} &= U_4 - U_3\cos(4\theta) \\
\overline{Q_{22}} &= U_1 - U_2\cos(2\theta) + U_3\cos(4\theta) \\
\overline{Q_{16}} &= -\frac{1}{2}U_2\sin(2\theta) - U_3\sin(4\theta) \\
\overline{Q_{26}} &= -\frac{1}{2}U_2\sin(2\theta) + U_3\sin(4\theta) \\
\overline{Q_{66}} &= U_5 - U_3\cos(4\theta)
\end{aligned}$$

where

$$
\begin{aligned}
U_1 &= \frac{1}{8}[3Q_{11} + 3Q_{22} + 2Q_{12} + 4Q_{66}] \\
U_2 &= \frac{1}{2}[Q_{11} - Q_{22}] \\
U_3 &= \frac{1}{8}[Q_{11} + Q_{22} - 2Q_{12} - 4Q_{66}] \\
U_4 &= \frac{1}{8}[Q_{11} + Q_{22} + 6Q_{12} - 4Q_{66}] \\
U_5 &= \frac{1}{8}[Q_{11} + Q_{22} - 2Q_{12} + 4Q_{66}]
\end{aligned}
\tag{2.18}
$$

Note that the terms U_1, U_4 and U_5 are invariant to a rotation relative to the three axes (perpendicular to the 1–2 planes).

2.5 The classical lamination theory

Now we shall present the addition of the properties of each lamina to form a laminate, which is the structure to be investigated when we apply loads, using the classical lamination theory (CLT), which is based on Kirchhoff–Love plate theory [3–5].

Referring to Fig. 2.5, one can write the displacement in the x-direction of a point at a z-distance from the mid-plane as follows (where w is the displacement in the z direction):

$$u = u_0 - z\frac{\partial w}{\partial x} \tag{2.19}$$

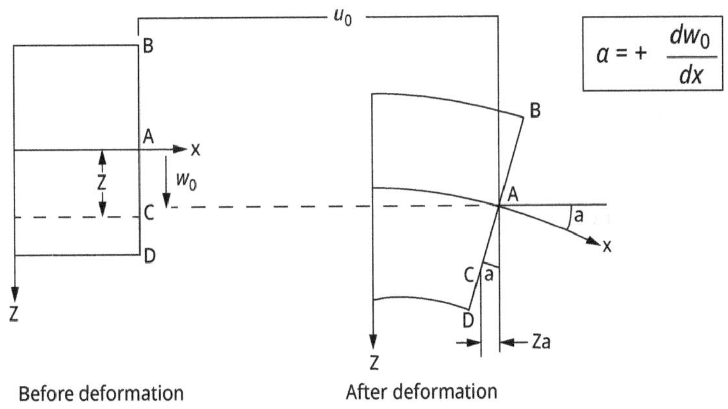

Fig. 2.5: The plate cross section before and after the deformation (CLT approach).

Similarly, the displacement in the y direction will be

$$v = v_0 - z\frac{\partial w}{\partial y} \tag{2.20}$$

Then the strains (ε_x, ε_y and γ_{xy}) and the curvatures (κ_x, κ_y and κ_{xy}) can be written as

$$\begin{aligned}
\varepsilon_x &\equiv \frac{\partial u}{\partial x} = \frac{\partial u_0}{\partial x} - z\frac{\partial^2 w_0}{\partial x^2} = \varepsilon_x^0 + z\kappa_x \\
\varepsilon_y &\equiv \frac{\partial v}{\partial y} = \frac{\partial v_0}{\partial y} - z\frac{\partial^2 w_0}{\partial y^2} = \varepsilon_y^0 + z\kappa_y \\
\gamma_{xy} &\equiv \frac{\partial u}{\partial y} + \frac{\partial v}{\partial x} = \frac{\partial u_0}{\partial y} + \frac{\partial v_0}{\partial x} - 2z\frac{\partial^2 w}{\partial x \partial y} = \gamma_{xy}^0 + z\kappa_{xy}
\end{aligned} \tag{2.21}$$

where ε_x^0, ε_y^0, γ_{xy}^0 are the strains at the neutral plane. In matrix notation, eq. (2.21) can be presented as

$$\begin{Bmatrix} \varepsilon_x \\ \varepsilon_y \\ \gamma_{xy} \end{Bmatrix} = \begin{Bmatrix} \varepsilon_x^0 \\ \varepsilon_y^0 \\ \gamma_{xy}^0 \end{Bmatrix} + z\begin{Bmatrix} \kappa_x \\ \kappa_y \\ \kappa_{xy} \end{Bmatrix} \Rightarrow \quad \{\varepsilon\} = \{\varepsilon^0\} + z\{\kappa\} \tag{2.22}$$

Then the stresses at the lamina level will be written as

$$\{\sigma\}^k = [\bar{Q}]^k\{\varepsilon^0\} + z[\bar{Q}]^k[\kappa] \tag{2.23}$$

Now we shall deal with force (N_x, N_y, N_{xy}) and moment (M_x, M_y, M_{xy}) resultants, per unit width (b in Fig. 2.6). Their definitions are given as (h is the total thickness of the laminate) follows:

$$\begin{aligned}
N_x &\equiv \int_{-h/2}^{h/2} \sigma_x dz, \quad N_y \equiv \int_{-h/2}^{h/2} \sigma_y dz, \quad N_{xy} \equiv \int_{-h/2}^{h/2} \tau_{xy} dz \\
M_x &\equiv \int_{-h/2}^{h/2} \sigma_x z dz, \quad M_y \equiv \int_{-h/2}^{h/2} \sigma_y z dz, \quad M_{xy} \equiv \int_{-h/2}^{h/2} \tau_{xy} z dz
\end{aligned} \tag{2.24}$$

Substituting the expressions of the stresses, one obtains expressions for the force and moment resultants as a function of the strain on the mid-plane, ε^0, and the curvature κ (see also [2]). The short written expressions are

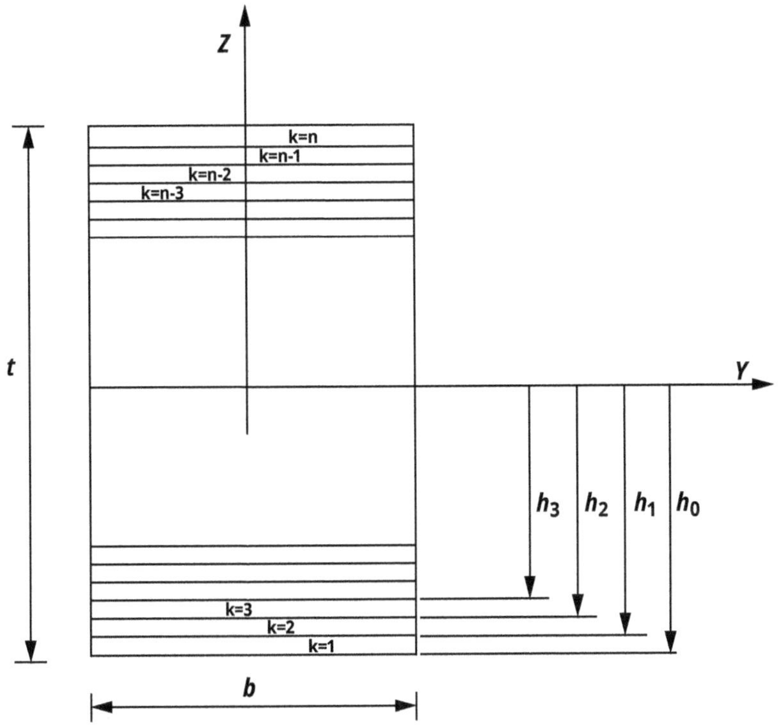

Fig. 2.6: Lamina notations within a given laminate.

$$\begin{Bmatrix} \{N\} \\ \{M\} \end{Bmatrix} = \begin{bmatrix} [A] & [B] \\ [B] & [D] \end{bmatrix} \begin{Bmatrix} \varepsilon^0 \\ \kappa \end{Bmatrix}$$

or

$$\begin{Bmatrix} \begin{Bmatrix} N_x \\ N_y \\ N_{xy} \end{Bmatrix} \\ \begin{Bmatrix} M_x \\ M_y \\ M_{xy} \end{Bmatrix} \end{Bmatrix} = \begin{bmatrix} \begin{bmatrix} A_{11} & A_{12} & A_{16} \\ A_{12} & A_{22} & A_{26} \\ A_{16} & A_{26} & A_{66} \end{bmatrix} & \begin{bmatrix} B_{11} & B_{12} & B_{16} \\ B_{12} & B_{22} & B_{26} \\ B_{16} & B_{26} & B_{66} \end{bmatrix} \\ \begin{bmatrix} B_{11} & B_{12} & B_{16} \\ B_{12} & B_{22} & B_{26} \\ B_{16} & B_{26} & B_{66} \end{bmatrix} & \begin{bmatrix} D_{11} & D_{12} & D_{16} \\ D_{12} & D_{22} & D_{26} \\ D_{16} & D_{26} & D_{66} \end{bmatrix} \end{bmatrix} \begin{Bmatrix} \begin{Bmatrix} \varepsilon_x^0 \\ \varepsilon_y^0 \\ \gamma_{xy}^0 \end{Bmatrix} \\ \begin{Bmatrix} \kappa_x \\ \kappa_y \\ \kappa_{xy} \end{Bmatrix} \end{Bmatrix} \tag{2.25}$$

where the various constants are defined as

$$A_{ij} \equiv \int_{-h/2}^{h/2} \bar{Q}_{ij}^k dz = \sum_{k=1}^{n} \bar{Q}_{ij}^k (h_k - h_{k-1})$$
$$B_{ij} \equiv \int_{-h/2}^{h/2} \bar{Q}_{ij}^k z dz = \frac{1}{2}\sum_{k=1}^{n} \bar{Q}_{ij}^k (h_k^2 - h_{k-1}^2) \tag{2.26}$$
$$D_{ij} \equiv \int_{-h/2}^{h/2} \bar{Q}_{ij}^k z^2 dz = \frac{1}{3}\sum_{k=1}^{n} \bar{Q}_{ij}^k (h_k^3 - h_{k-1}^3)$$

where $i, j = 1, 1; 1, 2; 2, 2; 1, 6; 2, 6; 6, 6$.

The way the sum is performed in eq. (2.26) is according to the notations in 2.5. The passage from integral over the thickness of the laminate to the sum over the thickness is dictated by the fact that the individual plies are very thin and the properties within each lamina are assumed constant in the thickness direction.

Finally, the equations of motion for a general case, applied to a thin plate made of laminated composite plies, using the CLT, are given as [1, 2] follows:

$$\frac{\partial N_x}{\partial x} + \frac{\partial N_{xy}}{\partial y} = I_1 \frac{\partial^2 u_0}{\partial t^2} - I_2 \frac{\partial^2}{\partial t^2}\left(\frac{\partial w_0}{\partial x}\right)$$
$$\frac{\partial N_{xy}}{\partial x} + \frac{\partial N_y}{\partial y} = I_1 \frac{\partial^2 v_0}{\partial t^2} - I_2 \frac{\partial^2}{\partial t^2}\left(\frac{\partial w_0}{\partial y}\right)$$
$$\frac{\partial^2 M_x}{\partial x^2} + 2\frac{\partial^2 M_{xy}}{\partial x \partial y} + \frac{\partial^2 M_y}{\partial y^2} + \frac{\partial}{\partial x}\left[N_{xx}\frac{\partial w_0}{\partial x} + N_{xy}\frac{\partial w_0}{\partial y}\right] \tag{2.27}$$
$$+ \frac{\partial}{\partial y}\left[N_{yy}\frac{\partial w_0}{\partial y} + N_{xy}\frac{\partial w_0}{\partial x}\right] = -p_z + I_1 \frac{\partial^2 w_0}{\partial t^2}$$
$$- I_3 \frac{\partial^2}{\partial t^2}\left(\frac{\partial^2 w_0}{\partial x^2} + \frac{\partial^2 w_0}{\partial y^2}\right) + I_2 \frac{\partial^2}{\partial t^2}\left(\frac{\partial w_0}{\partial x} + \frac{\partial w_0}{\partial y}\right)$$

where p_z is the load per unit area in the z direction[4] and the subscript 0 represents the values at the mid-plane of the cross section. N represents the in-plane loads, and the various moments of inertia, I_1, I_2 and I_3 are given by (ρ is the mass/unit length)

$$I_j = \int_{-h/2}^{h/2} \rho z^{j-1} dz; \quad j = 1, 2, 3 \tag{2.28}$$

4 Note that the z coordinate is normally used for the thickness direction, while x and y coordinates define the plate area.

To obtain the equations for a beam, one can use eq. (2.27), while all the derivations with respect to y are identically zero. This yields a 1D equation in the following form:

$$\frac{\partial^2 M_x}{\partial x^2} + \frac{\partial}{\partial x}\left(N_{xx}\frac{\partial w_0}{\partial x}\right) = -p_z + I_1\frac{\partial^2 w_0}{\partial t^2} - I_3\frac{\partial^4 w_0}{\partial t^2 \partial x^2} + I_2\frac{\partial^3 w_0}{\partial t^2 \partial x} \tag{2.29}$$

where N_x is the axial (in-plane, in the direction of the length of the beam) load. Remembering the relationship between transverse deflection, w, and the bending moment, we can rewrite eq. (2.29) in terms of w_0 only to yield

$$\begin{gathered} -D_{11}\frac{\partial^2 w_0}{\partial x^2} = M_x \Rightarrow \\ -\frac{\partial^2}{\partial x^2}\left(D_{11}\frac{\partial^2 w_0}{\partial x^2}\right) + \frac{\partial}{\partial x}\left(N_{xx}\frac{\partial w_0}{\partial x}\right) = -p_z + I_1\frac{\partial^2 w_0}{\partial t^2} - I_3\frac{\partial^4 w_0}{\partial t^2 \partial x^2} + I_2\frac{\partial^3 w_0}{\partial t^2 \partial x} \end{gathered} \tag{2.30}$$

with its associated boundary conditions:

$$\begin{gathered} \text{Geometric: specify either } w_0 \text{ or } \frac{\partial w_0}{\partial x} \\ \text{Natural: specify either } Q \equiv \frac{\partial M}{\partial x} \text{ or } M \end{gathered} \tag{2.31}$$

Typical boundary conditions normally used in the literature are in the following form:

$$\begin{gathered} \text{Simply supported: } w = 0 \text{ and } M = 0 \\ \text{Clamped: } w = 0 \text{ and } \frac{\partial w}{\partial x} = 0 \\ \text{Free: } Q \equiv \frac{\partial M}{\partial x} = 0 \text{ and } M = 0 \end{gathered} \tag{2.32}$$

The reader should be aware of thermal issues associated with the manufacturing of composite structures due to the differential thermal contraction during the post-curing phase and as a consequence of any temperature changes during the service life of the structure. This issue is caused by the relatively small axial thermal expansion coefficient of the modern reinforcing fibers (for carbon fibers it is even slightly negative), while the resin matrix has a large thermal coefficient. When cooling from a typical curing temperature, like 140 °C to room temperature, the fibers of the laminate composite will be in compression, while the matrix will show tension stresses [1]. Typical residual stresses due to thermal mismatch between the two components of the laminate are presented in Tab. 2.4.

Another important data for design are the experimental tension and compression strength as measured during various laboratory tests, as presented in Tab. 2.5 [1].

Finally, a table with a list of the main manufacturers of various composite materials is presented in Tab. 2.6.

Tab. 2.4: Typical thermal stresses in some common unidirectional composites (from [1]).

Matrix	Fiber	% Fiber volume, V_f	Temperature range ΔT (K)	Fiber residual stress (MPa)	Matrix residual stress (MPa)
Epoxy (high *T* cure)	T300 carbon	65	120	−19	36
Epoxy (low *T* cure)	E glass	65	100	−15	28
Epoxy (low *T* cure)	Kevlar-49	65	100	−16	30
Borosilicate glass	T300 carbon	50	520	−93	93
CAS* glass-ceramic	Nicalon SiC	40	1,000	−186	124

*CAS, $CaO–Al_2O_3–SiO_2$.

Tab. 2.5: Typical experimental tension and compression strengths for common composite materials (from [1]).

Material	Lay-up	% Fiber volume, V_f	Tensile strength, σ_t (GPa)	Compression strength, σ_c (GPa)	Ratio σ_c/σ_t
GRP	ud (unidirectional)	60	1.3	1.1	0.85
CFRP	ud (unidirectional)	60	2.0	1.1	0.55
KFRP	ud (unidirectional)	60	1.0	0.4	0.40
HTA/913 (CFRP)	$\left[\left(\pm45^0, 0_2^0\right)_2\right]_S$	65	1.27	0.97	0.77
T800/924 (CFRP)	$\left[\left(\pm45^0, 0_2^0\right)_2\right]_S$	65	1.42	0.90	0.63
T800/5245(CFRP)	$\left[\left(\pm45^0, 0_2^0\right)_2\right]_S$	65	1.67	0.88	0.53
SiC/CAS (CMC*)	ud (unidirectional)	37	334	1,360	4.07
SiC/CAS (CMC)	$\left[0^0, 90^0\right]_{3S}$	37	210	463	2.20

*CMC, ceramic matrix composites.

Tab. 2.6: List of the main manufacturers of various composite materials and resins.

Types of composite	Company	Company website
Thermoplastic composites	Milliken Tegris	tegris.milliken.com
Thermoplastic composites	Polystrand, Inc.	www.polystrand.com
Nonwoven fabrics (PolyWeb™) and foam	Wm. T. Burnett & Co.	www.williamtburnett.com
Thermoplastic composites	Schappe Techniques	www.schappe.com
Thermoplastic composites	TechFiber	www.fiber-tech.net
Thermoplastic composites	TenCate	www.tencate.com
Thermoplastic composites	Thercom	www.thercom.com
Thermoplastic composites	Vectorply	www.vectorply.com

Tab. 2.6 (continued)

Types of composite	Company	Company website
Composite materials: resins and fibers	SF composites	www.sf-composites.com
Formulation and manufacture of epoxy-based systems	SICOMIN	www.sicomin.com
Composite materials + resin, composite laminates	Lamiflex SPA	www.lamiflex.il
Composite materials + polyester	AMP Composite	www.amp-composite.il
Infusion, pultrusion, wet lay-up, prepreg, filament winding	Applied Poleramic Inc.	www.appliedpoleramic.com
Epoxy and polyurethane	Endurance Technologies	www.epoxi.com
Composite materials	Gurit	www.gurit.com
Advanced thermoset resins	Huntsman Advanced Materials	www.huntsman.com/advanced_materials/a/Home
Advanced thermoset resins	Lattice Composites	www.latticecomposites.com
Kevlar	DuPont™ Kevlar®	www.dupont.com/products-and-services/fabrics-fibers-nonwovens/fibers/brands/kevlar.html
UHMWPE – ultra-high-molecular-weight, high-performance polyethylene material	DuPont™ Tenslyon™	www.dupont.com
Innegra™ HMPP (polypropylene), high-performance fiber	Innegra Technologies	www.innegratech.com
Spread tow fabrics	TeXtreme	www.textreme.com/b2b
Adhesives and sealants	3M	solutions.3m.com
Prepreg and resins	Axiom Materials Inc.	www.axiommaterials.com
Fabrics, resins, and composite materials	Barrday Advanced Materials Solutions	www.barrday.com
Carbon prepreg	Hankuk Carbon Co., Ltd.	www.hcarbon.com/eng/product/overview.asp
Carbon fibers and prepregs	Hexcel®	www.hexcel.com/Products/Industries/ICarbon-Fiber

Tab. 2.6 (continued)

Types of composite	Company	Company website
Prepregs and compounds	Pacific Coast Composites	www.pccomposites.com
Prepregs and compounds	Quantum Composites	www.quantumcomposites.com

2.6 First-order shear deformation theory

Unlike isotropic materials, the ratio between the shear and the bending in orthotropic material, like laminated composite structures, is not negligible. The need to remove the somehow restricting assumptions from the CLT which, as stated above, are based on Kirchhoff–Love plate theory [3–5], like neglecting the influence of the shear strains and the fact that a plane before deformation remains plane after the deformation, led to derivation of more advanced bending theories for plates, like Mindlin theory of plates [6–8], which includes in-plane shear strains and is an extension of Kirchhoff–Love plate theory incorporating first-order shear effects.

Mindlin's theory assumes that there is a linear variation of displacement across the plate thickness, but the plate thickness does not change during deformation. An additional assumption is that the normal stress through the thickness is ignored, an assumption that is also called the plane stress condition. The Mindlin theory is often called the FOSDT (first-order shear deformation theory) of plates, and its application to composite materials is next presented. Under the assumptions and restrictions of Mindlin's theory (which has a similarity to Timoshenko's theory for beams [9–12]), the displacement field has five unknowns (u_0, v_0, w_0 – the displacements of the mid-plane in the x, y and z directions, respectively, and ϕ_x, ϕ_y – the rotations due to shear about x and y directions, respectively) and is given by (see also Fig. 2.7, which is similar to 2.4, but for FOSDT approach)

$$\begin{aligned} u(x,y,z,t) &= u_0(x,y,t) + z\phi_x(x,y,t) \\ v(x,y,z,t) &= v_0(x,y,t) + z\phi_y(x,y,t) \\ w(x,y,z,t) &= w_0(x,y,t) \end{aligned} \tag{2.33}$$

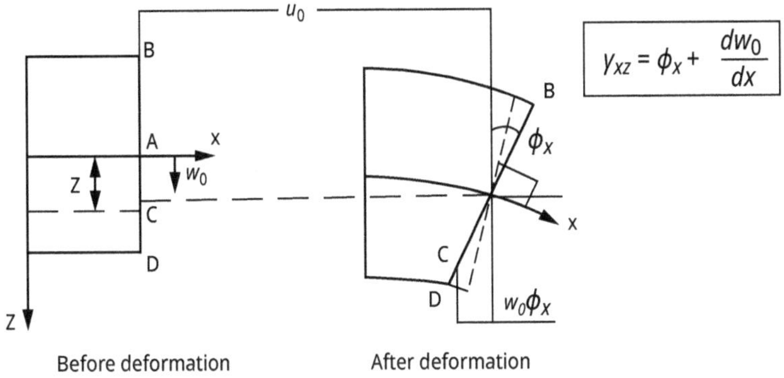

Fig. 2.7: The plate cross section before and after the deformation (FOSDT approach).

The associated strains [13][5] assuming nonlinear terms are

$$
\begin{aligned}
\varepsilon_x &= \frac{\partial u_0}{\partial x} + \frac{1}{2}\left(\frac{\partial w_0}{\partial x}\right)^2 + z\frac{\partial \phi_x}{\partial x} \\
\varepsilon_x &= \frac{\partial v_0}{\partial y} + \frac{1}{2}\left(\frac{\partial w_0}{\partial y}\right)^2 + z\frac{\partial \phi_y}{\partial y} \\
\varepsilon_z &= 0, \qquad \gamma_{xz} = \frac{\partial w_0}{\partial x} + \phi_x, \qquad \gamma_{yz} = \frac{\partial w_0}{\partial y} + \phi_y \\
\gamma_{xy} &= \left(\frac{\partial u_0}{\partial y} + \frac{\partial v_0}{\partial x} + \frac{\partial^2 w_0}{\partial x \partial y}\right) + z\left(\frac{\partial \phi_x}{\partial y} + \frac{\partial \phi_y}{\partial x}\right)
\end{aligned}
\tag{2.34}
$$

Multiplying eq. (2.34) by the stiffness matrix $[\bar{Q}]$ and integrating through the thickness of the laminate yield the force and moment resultants (as in eq. (2.23)), with two additional terms, the shear resultants, being defined by

$$
\begin{Bmatrix} Q_x \\ Q_y \end{Bmatrix} = k \int_{-h/2}^{h/2} \begin{Bmatrix} \tau_{xz} \\ \tau_{yz} \end{Bmatrix} dz \tag{2.35}
$$

where κ is called the shear correction coefficient and is defined by the ratio between the shear strain energies calculated by the actual shear distribution and the constant

5 Note that the assumption of constant shear strains across the height of the laminate is a rough approximation of the true strain distribution, which is at least quadratic through the thickness. However, although the rough approximation, the results of the application of Mindlin's plate theory present very good results when compared with experimental ones.

distribution assumed in the FOSDT theory. The value of κ is taken as 5/6 for a rectangular cross section.[6]

The equations of motion will then have the following form:

$$\begin{aligned}
&\frac{\partial N_x}{\partial x}+\frac{\partial N_{xy}}{\partial y}=I_1\frac{\partial^2 u_0}{\partial t^2}+I_2\frac{\partial^2 \phi_x}{\partial t^2}\\
&\frac{\partial N_{xy}}{\partial x}+\frac{\partial N_y}{\partial y}=I_1\frac{\partial^2 v_0}{\partial t^2}+I_2\frac{\partial^2 \phi_y}{\partial t^2}\\
&\frac{\partial Q_x}{\partial x}+\frac{\partial Q_y}{\partial y}+\frac{\partial}{\partial x}\left[N_{xx}\frac{\partial w_0}{\partial x}+N_{xy}\frac{\partial w_0}{\partial y}\right]\\
&\quad+\frac{\partial}{\partial y}\left[N_{yy}\frac{\partial w_0}{\partial y}+N_{xy}\frac{\partial w_0}{\partial x}\right]=-p_z+I_1\frac{\partial^2 w_0}{\partial t^2}\\
&\frac{\partial M_x}{\partial x}+\frac{\partial M_{xy}}{\partial y}-Q_x=I_3\frac{\partial^2 \phi_x}{\partial t^2}+I_1\frac{\partial^2 u_0}{\partial t^2}\\
&\frac{\partial M_{xy}}{\partial x}+\frac{\partial M_y}{\partial y}-Q_y=I_3\frac{\partial^2 \phi_y}{\partial t^2}+I_1\frac{\partial^2 v_0}{\partial t^2}
\end{aligned}\tag{2.36}$$

where p_z is the load per unit area in the z direction.[7] N_{xx}, N_{yy} and N_{xy} represent the in-plane loads, and the various moments of inertia, I_1, I_2 and I_3, are given by (ρ is the mass/unit length) eq. (2.28).

One should note that in addition to eq. (2.25), which describes the resultants of the force and the moment as a function of the stiffness coefficients A_{ij}, B_{ij} and D_{ij}, the shear resultants Q_x and Q_y are defined as

$$\begin{Bmatrix} Q_y \\ Q_x \end{Bmatrix}=\kappa\begin{bmatrix} A_{44} & A_{45} \\ A_{45} & A_{55} \end{bmatrix}\begin{Bmatrix} \dfrac{\partial w_0}{\partial y}+\phi_y \\ \dfrac{\partial w_0}{\partial x}+\phi_x \end{Bmatrix}\tag{2.37}$$

6 The accurate value is $\kappa=\frac{10(1+\upsilon)}{12+11\upsilon}$ for a rectangular cross section and $\kappa=\frac{6(1+\upsilon)}{7+6\upsilon}$ for a solid circular cross section.

7 Note that the coordinate z is normally used for the thickness direction, while x and y coordinates define the plate area.

where

$$A_{44} \equiv \kappa \int_{-h/2}^{h/2} \bar{Q}_{44}^{k} dz = \kappa \sum_{k=1}^{n} \bar{Q}_{44}^{k} (h_k - h_{k-1})$$
$$A_{45} \equiv \kappa \int_{-h/2}^{h/2} \bar{Q}_{45}^{k} dz = \kappa \sum_{k=1}^{n} \bar{Q}_{45}^{k} (h_k - h_{k-1}) \tag{2.38}$$
$$A_{55} \equiv \kappa \int_{-h/2}^{h/2} \bar{Q}_{55}^{k} dz = \kappa \sum_{k=1}^{n} \bar{Q}_{55}^{k} (h_k - h_{k-1})$$

and

$$\bar{Q}_{44} = Q_{44} \cos^2\theta + Q_{55} \sin^2\theta$$
$$\bar{Q}_{45} = (Q_{55} - Q_{44}) \cos\theta \sin\theta \tag{2.39}$$
$$\bar{Q}_{55} = Q_{44} \sin^2\theta + Q_{55} \cos^2\theta$$
$$\text{where } Q_{44} = G_{23} \text{ and } Q_{55} = G_{13}$$

Substituting the resultants defined in terms of the five unknown displacements (u_0, v_0, w_0, ϕ_x, ϕ_y), we get [13] five differential equations for the five unknown displacements:

$$\begin{aligned}
&A_{11}\left[\frac{\partial^2 u_0}{\partial x^2} + \frac{\partial^3 w_0}{\partial x^3}\right] + A_{12}\left[\frac{\partial^2 v_0}{\partial x \partial y} + \frac{\partial^3 w_0}{\partial x \partial y^2}\right] + A_{16}\left[2\frac{\partial^2 u_0}{\partial x \partial y} + \frac{\partial^2 v_0}{\partial x^2} + 3\frac{\partial^3 w_0}{\partial x^2 \partial y}\right] + \\
&A_{26}\left[\frac{\partial^2 v_0}{\partial y^2} + \frac{\partial^3 w_0}{\partial y^3}\right] + A_{66}\left[\frac{\partial^2 u_0}{\partial y^2} + \frac{\partial^2 v_0}{\partial x \partial y} + 2\frac{\partial^3 w_0}{\partial x \partial y^2}\right] + \\
&B_{11}\frac{\partial^2 \phi_x}{\partial x^2} + B_{12}\frac{\partial^2 \phi_y}{\partial x \partial y} + B_{16}\left[2\frac{\partial^2 \phi_x}{\partial x \partial y} + \frac{\partial^2 \phi_y}{\partial x^2}\right] + B_{26}\frac{\partial^2 \phi_y}{\partial y^2} + \\
&B_{66}\left[\frac{\partial^2 \phi_x}{\partial y^2} + \frac{\partial^2 \phi_y}{\partial x \partial y}\right] = I_1\frac{\partial^2 u_0}{\partial t^2} + I_2\frac{\partial^2 \phi_x}{\partial t^2}
\end{aligned} \tag{2.40}$$

$$\begin{aligned}
&A_{22}\left[\frac{\partial^2 v_0}{\partial y^2} + \frac{\partial^3 w_0}{\partial y^3}\right] + A_{12}\left[\frac{\partial^2 u_0}{\partial x \partial y} + \frac{\partial^3 w_0}{\partial x^2 \partial y}\right] + A_{16}\left[\frac{\partial^2 u_0}{\partial x^2} + \frac{\partial^3 w_0}{\partial x^3}\right] + \\
&A_{26}\left[\frac{\partial^2 u_0}{\partial y^2} + 2\frac{\partial^2 v_0}{\partial x \partial y} + 3\frac{\partial^3 w_0}{\partial x \partial y^2}\right] + A_{66}\left[\frac{\partial^2 u_0}{\partial x \partial y} + \frac{\partial^2 v_0}{\partial x^2} + 2\frac{\partial^3 w_0}{\partial x^2 \partial y}\right] + \\
&B_{22}\frac{\partial^2 \phi_y}{\partial y^2} + B_{12}\frac{\partial^2 \phi_x}{\partial x \partial y} + B_{16}\frac{\partial^2 \phi_x}{\partial x^2} + B_{26}\left[\frac{\partial^2 \phi_x}{\partial y^2} + 2\frac{\partial^2 \phi_y}{\partial x \partial y}\right] + \\
&B_{66}\left[\frac{\partial^2 \phi_y}{\partial x^2} + \frac{\partial^2 \phi_x}{\partial x \partial y}\right] = I_1\frac{\partial^2 v_0}{\partial t^2} + I_2\frac{\partial^2 \phi_y}{\partial t^2}
\end{aligned} \tag{2.41}$$

$$\kappa A_{55}\left[\frac{\partial^2 w_0}{\partial x^2}+\frac{\partial \phi_x}{\partial x}\right]+\kappa A_{45}\left[2\frac{\partial^2 w_0}{\partial x\partial y}+\frac{\partial \phi_y}{\partial x}+\frac{\partial \phi_x}{\partial y}\right]+\kappa A_{44}\left[\frac{\partial^2 w_0}{\partial y^2}+\frac{\partial \phi_y}{\partial y}\right]+$$
$$\frac{\partial}{\partial x}\left[N_{xx}\frac{\partial w_0}{\partial x}+N_{xy}\frac{\partial w_0}{\partial y}\right]+\frac{\partial}{\partial y}\left[N_{yy}\frac{\partial w_0}{\partial y}+N_{xy}\frac{\partial w_0}{\partial x}\right]=-p_z+I_1\frac{\partial^2 w_0}{\partial t^2} \tag{2.42}$$

$$\begin{aligned}
&B_{11}\left[\frac{\partial^2 u_0}{\partial x^2}+\frac{\partial^3 w_0}{\partial x^3}\right]+B_{12}\left[\frac{\partial^2 v_0}{\partial x\partial y}+\frac{\partial^3 w_0}{\partial x\partial y^2}\right]+\\
&B_{16}\left[2\frac{\partial^2 u_0}{\partial x\partial y}+\frac{\partial^2 v_0}{\partial x^2}+3\frac{\partial^3 w_0}{\partial x^2\partial y}\right]+B_{26}\left[\frac{\partial^2 v_0}{\partial y^2}+\frac{\partial^3 w_0}{\partial y^3}\right]+\\
&B_{66}\left[\frac{\partial^2 u_0}{\partial y^2}+\frac{\partial^2 v_0}{\partial x\partial y}+2\frac{\partial^3 w_0}{\partial x\partial y^2}\right]+D_{11}\frac{\partial^2 \phi_x}{\partial x^2}+D_{12}\frac{\partial^2 \phi_y}{\partial x\partial y}+\\
&D_{16}\left[2\frac{\partial^2 \phi_x}{\partial x\partial y}+\frac{\partial^2 \phi_y}{\partial x^2}\right]+D_{26}\frac{\partial^2 \phi_y}{\partial y^2}+D_{66}\left[\frac{\partial^2 \phi_x}{\partial y^2}+\frac{\partial^2 \phi_y}{\partial x\partial y}\right]-\\
&\kappa A_{55}\left[\frac{\partial w_0}{\partial x}+\phi_x\right]-\kappa A_{45}\left[\frac{\partial w_0}{\partial y}+\phi_y\right]=I_3\frac{\partial^2 u_0}{\partial t^2}+I_3\frac{\partial^2 \phi_x}{\partial t^2}
\end{aligned} \tag{2.43}$$

$$\begin{aligned}
&B_{22}\left[\frac{\partial^2 v_0}{\partial y^2}+\frac{\partial^3 w_0}{\partial y^3}\right]+B_{12}\left[\frac{\partial^2 u_0}{\partial x\partial y}+\frac{\partial^3 w_0}{\partial x^2\partial y}\right]+\\
&B_{16}\left[\frac{\partial^2 u_0}{\partial x^2}+\frac{\partial^3 w_0}{\partial x^3}\right]+B_{26}\left[\frac{\partial^2 u_0}{\partial y^2}+2\frac{\partial^2 v_0}{\partial x\partial y}+3\frac{\partial^3 w_0}{\partial x\partial y^2}\right]\\
&B_{66}\left[\frac{\partial^2 u_0}{\partial x\partial y}+\frac{\partial^2 v_0}{\partial x^2}+2\frac{\partial^3 w_0}{\partial x^2\partial y}\right]+D_{22}\frac{\partial^2 \phi_y}{\partial y^2}+D_{12}\frac{\partial^2 \phi_x}{\partial x\partial y}+\\
&D_{16}\frac{\partial^2 \phi_x}{\partial x^2}+D_{26}\left[\frac{\partial^2 \phi_x}{\partial y^2}+2\frac{\partial^2 \phi_y}{\partial x\partial y}\right]+D_{66}\left[\frac{\partial^2 \phi_y}{\partial x^2}+\frac{\partial^2 \phi_x}{\partial x\partial y}\right]-\\
&\kappa A_{44}\left[\frac{\partial w_0}{\partial y}+\phi_y\right]-\kappa A_{45}\left[\frac{\partial w_0}{\partial x}+\phi_x\right]=I_1\frac{\partial^2 v_0}{\partial t^2}+I_2\frac{\partial^2 \phi_y}{\partial t^2}
\end{aligned} \tag{2.44}$$

To solve the five differential equations, 10 boundary conditions should be supplied in the form of geometric and natural boundary conditions.[8]

To obtain the equations of motion for a beam, using FSDT, presented before for a plate, one should assume that all the derivations in the y direction should vanish, and v and ϕ_y should be identically zero. This yields, for the general case, three coupled equations of motion having the following form (assuming constant properties along the beam):

8 For further discussion about the types of boundary conditions to be imposed, the reader is referred to Ref. [13].

$$A_{11}\frac{\partial^2 u_0}{\partial x^2}+B_{11}\frac{\partial^2 \phi_x}{\partial x^2}=I_1\frac{\partial^2 u_0}{\partial t^2}+I_2\frac{\partial^2 \phi_x}{\partial t^2}$$
$$B_{11}\frac{\partial^2 u_0}{\partial x^2}+D_{11}\frac{\partial^2 \phi_x}{\partial x^2}-\kappa A_{55}\left[\frac{\partial w_0}{\partial x}+\phi_x\right]=I_3\frac{\partial^2 u_0}{\partial t^2}+I_3\frac{\partial^2 \phi_x}{\partial t^2} \tag{2.45}$$
$$\kappa A_{55}\left[\frac{\partial^2 w_0}{\partial x^2}+\frac{\partial \phi_x}{\partial x}\right]+N_{xx}\frac{\partial^2 w_0}{\partial x}=-p_z+I_1\frac{\partial^2 w_0}{\partial t^2}$$

with the following boundary conditions:

$$A_{11}\frac{\partial u_0}{\partial x}+B_{11}\frac{\partial \phi_x}{\partial x}=-N_{xx}\quad \underline{\text{or}}\quad u_0=0$$
$$A_{55}\left(\phi_x+\frac{\partial w_0}{\partial x}\right)-N_{xx}\frac{\partial w_0}{\partial x}=0\quad \underline{\text{or}}\quad w_0=0 \tag{2.46}$$
$$B_{11}\frac{\partial u_0}{\partial x}+D_{11}\frac{\partial \phi_x}{\partial x}=0\quad \underline{\text{or}}\quad \phi_x=0$$

The reader is referred to Refs. [14–20], as typical sources for solving buckling and natural frequencies of beams using the FSDT approach.

2.7 Higher-order theories

The need for higher-order shear deformation theories stems from the fact that although the FOSDT approach presents accurate results, it does not fulfill the shear-free boundary conditions on the top and bottom sides of the beam or a plate. One should remember that the FOSDT approach would need far less computer efforts as compared with any higher-order theories, which are known as demanding large computer memory.

One of the higher-order shear deformation theories was proposed by Reddy [13] and involves a third-order shear deformation theory (TOSDT), which will be presented next.

Based on the kinematics of the problem (Fig. 2.8), the displacement field has the following components:

$$u(x,y,z,t)=u_0(x,y,t)+z\alpha_x(x,y,t)+z^2\beta_x(x,y,t)+z^3\delta_x(x,y,t)$$
$$v(x,y,z,t)=v_0(x,y,t)+z\alpha_y(x,y,t)+z^2\beta_y(x,y,t)+z^3\delta_y(x,y,t) \tag{2.47}$$
$$w(x,y,z,t)=w_0(x,y,t)$$

The various terms in eq. (2.47), like α_x, α_y, β_x, β_y, δ_x and δ_y, are functions to be determined, having the following values at $z = 0$:

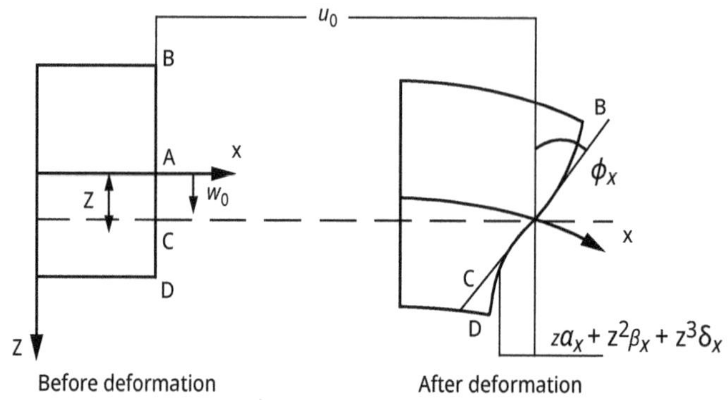

Fig. 2.8: The plate cross section before and after the deformation (TOSDT approach).

$$\alpha_x = \left[\frac{\partial u}{\partial z}\right], \qquad \alpha_y = \left[\frac{\partial v}{\partial z}\right], \qquad \beta_x = \frac{1}{2}\left[\frac{\partial^2 u}{\partial z^2}\right]$$
$$\beta_y = \frac{1}{2}\left[\frac{\partial^2 v}{\partial z^2}\right], \qquad \delta_x = \frac{1}{6}\left[\frac{\partial^3 u}{\partial z^3}\right], \qquad \delta_y = \frac{1}{6}\left[\frac{\partial^3 v}{\partial z^3}\right] \tag{2.48}$$

Accordingly, nine independent unknowns are to be found from nine second-order partial differential equations. However, the number of independent unknowns can be reduced for certain boundary conditions, like traction-free boundary conditions on the top and bottom layers of the plate [13], namely[9]

$$\tau_{xz}\left(x, y, +\frac{h}{2}, t\right) = 0, \quad \tau_{xz}\left(x, y, -\frac{h}{2}, t\right) = 0$$
$$\tau_{yz}\left(x, y, +\frac{h}{2}, t\right) = 0, \quad \tau_{yz}\left(x, y, -\frac{h}{2}, t\right) = 0 \tag{2.49}$$

For arbitrary values of Q_{55}, Q_{45} and Q_{44} (see also eqs. (2.37)–(2.39)), one can show that the boundary equations in eq. (2.49) can be written in terms of the shear strains (see also [13]), using the following expressions:

$$\gamma_{xz}\left(x, y, +\frac{h}{2}, t\right) = 0; \quad \gamma_{xz}\left(x, y, -\frac{h}{2}, t\right) = 0$$
$$\gamma_{yz}\left(x, y, +\frac{h}{2}, t\right) = 0; \quad \gamma_{yz}\left(x, y, -\frac{h}{2}, t\right) = 0 \tag{2.50}$$

Performing the necessary derivations to express the shear strains using the assumed strain field (eq. (2.47)), one gets

9 h being the total height of the beam (or plate), x-axis is along the beam's (or plate's) length, while y and z are perpendicular to it.

$$\begin{aligned}\gamma_{xz} &\equiv \frac{\partial w}{\partial x} + \frac{\partial u}{\partial z} = \frac{\partial w_0}{\partial x} + \alpha_x + 2z\beta_x + 3z^2\delta_x \\ \gamma_{yz} &\equiv \frac{\partial w}{\partial y} + \frac{\partial v}{\partial z} = \frac{\partial w_0}{\partial y} + \alpha_y + 2z\beta_y + 3z^2\delta_y\end{aligned} \tag{2.51}$$

Substituting the boundary conditions from eq. (2.50) into eq. (2.51), we get the following four equations:

$$\begin{aligned}@z = +\frac{h}{2}\ \gamma_{xz} &= \frac{\partial w_0}{\partial x} + \alpha_x + h\beta_x + \frac{3h^2}{4}\delta_x = 0, \quad \gamma_{yz} = \frac{\partial w_0}{\partial y} + \alpha_y + h\beta_y + \frac{3h^2}{4}\delta_y = 0 \\ @z = -\frac{h}{2}\ \gamma_{xz} &= \frac{\partial w_0}{\partial x} + \alpha_x - h\beta_x + \frac{3h^2}{4}\delta_x = 0, \quad \gamma_{yz} = \frac{\partial w_0}{\partial y} + \alpha_y - h\beta_y + \frac{3h^2}{4}\delta_y = 0\end{aligned} \tag{2.52}$$

Solving eq. (2.52) yields the following expressions:

$$\begin{aligned}\beta_x &= 0, \qquad \delta_x = -\frac{4}{3h^2}\left(\alpha_x + \frac{\partial w_o}{\partial x}\right) \\ \beta_y &= 0, \qquad \delta_y = -\frac{4}{3h^2}\left(\alpha_y + \frac{\partial w_o}{\partial y}\right)\end{aligned} \tag{2.53}$$

Back-substituting the expression in eq. (2.53) into eq. (2.47) provides the strain field for the posed problem:

$$\begin{aligned}u(x,y,z,t) &= u_0(x,y,t) + z\alpha_x(x,y,t) - z^3\frac{4}{3h^2}\left[\alpha_x(x,y,t) + \frac{\partial w_0(x,y,t)}{\partial x}\right] \\ v(x,y,z,t) &= v_0(x,y,t) + z\alpha_y(x,y,t) - z^3\frac{4}{3h^2}\left[\alpha_y(x,y,t) + \frac{\partial w_0(x,y,t)}{\partial y}\right] \\ w(x,y,z,t) &= w_0(x,y,t)\end{aligned} \tag{2.54}$$

One should notice that the number of unknowns had been reduced to only 5, similar to the case of FSTD approach being applied to a rectangular plate.

For the case of small strains and moderate rotations, the strain–displacement relations can be shown to be

$$\begin{aligned}\varepsilon_{xx} &= \frac{\partial u}{\partial x} + \frac{1}{2}\left(\frac{\partial w}{\partial x}\right)^2, \quad \varepsilon_{yy} = \frac{\partial v}{\partial y} + \frac{1}{2}\left(\frac{\partial w}{\partial y}\right)^2, \quad \varepsilon_{zz} = \frac{\partial w}{\partial z} \\ \varepsilon_{xy} &= \frac{1}{2}\left(\frac{\partial u}{\partial y} + \frac{\partial v}{\partial x} + \frac{\partial^2 w}{\partial x \partial y}\right), \quad \varepsilon_{xz} = \frac{1}{2}\left(\frac{\partial u}{\partial z} + \frac{\partial w}{\partial x}\right), \quad \varepsilon_{yz} = \frac{1}{2}\left(\frac{\partial v}{\partial z} + \frac{\partial w}{\partial y}\right)\end{aligned} \tag{2.55}$$

Substituting the displacement field, defined by eq. (2.54), into the nonlinear strain–displacement equations, eq. (2.55), yields the following expressions:

$$\begin{Bmatrix} \varepsilon_{xx} \\ \varepsilon_{yy} \\ \gamma_{xy} \end{Bmatrix} = \begin{Bmatrix} \dfrac{\partial u_0}{\partial x} + \dfrac{1}{2}\left(\dfrac{\partial w_0}{\partial x}\right)^2 \\ \dfrac{\partial v_0}{\partial y} + \dfrac{1}{2}\left(\dfrac{\partial w_0}{\partial y}\right)^2 \\ \dfrac{\partial u_0}{\partial y} + \dfrac{\partial v_0}{\partial x} + \dfrac{\partial^2 w_0}{\partial x \partial y} \end{Bmatrix} + z \begin{Bmatrix} \dfrac{\partial \alpha_x}{\partial x} \\ \dfrac{\partial \alpha_y}{\partial y} \\ \dfrac{\partial \alpha_x}{\partial y} + \dfrac{\partial \alpha_y}{\partial x} \end{Bmatrix} - z^3 \left(\dfrac{4}{3h^2}\right) \begin{Bmatrix} \dfrac{\partial \alpha_x}{\partial x} + \dfrac{\partial^2 w_0}{\partial x^2} \\ \dfrac{\partial \alpha_y}{\partial y} + \dfrac{\partial^2 w_0}{\partial y^2} \\ \dfrac{\partial \alpha_x}{\partial y} + \dfrac{\partial \alpha_y}{\partial x} + 2\dfrac{\partial^2 w_0}{\partial x \partial y} \end{Bmatrix}$$

$$\begin{Bmatrix} \gamma_{xz} \\ \gamma_{yz} \end{Bmatrix} = \begin{Bmatrix} \alpha_x + \dfrac{\partial w_0}{\partial x} \\ \alpha_y + \dfrac{\partial w_0}{\partial y} \end{Bmatrix} - z^2 \left(\dfrac{4}{h^2}\right) \begin{Bmatrix} \alpha_x + \dfrac{\partial w_0}{\partial x} \\ \alpha_y + \dfrac{\partial w_0}{\partial y} \end{Bmatrix} \quad \text{and} \quad \varepsilon_{zz} = 0 \tag{2.56}$$

where

$$\gamma_{xy} \equiv 2\varepsilon_{xy}, \quad \gamma_{xz} \equiv 2\varepsilon_{xz}, \quad \gamma_{yz} \equiv 2\varepsilon_{yz} \tag{2.57}$$

The stress resultant–strain relationship can be shown to have the following form (see also [13]):

$$\begin{Bmatrix} \{N\} \\ \{M\} \\ \{T\} \end{Bmatrix} = \begin{bmatrix} [A] & [B] & [E] \\ [B] & [D] & [F] \\ [E] & [F] & [G] \end{bmatrix} \begin{Bmatrix} \{\varepsilon^{(a)}\} \\ \{\varepsilon^{(b)}\} \\ \{\varepsilon^{(c)}\} \end{Bmatrix} \tag{2.58}$$

$$\begin{Bmatrix} \{Q\} \\ \{S\} \end{Bmatrix} = \begin{bmatrix} [A] & [D] \\ [D] & [F] \end{bmatrix} \begin{Bmatrix} \{\gamma^{(a)}\} \\ \{\gamma^{(b)}\} \end{Bmatrix} \tag{2.59}$$

where the terms $\{T\}$ and $\{S\}$ are higher-order stress resultants, stemmed from the present TOSDT. The various matrix terms for eq. (2.58) are defined in the following way:

$$A_{ij} = \sum_{k=1}^{N} \int_{z_k}^{z_{k+1}} \bar{Q}_{ij} dz = \sum_{k=1}^{N} \bar{Q}_{ij}^{(k)} [z_{k+1} - z_k]$$

$$B_{ij} = \sum_{k=1}^{N} \int_{z_k}^{z_{k+1}} \bar{Q}_{ij} z dz = \frac{1}{2} \sum_{k=1}^{N} \bar{Q}_{ij}^{(k)} \left[(z_{k+1})^2 - (z_k)^2 \right]$$

$$D_{ij} = \sum_{k=1}^{N} \int_{z_k}^{z_{k+1}} \bar{Q}_{ij} z^2 dz = \frac{1}{3} \sum_{k=1}^{N} \bar{Q}_{ij}^{(k)} \left[(z_{k+1})^3 - (z_k)^3 \right] \quad \text{while} \quad i,j = 1,2,6$$

$$E_{ij} = \sum_{k=1}^{N} \int_{z_k}^{z_{k+1}} \bar{Q}_{ij} z^3 dz = \frac{1}{4} \sum_{k=1}^{N} \bar{Q}_{ij}^{(k)} \left[(z_{k+1})^4 - (z_k)^4 \right]$$

$$F_{ij} = \sum_{k=1}^{N} \int_{z_k}^{z_{k+1}} \bar{Q}_{ij} z^4 dz = \frac{1}{5} \sum_{k=1}^{N} \bar{Q}_{ij}^{(k)} \left[(z_{k+1})^5 - (z_k)^5 \right]$$

$$G_{ij} = \sum_{k=1}^{N} \int_{z_k}^{z_{k+1}} \bar{Q}_{ij} z^6 dz = \frac{1}{7} \sum_{k=1}^{N} \bar{Q}_{ij}^{(k)} \left[(z_{k+1})^7 - (z_k)^7 \right] \tag{2.60}$$

while for eq. (2.59) their definition is

$$A_{lm} = \sum_{k=1}^{N} \int_{z_k}^{z_{k+1}} \bar{Q}_{lm} dz = \sum_{k=1}^{N} \bar{Q}_{lm}^{(k)} [z_{k+1} - z_k]$$

$$D_{lm} = \sum_{k=1}^{N} \int_{z_k}^{z_{k+1}} \bar{Q}_{lm} z^2 dz = \frac{1}{3} \sum_{k=1}^{N} \bar{Q}_{lm}^{(k)} \left[(z_{k+1})^3 - (z_k)^3 \right] \quad \text{while} \quad l,m = 4,5 \tag{2.61}$$

$$F_{lm} = \sum_{k=1}^{N} \int_{z_k}^{z_{k+1}} \bar{Q}_{lm} z^4 dz = \frac{1}{5} \sum_{k=1}^{N} \bar{Q}_{lm}^{(k)} \left[(z_{k+1})^5 - (z_k)^5 \right]$$

where

$$\left\{\varepsilon^{(a)}\right\} \equiv \begin{Bmatrix} \dfrac{\partial u_0}{\partial x} + \dfrac{1}{2}\left(\dfrac{\partial w_0}{\partial x}\right)^2 \\ \dfrac{\partial v_0}{\partial y} + \dfrac{1}{2}\left(\dfrac{\partial w_0}{\partial y}\right)^2 \\ \dfrac{\partial u_0}{\partial y} + \dfrac{\partial v_0}{\partial x} + \dfrac{\partial^2 w_0}{\partial x \partial y} \end{Bmatrix}, \left\{\varepsilon^{(b)}\right\} \equiv \begin{Bmatrix} \dfrac{\partial \alpha_x}{\partial x} \\ \dfrac{\partial \alpha_y}{\partial y} \\ \dfrac{\partial \alpha_x}{\partial y} + \dfrac{\partial \alpha_y}{\partial x} \end{Bmatrix}$$

$$\left\{\varepsilon^{(c)}\right\} \equiv -\left(\frac{4}{3h^2}\right) \begin{Bmatrix} \dfrac{\partial \alpha_x}{\partial x} + \dfrac{\partial^2 w_0}{\partial x^2} \\ \dfrac{\partial \alpha_y}{\partial y} + \dfrac{\partial^2 w_0}{\partial y^2} \\ \dfrac{\partial \alpha_x}{\partial y} + \dfrac{\partial \alpha_y}{\partial x} + 2\dfrac{\partial^2 w_0}{\partial x \partial y} \end{Bmatrix} \tag{2.62}$$

$$\left\{\gamma^{(a)}\right\} \equiv \begin{Bmatrix} \alpha_x + \dfrac{\partial w_0}{\partial x} \\ \alpha_y + \dfrac{\partial w_0}{\partial y} \end{Bmatrix}, \left\{\gamma^{(b)}\right\} \equiv -\left(\frac{4}{h^2}\right) \begin{Bmatrix} \alpha_x + \dfrac{\partial w_0}{\partial x} \\ \alpha_y + \dfrac{\partial w_0}{\partial y} \end{Bmatrix}$$

Finally, the equations of motion, expressed in stress resultants, for a TOSDT can be written as [13]

$$\frac{\partial N_{xx}}{\partial x} + \frac{\partial N_{xy}}{\partial y} = I_0 \ddot{u}_0 + \left(I_1 - \frac{4}{3h^2} I_3\right)\ddot{\alpha}_x - \frac{4}{3h^2} I_3 \frac{\partial \ddot{w}_0}{\partial x} \tag{2.63a}$$

$$\frac{\partial N_{xy}}{\partial x} + \frac{\partial N_{yy}}{\partial y} = I_0 \ddot{v}_0 + \left(I_1 - \frac{4}{3h^2} I_3\right)\ddot{\alpha}_y - \frac{4}{3h^2} I_3 \frac{\partial \ddot{w}_0}{\partial y} \tag{2.63b}$$

$$\begin{aligned} &\frac{\partial \bar{Q}_x}{\partial x} + \frac{\partial \bar{Q}_y}{\partial y} + \frac{\partial}{\partial x}\left(N_{xx}\frac{\partial w_0}{\partial x} + N_{xy}\frac{\partial w_0}{\partial y}\right) + \frac{\partial}{\partial y}\left(N_{xy}\frac{\partial w_0}{\partial y} + N_{yy}\frac{\partial w_0}{\partial y}\right) \\ &+ \frac{4}{3h^2}\left(\frac{\partial^2 T_{xx}}{\partial x^2} + 2\frac{\partial^2 T_{xy}}{\partial x \partial y} + \frac{\partial^2 T_{yy}}{\partial y^2}\right) + q = I_0 \ddot{w}_0 - \frac{16}{9h^4} I_6 \left(\frac{\partial^2 \ddot{w}_0}{\partial x^2} + \frac{\partial^2 \ddot{w}_0}{\partial y^2}\right) \\ &+ \frac{4}{3h^2}\left[I_3\left(\frac{\partial \ddot{w}_0}{\partial x} + \frac{\partial \ddot{w}_0}{\partial y}\right) + \left(I_4 - \frac{4}{3h^2} I_6\right)\left(\frac{\partial \ddot{\alpha}_x}{\partial x} + \frac{\partial \ddot{\alpha}_y}{\partial y}\right)\right] \end{aligned} \tag{2.63c}$$

$$\frac{\partial \bar{M}_{xx}}{\partial x} + \frac{\partial \bar{M}_{xy}}{\partial y} - \bar{Q}_x = \left(I_1 - \frac{4}{3h^2} I_3\right)\ddot{u}_0 + \left(I_2 - \frac{8}{3h^2} I_4 + \frac{16}{9h^4} I_6\right)\ddot{\alpha}_x - \frac{4}{3h^2}\left(I_4 - \frac{4}{3h^2} I_6\right)\frac{\partial \ddot{w}_0}{\partial x} \tag{2.63d}$$

$$\frac{\partial \bar{M}_{xy}}{\partial x} + \frac{\partial \bar{M}_{yy}}{\partial y} - \bar{Q}_y = \left(I_1 - \frac{4}{3h^2} I_3\right)\ddot{v}_0 + \left(I_2 - \frac{8}{3h^2} I_4 + \frac{16}{9h^4} I_6\right)\ddot{\alpha}_y - \frac{4}{3h^2}\left(I_4 - \frac{4}{3h^2} I_6\right)\frac{\partial \ddot{w}_0}{\partial y} \tag{2.63e}$$

where

$$I_0 = \sum_{k=1}^{N} \int_{z_k}^{z_{k+1}} \rho^{(k)} dz, \quad I_1 = \sum_{k=1}^{N} \int_{z_k}^{z_{k+1}} \rho^{(k)} z dz, \quad I_2 = \sum_{k=1}^{N} \int_{z_k}^{z_{k+1}} \rho^{(k)} z^2 dz$$
$$I_3 = \sum_{k=1}^{N} \int_{z_k}^{z_{k+1}} \rho^{(k)} z^3 dz, \quad I_4 = \sum_{k=1}^{N} \int_{z_k}^{z_{k+1}} \rho^{(k)} z^4 dz \tag{2.64}$$
$$I_5 = \sum_{k=1}^{N} \int_{z_k}^{z_{k+1}} \rho^{(k)} z^5 dz, \quad I_6 = \sum_{k=1}^{N} \int_{z_k}^{z_{k+1}} \rho^{(k)} z^6 dz$$

and

$$\bar{M}_{xx} = M_{xx} - \frac{4}{3h^2} T_{xx}, \quad \bar{M}_{xy} = M_{xy} - \frac{4}{3h^2} T_{xy}, \quad \bar{M}_{yy} = M_{yy} - \frac{4}{3h^2} T_{yy}$$
$$\bar{Q}_x = Q_4 - \frac{4}{h^2} S_4, \quad \bar{Q}_y = Q_5 - \frac{4}{h^2} S_5 \tag{2.65}$$

The solution of eqs. (2.63a)–(2.63e) will lead to the finding of the five unknowns, namely displacements u_0, v_0, w_0 and rotations α_x, α_y. Reddy [13] presents simplified results of the third-order theory, by neglecting the higher-order terms (T_{xx}, T_{xy}, T_{yy}), while keeping the other higher-order terms (S_x, S_y); however, the resulting theory comes out to be inconsistent in the energy sense.

A simpler high-order theory, a second-order shear deformation, was developed and presented by Khdeir and Reddy [21]. They propose the following displacement field:

$$u(x,y,z,t) = u_0(x,y,t) + z\alpha_x(x,y,t) + z^2\beta_x(x,y,t)$$
$$v(x,y,z,t) = v_0(x,y,t) + z\alpha_y(x,y,t) + z^2\beta_y(x,y,t) \tag{2.66}$$
$$w(x,y,z,t) = w_0(x,y,t)$$

For the case of linear strains, the strain–displacement relations can be shown to be written as

$$\varepsilon_{xx} \equiv \frac{\partial u}{\partial x} = \frac{\partial u_0}{\partial x} + z\frac{\partial \alpha_x}{\partial x} + z^2\frac{\partial \beta_x}{\partial x}, \varepsilon_{yy} \equiv \frac{\partial v_0}{\partial y} + z\frac{\partial \alpha_y}{\partial y} + z^2\frac{\partial \beta_y}{\partial y}, \varepsilon_{zz} \equiv \frac{\partial w}{\partial z} = 0$$

$$\varepsilon_{xy} \equiv \frac{1}{2}\left(\frac{\partial u}{\partial y} + \frac{\partial v}{\partial x}\right) = \frac{1}{2}\left[\left(\frac{\partial u_0}{\partial y} + \frac{\partial v_0}{\partial x}\right) + z\left(\frac{\partial \alpha_x}{\partial y} + \frac{\partial \alpha_y}{\partial x}\right) + z^2\left(\frac{\partial \beta_x}{\partial y} + \frac{\partial \beta_y}{\partial x}\right)\right]$$

$$\varepsilon_{xz} \equiv \frac{1}{2}\left(\frac{\partial u}{\partial z} + \frac{\partial w}{\partial x}\right) = \frac{1}{2}\left[\left(\alpha_x + \frac{\partial w_0}{\partial x}\right) + 2z\beta_x\right] \tag{2.67}$$

$$\varepsilon_{yz} \equiv \frac{1}{2}\left(\frac{\partial v}{\partial z} + \frac{\partial w}{\partial y}\right) = \frac{1}{2}\left[\left(\alpha_y + \frac{\partial w_0}{\partial x}\right) + 2z\beta_y\right]$$

The seven equations of motion presented in stress resultant terms, for the seven unknown displacements (u_0, v_0, w_0, α_x, α_y, β_x and β_y), assumed for the second-order shear deformation theory can be written as (see also [21])

$$\frac{\partial N_{xx}}{\partial x} + \frac{\partial N_{xy}}{\partial y} = I_1\ddot{u}_0 + I_2\ddot{\alpha}_x + I_3\ddot{\alpha}_y \tag{2.68a}$$

$$\frac{\partial N_{xy}}{\partial x} + \frac{\partial N_{yy}}{\partial y} = I_1\ddot{v}_0 + I_2\ddot{\beta}_x + I_3\ddot{\beta}_y \tag{2.68b}$$

$$\frac{\partial Q_x}{\partial x} + \frac{\partial Q_y}{\partial y} + q = I_1\ddot{w}_0 \tag{2.68c}$$

$$\frac{\partial M_{xx}}{\partial x} + \frac{\partial M_{xy}}{\partial y} - Q_x = I_2\ddot{u}_0 + I_3\ddot{\alpha}_x + I_4\ddot{\alpha}_y \tag{2.68d}$$

$$\frac{\partial M_{xy}}{\partial x} + \frac{\partial M_{yy}}{\partial y} - Q_y = I_2\ddot{v}_0 + I_3\ddot{\beta}_x + I_4\ddot{\beta}_y \tag{2.68e}$$

$$\frac{\partial T_{xx}}{\partial x} + \frac{\partial T_{xy}}{\partial y} - 2S_x = I_3\ddot{u}_0 + I_4\ddot{\alpha}_x + I_5\ddot{\alpha}_y \tag{2.68f}$$

$$\frac{\partial T_{xy}}{\partial x} + \frac{\partial T_{yy}}{\partial y} - 2S_y = I_3\ddot{v}_0 + I_4\ddot{\beta}_x + I_5\ddot{\beta}_y \tag{2.68g}$$

where

$$I_1 = \sum_{k=1}^{N}\int_{z_k}^{z_{k+1}} \rho^{(k)}dz, \quad I_2 = \sum_{k=1}^{N}\int_{z_k}^{z_{k+1}} \rho^{(k)}z\,dz, \quad I_3 = \sum_{k=1}^{N}\int_{z_k}^{z_{k+1}} \rho^{(k)}z^2dz$$
$$I_4 = \sum_{k=1}^{N}\int_{z_k}^{z_{k+1}} \rho^{(k)}z^3dz, \quad I_5 = \sum_{k=1}^{N}\int_{z_k}^{z_{k+1}} \rho^{(k)}z^4dz \tag{2.69}$$

The stress resultant–strain relationships, based on the above second-order displacement field, can be written as

$$\begin{Bmatrix} \{N\} \\ \{M\} \\ \{T\} \end{Bmatrix} = \begin{bmatrix} [A] & [B] & [D] \\ [B] & [D] & [E] \\ [D] & [E] & [F] \end{bmatrix} \begin{Bmatrix} \{\varepsilon^{(a)}\} \\ \{\varepsilon^{(b)}\} \\ \{\varepsilon^{(c)}\} \end{Bmatrix} \tag{2.70}$$

$$\begin{Bmatrix} \{Q\} \\ \{S\} \end{Bmatrix} = \begin{bmatrix} [A] & [B] \\ [B] & [D] \end{bmatrix} \begin{Bmatrix} \{\gamma^{(a)}\} \\ \{\gamma^{(b)}\} \end{Bmatrix} \tag{2.71}$$

where for eq. (2.70), the components are defined as

$$\begin{aligned} A_{ij} &= \sum_{k=1}^{N} \int_{z_k}^{z_{k+1}} \bar{Q}_{ij} dz = \sum_{k=1}^{N} \bar{Q}_{ij}^{(k)} [z_{k+1} - z_k] \\ B_{ij} &= \sum_{k=1}^{N} \int_{z_k}^{z_{k+1}} \bar{Q}_{ij} z dz = \frac{1}{2} \sum_{k=1}^{N} \bar{Q}_{ij}^{(k)} \left[(z_{k+1})^2 - (z_k)^2 \right] \\ D_{ij} &= \sum_{k=1}^{N} \int_{z_k}^{z_{k+1}} \bar{Q}_{ij} z^2 dz = \frac{1}{3} \sum_{k=1}^{N} \bar{Q}_{ij}^{(k)} \left[(z_{k+1})^3 - (z_k)^3 \right] \quad \text{while} \quad i,j = 1, 2, 6 \\ E_{ij} &= \sum_{k=1}^{N} \int_{z_k}^{z_{k+1}} \bar{Q}_{ij} z^3 dz = \frac{1}{4} \sum_{k=1}^{N} \bar{Q}_{ij}^{(k)} \left[(z_{k+1})^4 - (z_k)^4 \right] \\ F_{ij} &= \sum_{k=1}^{N} \int_{z_k}^{z_{k+1}} \bar{Q}_{ij} z^4 dz = \frac{1}{5} \sum_{k=1}^{N} \bar{Q}_{ij}^{(k)} \left[(z_{k+1})^5 - (z_k)^5 \right] \end{aligned} \tag{2.72}$$

while for eq. (2.71) their definition is

$$\begin{aligned} A_{lm} &= \sum_{k=1}^{N} \int_{z_k}^{z_{k+1}} \bar{Q}_{lm} dz = \sum_{k=1}^{N} \bar{Q}_{lm}^{(k)} [z_{k+1} - z_k] \\ B_{lm} &= \sum_{k=1}^{N} \int_{z_k}^{z_{k+1}} \bar{Q}_{lm} z dz = \frac{1}{2} \sum_{k=1}^{N} \bar{Q}_{lm}^{(k)} \left[(z_{k+1})^2 - (z_k)^2 \right] \quad \text{while} \quad l, m = 4, 5 \\ D_{lm} &= \sum_{k=1}^{N} \int_{z_k}^{z_{k+1}} \bar{Q}_{lm} z^2 dz = \frac{1}{3} \sum_{k=1}^{N} \bar{Q}_{lm}^{(k)} \left[(z_{k+1})^3 - (z_k)^3 \right] \end{aligned} \tag{2.73}$$

where

$$\left\{\varepsilon^{(a)}\right\} \equiv \begin{Bmatrix} \dfrac{\partial u_0}{\partial x} \\ \dfrac{\partial v_0}{\partial y} \\ \dfrac{\partial u_0}{\partial y} + \dfrac{\partial v_0}{\partial x} \end{Bmatrix}, \quad \left\{\varepsilon^{(b)}\right\} \equiv \begin{Bmatrix} \dfrac{\partial \alpha_x}{\partial x} \\ \dfrac{\partial \alpha_y}{\partial y} \\ \dfrac{\partial \alpha_x}{\partial y} + \dfrac{\partial \alpha_y}{\partial x} \end{Bmatrix}, \quad \left\{\varepsilon^{(c)}\right\} \equiv \begin{Bmatrix} \dfrac{\partial \beta_x}{\partial x} \\ \dfrac{\partial \beta_y}{\partial y} \\ \dfrac{\partial \beta_x}{\partial y} + \dfrac{\partial \beta_y}{\partial x} \end{Bmatrix} \tag{2.74}$$

$$\left\{\gamma^{(a)}\right\} \equiv \begin{Bmatrix} \alpha_x + \dfrac{\partial w_0}{\partial x} \\ \alpha_y + \dfrac{\partial w_0}{\partial y} \end{Bmatrix}, \quad \left\{\gamma^{(b)}\right\} \equiv 2\begin{Bmatrix} \beta_x \\ \beta_y \end{Bmatrix}$$

and

$$\gamma_{xy} \equiv 2\varepsilon_{xy}, \quad \gamma_{xz} \equiv 2\varepsilon_{xz}, \quad \gamma_{yz} \equiv 2\varepsilon_{yz} \tag{2.75}$$

The equations of motion (2.68a)–(2.68g) can be expressed using the seven unknown displacements to yield [21]:

$$\begin{aligned} &A_{11}\frac{\partial^2 u_0}{\partial x^2} + A_{66}\frac{\partial^2 u_0}{\partial y^2} + (A_{12}+A_{66})\frac{\partial^2 v_0}{\partial x \partial y} + 2B_{16}\frac{\partial^2 \alpha_x}{\partial x \partial y} + B_{16}\frac{\partial^2 \beta_x}{\partial x^2} \\ &B_{26}\frac{\partial^2 \beta_x}{\partial y^2} + D_{11}\frac{\partial^2 \alpha_y}{\partial x^2} + D_{66}\frac{\partial^2 \alpha_y}{\partial y^2} + (D_{12}+D_{66})\frac{\partial^2 \beta_y}{\partial x \partial y} = I_1\ddot{u}_0 + I_3\ddot{\alpha}_y \end{aligned} \tag{2.76a}$$

$$\begin{aligned} &A_{22}\frac{\partial^2 v_0}{\partial y^2} + A_{66}\frac{\partial^2 v_0}{\partial x^2} + (A_{12}+A_{66})\frac{\partial^2 u_0}{\partial x \partial y} + B_{16}\frac{\partial^2 \alpha_x}{\partial x^2} + B_{26}\frac{\partial^2 \alpha_x}{\partial y^2} \\ &B_{26}\frac{\partial^2 \beta_x}{\partial x \partial y} + D_{22}\frac{\partial^2 \beta_y}{\partial y^2} + D_{66}\frac{\partial^2 \beta_y}{\partial x^2} + (D_{12}+D_{66})\frac{\partial^2 \alpha_y}{\partial x \partial y} = I_1\ddot{v}_0 + I_3\ddot{\beta}_y \end{aligned} \tag{2.76b}$$

$$A_{55}\left(\frac{\partial \alpha_x}{\partial x} + \frac{\partial^2 w_0}{\partial x^2}\right) + A_{44}\left(\frac{\partial \beta_x}{\partial y} + \frac{\partial^2 w_0}{\partial y^2}\right) + 2B_{45}\left(\frac{\partial \beta_y}{\partial x} + \frac{\partial \alpha_y}{\partial y}\right) + q = I_1\,\ddot{w}_0 \tag{2.76c}$$

$$\begin{aligned} &2B_{16}\frac{\partial^2 u_0}{\partial x \partial y} + B_{16}\frac{\partial^2 v_0}{\partial x^2} + B_{26}\frac{\partial^2 v_0}{\partial y^2} + D_{11}\frac{\partial^2 \alpha_x}{\partial x^2} + D_{66}\frac{\partial^2 \alpha_x}{\partial y^2} + (D_{12}+D_{66})\frac{\partial^2 \beta_x}{\partial x \partial y} \\ &\quad + 2E_{16}\frac{\partial^2 \alpha_y}{\partial x \partial y} + E_{16}\frac{\partial^2 \beta_y}{\partial x^2} + E_{26}\frac{\partial^2 \beta_y}{\partial y^2} - A_{55}\left(\alpha_x + \frac{\partial w_0}{\partial x}\right) - 2B_{45}\beta_y = I_3\,\ddot{\alpha}_x \end{aligned} \tag{2.76d}$$

$$
\begin{aligned}
&2B_{26}\frac{\partial^2 v_0}{\partial x\partial y}+B_{16}\frac{\partial^2 u_0}{\partial x^2}+B_{26}\frac{\partial^2 u_0}{\partial y^2}+D_{22}\frac{\partial^2 \beta_x}{\partial y^2}+D_{66}\frac{\partial^2 \beta_x}{\partial x^2}+(D_{12}+D_{66})\frac{\partial^2 \alpha_x}{\partial x\partial y}\\
&\quad+E_{26}\frac{\partial^2 \alpha_y}{\partial y^2}+2E_{26}\frac{\partial^2 \beta_y}{\partial x\partial y}+E_{16}\frac{\partial^2 \alpha_y}{\partial x^2}-A_{44}\left(\alpha_x+\frac{\partial w_0}{\partial y}\right)-2B_{45}\alpha_y=I_3\ddot{\beta}_x
\end{aligned} \tag{2.76e}
$$

$$
\begin{aligned}
&D_{11}\frac{\partial^2 u_0}{\partial x^2}+D_{66}\frac{\partial^2 u_0}{\partial y^2}+(D_{12}+D_{66})\frac{\partial^2 v_0}{\partial x\partial y}+2E_{16}\frac{\partial^2 \alpha_x}{\partial x\partial y}+E_{16}\frac{\partial^2 \beta_x}{\partial x^2}+E_{26}\frac{\partial^2 \beta_x}{\partial y^2}\\
&\quad+F_{11}\frac{\partial^2 \alpha_y}{\partial x^2}+F_{66}\frac{\partial^2 \alpha_y}{\partial y^2}+(F_{12}+F_{66})\frac{\partial^2 \beta_y}{\partial x\partial y}-2B_{45}\left(\beta_x+\frac{\partial w_0}{\partial y}\right)-4D_{55}\alpha_y=I_3\ddot{u}_0+I_5\ddot{\alpha}_y
\end{aligned} \tag{2.76f}
$$

$$
\begin{aligned}
&D_{22}\frac{\partial^2 v_0}{\partial y^2}+D_{66}\frac{\partial^2 v_0}{\partial x^2}+(D_{12}+D_{66})\frac{\partial^2 u_0}{\partial x\partial y}+2E_{26}\frac{\partial^2 \beta_x}{\partial x\partial y}+E_{16}\frac{\partial^2 \alpha_x}{\partial x^2}+E_{26}\frac{\partial^2 \alpha_x}{\partial y^2}\\
&\quad+F_{221}\frac{\partial^2 \beta_y}{\partial y^2}+F_{66}\frac{\partial^2 \beta_y}{\partial x^2}+(F_{12}+F_{66})\frac{\partial^2 \alpha_y}{\partial x\partial y}-2B_{45}\left(\alpha_x+\frac{\partial w_0}{\partial x}\right)-4D_{44}\beta_y=I_3\ddot{v}_0+I_5\ddot{\beta}_y
\end{aligned} \tag{2.76g}
$$

The solution in [21] is given for a case of an antisymmetric angle-ply-laminated rectangular plate, with two opposite edges having simply supported boundary conditions (in the x direction), while the other two (in the y direction) may have any arbitrary combinations of free, clamped or simply supported edge conditions. The natural frequencies of such a plate are calculated using a generalized Lévy-type solution, namely selecting the assumed deflections and rotations to fulfill in the x direction, the simply supported boundary conditions (these functions will be dependent only on x), while in the perpendicular direction, having arbitrary boundary conditions, the functions will be dependent only on y. These functions (in the y direction) will now be the seven unknowns of the problem, after fulfillment of the prescribed boundary conditions in the y direction.

Another approach for using high-order approximation is known in the literature as “the Zig-Zag” (ZZ) theory, as in [22–35]. This approach is best summarized and reviewed by Carrera in [25] The idea behind this approach is to describe a piecewise continuous displacement field for basic elements like beams, plates and shells in the thickness direction and fulfill interlaminar continuity (IC) of the transverse stresses at each layer interface. One should note that in the literature the piecewise assumption of stresses and displacement fields is often referred to as ZZ and IC, respectively, while the theories applying both assumptions are named as ZZ theories. As Carrera [25] pointed out in his review, Lekhnitskii [36] was the first to suggest a theory for multilayered structures, followed by other fundamental contributors like Ambartsumian [37, 38] and Reisner [39, 40]. In the present context, we shall present only Lekhnitskii’s approach to highlight the ZZ theory.

Figure 2.9 presents the multilayered beam model as suggested by Lekhnitskii. It consists of a beam having the length L, total height H and width b. The cantilevered beam is loaded by a concentrated load P and a bending moment M. The beam consists of N_L layers. The coordinate system, as presented in Fig. 2.9, shows the Y coordinate along the length of the beam, with the Y–Z plane being perpendicular to it. The various Z coordinates of each layer are denoted by h_i, while the thickness of each layer is given by t_i. The black dots on the drawing stand for the interface surfaces.

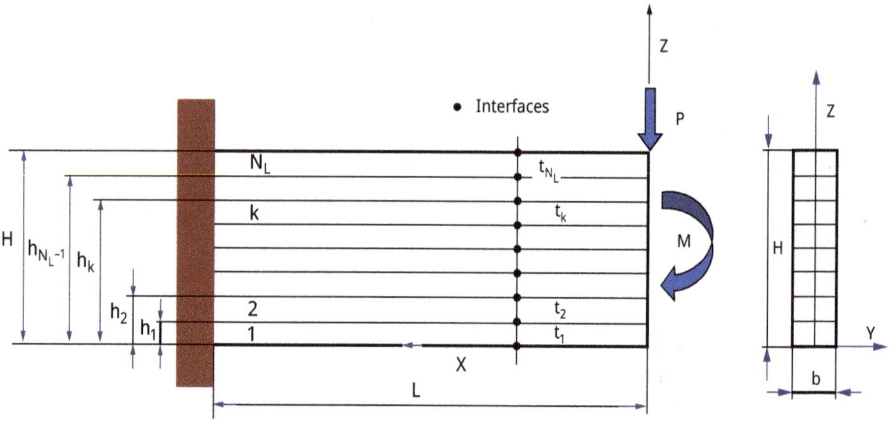

Fig. 2.9: Lekhnitskii's cantilever multilayered beam model – geometry and notations.

We shall denote the stresses for the k layer as σ_{xx}^k, τ_{xz}^k, σ_{zz}^k, while its displacements are written as u^k, w^k, with E^k, ν^k, G^k being the kth-layer Young's modulus, the Poisson's ratio and the shear modulus, respectively. As can be seen, the problem involves the X–Z plane, and it is a plane stress problem, which can be solved using stress functions φ^k for each layer k, defined as

$$\sigma_{xx}^k = \frac{\partial^2 \varphi^k}{\partial z^2}, \quad \sigma_{zz}^k = \frac{\partial^2 \varphi^k}{\partial x^2}, \quad \tau_{xz}^k = -\frac{\partial^2 \varphi^k}{\partial x \partial z} \tag{2.77}$$

The problem is then defined by demanding the compatibility of the resulted strains due to the applied loading, which can be written as

$$\frac{\partial^4 \varphi^k}{\partial x^4} + \frac{\partial^4 \varphi^k}{\partial x^2 \partial z^2} + \frac{\partial^4 \varphi^k}{\partial z^4} = 0 \tag{2.78}$$

Now Lekhnitskii assumed the following stress function:

$$\varphi^k(x,z) = \frac{\alpha^k}{6} z^3 + \frac{\beta^k}{2} z^2 + x\left[\frac{A^k}{3} z^3 + \frac{B^k}{2} z^2 + X^k z\right] \tag{2.79}$$

with α^k, β^k, A^k, B^k, X^k being constants to be latter determined from the boundary conditions and the continuity demands at the interface between two layers.

Using eq. (2.77), one can find the three plane stresses as

$$\sigma_{xx}^k \equiv \frac{\partial^2 \varphi^k}{\partial z^2} = \alpha^k z + \beta^k + x\left[2A^k z + B^k\right] \tag{2.80}$$

$$\sigma_{zz}^k \equiv \frac{\partial^2 \varphi^k}{\partial x^2} = 0 \tag{2.81}$$

$$\tau_{xz}^k \equiv -\frac{\partial^2 \varphi^k}{\partial x \partial z} = -A^k z^2 - B^k z - X^k \tag{2.82}$$

As shown in eqs. (2.80)–(2.82), the longitudinal stresses σ_{xx}^k are linear in both the longitudinal (x direction) and thickness (z direction) directions, and the stresses in the thickness direction, σ_{zz}^k, are zero throughout the X–Z plane, while the shear stresses τ_{xz}^k, have a parabolic distribution in the z direction. The above results, stemming from Lekhnitskii's assumption of the stress function, $\varphi^k(x,z)$ (eq. (2.79)), lead to a stress distribution that is acceptable for a beam theory.

To obtain the $5xN_L$ unknown constants, α^k, β^k, A^k, B^k, X^k, the following relations must be satisfied and then solved:

The strain–stress relation, expressed using the two displacements, u^k, w^k, can be written as (using the Hook's law)

$$\begin{aligned}
\varepsilon_{xx}^k &\equiv \frac{\partial u^k}{\partial x} = \frac{\sigma_{xx}^k}{E^k} - \nu^k \frac{\sigma_{zz}^k}{E^k} \\
\varepsilon_{zz}^k &\equiv \frac{\partial w^k}{\partial z} = \frac{\sigma_{zz}^k}{E^k} - \nu^k \frac{\sigma_{xx}^k}{E^k} \\
\gamma_{xz}^k &\equiv \frac{\partial u^k}{\partial z} + \frac{\partial w^k}{\partial x} = \frac{\tau_{xz}^k}{G^k}
\end{aligned} \tag{2.83}$$

The continuity (or compatibility) conditions for the displacements and the transverse stresses must be satisfied at the laminar interfaces (namely the ZZ phenomenon) according to the following equations:

$$u^k = u^{k-1} \text{ and } w^k = w^{k-1} \text{ for } k = 2, \ldots, N_L \tag{2.84}$$

$$\sigma_{yy}^k = \sigma_{yy}^{k-1} \text{ and } \sigma_{xy}^k = \sigma_{xy}^{k-1} \text{ for } k = 2, \ldots, N_L \tag{2.85}$$

Next, the various boundary conditions must be fulfilled. The upper and lower planes of the beam are free of stress. This can be written as

$$\text{At } Z = 0, H \quad \sigma_{yy}^1 = 0; \sigma_{yy}^{N_L} = 0 \text{ and } \sigma_{xy}^1 = 0; \sigma_{xy}^{N_L} = 0 \tag{2.86}$$

Finally, the equilibrium between the applied load (P and M) and the stresses should also be demanded, namely

$$b\sum_{1}^{N_L}\int_{h_{k-1}}^{h_k}\sigma_{zz}^k dz = 0$$
$$b\sum_{1}^{N_L}\int_{h_{k-1}}^{h_k}\sigma_{xx}^k z dz = \frac{M - Px}{H} \tag{2.87}$$
$$b\sum_{1}^{N_L}\int_{h_{k-1}}^{h_k}\sigma_{xz}^k dz = -\frac{P}{H}$$

Using expressions for stresses, as presented in eqs. (2.80)–(2.82), substituting it in eq. (2.83) and then integrating in the X and Y directions yields the following expressions for the displacements:

$$u^k = \int\frac{\sigma_{xx}^k}{E^k}dx = \frac{1}{E^k}\left[\left(\alpha^k z + \beta^k\right)x + \frac{x^2}{2}\left(2A^k z + B^k\right)\right] + f(z)$$
$$w^k = -\int\frac{v^k\sigma_{xx}^k}{E^k}dz = -\frac{v^k}{E^k}\left[\left(\alpha^k\frac{z^2}{2} + \beta^k z\right) + x\left(A^k z^2 + B^k z\right)\right] + f_1(x) \tag{2.88}$$

where $f(z)$ and $f_1(x)$ are functions to be determined using the last equation in eqs. (2.83), namely

$$\gamma_{xz} \equiv \frac{\partial u^k}{\partial z} + \frac{\partial w^k}{\partial z} = \tau_{xz} = -\frac{1}{G^k}\left[A^k z^2 + B^k z + X^k\right] \tag{2.89}$$

Substituting the expressions for u^k and w^k into eq. (2.89) yields

$$\frac{1}{E^k}\left[\alpha^k x + x^2 A^k\right] + \frac{\partial f(z)}{\partial z} - \frac{v^k}{E^k}\left[A^k z^2 + B^k z\right] + \frac{\partial f_1(x)}{\partial x} = -\frac{1}{G^k}\left[A^k z^2 + B^k z + X^k\right] \tag{2.90}$$

Equation (2.90) can be rearranged to yield the following form:

$$F(x) = \frac{1}{E^k}\left[\alpha^k x + x^2 A^k\right] + \frac{\partial f_1(x)}{\partial x} \equiv e^k$$
$$G(z) = \left[\frac{1}{G_{xz}} - \frac{v^k}{E^k}\right]\left[A^k z^2 + B^k z\right] + \frac{\partial f(z)}{\partial z} \equiv d^k \tag{2.91}$$
$$K = -\frac{X^k}{G_{xz}}$$

and

$$e^k + d^k = K = -\frac{X^k}{G^k} \Rightarrow e^k = -\left(\frac{X^k}{G^k} + d^k\right) \tag{2.92}$$

It is clear from eq. (2.90) that if the sum of the two functions $F(x)$ and $G(z)$ is equal to a constant, K, for every x and z, then each of these functions must be a constant, e^k and d^k, respectively (see eqs. (2.91) and (2.92)). Now we can proceed to integrate the functions presented in eq. (2.91) to yield the two functions $f(z)$ and $f_1(x)$. From eq. (2.92) we can see that

$$\begin{aligned} \frac{df(z)}{dz} &= d^k - \left(\frac{1}{G^k} - \frac{\nu^k}{E^k}\right)\left[A^k z^2 + B^k z\right] \\ \frac{df_1(x)}{dx} &= e^k - \frac{1}{E^k}\left[x\alpha^k + x^2 A^k\right] \\ &= -\left(\frac{X^k}{G^k} + d^k\right) - \frac{1}{E^k}\left[x\alpha^k + x^2 A^k\right] \end{aligned} \tag{2.93}$$

Integrating the equations presented in eq. (2.93) yields

$$\begin{aligned} f(z) &= d^k \cdot z - \left(\frac{1}{G^k} - \frac{\nu^k}{E^k}\right)\left[\frac{A^k z^3}{3} + B^k \frac{z^2}{2}\right] + d_1^k \\ f_1(x) &= -\left(\frac{X^k}{G^k} + d^k\right)x - \frac{1}{E^k}\left[\frac{x^2}{2}\alpha^k + \frac{x^3}{3}A^k\right] + e_1^k \end{aligned} \tag{2.94}$$

where d_1^k and e_1^k, e_1 are additional constants besides the constant d^k (see eq. (2.93)) defined earlier. Substituting the two functions from eq. (2.94), $f(z)$ and $f_1(x)$, into eq. (2.88) provides the expression for the two displacements u^k, w^k:

$$\begin{aligned} u^k &= \frac{1}{E^k}\left[\left(\alpha^k z + \beta^k\right)x + \frac{x^2}{2}\left(2A^k z + B^k\right)\right] + d^k \cdot z - \left(\frac{1}{G^k} - \frac{\nu^k}{E^k}\right)\left[\frac{A^k z^3}{3} + B^k \frac{z^2}{2}\right] + d_1^k \\ w^k &= -\frac{\nu^k}{E^k}\left[\left(\alpha^k \frac{z^2}{2} + \beta^k z\right) + x\left(A^k z^2 + B^k z\right)\right] - \left(\frac{X^k}{G^k} + d\right)x - \frac{1}{E^k}\left[\frac{x^2}{2}\alpha^k + \frac{x^3}{3}A^k\right] + e_1^k \end{aligned} \tag{2.95}$$

Having the two expressions for the displacements u^k, w^k, the requirements listed in eqs. (2.84)–(2.87) are used to obtain all the unknowns, including the constants d, d_1 and e_1. This results in the following relations [25]:

$$
\begin{aligned}
&\frac{\alpha^k}{E^k}=\frac{\alpha^{k-1}}{E^{k-1}},\quad \frac{A^k}{E^k}=\frac{A^{k-1}}{E^{k-1}}\\
&\frac{A^k}{E^k}h_{k-1}+\frac{B^k}{2E^k}h_{k-1}=\frac{A^{k-1}}{E^{k-1}}h_{k-1}+\frac{B^{k-1}}{2E^{k-1}}h_{k-1}\\
&\frac{\alpha^k h_{k-1}+\beta^k}{E^k}=\frac{\alpha^{k-1}h_{k-1}+\beta^{k-1}}{E^k}\\
&X^1=0, A^{N_L}h_{N_L}^2+B^{N_L}h_{N_L}+X^{N_L}=0\\
&A^k h_{k-1}^2+B^k h_{k-1}+X^k=A^{k-1}h_{k-1}^2+B^{k-1}h_{k-1}+X^{k-1}\\
&-\left(\frac{1}{G^{k-1}}-\frac{\nu^k}{E^{k-1}}\right)\left[\frac{A^k h_{k-1}^3}{3}+B^k\frac{h_{k-1}^2}{2}\right]-d^k h_{k-1}+d_1^k=\\
&=\left(\frac{1}{G^k}-\frac{\nu^{k-1}}{E^k}\right)\left[\frac{A^{k-1}h_{k-1}^3}{3}+B^{k-1}\frac{h_{k-1}^2}{2}\right]-d^{k-1}h_{k-1}+d_1^{k-1}\\
&\frac{\nu^k}{E^k}\left(A^k h_{k-1}^2+B^k h_{k-1}\right)-\frac{X^k}{G^k}+d^k=\frac{\nu^{k-1}}{E^{k-1}}\left(A^{k-1}h_{k-1}^2+B^{k-1}h_{k-1}\right)-\frac{X^{k-1}}{G^{k-1}}+d^{k-1}\\
&-\frac{\nu^k}{E^k}\left(\frac{\alpha^k}{2}h_{k-1}^2+\beta^k h_{k-1}\right)+e_1^k=-\frac{\nu^{k-1}}{E^{k-1}}\left(\frac{\alpha^{k-1}}{2}h_{k-1}^2+\beta^{k-1}h_{k-1}\right)+e_1^{k-1}
\end{aligned}
\tag{2.96}
$$

As a result of the expressions presented in eq. (2.96), the following recursive equations can be obtained:

$$
\begin{aligned}
&A^k=\frac{A^1E^k}{E^1},\quad B^k=\frac{B^1E^k}{E^1},\quad \alpha^k=\frac{\alpha^1E^k}{E^1},\quad \beta^k=\frac{\beta^1E^k}{E^1}\\
&X^k=\frac{1}{E^1}\left\{A^1\left[\sum_{t=1}^{k-1}(h_t^2-h_{t-1}^2)E^t-h_{k-1}E^k\right]+B^1\left[\sum_{t=1}^{k-1}(h_t-h_{t-1})E^t-h_{k-1}E^k\right]\right\}\\
&k=2,3,N_{L-1}
\end{aligned}
\tag{2.97}
$$

with

$$
X^1=1 \quad \text{and} \quad C_{N_L}=\frac{E^k}{E^1}\left[A^1h_{N_L}^2+B^1h_{N_L}\right]
\tag{2.98}
$$

The other constants can be written as

$$
A^1=\frac{6PE^1}{\delta b},\quad B^1=-\frac{6PE^1}{\delta b}\delta_2,\quad \alpha^1=-\frac{12ME^1}{\delta b}\delta_1,\quad \beta^1=-\frac{12ME^1}{\delta b}\delta_2
$$

where

$$\delta = 4\sum_{i=1}^{N_L}\left[h_i^3 - h_{i-1}^3\right]E^i + \sum_{i=1}^{N_L}\left[h_i - h_{i-1}\right]E^i - 3\left[\sum_{i=1}^{N_L}\left[h_i^2 - h_{i-1}^2\right]E^i\right]^2$$
$$\delta_1 = \sum_{i=1}^{N_L}\left[h_i^3 - h_{i-1}^3\right]E^i, \delta_2 = \sum_{i=1}^{N_L}\left[h_i^2 - h_{i-1}^2\right]E^i \tag{2.99}$$

One should remember that $h_0 = 0$ and $h_{N_L} = H$.

Finally, the expressions for the stresses can be written as

$$\sigma_{xx}^k = \frac{6E^k}{\delta b}[Px + M)][2\delta_1 z - \delta_2], \quad k = 1, 2, \ldots, N_L$$
$$\sigma_{zz}^k = 0, \quad k = 1, 2, \ldots, N_L$$
$$\sigma_{xz}^k = \frac{6P}{\delta b}\left\{\delta_1\left[\sum_{t=1}^{k-1}\left(h_t^2 - h_{t-1}^2\right)E^t + \left(z^2 - h_{t-1}^2\right)E^k\right]\right\} \tag{2.100}$$
$$-\left\{\delta_2\left[\sum_{t=1}^{k-1}(h_t - h_{t-1})E^t + (z - h_{t-1})E^k\right]\right\}, \quad k = 2, 3, \ldots, N_{L-1}$$

with

$$\sigma_{xz}^1 = -\frac{6PE^1}{\delta b}z[\delta_1 z - \delta_2], \quad \sigma_{xz}^{N_L} = -\frac{6PE^{N_L}}{\delta b}\left(h_{N_L} - z\right)\left[\delta_2 - \left(h_{N_L} + z\right)\delta_1\right]$$

With all the constants being determined, the expressions for the two displacements u^k, w^k can now be evaluated using eq. (2.95).

For other formulations of the various ZZ theories and the assumed stress functions, one can address references [22–27, 29–34] and similar available studies in the open literature.

References

[1] Harris, B. Engineering composite materials, The Institute of Materials, London, UK, 1999, 193.

[2] Jones, R.M., Mechanics of composite materials, 2nd edition, Taylor & Francis, Philadelphia, PA 19106, USA, 1999, 519.

[3] Love, A.E.H. On the small free vibrations and deformations of elastic shells, Philosophical Transactions of the Royal Society (London), 1888, série A, N°, 17, 491–549.

[4] Reddy, J.N. Theory and analysis of elastic plates and shells, CRC Press, Taylor and Francis, 2007.

[5] Timoshenko, S. and Woinowsky-Krieger, S. Theory of plates and shells, McGraw-Hill, New York, 1959.

[6] Mindlin, R.D. Influence of rotatory inertia and shear on flexural motions of isotropic, elastic plates, ASME Journal of Applied Mechanics, 18, 1951, 31–38.

[7] Reddy, J.N. Theory and analysis of elastic plates, Taylor and Francis, Philadelphia, 1999.

[8] Lim, G.T. and Reddy, J.N. On canonical bending relationships for plates, International Journal of Solids and Structures, 40, 2003, 3039–3067.

[9] Timoshenko, S.P. On the correction factor for shear of the differential equation for transverse vibrations of bars of uniform cross-section, Philosophical Magazine, 1921, 744.
[10] Timoshenko, S.P. On the transverse vibrations of bars of uniform cross-section, Philosophical Magazine, 1922, 125.
[11] Rosinger, H.E. and Ritchie, I.G. On Timoshenko's correction for shear in vibrating isotropic beams, Journal of Physics D: Applied Physics, 10, 1977, 1461–1466.
[12] Timoshenko, S.P. and Gere, J.M. Mechanics of materials, Van Nostrand Reinhold Co, 1972.
[13] Reddy, J.N. Mechanics of laminated composite plates and shells-theory and analysis, 2nd Edition, CRC Press LLC, Boca Raton, Florida, 33431 USA, 2004.
[14] Chandrashekhara, K., Krishnamurthy, K. and Roy, S. Free vibration of composite beams including rotary inertia and shear deformation, Composite Structures, 14(4), 1990, 269–279.
[15] Singh, G. and Venkateswara, R. Analysis of the nonlinear vibrations of unsymmetrically laminated composite beams, AAIA Journals, 29(10), October 1991.
[16] Abramovich, H. and Livshits, A. Free vibrations of non-symmetric cross-ply laminated composite beams, Journal of Sound and Vibration, 176(5), 6 October 1994, 597–612.
[17] Abramovich, H., Eisenberger, M. and Shulepov, O. Vibrations and buckling of non-symmetric laminated composite beams via the exact element method, AAIA Journals, 1995. http://arc.aiaa.org/doi/abs/10.2514/6.1995-1459.
[18] Abramovich, H., Eisenberger, M. and Shulepov, O. Vibrations and buckling of cross-ply nonsymmetric laminated composite beams, AAIA Journals, 34(5), May 1996.
[19] Li, J., Wu, Z., Kong, X., Li, X. and Wu, W. Comparison of various shear deformation theories for free vibration of laminated composite beams with general lay-ups, Composite Structures, 108, February 2014, 767–778.
[20] Abadi, M.M. and Daneshmehr, A.R. An investigation of modified couple stress theory in buckling analysis of micro composite laminated Euler–Bernoulli and Timoshenko beams, International Journal of Engineering Science, 75, February 2014, 40–53.
[21] Khdeir, A.A. and Reddy, J.N. Free vibrations of laminated composite plates using second-order shear deformation theory, Computers and Structures, 71(6), June 1999, 617–626.
[22] Li, X. and Liu, D. Zigzag theory for composite laminates, AIAA Journal, 33(6), 1995, 1163–1165.
[23] Cho, Y.B. and Averill, R.C. First-order zig-zag sublaminate plate theory and finite element model for laminated composite and sandwich panels, Composite Structures, 50, 2000, 1–15.
[24] Fares, M.E. and Elmarghany, M.Kh. A refined zigzag nonlinear first-order shear deformation theory of composite laminated plates, Composite Structures, 82, 2008, 71–83.
[25] Carrera, E. Historical review of Zig-Zag theories for multilayered plates and shells, Applied Mechanics Reviews, 56(3), May 2003, 287–308.
[26] Demasi, L. ∞6 mixed plate theories based on the generalized unified formulation. Part I: governing equations, Composite Structures, 87, 2009, 1–11.
[27] Demasi, L. ∞6 mixed plate theories based on the generalized unified formulation. Part II: layerwise theories, Composite Structures, 87, 2009, 12–22.
[28] Demasi, L. ∞6 mixed plate theories based on the generalized unified formulation. Part III: advanced mixed high order shear deformation theories, Composite Structures, 87, 2009, 183–194.
[29] Demasi, L. ∞6 mixed plate theories based on the generalized unified formulation. Part IV: Zig-zag theories, Composite Structures, 87, 2009, 195–205.
[30] Demasi, L. ∞6 mixed plate theories based on the generalized unified formulation, Part V: Results, Composite Structures, 88, 2009, 1–16.
[31] Sahoo, R. and Singh, B.N. A new inverse hyperbolic zigzag theory for the static analysis of composite and sandwich plates, Composite Structures, 105, 2013, 385–397.
[32] Tessler, A. Refined zigzag theory for homogenous, laminated composite, and sandwich plates: A homogenous limit methodology for zigzag function selection, NASA/TP-2010-216214, March 2010.

[33] Barut, A., Madenci, E. and Tessler, A. C°- continuous triangular plate element for laminated composite and sandwich plates using the {2,2}-, Refined Zigzag Theory, Composite Structures, 106, 2013, 835–853.

[34] Sahoo, R. and Singh, B.N. A new trigonometric zigzag theory for buckling and free vibration analysis of laminated composite and sandwich plates, Composite Structures, 117, 2014, 316–332.

[35] Sahoo, R. and Singh, B.N. A new trigonometric zigzag theory for static analysis of laminated composite and sandwich plates, Aerospace Science and Technology, 35, 2014, 15–28.

[36] Lekhnitskii, S.G. Strength calculation of composite beams, Vestnik inzhen i tekhnikov, 1935, 9.

[37] Ambartsumian, S.A. Analysis of two-layer orthotropic shells, Investiia Akad Nauk SSSR, Ot Tekh Nauk, 7, 1957.

[38] Ambartsumian, S.A. Two analysis method for two-layer orthotropic shells, Izv An Arm SSR Seiya Fiz-Matem nauk X(2), 1957.

[39] Reissner, E. On a certain mixed variational theory and a proposed application, International Journal of Numerical Methods in Engineering, 20, 1984, 1366–1368.

[40] Reissner, E. On a mixed variational theorem and on a shear deformable plate theory, International Journal of Numerical Methods in Engineering, 23, 1986, 193–198.

3 Design formulas

3.1 Introduction

This chapter aims to present the reader with useful design formulas to analyze aerospace structures and predict their structural behavior. First, some Airy functions will be presented, followed by the strength of materials' formulas and natural frequencies for various structures.

3.2 Airy functions

To solve directly the equations of elasticity for a posed problem is not an easy task. Therefore, the use of the Airy solutions (described in detail in Chapter 1 and in [1]) might provide a way to obtain elasticity solutions. One of these functions can be obtained by a combination of polynomials, which must satisfy the biharmonic equation (see eqs. (1.57) and (1.58)) and having at least a second degree to prevent obtaining a zero-stress solution. Some typical examples are presented in Tab. 3.1.

Tab. 3.1: Typical Airy functions and their resulting stresses.

Case no.	Airy function	σ_x or σ_r	σ_y or σ_θ	τ_{xy} or $\sigma_{r\theta}$
1	$\phi(x, y) = ax^2 + bxy + cy^2$	$2c$	$2a$	$-b$
2	$\phi(x, y) = ax^3 + bx^2y + cxy^2 + dy^3$	$2cx + 6dy$	$6ax + 2by$	$-2(b + c)$
3	$\phi(x, y) = ax^4 + bx^3y + cx^2y^2 + dxy^3 + ey^4$*	$-6(a+e)x^2 + 6dxy + 12ey^2$	$12ax^2 + 6bxy - 6(a+e)y^2$	$-3bx^2 - 3dy^2 + 12(a+e)xy$
4	$\phi(r, \theta) = Ar^2(1 - \cos 2\theta)$	$2A(1 + \cos 2\theta)$	$2A(1 - \cos 2\theta)$	$-A \sin 2\theta$

*If the relation $3(a + e) = -c$ holds, then eq. (1.57) in Chapter 1 is fulfilled.

One should note that for Case no. 1 in Tab. 3.1, assuming a rectangular plate, if $a = b = 0$, one gets simple constant tension in the x direction, while for $b = 0$, we would get constant tension in both x and y directions, and for $a = c = 0$, the result would represent a constant pure shear stress. For Case no. 2, if one assumes, for example, $a = b = c = 0$, the resulting stress $\sigma_x = 6dy$ would be linear, namely representing a pure bending of a rectangular plate.

https://doi.org/10.1515/9783111621104-003

3.3 Distribution of the shear forces, moments, deflections and slopes for beams

One of the basic elements to be used in any complex structure is a beam. Beam may be straight or curved, according to the required design. Loading a beam in bending and finding the appropriate reactions at its boundaries can be obtained by demanding equilibrium for all the forces and moments acting on it [2]. By the distribution of shear forces, V, and the moments, M, along a beam, loaded in bending, one can use either the section method [2] or the summation method, based on the next relations

$$\frac{dV}{dx} = -p \quad \frac{dM}{dx} = -V \Rightarrow \frac{d^2M}{dx^2} = +p \tag{3.1}$$

where p is the load per unit length, acting on the beam (see Fig. 3.1). One should note the definitions of positive external loading, as depicted in Fig. 3.1.

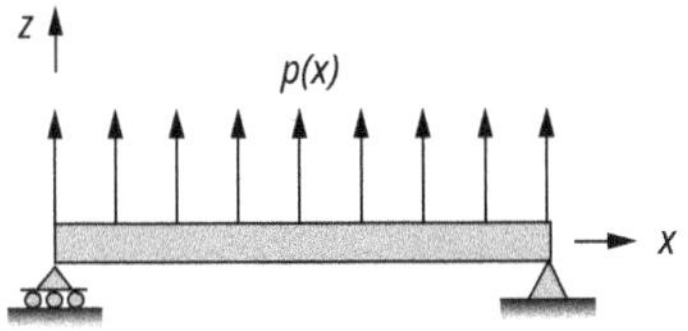

Fig. 3.1: A schematic beam on simply supported ends under load per unit length $p(x)$.

Establishing the relation in eq. (3.1), one can write the relations among the bending moment, M, the shear force, V, and the load per unit length, p, and the deflection, W, for the case of small deflections, to be

$$EI\frac{d^2W(x)}{dx^2} = M(x), \quad EI\frac{d^3W(x)}{dx^3} = -V(x), \quad EI\frac{d^4W(x)}{dx^4} = +p(x) \tag{3.2}$$

Solutions for deflections, slopes and reactions for beams on various boundary conditions and loadings are presented in Tab. 3.2.

3.4 Natural frequencies for common basic structures

For a single degree of freedom system, consisting of a mass, m, and a spring, k, the equation of motion solving the natural frequency is known to be [3]

$$m\ddot{x} + kx = 0 \tag{3.3}$$

where x is a coordinate and $\ddot{x}$ is the acceleration of the mass. Then the single natural frequency can be written as

Tab. 3.2: Reactions, deflections and slopes for typical loaded beams.

y	Loading	Boundary conditions		Reactions	Deflection and slope
		Left A	Right B		
1	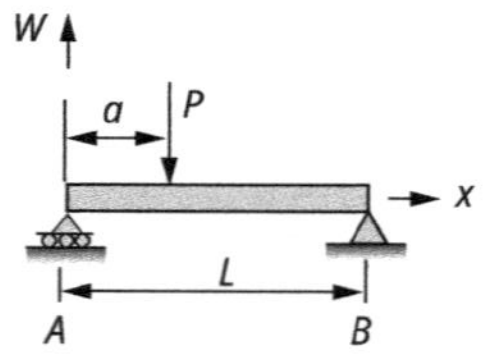	Simply supported	Simply supported	$R_A = P\left(1 - \frac{a}{L}\right);\ M_A = 0$ $R_B = P \cdot \frac{a}{L};\ M_B = 0$	For $0 \le x \le a$ $W(x) = \frac{1}{EI}\left[P\left(1 - \frac{a}{L}\right)\frac{x^3}{6} + A_1 x\right]$ $\theta(x) = \frac{1}{EI}\left[P\left(1 - \frac{a}{L}\right)\frac{x^2}{2} + A_1\right]$ For $a \le x \le L$ $W(x) = \frac{1}{EI}\left[P\frac{ax^2}{2}\left(1 - \frac{x}{3L}\right) + B_1 x + B_2\right]$ $\theta(x) = \frac{1}{EI}\left[Pax - P\frac{a}{L}\frac{x^2}{2} + B_1\right]$ where $A_1 = \frac{Pa^2}{6}\left\{3 - 2\frac{L}{a}\left[1 + \left(\frac{a}{L}\right)^2\left(1 + \frac{a}{L}\right)\right]\right\}$ $B_1 = -\frac{PaL}{3}\left[1 + \left(\frac{a}{L}\right)^2\left(1 + \frac{a}{L}\right)\right]$ $B_2 = \frac{Pa^3}{3}\left(1 + \frac{a}{L}\right)$

2	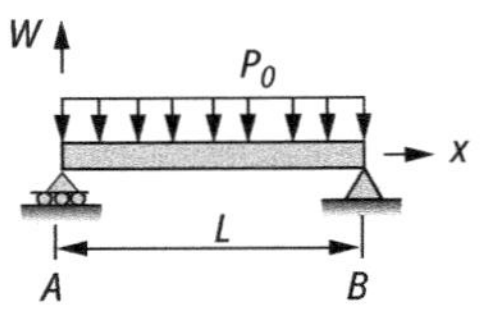	Simply supported	Simply supported	$R_A = \frac{p_0 L}{2};\ M_A = 0$ $R_B = \frac{p_0 L}{2};\ M_B = 0$	For $0 \le x \le L$ $W(x) = \frac{p_0}{24EI}\left[2Lx^3 - x^4 - xL^3\right]$ $\theta(x) = \frac{p_0}{24EI}\left[6Lx^2 - 4x^3 - L^3\right]$
3	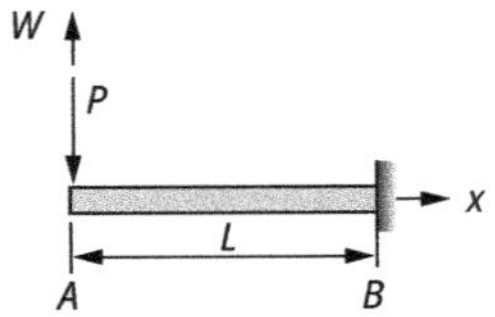	Free	Clamped	$R_A = 0;\ M_A = 0$ $R_B = P;\ M_B = -PL$	For $0 \le x \le L$ $W(x) = \frac{P}{6EI}\left[3L^2x - x^3 - 2L^3\right]$ $\theta(x) = \frac{P}{6EI}\left[3L^2 - 3x^2\right]$
4	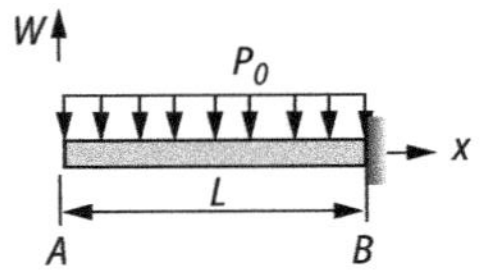	Free	Clamped	$R_A = 0;\ M_A = 0$ $R_B = p_0 L;\ M_B = -p_0 \frac{L^2}{2}$	For $0 \le x \le L$ $W(x) = \frac{p_0}{24EI}\left[4L^3x - x^4 - 3L^4\right]$ $\theta(x) = \frac{p_0}{6EI}\left[L^3 - x^3\right]$

(continued)

Tab. 3.2 (continued)

y	Loading	Boundary conditions		Reactions	Deflection and slope
		Left A	Right B		
5	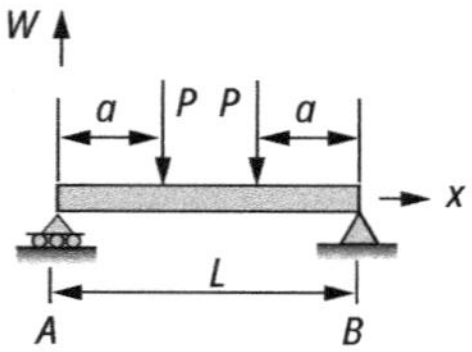	Simply supported	Simply supported	$R_A = P;\ M_A = 0 -$ $R_B = P;\ M_B = 0$	For $0 \le x \le a$ $W(x) = \frac{1}{EI}\left[\frac{Px^3}{6} + A_1 x\right]$ $\theta(x) = \frac{1}{EI}\left[\frac{Px^2}{2} + A_1\right]$ For $a \le x \le (L-a)$ $W(x) = \frac{1}{EI}\left[\frac{Pax^2}{2} + B_1 x + B_2\right]$ $\theta(x) = \frac{1}{EI}[Pax + B_1]$ For $(L-a) \le x \le L$ $W(x) = \frac{1}{EI}\left[P\frac{ax^2}{2} - P\frac{x^3}{6} + P(L-a)\frac{x^2}{2} + C_1 x + C_2\right]$ $\theta(x) = \frac{1}{EI}\left[Pax - P\frac{x^2}{2} + P(L-a)x + C_1\right]$ where $A_1 = \frac{P}{6}\left(\frac{a}{L}\right)\left[3aL + 2\left(a^2 - L^2\right)\right]$ $B_1 = \frac{Pa}{3L}\left(a^2 - L^2\right) \quad B_2 = -\frac{Pa^3}{6}$ $C_1 = \frac{Pa^3}{3L} - \frac{P}{2}\left(a^2 + L^2\right)$ $C_2 = \frac{P(L-a)^3}{6} - \frac{Pa^3}{6}$

6	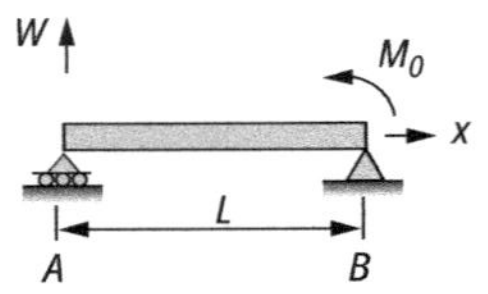	Simply supported	Simply supported	$R_A = \frac{M_0}{L};\ M_A = 0$ $R_B = -\frac{M_0}{L};\ M_B = M_0$	For $0 \le x \le L$ $W(x) = \frac{M_0 x}{6EIL}\left[x^2 - L^2\right]$ $\theta(x) = \frac{M_0}{6EIL}\left[3x^2 - L^2\right]$
7	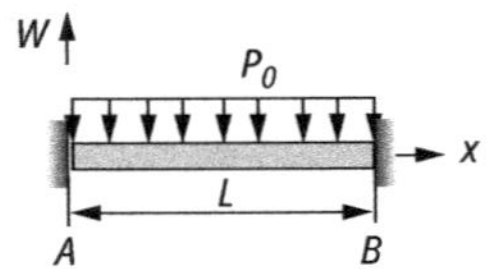	Clamped	Clamped	$R_A = \frac{p_0 L}{2};\ M_A = -\frac{p_0 L^2}{12}$ $R_B = \frac{p_0 L}{2};\ M_B = -\frac{p_0 L^2}{12}$	For $0 \le x \le L$ $W(x) = \frac{p_0 x^2}{24EI}\left[2Lx - x^2 - L^2\right]$ $\theta(x) = \frac{p_0 x}{12EIL}\left[3Lx - 2x^2 - L^2\right]$
8	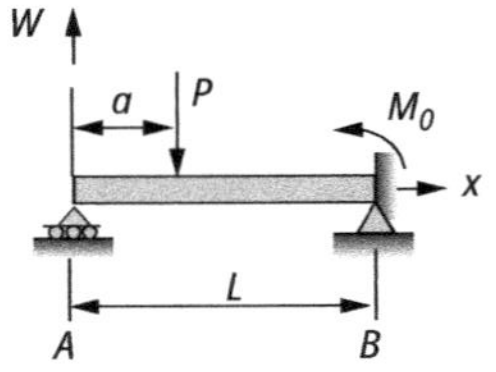	Simply supported	Clamped	$R_A = P - \frac{P}{2}\left(\frac{a}{L}\right)\left[\left(\frac{a}{L}\right)^2 + 3\right];$ $M_A = 0$ $R_B = \frac{P}{2}\left(\frac{a}{L}\right)\left[\left(\frac{a}{L}\right)^2 + 3\right];$ $M_B = \frac{Pa}{2}\left[\left(\frac{a}{L}\right)^2 + 1\right]$	For $01 \le x \le a$ $W(x) = \frac{1}{EI}\left[R_A \frac{x^3}{6} + A_1 X\right]$ $\theta(x) = \frac{1}{EI}\left[R_A \frac{x^2}{2} + A_1\right]$ For $a \le x \le L$ $W(x) = \frac{1}{EI}\left[R_A \frac{x^3}{6} - \frac{Px^3}{6} + \frac{Pax^2}{2} + B_1 x + B_2\right]$ $\theta(x) = \frac{1}{EI}\left[R_A \frac{x^2}{2} - \frac{Px^2}{2} + Pax + B_1\right]$ where $A_1 = B_1 + \frac{Pa^2}{2}$ $B_1 = \frac{PL^2}{4}\left(\frac{a}{L}\right)\left[\left(\frac{a}{L}\right)^2 + 3\right] - PaL$ $B_2 = +\frac{Pa^3}{6}$

(continued)

Tab. 3.2 (continued)

y	Loading	Boundary conditions: Left A	Boundary conditions: Right B	Reactions	Deflection and slope
9	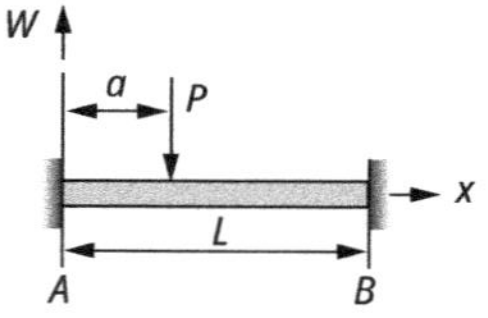	Clamped	Clamped	$R_A = P + 3P\left(\frac{a}{L}\right)^2\left(1 - \frac{3a}{2L}\right);$ $M_A = Pa + P(2L - 3a)\left(\frac{a}{L}\right)^2$ $R_B = 3P\left(\frac{a}{L}\right)^2\left(\frac{3a}{2L} - 1\right);$ $M_B = P\left(\frac{a}{L}\right)^2\left(\frac{3a}{2} - L\right)$	For $0 \le x \le a$ $W(x) = \frac{1}{EI}\left[R_A\frac{x^3}{6} - M_A\frac{x^2}{2}\right]$ $\theta(x) = \frac{1}{EI}\left[R_A\frac{x^2}{2} - M_A x\right]$ For $a \le x \le L$ $W(x) = \frac{1}{EI}\left[-R_B\frac{x^3}{6} - (M_A - Pa)\frac{x^2}{2} + B_1 x + B_2\right]$ $\theta(x) = \frac{1}{EI}\left[-R_B\frac{x^2}{2} - (M_A - Pa)x + B_1\right]$ where $B_1 = -\frac{Pa^2}{2},\ B_2 = +\frac{Pa^3}{2}$
10	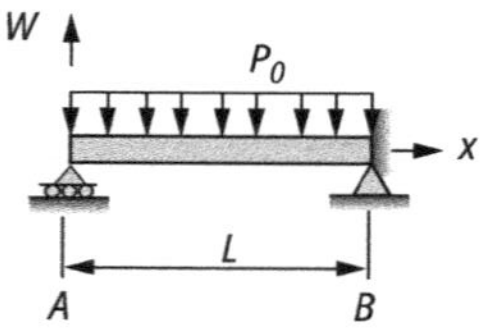	Simply supported	Clamped	$R_A = \frac{3p_0 L}{8};\ M_A = 0$ $R_B = \frac{5p_0 L}{8};\ M_B = -\frac{p_0 L^2}{8}$	For $0 \le x \le L$ $W(x) = \frac{p_0 x}{48EI}\left[3Lx^2 - 2x^3 - L^3\right]$ $\theta(x) = \frac{p_0}{48EIL}\left[9Lx^2 - 8x^3 - L^3\right]$

$$\omega_n = \sqrt{\frac{k}{m}} \Rightarrow f_\text{n} = \frac{1}{2\pi}\sqrt{\frac{k}{m}} \tag{3.4}$$

where f_n is the natural frequency in Hz, and $\omega_\text{n} = 2\pi f_\text{n}$ (the circular frequency expressed in rad/s).

For cases where the spring is provided by the deflection of a beam, the resulting natural frequencies are presented in Tab. 3.3, while the mass of the beam is negligible compared to the mass m, and in Tab. 3.5, when this assumption is removed.

Tab. 3.3: Natural frequencies for a concentrated mass resting on a massless beam.

No.	Case	Boundary conditions		Natural frequency, f_n (Hz)
		Left A	Right B	
1	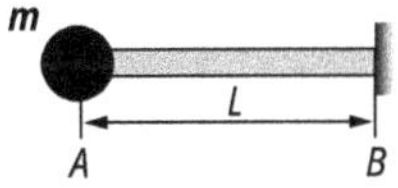	Free	Clamped	$\frac{1}{2\pi}\sqrt{\frac{3EI}{mL^3}}$
2	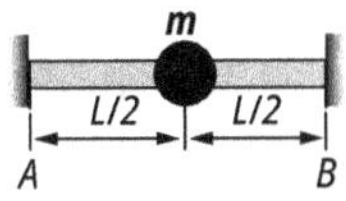	Clamped	Clamped	$\frac{1}{2\pi}\sqrt{\frac{192EI}{mL^3}}$
3	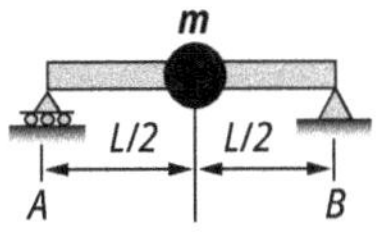	Simply supported	Simply supported	$\frac{1}{2\pi}\sqrt{\frac{48EI}{mL^3}}$
4	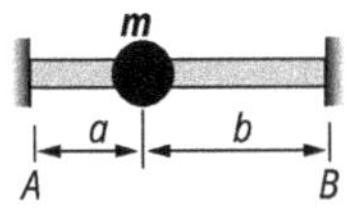	Clamped	Clamped	$\frac{1}{2\pi}\sqrt{\frac{3EIL^3}{ma^3b^3}}$
5	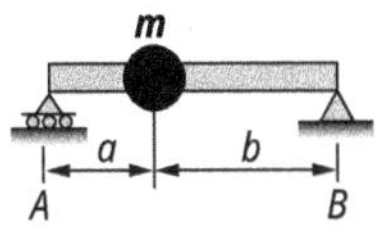	Simply supported	Simply supported	$\frac{1}{2\pi}\sqrt{\frac{3EIL}{ma^2b^2}}$

Note that in Tab. 3.3, E represents the Young's modulus of the beam, I is the cross-sectional area moment of inertia, L is the beam's length, and m is the concentrated mass.

For the case of an isotropic continuous beam, the equation of motion to find its natural frequencies has the following form:

$$\frac{\partial^2}{\partial x^2}\left(EI\frac{\partial^2 w(x,t)}{\partial x^2}\right) + \rho A\frac{\partial^2 w(x,t)}{\partial t^2} = 0 \tag{3.5}$$

Assuming constant structural properties along the beam and using the following expression

$$w(x,t) = W(x)e^{i\omega t} \tag{3.6}$$

Equation (3.5) transforms into the following regular differential equation

$$EI\frac{d^4W(x)}{dx^4} - \rho A\omega^2 W(x) = 0 \tag{3.7}$$

having the following solution:

$$W(x) = C_1\cos(\lambda x) + C_2\sin(\lambda x) + C_3\cosh(\lambda x) + C_4\sinh(\lambda x)$$

where

$$\lambda^4 = \frac{\rho A\omega^2}{EI} \Rightarrow \omega = \lambda^2\sqrt{\frac{EI}{\rho A}} \tag{3.8}$$

where (ρA) is the mass per unit length of the beam, A being the cross section of the beam and ρ its density, EI is the bending stiffness of the beam, and C_1, C_2, C_3 and C_4 are constants, to be determined by imposing boundary conditions at both ends of the beam. This leads to an eigenvalue problem, from which the natural frequencies of the beam are calculated together with their associated mode shapes. The values for the expression $(\lambda L)^2$, where L is the beam's length, for typical configurations of a beam are given in Tab. 3.4.

Tab. 3.4: Natural frequencies for typical beams.

No.	Case	Boundary conditions		$(\lambda L)^2$		
		Left A	Right B	First mode	Second mode	Third mode
1	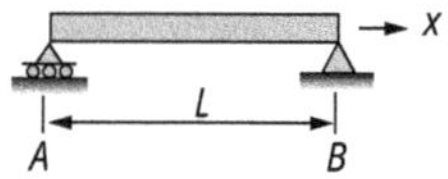	Simply supported	Simply supported	9.87	39.5	88.9
2	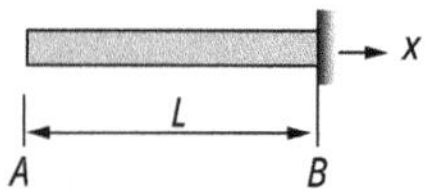	Free	Clamped	3.52	22.4	61.7
3	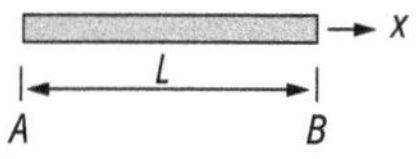	Free	Free	0	22.40	61.7

Tab. 3.4 (continued)

No.	Case	Boundary conditions		$(\lambda L)^2$		
		Left A	**Right B**	**First mode**	**Second mode**	**Third mode**
4	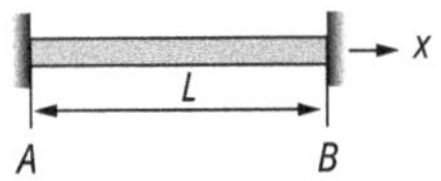	Clamped	Clamped	22.40	61.7	121.0
5	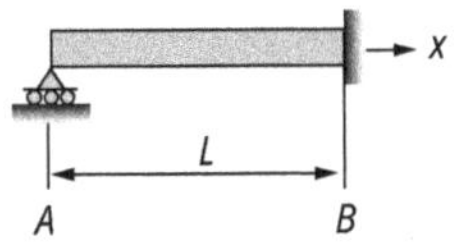	Simply supported	Clamped	15.40	50.0	104.0
6	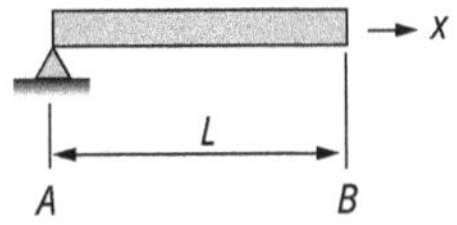	Simply supported	Free	0	15.4	50.0

When a concentrated mass is added to a beam, its first natural frequency is lowered. Typical values are presented in Tab. 3.5.

Tab. 3.5: Natural frequencies for a concentrated mass resting on a beam with a mass of M_b.

No.	Case	Boundary conditions		Natural frequency, f_n (Hz)
		Left A	**Right B**	
1	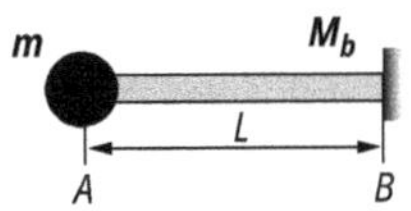	Free	Clamped	$\frac{1}{2\pi}\sqrt{\frac{3EI}{(m+0.23M_b)L^3}}$
2	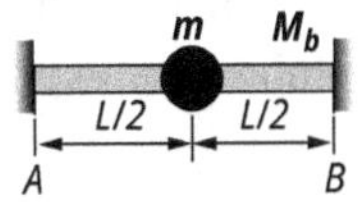	Clamped	Clamped	$\frac{1}{2\pi}\sqrt{\frac{196EI}{(m+0.35M_b)L^3}}$
3	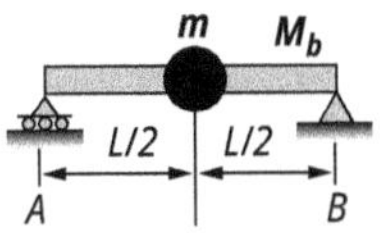	Simply supported	Simply supported	$\frac{1}{2\pi}\sqrt{\frac{48EI}{(m+0.5M_b)L^3}}$

The classical equation of motion for the transverse deflection, w, of thin plates [4] can be written as

$$D\nabla^4 w + \bar{m}\frac{\partial^2 w}{\partial t^2} = 0 \tag{3.9}$$

where the expression for D, the flexural rigidity of the plate, is given as

$$D = \frac{Eh^3}{12(1-\upsilon^2)} \tag{3.10}$$

where $\bar{m}$ is the mass density per unit area of the plate, E the Young's modulus, h the plate thickness, υ the Poisson's ratio, t the time and ∇^4 the Laplacian (∇^2) squared, defined in Cartesian coordinates as

$$\nabla^4 \equiv \left[\frac{\partial^2}{\partial x^2} + \frac{\partial^2}{\partial y^2}\right]^2 = \frac{\partial^4}{\partial x^4} + 2\frac{\partial^4}{\partial x^2 \partial y^2} + \frac{\partial^4}{\partial y^4} \tag{3.11}$$

In polar coordinates (r, θ), the expression for ∇^4 is written as

$$\nabla^4 \equiv \left[\frac{\partial^2}{\partial r^2} + \frac{1}{r}\frac{\partial}{\partial r} + \frac{1}{r^2}\frac{\partial^2}{\partial \theta^2}\right]^2 \tag{3.12}$$

while for skew coordinates (ξ, η) that are related to x, y Cartesian coordinates (where α is the angle between y and η coordinates)

$$\xi = x - y\tan\alpha \qquad \eta = \frac{y}{\cos\alpha} \tag{3.13}$$

the expression for ∇^4 is given as

$$\nabla^4 \equiv \left[\frac{1}{\cos^2\alpha}\left(\frac{\partial^2}{\partial \xi^2} - 2\sin\alpha\frac{\partial^2}{\partial \xi \partial \eta} + \frac{\partial^2}{\partial \eta^2}\right)\right]^2 \tag{3.14}$$

For transforming the partial differential equation (3.9) into an ordinary differential equation as for free vibration of beams, presented above, the following expression is used:

$$w = We^{i\omega t} \tag{3.15}$$

which leads to

$$D\nabla^4 W - \bar{m}\omega^2 W = 0 \tag{3.16}$$

where ω is the circular frequency (in rad/s) and W is a function only of the position coordinates.

Typical values for the eigenvalues, λ, of various circular plates with different boundary conditions are presented in Tab. 3.6 (from [4]).

Tab. 3.6: Values of $\lambda^2 = \omega R^2 \sqrt{m/D}$ for a circular plate with a radius R.

s n	0	1	2	3	4
All around clamped circular plate at $r = R$					
0	10.2168	21.26	34.88	51.04	69.6659
1	39.771	60.82	84.68	111.01	140.1079
2	89.104	120.08	163.81	190.30	229.5186
3	168.183	199.06	242.71	289.17	338.4113
All around simply supported circular plate at $r = R$, $v = 0.3$					
0	4.977	13.94	25.65	–	–
1	29.76	48.51	70.14	–	–
2	74.20	102.80	134.33	–	–
3	138.34	176.84	218.24	–	–
Completely free circular plate, $v = 0.33$					
0	–	–	5.253	12.23	21.6
1	9.084	20.52	35.25	52.91	73.1
2	38.55	59.86	83.9	111.3	142.8
3	87.80	119.0	154.0	192.1	232.3

For rectangular plates, the natural frequencies can be obtained analytically only for a plate on simply supported boundary conditions for all four edges. This expression is given as

$$\omega_{m,n} = \sqrt{\frac{D}{\bar{m}}}\left[\left(\frac{m\pi}{a}\right)^2 + \left(\frac{n\pi}{b}\right)^2\right] \tag{3.17}$$

where a and b are the length and width of the plate, respectively, and m and n are half-waves in the longitudinal and lateral directions, respectively. For other boundary conditions, various approximate methods have been used to determine the natural frequencies. Table 3.7 presents the results obtained using the Rayleigh method, where the assumed function is given in Tab. 3.7, and the natural frequency is calculated assuming the following expression (for $\upsilon = 0.25$, see [4]):

$$\omega = \frac{\pi^2}{a^2}\sqrt{\frac{D\alpha}{\bar{m}\beta}} \tag{3.18}$$

Tab. 3.7: Frequency coefficients for eq. (3.18) and various mode shapes at ν = 0.25.

<table>
<tr><th>Boundary conditions</th><th>Deflection function or mode shape</th><th>α</th><th>β</th></tr>
<tr><td>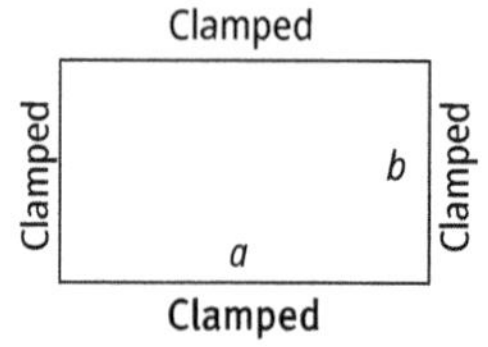
</td><td>$\left(\cos\frac{2\pi x}{a}-1\right)$ $\left(\cos\frac{2\pi y}{b}-1\right)$</td><td>$12+8\left(\frac{a}{b}\right)^2+12\left(\frac{a}{b}\right)^4$</td><td>2.25</td></tr>
<tr><td>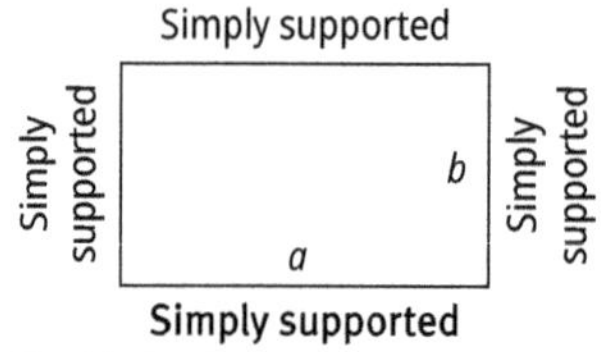
</td><td>$\sin\frac{\pi x}{a}\sin\frac{\pi y}{b}$</td><td>$\frac{1}{4}+\frac{1}{2}\left(\frac{a}{b}\right)^2+\frac{1}{4}\left(\frac{a}{b}\right)^4$</td><td>0.25</td></tr>
<tr><td>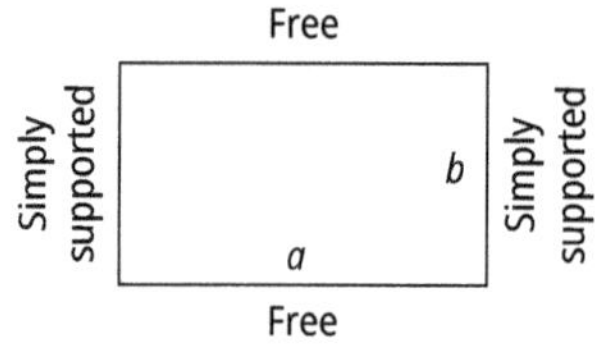
</td><td>$\sin\frac{\pi x}{a}$</td><td>$\frac{1}{2}$</td><td>0.50</td></tr>
<tr><td>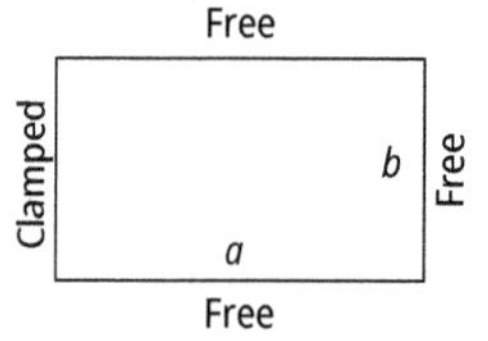
</td><td>$1-\cos\frac{\pi x}{2a}$</td><td>0.0313</td><td>0.2268</td></tr>
<tr><td>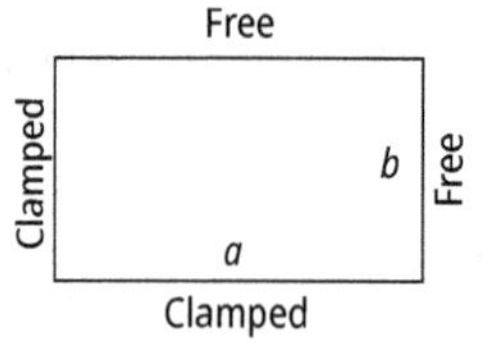
</td><td>$\left(1-\cos\frac{\pi x}{2a}\right)$ $\left(1-\cos\frac{\pi y}{2b}\right)$</td><td>$0.0071+0.024\left(\frac{a}{b}\right)^2$ $+0.071\left(\frac{a}{b}\right)^4$</td><td>0.0514</td></tr>
<tr><td>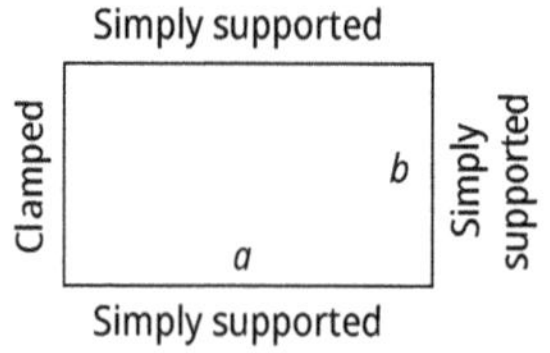
</td><td>$\left(\cos\frac{3\pi x}{2a}-\cos\frac{\pi x}{2a}\right)$ $\sin\frac{\pi y}{b}$</td><td>$1.28+\frac{5}{4}\left(\frac{a}{b}\right)^2+\frac{1}{2}\left(\frac{a}{b}\right)^4$</td><td>0.50</td></tr>
</table>

Tab. 3.7 (continued)

Boundary conditions	Deflection function or mode shape	α	β
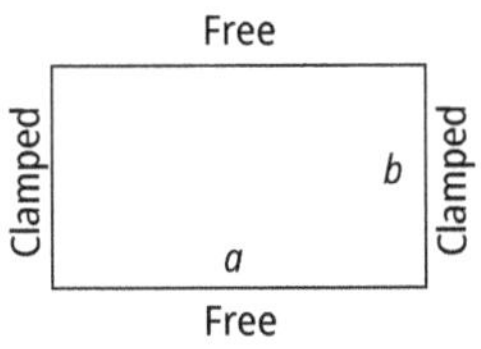	$\cos\frac{2\pi x}{a}-1$	1.5	1.50
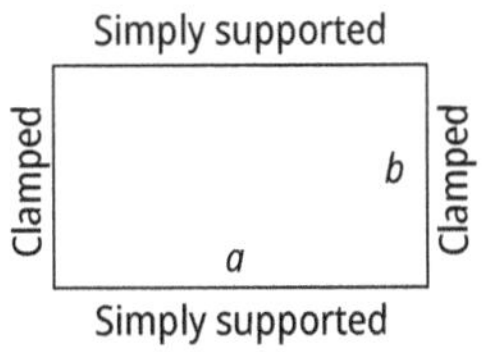	$\left(\cos\frac{2\pi x}{a}-1\right)\sin\frac{\pi y}{b}$	$4+2\left(\frac{a}{b}\right)^2+\frac{3}{4}\left(\frac{a}{b}\right)^4$	0.75
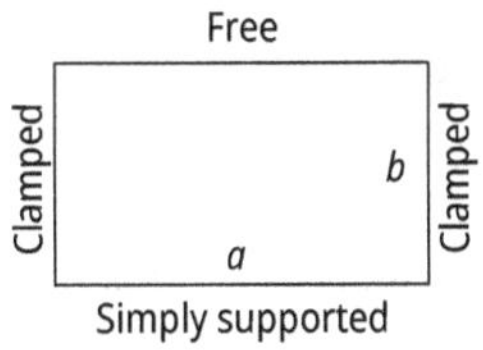	$\left(\cos\frac{2\pi x}{a}-1\right)\frac{y}{b}$	$2.67+0.304\left(\frac{a}{b}\right)^2$	0.50

3.5 Torsion of bars

When a bar is held at one side and being rotated at the other side by a torque T (see Fig. 3.2), the bar will be under torsion, leading to shear stresses. The calculation of the stresses for bars having circular cross section assumes that the cross section rotates and remains planar, leading to the following expression:

$$\tau=\frac{Tr}{J}\Rightarrow\tau_{\max}=\frac{TR}{J} \tag{3.19}$$

where τ is the shear stress at an intermediate radius r, T is the applied torque and J is the polar moment of inertia expressed in m^4. The maximal shear stress $\tau_{\max}$ will appear at $r=R$. The angle of rotation, θ, is calculated according to the following equation:

$$\theta=\frac{TL}{GJ} \tag{3.20}$$

where G is the shear modulus and L is the length of the bar. Note that for isotropic materials G can be expressed using the Young's modulus, E, and the Poisson's ratio, υ, to yield

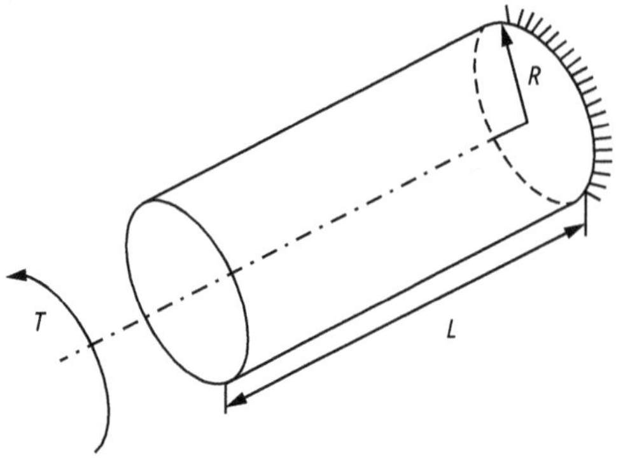

Fig. 3.2: A circular bar with a radius *R* and length *L* under a torque *T*.

$$G = \frac{E}{2(1+\upsilon)} \tag{3.21}$$

For noncircular cross sections, the behavior of the bars is nonsymmetric when a torque is applied and the plane sections do not remain plane. Note that the distribution of stress in a noncircular cross section is not necessarily linear as for circular cross sections. Typical expressions for the polar moment of inertia and the associated maximal shear stresses are presented in Tab. 3.8. For additional cases, the reader is addressed to Ref. [5].

Tab. 3.8: Polar moment of inertia and maximal shear stresses.

Type of cross section	Polar moment of inertia, J	Maximal shear stress, τ_{max}
	$\frac{\pi R^4}{2}$	$\frac{2T}{\pi R^3}$
	$\frac{\pi a^4}{2} - \frac{\pi b^4}{2}$	$\frac{2Ta}{\pi(a^4 - b^4)}$
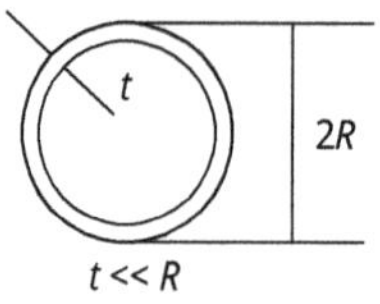	$2\pi R^3 t$	$\frac{T}{2\pi R^2 t}$

Tab. 3.8 (continued)

Type of cross section	Polar moment of inertia, J	Maximal shear stress, τ_{max}
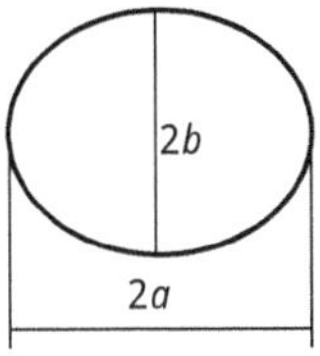	$\frac{\pi a^3 b^3}{(a^2+b^2)}$	$\frac{2T}{\pi a b^2}$ @ ends of minor axis
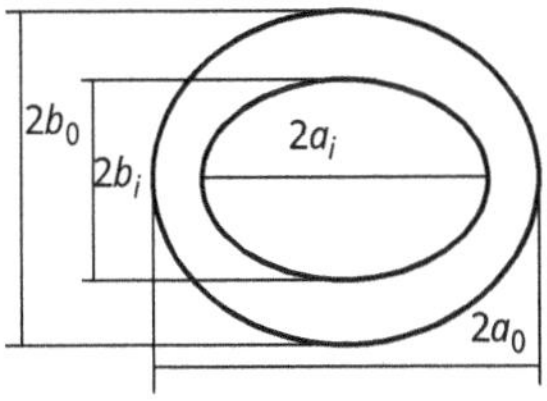	$\frac{\pi a^3 b^3}{(a^2+b^2)}\left(1-\kappa^4\right)$ $\kappa \equiv \frac{a_i}{a_o}=\frac{b_i}{b_o}$	$\frac{2T}{\pi a b^2\left(1-\kappa^4\right)}$
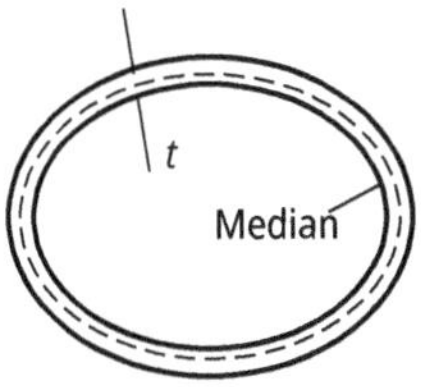	$\frac{4A^2t}{L'}$, A = area enclosed by the median, L' = length of median	$\frac{T}{2tA}$
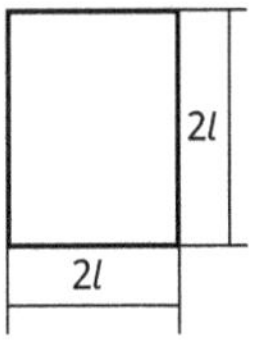	$\frac{9l^4}{4}$	$\frac{0.601T}{l^3}$ @ mid of each edge
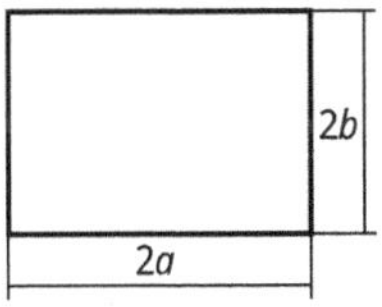	$ab^3\left[\frac{16}{3}-3.36\frac{b}{a}\left(1-\frac{1}{12}\left(\frac{b}{a}\right)^4\right)\right]$	$\frac{3T}{8ab^2}\{1+0.6095\left(\frac{b}{a}\right)$ $+0.8865\left(\frac{b}{a}\right)^2-1.8023\left(\frac{b}{a}\right)^3$ $+0.9100\left(\frac{b}{a}\right)^4$ @ mid of each edge a
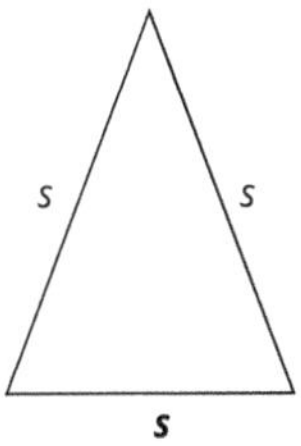	$\frac{\sqrt{3}s^4}{80}$	$\frac{20T}{s^3}$ @ mid of each edge s

Tab. 3.8 (continued)

Type of cross section	Polar moment of inertia, J	Maximal shear stress, τ_{max}
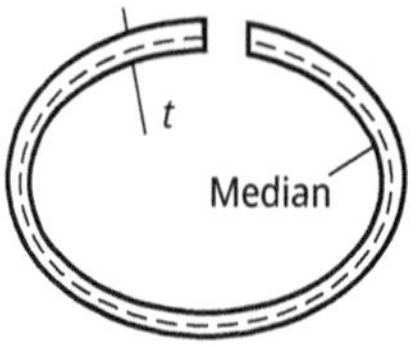	$\frac{L't^2}{3}$, L' = length of median	$\frac{T(3L'+1.8t)}{t^2(L')^2}$ @ mid of each edge s
Th in-walled profiles $L_i >> t_i$ 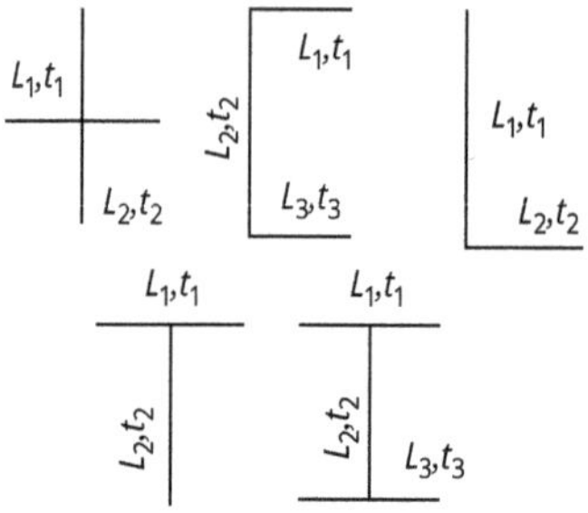 L_i = length of element t_i = thickness of element	$\frac{1}{3}\sum_{i=1}^{n}(L_i)^3 t_i$	$\frac{TL_{max}}{\frac{1}{3}\sum_{i=1}^{n}(L_i)^3 t_i}$ @ mid-way along t_{max} element

References

[1] Timoshenko, S.P. and Goodier, J.N. Theory of elasticity, 3rd Ed., International Student Edition, McGraw-Hill Book Company, Kögakusha Company, Ltd., Tokyo, Japan, ©, McGraw-Hill Inc., 1970, 567.

[2] Popov, E.P. Introduction to mechanics of solids, Prentice-Hall, Inc., Englewood Cliffs, N.J., 1968, 571.

[3] Thompson, W.T. Vibration theory and applications, George Allen & Unwin Ltd., Ruskin House, 40 Museum Street, W. C. 1, London, UK, 1965, 384.

[4] Leissa, A.W. Vibration of Plates, NASA SP-160, 1969, 350.

[5] Young, W.C. and Budynas, R.G. Roark's formulas for stress and strain, 7th Edition, Mc-Graw Hill, New York, NY 10121-2298, USA., 2002, 852.

4 Introduction to fatigue

4.1 Introduction

This chapter presents the behavior of structures under fatigue. The structures can be made of metals or laminated composite materials. The issue of crack propagation, normally associated with the fatigue phenomenon, will be presented in Chapter 5.

4.2 Definition of fatigue

Fatigue (coming from the French word fatigué = tired) of metal structures had been identified already in 1840 in the English railway industry, mainly on train axle failures, as people thought that the materials became "tired" and weak due to cyclic repeated loading [1]. The name *fatigue* was then introduced to describe these structural failures due to repeated stresses and continued to be used until today to describe failures, mainly caused by fracture of the structure due to cycling loadings (see Fig. 4.1 for typical fatigue failures performed at the Aircraft Structures Laboratory, Faculty of Aerospace Engineering, Technion, I.I.T., 32000 Haifa, Israel, by the author of this book).

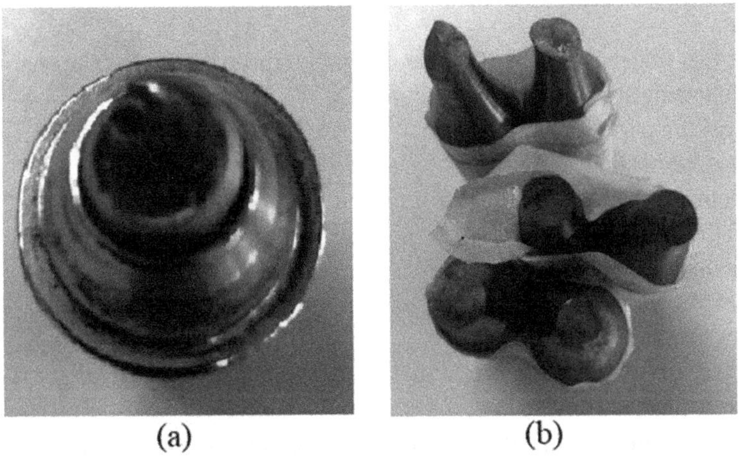

(a) (b)

Fig. 4.1: Cross section after fatigue failure: (a) wheel car bolt and (b) test specimens.

Most of the structural parts of movable machines are susceptible to fatigue; therefore, it is very important to assure that these parts will not fail due to repetitive stresses, which might be expected to reach at least a few million load cycles. The aerospace sector, the automotive and train sector, the wind-power turbines sector and the mari-

https://doi.org/10.1515/9783111621104-004

time sector are the main domains of engineering that are heavily influenced by the fatigue. Fatigue of various aerospace, mechanical and civil structures had been investigated in depth during the past years, with numerous books trying to address the issue (see typical books in [1–19]). Apart from books, the literature is full of dedicated manuscripts covering all the aspects of fatigue and the ways to prevent failures due to it [20–34].

4.2.1 Basic fatigue concepts

The various factors causing fatigue failure of a given structural member can be listed as
- High value of tensile stresses
- A large number of variations and changes of the applied stresses
- A sufficient large number of the applied stress cycles
- The metallurgical structure or material characteristics
- Stress concentration
- Corrosion
- Residual stress
- Combined stresses
- Temperature
- Overloading the structure

One should note that under the point "The metallurgical structure and material characteristics" the following attributes could be found:[1]

a. Due to the way the part was fabricated:
   - Method of fabrication
     - Shaping method: ground, turned, broached, milled, EDM,[2] chemical milling
     - Bending, spinning, stretching
     - Welding
     - Heat treatment
   - Finishing process
     - Surface finish: polished, as-cast surface, as-machined surface (with maximal surface roughness)
     - Special treatments like cold working, case hardening, plating, anodized within specified thickness

1 Note that the material attributes are normally controlled by the material specifications such as *SAE/AMS* (*S*ociety of *A*utomotive *E*ngineering/*A*erospace *M*aterial *S*pecification) number, part drawing, the manufacturing company standards and by the way the part is fabricated.

2 *EDM*, *e*lectrical *d*ischarge *m*achining.

b. Due to the SAE/AMS number, part drawing and company standards:
   - The material itself
     - The chemical composition
     - The cleanliness of the process (like commercial quality), the melting method (in air vs. in vacuum) and so on
   - Forming the material and its processing
     - Casting, un-HIP[3]'d or HIP'd (location in the casting form)
     - Forging, bar, extrusion, plate sheet (its texture and cold working process)
     - Type of the metal powder
     - Microstructure (the size of the grain)
     - The overall size of the part
     - Heat treatment
     - Hardness

The fluctuating stresses acting on the structure can be either sinusoidal above the time axis, as presented in Fig. 4.2, meaning the stresses are only tensile, or completely reversed cycle of stress, in which the sinusoidal curve passes the time axis, leading to alternating tensile and compressive stresses. The alternating stresses can also appear in irregular or random form. One should note that tensile stresses are positive, with negative values being attributed to the compression ones.

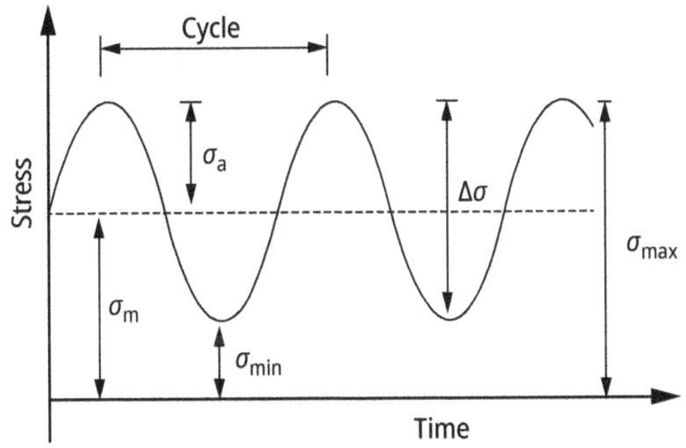

Fig. 4.2: Schematic definition of various expressions for cyclic stress.

The definition of the various expressions is displayed in Fig. 4.2, and the additional ones are depicted in Tab. 4.1.

One should note that a structure undergoing cyclic or fluctuating stresses might fail at loads lower than the static ones. It was noted that almost 90% of all the failure of metallic components, like aircraft, bridges and machine parts, are attributed to

3 *HIP*, *h*ot *i*sostatic *p*ressing.

fatigue. The fatigue failure is sudden and catastrophic and has the nature of brittle-type failure, even in normally ductile-type materials.

Tab. 4.1: Definition of various expressions connected with cyclic stresses.

No.	Name	Equation/notation
1	Maximum stress	σ_{max}
2	Minimum stress	σ_{min}
3	Stress range	$\Delta\sigma \equiv \sigma_r = \sigma_{max} - \sigma_{min}$
4	Fluctuating stress	$\sigma_a \equiv \frac{\Delta\sigma}{2} = \frac{\sigma_{max} - \sigma_{min}}{2}$
5	Average stress	$\sigma_m \equiv \frac{\sigma_{max} + \sigma_{min}}{2}$
6	Stress ratio	$R \equiv \frac{\sigma_{min}}{\sigma_{max}}$
7	Amplitude ratio	$A \equiv \frac{\sigma_a}{\sigma_m} = \frac{1-R}{1+R}$

The behavior of the two types of materials, brittle versus ductile ones, is presented in Fig. 4.3, where the relevant curves of the stress versus the strain are plotted. It is evident that a brittle material presents no yield point and its failure is a sudden breakage into pieces due to the application of tensile stresses. A ductile material exhibits a well-defined yield point with failure in the plastic strain regime.

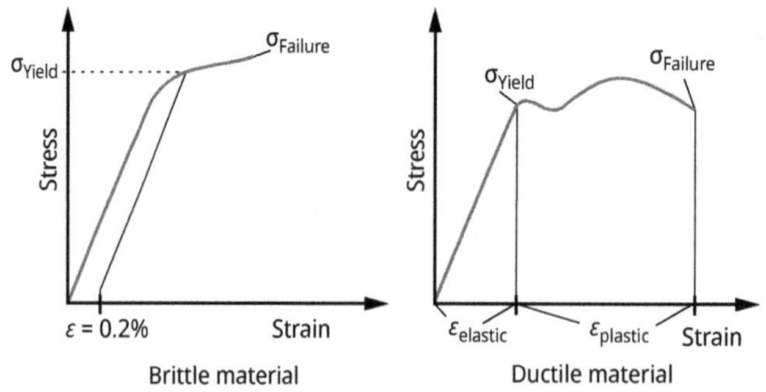

Fig. 4.3: Behavior of brittle and ductile materials under tensile loading.

4.2.2 The four steps of fatigue failure

One of the basic questions related to fatigue is "why the structural component would fatigue fail if the working stress is below the yield stress?" The answer to this query is microscopic plasticity behavior at fatigue failure, occurring below the yield stress, and another important factor connected to damage accumulation on the time axis is the many load cycles being applied on the structural component. Roughly, four steps can be identified up to the failure:

1. Crack initiation on external and internal surfaces of the specimen due to slip process. This will take part within the first 10% of the total component life.
2. The first stage of the crack growth along planes under high shear stresses, a continuation of the slip process encountered in the first step. This causes the deepening of the crack.
3. The second stage of the crack growth, which is characterized by a crack growth along high tensile stresses. This is called transgranular crack propagation.
4. Ductile failure of the component due to the crack propagation, which leads to the reduction in the load-bearing area.

4.3 The *S–N* curve

The well-known curve, the *S–N* curve, known also as Wöhller[4] curve, is the basic representation of engineering fatigue data, in which the stress, *S*, versus the number of cycles, *N*, is drawn (see Fig. 4.4). The *y*-axis, the stress, can be one of the following values (see Tab. 4.1), σ_{max} or σ_{min}, while mentioning the σ_m, *R* or *A*.

On the graph presented in Fig. 4.4, material "A" displays a threshold value called "*fatigue limit*" or "*endurance limit*" in the range of 10^7 or 10^8 cycles. Below this limit, the material presumably can endure an infinite number of cycles before failure. Materials like mild steel (St) and titanium (Ti) show this property. The second graph, in Fig. 4.4, material "B," shows no fatigue limit, a property attributed to nonferrous materials like aluminum (Al), magnesium (Mg) and copper (Cu), for which the fatigue strength is defined at 10^8 cycles. Another interesting property appearing in the *S–N* curve is the regime named *fatigue failure at high numbers of cycles* or *high cycle fatigue (HCF)* defined as $N > 10^5$ cycles, while the region $N < 10^5$ cycles is called *low cycle fatigue (LCF)*. Note that the value of *N* would increase with the decrease in the stress level. In addition, the HCF is characterized by low strains, while LCF would display high strains. The *S–N* curve for the HCF region is sometimes analytically described by the Basquin equation [35], having the following form:

4 August Wöhller (1819–1914) was a German railway engineer who investigated in-depth metal fatigue failures of railway axles.

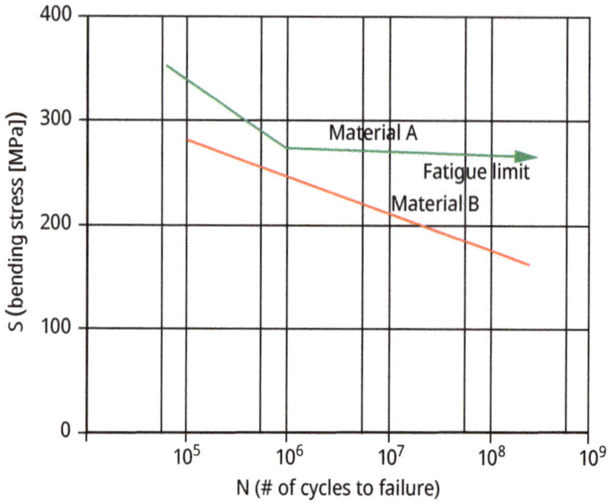

Fig. 4.4: A schematic *S–N* curve.

$$\sigma_{\max} = \alpha N_f^{\beta} \tag{4.1}$$

where α and β are model fitting constants, $\sigma_{\max}$ is the stress amplitude at $N < 10^6$, and N_f is the number of cycles at failure. To determine the constants, the following procedure is performed: taking the log of eq. (4.1) yields

$$\log(\sigma_{\max}) = \log(\alpha) + \beta(N_f) \tag{4.2}$$

Assume that $\sigma_{\max}$ at low number of cycles, let say $N = 1{,}000$, is 90% of the ultimate strength of the tested material, σ_{ul}, and that for $N = 10^6$ $\sigma_{\max} = \sigma_\infty$ (where σ_∞ stands for stress at fatigue limit), thus yielding indefinite fatigue life, which leads to the following equations:

$$\begin{aligned} \log(0.9\sigma_{ul}) &= \log(\alpha) + \beta(1,000) = \log(\alpha) + 3\beta \\ \log(\sigma_\infty) &= \log(\alpha) + \beta(1,000,000) = \log(\alpha) + 6\beta \end{aligned} \tag{4.3}$$

from which the two constants, α and β, can be determined to yield

$$\alpha = \frac{\sigma_{F.L.}}{10^{6\beta}}; \quad \beta = \frac{\log(\sigma_\infty) - \log(0.9S_{ul})}{3}. \tag{4.4}$$

To determine the N_f, for a given stress $\sigma_{\max}$, once the constants α and β were calculated (see eq. (4.4)), eq. (4.1) is changed to yield

$$N_f = \left[\frac{\sigma_{\max}}{\alpha}\right]^{\frac{1}{\beta}} \tag{4.5}$$

Another well-known description of the S–N curve carries the name of Weibull distribution [36], which suggested the following expression:

$$\frac{\sigma_{\max} - \sigma_{\infty}}{\sigma_{\mathrm{ul}} - \sigma_{\infty}} = 1 - \varphi[\alpha \log N_{\mathrm{f}} + \beta] \tag{4.6}$$

where $\sigma_{\max}$ is the maximal applied stress, σ_{ul} is the ultimate tensile strength stress, σ_{∞} stands for the fatigue limit stress, α, β are model fitting parameters, N_{f} is the number of cycles at failure, and φ is a Gaussian error function. Based on eq. (4.6), Weibull proposed an S–N curve model having the following expression:

$$\sigma_{\max} = (\sigma_{\mathrm{ul}} - \sigma_{\infty})e^{\left[-\alpha(\log N_{\mathrm{f}})^{\beta}\right]} + \sigma_{\infty} \tag{4.7}$$

which includes both the ultimate strength stress (σ_{ul}) and the fatigue limit stress (σ_{∞}). The two fitting parameters (α and β) in eq. (4.7) can be determined by taking log on both sides of the equation, namely

$$\frac{\sigma_{\max} - \sigma_{\infty}}{\sigma_{\mathrm{ul}} - \sigma_{\infty}} = e^{\left[-\alpha(\log N_{\mathrm{f}})^{\beta}\right]} \tag{4.8}$$

One should note that based on the characterization of the materials using the S–N curves, some designed philosophies were used. The old design standard, also known as infinite life design, was based on the empirical information of the fatigue life presented by the S–N curves, adding to it a large safety factor, and requesting the retiring of parts and components at the preset life limit value of $N_{\mathrm{failure}} = 10^7$. Today, the American air force (but not the navy) adopted a standard based on the damage-tolerant design, which accepts the presence of cracks in the various components of a structure. The determination of the structural life is based on the prediction of the crack growth rate (see Chapter 5).

The experimental construction of the S–N is usually done using 8–12 test specimens. Normally, the tests would start at a stress $\sigma = 2/3\sigma_{\text{static tensile strength}}$ of the tested material and the stress is lowered until the specimens do not fail at about 10^7 cycles. As expected, a considerable scatter in the experimental data can be expected, with adoption of statistical approach to define the fatigue limit. Usually, a logarithmic normal distribution of the fatigue life is assumed for the region with failure probability of $0.10 < P < 0.90$ (P = probability).

Note also that the increase in the mean stress applied to a component would decrease its fatigue strength.

One should be aware of the great influence of the stress ratio, R, as depicted in Fig. 4.5, on the fatigue life of the tested specimens. The largest fatigue life would be obtained at $R = -1$, while the minimal one at $R = +1$, with a large drop in the interval $-1 < R < +1$.

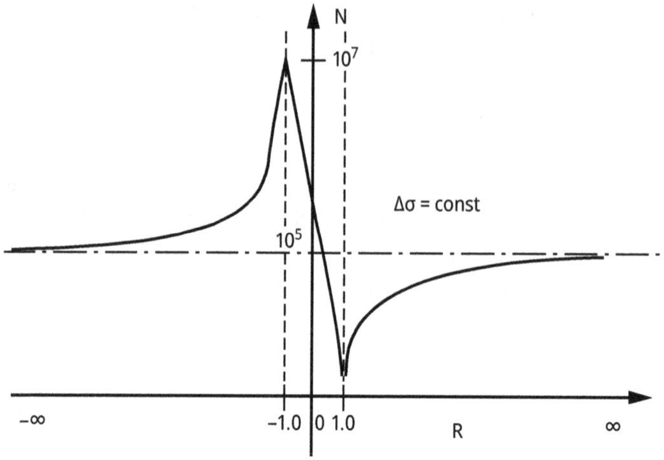

Fig. 4.5: Schematic drawing of the influence of stress ratio, *R*, on the fatigue life of samples.

4.3.1 The probability distributions for fatigue life

The log-normal distribution [37] is one of the most useful probability distributions to model failure times of materials and structures undergoing fatigue. Its probability density function (PDF) has the following form:

$$f(x) = \frac{1}{\sqrt{2\pi} \cdot \sigma} e^{\left[-\frac{1(x-\mu)^2}{2\ \sigma^2}\right]}, \quad -\infty < x < +\infty \tag{4.9}$$

where $x = \log N_f$.

Note that if the fatigue life N_f has a log normal, the logarithm of N_f, namely x, has a normal distribution. This conclusion is very important as one can use the theoretical results and the equation of normal distribution, while applying the log-normal distribution. The cumulative distribution function (CDF) of the log-normal distribution has the following form:

$$F(x) = \int_{-\infty}^{x} f(t)dt \tag{4.10}$$

where the variable t is the integration one, and $f(t) = f(x)$ is given by eq. (4.9). Note that the integral in eq. (4.10) has no closed-form solution and a numerical algorithm is needed to evaluate it. Next, the evaluation of the two variables has to be estimated using the following relationships:

$$\mu = \frac{1}{n}\sum_{i=1}^{n} x_i; \qquad \sigma = \sqrt{\frac{\sum (x_i - \mu)^2}{n-1}}. \tag{4.11}$$

For a fatigue data obtained with n samples with $x_i = \log(N_i)$, $i = 1, \ldots, n$.

Another very popular model, successfully applied for fatigue issues, is the Weibull distribution [37], having the following PDF form:

$$f(x) = \frac{\beta}{\Omega}\left(\frac{x - x_0}{\Omega}\right)^{\beta-1} e^{-\left(\frac{x - x_0}{\Omega}\right)^{\beta}} \tag{4.12}$$

where β is the shape parameter changing the shape of the PDF curve and it can have the following values:

a. If $\beta < 1$, PDF shape is similar to an exponential distribution.
b. If $1 < \beta < 3$, the PDF shape is positively skewed.
c. If $\beta > 3$, the PDF is symmetrical.

The symbol Ω in eq. (4.12) is the scale parameter or the dispersion of the distribution, affecting both the mean and standard deviation. The third parameter, x_0, the location parameter for the Weibull distribution, is defined as $x_0 = \log N_0$, where N_0 is a lower fatigue life, namely no fatigue failure will happen under this value. If $x_0 = 0$, the three-parameter Weibull distribution will be reduced to two-parameter Weibull distribution, with the previous one providing the best fitting capacity. The CDF of Weibull distribution has an explicit expression with the following form:

$$F(x) = 1 - e\left(-\left(\frac{x - x_0}{\Omega}\right)^{\beta}\right), \quad x \geq x_0 \tag{4.13}$$

To estimate the three parameters in the Weibull distribution, the most used method is to fit the experimental points in a linear regression line in the form $y = a + bx$, using the least square method. Using eq. (4.13), one obtains

$$\log\left[\log\left(\frac{1}{1 - F(x)}\right)\right] = -\beta\log(\Omega) + \beta\log(x_i - x_0) \tag{4.14}$$

Applying the least-square fit yields

$$\begin{aligned} y_i &= \log\left[\log\left(\frac{1}{1 - F(x_i)}\right)\right] \\ x_i &= \log(N_i) \\ a &= \beta \\ b &= -\beta\log(\Omega) \end{aligned} \tag{4.15}$$

For a Weibull distribution, an approximated straight line could be drawn, with the parameter β being the slope of the graph, and the parameter Ω is estimated for a point of the straight line corresponding to 63.2% failure, namely $F(x_i) = 0.632$. The third parameter, x_0, is estimated using an iteration procedure (no closed form is available).

4.3.2 Fatigue life for various combinations of alternating and mean stresses

When plotting the alternating stress σ_a versus the mean stress σ_m, as shown in Fig. 4.6, various lines can be achieved and compared with test results. These lines have the following expressions (note that σ_e is the fatigue limit stress for a completely reversible loading ($R = -1$)):

a. Goodman line – a linear approximation

$$\left(\frac{\sigma_m}{\sigma_{ul}}\right) + \left(\frac{\sigma_a}{\sigma_e}\right) = 1 \tag{4.16}$$

b. Soderberg line – a linear approximation as in eq. (4.16) with the σ_{ul} being replaced by σ_y, the yield strength of the material to give

$$\left(\frac{\sigma_m}{\sigma_y}\right) + \left(\frac{\sigma_a}{\sigma_e}\right) = 1 \tag{4.17}$$

c. Gerber's parabola line – a nonlinear approximation with the following form:

$$\left(\frac{\sigma_m}{\sigma_{ul}}\right)^2 + \left(\frac{\sigma_a}{\sigma_e}\right) = 1 \tag{4.18}$$

d. Modified Goodman line – starts at the yield strength σ_y for the mean stress σ_m and continues along the yield line $[(\sigma_m = \sigma_y, 0), (0, \sigma_a = \sigma_y)]$ (see Fig. 4.6) till it intercepts the Goodman line. From that point, it continues along the Goodman line.

One should note that any combination of mean and alternating stresses under the corresponding lines (depending on which theory is chosen) would lead to the required number of cycles, and all these points above the lines would expect to fail earlier.

Appendix A provides an example for the use of Fig. 4.6 and its associated eqs. (4.16), (4.17) and (4.18).

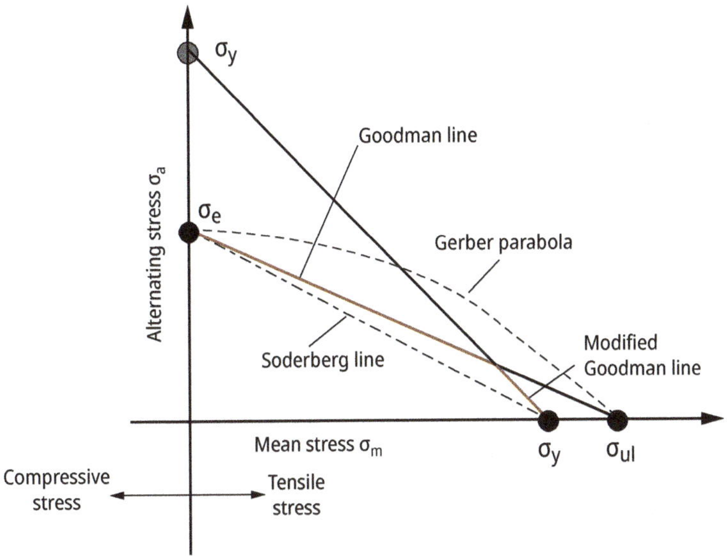

Fig. 4.6: Constant life diagrams: combinations of σ_a and σ_m leading to the same fatigue life.

4.4 Miner rule – the cumulative damage

In real life, a component or a structure might experience loads at variable amplitudes. Designing of fatigue for these structures is performed by the famous approach called Miner's rule or Miner's linear damage rule [38]. According to it, the fatigue damage D is found using the following equation:

$$D = \frac{n_1}{N_1} + \frac{n_2}{N_2} + \frac{n_3}{N_3} + \cdots + \frac{n_i}{N_i} = \sum_{i=1}^{k} \frac{n_i}{N_i} \leq 1.0 \tag{4.19}$$

where $n_1, n_2, n_3, \ldots, n_i$ are the number of cycles at stresses $S_1, S_2, S_3, \ldots, S_i$ expected during the life of the component, while $N_1, N_2, N_3, \ldots, N_i$ are the number of cycles at failure under constant load amplitude (see Fig. 4.7 for a typical example). For values less and/or equal to unity, no failure is expected. Values above unity will mean failure due to fatigue. Note that the failure cycle N_i should be obtained from a relevant component design S–N curve.

One should be aware that the Miner's rule does not account for the sequence of the stress ranges, but only for the number of stress ranges. To take this issue into account, the design of the component should be more conservative, thus reducing the limiting damage to a value less than unity.

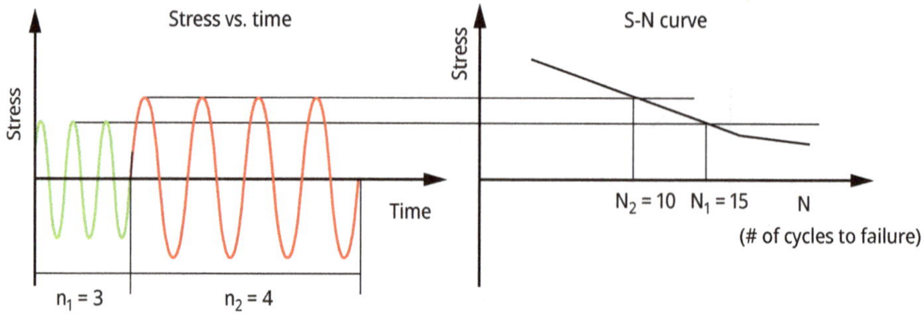

$$D = \frac{n_1}{N_1} + \frac{n_2}{N_2} = \frac{3}{15} + \frac{4}{10} = 0.6 < 1.0 \Longrightarrow \text{No failure !!!}$$

Fig. 4.7: Schematic application of the Miner's rule.

4.5 Fatigue of composite materials

Although laminated composite materials experience high static strength and stiffness together with a low specific mass (in the order of 15–1.6 g/cm^3), showing an almost fully elastic behavior up to failure, as described by Schijve [11], it seems that fatigue problems might appear also for this type of material. This is reflected in the numerous articles dealing with fatigue of composite materials. References [27–33] are only typical for those papers dealing with the behavior of fiberglass-type materials under fatigue, while authors in [39–45] address the way the graphite/epoxy-type materials (used mainly in the aerospace sector) behave under fatigue loads.

In considering fatigue damage and failure, it is generally agreed that fatigue damage consists of various combinations of matrix cracking, fiber–matrix debonding, delamination, void growth and local fiber breakage. Moreover, mechanism, type and distribution of damage would depend upon the material system (combination of fiber and matrix material), stacking sequence of plies, fabrication techniques, geometry, stress state and loading history. Finally, it can be stated that these mechanisms are sensitive to one or more parameters, including type of loading, frequency of cyclic loading, temperature and moisture.

Already in 1979, Rosenfeld and Huang [39] investigated the fatigue behavior of graphite/epoxy laminates under compression for $R = 0, -\infty, -1$. They reported a significant life reduction for the load cases $R = -1$ and $R = -\infty$. The mechanism causing failure is reported to be the failure of the matrix near a stress riser, leading to fiber split that caused a progressive delamination yielding the buckling of the fibers followed by the laminate failure. Rotem [40] reports the behavior of orthotropic laminates under applying tension–compression loading. He investigated two types of laminates: $L_1 = [0^0, \pm 45^0, 0^0]_{2\ s}$ and $L_2 = [90^0, \pm 45^0, 90^0]_{2\ s}$. Both laminates started their failure due to delamination, with the L_1 laminate failing in compression due to delamination between

0^0 and 45^0 laminae, while L_2 failed under tension caused by the delamination between $+45^0$ and -45^0 laminae. Typical results for both laminates are shown in Fig. 4.8.

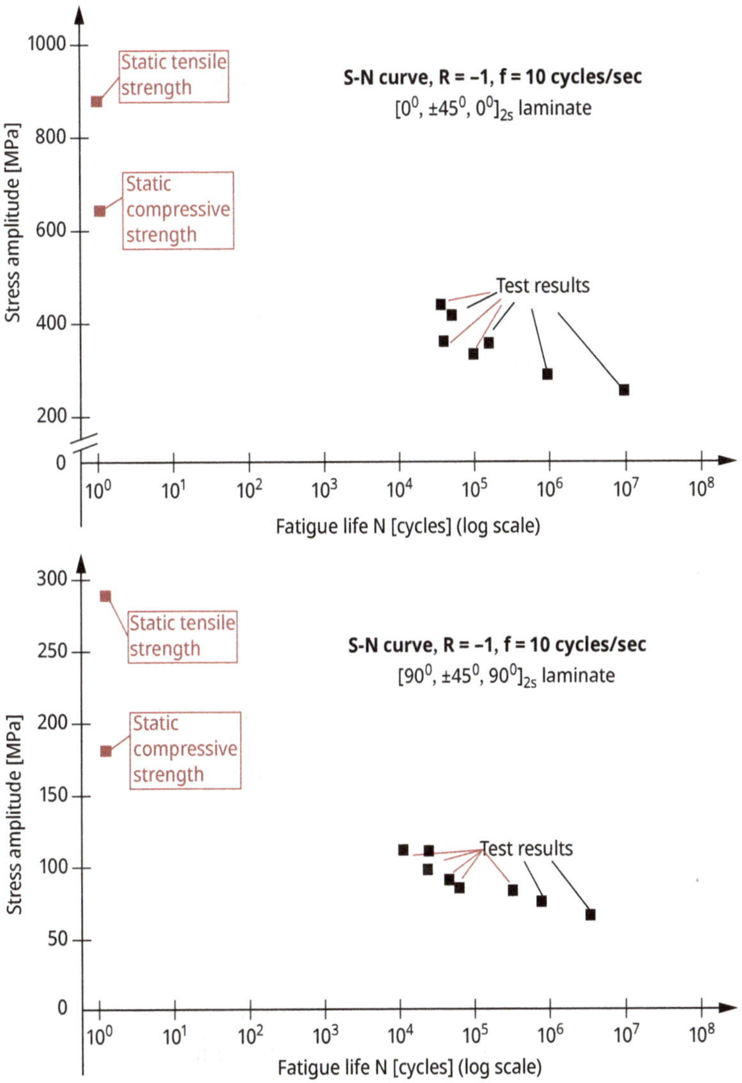

Fig. 4.8: Typical test results (adapted from [40]).

Another team of researchers, Caprino and D'Amore [41], also investigated the fatigue life of graphite/epoxy laminates subjected to tension–compression loading, using the two-parameter model they developed earlier for random glass fiber-reinforced plastics loaded in tension. This model assumes that a continuous decrease in the material strength will be experienced with the increasing number of cycles N, according to the following equation [41]:

$$\frac{d\sigma_{tN}}{dN} = -a_t \cdot N^{-b_t} \tag{4.20}$$

where σ_{tN} is the residual tensile materials strength after N cycles; a_t and b_t are two positive constants to be determined from tests. It was further assumed that the constant a_t can be written as

$$a_t = a_{0t} \cdot (\sigma_{max} - \sigma_{min}) \tag{4.21}$$

with a_{0t} being a constant. Integrating eq. (4.20) using the boundary condition $N = 1 \rightarrow \sigma_{tN} = \sigma_{t0}$, where σ_{t0} is the monotonic tensile strength of the virgin material, one obtains

$$\sigma_{tN} = \sigma_{t0} - \alpha_t \cdot \sigma_{max} \cdot (1 - R) \cdot \left(N^{\beta_t} - 1\right)$$

where

$$\alpha_t = \frac{a_{0t}}{1 - b_t}, \quad \beta_t = 1 - b_t \tag{4.22}$$

Finally, the critical number of cycles to failure, N_t, is calculated using the following expression:

$$N_t = \left[1 + \frac{\sigma_{t0} - \sigma_{max}}{\alpha_t \cdot (\sigma_{max} - \sigma_{min})}\right]^{\frac{1}{\beta_t}} \tag{4.23}$$

To enable the use of the above-developed expressions also for the compression case, eqs. (4.20) and (4.23) were modified by Caprino and D'Amore [41] to yield

$$\frac{d\sigma_{cN}}{dN} = +a_c \cdot N^{-b_c} \tag{4.24}$$

$$N_c = \left[1 + \frac{\sigma_{min} - \sigma_{c0}}{\alpha_c \cdot (\sigma_{max} - \sigma_{min})}\right]^{\frac{1}{\beta_c}} \tag{4.25}$$

The main conclusions from the research presented in [41] claim that the two-constant model is capable of describing accurately the classical S–N curve for composites. The experimental calculation of the two constants seems to be sensitive to the loads applied and yields different values for tension–tension and tension–compression cases. Also, a reasonable agreement was found between the prediction and the experimental results.

Minak et al. [42] present another interesting research addressing the fatigue residual strength of graphite/epoxy circular plates damaged by low velocity impact and undamaged specimens. Good experimental data are presented and discussed. Williams et al. [43] investigated Hercules As/3501-6 graphite fiber/epoxy composite using an eight-ply $[0^0, \pm45^0, 0^0]_2$ laminate. The specimens were subjected to a sinusoidal flexural fatigue in a cantilever mode, with the load varying between + 73.5 and −73.5 N at 30 Hz at room temperature. The structural state of the specimens was measured using ultrasonic 4.0 MHz attenuation method. The method could measure the void volume fraction. They report that beyond 10,000 cycles, the flexural stiffness of the tested specimens decreased with the number of fatigue cycles and the ultrasonic through transmission attenuation increased with the number of fatigue cycles. Minak [44] presents an interesting research on the determination of fatigue life of specimens manufactured from graphite/epoxy, by measuring the temperature variation of the tested specimens throughout the alternating loads. Two types of laminates were tested: type A $[0^0, \pm45^0, 90^0, \pm45^0, 0^0]$ and type B $[90^0, \pm45^0, 0^0, \pm45^0, 90^0]_2$. The load ratio applied was $R = 0.1$, with three levels of frequencies 5, 10 and 15 Hz being used throughout the tests. Measuring the temperature of the external ply and in parallel the variation of the specimen's variation enabled the author to conclude, "dissipated energy-life relation could be used as an alternative method for CFRP components life prediction" [44].

Finally, Jamison and Reifsnider [45] performed a very comprehensive research to establish the character and sequence of development of advanced damage resulting from tension–tension cyclic loading of graphite/epoxy laminates. They defined the advanced damage development as the damage produced by the part of the loading history, which is applied subsequent to the development of the characteristic damage state for matrix cracking.

Three laminate types, $\left[0^0, 90_2^0\right]_s$, $[0^0, \pm45^0]_s$ and $[0^0, 90^0, \pm45^0]_s$, manufactured from T300/5208 graphite/epoxy material, representing a broad range of intralaminar and interlaminar stress conditions were subjected to tensile fatigue loading. Post-fatigue analysis was performed by both nondestructive and destructive (microscopic) means. During the tests, the dynamic secant modulus was continuously computed with any changes in this quantity serving as an indication of damage in the specimen. The work presents a large data of experimental results for graphite/epoxy specimens.

The authors discovered local delamination regions near interior matrix cracks, and they claim that internal stress redistributions are much larger than previously suspected and can be large enough to cause strength reductions of as much as 30–50%, levels that are commonly observed in long-term cyclic loading. It appears that fiber failure is much less consequential in these large strength reductions than has been suggested in the literature, and that the acceleration of damage quite near the end of life for laminates of this type is caused by a localization of damage, primarily secondary matrix cracking and local delamination. No evidence of accelerating fiber fracture in that region was found. It seems that the most important strength reduction mechanism for the tested laminated composite material is the redistribution of the internal stresses.

Another important conclusion states, "matrix cracks in off-axis plies of angle-ply laminates seem to be important for long-term fatigue behavior primarily in the sense that they act as initiation points of fiber failure and local delamination." This means that while matrix cracks alone do not reduce the residual strength during cyclic loading, the events associated with and nucleated by their presence are the main drive to the development of subsequent damage that does reduce the strength, stiffness and life of composite laminates.

Moreover, the longitudinal engineering modulus of composite laminates seems to change during long-term cyclic loading, in large, systematic and reproducible ways, which are directly and quantitatively related to the details of the micro-events that influence the residual properties of such laminates. This change can be measured using nondestructive techniques to characterize the internal damage for a specific specimen. Using the measured internal damage, one can apply adequate models to predict actual longitudinal engineering stiffness, normally yielding an excellent agreement (within a few percent) with experimentally measured values.

References

[1] Stephens, R.I., Fatemi, A., Stephens, R.R. and Fuchs, H.O. Metal fatigue in engineering, John Wiley and Sons, 2000, 496.
[2] Rolfe, S.T. and Barsom, J.M. Fracture and fatigue control in structures: Applications of fracture mechanics, ASTM international, Technology & Engineering, 1977, 562.
[3] Almar-Naess, A. Fatigue handbook: Offshore steel structures, Tapir, 1985, 520.
[4] Mann, J.Y. Bibliography on the fatigue of materials, components and structures, science direct, Elsevier Ltd., 1990, 509.
[5] Radaj, D. Design and analysis of fatigue resistant welded structures, Woodhead Publishing Series in Welding and Other Joints Technologies, 1990, 378.
[6] Reifsnider, K.L. Ed. Fatigue of composite materials, vol. 4, composite materials series, 1st, Elsevier Science, 1991, 518.
[7] Maddox, S.J. Fatigue strength of welded structures, 2nd, Woodhead Publishing, 1991, 208.
[8] Gurney, T.R. Fatigue of thin walled joints under complex loading, 1st, Abington Publishing, 1997, 222.
[9] Etube, L. Fatigue and fracture mechanics of offshore structures, Wiley, 2000, 164.
[10] Lassen, T. and Recho, N. Fatigue life analyses of welded structures: Flaws, Wiley ISTE, 2006, 407.
[11] Schijve, J. Fatigue of structures and materials, Springer, 2009, 623.
[12] Maranian, P. Reducing brittle and fatigue failures in steel structures, American Society of Civil Engineers, 2010, 196.
[13] Nussbaumer, A., Borges, L. and Davaine, L. Fatigue design of steel and composite structures: Eurocode 3: Design of steel structures, part 1–9 fatigue; design of composite steel and concrete structures, Wiley, 2011, 334.
[14] Campbell, F.C. Fatigue and fracture: Understanding the basics, ASM International, 2012, 685.
[15] Radaj, D. and Vormwald, M. Advanced methods of fatigue assessment, Springer, 2013, 481.
[16] Dowling, N.E. Mechanical behavior of materials: Engineering methods for deformation, fracture, and fatigue, Pearson Education Ltd., 2013, 954.
[17] Bathias, C. and Pineau, A. Eds. Fatigue of materials and structures: Fundamentals, Wiley, 2013, 511.
[18] Bathias, C. and Pineau, A. Eds. Fatigue of materials and structures: Application to design and analysis, Wiley, 2013, 344.

[19] Lotsberg, I. Fatigue design of marine structures, Cambridge University Press, March 2016.
[20] Mikhailov, S.E. and Namestnikova, I.V. Fatigue strength and durability analysis by normalized equivalent stress functionals, in Proceedings of the 9th international conference on the mechanical behavior of materials, ICM9, Geneva, Switzerland, 2003, 10.
[21] Ritchie, R.O. Mechanisms of fatigue crack propagation in metals, ceramics and composites: role of crack tip shielding, Material Sciences and Engineering A, 103, 1988, 15–28.
[22] DET NORSKE VERITAS AS. Fatigue design of offshore steel structures, Recommended Practice DNV-RP-C203, 2012, 178.
[23] Azeez, A.A. Fatigue failure and testing method, M.Sc. thesis, Mechanical Engineering and Production technology, Mechatronics, HAMK University of Applied Sciences, Riihimäki, Finland, 2013, 32.
[24] Santecchia, E., Hamouda, A.M.S., Musharavati, F., Zalnezhad, E., Cabibbo, M., El Mehmedi, M. and Spigarelli, S. A review on fatigue life predictions methods for metals, Advances in Materials Science and Engineering, 2016, 2016, Article ID 9573524. 26. doi: 10.1155/2016/9573524.
[25] Özdeş, H., Tiryakioğlu, M. and Eason, P.D. On estimating axial high cycle fatigue behavior by rotating beam fatigue testing: application to A356 Aluminum alloy castings, Materials Science and Engineering A, 697, 2017, 95–100. doi: 10.1016/j.msea.2017.05.008.
[26] Zhang, Z., Ma, H., Zheng, R., Hu, Q., Nakatani, M., Ota, M., Chen, G., Chen, X., Ma, C. and Ameyama, K. Fatigue behavior of a harmonic structure designed austenitic stainless steel under uniaxial stress loading, Materials Science and Engineering A, 707, 2017, 287–294. doi: 10.1016/j.msea.2017.05.008.
[27] Freire, R.C.S., Jr. and de Aquino, E.M.F. Fatigue damage mechanism and failure prevention in fiberglass reinforce plastic, Materials Research, 8(1), 2005, 45–49.
[28] Ellyin, F., Xia, Z. and Li, C.-S. Fatigue damage of particle reinforce metal matrix composites, WIT Transaction on State of the Art in Science and Engineering, 21(4), 2005, 73–103.
[29] Zhou, A., Post, N., Pingry, R., Cain, J., Lesko, J.J. and Case, S.W. Durability of composite under fatigue loads, Durability of composites for civil structural applications, Vol. 7, Karbhari, V.M. Ed. Woodhead Publishing, Boca Raton, FL., 2007, 126–149.
[30] Brighenti, R., Carpinteri, A. and Scorza, D. Fatigue crack propagation simulating fiber debonding in cyclically loaded composites, Procedia Materials Science, 3, 2014, 357–362.
[31] Mouhoubi, S. and Azouaoui, K. Residual properties of composite plates subjected to impact fatigue, Journal of Composite Materials, 53(6), 2018, 799–817. doi: 10.1177/0021998318791324.
[32] Deveci, H.A. and Artem, H.S. On the estimation and optimization capabilities of the fatigue life prediction models in composite laminates, Reinforced Plastics and Composites, 37(21), 2018, 1304–1321. doi: 10.1177/0731684418791231.
[33] Burhan, I. and Kim, H.S. S-N curve models for composite materials characterization: An evaluative review, Journal of Composite Science, 2(38), 2018, 29. doi: 10.3390/jcs2030038.
[34] Klevtsov, G.V. and Klevtsova, N.A. Influence of stress ratio R on the fatigue strength and fatigue crack path in metal materials, in Proceedings of the international conference on Crack Paths (CP2009), University of Padua, Vicenza, Italy, 23–25 Sept. 2009, 359–365.
[35] Basquin, O.H. The exponential law of endurance test, Proceedings of the American Society for Testing and Materials, 10, 1910, 625–630.
[36] Weibull, W. The statistical aspect of fatigue failures and its consequences, fatigue and fracture of metals, a symposium held at the Massachusetts institute of technology, Vol. 1952, Murray, W.M. Ed. John Wiley & Sons, New York, NY, USA, June 19–22 1950, 182–196.
[37] Li, H., Wen, D., Lu, Z., Wang, Y. and Deng, F. Identifying the probability distribution on fatigue using the maximum entropy principle, Entropy, 18(111), 2016, 19. doi: 10.3390/e18040111.
[38] Ballio, G. and Castiglioni, C.A. A unified approach for the design of steel structures under low and/or high cycle fatigue, Journal of Constructional Steel Research, 34, 1995, 75–101.
[39] Rosenfeld, M.S. and Huang, S.L. Fatigue characteristics of graphite/epoxy laminates under compression loading, Journal of Aircraft, 15(5), 1978, 264–268.

[40] Rotem, A. The fatigue behavior of orthotropic laminates under-tension-compression loading, International Journal of Fatigue, 13(3), 1991, 209–215.
[41] Caprino, G. and D'Amore, A. Fatigue life of graphite/epoxy laminates subjected to tension-compression loading, Mechanics of Time-dependent Materials, 4, 2000, 139–154.
[42] Minak, G., Morreli, P. and Zucchelli, A. Fatigue residual strength of circular laminate graphite-epoxy composite plates damaged by transverse load, Composite Science and Technology, 69, 2009, 1358–1363.
[43] Williams, J.H., Jr., Yuce, H. and Lee, S.S. Ultrasonic and mechanical characterizations of fatigue states of graphite epoxy composite laminates, NASA CR, 3504, January 1982, 26.
[44] Minak, G. On the determination of the fatigue life of laminated graphite-epoxy composite by means of temperature measurement, Journal of Composite Materials, 44(14), 2010, 1739–1752. doi: 10.1177/0021998309359815.
[45] Jamison, R.D. and Reifsnider, K.L. Advanced fatigue damage development in graphite-epoxy laminates, ADA130190, AFWAL TR-82-3130, Flight Dynamic Laboratory, Air Force Wright Aeronautical Laboratories, Air Force Systems Command, Wright Patterson Air Force Base, Ohio, 45433, USA, December 1982, 222.

Appendix A: Application of Fig. 4.6 and its associated equations

Determine the diameter of a 4,340 steel rod undergoing alternating axial load, which varies from a maximal value of 350 kN tension to a minimal compression of 150 kN, taking into account a safety factor (S.F.) = 2.5.

The mechanical properties of 4,340 steel are:

σ_{ul} = 1,092 MPa; σ_y = 1,005 MPa; σ_e = 520 MPa (for the safety factor of 1.0)

Solution:

Assuming that the cross-sectional area of the road is given by the variable A, with d standing for the rod diameter:

$$A = \frac{\pi}{4} d^2 \Rightarrow d = \sqrt{\frac{4A}{\pi}}$$

we can write the following expressions:

$$\sigma_{max} = \frac{0.35}{A} \text{ [MPa]}, \quad \sigma_{min} = -\frac{0.15}{A} \text{ [MPa]}$$

$$\Rightarrow \sigma_{mean} \equiv \sigma_m = \frac{\sigma_{max} + \sigma_{min}}{2} = \frac{0.35 - 0.15}{2} = \frac{0.1}{A} \text{ [MPa]}$$

$$\sigma_a = \frac{\sigma_{max} - \sigma_{min}}{2} = \frac{0.35 + 0.15}{2A} = \frac{0.25}{A} \text{ [MPa]}$$

Using eq. (4.16), namely

$$\left(\frac{\sigma_m}{\sigma_{ul}}\right) + \left(\frac{\sigma_a}{\sigma_e}\right) = 1$$

and substituting for σ_m and σ_a, taking into account the safety factor (S.F.) = 2.5 yields

$$\sigma_e = \tfrac{520}{2.5} = 208\ [\text{MPa}]$$

$$\left(\frac{0.1}{A \cdot 1,092}\right) + \left(\frac{0.25}{A \cdot 208}\right) = 1 \Rightarrow A = \left(\frac{0.1}{1,092}\right) + \left(\frac{0.25}{208}\right) = 1,293(\text{mm})^2$$

Therefore, the diameter of the rod will be

$$\Rightarrow d = \sqrt{\frac{4A}{\pi}} = \sqrt{\frac{4 \cdot 1293}{\pi}} = 40.58[\text{mm}]$$

5 Introduction to crack propagation analysis

5.1 Introduction

This chapter presents the crack behavior of various materials under repeated loading. This is a continuation of Chapter 4 and presents the crack propagation philosophy associated with fatigue behavior of components.

5.2 Foundations of fracture mechanics

5.2.1 Introductory concepts

Fracture mechanics deals with the appearance of a crack in an engineering material, be it a metal or composite, trying to provide answers to two key issues:

a. The failure criteria
b. Determination of strain and stress fields and the deformation in the vicinity of the geometrical singularity of the crack tip

Griffith [1] was the first to lay the foundations of fracture mechanics already in 1921. He suggested using the following relation to find the σ_f, the stress at the fracture and the length of the crack (see Fig. 5.1):

$$\sigma_f \cdot \sqrt{a} \approx \text{Const} = \sqrt{\frac{2E\Psi}{\pi}} \tag{5.1}$$

where E is the Young's modulus of the material and Ψ is its surface energy density (having the value of 1 J/m^2 for glass).

In 1939, Westergaard [2] presented a solution for the stress field around a crack in the form (Fig. 5.2)

$$\sigma_{xx} = \sigma_{yy} = \frac{\sigma_{\text{inf.}}}{\sqrt{1 - \left(\frac{a}{x}\right)^2}} \tag{5.2}$$

It is clear that the stresses would tend to infinity at the crack tip and would get the value $\sigma_{\text{inf.}}$ at a far distance from it.

Irwin [3] started the extension of the discipline in 1957, by modifying Griffith's equation (eq. (5.1)), and suggested to include also a plastic energy dissipation, leading to the following relation:

https://doi.org/10.1515/9783111621104-005

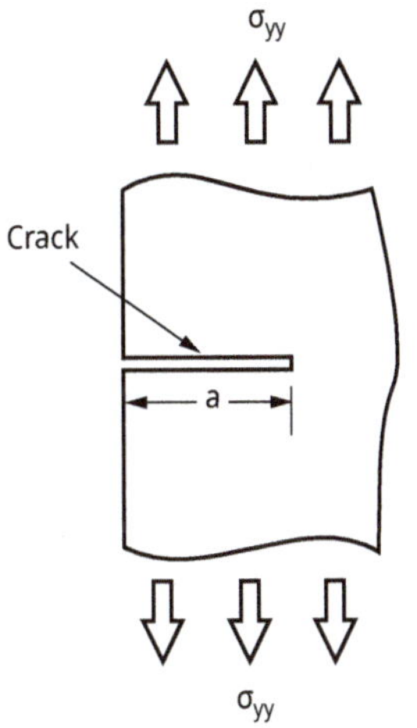

Fig. 5.1: A schematic drawing of a crack (Griffith's model).

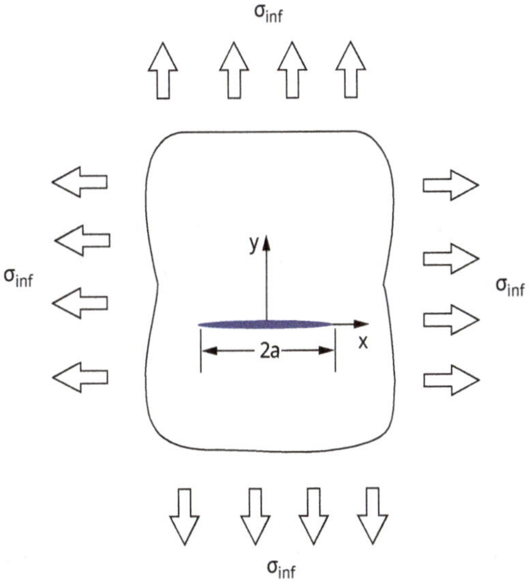

Fig. 5.2: A schematic drawing of a crack having the length $2a$ under tensile stresses at infinity, σ_{inf}.

$$\sigma_{\mathrm{f}} \cdot \sqrt{a} = \sqrt{\frac{EG}{\pi}}, \quad G = 2\Psi + G_{\mathrm{p}} \tag{5.3}$$

where G is the total energy dissipation and G_{p} is the plastic part of it. One should note that for glass (a brittle material), the term Ψ would dominate leading to $G \approx 2\Psi = 2\ \mathrm{J/m^2}$, while for ductile materials (metals like steel and some aluminum alloys) we will obtain $G \approx G_p = 1{,}000$ J/m^2.

As can be deducted from eq. (5.2), in the vicinity of the crack tip, a stress concentration is observed. To visualize this phenomenon, flow lines are depicted in Fig. 5.3 for a virgin specimen in comparison with the same specimen with a central crack, which experiences a concentration of the flow lines at its tips.

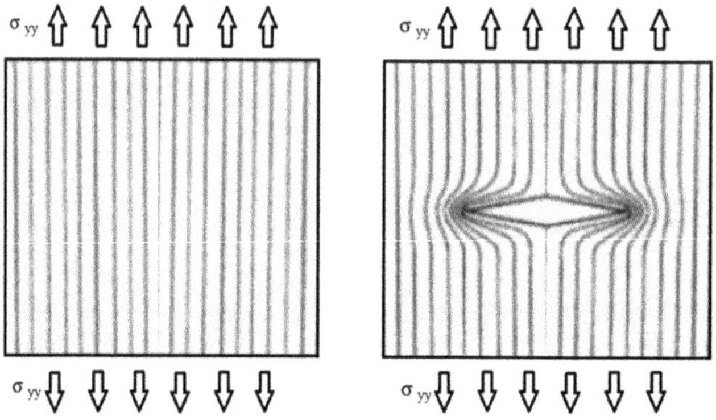

Fig. 5.3: Concentration of flow lines in the vicinity of the crack tips compared with a virgin specimen – a schematic drawing.

Today the fracture mechanics includes nonlinear problems due to their geometric and material variations. The topic of crack bifurcation in mixed loading modes has also been extensively researched over the last decade. In the last decade, the fracture mechanics had been applied to composites using numerical algorithms and FE (finite element) codes (see typical references in [4–24]).

5.2.2 Basic failure modes

The basic failure modes due to cracking are presented in Fig. 5.4. The first failure mode called in the literature *Mode I (opening mode)* is presenting the opening of the crack lips in a perpendicular direction to its propagation. The second mode, *Mode II* (*sliding mode*), is an in-plane shear with the crack propagation being parallel to the crack lip displacements. The third and last mode, *Mode III* (*tearing mode*), is an out-of-plane shear (torsion type) in which the crack lip displacements are perpendicular to its propagation. Note that in real cases, a combination of more than one type of crack failure might occur. To determine, after fracture, what type of failure mode occurred, one should carefully examine the topography of the failure. In general cases, a smooth zone characteristically to crack propagation by fatigue may be observed in ductile materials, while brittle one would show crystalline and/or apparent grains along the fracture boundary.

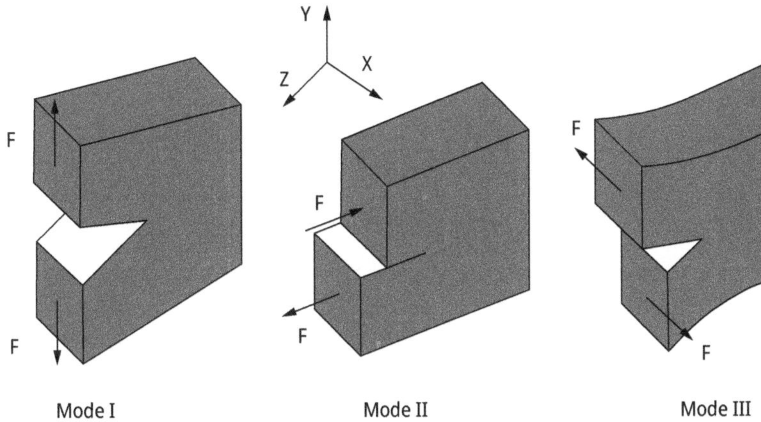

Fig. 5.4: The three crack modes.

5.2.3 The stress intensity factor, *K*

Irwin [3] presented expressions for the displacements u, v along the X, Y axes (see Fig. 5.4) and the various stresses in the singular zone of the crack tip (which presents a singularity in the form of $1/\sqrt{r}$) using parameters called stress intensity factors, K_{I}, K_{II} and K_{III}, for the three modes of failure presented earlier. These expressions are

$$\begin{aligned} u &= \frac{K_I}{2\mathrm{H}}\sqrt{\frac{r}{2\pi}}\cos\frac{\theta(k-\cos\theta)}{2} + \frac{K_{II}}{2\mathrm{H}}\sqrt{\frac{r}{2\pi}}\sin\frac{\theta(k+\cos\theta+2)}{2} \\ v &= \frac{K_I}{2\mathrm{H}}\sqrt{\frac{r}{2\pi}}\sin\frac{\theta(k-\cos\theta)}{2} - \frac{K_{II}}{2\mathrm{H}}\sqrt{\frac{r}{2\pi}}\cos\frac{\theta(k+\cos\theta-2)}{2} \end{aligned} \tag{5.4}$$

$$\begin{aligned} \sigma_x &= \frac{K_I}{\sqrt{2\pi r}}\cos\frac{\theta}{2}\left(1-\sin\frac{\theta}{2}\sin\frac{3\theta}{2}\right) - \frac{K_{II}}{\sqrt{2\pi r}}\sin\frac{\theta}{2}\left(2+\cos\frac{\theta}{2}\cos\frac{3\theta}{2}\right) \\ \sigma_y &= \frac{K_I}{\sqrt{2\pi r}}\cos\frac{\theta}{2}\left(1+\sin\frac{\theta}{2}\sin\frac{3\theta}{2}\right) + \frac{K_{II}}{\sqrt{2\pi r}}\sin\frac{\theta}{2}\cos\frac{\theta}{2}\cos\frac{3\theta}{2} \\ \tau_{xy} &= \frac{K_I}{\sqrt{2\pi r}}\cos\frac{\theta}{2}\sin\frac{\theta}{2}\cos\frac{3\theta}{2} + \frac{K_{II}}{\sqrt{2\pi r}}\cos\frac{\theta}{2}\left(1-\sin\frac{\theta}{2}\sin\frac{3\theta}{2}\right) \end{aligned} \tag{5.5}$$

where

$k = 3 - 4\upsilon$ for plane strain; $k = \frac{3-\upsilon}{1+\upsilon}$ for plane stress; $H \equiv \frac{E}{2(1+\upsilon)}$ is the shear modulus, E is the Young's modulus, υ is the Poisson's ratio, and r, θ are the radius and angle for polar coordinates (see Fig. 5.5).

Note that for out-of-plane tractions, the Z displacement component, w, is considered and has the following form:

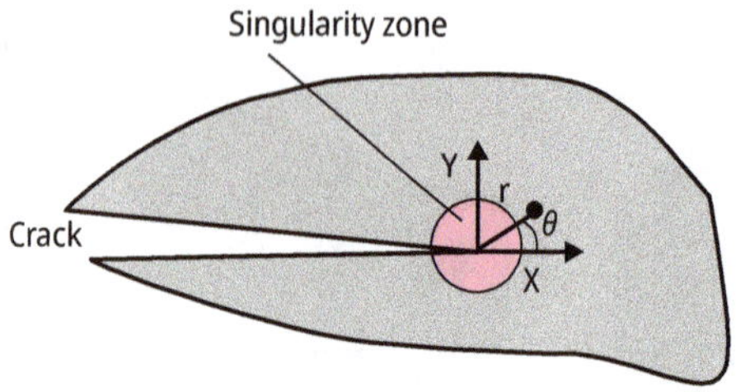

Fig. 5.5: A schematic drawing of the crack and its singularity zone.

$$w = \frac{2K_{\mathrm{III}}}{H}\sqrt{\frac{r}{2\pi}}\sin\frac{\theta}{2} \tag{5.6}$$

The associated stress components are

$$\begin{aligned} \tau_{xz} &= -\frac{2K_{\mathrm{III}}}{\sqrt{2\pi r}}\sin\frac{\theta}{2}: \\ \tau_{yz} &= -\frac{2K_{\mathrm{III}}}{\sqrt{2\pi r}}\cos\frac{\theta}{2} \end{aligned} \tag{5.7}$$

The definitions of the stress intensity factors, K_{I}, K_{II} and K_{III}, are

$$\begin{aligned} K_{\mathrm{I}} &\equiv \lim_{r\to 0}\left[\frac{E}{8\alpha}\sqrt{\frac{2\pi}{r}}(v)\right] \\ K_{\mathrm{II}} &\equiv \lim_{r\to 0}\left[\frac{E}{8\alpha}\sqrt{\frac{2\pi}{r}}(u)\right] \\ K_{\mathrm{III}} &\equiv \lim_{r\to 0}\left[\frac{E}{8(1+\upsilon)}\sqrt{\frac{2\pi}{r}}(w)\right] \end{aligned} \tag{5.8}$$

with $\alpha = 1$ for in-plane stress, $\alpha = 1 - \upsilon^2$ for in-plane strain, and u, v and w are the expressions for the displacements of the crack lips. As can be deducted from eq. (5.8), the three stress concentration factors independent of the coordinates r, θ, depend only on the applied tractions and the crack geometry and proportional to the displacements of the crack lips.

5.2.4 The energy release rate, *G*

Griffith [1] presented in its seminal work the issue of bodies containing cracks using an energy approach and introducing a variable G, the energy release rate, defined as

$$G \equiv -\frac{dW}{d(a)} = -\frac{W_{ext.} + W_{elastic}}{d(a)}$$

where (5.9)

$$W_{elastic} = \int_{Volume} \left(\int \sigma_{ij} d\varepsilon_{ij} \right) dV, \quad i,j = x,y,z$$

while $W_{ext.}$ is the potential energy of the external tractions and a is half of the crack length. The formal wording definition for G is *the energy needed to extend the front (tip) of the crack by a unit length da*, and from the energy point of view *it corresponds to the reduction (the reason for the minus sign of G) of the total energy W of a body containing a crack passing from an initial crack length 2a* to *(2a + da)*.

The energy release rate G can also be defined in the following way:

$$G \equiv \frac{(K_I^2 + K_{II}^2)}{\Gamma} + \frac{K_{III}^2}{2H} \tag{5.10}$$

where $\Gamma = E$ for an in-plane stress case; $\Gamma = \frac{E}{1-\upsilon^2}$ for an in-plane strain case.

For a simple case of a crack in a thin rectangular plate, with the traction being perpendicular to the notch, we can get the criterion for a crack propagation as

If $G \geq G_c$ the crack will propagate

where (5.11)

$$G = \frac{\pi\sigma^2 a}{E}, \quad G_c = \frac{\pi\sigma_f^2 a}{E}$$

Note that eq. (5.11) is true for in-plane stress, while for the case of in-plane strain instead of E one would use the expression $E/(1-\upsilon^2)$.

For mode I, one can write the fracture toughness K_I, for a center-cracked infinite plate as

$$K_I = \sigma\sqrt{\pi a};$$

$$K_{Ic} = \sqrt{EG_c} \text{ in-plane stress} \tag{5.12}$$

$$K_{Ic} = \sqrt{\frac{EG_c}{1-\upsilon^2}} \text{ in-plane strain}$$

Note that fracture will occur if $K_I \geq K_{Ic}$. To take into account the geometry of the specimen, eq. (5.12) is multiplied by a dimensionless correction factor f which is a function of the crack length and width, b, of the specimen. In the case of a plate with finite width denoted by $\bar{W}$ and a through-thickness crack of length $2a$, this factor will have the following form:

$$f\left(\frac{a}{\overline{W}}\right) = \sqrt{\sec\left(\frac{\pi a}{\overline{W}}\right)} \tag{5.13}$$

while for a plate with a finite width $\overline{W}$ and a through-thickness edge crack of length a the correction factor will be

$$f\left(\frac{a}{\overline{W}}\right) = 1.12 - \frac{0.41}{\sqrt{\pi}}\frac{a}{\overline{W}} + \frac{18.7}{\sqrt{\pi}}\left(\frac{a}{\overline{W}}\right)^2 - \cdots \tag{5.14}$$

5.2.5 The *J*-integral

A new interpretation to characterize the singularity of the stress field in the vicinity of a notch (crack) was proposed in 1968 by Rice [23] by defining an integral of a contour defined as

$$J \equiv \int_{\Gamma} \left(W dy - T \frac{\partial \vec{u}}{\partial x} ds \right)$$

where (5.15)

$$W = W(x, y) = W(\vec{e}) = \int_{\varepsilon}^{0} \sigma_{ij} d\varepsilon_{ij}$$

where $\vec{e} = [\varepsilon_{ij}]$ is an infinitesimal strain tensor, Γ (see Fig. 5.6) is a closed curve surrounding the tip of the crack, and the integral should be evaluated in a counterclockwise direction starting from the lower flat part of the crack surface and continuing along the path Γ to the upper flat surface; T is the traction vector defined according to the outward normal $\vec{n}$ along Γ; $T_i = \sigma_{ij} n_j$; $\vec{u}$ is the displacement vector and ds is an arc length element along Γ. Note that the *J*-integral is path independent [23].

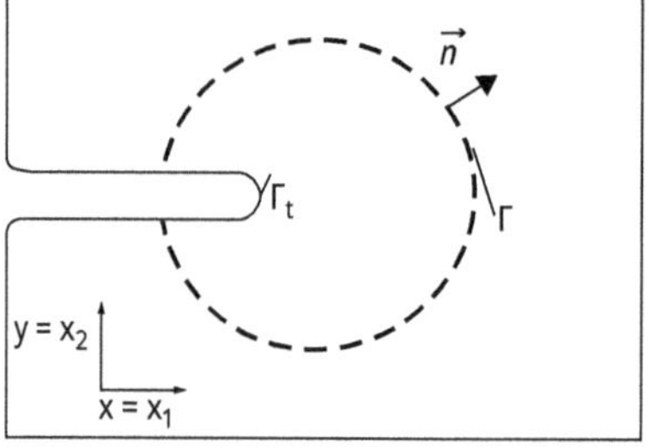

Fig. 5.6: The integral path for the *J*-integral-flat 2D-surfaced notch (crack) deformation field. Γ is any curve surrounding the crack tip, while Γ_t denotes the curved crack tip.

Rice interpreted the J-integral as the difference in potential energy W of two cracked bodies submitted to the same boundary conditions, with lengths of crack differing by a length Δa, expressing it by the following relation:

$$J \equiv -\lim_{\Delta a \to 0} \frac{W(a + \Delta a) - W(a)}{\Delta a} = -\frac{dW}{\Delta a} \tag{5.16}$$

Assuming a case of an elastic material near the crack zone, we can write

$$J = G = -\frac{dW}{\Delta a} \tag{5.17}$$

5.2.6 The crack opening displacement

The COD is a variable used to model crack propagation. It is the vector for the displacement at the crack tip. It represents the displacement that makes the crack propagate. Normally, measurements of the COD are limited to the crack of mode I, which propagates along a fixed direction, which corresponds to the axis of the initial crack.

When treating a mixed mode, these measurements would be based on the vector CTD (crack tip displacement), which is defined in the literature as a combination of CTOD (crack tip opening displacement) at loading in mode I and of CTSD (crack tip sliding displacement) which is the vector of displacement due to the slip of crack lips corresponding to the loading in mode II.

5.2.7 Some closure notes

Note that the above-described parameters of fracture mechanics K, G, J and COD are used to predict the direction of crack propagation under loadings in mixed mode, being limited to the failure in linear elastic medium. Their applications lead to very good predictions for fragile elastic materials containing a real crack.

However, for cases with ductile material, or when the load is alternating like in fatigue and/or in the presence of overloading or residual stresses, the applications of these criteria would lead to less good predictions.

5.3 Fatigue crack propagation

5.3.1 Introductory concepts

In Chapter 4, the fatigue process till failure of the component was described as being composed of four steps all connected to cracks, their initiation, growth, propagation and failure. Once the crack has been formed, its growth can be divided into two steps:

1. The first step in which growth is done along slip bands due to shear stresses, leading to formation of intrusions or crack deepening. During this step, the crack would extend by only a few grain diameters at a rate of a few nanometers per cycle.
2. The second step is associated with a much faster crack growth at a rate of microns per cycle, being dictated by the normal stress applied on the specimen. This stage would show also fatigue striations[1] produced by a cycle of stress. The formation of the striations is explained in the literature by three consecutive laps, starting with formation of double notch concentrating slip at 45° due to a tensile stress, followed by a crack widening, with the third lap being crack tip extension and its blunting. From the fatigue failure point of view, the second step is the most important, as at this step the component can be replaced before the crack would reach its critical value.

5.3.2 The Paris law

In 1963, Paris and Erdogan [4] proposed the following relation, to be known in the literature as Paris–Erdogan rule, or in short Paris law:

$$\frac{da}{dN} = C \cdot (\Delta K)^{\alpha} \tag{5.18}$$

where a is the length of the crack, C is a constant in region II (see Fig. 5.7) of the graph and α is a constant, both to be empirically determined (usually, $\alpha \approx 3$ for steel-type materials and $\alpha \approx 3-4$ for aluminum alloys). ΔK has the following form[2]:

$$\Delta K \equiv K_{\max} - K_{\min} = \sigma_{\max}\sqrt{\pi a} - \sigma_{\min}\sqrt{\pi a} = \Delta\sigma\sqrt{\pi a} \tag{5.19}$$

As shown in Fig. 5.7, the graph has three distinct regions. Region I is characterized by a slow and negligible crack growth, with a large influence of microstructure mean

1 Striations: a word coming from geology, meaning ridges, furrows or linear marks due to the movement of parts.

2 One should use a range of ΔK due to the fatigue-loading mode and this makes the link between fatigue and fracture mechanics. Note that K is not defined for compression mode and it is taken as zero.

stress environment [11], with ΔK_{incep} being the threshold of ΔK beyond which the crack will start to grow considerably. The second region, II, also called the Paris domain, due to the Paris law, presents stable crack growth with a linear behavior between crack growth rate and log of the stress intensity factor range (log(ΔK)). This region shows the primary mechanism of the crack, the development of striations, very small influence of the microstructure thickness and the large influence of mean stress, fatigue frequency and certain combinations of environment [11]. Region III, the third and the last region, presents an unstable growth of the crack leading to failure, as $K_{max} > K_c$ of the certain material. Ritchie [11] calls this region "the static mode mechanisms" as it shows cleavage, intergranular and fibrous mechanisms. The region presents a large influence of the microstructure, mean stress and thickness, with very little influence of the environment [11].

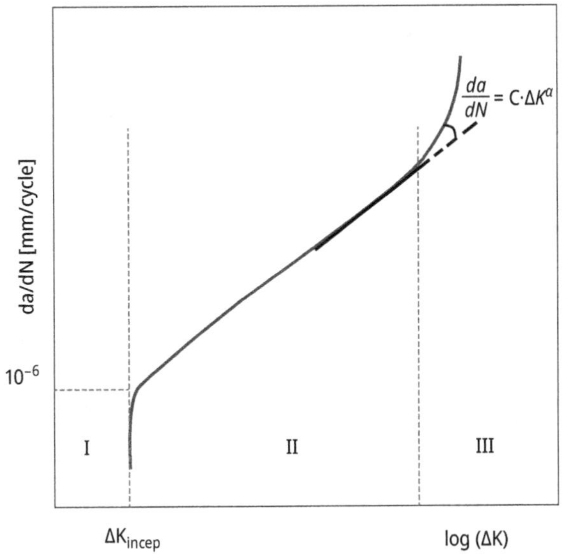

Fig. 5.7: Schematic description of the Paris law.

Note that the fatigue life of a specimen can be obtained by integrating Paris law (eq. (5.18)) to yield the following expression:

$$\frac{da}{dN} = C \cdot (\Delta K)^a \Rightarrow N_{final} = \int_0^{N_{final}} dN = \int_{a_{start}}^{a_{final}} \frac{da}{C \cdot (\Delta K)^a} \tag{5.20}$$

Introducing the relation between the stress intensity factor and the crack length + stress, together with the function f (see eq. (5.14)), one obtains the following expression:

$$N_{final} = \frac{1}{C(f\Delta\sigma\sqrt{\pi})^{\alpha}}\left(\frac{1}{0.5\alpha - 1}\right)\left[\frac{1}{a_{start}^{0.5\alpha-1}} - \frac{1}{a_{final}^{0.5\alpha-1}}\right] \tag{5.21}$$

Note that the correction factor f was introduced in eq. (5.21) after integration as it is an empirical value.

5.3.3 Experimental fracture mechanics

5.3.3a Experimental determination of the Paris law

The geometry of the samples is according to ASTM E399 [25] with the variables w and B (Fig. 5.8) being $W = 1''$ (25.4 mm) and $B = ½''$ (12.7 mm) (more details in [24]).

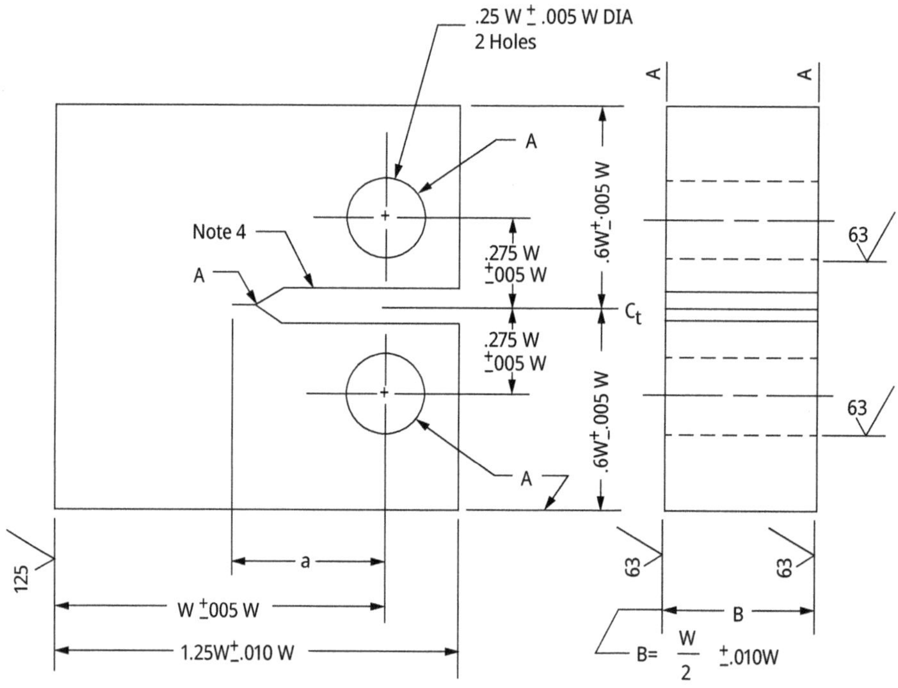

Fig. 5.8: Schematic drawing of the crack propagation specimen.

The tests were performed on an MTS hydraulic fatigue tensile machine equipped with a 100 kN load cell. At the end of the crack, an MTS COD (crack opening displacement) gage was placed and it measured the crack opening value at each load cycle (Fig. 5.9). For tests on solid specimens, a crack length gage Vishay TK-09-CPA01-005/DP was bonded to the specimen (Fig. 5.10). The gage has 20 wires at 0.25 mm distance between two adjacent ones, and as the crack propagates, the wires tear, thus indicating the crack length.

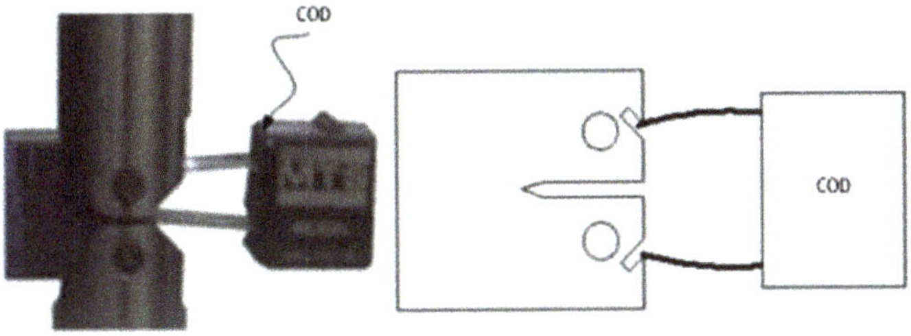

Fig. 5.9: COD implementation.

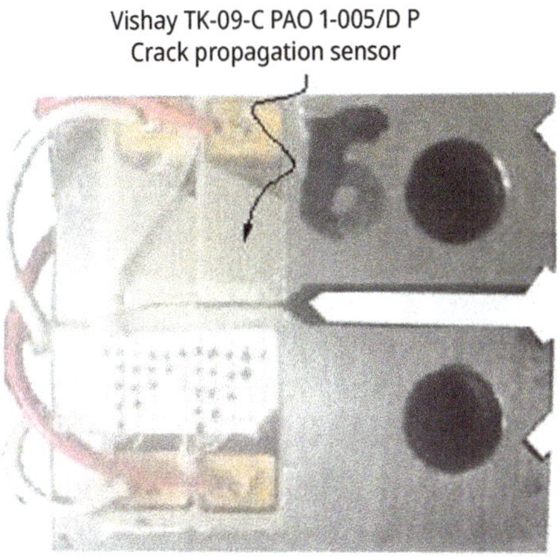

Fig. 5.10: Crack propagation sensor implementation.

5.3.3b Experimental determination of the Paris law constants

The sample was installed on the loading machine, which applied a 10 Hz sinusoidal load at $R = 0.1$ (i.e., the machine applies only tensile force). During the test, every 10 cycles, a sample of the crack length gage and a sample of the COD gage were taken and stored. The output measurements include the maximum and minimum forces at each cycle and the amplitude of the COD gage and the readings of the crack propagation sensor. The large database is then filtered to reduce its size for further processing, by keeping only the data when the crack length changes its value by a unit.

For each cycle, the stiffness of the tested specimen is determined using the following relation:

$$k_s = \frac{\text{Force}_{\text{max}} - \text{Force}_{\text{min}}}{\text{COD}_{\text{length}}} \tag{5.22}$$

After calculating the stiffness of the specimen with respect to the crack length, a polynomial expression is curve-fitted between the crack length and the stiffness defined in eq. (5.22). A typical graph of the crack length versus the specimen's stiffness is presented in Fig. 5.11.

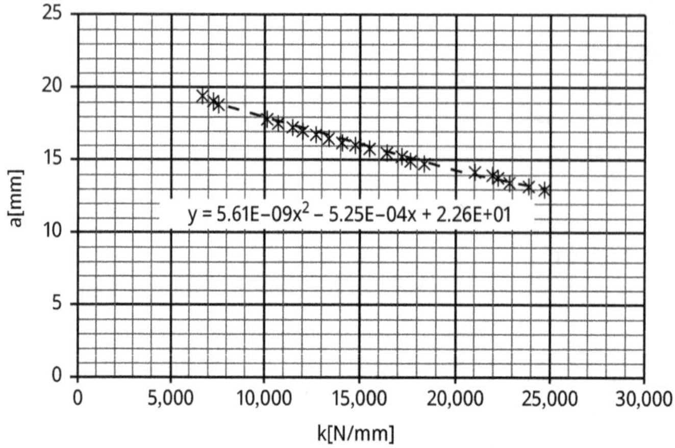

Fig. 5.11: Typical experimental graph of crack length (*a*) versus specimen's stiffness (*k*).

Knowing the crack length and the applied load yields, the ΔK_1 values for the entire data as well as the crack length rate *da/dN* (*a* is the crack length and *N* stands for the number of cycles) lead to a graph of the crack length rate versus ΔK_1 (Fig. 5.12). ΔK_1 is calculated using the following expression:

$$\Delta K_1 = \frac{\Delta P}{B \cdot \sqrt{W}} \cdot f\left(\frac{a}{W}\right) \tag{5.23}$$

where

$$f(\chi) = \frac{2+\chi}{(1-\chi)^{3/.2}}\Gamma$$
$$\Gamma \equiv \left[0.886 + 4.64(\chi) - 13.32(\chi)^2 + 14.72(\chi)^3 - 5.6(\chi)^4\right] \tag{5.24}$$
$$\chi \equiv \frac{a}{W}$$

ΔP is the load amplitude, a is the crack length, and W and B are dimensions as presented in Fig. 5.8. The graph depicted in Fig. 5.8 enables to find the Paris coefficients C and α appearing in Paris law (see eqs. (5.18) and (5.19)) as expressed by the following equations:

$$\frac{da}{dN} = C \cdot \Delta K^{\alpha} \tag{5.25}$$

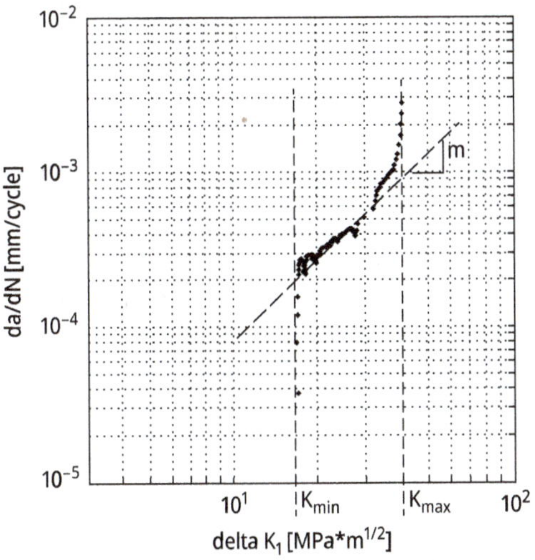

Fig. 5.12: Typical experimental graph of (da/dN) versus ΔK_1.

where

$$\Delta K = K_{max} - K_{min} \tag{5.26}$$

One should note that for the typical graph shown in Fig. 5.12, the following values were obtained: $a = 1.86$, $K_{min} = 17\left[\mathrm{MPa}\cdot\sqrt{m}\right]$, $K_{max} = 40\left[\mathrm{MPa}\cdot\sqrt{m}\right]$ and $C = 6.92 \cdot 10^{-21}[m^{0.5}/(\text{cycle} \cdot \mathrm{MPa})]$. For the case $da/dN \to \infty$, $K_{max} \to K_{1C}$ and sometimes K_{min} is also named ΔK_{th}, occurred at lower threshold value of da/dN, below which no crack propagation would occur (see also Fig. 5.7).

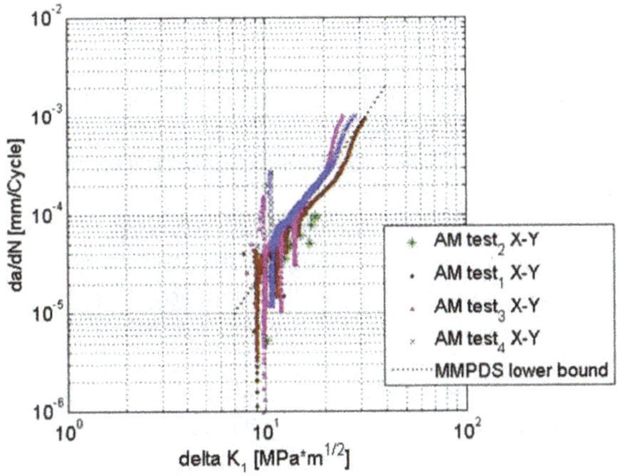

Fig. 5.13: Experimental results for four additive manufactured specimens printed in the x–y plane direction [24].

It seems that the results in Fig. 5.13 are in close agreement with the results presented in the Metallic Materials Properties Development and Standardization Scientific Report [25].

References

[1] Griffith, A.A. The phenomena of rupture and flow in solids, Philosophical Transactions of the Royal Society, Series A, 221, 1921, 163–197. doi: 10.1098/rsta.1921.0006.

[2] Westergaard, H.M. Bearing pressures and cracks, Journal of Applied Mechanics, 6, 1939, A49–53.

[3] Irwin, G. Analysis of stresses and strains near the end of a crack traversing a plate, Journal of Applied Mechanics, Transactions of ASME, 24, 1957, 361–364.

[4] Paris, P. and Erdogan, F. A critical analysis of crack propagation laws, Journal Basic Engineering, 85(4), 1963, 528–533.

[5] Tomkins, B. Fatigue crack propagation – an analysis, Journal Philosophical Magazine, 18(155), 1968, 1041–1066.

[6] Engle, R.M., Jr. CRACKS, A Fortran IV digital computer program for crack propagation analysis, Technical Report AFFDL-TR-70-107, Air force Flight Dynamics Laboratory, Air Force Systems Command, Wright-Patterson Air Force Base, Ohio, USA, 1970, 61.

[7] Jaske, C.E., Feddersen, C.E., Davies, K.B. and Rice, R.C. Analysis of fatigue, fatigue-crack propagation and fracture data, NASA CR-132332, 1973, 188.

[8] Engelder, T. Brittle crack propagation, Chapter 3 in *Continental Tectonics*, Hancock, P. Ed. Pergamon Press, Oxford, 1994, 43–52.

[9] Ortiz, M. and Pandolfi, A. Finite-deformation irreversible cohesive elements for three-dimensional crack-propagation analysis, International Journal for Numerical Methods in Engineering, 444, 1999, 1267–1282.

[10] Kerans, R.J. and Parthasarathy, T.A. Crack deflection in ceramic composites and fiber coating design criteria, Composites Part A: Applied Science and Manufacturing, 30(4), 1999, 521–524.

[11] Ritchie, R.O. Mechanisms of fatigue-crack propagation in ductile and brittle solids, International Journal of Fracture, 100, 1999, 55–83.

[12] Scheider, I. Cohesive model for crack propagation analyses of structures with elastic-plastic material behavior-Foundations and implementation, Technical report, GKSS Research Center, Internal report No. WMS/2000/19, 2001, 41.

[13] Scheider, I., Schödel, M., Brocks, W. and Schönfeld, W. Crack propagation analyses with CTOA and cohesive model: comparison and experimental validation, Engineering Fracture Mechanics, 73, 2006, 252–263.

[14] Patrício, M. and Mattheij, R.M.M. Crack propagation analysis, CASA-Report, Vol. 0723, 2007, Eindhoven, Technische Universiteit Eindhoven, The Netherlands, 25.

[15] Souiyah, M., Muchtar, A., Alshoaibi, A. and Ariffin, A.K. Finite element analysis of the crack propagation for solid materials, American Journal of Applied Sciences, 6(7), 2009, 1396–1402.

[16] Kostson, E. and Fromme, P. Fatigue crack monitoring in multi-layered aircraft structures using guided ultrasonic waves, Journal of Physics: Conference Series, 195, 2009, Article Id: 012003, 10. doi: 10.1088/1742-6596/195/1/012003.

[17] Chin, P.L. Stress analysis, crack propagation and stress intensity factor computation of a Ti-6Ai-4V aerospace bracket using ANSYS and FRANC3D, Master Thesis, Rensselaer Polytechnic Institute, Hartford, Connecticut, USA, 2011, 65.

[18] Okada, H., Kawai, H., Tokuda, T. and Fukui, Y. Fully automated mixed mode crack propagation analyses based on tetrahedral finite element and VCCM (virtual crack closure-integral method), International Journal of Fatigue, 50, 2013, 33–39.

[19] Dassios, K.G., Kordatos, E.Z., Aggelis, D.G. and Matikas, T.E. Crack growth monitoring in ceramic matrix composites by combined infrared thermography and acoustic emission, Journal of the American Ceramic Society, 97(1), 2014, 251–257. doi: 10.1111/jace.12592.

[20] Dündar, H. and Ayhan, A.O. Three-dimensional fracture and fatigue crack propagation analysis in structures with multiple cracks, Computers and Structures, 158, 2015, 259–273.

[21] Jensen, B.E.W. Numerical analysis of crack propagation and lifetime estimation, Master thesis, Aalborg University Esbjerg, Denmark, 2015, 94.

[22] Hashtroodi, S. Crack propagation analysis of a pre-stressed L-shaped spandrel parking garage beam, Master thesis in Civil Engineering, The University of Toledo, Toledo, Ohio, US, 2015, 120.

[23] Rice, J.R. A path independent integral and the approximate analysis of strain concentrations by notches and cracks, Journal of Applied Mechanics, 35, 1968, 379–386.

[24] Abramovich, H., Broitman, N. and Shirizly, A. Investigation of crack propagation properties of additive manufacturing products, 31st Congress of the International Council of the Aeronautical Sciences, Belo Horizonte, Brazil, Sept. 9–14 2018, 9.

[25] Standard Test Methods for Measurement of Fatigue. Crack Growth Rates. ASTM E647. 100 Barr Harbor Dr. West Conshohocken, PA 19428, USA, 1999.

6 Buckling of thin-walled structures

6.1 Introduction

This chapter presents the behavior of thin-walled structures like columns, plates and shells. The thin-walled structures can be either isotropic or orthotropic, and their critical loads will be analytically evaluated for applicable cases.

6.2 Buckling of columns

6.2.1 Euler buckling

To find the buckling[1] load of a column, having the length L, and stiffness EI[2] resting on simply supported, or hinged–hinged[3] boundary conditions and being compressed by a compressive load P (see Fig. 6.1), one needs to write the moment at a chosen axial coordinate, x, and equate it to the known expression of EId^2z/dx^2. One should note that in Fig. 6.1, the column is described after its deformation in an exaggerated manner. The result is [1, 2]

$$EI\frac{d^2z(x)}{dx^2} = -Pz(x) \tag{6.1}$$

Rearranging eq. (6.1) and dividing by EI, one obtains

$$\frac{d^2z(x)}{dx^2} + k^2z(x) = 0, \quad k^2 = \frac{P}{EI} \tag{6.2}$$

The general solution for eq. (6.2) is given by

$$z(x) = A\sin(kx) + B\cos(kx) \tag{6.3}$$

while A and B are constants to be determined from the boundary conditions of the problem. For the hinged–hinged column case, the boundary conditions would require the lateral displacement and the moment to vanish at its both ends. One should note that for the equation of motion displayed in eq. (6.1), the natural boundary conditions,

1 "Buckling" phenomenon has various definitions. One of these definitions *describes* it as the state of a compressed column, when a slight increase in the compressive load would induce sudden large lateral (unbounded) deformation of structure, whereas before that compressive load (critical load), the column had experienced little, if any, lateral deformation.

2 E, Young's modulus; I, moment of inertia of the cross section (for a rectangular cross, having a height of h and width of b, $I = bh^3/12$).

3 In some of the books and manuscripts, these boundary conditions are referred to as *pinned–pinned* boundary conditions.

https://doi.org/10.1515/9783111621104-006

namely the moment is inherently fulfilled. Therefore, we are left with only two geometric boundary conditions, which lead to

$$\begin{aligned} &z(0)=0 \Rightarrow B=0 \\ &z(L)=0 \Rightarrow A\sin(kL)=0 \\ &A\neq 0 \text{ and } \sin(kL)=0 \Rightarrow kL=n\pi, \quad n=1,2,3,\ \ldots \end{aligned} \tag{6.4}$$

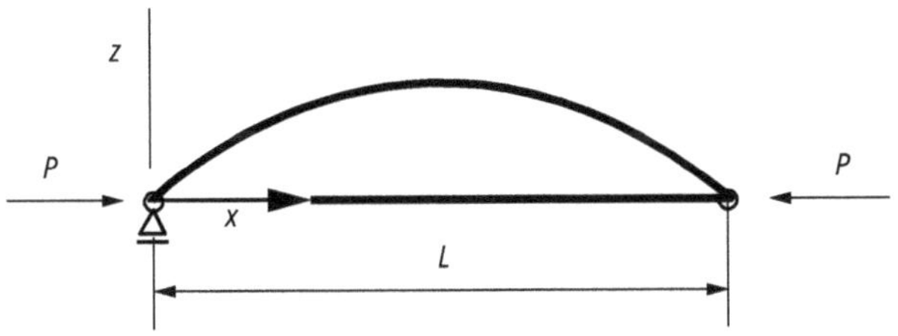

Fig. 6.1: A hinged–hinged column under compressive loads P.

The mathematical result from eq. (6.4) is that the calculation of the buckling of a column leads to an eigenvalue problem. Substituting the expression for k (see eq. (6.2)) and taking $n=1$ leads to the expression for what it is called the *Euler buckling load*:

$$k^2 \equiv \frac{P}{EI} = \left(\frac{n\pi}{L}\right)^2 \Rightarrow P = \left(\frac{n\pi}{L}\right)^2 EI, \quad P_{n=1} \equiv P_{\text{Euler}} = \frac{\pi^2 EI}{L^2} \tag{6.5}$$

For other boundary conditions (besides hinged–hinged), incorporating both geometrical and natural boundary conditions, the equation of motion would have the following form:[4]

$$\frac{d^4 z(x)}{dx^4} + k^2 \frac{d^2 z}{dx^2}(x) = 0, \quad k^2 = \frac{P}{EI} \tag{6.6}$$

The general solution for eq. (6.6) is then given by

$$z(x) = A\sin(kx) + B\cos(kx) + Cx + D \tag{6.7}$$

Application of various boundary conditions (see Tab. 6.1 for the most used one) would provide the relevant buckling load and its associated mode. For an enhanced understanding of the buckling phenomenon, which is a stability issue, the reader is referred to Refs. [1–6].

4 The theory leading to eq. (6.6) is usually denoted in the literature as the *Bernoulli–Euler* theory.

Tab. 6.1: Buckling factors and associated modes.

Hinged–hinged	Clamped–clamped	Clamped–free	Hinged–guided	Clamped–guided	Clamped–hinged
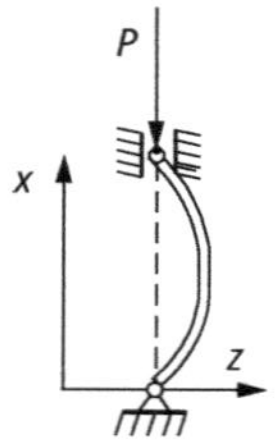	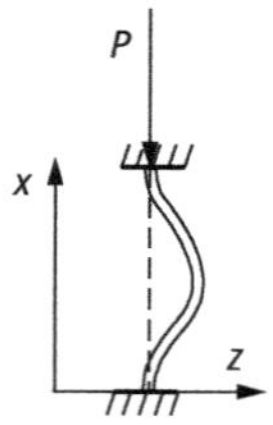	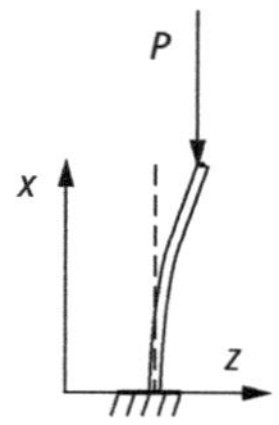	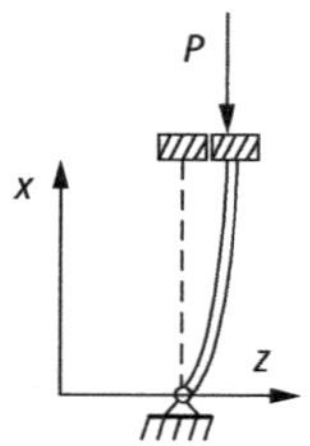	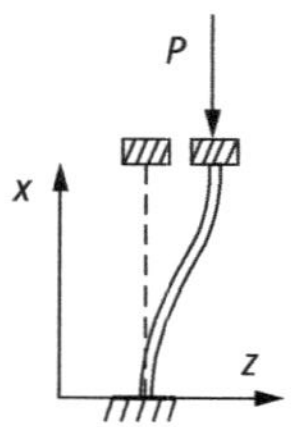	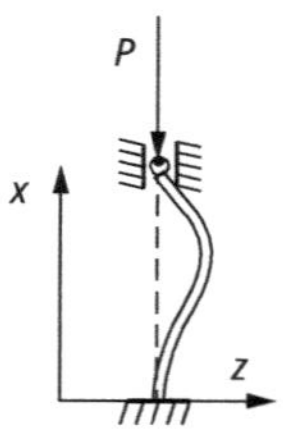
$z(0)=z(L)=0$ $z''(0)=z''(L)=0$	$z(0)=z(L)=0$ $z'(0)=z'(L)=0$	$z(0)=z'(0)=0$ $z''(L)=0$ $z'''(L)+k^2z'(L)=0$	$z(0)=z''(0)=0$ $z'(L)=0$ $z'''(L)+k^2z'(L)=0$	$z(0)=z'(0)=0$ $z'(L)=0$ $z'''(L)+k^2z'(L)=0$	$z(0)=z(L)=0$ $z'(0)=z''(L)0$
K = 1	K = 4	K = 0.25	K = 0.25	K = 1	K = 2.045
$z(x)=\sin\left(\frac{\pi x}{L}\right)$	$z(x)=1-\cos\left(\frac{2\pi x}{L}\right)$	$z(x)=1-\cos\left(\frac{\pi x}{2L}\right)$	$z(x)=1-\cos\left(\frac{\pi x}{2L}\right)$	$z(x)=1-\cos\left(\frac{\pi x}{2L}\right)$	$z(x)=\sin(\alpha x)+\alpha L\left[1-\cos(\alpha x)-\frac{x}{L}\right]$ $\alpha=1.4318\frac{\pi}{L}$
$L_{eff}=L$	$L_{eff}=0.5L$	$L_{eff}=2L$	$L_{eff}=2L$	$L_{eff}=L$	$L_{eff}=0.7L$

Note that $z'(x)\equiv\frac{dz(x)}{dx}$; $z''(x)\equiv\frac{d^2z(x)}{dx^2}$; $z'''(x)\equiv\frac{d^3z(x)}{dx^3}$. $P_{(\text{any BC})}=KP_{(\text{Euler})}=K\frac{\pi^2EI}{L^2}$, $K\leq 4$.
L_{eff} (*effective length*) of a *column* is the distance between two successive inflection points.

6.2.2 Rankin–Gordon formula

The buckling formulas derived in the previous section dealt with what is called “slender” columns. Slenderness ratio of a column (λ) is defined as the ratio of the effective length of a column (L_{eff} – see last row in Tab. 6.1) and the least radius of gyration (r) about the axis under consideration. Consequently, this can be expressed as

$$\lambda \equiv \frac{L_{eff}}{r}, \quad r = \sqrt{\frac{I}{A}} \tag{6.8}$$

where A is the cross-sectional area of the column and I is its moment of inertia.

The Euler buckling formula, developed in the previous section, is not suitable for short stubby columns. There are a few formulas to be used for relatively low slenderness ratio, with the Johnson’s parabola [7] and the Rankine–Gordon[5] semiempirical formula, being the most known ones.

Johnson’s parabola [7] is given by the following expression (see also Fig. 6.2):

$$\sigma \equiv \sigma_y - \frac{1}{E}\left(\frac{\sigma_y}{2\pi}\right)^2\left(\frac{L_{eff}}{r}\right)^2 = \sigma_y - \frac{1}{E}\left(\frac{\sigma_y}{2\pi}\right)^2\lambda^2 \tag{6.9}$$

where σ_y is the yield stress for a given material.

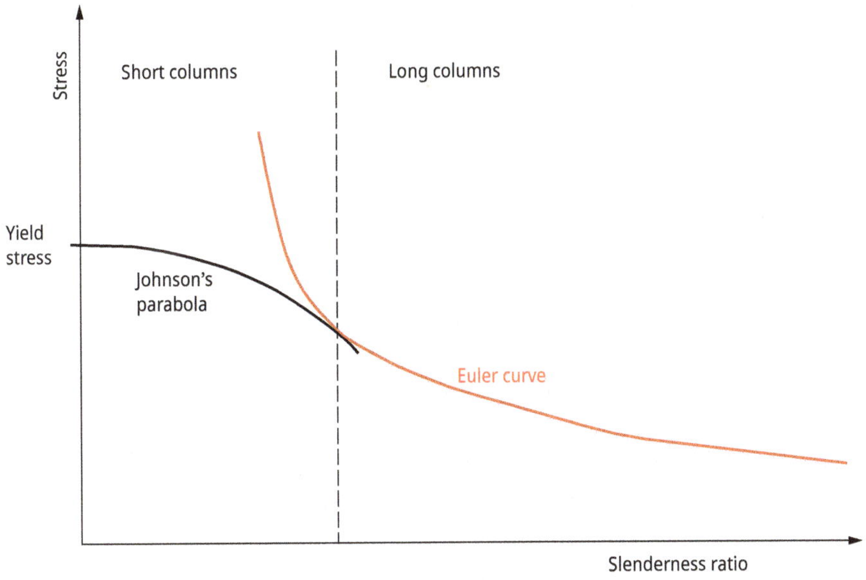

Fig. 6.2: Schematic diagram presenting Euler’s curve and Johnson’s parabola.

5 William John Macquorn Rankine (1820–1872) and Perry Hugesworth Gordon (1894–1966).

Rankine–Godman's semiempirical formula can be written as

$$\frac{1}{P_{\mathrm{R-G}}} \equiv \frac{1}{P_{\mathrm{Euler}}} + \frac{1}{P_{\mathrm{Crush}}} \tag{6.10}$$

where $P_{\mathrm{Crush}} \equiv \sigma_{\mathrm{y}} \cdot A$ is the crushing load or yield point load in compression for a given material. Equation (6.10) can be rearranged to show the stress in the column according to Rankine–Godman approach to yield

$$\sigma_{\mathrm{R-G}} = \frac{\sigma_{\mathrm{y}}}{1 + \frac{\sigma_{\mathrm{y}}}{\pi^2 E}\left(\frac{L_{\mathrm{eff}}}{r}\right)^2} = \frac{\sigma_{\mathrm{y}}}{1 + \frac{\sigma_{\mathrm{y}}}{\pi^2 E}\lambda^2} \tag{6.11}$$

A schematic diagram presenting the Euler curve and the Rankine–Godman curve is shown in Fig. 6.3.

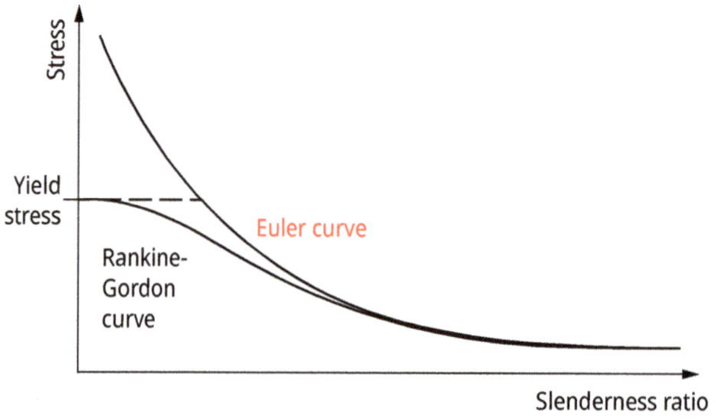

Fig. 6.3: Schematic diagram presenting Euler's curve and Rankine–Gordon's curve.

One can see from Fig. 6.3 that for short columns the stress predicted by the Rankine–Gordon's curve is tending to the yield stress, while for long columns, both curves would coincide.

6.2.3 Composite columns – CLT approach

Buckling of laminated composite columns (see Chapter 2), which are considered as 1D elements (see Fig. 6.4), can be derived for a general laminate using the classical lamination theory (CLT), to yield the following equations of motion:

$$A_{11}\left(\frac{d^2u_0}{dx^2} + \frac{d^3w_0}{dx^3}\right) - B_{11}\frac{d^3w_0}{dx^3} = 0 \tag{6.12}$$

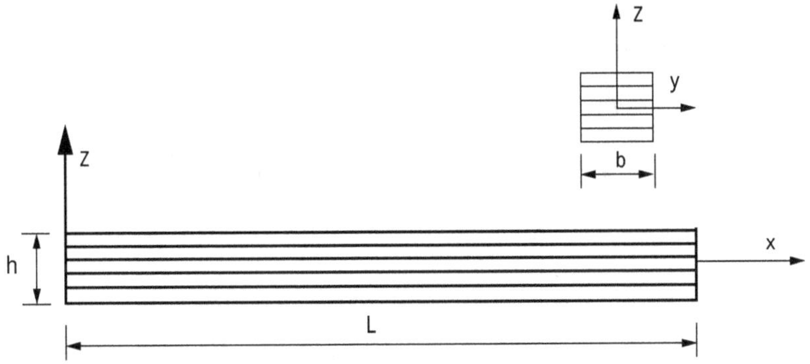

Fig. 6.4: The laminated composite column model.

$$-B_{11}\left(\frac{d^3u_0}{dx^3}+\frac{d^4w_0}{dx^4}\right)+D_{11}\frac{d^4w_0}{dx^4}+\bar{N}_{xx}\frac{d^2w_0}{dx^2}=0 \tag{6.13}$$

while A_{11}, B_{11} and D_{11} are the stretching, coupled stretching–bending and bending stiffness, accordingly, and stands for the axial compression per unit width (b), and it is assumed to be independent of the coordinate x. The two coupled equations of motion (eqs. (6.12) and (6.13)) can be decoupled to yield one single equation having the following form:

$$\left(D_{11}-\frac{B_{11}^2}{A_{11}}\right)\frac{d^4w_0}{dx^4}+\bar{N}_{xx}\frac{d^2w_0}{dx^2}=0 \tag{6.14}$$

or

$$\frac{d^4w_0}{dx^4}+\frac{\bar{N}_{xx}}{\left(D_{11}-\frac{B_{11}^2}{A_{11}}\right)}\frac{d^2w_0}{dx^2}=0 \tag{6.15}$$

6.2.3.1 Symmetric laminate ($B_{11} = 0$)

For the case of symmetric laminate, $B_{11} = 0$, and where w_b is the buckling deflection, one obtains

$$\frac{d^4w_b}{dx^4}+\frac{P}{E_{xx}\cdot I_{yy}}\frac{d^2w_b}{dx^2}=0 \tag{6.16}$$

where

$$b\cdot D_{11}=E_{xx}\cdot I_{yy} \quad \text{and} \quad b\cdot\bar{N}_{xx}=P \tag{6.17}$$

One should note that eq. (6.16) is exactly the Bernoulli–Euler equation of buckling for columns (see eq. (6.6)).

Now we shall present the general solution for eq. (6.16), applicable to the buckling of symmetric laminated composite columns ($B_{11} = 0$). Rearranging eq. (6.16), we obtain

$$\frac{d^4 w_b}{dx^4} + \lambda^2 \frac{d^2 w_b}{dx^2} = 0, \quad \lambda^2 = \frac{\bar{N}_{xx}}{D_{11}} \tag{6.18}$$

The solution of eq. (6.18) has the following general form (which is identical to eq. (6.7) applied for Bernoulli–Euler columns):

$$w_b(x) = C_1 \sin(\lambda x) + C_2 \cos(\lambda x) + C_3 x + C_4 \tag{6.19}$$

where C_1, C_2, C_3 and C_4 are constants to be determined using four boundary conditions of a given problem. For example, assuming a simply supported column having a length L, one can write its boundary conditions as

$$\begin{aligned} w_b(0) = 0; \quad M_{yy}(0) = 0 \Rightarrow \frac{d^2 w_b(0)}{dx^2} \equiv w_b''(0) = 0 \\ w_b(L) = 0; \quad M_{yy}(L) = 0 \Rightarrow \frac{d^2 w_b(L)}{dx^2} \equiv w_b''(L) = 0 \end{aligned} \tag{6.20}$$

Substituting the boundary conditions in eq. (6.19) yields a set of four equations with four unknowns, which can be written in a 4×4 matrix, while the right-hand sides of the four equations are identically zero. To obtain a unique solution, the determinant of the matrix must vanish. This leads to the following characteristic equation:

$$\sin(\lambda L) = 0 \quad \Rightarrow \quad \lambda L = n\pi, \quad n = 1, 2, 3, 4, \ldots \tag{6.21}$$

The critical buckling load of the column (the lowest one, $n = 1$) will then be

$$(\bar{N}_{xx})_{cr} = \frac{\pi^2}{L^2} D_{11} \tag{6.22}$$

One should remember that the term in eq. (6.22) should be multiplied by the width of the beam, b, to obtain P_{cr} [N]. One can see that the buckling load for a laminated symmetric composite column (eq. (6.22)) is exactly the same as for the Bernoulli–Euler column presented in eq. (6.5).

The column buckling shape (the eigenfunction) has the following form (this is obtained by back-substituting the eigenvalue from eq. (6.21) into the matrix form):

$$w_b(x) = C_1 \sin(\lambda x) = C_1 \sin\left(\frac{\pi x}{L}\right) \tag{6.23}$$

Tables 6.2 and 6.3[6] present some of the most encountered column cases, while the schematic drawings of the various out-of-plane boundary conditions are presented in

6 Some of the boundary conditions and their relevant buckling loads are also presented in Tab. 6.1.

Fig. 6.5. One should note that only the out-of-plane boundary conditions can influence the critical buckling load and the shape at buckling (see also Appendix A).

Tab. 6.2: Buckling of laminated composite columns – out-of-plane boundary conditions – CLT approach.

No.	B.C. name	Out-of-plane boundary conditions	
		$x = 0$	$x = L$
1	SS-SS*	$w_b = 0; \frac{d^2 w_b}{dx^2} = 0$	$w_b = 0; \frac{d^2 w_b}{dx^2} = 0$
2	C-C**	$w_b = 0; \frac{dw_b}{dx} = 0$	$w_b = 0; \frac{dw_b}{dx} = 0$
3	C-F***	$w_b = 0; \frac{dw_b}{dx} = 0$	$\frac{d^2 w_b}{dx^2} = 0; \frac{d^3 w_b}{dx^3} + \lambda^2 \frac{dw_b}{dx} = 0$
4	F-F	$\frac{d^2 w_b}{dx^2} = 0; \frac{d^3 w_b}{dx^3} + \lambda^2 \frac{dw_b}{dx} = 0$	$\frac{d^2 w_b}{dx^2} = 0; \frac{d^3 w_b}{dx^3} + \lambda^2 \frac{dw_b}{dx} = 0$
5	SS-C	$w_b = 0; \frac{d^2 w_b}{dx^2} = 0$	$w_b = 0; \frac{dw_b}{dx} = 0$
6	SS-F	$w_b = 0; \frac{d^2 w_b}{dx^2} = 0$	$\frac{d^2 w_b}{dx^2} = 0; \frac{d^3 w_b}{dx^3} + \lambda^2 \frac{dw_b}{dx} = 0$
7	G****-F	$\frac{dw_b}{dx} = 0; \frac{d^3 w_b}{dx^3} + \lambda^2 \frac{dw_b}{dx} = 0$	$\frac{d^2 w_b}{dx^2} = 0; \frac{d^3 w_b}{dx^3} + \lambda^2 \frac{dw_b}{dx} = 0$
8	G-SS	$\frac{dw_b}{dx} = 0; \frac{d^3 w_b}{dx^3} + \lambda^2 \frac{dw_b}{dx} = 0$	$w_b = 0; \frac{d^2 w_b}{dx^2} = 0$
9	G-G	$\frac{dw_b}{dx} = 0; \frac{d^3 w_b}{dx^3} + \lambda^2 \frac{dw_b}{dx} = 0$	$\frac{dw_b}{dx} = 0; \frac{d^3 w_b}{dx^3} + \lambda^2 \frac{dw_b}{dx} = 0$
10	G-C	$\frac{dw_b}{dx} = 0; \frac{d^3 w_b}{dx^3} + \lambda^2 \frac{dw_b}{dx} = 0$	$w_b = 0; \frac{dw_b}{dx} = 0$

*SS, simply supported or hinged–hinged or pinned–pinned.
**C, clamped.
***F, free.
****G, guided.

To conclude this section, a column with general boundary conditions is presented in Fig. 6.6.

Defining the following variables as

$$\bar{k}_1 = \frac{k_1}{\bar{D}_{11}}; \bar{k}_2 = \frac{k_2}{\bar{D}_{11}}; \bar{k}_{\theta_1} = \frac{k_{\theta_1}}{\bar{D}_{11}}; \bar{k}_{\theta_2} = \frac{k_{\theta_2}}{\bar{D}_{11}} \tag{6.24}$$

where $\bar{D}_{11} \equiv D_{11} - \frac{B_{11}^2}{A_{11}}$ and k_1 and k_2 are linear springs, while $\bar{k}_{\theta_1}$ and $\bar{k}_{\theta_2}$ are torsion springs. The boundary conditions of the problem depicted in Fig. 6.6 are given by

Tab. 6.3: Buckling loads and relevant buckling modes of laminated composite columns using CLT approach.

No.	Name	Characteristic equation	Critical buckling load	Mode shape
1	SS-SS	$\sin(\lambda L)=0$ $\lambda L=n\pi$	$(\bar{N}_{xx})_{cr}=\frac{\pi^2}{L^2}D_{11}$	$\sin\left(\frac{\pi x}{L}\right)$
2	C-C	$\lambda L\sin(\lambda L)=2[1-\cos(\lambda L)]$ $\lambda L=2\pi,\ 8.987, 4\pi,\ \ldots$	$(\bar{N}_{xx})_{cr}=4\frac{\pi^2}{L^2}D_{11}$	$1-\cos\left(\frac{2\pi x}{L}\right)$
3	C-F	$\cos(\lambda L)=0$ $\lambda L=(2n-1)\pi/2$	$(\bar{N}_{xx})_{cr}=\frac{\pi^2}{4L^2}D_{11}$	$1-\cos\left(\frac{\pi x}{2L}\right)$
4	F-F	$\sin(\lambda L)=0$ $\lambda L=n\pi$	$(\bar{N}_{xx})_{cr}=\frac{\pi^2}{L^2}D_{11}$	$\sin\left(\frac{\pi x}{L}\right)$
5	SS-C	$\tan(\lambda L)=\lambda L$ $\lambda L=1.430\pi, 2.459\pi,\ \ldots$	$(\bar{N}_{xx})_{cr}=2.045\frac{\pi^2}{L^2}D_{11}$	$\sin(ax)+aL\left[1-\cos(ax)-\frac{x}{L}\right]$ while $a=1.4318\frac{\pi}{L}$
6	SS-F	$\sin(\lambda L)=0$ $\lambda L=n\pi$	$(\bar{N}_{xx})_{cr}=\frac{\pi^2}{L^2}D_{11}$	$\sin\left(\frac{\pi x}{L}\right)$
7	G-F	$\cos(\lambda L)=0$ $\lambda L=(2n-1)\pi/2$	$(\bar{N}_{xx})_{cr}=\frac{\pi^2}{4L^2}D_{11}$	$\cos\left(\frac{\pi x}{2L}\right)$
8	G-SS	$\cos(\lambda L)=0$ $\lambda L=(2n-1)\pi/2$	$(\bar{N}_{xx})_{cr}=\frac{\pi^2}{4L^2}D_{11}$	$\cos\left(\frac{\pi x}{2L}\right)$
9	G-G	$\sin(\lambda L)=0$ $\lambda L=n\pi$	$(\bar{N}_{xx})_{cr}=\frac{\pi^2}{L^2}D_{11}$	$\cos\left(\frac{\pi x}{L}\right)$
10	G-C	$\sin(\lambda L)=0$ $\lambda L=n\pi$	$(\bar{N}_{xx})_{cr}=\frac{\pi^2}{L^2}D_{11}$	$1-\cos\left(\frac{\pi x}{L}\right)$

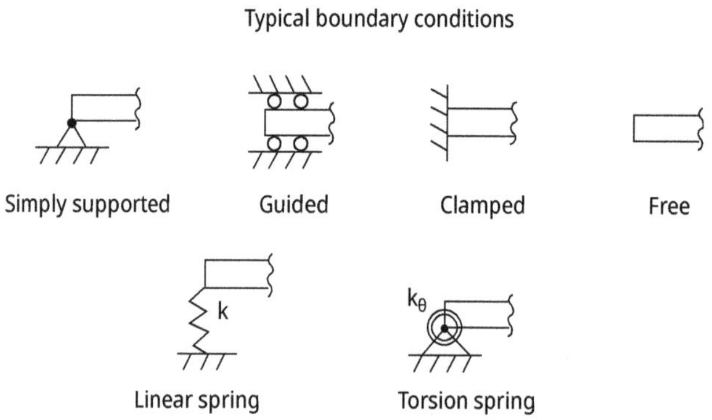

Fig. 6.5: Typical out-of-plane boundary conditions for laminated composite columns (and isotropic columns) under axial compression.

Fig. 6.6: Typical laminated composite columns under axial compression having spring-based boundary conditions.[7]

$$\begin{aligned}
&\frac{d^3 w_b(0)}{dx^3} + \lambda^2 \frac{dw_b(0)}{dx} = -\bar{k}_1 \cdot w_b(0) \qquad \frac{d^2 w_b(0)}{dx^2} = \bar{k}_{\theta_1} \frac{dw_b(0)}{dx} \\
&\frac{d^3 w_b(L)}{dx^3} + \lambda^2 \frac{dw_b(L)}{dx} = \bar{k}_2 \cdot w_b(L) \qquad \frac{d^2 w_b(L)}{dx^2} = -\bar{k}_{\theta_2} \frac{dw_b(L)}{dx}
\end{aligned} \tag{6.25}$$

Application of the boundary conditions to the general solution presented by eq. (6.19) yields the following characteristic equation (see details in [3]):

$$\begin{aligned}
&\left\{-(\bar{k}_1+\bar{k}_2)\lambda^6 + \left[\bar{k}_{\theta_1}\cdot\bar{k}_{\theta_2}(\bar{k}_1+\bar{k}_2)+\bar{k}_1\cdot\bar{k}_2\cdot L\right]\lambda^4 + \bar{k}_{\theta_1}\cdot\bar{k}_{\theta_2}(\bar{k}_{\theta_1}+\bar{k}_{\theta_2}-\bar{k}_{\theta_1}\cdot\bar{k}_{\theta_2}\cdot L)\lambda^2\right\}\sin(\lambda L) \\
&+\left\{(\bar{k}_1+\bar{k}_2)(\bar{k}_{\theta_1}+\bar{k}_{\theta_2})\lambda^3 - \bar{k}_1\cdot\bar{k}_2\cdot L(\bar{k}_{\theta_1}+\bar{k}_{\theta_2})\lambda^3 - 2\cdot\bar{k}_1\cdot\bar{k}_2\cdot\bar{k}_{\theta_1}\cdot\bar{k}_{\theta_2}\cdot\lambda\right\}\cos(\lambda L) \\
&+\bar{k}_1\cdot\bar{k}_2\cdot\bar{k}_{\theta_1}\cdot\bar{k}_{\theta_2}\cdot\lambda = 0
\end{aligned} \tag{6.26}$$

Equation (6.26) can be used to solve not only problems involving boundary conditions with springs but also classical boundary conditions. Letting the linear springs k_1 and k_2 tend to infinity (∞) while the torsional ones, $\bar{k}_{\theta_1}$ and $\bar{k}_{\theta_2}$, are set to zero would lead to the classical hinged–hinged boundary conditions at both ends of the compressed beam. Similarly, setting all the springs to zero would yield free–free boundary conditions. Clamped boundary conditions can be obtained when both types of springs (linear and torsion) would tend to infinity.

6.2.3.2 Nonsymmetric laminate ($B_{11} \neq 0$)

To solve the case of a nonsymmetric laminate ($B_{11} \neq 0$), we can use eq. (6.15). The general solution can be written as

$$w_0(x) = C_1 \sin\left(\hat{\lambda} x\right) + C_2 \cos\left(\hat{\lambda} x\right) + C_3 x + C_4 \tag{6.27}$$

7 The drawing presented can also be suitable for a Bernoulli–Euler column.

where

$$\widehat{\lambda}^2 = \frac{\bar{N}_{xx}}{\left(D_{11} - \frac{B_{11}^2}{A_{11}}\right)} \tag{6.28}$$

The in-plane displacement will then be given as (see Appendix A)

$$u_0(x) = C_5 \sin\left(\widehat{\lambda} x\right) + C_6 \cos\left(\widehat{\lambda} x\right) + C_7 x + C_8 \tag{6.29}$$

where

$$\begin{aligned} C_5 &= -\frac{B_{11}}{A_{11}} \cdot \widehat{\lambda} \cdot C_2 \\ C_6 &= +\frac{B_{11}}{A_{11}} \cdot \widehat{\lambda} \cdot C_1 \end{aligned} \tag{6.30}$$

Solving for a hinged–hinged column (or pinned–pinned, or simply supported at both ends), with the following boundary conditions, two for the in-plane and four for the out-of-plane (see also Appendix A)

$$w_0(0) = 0; \quad M_{xx}(0) = 0 \quad \Rightarrow \quad -B_{11}\frac{du_0(0)}{dx} + D_{11}\frac{d^2 w_0(0)}{dx^2} = 0 \tag{6.31}$$

$$w_0(L) = 0; \quad M_{xx}(L) = 0 \quad \Rightarrow \quad -B_{11}\frac{du_0(L)}{dx} + D_{11}\frac{d^2 w_0(L)}{dx^2} = 0 \tag{6.32}$$

$$u_0(0) = 0 \tag{6.33}$$

$$A_{11}\frac{du_0(L)}{dx} - B_{11}\frac{d^2 w_0(L)}{dx^2} = -\bar{N}_{xx} \tag{6.34}$$

yields the following eigenvalue:

$$\sin\left(\widehat{\lambda} L\right) = 0 \quad \Rightarrow \quad \widehat{\lambda} L = n\pi, \quad n = 1, 2, 3, 4, \ldots \tag{6.35}$$

which gives the critical buckling load per unit width having the following form:

$$(\bar{N}_{xx})_{\text{cr}} = \frac{\pi^2}{L^2}\left(D_{11} - \frac{B_{11}^2}{A_{11}}\right) \tag{6.36}$$

The buckling shape will be like the symmetric case presented in Tab. 6.2, namely $\sin\left(\frac{\pi x}{L}\right)$. Note that as expected, the buckling load for a symmetric laminate will be higher than a nonsymmetric one, having the same number of layers. Tables 6.4 and 6.5 present the boundary conditions for the nonsymmetric case, while Tab. 6.6 gives the buckling loads and the associated buckling modes for the most used cases.

Tab. 6.4: Buckling of nonsymmetric laminated composite columns – out-of-plane boundary conditions using CLT approach.

No.	Name	Out-of-plane boundary conditions	
		$x=0$	$x=L$
1	SS-SS*	$w_0=0;\ -B_{11}\frac{du_0}{dx}+D_{11}\frac{d^2w_0}{dx^2}=0$	$w_0=0;\ -B_{11}\frac{du_0}{dx}+D_{11}\frac{d^2w_0}{dx^2}=0$
2	C-C**	$w_0=0;\ \frac{dw_0}{dx}=0$	$w_0=0;\ \frac{dw_0}{dx}=0$
3	C-F***	$w_0=0;\ \frac{dw_0}{dx}=0$	$-B_{11}\frac{u_0}{dx}+D_{11}\frac{d^2w_0}{dx^2}=0$ $-B_{11}\frac{d^2u_0}{dx^2}+D_{11}\frac{d^3w_0}{dx^3}+\bar{N}_{xx}\frac{dw_0}{dx}=0$
4	F-F	$-B_{11}\frac{du_0}{dx}+D_{11}\frac{d^2w_0}{dx^2}=0$ $-B_{11}\frac{d^2u_0}{dx^2}+D_{11}\frac{d^3w_0}{dx^3}+\bar{N}_{xx}\frac{dw_0}{dx}=0$	$-B_{11}\frac{du_0}{dx}+D_{11}\frac{d^2w_0}{dx^2}=0$ $-B_{11}\frac{d^2u_0}{dx^2}+D_{11}\frac{d^3w_0}{dx^3}+\bar{N}_{xx}\frac{dw_0}{dx}=0$
5	SS-C	$w_0=0$ $-B_{11}\frac{du_0}{dx}+D_{11}\frac{d^2w_0}{dx^2}=0$	$w_0=0;\ \frac{dw_0}{dx}=0$
6	SS-F	$w_0=0;\ -B_{11}\frac{du_0}{dx}+D_{11}\frac{d^2w_0}{dx^2}=0$	$-B_{11}\frac{du_0}{dx}+D_{11}\frac{d^2w_0}{dx^2}=0$ $-B_{11}\frac{d^2u_0}{dx^2}+D_{11}\frac{d^3w_0}{dx^3}+\bar{N}_{xx}\frac{dw_0}{dx}=0$
7	G***-F	$\frac{dw_0}{dx}=0$ $-B_{11}\frac{d^2u_0}{dx^2}+D_{11}\frac{d^3w_0}{dx^3}+\bar{N}_{xx}\frac{dw_0}{dx}=0$	$-B_{11}\frac{du_0}{dx}+D_{11}\frac{d^2w_0}{dx^2}=0$ $-B_{11}\frac{d^2u_0}{dx^2}+D_{11}\frac{d^3w_0}{dx^3}+\bar{N}_{xx}\frac{dw_0}{dx}=0$
8	G-SS	$\frac{dw_0}{dx}=0$ $-B_{11}\frac{d^2u_0}{dx^2}+D_{11}\frac{d^3w_0}{dx^3}+\bar{N}_{xx}\frac{dw_0}{dx}=0$	$w_0=0;\ -B_{11}\frac{du_0}{dx}+D_{11}\frac{d^2w_0}{dx^2}=0$
9	G-G	$\frac{dw_0}{dx}=0$ $-B_{11}\frac{d^2u_0}{dx^2}+D_{11}\frac{d^3w_0}{dx^3}+\bar{N}_{xx}\frac{dw_0}{dx}=0$	$\frac{dw_0}{dx}=0$ $-B_{11}\frac{d^2u_0}{dx^2}+D_{11}\frac{d^3w_0}{dx^3}+\bar{N}_{xx}\frac{dw_0}{dx}=0$
10	G-C	$\frac{dw_0}{dx}=0$ $-B_{11}\frac{d^2u_0}{dx^2}+D_{11}\frac{d^3w_0}{dx^3}+\bar{N}_{xx}\frac{dw_0}{dx}=0$	$w_0=0;\ \frac{dw_0}{dx}=0$

*SS, simply supported, or hinged–hinged or pinned–pinned.
**C, clamped.
***F, free.
***G, guided.

Tab. 6.5: Buckling of nonsymmetric laminated composite columns – in-plane boundary conditions using CLT approach.

No.	Name	In-plane boundary conditions	
		$x = 0$	**$x = L$**
1	SS-SS	$u_0 = 0$	$A_{11}\frac{du_0}{dx} - B_{11}\frac{d^2w_0}{dx^2} = -\bar{N}_{xx}$
2	C-C	$u_0 = 0$	$A_{11}\frac{du_0}{dx} - B_{11}\frac{d^2w_0}{dx^2} = -\bar{N}_{xx}$
3	C-F	$u_0 = 0$	$A_{11}\frac{du_0}{dx} - B_{11}\frac{d^2w_0}{dx^2} = -\bar{N}_{xx}$
4	F-F	$A_{11}\frac{du_0}{dx} - B_{11}\frac{d^2w_0}{dx^2} = -\bar{N}_{xx}$	$A_{11}\frac{du_0}{dx} - B_{11}\frac{d^2w_0}{dx^2} = -\bar{N}_{xx}$
5	SS-C	$u_0 = 0$	$A_{11}\frac{du_0}{dx} - B_{11}\frac{d^2w_0}{dx^2} = -\bar{N}_{xx}$
6	SS-F	$u_0 = 0$	$A_{11}\frac{du_0}{dx} - B_{11}\frac{d^2w_0}{dx^2} = -\bar{N}_{xx}$
7	G-F	$u_0 = 0$	$A_{11}\frac{du_0}{dx} - B_{11}\frac{d^2w_0}{dx^2} = -\bar{N}_{xx}$
8	G-SS	$u_0 = 0$	$A_{11}\frac{du_0}{dx} - B_{11}\frac{d^2w_0}{dx^2} = -\bar{N}_{xx}$
9	G-G	$u_0 = 0$	$A_{11}\frac{du_0}{dx} - B_{11}\frac{d^2w_0}{dx^2} = -\bar{N}_{xx}$
10	G-C	$u_0 = 0$	$A_{11}\frac{du_0}{dx} - B_{11}\frac{d^2w_0}{dx^2} = -\bar{N}_{xx}$

Tab. 6.6: Buckling loads and relevant buckling modes of laminated composite columns using CLT approach.

No.	Name	Characteristic equation	Critical buckling load per unit width	Mode shape
1	SS-SS	$\sin\left(\widehat{\lambda L}\right) = 0;\quad \widehat{\lambda L} = n\pi$	$\left(\bar{N}_{xx}\right)_{cr} = \frac{\pi^2}{L^2}\bar{D}_{11}{}^*$	$\sin\left(\frac{\pi x}{L}\right)$
2	C-C	$\widehat{\lambda L}\sin\left(\widehat{\lambda L}\right) = 2\left[1 - \cos\left(\widehat{\lambda L}\right)\right]$ $\widehat{\lambda L} = 2\pi, 8.987, 4\pi, .$	$\left(\bar{N}_{xx}\right)_{cr} = 4\frac{\pi^2}{L^2}\bar{D}_{11}$	$1 - \cos\left(\frac{\pi x}{L}\right)$
3	C-F	$\cos\left(\widehat{\lambda L}\right) = 0\ \widehat{\lambda L} = (2n-1)\frac{\pi}{2}$	$\left(\bar{N}_{xx}\right)_{cr} = \frac{\pi^2}{4L^2}\bar{D}_{11}$	$1 - \cos\left(\frac{\pi x}{2L}\right)$
4	F-F	$\sin\left(\widehat{\lambda L}\right) = 0;\quad \widehat{\lambda L} = n\pi$	$\left(\bar{N}_{xx}\right)_{cr} = \frac{\pi^2}{L^2}\bar{D}_{11}$	$\sin\left(\frac{\pi x}{L}\right)$

Tab. 6.6 (continued)

No.	Name	Characteristic equation	Critical buckling load per unit width	Mode shape
5	SS-C	$\tan\left(\widehat{\lambda}L\right)=\widehat{\lambda}L$ $\widehat{\lambda}L=1.430\pi, 2.459\pi, \ldots$	$\left(\bar{N}_{xx}\right)_{cr}=2.045\frac{\pi^2}{L^2}\bar{D}_{11}$	$\sin(\alpha x)+\alpha L\left[1-\cos(\alpha x)-\frac{x}{L}\right]$ while $\alpha=1.4318\frac{\pi}{L}$
6	SS-F	$\sin\left(\widehat{\lambda}L\right)=0$ $\widehat{\lambda}L=n\pi$	$\left(\bar{N}_{xx}\right)_{cr}=\frac{\pi^2}{L^2}\bar{D}_{11}$	$\sin\left(\frac{\pi x}{L}\right)$
7	G-F	$\cos\left(\widehat{\lambda}L\right)=0;\quad \widehat{\lambda}L=(2n-1)\frac{\pi}{2}$	$\left(\bar{N}_{xx}\right)_{cr}=\frac{\pi^2}{4L^2}\bar{D}_{11}$	$\cos\left(\frac{\pi x}{2L}\right)$
8	G-SS	$\cos\left(\widehat{\lambda}L\right)=0\ \widehat{\lambda}L=(2n-1)\pi/2$	$\left(\bar{N}_{xx}\right)_{cr}=\frac{\pi^2}{4L^2}\bar{D}_{11}$	$\cos\left(\frac{\pi x}{2L}\right)$
9	G-G	$\sin\left(\widehat{\lambda}L\right)=0$ $\widehat{\lambda}L=n\pi$	$\left(\bar{N}_{xx}\right)_{cr}=\frac{\pi^2}{L^2}\bar{D}_{11}$	$\cos\left(\frac{\pi x}{2L}\right)$
10	G-C	$\sin\left(\widehat{\lambda}L\right)=0$ $\widehat{\lambda}L=n\pi$	$\left(\bar{N}_{xx}\right)_{cr}=\frac{\pi^2}{L^2}\bar{D}_{11}$	$1-\cos\left(\frac{\pi x}{L}\right)$

$* \bar{D}_{11}=\left(D_{11}-\frac{B_{11}^2}{A_{11}}\right).$

6.3 Buckling of columns – FOSDT approach

Buckling of columns using the first-order shear deformation theory (FOSDT) (see also [2, 5, 8–19]) can be obtained using the equations presented in Chapter 2:[8]

$$A_{11}\left(\frac{d^2u_0}{dx^2}+\frac{d^3w_b}{dx^3}\right)+B_{11}\frac{d^2\phi_x}{dx^2}=0 \tag{6.37}$$

$$KA_{55}\left(\frac{d^2w_b}{dx^2}+\frac{d\phi_x}{dx}\right)-\bar{N}_{xx}\frac{d^2w_b}{dx^2}=0 \tag{6.38}$$

$$B_{11}\left(\frac{d^2u_0}{dx^2}+\frac{\partial^3w_b}{dx^3}\right)+D_{11}\frac{d^2\phi_x}{dx^2}-KA_{55}\left(\frac{dw_b}{dx}+\phi_x\right)=0 \tag{6.39}$$

Decoupling the equations leads to the following uncoupled ones[9] (see also [12, 13]):

8 $w_0=w_0^p+w_b$ (w_0^p is the prebuckling deflection and w_b is the buckling deflection).
9 Note that to uncouple the equations one needs to neglect high-order terms like $\frac{d^3w_b}{dx^3}$.

$$\frac{d^4w_b}{dx^4}+\Gamma^2\frac{d^2w_b}{dx^2}=0 \tag{6.40}$$

$$\frac{d^3\phi_x}{dx^3}+\Gamma^2\frac{d^2\phi_x}{dx}=0 \tag{6.41}$$

$$\frac{d^4u_0}{dx^4}+\Gamma^2\frac{d^2u_0}{dx^2}=0 \tag{6.42}$$

where

$$\Gamma^2=\frac{\bar{N}_{xx}}{\left(D_{11}-\frac{B_{11}^2}{A_{11}}\right)\left(1-\frac{\bar{N}_{xx}}{KA_{55}}\right)} \quad\Rightarrow\quad \bar{N}_{xx}=\frac{\Gamma^2\left(D_{11}-\frac{B_{11}^2}{A_{11}}\right)}{1+\frac{\Gamma^2\left(D_{11}-\frac{B_{11}^2}{A_{11}}\right)}{KA_{55}}} \tag{6.43}$$

Accordingly, the solutions for eqs. (6.40)–(6.42) have a form similar to that presented before, namely

$$w_b(x)=C_1\sin(\Gamma x)+C_2\cos(\Gamma x)+C_3x+C_4 \tag{6.44}$$

$$\phi_x(x)=C_5\sin(\Gamma x)+C_6\cos(\Gamma x)+C_7 \tag{6.45}$$

$$u_0(x)=C_8\sin(\Gamma x)+C_9\cos(\Gamma x)+C_{10}x+C_{11} \tag{6.46}$$

One should note that the 11 constants (C_1–C_{11}) are not independent. Their dependency can be obtained by back-substituting the solutions (eqs. (6.44)–(6.46)) in the coupled equations (6.37) and (6.39) to yield the following five relations:

$$A_{11}\cdot C_8+B_{11}\cdot C_5=0 \tag{6.47}$$

$$A_{11}\cdot C_9+B_{11}\cdot C_6=0 \tag{6.48}$$

$$-B_{11}\cdot C_8\cdot\hat{\lambda}^2-D_{11}\cdot C_5\cdot\hat{\lambda}^2-KA_{55}\left(-C_2\cdot\hat{\lambda}+C_5\right)=0 \tag{6.49}$$

$$-B_{11}\cdot C_9\cdot\hat{\lambda}^2-D_{11}\cdot C_6\cdot\hat{\lambda}^2-KA_{55}\left(C_1\cdot\hat{\lambda}+C_6\right)=0 \tag{6.50}$$

$$-KA_{55}\cdot(C_3+C_7)=0 \quad\Rightarrow\quad C_3+C_7=0 \tag{6.51}$$

These five relationships together with the six boundary conditions of the problem will provide the needed 11 equations with 11 unknowns, C_1–C_{11}. The six boundary conditions of the problem are (see also Chapter 2):

$$A_{11}\frac{du_0}{dx}+B_{11}\frac{d\phi_x}{dx}=-\bar{N}_{xx} \quad \text{or} \quad u_0=0 \tag{6.52}$$

$$KA_{55}\left(\frac{dw_b}{dx}+\phi_x\right)-\bar{N}_{xx}\frac{dw_b}{dx}=0 \quad \text{or} \quad w_b=0 \tag{6.53}$$

$$B_{11}\frac{du_0}{dx}+D_{11}\frac{d\phi_x}{dx}=0 \quad \text{or} \quad \phi_x=0 \tag{6.54}$$

Let us present the general case of a simply supported laminated composite beam having the following boundary conditions:

$$u_0(0)=0, \quad A_{11}\frac{du_0(L)}{dx}+B_{11}\frac{d\phi_x(L)}{dx}=-\bar{N}_{xx} \tag{6.55}$$

$$w_b(0)=0, \quad w_b(L)=0 \tag{6.56}$$

$$B_{11}\frac{du_0(0)}{dx}+D_{11}\frac{d\phi_x(0)}{dx}=0, \quad B_{11}\frac{du_0(L)}{dx}+D_{11}\frac{d\phi_x(L)}{dx}=0 \tag{6.57}$$

Solving for the 11 unknowns, C_1–C_{11}, we get the following results for the first four constants:

$$C_1=\frac{B_{11}}{A_{11}}\frac{[1-\cos(\Gamma L)]}{\sin(\Gamma L)}, \quad C_2=\frac{B_{11}}{A_{11}}, \quad C_3=0, \quad C_4=-\frac{B_{11}}{A_{11}}. \tag{6.58}$$

To obtain the buckling load, one should demand that $w_0(x)\to\infty$, leading to the same procedure applied for a general beam using the CLT approach. Observe that the constant C_1 leads to the following eigenvalue solution:

$$\sin(\Gamma L)=0 \quad \Rightarrow \quad \Gamma L=n\pi, \quad n=1,2,3,4,\ldots \tag{6.59}$$

which leads to the critical buckling load per unit width having the following form:

$$(\bar{N}_{xx})_{cr}=\frac{\left(\frac{\pi}{L}\right)^2\left(D_{11}-\frac{B_{11}^2}{A_{11}}\right)}{1+\frac{\left(\frac{\pi}{L}\right)^2\left(D_{11}-\frac{B_{11}^2}{A_{11}}\right)}{KA_{55}}}=\frac{\left[(\bar{N}_{xx})_{cr}\right]_{CLT}}{1+\frac{\left[(\bar{N}_{xx})_{cr}\right]_{CLT}}{KA_{55}}} \tag{6.60}$$

If one defines the following term:

$$H\equiv\frac{\left[(\bar{N}_{xx})_{cr}\right]_{CLT}}{KA_{55}} \tag{6.61}$$

then eq. (6.60) can be written as

$$(\bar{N}_{xx})_{cr}=\frac{\left[(\bar{N}_{xx})_{cr}\right]_{CLT}}{1+H} \tag{6.62}$$

The shape of the buckling will be $\sin\left(\frac{\pi x}{L}\right)$ like for the symmetric case, calculated using the CLT approach. One should notice that the buckling load calculated using FOSDT approach is lower than the one calculated according to CLT. Like before, a nonsymmetric laminate would yield an even lower buckling load as compared with a symmetric one having the same layers. Some typical buckling loads are presented in Tab. 6.7.

Tab. 6.7: Buckling loads and relevant buckling modes of laminated composite columns using FOSDT approach.

No.	Name	$\left[(\bar{N}_{xx})_{cr}\right]_{CLT}$	Critical buckling load per unit width	Mode shape
1	SS-SS	$(\bar{N}_{xx})_{cr} = \frac{\pi^2}{L^2}\bar{D}_{11}{}^*$	$(\bar{N}_{xx})_{cr} = \frac{\left[(\bar{N}_{xx})_{cr}\right]_{CLT}}{1+H}{}^{**}$	$\sin\left(\frac{\pi x}{L}\right)$
2	C-C	$(\bar{N}_{xx})_{cr} = 4\frac{\pi^2}{L^2}\bar{D}_{11}$	$(\bar{N}_{xx})_{cr} = \frac{\left[(\bar{N}_{xx})_{cr}\right]_{CLT}}{1+H}$	$1-\cos\left(\frac{2\pi x}{L}\right)$
3	C-F	$(\bar{N}_{xx})_{cr} = \frac{\pi^2}{4L^2}\bar{D}_{11}$	$(\bar{N}_{xx})_{cr} = \frac{\left[(\bar{N}_{xx})_{cr}\right]_{CLT}}{1+H}$	$1-\cos\left(\frac{\pi x}{2L}\right)$
4	F-F	$(\bar{N}_{xx})_{cr} = \frac{\pi^2}{L^2}\bar{D}_{11}$	$(\bar{N}_{xx})_{cr} = \frac{\left[(\bar{N}_{xx})_{cr}\right]_{CLT}}{1+H}$	$\sin\left(\frac{\pi x}{L}\right)$
5	SS-C	$(\bar{N}_{xx})_{cr} = 2.045\frac{\pi^2}{L^2}\bar{D}_{11}$	$(\bar{N}_{xx})_{cr} = \frac{\left[(\bar{N}_{xx})_{cr}\right]_{CLT}}{1+H}$	$\sin(\alpha x) + \alpha L\left[1-\cos(\alpha x) - \frac{x}{L}\right]$ while $\alpha = 1.4318\frac{\pi}{L}$
6	SS-F	$(\bar{N}_{xx})_{cr} = \frac{\pi^2}{L^2}\bar{D}_{11}$	$(\bar{N}_{xx})_{cr} = \frac{\left[(\bar{N}_{xx})_{cr}\right]_{CLT}}{1+H}$	$\sin\left(\frac{\pi x}{L}\right)$
7	G-F	$(\bar{N}_{xx})_{cr} = \frac{\pi^2}{4L^2}\bar{D}_{11}$	$(\bar{N}_{xx})_{cr} = \frac{\left[(\bar{N}_{xx})_{cr}\right]_{CLT}}{1+H}$	$\cos\left(\frac{\pi x}{2L}\right)$
8	G-SS	$(\bar{N}_{xx})_{cr} = \frac{\pi^2}{4L^2}\bar{D}_{11}$	$(\bar{N}_{xx})_{cr} = \frac{\left[(\bar{N}_{xx})_{cr}\right]_{CLT}}{1+H}$	$\cos\left(\frac{\pi x}{2L}\right)$
9	G-G	$(\bar{N}_{xx})_{cr} = \frac{\pi^2}{L^2}\bar{D}_{11}$	$(\bar{N}_{xx})_{cr} = \frac{\left[(\bar{N}_{xx})_{cr}\right]_{CLT}}{1+H}$	$\cos\left(\frac{\pi x}{L}\right)$
10	G-C	$(\bar{N}_{xx})_{cr} = \frac{\pi^2}{L^2}\bar{D}_{11}$	$(\bar{N}_{xx})_{cr} = \frac{\left[(\bar{N}_{xx})_{cr}\right]_{CLT}}{1+H}$	$1-\cos\left(\frac{\pi x}{2L}\right)$

$^*\bar{D}_{11} = \left(D_{11} - \frac{B_{11}^2}{A_{11}}\right)$, $^{**}H \equiv \frac{\left[(\bar{N}_{xx})_{cr}\right]_{CLT}}{KA_{55}}$.

6.4 Buckling of plates

In contrast to the previous sections, which were dealing with a 1D structure, like columns or beams, plates are considered to be 2D entities, with the third dimension, the thickness, being much smaller than the other planar dimensions. Although the issue of stability of plates had been treated intensively in the literature (see, for example, [1, 4, 5, 20–37]), few closed-form solutions are available. Most of the buckling solutions are based on approximate methods, like Rayleigh–Ritz, Galerkin–Bubnov or extended Kantorovich methods (which will be presented in this chapter).

6.4.1 Buckling of isotropic plates

The results for isotropic plates to be presented in this section are based on the seminal work of Timoshenko and Gere [1]. The equation of motion for a thin isotropic plate (thickness, h) under in-plane loads (assuming the plate has some initial curvature due to those loads) is given by the following expression:

$$\frac{\partial^4 w(x,y)}{\partial x^4}+2\frac{\partial^4 w(x,y)}{\partial x^2 \partial y^2}+\frac{\partial^4 w(x,y)}{\partial y^4}=\frac{1}{D}\left(N_x\frac{\partial^2 w(x,y)}{\partial x^2}+2N_{xy}\frac{\partial^2 w(x,y)}{\partial x \partial y}+N_y\frac{\partial^2 w(x,y)}{\partial y^2}\right) \tag{6.63}$$

where N_x and N_y are compressive loads per unit length in the x and y directions of the plate (acting at the midplane), N_{xy} is the shear load per unit length and $D=\frac{Eh^3}{12(1-\nu^2)}$ is the bending rigidity of the plate, while E and ν are its Young's modulus and Poisson's ratio, respectively.

Let us present the solution for simply supported boundary conditions around all four sides of a thin plate having the length a, width b and thickness h, subjected to in-plane compression load N_x, while $N_y = N_{xy} = 0$, also known as the Navier method.[10] The out-of-plane displacement can be assumed to be

$$w(x,y)=\sum_{m=1}^{\infty}\sum_{n=1}^{\infty}\sin\frac{m\pi}{a}\sin\frac{n\pi}{b} \tag{6.64}$$

Note that the double series, representing the lateral deflection, satisfy both eq. (6.63) and the boundary conditions of the case, namely

$$\begin{aligned} &w_0(x,0)=0, \quad w_0(x,b)=0, \quad w_0(0,y)=0, \quad w_0(a,y)=0 \\ &M_{xx}(0,y)=0, \quad M_{xx}(a,y)=0, \quad M_{yy}(x,0)=0, \quad M_{yy}(x,b)=0 \end{aligned} \tag{6.65}$$

10 Claude-Louis Navier (1785–1836).

where

$$M_{xx}=-D\left[\frac{\partial^2 w}{\partial x^2}+\nu\frac{\partial^2 w}{\partial y^2}\right],\quad M_{yy}=-D\left[\frac{\partial^2 w}{\partial y^2}+\nu\frac{\partial^2 w}{\partial x^2}\right]\tag{6.66}$$

Substituting eq. (6.64) into eq. (6.63) leads to the following expression:

$$N_x=\frac{\pi^2 a^2}{m^2}D\left[\left(\frac{m}{a}\right)^2+\left(\frac{n}{b}\right)^2\right]^2\tag{6.67}$$

To obtain the critical load, one should find the smallest value of N_x in eq. (6.67), which is easy to see that it will happen for $n=1$, namely half-wave in the width direction (perpendicular to the applied in-plane load N_x). Therefore, the critical buckling load can be written as

$$\begin{gathered}(N_x)_{\text{cr}}=\frac{\pi^2}{m^2b^2}D\left[\left(\frac{b}{a}\right)m^2+\left(\frac{a}{b}\right)\right]^2=k\frac{\pi^2}{b^2}D\\ k\equiv\frac{1}{m^2}\left[\left(\frac{b}{a}\right)m^2+\left(\frac{a}{b}\right)\right]^2\end{gathered}\tag{6.68}$$

The variation of k with the aspect ratio a/b and the number of half-waves in the longitudinal direction, m, is presented in Fig. 6.7. The minimal value of k is 4, and it occurs at an integer value of the aspect ratio, a/b, with the critical buckling waves following the aspect ratio, in the longitudinal direction.

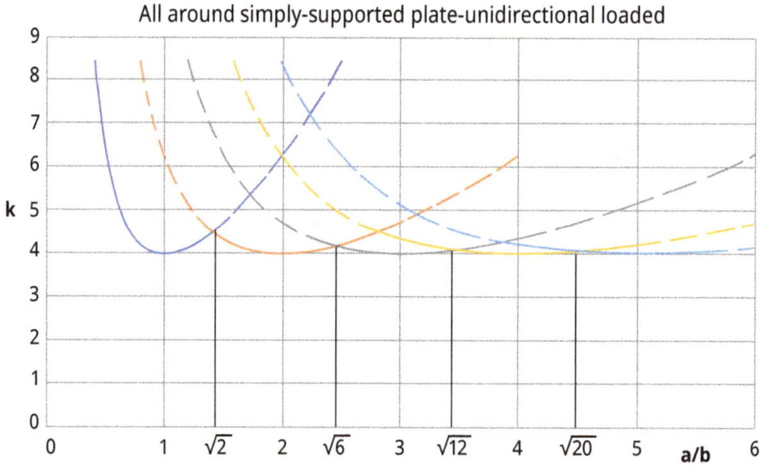

Fig. 6.7: All-around simply supported thin plate loaded in the longitudinal direction – buckling factor, *k*, versus aspect ratio, *a*/*b*.

Other cases of interest would include plates with two opposite sides on simply supported boundary conditions, while the other two sides have any combination of boundary conditions, with the axial compression load being applied on the simply supported edges, also known as the Lévy method.[11] The general solution has the following form [1]:

$$w(x,y) = \sin\frac{m\pi}{a}\left[A_1 e^{-\alpha y} + A_2 e^{\alpha y} + A_3\cos(\beta y) + A_4\sin(\beta y)\right]$$

where

$$\alpha \equiv \sqrt{\left(\frac{m\pi}{a}\right)^2 + \sqrt{\frac{N_x}{D}}\left(\frac{m\pi}{a}\right)^2}, \quad \beta \equiv \sqrt{-\left(\frac{m\pi}{a}\right)^2 + \sqrt{\frac{N_x}{D}}\left(\frac{m\pi}{a}\right)^2} \tag{6.69}$$

The constants A_1–A_4 are to be determined by the application of proper boundary conditions, leading to the eigenvalue problem, which determines the critical buckling load (per unit width). Typical boundary condition for the unloaded edges of the plate is next given for clamped–clamped and free–free cases.

a. Clamped–clamped

$$w(x,0) = 0, \quad \frac{\partial w(x,0)}{\partial x} = 0, \quad w(x,b) = 0, \quad \frac{\partial w(x,0)}{\partial x} = 0 \tag{6.70}$$

b. Free–free

$$\begin{aligned} &\frac{\partial^2 w(x,0)}{\partial y^2} + \nu\frac{\partial^2 w(x,0)}{\partial x^2} = 0, \quad \frac{\partial^3 w(x,0)}{\partial y^3} + (2-\nu)\frac{\partial^3 w(x,0)}{\partial x^2 \partial y} = 0 \\ &\frac{\partial^2 w(x,b)}{\partial y^2} + \nu\frac{\partial^2 w(x,b)}{\partial x^2} = 0, \quad \frac{\partial^3 w(x,b)}{\partial y^3} + (2-\nu)\frac{\partial^3 w(x,b)}{\partial x^2 \partial y} = 0 \end{aligned} \tag{6.71}$$

Typical results for the common boundary conditions are presented in Fig. 6.8 (adopted from [38]).

Timoshenko and Gere [1] presented an expression for the case E (see Fig. 6.8) for long plates, having the following form:

$$k_c = 0.456 + \left(\frac{b}{a}\right)^2 \tag{6.72}$$

More details about the buckling of flat isotropic plates, involving other in-plane loadings, bending and buckling of plates having elastic boundary conditions, can be found in [1, 38].

To conclude this section, we shall present the case of a rectangular flat isotropic plate under pure shear ($N_x = N_y = 0$, $N_{xy} \neq 0$), which was treated in [1, 38–42]. The equation of motion for this case is given by

11 Maurice Lévy (1838–1910), *Comptes Rendus*, vol. 129, 1899, pp. 535–539.

$$\frac{\partial^4 w(x,y)}{\partial x^4} + 2\frac{\partial^4 w(x,y)}{\partial x^2 \partial y^2} + \frac{\partial^4 w(x,y)}{\partial y^4} = \frac{2N_{xy}}{D}\frac{\partial^2 w(x,y)}{\partial x \partial y} \tag{6.73}$$

Due to the mixed derivation on the right-hand side of eq. (6.73), no trigonometric solution for the deflection, w, is possible. The approximate solutions had been obtained using energy methods [1, 40]. Figure 6.9 presents the buckling coefficient, k_s, for a rectangular flat plate resting on simply supported boundary conditions under shear stresses along its perimeter.

From Fig. 6.9 one can obtain the buckling coefficient for infinitely long flat plates on simply supported boundary conditions to yield

$$(\tau_{cr})_{SS} = k_s \frac{\pi^2 D}{b^2 h} = 5.35 \frac{\pi^2 D}{b^2 h} \tag{6.74}$$

A similar expression is presented in [1] for infinitely long flat plates on clamped boundary conditions, namely

$$(\tau_{cr})_{\text{clamped}} = k_s \frac{\pi^2 D}{b^2 h} = 8.98 \frac{\pi^2 D}{b^2 h} \tag{6.75}$$

In the same reference [1], a general expression for the shear buckling coefficient, k_s, is presented for rectangular plates on simply supported boundary conditions, starting with a $a/b = 1$ up to long infinite one, $a/b \to \infty$,

$$k_s = 5.35 + 4\left(\frac{b}{a}\right)^2 \tag{6.76}$$

6.4.2 Buckling of orthotropic plates

Finding the buckling of orthotropic plates is more complicated as compared to the isotropic case, and only few closed-form solutions are available for certain boundary conditions and layups. Other rigorous plate solutions can be found based on the Navier method, which treats rectangular plates on simply supported boundary conditions all around it or the Lévy method suitable for plates with two opposite simply supported edges, while the other two can have any combination of boundary conditions. Approximate solutions can be obtained using Rayleigh–Ritz, Galerkin–Bubnov or extended Kantorovich methods, all being based on energy approaches (see details in [67]).

The first case to be presented is also called *special orthotropic plates* for which the bending–stretching coupling terms B_{ij} and the bending–twisting coefficients D_{16} and D_{26} are set to zero. This leads to the following equation, assuming constant in-plane compressive loads per unit width in the x and y directions, $\bar{N}_{xx}$ and $\bar{N}_{yy}$, respectively, and shear loads $\bar{N}_{xy}$:

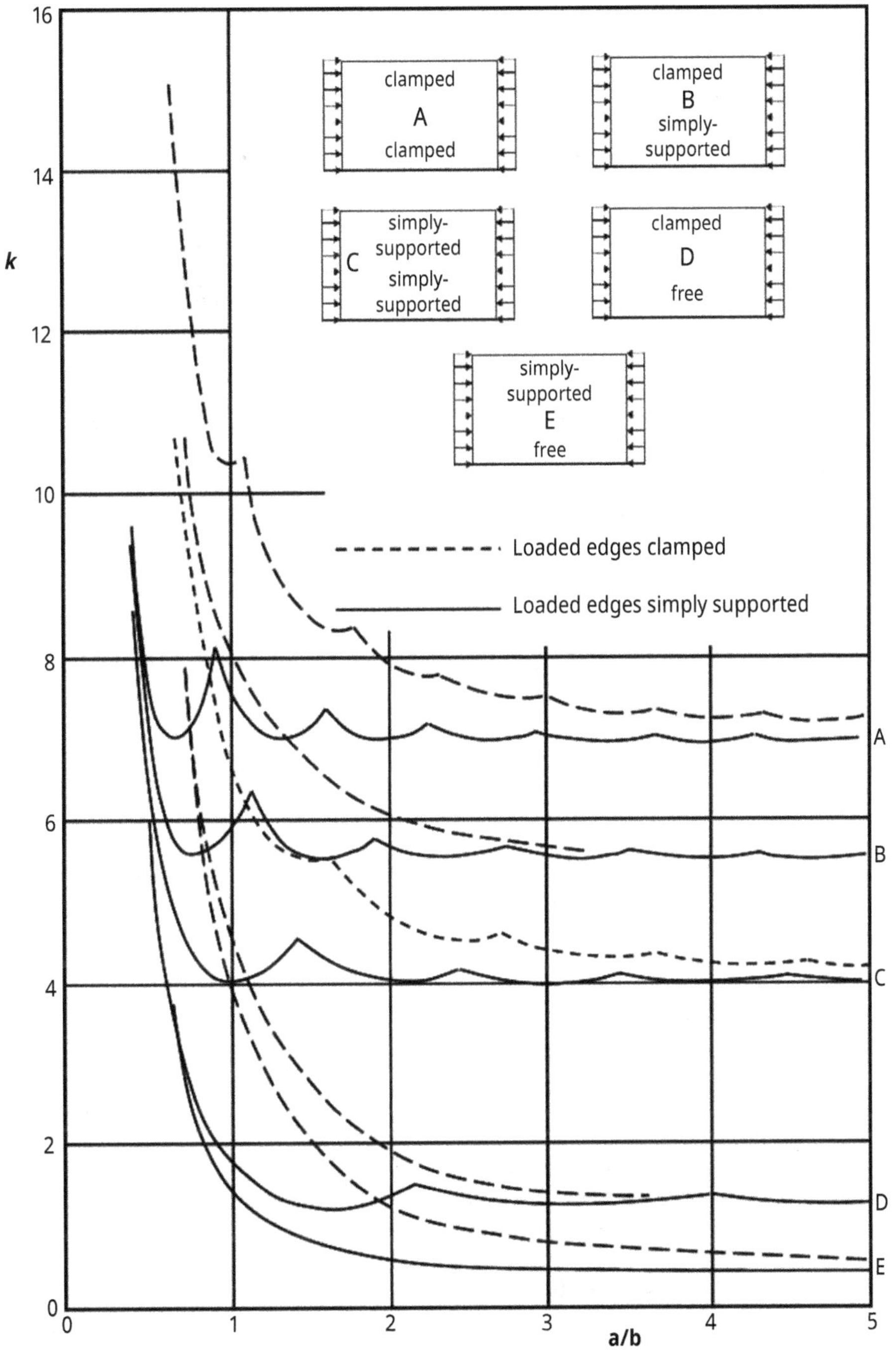

Fig. 6.8: Rectangular thin plates loaded in the longitudinal direction – buckling factor, *k*, versus aspect ratio, *a*/*b*, for various boundary conditions (adapted from [38]).

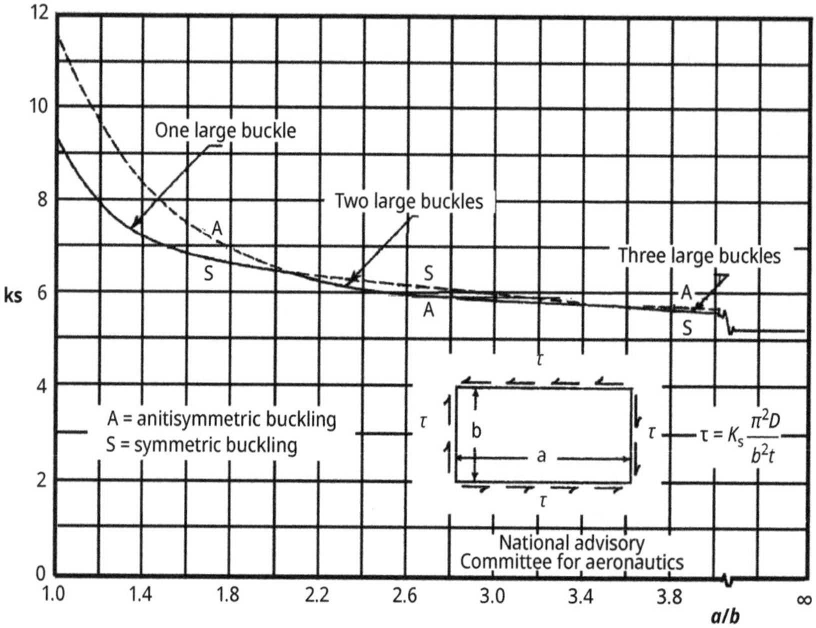

Fig. 6.9: Rectangular thin plates loaded in pure shear-buckling factor, *k*, versus aspect ratio, *a*/*b*, for simply supported boundary conditions (adapted from [40]).

$$D_{11}\frac{\partial^4 w_0}{\partial x^4}+2(D_{12}+2D_{66})\frac{\partial^4 w_0}{\partial x^2\partial y^2}+D_{22}\frac{\partial^4 w_0}{\partial y^4}=\bar{N}_{xx}\frac{\partial^2 w_0}{\partial x^2}+2\bar{N}_{xy}\frac{\partial^2 w_0}{\partial x\partial y}+\bar{N}_{yy}\frac{\partial^2 w_0}{\partial y^2} \tag{6.77}$$

Note that the simply supported boundary conditions presented in eq. (6.65), for the isotropic case, still hold; however, the expressions for the moments per unit width are for the orthotropic case:

$$M_{xx}=-\left[D_{11}\frac{\partial^2 w_0}{\partial x^2}+D_{12}\frac{\partial^2 w_0}{\partial y^2}\right],\quad M_{yy}=-\left[D_{12}\frac{\partial^2 w_0}{\partial x^2}+D_{22}\frac{\partial^2 w_0}{\partial y^2}\right],\quad M_{xy}=-2D_{66}\frac{\partial^2 w_0}{\partial x\partial y} \tag{6.78}$$

For the clamped boundary conditions along the plate's perimeter, eq. (6.70) presented for the isotropic case is valid also for the present orthotropic case. The issue of free boundary conditions on all four sides of the plate, presented in eq. (6.71) for the isotropic case, is somehow more complicated and is presented as

$$\begin{aligned} &M_{xx}(0,y)=0,\quad M_{xx}(a,y)=0,\quad M_{yy}(x,0)=0,\quad M_{yy}(x,b)=0\\ &V_y(0,y)=0,\quad V_y(a,y)=0,\quad V_x(x,0)=0,\quad V_y(x,b)=0 \end{aligned} \tag{6.79}$$

where

$$V_x = \frac{\partial M_{xx}}{\partial x} + \frac{\partial M_{xy}}{\partial y} + \frac{\partial}{\partial x}\left(\bar{N}_{xy}\frac{\partial w_0}{\partial y} - \bar{N}_{xx}\frac{\partial w_0}{\partial x}\right) + \frac{\partial M_{xy}}{\partial y}$$
$$V_y = \frac{\partial M_{yy}}{\partial y} + \frac{\partial M_{xy}}{\partial x} + \frac{\partial}{\partial y}\left(\bar{N}_{xy}\frac{\partial w_0}{\partial x} - \bar{N}_{yy}\frac{\partial w_0}{\partial y}\right) + \frac{\partial M_{xy}}{\partial x} \tag{6.80}$$

Solving eq. (6.77) for a rectangular plate loaded in the x direction only ($\bar{N}_{xy} = \bar{N}_{yy} = 0$) for all-around simply supported boundary conditions (eqs. (6.65) and (6.78)), which can be shown to be simplified yielding the following expressions:

$$w_0(x,0) = \frac{\partial^2 w_0(x,0)}{\partial x^2} = 0, \quad w_0(x,b) = \frac{\partial^2 w_0(x,b)}{\partial x^2} = 0$$
$$w_0(0,y) = \frac{\partial^2 w_0(0,y)}{\partial y^2} = 0, \quad w_0(a,y) = \frac{\partial^2 w_0(a,y)}{\partial y^2} = 0 \tag{6.81}$$

and substituting the following out-of-plane deflection w_0, which satisfies the boundary conditions (eq. (6.81)), into eq. (6.77)

$$w_0 = A_{mn}\sin\frac{m\pi x}{a}\sin\frac{n\pi y}{b}, \quad m,n = 1,2,3,\ldots \tag{6.82}$$

where A_{mn} is a small arbitrary amplitude coefficient, which yields the following critical load per unit width (b being the width of the plate)

$$(\bar{N}_{xx})_{cr} = \pi^2\left[D_{11}\left(\frac{m}{a}\right)^2 + 2(D_{12} + 2D_{66})\left(\frac{n}{b}\right)^2 + D_{22}\left(\frac{n}{b}\right)^4\left(\frac{a}{m}\right)^2\right] \tag{6.83}$$

Equation (6.83) is a function of both m and n, which are the number of half-waves in the x and y directions, respectively. It is clear that the minimum can be obtained for $n = 1$. Rearranging eq. (6.83) leads to the following expression:

$$\frac{K_x}{\pi^2} \equiv \frac{\bar{N}_{xx}\cdot b^2}{\pi^2 D_{22}} = \left[\frac{D_{11}}{D_{22}}\left(\frac{b}{a}\right)^2 m^2 + 2\left(\frac{D_{12}}{D_{22}} + 2\frac{D_{66}}{D_{22}}\right) + \left(\frac{a}{b}\right)^2\frac{1}{m^2}\right] \tag{6.84}$$

Equation (6.84) enables to find the lowest values for K_x as a function of the aspect ratio a/b and m for given values of D_{11}, D_{22}, D_{12} and D_{66}. Figure 6.10 presents the behavior of eq. (6.84) for three values of D_{11}/D_{22} assuming $(D_{12} + 2D_{66})/D_{22} = 1$. Similar curves would be achieved for other values of those parameters.

One should note that the minimum value for each curve generated by eq. (6.84) is found for a certain aspect ratio, thus obtaining $(K_x/\pi^2)_{\min}$

$$\frac{a}{b} = m\cdot\sqrt[4]{\frac{D_{11}}{D_{22}}} \Rightarrow \left(\frac{K_x}{\pi^2}\right)_{\min} = 2\left(\sqrt{\frac{D_{11}}{D_{22}}} + \frac{D_{12}}{D_{22}} + 2\frac{D_{66}}{D_{22}}\right) \tag{6.85}$$

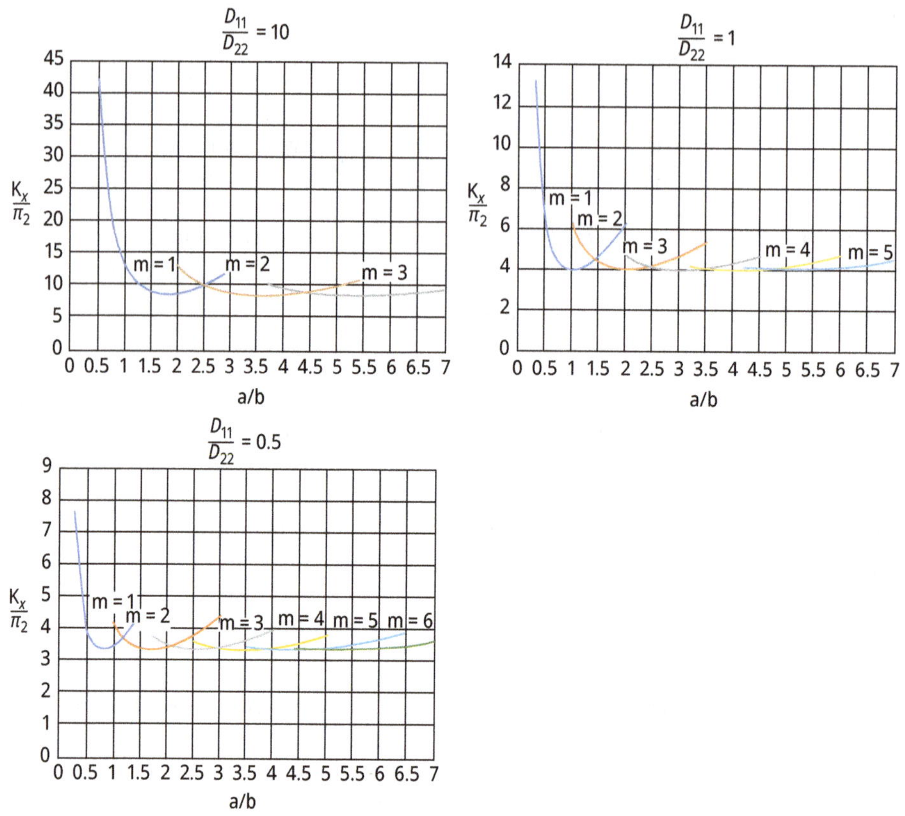

Fig. 6.10: Uniaxial buckling of a rectangular special orthotropic plate simply supported on all its sides for three values of D_{11}/D_{22} while $(D_{12} + 2D_{66})/D_{22} = 1$.

The biaxial case, where only in-plane compressive forces ($\bar{N}_{xy} = 0$) are applied in the x and y directions of an all-around simply supported plate, can also be solved, provided one can write $\kappa \equiv \bar{N}_{yy}/\bar{N}_{xx}$ (the y compression load is proportional to the x compression load). Then using eqs. (6.77) and (6.82) one obtains the expression for the critical buckling load, as

$$(\bar{N}_{xx})_{cr} = \pi^2 \frac{\left[D_{11}\left(\frac{m}{a}\right)^2 + 2(D_{12} + 2D_{66})\left(\frac{n}{b}\right)^2 + D_{22}\left(\frac{n}{b}\right)^4\left(\frac{a}{m}\right)^2\right]}{1 + \kappa\left(\frac{a}{b}\right)^2\left(\frac{n}{m}\right)^2} \tag{6.86}$$

yielding as before the value of $\left(K_x/\pi^2\right)$ as a function of the aspect ratio (a/b), stiffness D_{11}, D_{12}, D_{22}, D_{66} and the half-waves in the x- and y-plate directions, m and n, respectively

$$\frac{K_x}{\pi^2} = \frac{\left[D_{11}\left(\frac{m}{a}\right)^2 + 2(D_{12} + 2D_{66})\left(\frac{n}{b}\right)^2 + D_{22}\left(\frac{n}{b}\right)^4\left(\frac{a}{m}\right)^2\right]}{1 + \kappa\left(\frac{a}{b}\right)^2\left(\frac{n}{m}\right)^2} \tag{6.87}$$

Note that in contrast to the isotropic case, the assumption of $n = 1$ for the special orthotropic case of a plate under biaxial loads will not necessarily lead to the lowest value of the critical compression load, and therefore, its lowest value will be found after calculating the critical buckling loads for a few combinations of m and n.

The case of a special orthotropic rectangular plate on simply supported boundary conditions under pure shear ($\bar{N}_{xy} \neq 0$, $\bar{N}_{xx} \neq 0$, $\bar{N}_{yy} \neq 0$) case is more complicated than the previous two ones. The equation to be solved is given by

$$D_{11}\frac{\partial^4 w_0}{\partial x^4} + 2(D_{12} + 2D_{66})\frac{\partial^4 w_0}{\partial x^2 \partial y^2} + D_{22}\frac{\partial^4 w_0}{\partial y^4} = 2\bar{N}_{xy}\frac{\partial^2 w_0}{\partial x \partial y} \tag{6.88}$$

There are no analytical solutions for this case, even for an all-around simply supported rectangular plate. Bergmann and Reisner [30] assumed an infinitely long plate in the x-direction ($a/b \to \infty$) with the D_{11} bending rigidity being neglected. The exact solution for this case, where simply supported boundary conditions were assumed along the long edges, has the following form:

$$\frac{\bar{N}_{xy} \cdot b^2}{4\sqrt{[D_{22}(D_{12} + 2D_{66})]}} = 11.71 \tag{6.89}$$

Another solution for eq. (6.88) was given by Leissa [29], which has the following form:

$$w(x,y) = f(y)\exp\left(i\kappa\frac{x}{b}\right) \tag{6.90}$$

where $i = \sqrt{-1}$, b is the plate's width and κ is a wavelength constant to be determined. Substituting eq. (6.90) into eq. (6.88) yields

$$D_{11}\left(\frac{\kappa}{b}\right)^4 f(y) - 2(D_{12} + 2D_{66})\left(\frac{\kappa}{b}\right)^2\frac{d^2 f(y)}{dy^2} + D_{22}\frac{d^4 f(y)}{dy^4} - 2i\bar{N}_{xy}\left(\frac{\kappa}{b}\right)\frac{df(y)}{dy} = 0 \tag{6.91}$$

The solution of eq. (6.91) has the following form:

$$f(y) = A_1\exp\left(i\beta_1\frac{y}{b}\right) + A_2\exp\left(i\beta_2\frac{y}{b}\right) + A_3\exp\left(i\beta_3\frac{y}{b}\right) + A_4\exp\left(i\beta_4\frac{y}{b}\right) \tag{6.92}$$

where β_1, β_2, β_3 and β_4 are the roots of the fourth-degree polynomial equation presented by eq. (6.91). Then the simply supported boundary conditions are substituted at $y = 0$, b into eq. (6.92), leading to a fourth-order characteristic determinant, with κ and $\bar{N}_{xy}$ being free parameters. It was shown in [29] that the following parameter, $\sqrt{D_{11}/D_{22}}/[D_{12} + 2D_{66}/D_{22}]$, determines the solution.

When only two opposite edges of a plate are on simply supported boundary conditions, while the other two sides can be either clamped, free or any other combination of boundaries, the Lévy method can be applied to solve the buckling problem. Assuming the out-of-plane displacement has the following form

$$w_0(x,y) = G_n(x)\sin\frac{n\pi y}{b}, \quad n = 1,2,3 \tag{6.93}$$

and substituting into eq. (6.77), for constant in-plane loads, without shear-type ones ($\bar{N}_{xx} \neq 0$, $\bar{N}_{yy} \neq 0$, $\bar{N}_{xy} \neq 0$) yields

$$D_{11}\frac{d^4G_n(x)}{dx^4} - \left[2(D_{12}+2D_{66})\left(\frac{n\pi}{b}\right)^2 + \bar{N}_{xx}\right]\frac{d^2G_n(x)}{dx^2} + \left(\frac{n\pi}{b}\right)^2\left[D_{22}\left(\frac{n\pi}{b}\right)^2 + \bar{N}_{yy}\right]G_n(x) = 0 \tag{6.94}$$

The general solution for eq. (6.94) can be assumed to have the following form:

$$G_n(x) = A_1\sinh(\lambda_1 x) + A_2\cosh(\lambda_1 x) + A_3\sin(\lambda_2 x) + A_4\cos(\lambda_2 x) \tag{6.95}$$

where λ_1 and λ_2 are the roots of the following characteristic equation:

$$D_{11}\lambda^4 - \left[2(D_{12}+2D_{66})\left(\frac{n\pi}{b}\right)^2 + \bar{N}_{xx}\right]\lambda^2 + \left[D_{22}\left(\frac{n\pi}{b}\right)^4 + \bar{N}_{yy}\left(\frac{n\pi}{b}\right)^2\right] = 0 \tag{6.96}$$

The roots of eq. (6.96) are given by

$$\begin{aligned}(\lambda_1)^2 &= \frac{1}{D_{11}}\left[B + \sqrt{B^2 - D_{11}C}\right]\\ (\lambda_2)^2 &= \frac{1}{D_{11}}\left[-B + \sqrt{B^2 - D_{11}C}\right]\end{aligned} \tag{6.97}$$

where

$$B \equiv \left[(D_{12}+2D_{66})\left(\frac{n\pi}{b}\right)^2 + \frac{\bar{N}_{xx}}{2}\right], \quad C \equiv \left[D_{22}\left(\frac{n\pi}{b}\right)^4 + \bar{N}_{yy}\left(\frac{n\pi}{b}\right)^2\right] \tag{6.98}$$

The constants A_1–A_4 are to be determined using the appropriate boundary conditions at $x = 0, a$, yielding the buckling load of the plate.

6.5 Buckling of shells

In contrast to the previous sections, which were dealing with 1D structures, like columns or beams, and 2D structures like plates or panels, shells are considered to be 3D entities, with the third dimension, the thickness, being much smaller than the other length and radius dimensions. Although the issue of stability of shells had been treated intensively in the literature (see, e.g., [43–66]), few closed-form solutions are available.

To become aware of the complexity of this thin-walled structure type, a short introduction will be presented, followed by their structural behavior according to various theories available in the literature.

6.5.1 Buckling of isotropic shells

Shells are considered to be thin-walled structures to be used as structural elements in aerospace, marine, civil, architecture, mechanical engineering and other applications. Typical examples would be aircraft fuselage, missiles, rockets, ships and submarines or water tanks, silos, piping systems and pressure vessels. A shell structure displays some important advantages like high stiffness, high strength-to-weight ratio, good efficiency of load-carrying behavior and structural integrity.

In general, the shells are classified according to its thickness, namely thin-type shell for which $t/R \ll 1$ (t is the shell thickness and R is its radius) and thick-type shells. It is customary to define a thin shell by the following expression:

$$\left(\frac{t}{R}\right)_{\max} \leq \frac{1}{20} \quad \text{or} \quad \frac{1}{1{,}000} \leq \left(\frac{t}{R}\right) \leq \frac{1}{20} \tag{6.99}$$

while for a thick shell, its thickness ratio would be

$$\left(\frac{t}{R}\right)_{\max} \geq \frac{1}{20} \tag{6.100}$$

Another important property of the shell is its curvature leading to a variety of shapes, like cylindrical, conical paraboloid and others as defined and presented in Tab. 6.8.

Figure 6.11 displays some of the shells being tested by the author of this book at his Aerospace Structures Laboratory, Technion, I.I.T., 32000, Haifa, Israel.

Depending on the shape of the shell, various loads can be applied like axial, lateral, a combination of lateral and axial loads, external and internal pressure, bending and torsion.

This chapter deals only with the structural behavior of circular shells, which is displayed in Fig. 6.12 for axial compression. It clearly represents the difference between a perfect ideal shell (λ_{perf}) and a real imperfect shell ($\lambda_{\text{LL-imp.}}$), with the bifurcation load ($\lambda_{\text{Bif.}}$) as a reference.

Browsing Fig. 6.12 reveals that $\lambda_{\text{perf}} > \lambda_{\text{Bif}}$, which means that the test results are well beyond the theoretical predictions. The reasons for those discrepancies might be one or a combination of the following topics:

a. Boundary conditions
b. Initial geometric imperfections
c. Load eccentricity
d. Shell geometry (length and the ratio R/t)

e. Plasticity and residual stresses from welding and rolling
f. Uneven load distribution
g. Variation of temperatures

Tab. 6.8: Shells and their respective shape equations.

Shell name	Shell shape	Shape expression
Circular cylinder		$x^2 + y^2 = R^2$
Elliptical cylinder		$\frac{x^2}{a^2} + \frac{y^2}{b^2} = 1$
Sphere		$x^2 + y^2 + z^2 = R^2$
General cone		$\frac{x^2}{a^2} + \frac{y^2}{b^2} = r^2 = \frac{z^2}{c^2}$ $1/c^2 = \tan^2\theta = r^2/z^2$
Truncated general cone		$\frac{x^2}{a^2} + \frac{y^2}{b^2} = r^2 = \frac{z^2}{c^2}$ $1/c^2 = \tan^2\theta = r^2/z^2$
Ellipsoid		$\frac{x^2}{a^2} + \frac{y^2}{b^2} + \frac{z^2}{c^2} = 1$
One-sheet hyperboloid	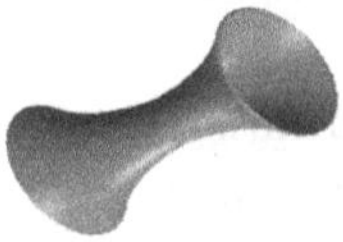	$\frac{x^2}{a^2} + \frac{y^2}{b^2} - \frac{z^2}{c^2} = 1$
Two-sheet hyperboloid	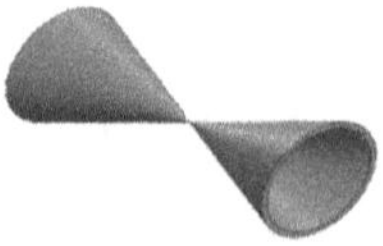	$-\frac{x^2}{a^2} - \frac{y^2}{b^2} + \frac{z^2}{c^2} = 1$

Tab. 6.8 (continued)

Shell name	Shell shape	Shape expression
Circular paraboloid		$\frac{x^2}{a^2} + \frac{y^2}{b^2} = \frac{z}{c}$
Elliptical paraboloid	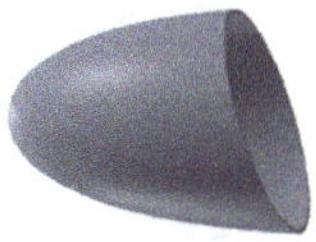	$\frac{x^2}{a^2} + \frac{y^2}{b^2} = \frac{z}{c}$
Torus		$\left[R - \sqrt{x^2 - y^2}\right]^2 + z^2 = r^2$ R is the distance from the hole center to the tube center; r is the tube radius

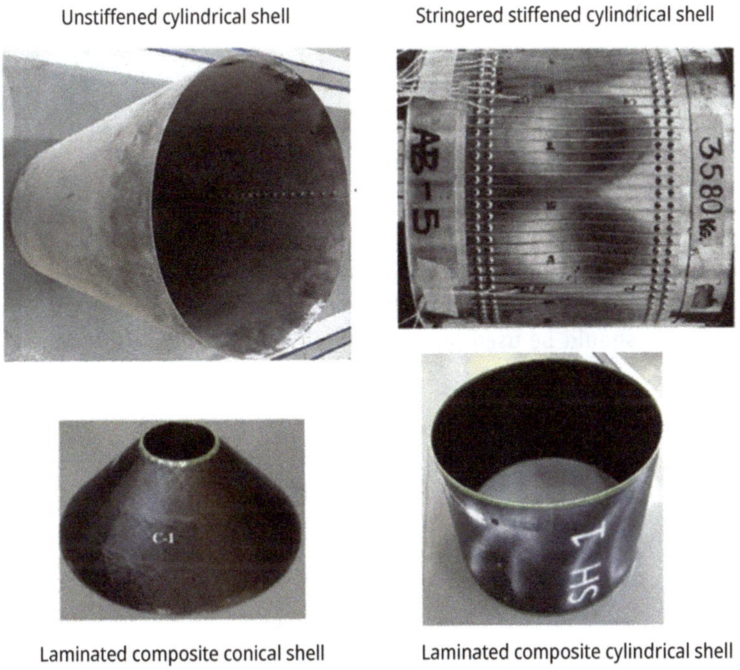

Fig. 6.11: Some of the typical shells being tested by the author of the present book.

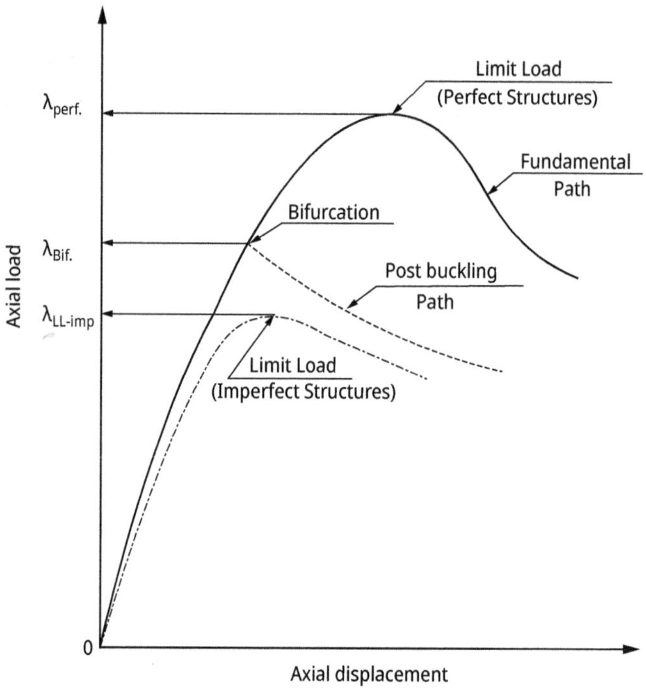

Fig. 6.12: Typical behavior of a compressed circular cylindrical shell.

The first two factors, boundary conditions and initial geometric imperfections, are the most significant factors in reducing the buckling load of axially compressed cylindrical shells. Therefore, to be able to design shell-type structures in a safe way, the following design equation should be used (according to NASA SP-8007 report as presented in [65]):

$$\left(\frac{t}{R}\right)_{\max} \geq \frac{1}{20}$$

$$F_{\text{design}} = F_{\text{theory}} \cdot \kappa$$

where (6.101)

$$\kappa = \frac{P}{P_{\text{CL}}} = 1 - 0.901\left[1 - e^{-\frac{1}{16}\sqrt{\frac{R}{t}}}\right]$$

$$\Rightarrow \kappa_{\frac{R}{t}=500} = 0.321; \quad \kappa_{\frac{R}{t}=1,000} = 0.223; \quad \kappa_{\frac{R}{t}=1,500} = 0.178; \quad \kappa_{\frac{R}{t}=2,000} = 0.153$$

One can see from eq. (6.101) that the knock-down factor, κ, is reducing with the increase of the radius to the thickness ratio R/t. This means that for a shell subjected to axial compression, the design load would be in the range of $(0.321\text{–}0.153)F_{\text{theory}}$.

6.5.2 Buckling of isotropic cylindrical shells

6.5.2.1 Axial compression

Batdorf [45] presented already in 1947 an approximate equation to calculate the buckling load of isotropic cylindrical shells loaded in compression. Its form is

$$N_x = k_x \frac{\pi^2 D}{L^2}, \quad \text{where} \quad D = \frac{Et^3}{12(1-\upsilon^2)} \tag{6.102}$$

Note that N_x is the axial compression load per unit shell circumference, namely $N_x = F_{\text{axial}}/2\pi R$, R is the shell radius, t is its thickness, and E and υ are the Young's modulus and the Poisson's ratio, respectively.

The buckling factor k in eq. (6.102) is defined (see also [65]) as

$$k_x = m^2(1+\beta^2)^2 + \frac{12}{\pi^4}\frac{k^2 Z^2}{m^2(1+\beta^2)^2}$$

where (6.103)

$$\beta = \frac{nL}{\pi R m} \quad Z = \frac{L^2}{Rt}\sqrt{1-\upsilon^2}$$

n and m appearing in eq. (6.103) represent the number of buckle half waves in the circumferential and axial directions, respectively, and L is the shell length.

It is stated in [65] that for $\kappa Z > (\sqrt{3}\pi^2)/6$ (moderately long cylinders) the buckling factor can be approximated to the following expression:

$$k_x = \frac{4\sqrt{3}}{\pi^2}\kappa Z \tag{6.104}$$

Substituting eq. (6.1054) into (6.102) and remembering the expressions for the variables D and Z yield the critical buckling stress for an isotropic cylindrical shell under axial compression and its respective buckling load:

$$\sigma_x = \sigma_{cr} = \frac{N_x}{t} = \frac{\kappa E}{\sqrt{3(1-\upsilon^2)}}\frac{t}{R} \Rightarrow \text{for} \quad \upsilon = 0.316, \quad \sigma_{cr} = 0.608\kappa E\frac{t}{R}$$

or

$$P_{cr} = 2\pi R t \cdot \sigma_{cr} = 1.216\pi\kappa E t^2$$

while

$$\kappa = 1 - 0.731\left(1 - e^{-\frac{1}{16}\sqrt{\frac{R}{t}}}\right) \tag{6.105}$$

6.5.2.2 External pressure

When dealing with a cylindrical shell loaded by an external pressure, one has to distinguish between two terms: the term "lateral pressure" which stands for the application of external pressure only on the curved part of the shell and the term "hydrostatic pressure" for which the applied pressure is also on the ends of the cylinder. For the lateral pressure case, the load on the shell wall is given by

$$N_y = \sigma_y t = p \cdot R, \quad N_x = 0 \tag{6.106}$$

while for the hydrostatic pressure the load on the cylinder wall would be

$$N_y = \sigma_y t = p \cdot R \quad N_x = \sigma_x t = \frac{p \cdot R}{2} \tag{6.107}$$

where p is the pressure and R stands for the shell radius.

An approximate buckling equation for cylinders loaded by lateral pressure [45] is given by the following expression:

$$N_y = k_y \frac{\pi^2 D}{L^2} \tag{6.108}$$

The buckling factor k_y in eq. (3.108) is defined [65] as

$$k_y \equiv \frac{p \cdot R \cdot L^2}{\pi^2 D} = \frac{1}{\beta^2}\left[\left(1+\beta^2\right)^2 + \frac{12}{\pi^4}\frac{\kappa^2 Z^2}{\left(1+\beta^2\right)^2}\right]$$

where

$$\beta = \frac{nL}{\pi Rm} \quad Z = \frac{L^2}{Rt}\sqrt{1-\upsilon^2} \tag{6.109}$$

For the hydrostatic pressure case, we will have similar expressions, namely

$$N_y = k_{\text{h.p.}} \frac{\pi^2 D}{L^2} \tag{6.110}$$

The buckling factor $k_{\text{h.p.}}$ is defined [66] as

$$k_{\text{h.p.}} \equiv \frac{p \cdot R \cdot L^2}{\pi^2 D} = \frac{1}{\left(\beta^2+0.5\right)}\left[\left(1+\beta^2\right)^2 + \frac{12}{\pi^4}\frac{\kappa^2 Z^2}{\left(1+\beta^2\right)^2}\right] \tag{6.111}$$

Figure 6.13 presents the variation of the minimal values of the buckling coefficients as a function of κZ. It was obtained by allowing β (the buckle aspect ratio) to be a continuous variable along the region of κZ and by differentiation with respect to β to obtain minimal values for the buckling coefficients.

As can be observed in Fig. 6.13, the straight-line part of the curve for $\kappa Z > 100$ can be approximated by the following expression (see also [65]):

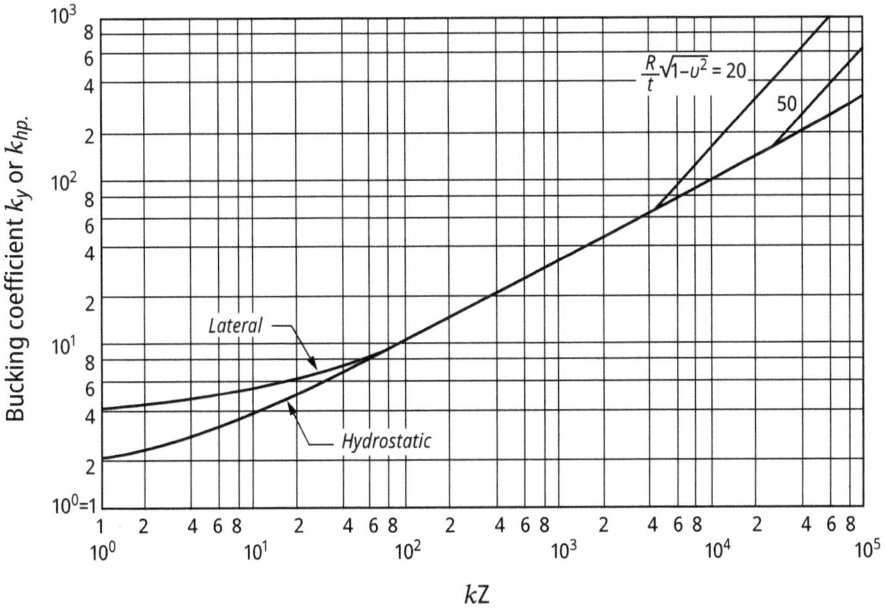

Fig. 6.13: Buckling coefficients for simply supported isotropic circular shell under external pressure (adapted from [65]).

$$k_y = 1.04\sqrt{\kappa Z} \tag{6.112}$$

This leads to the following expression for critical pressure:

$$p_{cr} = \frac{0.855}{(1-\upsilon^2)^{\frac{3}{4}}} \frac{E\sqrt{\kappa}}{\left(\frac{R}{t}\right)^{\frac{5}{2}}\left(\frac{L}{R}\right)} \Rightarrow p_{cr}(\upsilon = 0.3) = 0.926 \frac{E\sqrt{\kappa}}{\left(\frac{R}{t}\right)^{\frac{5}{2}}\left(\frac{L}{R}\right)} \tag{6.113}$$

For high values of κZ appearing in Fig. 6.13, the buckling coefficient would have the following form (corresponding to the buckling of the shell into an oval shape):

$$k_y = \frac{3}{\pi^2} \frac{\kappa Z}{\left(\frac{R}{t}\right)\sqrt{1-\upsilon^2}} \Rightarrow p_{cr} = \frac{\kappa E}{4(1-\upsilon^2)}\left(\frac{t}{R}\right)^3 \tag{6.114}$$

Note that changing the boundary conditions to no longitudinal movement and/or no rotation might increase the theoretical buckling pressure by 50%, as mentioned in [65].

6.5.2.3 Torsion

The application of a torsion moment would induce shear stresses in the wall of the shell. The variation of the buckling coefficient k_{xy} with the variable κZ is depicted in Fig. 6.14. The linear part of the curve, namely for the range $50 < \kappa Z < 78(R/t)^2(1-\upsilon^2)$, can be written as

$$k_{xy} = \frac{N_{xy}L^2}{\pi^2 D} = 0.85(\kappa Z)^{\frac{3}{4}} \Rightarrow \tau_{xy} = \frac{0.747(\kappa)^{\frac{3}{4}}E}{\left(\frac{R}{t}\right)^{\frac{5}{4}}\sqrt{\left(\frac{L}{R}\right)}} \tag{6.115}$$

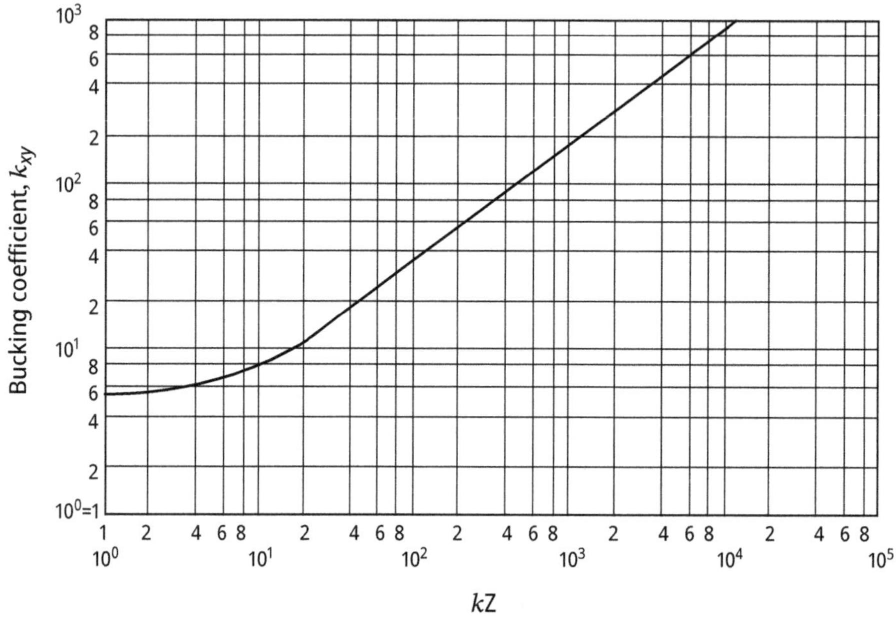

Fig. 6.14: Buckling coefficients for simply supported isotropic circular shell under torsion (adapted from [65]).

For cases in which $\kappa Z > 78(R/t)^2(1-\upsilon^2)$, the shell is expected to buckle with two circumferential half waves ($m = 2$), as reported in [65], and its corresponding buckling coefficient will have the following form:

$$k_{xy} = \frac{2\sqrt{2}\kappa Z}{\pi^2\sqrt{\left(\frac{R}{t}\right)}\sqrt[4]{(1-\upsilon^2)}} \Rightarrow \tau_{xy} = \frac{\kappa E}{3\sqrt{2}\cdot\sqrt[4]{(1-\upsilon^2)^3}}\left(\frac{t}{R}\right)^{\frac{3}{2}} \tag{6.116}$$

It is recommended for moderately long shells [65] to use the knockdown factor of

$$\kappa^{\frac{3}{4}} = 0.67 \Rightarrow \kappa = 0.67^{\frac{4}{3}} = 0.586 \tag{6.117}$$

6.5.2.4 Combined loadings

For cases with more than one type of loading, the following interaction curves are valid:

a. A combination of axial compression and bending:

$$\frac{\sigma_x}{\sigma_{x,\,\mathrm{cr}}} + \frac{\sigma_b}{\sigma_{b,\,\mathrm{cr}}} = 1 \tag{6.118}$$

where $\sigma_{x,\mathrm{cr}}$ and $\sigma_{b,\mathrm{cr}}$ are the critical stresses for axial compression and bending, respectively.

b. A combination of axial compression and external pressure:

$$\frac{\sigma_x}{\sigma_{x,\mathrm{cr}}} + \frac{\sigma_p}{\sigma_{p,\mathrm{cr}}} = 1 \tag{6.119}$$

where $\sigma_{x,\mathrm{cr}}$ and $\sigma_{p,\mathrm{cr}}$ are the critical stresses for axial compression and external pressure, respectively.

c. A combination of axial compression and torsion:

$$\frac{\sigma_x}{\sigma_{x,\mathrm{cr}}} + \frac{\sigma_t}{\sigma_{t,\mathrm{cr}}} = 1 \tag{6.120}$$

where $\sigma_{x,\mathrm{cr}}$ and $\sigma_{t,\mathrm{cr}}$ are the critical stresses for axial compression and torsion, respectively.

6.5.2.5 Buckling of orthotropic cylindrical shells under axial compression

Dealing with orthotropic shells makes the study more complicated as compared with the isotropic ones. Few works present detailed expressions as done by Jones [66] in 1968. He presented a detailed expression to calculate the buckling load of an orthotropic cylindrical shell in compression with more than four half-circumferential waves ($n \geq 4$). Its form is

$$N_x = \left(\frac{L}{\pi m}\right)^2 \frac{\det(\Delta)}{\det(\mathrm{E})},$$

where

$$\det(\Delta) = \begin{vmatrix} \Delta_{11} & \Delta_{12} & \Delta_{13} \\ \Delta_{21} & \Delta_{22} & \Delta_{23} \\ \Delta_{31} & \Delta_{32} & \Delta_{33} \end{vmatrix}; \ \det(\mathrm{E}) = \begin{vmatrix} \Delta_{11} & \Delta_{12} \\ \Delta_{21} & \Delta_{22} \end{vmatrix} \tag{6.121}$$

while n and m represent the number of buckle half waves in the circumferential and axial directions, respectively, and L is the shell length.

The general expressions for the various members of determinants in eq. (6.121) are defined as

$$\Delta_{11}=\widehat{E}_x\left[\frac{m\pi}{L}\right]^2+\widehat{G}_{xy}\left[\frac{n}{R}\right]^2,\quad \Delta_{12}=\Delta_{21}=\left[\widehat{E}_x+\widehat{G}_{xy}\right]\left[\frac{m\pi}{L}\right]\left[\frac{n}{R}\right]$$

$$\Delta_{22}=\widehat{E}_y\left[\frac{n}{R}\right]^2+\widehat{G}_{xy}\left[\frac{m\pi}{L}\right]^2,\quad \Delta_{31}=\Delta_{13}=\left[\widehat{C}_{xy}+2\widehat{K}_{xy}\right]\left[\frac{n}{R}\right]^2\left[\frac{m\pi}{L}\right]+\frac{\widehat{E}_{xy}}{R}\left[\frac{m\pi}{L}\right]+\widehat{C}_x\left[\frac{m\pi}{L}\right]^3$$

$$\Delta_{23}=\Delta_{32}=\left[\widehat{C}_{xy}+2\widehat{K}_{xy}\right]\left[\frac{m\pi}{L}\right]^2\left[\frac{n}{R}\right]+\frac{\{\widehat{E}\}_y}{R}\left[\frac{n}{R}\right]+\widehat{C}_y\left[\frac{n}{R}\right]^3$$

$$\Delta_{33}=\widehat{D}_x\left[\frac{m\pi}{L}\right]^4+\widehat{D}_{xy}\left[\frac{m\pi}{L}\right]^2\left[\frac{n}{R}\right]^2+\widehat{D}_y\left[\frac{n}{R}\right]^4+\frac{\widehat{E}_y}{R^2}+2\frac{\widehat{C}_y}{R}\left[\frac{n}{R}\right]^2+2\frac{\widehat{C}_{xy}}{R}\left[\frac{m\pi}{L}\right]^2 \tag{6.122}$$

The extensional stiffness of the shell's wall can be written as

$$\widehat{E}_x=\sum_{k=1}^{N}\left[\frac{E_x}{1-\upsilon_x\upsilon_y}\right]_k t_k,\quad \widehat{E}_y=\sum_{k=1}^{N}\left[\frac{E_y}{1-\upsilon_x\upsilon_y}\right]_k t_k$$
$$\widehat{E}_{xy}=\sum_{k=1}^{N}\left[\frac{\upsilon_yE_x}{1-\upsilon_x\upsilon_y}\right]_k t_k=\sum_{k=1}^{N}\left[\frac{\upsilon_xE_y}{1-\upsilon_x\upsilon_y}\right]_k t_k \tag{6.123}$$

The shear stiffness of the shell's wall is given by

$$\widehat{G}_{xy}=\sum_{k=1}^{N}\left[G_{xy}\right]_k t_k \tag{6.124}$$

The bending stiffness (in the directions x and y and for twisting) of the shell's wall (per unit wall width) is given as (see also Fig. 6.15)

$$\widehat{D}_x=\sum_{k=1}^{N}\left[\frac{E_x}{1-\upsilon_x\upsilon_y}\right]_k\left[\frac{t_k^3}{12}+t_k\bar{z}_k^2\right],\quad \widehat{D}_y=\sum_{k=1}^{N}\left[\frac{E_y}{1-\upsilon_x\upsilon_y}\right]_k\left[\frac{t_k^3}{12}+t_k\bar{z}_k^2\right]$$
$$\widehat{D}_{xy}=\sum_{k=1}^{N}\left[4G_{xy}+\frac{\upsilon_yE_x}{1-\upsilon_x\upsilon_y}+\frac{\upsilon_xE_y}{1-\upsilon_x\upsilon_y}\right]_k\left[\frac{t_k^3}{12}+t_k\bar{z}_k^2\right] \tag{6.125}$$

The expressions presented in eqs. (6.124)–(6.126) would simplify for the case of an isotropic shell yielding

$$\widehat{E}_x=\widehat{E}_y=\frac{E\cdot t}{1-\upsilon^2},\quad \widehat{E}_{xy}=\frac{\upsilon\cdot E\cdot t}{1-\upsilon^2}$$
$$\widehat{G}_{xy}=G\cdot t \tag{6.126}$$
$$\widehat{D}_x=\widehat{D}_y=D=\frac{E\cdot t^3}{1-\upsilon^2},\quad \widehat{D}_{xy}=2D=2\frac{E\cdot t^3}{1-\upsilon^2}$$

The coupling stiffness of the shell's wall is given by the following expressions (see also Fig. 6.15):

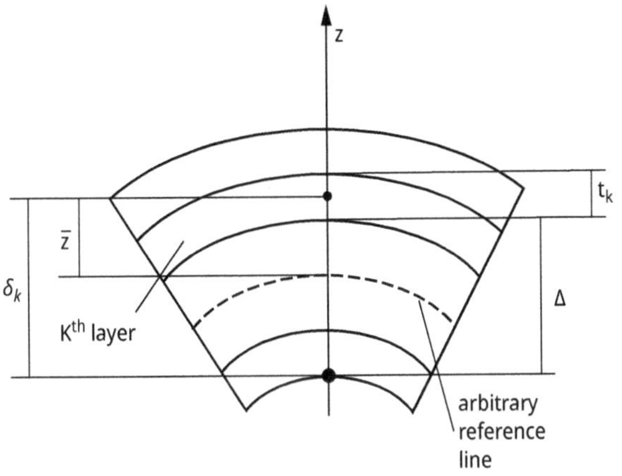

Fig. 6.15: The arrangement of the layers in an orthotropic shell wall (adapted from [65]).

$$\widehat{C}_x = \sum_{k=1}^{N} \left[\frac{E_x}{1-\upsilon_x\upsilon_y}\right]_k t_k\bar{z}_k, \quad \widehat{C}_y = \sum_{k=1}^{N} \left[\frac{E_y}{1-\upsilon_x\upsilon_y}\right]_k t_k\bar{z}_k$$
$$\widehat{C}_{xy} = \sum_{k=1}^{N} \left[\frac{\upsilon_y E_x}{1-\upsilon_x\upsilon_y}\right]_k t_k\bar{z}_k = \sum_{k=1}^{N} \left[\frac{\upsilon_x E_y}{1-\upsilon_x\upsilon_y}\right]_k t_k\bar{z}_k \tag{6.127}$$
$$\widehat{K}_{xy} = \sum_{k=1}^{N} \left[G_{xy}\right]_k t_k\bar{z}_k$$

where the subscript k refers to the properties (material and geometry) of the kth layer in an N-layered shell's wall.

Note that due to the large number of variables, no closed-form solution can be presented. For configuration for which the coupling coefficients can be neglected, the buckling load should be calculated using the following equation:

$$N_x = \frac{m^2\pi^2}{L^2}\widehat{D}_x\left[1+2\frac{\widehat{D}_{xy}}{\widehat{D}_x}\beta^2+\frac{\widehat{D}_y}{\widehat{D}_x}\beta^4\right]+ \frac{k^2L^4}{\pi^4 m^2\widehat{D}_x R^2}\frac{\tilde{E}_x\tilde{E}_y-\tilde{E}_{xy}^2}{\widehat{E}_x+\left[\frac{\widehat{E}_x\widehat{E}_y-\widehat{E}_{xy}^2}{\widehat{G}_{xy}}-2\widehat{E}_{xy}\right]\beta^2+\widehat{E}_y\beta^4} \tag{6.128}$$

The correlation factor should be taken as the same form used for isotropic shells, while the wall thickness should be replaced by the geometrical mean of the radii of gyrations for the axial and circumferential directions, namely

$$\kappa = 1 - 0.901\left[1 - e^{-\frac{1}{29.8}\sqrt{\frac{R}{\sqrt[4]{\frac{\hat{D}_x\hat{D}_y}{\hat{E}_x\hat{E}_y}}}}}\right] \tag{6.129}$$

Note that for the bending case, the correlation factor should be taken as [65]

$$\kappa = 1 - 0.731\left[1 - e^{-\frac{1}{29.8}\sqrt{\frac{R}{\sqrt[4]{\frac{\hat{D}_x\hat{D}_y}{\hat{E}_x\hat{E}_y}}}}}\right] \tag{6.130}$$

6.5.2.6 Buckling of orthotropic cylindrical shells under external pressure

For the case of an orthotropic cylindrical shell, the equation for the buckling under lateral pressure can be written in a similar way as given in eq. (6.122), namely

$$p_{\mathrm{cr}} = \frac{R}{n^2}\frac{\det(\Delta)}{\det(E)} \tag{6.131}$$

where $\det(\Delta)$ and $\det(E)$ are defined by eqs. (6.122) and (6.123).

For the hydrostatic case, we can write

$$p_{\mathrm{cr}} = \frac{R}{n^2 + 0.5\left(\frac{m\pi R}{L}\right)^2}\frac{\det(\Delta)}{\det(E)} \tag{6.132}$$

One should note that for the lateral pressure case, m would be unity, while n would vary to yield a minimal critical pressure (but not less than $n = 2$) in eq. (6.132). For the hydrostatic pressure case, one should vary also the value of m in addition to n. For long shells, eq. (6.132) is simplified to the following form:

$$p_{\mathrm{cr}} = \frac{3\left(\frac{\widehat{D}_y - \widehat{C}_y^2}{\widehat{E}_y}\right)}{R^3} \tag{6.133}$$

6.5.2.7 Buckling of orthotropic cylindrical shells under torsion

Weingarten et al. [65] report the torsional buckling of orthotropic cylindrical shells, for negligible coupling of bending and extension. The expression is given as

$$M_{\text{torsion, cr}} \approx 21.75 \widehat{D}_y^{\frac{5}{8}} \frac{R^{\frac{5}{4}}}{\sqrt{L}} \left[\frac{\widehat{E}_x \widehat{E}_y - \widehat{E}_{xy}^2}{\widehat{E}_y} \right]^{\frac{3}{8}} \cdot \kappa, \quad \text{with} \quad \kappa = 0.67$$

$$\text{for} \quad \sqrt{\frac{\widehat{E}_x \widehat{E}_y - \widehat{E}_{xy}^2}{12 \widehat{E}_y \widehat{D}_x}} \left[\frac{\widehat{D}_y}{\widehat{D}_x} \right]^{\frac{5}{6}} \frac{L^2}{R} \geq 500 \tag{6.134}$$

6.5.2.8 Combined bending and axial compression of orthotropic cylindrical shells

Due to a limited experimental data, Weingarten et al. [65] recommend to use the following expression:

$$\frac{\sigma_x}{\sigma_{x,cr}} + \frac{\sigma_b}{\sigma_{b,cr}} = 1 \tag{6.135}$$

where σ_x and σ_b are the applied axial and bending stresses, respectively, and $\sigma_{x,\text{cr}}$ and $\sigma_{b,\text{cr}}$ are the critical axial and bending stresses, respectively.

6.6 Miscellaneous stability problems

6.6.1 Case 1 – an axially compressed beam laterally loaded by a single force

Following Ref. [1], we shall investigate the case presented in Fig. 6.16. Let us denote the slopes at points A and B as θ_A and θ_B, respectively. Assuming the Bernoulli–Euler theorem (thus neglecting the influence of the shear deformation), two differential equations can be written for the two beam segments, AO and OB, namely

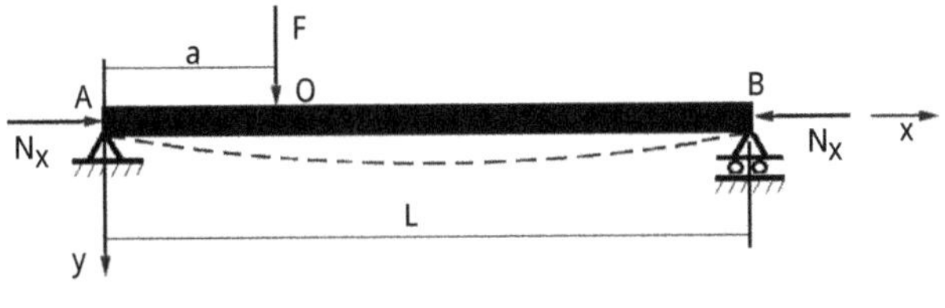

Fig. 6.16: An axially compressed beam laterally loaded by a single force.

$$\begin{aligned} EI\frac{d^2y}{dx^2} &= -M = -\frac{F(L-x)}{L}x - N_x y \\ EI\frac{d^2y}{dx^2} &= -M = -\frac{Fa(L-x)}{L} - N_x y \end{aligned} \tag{6.136}$$

where EI is the flexural bending of the uniform beam.

The general solutions for eqs. (6.136) can be found in the following way. Division of both equations by the factor EI and defining a new constant $\lambda^2 = N_x/EI$ yields

$$\frac{d^2y}{dx^2} + \lambda^2 y = -\frac{F(L-x)}{L}x$$
$$\frac{d^2y}{dx^2} + \lambda^2 y = -\frac{Fa(L-x)}{L} \tag{6.137}$$

The general solutions for eqs. (6.137) can be written as

$$\text{For segment } AO\text{: } y(x) = A_1 \cos \lambda x + A_2 \sin \lambda x - \frac{F(L-x)}{N_x L}x \tag{6.138a}$$

$$\text{For segment } OB\text{: } y(x) = B_1 \cos \lambda x + B_2 \sin \lambda x - \frac{Fa(L-x)}{N_x L} \tag{6.138b}$$

The four constants, A_1, A_2, B_1 and B_4, are to be determined using the boundary conditions of the case presented in Fig. 6.16, namely simply supported at both ends. Applying the two boundary conditions (zero vertical deflections at both ends) and two continuity conditions at point O (the deflections and their respective slopes should be equal) provides the expressions for the four constants, namely

$$y(0) = 0 \Rightarrow A_1 = 0$$
$$y(L) = 0 \Rightarrow \mathrm{B}_1 \cos \lambda L + \mathrm{B}_2 \sin \lambda L = 0 \rightarrow B_1 = -B_2 \tan \lambda L \tag{6.139}$$

The continuity requirements can be written as

$$A_2 \sin \lambda a - \frac{Fa(L-a)}{N_x L} = B_2[\sin \lambda a - \tan \lambda L \cos \lambda a] - \frac{Fa(L-a)}{N_x L}$$
$$A_2 \lambda \cos \lambda a - \frac{F(L-a)}{N_x L} = B_2 \lambda[\cos \lambda a + \tan \lambda L \sin \lambda a] + \frac{Fa}{N_x L} \tag{6.140}$$
$$\Rightarrow A_2 = \frac{F \sin[\lambda(L-a)]}{N_x \lambda \sin \lambda L}, \quad \mathrm{B}_2 = -\frac{F \sin \lambda a}{N_x \lambda \tan \lambda L}$$

Substituting the four constants, the equations for the deflection, the slope and the moment for each segment can be written as

For segment AO $0 \le x \le a$

$$y(x) = \frac{F\sin[\lambda(L-a)]}{N_x\lambda\sin\lambda L}\sin\lambda x - \frac{F(L-a)}{N_xL}x$$
$$\frac{dy(x)}{dx} = \frac{F\sin[\lambda(L-a)]}{N_x\sin\lambda L}\cos\lambda x - \frac{F(L-a)}{N_xL} \tag{6.141}$$
$$\frac{d^2y(x)}{dx^2} = \frac{F\lambda\sin[\lambda(L-a)]}{N_x\sin\lambda L}\sin\lambda x$$

For segment OB $a \le x \le L$

$$y(x) = \frac{F\sin\lambda a}{N_x\lambda\sin\lambda L}\sin\lambda(L-x) - \frac{Fa(L-x)}{N_xL}$$
$$\frac{dy(x)}{dx} = -\frac{F\sin\lambda a}{N_x\sin\lambda L}\cos\lambda(L-x) - \frac{Fa}{N_xL} \tag{6.142}$$
$$\frac{d^2y(x)}{dx^2} = -\frac{F\lambda\sin\lambda a}{N_x\sin\lambda L}\sin\lambda(L-x)$$

One should note that for the particular case in which the concentrated force F is applied at the mid-beam, the maximal deflection of the beam (at the beam center) can be written as

$$\text{Delta}_{\text{max}} = y(x = L/2) = \frac{F}{2N_x\lambda}\left(\tan\frac{\lambda L}{2} - \frac{\lambda L}{2}\right) \tag{6.143}$$

Let us investigate eq. (6.143) for some extreme cases. To do so, a new variable κ will be defined as

$$\kappa = \frac{\lambda L}{2} = \frac{L}{2}\sqrt{\frac{N_x}{EI}} \Rightarrow \kappa^2 = \frac{L^2}{4}\frac{N_x}{EI} \text{ or } N_x = \frac{4\kappa^2 EI}{L^2} \tag{6.144}$$

Substituting eq. (6.144) into eq. (6.143) yields the following expression:

$$\text{Delta}_{\text{max}} = y(x = L/2) = \frac{3FL^3}{48EI}(\tan - \kappa) \tag{6.145}$$

For small values of the variable κ, eq. (6.145) can be written as

$$\text{Delta}_{\text{max}} = y(x = L/2) = \delta_{\text{SS}} = \frac{FL^3}{48EI} \tag{6.146}$$

Note that δ_{SS} is the mid-beam deflection of a beam on simply supported conditions at both ends and loaded by a lateral load F. Also, the transition from eq. (6.145) to eq. (6.146) was enabled by the two equations presented in eq. (6.147):

$$\tan\kappa = \kappa + \frac{\kappa^3}{3} + \frac{2\kappa^5}{15} + o(\kappa^7)$$
$$\Rightarrow 3\frac{\tan - \kappa}{\kappa^3} = 3\frac{\kappa + \frac{\kappa^3}{3} - \kappa}{\kappa^3} = 1 \tag{6.147}$$

For the other extreme case, when $\kappa \to \pi/2$ the expression $\tan\kappa \to \infty$ in eq. (6.145) causes the value of $Delta_{max}$ to reach also infinity. Therefore, we obtain the Bernoulli–Euler buckling expression as presented in the following equation:

$$\kappa = \frac{\lambda L}{2} = \frac{\pi}{2} \Rightarrow \lambda^2 = \frac{N_x}{EI} = \frac{\pi^2}{L^2} \quad \to \quad N_{x_{cr}} = \frac{\pi^2 EI}{L^2} \tag{6.148}$$

Substituting back in eq. (6.144) yields the following expression:

$$\kappa = \frac{\pi}{2}\sqrt{\frac{N_x}{N_{x_{cr}}}} \tag{6.149}$$

meaning that the variable κ would depend only on the ratio N_x/N_{xcr}.

Note that eq. (6.145) can be easily calculated for any combination of the ratio N_x/N_{xcr} and the lateral load F.

To complete the investigation of the case, the slope of the beam and the maximal moment will be next derived.

Using the second expression in eqs. (6.141) for $x = L/2$ yields

$$\theta_A = \frac{dy(0)}{dx} = \frac{F\sin\lambda\frac{L}{2}}{N_x \sin\lambda L}\cos 0 - \frac{F\frac{L}{2}}{N_x L} = \frac{F}{2N_x}\frac{1}{\cos\frac{\lambda L}{2}} - \frac{F}{2N_x} =$$
$$= \frac{F}{2N_x}\left(\frac{1}{\cos\frac{\lambda L}{2}} - 1\right) \tag{6.150}$$

Using eq. (6.144), the expression for the slope at point A can be written as

$$\theta_A = \frac{FL^2}{16}\frac{2(1 - \cos\kappa)}{\kappa^2 \cos\kappa} \tag{6.151}$$

In a similar way, the slope at point B can be evaluated to be

$$\theta_B = \frac{dy(L)}{dx} = \frac{F\sin\lambda a}{N_x \sin\lambda L}\cos\lambda(L - L) + \frac{Fa}{N_x L}$$
$$\text{but } a = \frac{L}{2} \text{ and } \cos\lambda 0 = 1. \text{ Hence} \tag{6.152}$$
$$\theta_B = -\frac{F}{2N_x \cos\frac{\lambda L}{2}} + \frac{F}{2N_x} = \frac{F}{2N_x}\left(1 - \frac{1}{\cos\frac{\lambda L}{2}}\right) = \frac{FL^2}{16EI}\frac{(\cos\kappa - 1)}{\kappa^2\cos\kappa}$$

The expression for the maximal bending moment, which will occur for the present case at the mid-beam, would be

$$M_{\max} = M\left(\frac{L}{2}\right) = -EI\frac{d^2y\left(\frac{L}{2}\right)}{dx^2} = EI\frac{F\lambda \sin \lambda a}{N_x \sin \lambda L}\sin\lambda\left(L - \frac{L}{2}\right)$$

$$but\ a = \frac{L}{2}.\ \text{Therefore} \tag{6.153}$$

$$M_{\max} = \frac{EIF\lambda}{2N_x}\tan\frac{\lambda L}{2} = \frac{FL}{4}\frac{\tan \kappa}{\kappa}$$

6.6.2 Case 2 – an axially compressed beam under a bending moment

Based on the solution presented in Section 6.6.1, one can solve other important cases, like the one presented in Fig. 6.17.

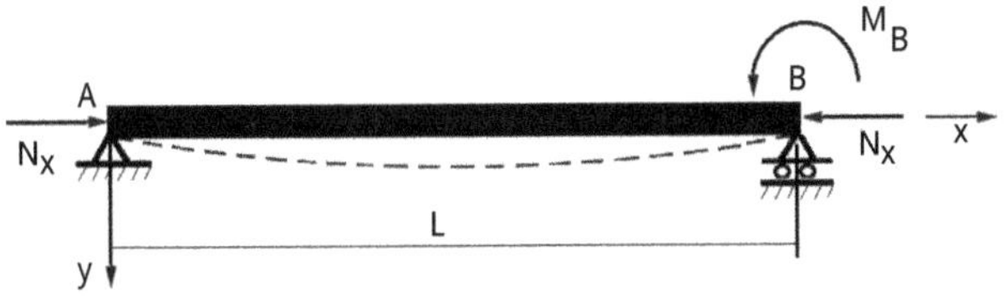

Fig. 6.17: An axially compressed beam under a single bending moment..

Like in the previous chapter, the slopes at points A and B will be denoted as θ_A and θ_B, respectively, and the Bernoulli–Euler theorem (thus neglecting the influence of the shear deformation) applies for this case.

If we assume that the distance ($L–a$) presented in Fig. 6.16 approaches zero while the applied lateral force F is increasing, the product $F(L–a) = M_B$ would lead to the case presented in Fig. 16.7, thus solving the present case using the various derivations presented for Case 1.

Using the first equation in eqs. (6.141), the deflection curve for the present case can be written as[12]

12 Note that the term $L–a \to 0$.

$$y(x) = \frac{F \sin[\lambda(L-a)]}{N_x \lambda \sin \lambda L} \sin\lambda x - \frac{F(L-a)}{N_x L} x =$$
$$= \frac{F\lambda(L-a)}{N_x \lambda \sin \lambda L} \sin\lambda x - \frac{F(L-a)}{N_x L} x = \frac{M_B}{N_x}\left(\frac{\sin \lambda x}{\sin \lambda L} - \frac{x}{L}\right) \tag{6.154}$$

where $M_B = F(L-a)$ and

$$\sin \lambda(L-a) \to \lambda(L-a) \text{ for small angles}$$

Next, the slopes of the beams,[13] θ_A and θ_B, can be evaluated using the following path.

Remembering the relation

$$\kappa = \frac{\lambda L}{2} = \frac{L}{2}\sqrt{\frac{N_x}{EI}}$$
$$\Rightarrow \kappa^2 = \frac{L^2}{4}\frac{N_x}{EI} \to N_x = \frac{4\kappa^2 EI}{L^2} \tag{6.155}$$

Equation (6.156) can be derived to be

$$\theta_A = \frac{dy(0)}{dx} = \frac{M_B}{N_x}\left(\frac{\lambda \cos 0}{\sin \lambda L} - \frac{1}{L}\right) = \frac{M_B L}{2\kappa EI}\left(\frac{1}{\sin 2\kappa} - \frac{1}{2\kappa}\right) \tag{6.156}$$

Similarly, the slope at point B will be

$$\theta_B = \frac{dy(L)}{dx} = -\frac{M_B}{N_x}\left(\frac{\lambda \cos \lambda L}{\sin \lambda L} - \frac{1}{L}\right) = -\frac{M_B L}{2\kappa EI}\left(\frac{1}{\tan 2\kappa} - \frac{1}{2\kappa}\right) \tag{6.157}$$

As shown for Case 1, the factor multiplying the term $M_B L/2EI\kappa$ would tend to unity for $\kappa \to 0$ (presenting the case where no compression is applied to the beam), and for $\kappa \to \pi/2$ it will tend to infinity (no bending moment is applied, and the beam will buckle as an Euler beam).

6.6.3 Case 3 – an axially compressed beam under bending moments at both ends

The drawing presented in Fig. 6.18a and b displays a case where the axial compression N_x is eccentrically applied at a distance ecc from the beam's neutral axis (Fig. 6.18a). The same phenomenon can be presented by the drawing in Fig. 6.18b, where bending moments are applied at both ends, namely $M_A = N_x \cdot ec_A$ and $M_B = N_x \cdot ec_B$.

Based on the derivation presented in Section 6.6.2, the deflection curve for the present case can be easily obtained using the superposition method. Using eq. (6.154) and adding the contribution of the bending moment M_A, where the term M_B in eq. (6.154) is replaced by the term M_A and x is changed to $(L-x)$, yields

13 A positive angle would be for a positive bending moment (downward in Fig. 6.17).

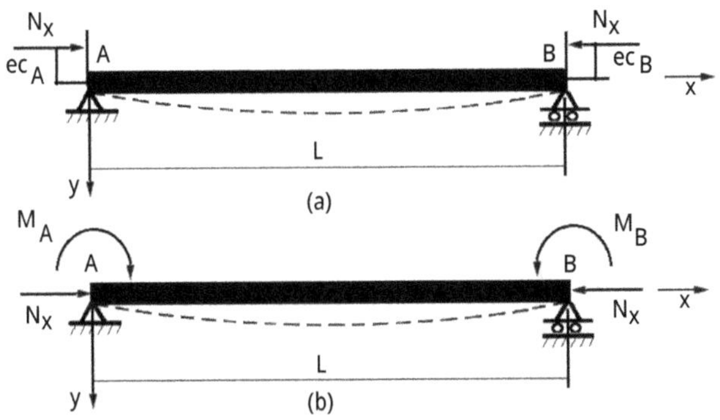

Fig. 6.18: An axially compressed beam: (a) eccentric compressive axial load and (b) under bending moments at both ends.

$$y(x) = \frac{M_B}{N_x}\left(\frac{\sin\lambda x}{\sin\lambda L} - \frac{x}{L}\right) + \frac{M_A}{N_x}\left[\frac{\sin\lambda(L-x)}{\sin\lambda L} - \frac{(L-x)}{L}\right] \tag{6.158}$$

Note that eq. (6.158) is also applicable for the problem presented in Fig. 6.18a. This is done by substituting the following terms $M_A = N_x \cdot ec_A$ and $M_B = N_x \cdot ec_B$ to yield

$$y(x) = ec_B\left(\frac{\sin\lambda x}{\sin\lambda L} - \frac{x}{L}\right) + ec_A\left[\frac{\sin\lambda(L-x)}{\sin\lambda L} - \frac{(L-x)}{L}\right] \tag{6.159}$$

The slopes at both ends of the beam can be shown to be for the case shown in Fig. 6.18b

$$\begin{aligned} \theta_A &= \frac{M_A L}{2EI\kappa}\left(\frac{1}{2\kappa} - \frac{1}{\tan 2\kappa}\right) + \frac{M_B L}{2EI\kappa}\left(\frac{1}{\sin 2\kappa} - \frac{1}{2\kappa}\right) \\ \theta_B &= \frac{M_A L}{2EI\kappa}\left(\frac{1}{\sin 2\kappa} - \frac{1}{2\kappa}\right) + \frac{M_B L}{2EI\kappa}\left(\frac{1}{2\kappa} - \frac{1}{\tan 2\kappa}\right) \end{aligned} \tag{6.160}$$

For a simpler case where $M_A = M_B = M$, we obtain from eq. (6.158) (using well-known trigonometric identities)

$$\begin{aligned} y(x) &= \frac{M}{N_x \cos\frac{\lambda L}{2}}\left[\cos\left(\frac{\lambda L}{2} - \lambda x\right) - \cos\frac{\lambda L}{2}\right] = \\ &= \frac{ML^2}{4\kappa^2 EI\cos\kappa}\left[\cos\left(\kappa - \frac{2\kappa x}{L}\right) - \cos\kappa\right] \end{aligned} \tag{6.161}$$

The beam mid-deflection can now be obtained by substituting $x = L/2$ in eq. (6.161) to yield

$$y\left(\frac{L}{2}\right) = \frac{ML^2}{4\kappa^2 EI\cos\kappa}\left[\cos\left(\kappa - \frac{2\kappa\frac{L}{2}}{L}\right) - \cos\kappa\right] = \frac{ML^2}{4\kappa^2 EI}\frac{1-\cos\kappa}{\cos\kappa} \tag{6.162}$$

The slopes at both ends of the beam are obtained from the first differentiation with respect to x of eq. (6.161) and setting $x = 0$ to yield

$$|\theta_A| = |\theta_B| = \frac{dy(0)}{dx} = \frac{ML}{2\kappa EI}\tan\kappa \tag{6.163}$$

Taking the second derivative with respect to x of eq. (6.161), multiplying by $-EI$ and inserting $x = L/2$ will yield the expression for M, the maximal bending moment, namely

$$M_{\max} = -EI\frac{d^2y\left(\frac{L}{2}\right)}{dx^2} = \frac{M}{\cos\kappa} = M\sec\kappa \tag{6.164}$$

6.6.4 Case 4 – an axially compressed beam laterally loaded by uniform pressure

The case presented in Fig. 6.19 can be shown to be equivalent to a system of infinitesimally concentrated forces laterally loading the given beam. Then the lateral deflection of the beam would be a superposition of all the deflections caused by the infinitesimally concentrated forces. Using the solution given in Section 6.6.1, the expression for the case presented in Fig. 6.19 can be written as (see also [1])

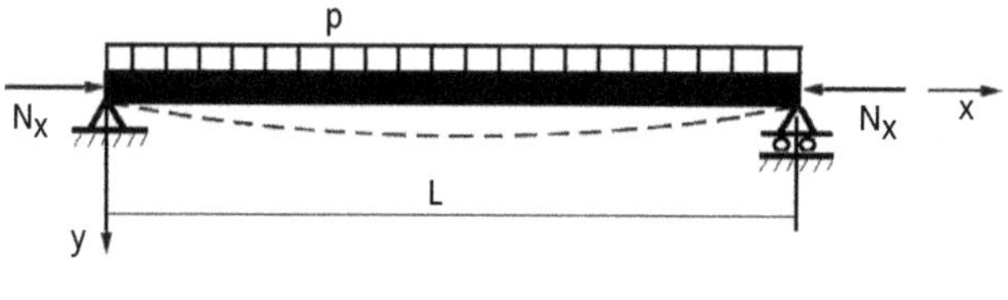

Fig. 6.19: An axially compressed beam laterally loaded by a uniform pressure p.

$$y(x) = \frac{pL^4}{16EI\kappa^4}\left[\frac{\cos\left(\kappa - \frac{2\kappa x}{L}\right)}{\cos\kappa} - 1\right] - \frac{pL^2}{8EI\kappa^2}x(L-x) \tag{6.165}$$

The beam's mid-deflection can be written as

$$\text{DELTA}_{\max} = y\left(\frac{L}{2}\right) = \frac{5pL^4}{384EI}\frac{12(\sec\kappa - 2 - \kappa^2)}{5\kappa^4} \tag{6.166}$$

Note that the first fraction on the right-hand side of the equation stays for the central deflection of a beam on simply supported boundary conditions and loaded by a uniform pressure p. The second fraction represents the influence of axial compression force on the lateral deflection of the beam. It can be shown by expanding[14] the sec κ in a form of series that the fraction would tend to unity when $\kappa \to 0$ and it would tend to infinity for $\kappa \to \pi/2$.

The slopes at both ends of the beam are obtained by differentiating the deflection equation [eq. (6.165)] to yield

$$|\theta_A| = |\theta_B| = \frac{dy(0)}{dx} = \frac{pL^3}{24EI}\frac{3(\tan\kappa - \kappa)}{\kappa^2} \tag{6.167}$$

As for the deflection equation (6.164), the first fraction on the right-hand side of the equation represents the slope of a beam on simply supported boundary conditions and loaded by uniform load, p. The second fraction stands for the influence of the compression load on the slope, and it will tend to unity when $\kappa \to 0$ and it would tend to infinity for $\kappa \to \pi/2$.

Twice differentiating eq. (6.165) and setting $x = L/2$ would represent the maximal moment having the following expression:

$$M_{\max} = -EI\frac{d^2y\left(\frac{L}{2}\right)}{dx^2} = \frac{pL^2}{8}\frac{2(1-\cos\kappa)}{\kappa^2\cos\kappa} \tag{6.168}$$

6.6.5 Approximate solutions for beam's lateral deflection

The four cases presented above have the same pattern for the beam's lateral deflection. It consists of an expression for the lateral deflection due to lateral loads only multiplied by another expression indicating the influence of the axial compressive load on the total deflection. This influence can be approximated by using the following expression:

$$\text{Factor} = \frac{1}{1 - {}^{N_x}/_{N_{x\mathrm{cr}}}} \tag{6.169}$$

The amplification factor given in eq. (6.169) can be used for ${}^{N_x}/_{N_{x\mathrm{cr}}} \le 0.6$ with 2% error (see [1]). The expression in eq. (6.169) is depicted in Fig. 6.20.

14 $\sec x = 1 + \frac{x^2}{2} + \frac{5x^4}{24} + \frac{61x^6}{720} + \frac{1{,}385x^8}{40{,}320} + \ldots$

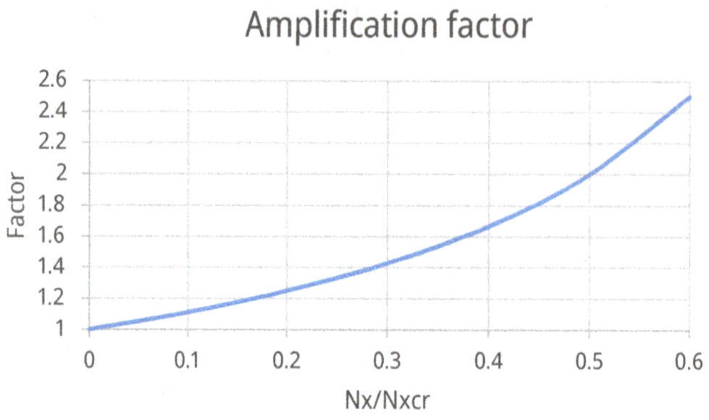

Fig. 6.20: The amplification factor versus $N_x/N_{x_{cr}} \leqq 0.6$.

6.6.6 Case 5 – an axially compressed beam laterally loaded by a uniform pressure on clamped boundary conditions

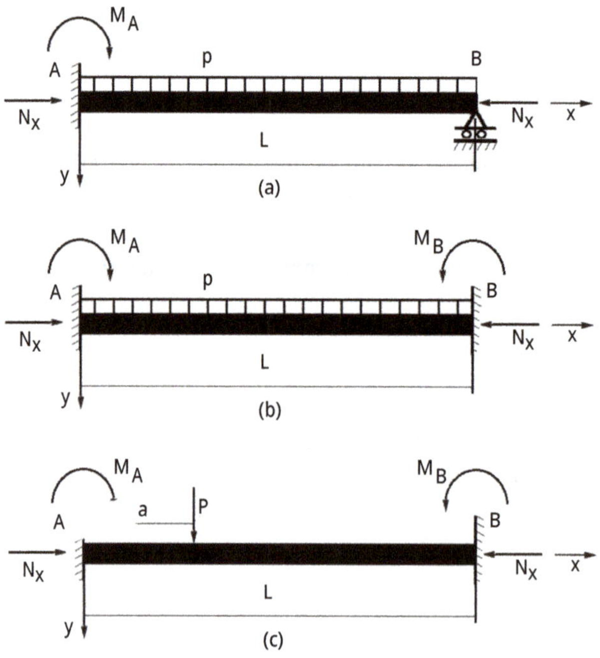

Fig. 6.21: An axially compressed beam under various lateral loading and boundary conditions: (a) continuous uniform pressure with clamped–simply supported boundaries, (b) continuous uniform pressure with clamped–clamped boundaries and (c) concentrated single force with clamped–clamped boundaries.

Using the various derivations presented in the previous chapters, one can calculate the behavior of axially loaded beams resting on clamped boundary conditions as depicted in Fig. 6.21a–c. First, let us concentrate on Fig. 6.21a. This is a static indeterminate case with the moment M_A to be found by demanding the slope at point A to be zero; this means that the slope due to lateral uniform continuous pressure [eq. (6.167) in Section 6.6.4] plus the slope due to the action of the bending moment M_A [eq. (6.157) in Section 6.6.2] must be zero. Therefore,

$$\frac{pL^3}{24EI}\frac{3(\tan\kappa-\kappa)}{\kappa^2}-\frac{M_A L}{2EI\kappa}\left(\frac{1}{\tan\kappa}-\frac{1}{2\kappa}\right)=0$$

$$\Rightarrow M_A=-\frac{\dfrac{pL^3}{24EI}\dfrac{3(\tan\kappa-\kappa)}{\kappa^2}}{\dfrac{L}{2EI\kappa}\left(\dfrac{1}{2\kappa}-\dfrac{1}{\tan\kappa}\right)}=-\frac{pL^2}{4\kappa}\frac{3(\tan\kappa-\kappa)}{\left(\dfrac{1}{2\kappa}-\dfrac{1}{\tan\kappa}\right)}= \tag{6.170}$$

$$=-\frac{pL^2}{2}\frac{3(\tan\kappa-\kappa)\tan\kappa}{(\tan\kappa-2\kappa)}$$

Note that the negative sign in eq. (6.170) shows that the direction of the moment at point A is opposite to what is depicted in Fig. 6.21a.

With moment M_A being found, the solution for the case depicted in Fig. 6.21a can be obtained using superposition of the solutions of Cases 2 and 4.

For the uniformly loaded beam case with two clamped boundary conditions (Fig. 6.21b), the following procedure is suggested. Let us assume that the moments at the boundaries are equal, $M_A=M_B=M$. Then the case is symmetric, and to find the moment at the boundaries, M, we use the same demand as done before; namely, the slopes due to the continuous load at both ends should be cancelled by the slopes created by the moments at the beam's boundaries. The following expression can be written, and the unknown moment, M, can be found:

$$\frac{pL^3}{24EI}\frac{3(\tan\kappa-\kappa)}{\kappa^2}+\frac{ML}{2EI\kappa}\frac{\tan\kappa}{\kappa}=0$$

$$\Rightarrow M=-\frac{\dfrac{pL^3}{24EI}\dfrac{3(\tan\kappa-\kappa)}{\kappa^2}}{\dfrac{L}{2EI\kappa}\dfrac{\tan\kappa}{\kappa}}=-\frac{pL^2}{4\kappa}\frac{3(\tan\kappa-\kappa)}{\tan\kappa} \tag{6.171}$$

As before, the minus sign indicates that the moment would act in the opposite direction to the one drawn in Fig. 6.21b. With the unknown term M being found, the deflection curve can be also found by the superposition of the deflection due to the lateral load and the deflection due to the two end moments. The moment at mid-beam will be the sum of the moment due to the lateral load and the moment due to the moments M acting at the boundaries to yield

$$M\left(\frac{l}{2}\right)=\frac{pL^2}{8}\frac{2(1-\cos\kappa)}{\kappa^2\cos\kappa}-\frac{pL^2}{4\kappa}\frac{3(\tan\kappa-\kappa)}{\tan\kappa}$$
$$\Rightarrow M\left(\frac{l}{2}\right)=\frac{pL^2}{24}\left[\frac{6(1-\cos\kappa)}{\kappa^2\cos\kappa}-\frac{18(\tan\kappa-\kappa)}{\kappa\tan\kappa}\right] \tag{6.172}$$

Finally, to solve the case presented in Fig. 6.21c for a nonsymmetric loaded beam, the slopes at both boundaries should be zero. Therefore, the sum of the angles at both beam's ends due to the concentrated load and the relevant slopes due to the concentrated bending moments, M_A and M_B, must be zero. Two algebraic equations in two unknowns (M_A and M_B) would be generated as presented in the following equation:

$$\theta_A=\frac{dy(0)}{dx}=\left[\frac{F\sin[\lambda(L-a)]}{N_x\sin\lambda L}\cos\lambda 0-\frac{F(L-a)}{N_xL}\right]+$$
$$+\left[\frac{M_AL}{2EI\kappa}\left(\frac{1}{2\kappa}-\frac{1}{\tan 2\kappa}\right)\right]+\left[\frac{M_BL}{2EI\kappa}\left(\frac{1}{\sin 2\kappa}-\frac{1}{2\kappa}\right)\right]$$
$$\theta_B=\frac{dy(L)}{dx}=\left[-\frac{F\sin\lambda a}{N_x\sin\lambda L}\cos\lambda(L-L)-\frac{Fa}{N_xL}\right]+$$
$$+\left[\frac{M_AL}{2EI\kappa}\left(\frac{1}{\sin 2\kappa}-\frac{1}{2\kappa}\right)\right]+\left[\frac{M_BL}{2EI\kappa}\left(\frac{1}{2\kappa}-\frac{1}{\tan 2\kappa}\right)\right]$$
$$\Rightarrow \theta_A=\left[\frac{F\sin[\lambda(L-a)]}{N_x\sin\lambda L}-\frac{F(L-a)}{N_xL}\right]+ \tag{6.173}$$
$$+\left[\frac{M_AL}{2EI\kappa}\left(\frac{1}{2\kappa}-\frac{1}{\tan 2\kappa}\right)\right]+\left[\frac{M_BL}{2EI\kappa}\left(\frac{1}{\sin 2\kappa}-\frac{1}{2\kappa}\right)\right]$$
$$\Rightarrow \theta_B=\left[-\frac{F\sin\lambda a}{N_x\sin\lambda L}-\frac{Fa}{N_xL}\right]+$$
$$+\left[\frac{M_AL}{2EI\kappa}\left(\frac{1}{\sin 2\kappa}-\frac{1}{2\kappa}\right)\right]+\left[\frac{M_BL}{2EI\kappa}\left(\frac{1}{2\kappa}-\frac{1}{\tan 2\kappa}\right)\right]$$

References

[1] Timoshenko, S.P. and Gere, J.M. Elastic stability, 2nd, McGraw-Hill, New York, 1963, 541.

[2] Abramovich, H. Ed. Stability and vibrations of thin-walled composite structures, Woodhead Publishing Series in Composites Science and Engineering, The Officers' Mess Business Centre, Royston Road, Duxford, CB22 4QH, Elsevier Ltd, United Kingdom, Copyright ©, 2017, 758.

[3] Simitses, G.J. An introduction to the elastic stability of structures, Prentice-Hall, Inc., Englewood Cliffs, New Jersey, 1976, 253.

[4] Reddy, J.N. Energy principles and variational methods in applied mechanics, 2nd, John Wiley, New York, 2002, 608.

[5] Brush, D.O. and Almroth, B.O. Buckling of bars, plates and shells, McGraw-Hill international student edition, New York (NY), USA, 1975, 379.

[6] Thompson, J.M.T. and Hunt, G.W. Elastic instability phenomena, John Wiley and Sons, Chichester, UK, 1984, 216.
[7] Johnson, J.B., Bryan, C.W., Turneaure, F.E. and Kinne, W.S. The theory and practice of modern framed structures-Designed for the use of schools and for engineers in professional practice, Part III: Design, John Wiley & Sons, New York, 1916, 565.
[8] Whitney, J.M. Cylindrical bending of unsymmetrically laminated plates, Journal of Composite Materials, 3(4), 1969, 715–719.
[9] Timoshenko, S.P. On the correction factor for shear of the differential equation for transverse vibrations of bars of uniform cross-section, Philosophical Magazine, 41(245), 1921, 744–746.
[10] Timoshenko, S.P. On the transverse vibrations of bars of uniform cross-section, Philosophical Magazine, 43(253), 1922, 125–131.
[11] Reissner, E. The effect of transverse shear deformation on the bending of elastic plates, Journal of Applied Mechanics, 12(2), 1945, A69–A77.
[12] Abramovich, H. Shear deformation and rotary inertia effects of vibrating composite beams, Composite Structures, 20, 1992, 165–173.
[13] Abramovich, H. and Hamburger, O. Vibration of a uniform cantilever Timoshenko beam with translational and rotational springs and with a tip mass, Journal of Sound and Vibration, 154(1), 1992, 67–80.
[14] Abramovich, H. A note on experimental investigation on a vibrating Timoshenko cantilever beam, Journal of Sound and Vibration, 160(1), 1993, 167–171.
[15] Abramovich, H. and Livshits, A. Dynamic behavior of cross-ply laminated beams with piezoelectric layers, Composite Structures, 25(1–4), 1993, 371–379.
[16] Abramovich, H. and Livshits, A. Free vibrations of non-symmetric cross-ply laminated composite beams, Journal of Sound and Vibration, 176(5), 1994, 597–612.
[17] Abramovich, H. Thermal buckling of cross-ply composite laminates using a first-order shear deformation theory, Composites Structures, 28, 1994, 201–213.
[18] Abramovich, H. Deflection control of laminated composite beams with piezoceramic layers- closed form solutions, Composite Structures, 43, 1998, 217–231.
[19] Abramovich, H., Eisenberger, M. and Shulepov, O. Vibrations and buckling of cross-ply non-symmetric laminated composite beams, AIAA Journal, 34(5), 1996, 1064–1069.
[20] Timoshenko, S.P. and Woinowsky-Krieger, S. Theory of plates and shells, McGraw-Hill, New York, 1959, 580.
[21] Reissner, E. and Stavsky, Y. Bending and stretching of certain types of heterogeneous aeolotropic elastic plates, Journal of Applied Mechanics, 28(3), 1961, 402–408.
[22] Dong, S.B., Pister, K.S. and Taylor, R.L. On the theory of laminated anisotropic shells and plates, Journal of Aerospace Sciences, 29(8), 1962, 969–975.
[23] Whitney, J.M. and Leissa, A.W. Analysis of heterogeneous anisotropic plates, Journal of Applied Mechanics, 36(2), 1969, 261–266.
[24] Reddy, J.N. Mechanics of laminated composite plates and shells- Theory and analysis, 2nd, CRC Press, Boca Raton, Florida, USA, 2004, 831.
[25] Mindlin, R.D. Influence of rotatory inertia and shear on flexural motions of isotropic, elastic plates, ASME Journal of Applied Mechanics, 18(1), 1951, 31–38.
[26] Whitney, J.M. The effect of transverse shear deformation in the bending of laminated plates, Journal of Composite Materials, 3(4), 1969, 534–547.
[27] Whitney, J.M. Shear correction factors for orthotropic laminates under static load, Journal of Applied Mechanics, 40(1), 1973, 302–304.
[28] Reissner, E. Note on the effect of transverse shear deformation in laminated anisotropic plates, Computer Methods in Applied Mechanics and Engineering, 20, 1979, 203–209.

[29] Leissa, A.W., Buckling of laminated composite plates and shell panels, AFWAL-TR-85-3069, June 1985.
[30] Bergmann, S. and Reissner, H. Neuere probleme aus der flugzeugstatik uber die knickung von wellblechstreifen bei schubbeanspruchung, Zeitschrift fur Flugtechnik und Motorluftschiffahrt (Z.F.M.), 20(18), 1929, 475–481.
[31] Jones, R.M. Buckling and vibration of unsymmetrically laminated cross-ply rectangular plates, AIAA Journal, 11(12), 1973, 1626–1632.
[32] Whitney, J.M. A study of the effects of coupling between bending and stretching on the mechanical behavior of layered anisotropic composite materials, Ph.D. Dissertation, Ohio State University, 1968, Also Tech. Rept. AFML-TR-68-330, April 1969, 80.
[33] Whitney, J.M. and Leissa, A.W. Analysis of heterogeneous anisotropic plates, Transactions ASME, Journal of Applied Mechanics, 36(2), 1969, 261–266.
[34] Whitney, J.M. Bending, vibrations, and buckling of laminated anisotropic rectangular plates, Wright Patterson AFBML, Ohio, Technical Report. AFML-TR-70–75, Aug. 1970, 35.
[35] Whitney, J.M. The effect of boundary conditions of the response of laminated composites, Journal of Composite Materials, 4, 1970, 192–203.
[36] Khdeir, A.A. Stability of antisymmetric angle-ply laminated plates, Journal of Engineering Mechanics, 115(5), 1989, 952–962.
[37] Reddy, J.N., Khdeir, A.A. and Librescu, L. Lévy type solutions for symmetrically laminated rectangular plates using first order shear deformation theory, Journal of Applied Mechanics, 54, 1987, 740–742.
[38] Gerard, G. and Becker, H. Handbook of structural stability, PART I -Buckling of flat plates, NACA TN 3781, 106.
[39] Johns, D.J. Shear buckling of isotropic and orthotropic plates- A review, Aeronautical Research Council, R&M No. 3677, London: Her Majesty's Stationery Office, UK, March 1972, 37.
[40] Stein, M. and Neff, J. Buckling stresses of simply supported rectangular flat plates in shear, NACA TN, 1222, March 1947, 14.
[41] Batdorf, S.B. and Stein, M. Critical combinations of shear and direct stress for simply supported rectangular flat plates, NACA TN, 1223, March 1947, 32.
[42] Whitney, J.M. and Pagano, N.J. Shear deformation in heterogeneous anisotropic plates, Journal of Applied Mechanics, 37(4), 1970, 1031–1036.
[43] Donnell, L.H. Stability of thin-walled tubes under torsion, NACA Report 479, 1933, 27.
[44] Donnell, L.H. A new theory for the buckling of thin cylinders under axial compression and bending, AER-56-12, Trans. Of the American Society of Mechanical Engineers, Aeronautical Engineering Division, Annual Meeting, New York N.Y., Dec. 3–7 1934, 796–806.
[45] Batdorf, S.B. A simplified method of elastic –stability analysis for thin cylindrical shells II- Modified equilibrium equation, NACA TN-1342, 1947, 36.
[46] Sanders, J.L., Jr. An improved first –approximation theory for thin shells, NASA TR R-24, 1959, 11.
[47] Weingarten, V.I., Morgan, E.J. and Seide, P. Elastic stability of thin-walled cylindrical and conical shells under axial compression, AIAA Journal, 3(3), 1965, 500–505.
[48] Block, D.L., Card, M.F. and Mikulas, M.M., Jr. Buckling of eccentrically stiffened orthotropic cylinders, NASA TN D-2960, NASA Langley Research Center, Hampton, Virginia, August 1965, 31.
[49] Jones, R.M. and Morgan, H.S. Buckling and vibration of cross-ply laminated circular cylindrical shells, AIAA Journal, 13(5), 1975, 664–671.
[50] Simitses, G.J. Buckling of moderately thick laminated cylindrical shells- a review, Composites Part B 27B, 1359–8368(95), 1996, 581–587.
[51] Simitses, G.J., Sheinman, I. and Shaw, D. The accuracy of Donnell's equations for axially-loaded, imperfect orthotropic cylinders, Computers and Structures, 20(6), 1985, 939–945.
[52] Simitses, G.J. Buckling and postbuckling of imperfect cylindrical shells- a review, Applied Mechanics Review, 39(10), 1986, 1517–1524.

[53] Hutchinson, J. Buckling of imperfect cylindrical shells under axial compression and external pressure, AIAA Journal, 3(10), 1965, 1969–1970.
[54] Teng, J.G. Buckling of thin shells: recent advances and trends, Applied Mechanics Review, 49(4), 1996, 263–274.
[55] Hunt, G.W., Lord, G.J. and Peletier, M.A. Cylindrical shell buckling: a characterization of localization and periodicity, Discrete and Continuous Dynamical Systems-Series B, 3(4), 2003, 505–518.
[56] Ventsel, E. and Krauthammer, T. Thin plates and shells, Theory analysis and application, Marcel Dekker, Inc., New York, Basel, 2001, 671.
[57] Aly, S.S.S. Buckling assessment of axially loaded cylindrical shells with random imperfections, Ph. D. thesis, IOWA State University, 1995, 154.
[58] Bushnell, D., Computerized buckling analysis of shells, AFWAL-TR-81-3049, Dec. 1981, 805.
[59] Godoy, L.A. and Sosa, E.M. Computational buckling analysis and practice, Mećanica Computacional, Vol. XXI, Idelsohn, S.R., Sonzogni, V.E. and Cardona, A. Eds., Santa Fe-Paraná, Argentina, 2002, 1652–1667.
[60] Takano, A. Simple closed-form solution for the buckling of moderately thick anisotropic cylinders, Transactions of the Japan Society for Aeronautical and Space Sciences, Aerospace Technology, 10, 2012, 17–26.
[61] Lvov, G., Pupazescu, A., Beschetnikov, D. and Zaharia, M. Buckling analysis of a thin –walled cylindrical shell strengthened by fiber-reinforced polymers, Materiale Plastice, 52(1), 2015, 28–31.
[62] Chung, S.W. and Hong, S.G. A comparison of membrane shell theories of hybrid anisotropic materials, European Journal of Engineering and Technology, 4(5), 2016, 83–106.
[63] Bushnell, D. Buckling of shells-pitfall for designers, AIAA Journal, 19(9), 1981, 1183–1226.
[64] Arbocz, J. and Starnes, J.H., Jr. Future directions and challenges in shell stability analysis, Thin-Walled Structures, 40(9), 2002, 729–754.
[65] Weingarten, V.I., Seide, P. and Peterson, J.P. Buckling of thin-walled circular cylinders, NASA SP-8007, Sept. 1965, revised version Aug. 1968, 49.
[66] Jones, R.M. Buckling of circular cylindrical shells with multiple orthotropic layers and eccentric stiffeners, AIAA Journal, 6(12), 1968, 2301–2305.
[67] Abramovich, H. Advanced topics of thin-walled structures, © 2021 World Scientific Publishing Co. Pte. Ltd., Singapore; Hackensack, NJ; London, 404.

Appendix A: Nonsymmetric laminated composite beam – CLT approach

The two coupled equations of motion for a nonsymmetric laminated composite beam under compressive loads based on the CLT approach are

$$A_{11}\frac{d^2u_0}{dx^2} - B_{11}\frac{d^3w_0}{dx^3} = 0 \tag{A.1}$$

$$-B_{11}\frac{d^3u_0}{dx^3} + D_{11}\frac{d^4w_0}{dx^4} + \bar{N}_{xx}\frac{d^2w_0}{dx^2} = 0 \tag{A.2}$$

Derivation with respect to x of eq. (A.1) and substituting in eq. (A.2) lead to the following decoupled equation:

$$-\frac{B_{11}}{A_{11}}\frac{d^4w_0}{dx^4}+D_{11}\frac{d^4w_0}{dx^4}+\bar{N}_{xx}\frac{d^2w_0}{dx^2}=0 \tag{A.3}$$

The solution of eq. (A.3) has the following general form:

$$w_0(x)=C_1\sin\left(\widehat{\lambda}x\right)+C_2\cos\left(\widehat{\lambda}x\right)+C_3x+C_4 \tag{A.4}$$

where

$$\widehat{\lambda}^2=\frac{\bar{N}_{xx}}{\left(D_{11}-\frac{B_{11}^2}{A_{11}}\right)} \tag{A.5}$$

Let us assume that the axial displacement $u_0(x)$ has the same form as the lateral displacement [8, 12–19, 26–28, 42], namely

$$u_0(x)=C_5\sin\left(\widehat{\lambda}x\right)+C_6\cos\left(\widehat{\lambda}x\right)+C_7x+C_8 \tag{A.6}$$

The four constants in eq. (A.6) are connected to the four constants in eq. (A.4). By substituting eqs. (A.4) and (A.6) into eq. (A.1), we get the following relationships:

$$\begin{aligned} C_5 &= -\frac{B_{11}}{A_{11}}\cdot\widehat{\lambda}\cdot C_2 \\ C_6 &= +\frac{B_{11}}{A_{11}}\cdot\widehat{\lambda}\cdot C_1 \end{aligned} \tag{A.7}$$

The rest of the constants C_1, C_2, C_3, C_4, C_7 and C_8 will be determined using the six boundary conditions of the problem. For the case of simply supported at both ends of the column, the out-of-plane boundary conditions have the following expressions:

$$w_0(0)=0,\quad M_{xx}(0)=0\ \Rightarrow\ -B_{11}\frac{du_0(0)}{dx}+D_{11}\frac{d^2w_0(0)}{dx^2}=0 \tag{A.8}$$

$$w_0(L)=0,\quad M_{xx}(L)=0\ \Rightarrow\ -B_{11}\frac{du_0(L)}{dx}+D_{11}\frac{d^2w_0(L)}{dx^2}=0 \tag{A.9}$$

and the in-plane boundary conditions can be written (assuming that at $x=0$ there is no axial displacement, while at the other end of the beam, at $x=L$, there is a compression load) as

$$u_0(0)=0 \tag{A.10}$$

$$A_{11}\frac{du_0(L)}{dx}-B_{11}\frac{d^2w_0(L)}{dx^2}=-\bar{N}_{xx} \tag{A.11}$$

Applying the four out-of-plane boundary conditions (eqs. (A.8) and (A.9)) leads to the following four equations:

$$
\begin{aligned}
&C_2 + C_4 = 0 \\
&\frac{B_{11}^2}{A_{11}} \cdot \hat{\lambda}^2 \cdot C_2 - B_{11} \cdot C_7 - D_{11} \cdot \hat{\lambda}^2 \cdot C_2 = 0 \\
&C_1 \sin\left(\hat{\lambda} L\right) + C_2 \cos\left(\hat{\lambda} L\right) + C_3 x + C_4 = 0 \\
&-\frac{B_{11}^2}{A_{11}} \cdot \hat{\lambda}^2 \cdot C_2 \cdot \cos\left(\hat{\lambda} L\right) - \frac{B_{11}^2}{A_{11}} \cdot \hat{\lambda}^2 \cdot C_1 \cdot \sin\left(\hat{\lambda} L\right) + B_{11} \cdot C_7 \\
&D_{11} \cdot \hat{\lambda}^2 \cdot C_1 \cdot \sin\left(\hat{\lambda} L\right) + D_{11} \cdot \hat{\lambda}^2 \cdot C_2 \cdot \cos\left(\hat{\lambda} L\right) = 0
\end{aligned}
\tag{A.12}
$$

Applying the two in-plane boundary conditions (eqs. (A.10) and (A.11)) yields

$$
\begin{aligned}
&\tfrac{B_{11}}{A_{11}} \cdot \hat{\lambda} \cdot C_1 + C_8 = 0 \\
&-B_{11} \cdot \hat{\lambda}^2 \cdot C_2 \cdot \cos\left(\hat{\lambda} L\right) - B_{11} \cdot \hat{\lambda}^2 \cdot C_1 \cdot \sin\left(\hat{\lambda} L\right) + A_{11} \cdot C_7 \\
&+ B_{11} \cdot \hat{\lambda}^2 \cdot C_1 \cdot \sin\left(\hat{\lambda} L\right) + - B_{11} \cdot \hat{\lambda}^2 \cdot C_2 \cdot \cos\left(\hat{\lambda} L\right) = -\bar{N}_{xx} \\
&\Rightarrow C_7 = -\frac{\bar{N}_{xx}}{A_{11}}
\end{aligned}
\tag{A.13}
$$

Solving for the six constants, we get the following results:

$$
\begin{aligned}
&C_1 = \frac{B_{11}}{A_{11}} \frac{\left[\cos\left(\hat{\lambda} L\right) - 1\right]}{\sin\left(\hat{\lambda} L\right)}, \quad C_2 = \frac{B_{11}}{A_{11}}, \quad C_3 = 0 \\
&C_4 = -\frac{B_{11}}{A_{11}}, \quad C_7 = -\frac{\bar{N}_{xx}}{A_{11}}, \quad C_8 = -\hat{\lambda} \frac{B_{11}^2}{A_{11}^2} \frac{\left[\cos\left(\hat{\lambda} L\right) - 1\right]}{\sin\left(\hat{\lambda} L\right)}.
\end{aligned}
\tag{A.14}
$$

To obtain the buckling load, one should demand that $w_0(x) \to \infty$, namely finding when one of the constants would tend to infinity. Observing the constant C_1 leads to the following eigenvalue solution

$$
\sin\left(\hat{\lambda} L\right) = 0 \quad \Rightarrow \quad \hat{\lambda} L = n\pi, \quad n = 1, 2, 3, 4, \ldots \tag{A.15}
$$

which leads to the critical buckling load per unit width, b, having the following form:

$$
\left(\bar{N}_{xx}\right)_{cr} = \frac{\pi^2}{L^2} \left(D_{11} - \frac{B_{11}^2}{A_{11}} \right) \tag{A.16}
$$

The shape of the buckling will be $\sin\left(\frac{\pi x}{L}\right)$ like for the symmetric case.

One should note that the critical buckling load per unit width $(\bar{N}_{xx})_{cr}$ is influenced by the coupling coefficient B_{11}, presenting for nonsymmetric cases, a buckling load below that obtained for the symmetric case, having the same number of laminates.

For the clamped–clamped case, the involved boundary conditions are

$$w_0(0) = 0, \quad \frac{dw_0(0)}{dx} = 0 \tag{A.17}$$

$$w_0(L) = 0, \quad \frac{dw_0(L)}{dx} = 0 \tag{A.18}$$

$$u_0(0) = 0, \quad A_{11}\frac{du_0(L)}{dx} - B_{11}\frac{d^2w_0(L)}{dx^2} = -\bar{N}_{xx} \tag{A.19}$$

Applying the boundary conditions yields the following characteristic equation:

$$\left(\widehat{\lambda L}\right)\cdot\sin\left(\widehat{\lambda L}\right) = 2\left[1-\cos\left(\widehat{\lambda L}\right)\right] \quad \Rightarrow \quad \widehat{\lambda L} = 2\pi, 8.987, 4\pi, \ldots \tag{A.20}$$

with $C_1 = C_3 = C_4 = C_8 = 0$, $C_2 \neq 0$ and $C_7 = -\frac{\bar{N}_{xx}}{A_{11}}$. The critical buckling load per unit width can then be written as

$$(\bar{N}_{xx})_{cr} = \frac{4\pi^2}{L^2}\left(D_{11} - \frac{B_{11}^2}{A_{11}}\right) \tag{A.21}$$

The buckling shape can be written as

$$w_0(x) = C_2\left[\cos\left(\frac{2\pi x}{L}\right) - 1\right] \tag{A.22}$$

7 Vibrations of thin-walled structures

7.1 Introduction

This chapter is the complementary chapter to Chapter 6, which dealt with stability issues for isotropic and composite columns and plates. It is aimed at presenting the vibration analysis of isotropic and composite columns and plates based on the classical lamination theory (CLT) and the first-order shear deformation theory (FOSDT) developed in Chapter 2. Some typical references dealing with 1D behavior characteristically to beams are given in Refs. [1–17].

7.1.1 CLPT approach

As described in the previous chapter, the classical lamination plate theory (CLPT) is an extension of the well-known Kirchhoff–Love classical plate theory to be applied to laminated composite plates. The equations of motion are

$$\begin{aligned}
&\frac{\partial N_{xx}}{\partial x}+\frac{\partial N_{xy}}{\partial y}=I_0\frac{\partial^2 u_0}{\partial t^2}-I_1\frac{\partial^2}{\partial t^2}\left(\frac{\partial w_0}{\partial x}\right)\\
&\frac{\partial N_{xy}}{\partial x}+\frac{\partial N_{yy}}{\partial y}=I_0\frac{\partial^2 v_0}{\partial t^2}-I_1\frac{\partial^2}{\partial t^2}\left(\frac{\partial w_0}{\partial y}\right)\\
&\frac{\partial^2 M_{xx}}{\partial x^2}+2\frac{\partial^2 M_{xy}}{\partial x\partial y}+\frac{\partial^2 M_{yy}}{\partial y^2}+\frac{\partial}{\partial x}\left(N_{xx}\frac{\partial w_0}{\partial x}+N_{xy}\frac{\partial w_0}{\partial y}\right)\\
&+\frac{\partial}{\partial y}\left(N_{xy}\frac{\partial w_0}{\partial x}+N_{yy}\frac{\partial w_0}{\partial y}\right)=q+I_0\frac{\partial^2 w_0}{\partial t^2}+I_1\frac{\partial^2}{\partial t^2}\left(\frac{\partial u_0}{\partial x}+\frac{\partial v_0}{\partial y}\right)\\
&-I_2\frac{\partial^2}{\partial t^2}\left(\frac{\partial^2 w_0}{\partial x^2}+\frac{\partial^2 w_0}{\partial y^2}\right)
\end{aligned}\tag{7.1}$$

where q is the distributed pressure on the surface of the plate and the mass moments of inertia I_0, I_1 and I_2 are defined as

$$\begin{Bmatrix} I_0\\ I_1\\ I_2\end{Bmatrix}=\int_{-h/2}^{+h/2}\begin{Bmatrix}1\\ z\\ z^2\end{Bmatrix}\cdot\rho_0\cdot dz \tag{7.2}$$

where h is the total thickness of the plate and ρ_0 is the relevant density.

In eq. (7.1), N_{xx}, N_{xy}, N_{yy} are the force resultants per unit length and M_{xx}, M_{xy}, M_{yy} are the moment resultants per unit length defined as

https://doi.org/10.1515/9783111621104-007

$$\begin{Bmatrix} N_{xx} \\ N_{yy} \\ N_{xy} \end{Bmatrix} = \int_{-h/2}^{+h/2} \begin{Bmatrix} \sigma_{xx} \\ \sigma_{yy} \\ \tau_{xy} \end{Bmatrix} dz; \quad \begin{Bmatrix} M_{xx} \\ M_{yy} \\ M_{xy} \end{Bmatrix} = - \int_{-h/2}^{+h/2} \begin{Bmatrix} \sigma_{xx} \\ \sigma_{yy} \\ \tau_{xy} \end{Bmatrix} \cdot z \cdot dz \tag{7.3}$$

In eq. (7.3), z is a coordinate normal to the surface of the plate, and σ_{xx} and σ_{yy} are the normal stresses in the x and y directions, respectively, while τ_{xy} is the shear stress.

The stress resultants N_{xx}, N_{xy}, N_{yy} and M_{xx}, M_{xy}, M_{yy} can be defined using the assumed displacements to yield:

$$\begin{Bmatrix} N_{xx} \\ N_{yy} \\ N_{xy} \end{Bmatrix} = \begin{bmatrix} A_{11} & A_{12} & A_{16} \\ A_{12} & A_{22} & A_{26} \\ A_{16} & A_{26} & A_{66} \end{bmatrix} \begin{Bmatrix} \frac{\partial u_0}{\partial x} + \frac{1}{2}\left(\frac{\partial w_0}{\partial x}\right)^2 \\ \frac{\partial v_0}{\partial y} + \frac{1}{2}\left(\frac{\partial w_0}{\partial y}\right)^2 \\ \frac{\partial u_0}{\partial y} + \frac{\partial v_0}{\partial x} + \frac{\partial^2 w_0}{\partial x \partial y} \end{Bmatrix} - \begin{bmatrix} B_{11} & B_{12} & B_{16} \\ B_{12} & B_{22} & B_{26} \\ B_{16} & B_{26} & B_{66} \end{bmatrix} \begin{Bmatrix} \frac{\partial^2 w_0}{\partial x^2} \\ \frac{\partial^2 w_0}{\partial y^2} \\ 2\frac{\partial^2 w_0}{\partial x \partial y} \end{Bmatrix} \tag{7.4}$$

$$\begin{Bmatrix} M_{xx} \\ M_{yy} \\ M_{xy} \end{Bmatrix} = \begin{bmatrix} B_{11} & B_{12} & B_{16} \\ B_{12} & B_{22} & B_{26} \\ B_{16} & B_{26} & B_{66} \end{bmatrix} \begin{Bmatrix} \frac{\partial u_0}{\partial x} + \frac{1}{2}\left(\frac{\partial w_0}{\partial x}\right)^2 \\ \frac{\partial v_0}{\partial y} + \frac{1}{2}\left(\frac{\partial w_0}{\partial y}\right)^2 \\ \frac{\partial u_0}{\partial y} + \frac{\partial v_0}{\partial x} + \frac{\partial^2 w_0}{\partial x \partial y} \end{Bmatrix} - \begin{bmatrix} D_{11} & D_{12} & D_{16} \\ D_{12} & D_{22} & D_{26} \\ D_{16} & D_{26} & D_{66} \end{bmatrix} \begin{Bmatrix} \frac{\partial^2 w_0}{\partial x^2} \\ \frac{\partial^2 w_0}{\partial y^2} \\ 2\frac{\partial^2 w_0}{\partial x \partial y} \end{Bmatrix} \tag{7.5}$$

where

$$A_{ij} = \sum_{k=1}^{N} \bar{Q}_{ij}^{(k)} (z_{k+1} - z_k); \quad B_{ij} = \frac{1}{2}\sum_{k=1}^{N} \bar{Q}_{ij}^{(k)} (z_{k+1}^2 - z_k^2); \quad D_{ij} = \frac{1}{3}\sum_{k=1}^{N} \bar{Q}_{ij}^{(k)} (z_{k+1}^3 - z_k^3) \tag{7.6}$$

with $\bar{Q}_{ij}^{(k)}$ being the lamina stiffness after transformation. Substituting eqs. (7.4) and (7.5) into eq. (7.1) leads to the equations of motion for a laminated composite plate expressed by the three assumed displacements (u_0, v_0 and w_0):

$$A_{11}\left(\frac{\partial^2 u_0}{\partial x^2}+\frac{\partial^3 w_0}{\partial x^3}\right)+A_{12}\left(\frac{\partial^2 v_0}{\partial x\partial y}+\frac{\partial^3 w_0}{\partial x\partial y^2}\right)+A_{16}\left(2\frac{\partial^2 u_0}{\partial x\partial y}+3\frac{\partial^3 w_0}{\partial x^2\partial y}+\frac{\partial^2 v_0}{\partial x^2}\right)$$
$$+A_{26}\left(\frac{\partial^2 v_0}{\partial y^2}+\frac{\partial^3 w_0}{\partial y^3}\right)+A_{66}\left(\frac{\partial^2 u_0}{\partial y^2}+\frac{\partial^2 v_0}{\partial x\partial y}+2\frac{\partial^3 w_0}{\partial x\partial y^2}\right)-B_{11}\frac{\partial^3 w_0}{\partial x^3}-B_{12}\frac{\partial^3 w_0}{\partial x\partial y^2} \tag{7.7}$$
$$-3B_{16}\frac{\partial^3 w_0}{\partial x^2\partial y}-B_{26}\frac{\partial^3 w_0}{\partial y^3}-2B_{66}\frac{\partial^3 w_0}{\partial x\partial y^2}=I_0\frac{\partial^2 u_0}{\partial t^2}-I_1\frac{\partial^3 w_0}{\partial x\partial t^2}$$

$$A_{22}\left(\frac{\partial^2 v_0}{\partial y^2}+\frac{\partial^3 w_0}{\partial y^3}\right)+A_{12}\left(\frac{\partial^2 u_0}{\partial x\partial y}+\frac{\partial^3 w_0}{\partial x^2\partial y}\right)+A_{16}\left(\frac{\partial^2 u_0}{\partial x^2}+\frac{\partial^3 w_0}{\partial x^3}\right)$$
$$+A_{26}\left(\frac{\partial^2 u_0}{\partial y^2}+2\frac{\partial^2 v_0}{\partial x\partial y}+3\frac{\partial^3 w_0}{\partial x\partial y^2}\right)+A_{66}\left(\frac{\partial^2 u_0}{\partial x\partial y}+\frac{\partial^2 v_0}{\partial x^2}+2\frac{\partial^3 w_0}{\partial x^2\partial y}\right)-B_{12}\frac{\partial^3 w_0}{\partial x^2\partial y} \tag{7.8}$$
$$-B_{22}\frac{\partial^3 w_0}{\partial y^3}-B_{16}\frac{\partial^3 w_0}{\partial x^3}-3B_{26}\frac{\partial^3 w_0}{\partial x\partial y^2}-2B_{66}\frac{\partial^3 w_0}{\partial x^2\partial y}=I_0\frac{\partial^2 v_0}{\partial t^2}-I_1\frac{\partial^3 w_0}{\partial y\partial t^2}$$

$$B_{11}\left(\frac{\partial^3 u_0}{\partial x^3}+\frac{\partial^4 w_0}{\partial x^4}\right)+B_{12}\left(\frac{\partial^3 v_0}{\partial x^2\partial y}+\frac{\partial^3 u_0}{\partial x\partial y^2}+4\frac{\partial^3 w_0}{\partial x^2\partial y^2}\right)+B_{16}\left(3\frac{\partial^3 u_0}{\partial x^2\partial y}+\frac{\partial^3 v_0}{\partial x^3}+8\frac{\partial^4 w_0}{\partial x^3\partial y}\right)$$
$$+B_{22}\left(\frac{\partial^3 v_0}{\partial y^3}+2\frac{\partial^4 w_0}{\partial y^4}\right)+B_{26}\left(\frac{\partial^3 u_0}{\partial y^3}+3\frac{\partial^3 v_0}{\partial x\partial y^2}+8\frac{\partial^4 w_0}{\partial x\partial y^3}\right)+2B_{66}\left(\frac{\partial^3 u_0}{\partial x\partial y^2}+\frac{\partial^3 v_0}{\partial x^2\partial y}+4\frac{\partial^4 w_0}{\partial x^2\partial y^2}\right)$$
$$-D_{11}\frac{\partial^4 w_0}{\partial x^4}-2D_{12}\frac{\partial^4 w_0}{\partial x^2\partial y^2}-D_{22}\frac{\partial^4 w_0}{\partial y^4}-4D_{16}\frac{\partial^4 w_0}{\partial x^3\partial y}-4D_{26}\frac{\partial^4 w_0}{\partial x\partial y^3}-4D_{66}\frac{\partial^4 w_0}{\partial x^2\partial y^2}+\mathrm{P}(w_0)$$
$$=q+I_0\frac{\partial^2 w_0}{\partial t^2}-I_2\frac{\partial^2}{\partial t^2}\left(\frac{\partial^2 w_0}{\partial x^2}+\frac{\partial^2 w_0}{\partial y^2}\right)+I_1\frac{\partial^2}{\partial t^2}\left(\frac{\partial u_0}{\partial x}+\frac{\partial v_0}{\partial y}\right) \tag{7.9}$$

where

$$P(w_0)=\frac{\partial}{\partial x}\left(N_{xx}\frac{\partial w_0}{\partial x}+N_{xy}\frac{\partial w_0}{\partial y}\right)+\frac{\partial}{\partial y}\left(N_{xy}\frac{\partial w_0}{\partial x}+N_{yy}\frac{\partial w_0}{\partial y}\right) \tag{7.10}$$

7.1.2 FOSDPT approach

The first-order shear deformation plate theory (FOSDPT) described in detail in Chapter 2 is an extension of the well-known Timoshenko beam theory and/or Mindlin–Reissner plate theory to be applied to laminated composite plates.

The equations of motion in terms of displacements for the FOSDPT approach can be written as

$$A_{11}\left(\frac{\partial^2 u_0}{\partial x^2}+\frac{\partial^3 w_0}{\partial x^3}\right)+A_{12}\left(\frac{\partial^2 v_0}{\partial x\partial y}+\frac{\partial^3 w_0}{\partial x\partial y^2}\right)+A_{16}\left(2\frac{\partial^2 u_0}{\partial x\partial y}+3\frac{\partial^3 w_0}{\partial x^2\partial y}+\frac{\partial^2 v_0}{\partial x^2}\right)$$
$$+A_{26}\left(\frac{\partial^2 v_0}{\partial y^2}+\frac{\partial^3 w_0}{\partial y^3}\right)+A_{66}\left(\frac{\partial^2 u_0}{\partial y^2}+\frac{\partial^2 v_0}{\partial x\partial y}+2\frac{\partial^3 w_0}{\partial x\partial y^2}\right)+B_{11}\frac{\partial^2\phi_x}{\partial x^2}+B_{12}\frac{\partial^2\phi_y}{\partial x\partial y} \quad (7.11)$$
$$+B_{16}\left(2\frac{\partial^2\phi_x}{\partial x\partial y}+\frac{\partial^2\phi_y}{\partial x^2}\right)+B_{26}\frac{\partial^2\phi_y}{\partial y^2}+B_{66}\left(\frac{\partial^2\phi_x}{\partial y^2}+\frac{\partial^2\phi_y}{\partial x\partial y}\right)=I_0\frac{\partial^2 u_0}{\partial t^2}+I_1\frac{\partial^2\phi_x}{\partial t^2}$$

$$A_{22}\left(\frac{\partial^2 v_0}{\partial y^2}+\frac{\partial^3 w_0}{\partial y^3}\right)+A_{12}\left(\frac{\partial^2 u_0}{\partial x\partial y}+\frac{\partial^3 w_0}{\partial x^2\partial y}\right)+A_{16}\left(\frac{\partial^2 u_0}{\partial x^2}+\frac{\partial^3 w_0}{\partial x^3}\right)$$
$$+A_{26}\left(\frac{\partial^2 u_0}{\partial y^2}+2\frac{\partial^2 v_0}{\partial x\partial y}+3\frac{\partial^3 w_0}{\partial x\partial y^2}\right)+A_{66}\left(\frac{\partial^2 u_0}{\partial x\partial y}+\frac{\partial^2 v_0}{\partial x^2}+2\frac{\partial^3 w_0}{\partial x^2\partial y}\right)+B_{12}\frac{\partial^2\phi_x}{\partial x\partial y}$$
$$+B_{22}\frac{\partial^2\phi_y}{\partial y^2}+B_{16}\frac{\partial^2\phi_x}{\partial x^2}+B_{26}\left(2\frac{\partial^2\phi_y}{\partial x\partial y}+\frac{\partial^2\phi_x}{\partial y^2}\right)+B_{66}\left(\frac{\partial^2\phi_x}{\partial x\partial y}+\frac{\partial^2\phi_y}{\partial x^2}\right)=I_0\frac{\partial^2 v_0}{\partial t^2}+I_1\frac{\partial^2\phi_y}{\partial t^2}$$
$$(7.12)$$

$$KA_{44}\left(\frac{\partial^2 w_0}{\partial y^2}+\frac{\partial\phi_y}{\partial y}\right)+KA_{45}\left(\frac{\partial^2 w_0}{\partial x\partial y}+\frac{\partial\phi_y}{\partial x}\right)+KA_{45}\left(\frac{\partial^2 w_0}{\partial x\partial y}+\frac{\partial\phi_x}{\partial y}\right) \quad (7.13)$$
$$+KA_{55}\left(\frac{\partial^2 w_0}{\partial x^2}+\frac{\partial\phi_x}{\partial x}\right)+\mathrm{P}(w_0)=q+I_0\frac{\partial^2 w_0}{\partial t^2}$$

$$B_{11}\left(\frac{\partial^2 u_0}{\partial x^2}+\frac{\partial^3 w_0}{\partial x^3}\right)+B_{12}\left(\frac{\partial^2 v_0}{\partial x\partial y}+\frac{\partial^3 w_0}{\partial x\partial y^2}\right)+B_{16}\left(2\frac{\partial^2 u_0}{\partial x\partial y}+\frac{\partial^2 v_0}{\partial x^2}+3\frac{\partial^4 w_0}{\partial x^3\partial y}\right)$$
$$+B_{26}\left(\frac{\partial^2 v_0}{\partial y^2}+\frac{\partial^3 w_0}{\partial y^3}\right)+B_{66}\left(\frac{\partial^2 u_0}{\partial y^2}+\frac{\partial^2 v_0}{\partial x\partial y}+2\frac{\partial^3 w_0}{\partial x\partial y^2}\right)+D_{11}\frac{\partial^2\phi_x}{\partial x^2}+D_{12}\frac{\partial^2\phi_y}{\partial x\partial y} \quad (7.14)$$
$$+D_{16}\left(2\frac{\partial^2\phi_x}{\partial x\partial y}+\frac{\partial^2\phi_y}{\partial x^2}\right)+D_{26}\frac{\partial^2\phi_y}{\partial y^2}+D_{66}\left(\frac{\partial^2\phi_x}{\partial y^2}+\frac{\partial^2\phi_y}{\partial x\partial y}\right)-KA_{55}\left(\frac{\partial w_0}{\partial x}+\phi_x\right)$$
$$-KA_{45}\left(\frac{\partial w_0}{\partial y}+\phi_y\right)=I_2\frac{\partial^2\phi_x}{\partial t^2}+I_1\frac{\partial^2 u_0}{\partial t^2}$$

$$B_{12}\left(\frac{\partial^2 u_0}{\partial x \partial y}+\frac{\partial^3 w_0}{\partial x \partial y^2}\right)+B_{16}\left(\frac{\partial^2 u_0}{\partial x^2}+\frac{\partial^3 w_0}{\partial x^3}\right)+B_{22}\left(\frac{\partial^2 v_0}{\partial y^2}+\frac{\partial^3 w_0}{\partial y^3}\right)$$
$$+B_{26}\left(\frac{\partial^2 u_0}{\partial y^2}+2\frac{\partial^2 v_0}{\partial x \partial y}+3\frac{\partial^3 w_0}{\partial x \partial y^2}\right)+B_{66}\left(\frac{\partial^2 v_0}{\partial x^2}+\frac{\partial^2 u_0}{\partial x \partial y}+2\frac{\partial^3 w_0}{\partial x \partial y^2}\right)$$
$$+D_{12}\frac{\partial^2 \phi_x}{\partial x \partial y}+D_{22}\frac{\partial^2 \phi_y}{\partial y^2}+D_{16}\frac{\partial^2 \phi_x}{\partial x^2}+D_{26}\left(\frac{\partial^2 \phi_x}{\partial y^2}+2\frac{\partial^2 \phi_y}{\partial x \partial y}\right)+D_{66}\left(\frac{\partial^2 \phi_y}{\partial x^2}+\frac{\partial^2 \phi_x}{\partial x \partial y}\right) \tag{7.15}$$
$$-KA_{45}\left(\frac{\partial w_0}{\partial x}+\phi_x\right)-KA_{44}\left(\frac{\partial w_0}{\partial y}+\phi_y\right)=I_2\frac{\partial^2 \phi_y}{\partial t^2}+I_1\frac{\partial^2 v_0}{\partial t^2}$$

where $P(w_0)$ is given by eq. (7.10); u_0, v_0, w_0 are the displacements in the directions x, y and z, respectively; ϕ_x, ϕ_y are rotations about the x and y axes, respectively, and the various terms A_{ij}, B_{ij} and D_{ij} are defined in detail in Chapter 6.

7.2 Vibrations of columns – CLT approach

The equations of motion for calculating a general laminate column can be derived from eq. (7.1) by assuming that the displacement of the model is only a function of the x coordinate, yielding the following equations (v_0 is assumed to be zero):

$$A_{11}\left(\frac{\partial^2 u_0}{\partial x^2}+\frac{\partial^3 w_0}{\partial x^3}\right)-B_{11}\frac{\partial^3 w_0}{\partial x^3}=I_0\frac{\partial^2 u_0}{\partial t^2}-I_1\frac{\partial^3 w_0}{\partial x \partial t^2} \tag{7.16}$$

$$B_{11}\left(\frac{\partial^3 u_0}{\partial x^3}+\frac{\partial^4 w_0}{\partial x^4}\right)-D_{11}\frac{\partial^4 w_0}{\partial x^4}=q+\bar{P}(w_0)+I_0\frac{\partial^2 w_0}{\partial t^2}-I_2\frac{\partial^2}{\partial t^2}\left(\frac{\partial^2 w_0}{\partial x^2}\right)+I_1\frac{\partial^2}{\partial t^2}\left(\frac{\partial u_0}{\partial x}\right) \tag{7.17}$$

where

$$\bar{P}(w_0)=\frac{\partial}{\partial x}\left(N_{xx}\frac{\partial w_0}{\partial x}\right) \tag{7.18}$$

$$\begin{Bmatrix} I_0 \\ I_1 \\ I_2 \end{Bmatrix}=\int_{-h/2}^{+h/2}\begin{Bmatrix} 1 \\ z \\ z^2 \end{Bmatrix}\cdot \rho_0 \cdot dz \tag{7.19}$$

and h is the total thickness of the plate and ρ_0 is the relevant density.

To analyze the vibration problem of the columns, the lateral load q will be assumed to be zero, while the axial compressive load, $\bar{P}(w_0)$, will be assumed to be nonzero. This leads to the following coupled equations:

$$A_{11}\left(\frac{\partial^2 u_0}{\partial x^2}+\frac{\partial^3 w_0}{\partial x^3}\right)-B_{11}\frac{\partial^3 w_0}{\partial x^3}=I_0\frac{\partial^2 u_0}{\partial t^2}-I_1\frac{\partial^2}{\partial t^2}\left(\frac{\partial w_0}{\partial x}\right) \tag{7.20}$$

$$B_{11}\left(\frac{\partial^3 u_0}{\partial x^3}+\frac{\partial^4 w_0}{\partial x^4}\right)-D_{11}\frac{\partial^4 w_0}{\partial x^4}=\bar{P}(w_0)+I_0\frac{\partial^2 w_0}{\partial t^2}-I_2\frac{\partial^2}{\partial t^2}\left(\frac{\partial^2 w_0}{\partial x^2}\right)+I_1\frac{\partial^2}{\partial t^2}\left(\frac{\partial u_0}{\partial x}\right) \tag{7.21}$$

7.2.1 Symmetric laminate ($B_{11} = 0, I_1 = 0$)

For the case of symmetric laminate, namely $B_{11} = 0$, and $I_1 = 0$, the following single equation is obtained, while eq. (7.20) (assuming the term $\frac{\partial^3 w_0}{\partial x^3}$ is neglected) is solved independently

$$-D_{11}\frac{\partial^4 w_0}{\partial x^4}-\bar{N}_{xx}\frac{\partial^2 w_0}{\partial x^2}=I_0\frac{\partial^2 w_0}{\partial t^2}-I_2\frac{\partial^2}{\partial t^2}\left(\frac{\partial^2 w_0}{\partial x^2}\right) \tag{7.22}$$

Assuming that $w_0(x,t) = W(x)e^{i\omega t}$, where ω is the circular natural frequency, and substituting into eq. (7.22) yields the following differential equation:

$$D_{11}\frac{d^4 W}{dx^4}+\left(\bar{N}_{xx}+\omega^2 I_2\right)\frac{d^2 W}{dx^2}-\omega^2 I_0\cdot W=0 \tag{7.23}$$

The solution of eq. (7.23) has the following general form:

$$W(x)=A_1\sinh(\alpha_1 x)+A_2\cosh(\alpha_1 x)+A_3\sin(\alpha_2 x)+A_4\cos(\alpha_2 x) \tag{7.24}$$

where A_1, A_2, A_3 and A_4 are constants to be determined using boundary conditions and the terms α_1 and α_2 are defined as

$$\begin{aligned}\alpha_1&=\sqrt{\frac{-\left(N_{xx}+\omega^2 I_2\right)+\sqrt{\left(\bar{N}_{xx}+\omega^2 I_2\right)^2+4D_{11}\omega^2 I_0}}{2D_{11}}}\\ \alpha_2&=\sqrt{\frac{\left(N_{xx}+\omega^2 I_2\right)+\sqrt{\left(\bar{N}_{xx}+\omega^2 I_2\right)^2+4D_{11}\omega^2 I_0}}{2D_{11}}}\end{aligned} \tag{7.25}$$

To investigate the influence of the compression load, $\bar{N}_{xx}$, and the rotary inertia, I_2, on the natural frequency, the circular natural frequency squared (ω^2) is rewritten using eq. (7.25) to yield

$$\omega^2 = \frac{D_{11}\alpha_1^4 + \bar{N}_{xx}\alpha_1^2}{I_0 - I_2\alpha_1^2} = \alpha_1^4 \frac{D_{11}}{I_0} \frac{\left(1 + \frac{\bar{N}_{xx}}{D_{11}\alpha_1^2}\right)}{\left(1 - \frac{I_2}{I_0}\alpha_1^2\right)}$$
$$\omega^2 = \frac{D_{11}\alpha_2^4 - \bar{N}_{xx}\alpha_2^2}{I_0 + I_2\alpha_2^2} = \alpha_2^4 \frac{D_{11}}{I_0} \frac{\left(1 - \frac{\bar{N}_{xx}}{D_{11}\alpha_2^2}\right)}{\left(1 + \frac{I_2}{I_0}\alpha_2^2\right)} \tag{7.26}$$

The expressions in eq. (7.26) are the same, and only one of the equations should be used, for instance, when α_2 is known. Also, it is clear from eq. (7.26) (the second expression) that the compressive load $\bar{N}_{xx}$ reduces the natural frequency, and the same tendency is obtained by including the rotary inertia, I_2. Assuming $\bar{N}_{xx} = I_2 = 0$, one obtains the known frequency equation used by Bernoulli–Euler theory, namely

$$\omega^2 = \alpha_2^4 \frac{D_{11}}{I_0} \tag{7.27}$$

which holds true also for an isotropic column, by letting the bending stiffness, to be written as $D_{11} \equiv D = Eh^3/[12(1-\upsilon^2)]$, where E is the Young's modulus, υ is the Poisson's ratio and h is the thickness of the column.

Let us show the application of boundary conditions in eq. (7.24) for the simply supported case. The boundary conditions for this case are

$$\begin{aligned} W(0) = 0; \quad M_{xx}(0) = 0 \quad &\Rightarrow \quad \frac{d^2W(0)}{dx^2} = 0 \\ W(L) = 0; \quad M_{xx}(L) = 0 \quad &\Rightarrow \quad \frac{d^2W(L)}{dx^2} = 0 \end{aligned} \tag{7.28}$$

Substituting the boundary conditions in eq. (7.24) yields a set of four equations with four unknowns, having the following matrix form:

$$\begin{bmatrix} 0 & 1 & 0 & 1 \\ 0 & \alpha_1^2 & 0 & -\alpha_2^2 \\ \sinh(\alpha_1 L) & \cosh(\alpha_2 L) & \sin(\alpha_2 L) & \cos(\alpha_2 L) \\ \alpha_1^2\sinh(\lambda L) & \alpha_1^2\cosh(\lambda L) & -\alpha_2^2\sin(\alpha_2 L) & -\alpha_2^2\cos(\alpha_2 L) \end{bmatrix} \begin{Bmatrix} C_1 \\ C_2 \\ C_3 \\ C_4 \end{Bmatrix} = \begin{Bmatrix} 0 \\ 0 \\ 0 \\ 0 \end{Bmatrix} \tag{7.29}$$

To obtain a unique solution, the determinant of the matrix appearing in eq. (7.29) must vanish. This leads to the following characteristic equation:

$$\sin(\alpha_2 L) = 0 \quad \Rightarrow \quad \alpha_2 L = n\pi,\ n = 1, 2, 3, 4, \ldots \quad \Rightarrow \quad \alpha_2^2 = \frac{n^2\pi^2}{L^2} \tag{7.30}$$

Substituting the result of eq. (7.30) into the second expression of eq. (7.26) leads to the natural frequencies of a symmetric layered column, having a length L, under compressive load, including rotary inertia on simply supported boundary conditions

$$\omega = \left(\frac{n\pi}{L}\right)^2 \sqrt{\frac{D_{11}}{I_0}} \sqrt{\frac{\left(1 - \frac{\bar{N}_{xx}}{D_{11}\left(\frac{n\pi}{L}\right)^2}\right)}{\left(1 + \frac{I_2}{I_0}\left(\frac{n\pi}{L}\right)^2\right)}}, \quad n = 1, 2, 3, 4, \ldots \tag{7.31}$$

Neglecting the rotary inertia leads to the following expression:

$$\omega = \left(\frac{n\pi}{L}\right)^2 \sqrt{\frac{D_{11}}{I_0}} \sqrt{\left(1 - \frac{\bar{N}_{xx}}{D_{11}\left(\frac{n\pi}{L}\right)^2}\right)}, \quad n = 1, 2, 3, 4, \ldots \tag{7.32}$$

showing that the axial compression load reduces the natural frequencies of a column, while for the case of no axial compression, but including the rotary inertia the expression has the following form:

$$\omega = \left(\frac{n\pi}{L}\right)^2 \sqrt{\frac{D_{11}}{I_0}} \sqrt{\frac{1}{\left(1 + \frac{I_2}{I_0}\left(\frac{n\pi}{L}\right)^2\right)}}, \quad n = 1, 2, 3, 4, \ldots \tag{7.33}$$

displaying the same tendency as before, namely the rotary inertial tends to reduce the natural frequencies of the column. For the simple case of a column, without axial compression and rotary inertia, the expression for the natural frequencies is like for an isotropic case using the classical beam theory.

Tab. 7.1: Characteristic equations and eigenvalues for natural vibrations of laminated composite columns using CLT approach.

No.	Name	Characteristic equation	Eigenvalues
1	SS-SS*	$\sin(\alpha_2 L) = 0$ $(\alpha_2 L)_n = n\pi, \quad n = 1, 2, 3, \ldots$	$\omega_n = \left(\frac{n\pi}{L}\right)^2 \sqrt{\frac{D_{11}}{I_0}}$
2	C-C**	$\cos(\alpha_2 L)\cosh(\alpha_2 L) = 1$ $(\alpha_2 L)_n = 4.73004, 7.85321, 10.9956, \ldots, \frac{(2n+1)\pi}{2}$	$\omega_1 = \left(\frac{4.73004}{L}\right)^2 \sqrt{\frac{D_{11}}{I_0}}$
3	C-F***	$\cos(\alpha_2 L)\cosh(\alpha_2 L) = -1$ $(\alpha_2 L)_n = 1.87510, 4.69409, 7.85340, \ldots, \frac{(2n-1)\pi}{2}$	$\omega_1 = \left(\frac{1.87351}{L}\right)^2 \sqrt{\frac{D_{11}}{I_0}}$
4	F-F	$\cos(\alpha_2 L)\cosh(\alpha_2 L) = 1$ $(\alpha_2 L)_n = 4.73004, 7.85321, 10.9956, \ldots, \frac{(2n+1)\pi}{2}$	$\omega_1 = \left(\frac{4.73004}{L}\right)^2 \sqrt{\frac{D_{11}}{I_0}}$
5	SS-C	$\tan(\alpha_2 L) = \tanh(\alpha_2 L)$ $(\alpha_2 L)_n = 3.9266, 7.0686, 10.2102, \ldots, \frac{(4n+1)\pi}{4}$	$\omega_1 = \left(\frac{3.9266}{L}\right)^2 \sqrt{\frac{D_{11}}{I_0}}$

Tab. 7.1 (continued)

No.	Name	Characteristic equation	Eigenvalues
6	SS-F	$\tan(a_2L) = \tanh(a_2L)$ $(a_2L)_n = 3.9266, 7.0686, 10.2102, \ldots, \frac{(4n+1)\pi}{4}$	$\omega_1 = \left(\frac{3.9266}{L}\right)^2 \sqrt{\frac{D_{11}}{I_0}}$
7	G****-F	$\tan(a_2L) = -\tanh(a_2L)$ $(a_2L)_n = 2.3650, 5.4978, 8.6394, \ldots, \frac{(4n-1)\pi}{4}$	$\omega_1 = \left(\frac{2.3650}{L}\right)^2 \sqrt{\frac{D_{11}}{I_0}}$
8	G-SS	$\cos(a_2L) = 0 (a_2L)_n = (2n-1)\frac{\pi}{2}, \quad n = 1, 2, 3, \ldots$	$\omega_n = \left[\frac{(2n-1)\pi}{2L}\right] \sqrt{\frac{D_{11}}{I_0}}$
9	G-G	$\sin(a_2L) = 0$ $(a_2L)_n = n\pi, \quad n = 1, 2, 3, \ldots$	$\omega_n = \left(\frac{n\pi}{L}\right)^2 \sqrt{\frac{D_{11}}{I_0}}$
10	G-C	$\tan(a_2L) = -\tanh(a_2L)$ $(a_2L)_n = 2.3650, 5.4978, 8.6394, \ldots, \frac{(4n-1)\pi}{4}$	$\omega_1 = \left(\frac{2.3650}{L}\right)^2 \sqrt{\frac{D_{11}}{I_0}}$

$$\omega_n = \left(\frac{n\pi}{L}\right)^2 \sqrt{\frac{D_{11}}{I_0}}, \quad n = 1, 2, 3, 4, \ldots \tag{7.34}$$

Tables 7.1 and 7.2 present some of the most encountered column cases having various boundary conditions (see similar tables in Chapter 6). The various expressions given in Tab. 7.1 are for the case without axial compression load and neglecting the rotary inertia term. For this case, eq. (7.24) simplifies into the following equation:

$$W(x) = A_1 \sinh(a_2x) + A_2 \cosh(a_2x) + A_3 \sin(a_2x) + A_4 \cos(a_2x) \tag{7.35}$$

Tab. 7.2: Mode shapes and their relevant eigenvalues for natural vibrations of laminated composite columns using CLT approach.

No.	Name	Mode shape	Eigenvalues
1	SS-SS	$W_n(x) = \sin\left(\frac{n\pi x}{L}\right)$	$\omega_n = \left(\frac{n\pi}{L}\right)^2 \sqrt{\frac{D_{11}}{I_0}}$
2[†]	C-C	$W_n(x) = \cosh\left[\frac{(a_2L)_n x}{L}\right] - \cos\left[\frac{(a_2L)_n x}{L}\right] - \frac{\cosh[(a_2L)_n] - \cos[(a_2L)_n]}{\sinh[(a_2L)_n] - \sin[(a_2L)_n]}\left\{\sinh\left[\frac{(a_2L)_n x}{L}\right] - \sin\left[\frac{(a_2L)_n x}{L}\right]\right\}$	$\omega_n = \left[\frac{(2n+1)\pi}{2L}\right]^2 \sqrt{\frac{D_{11}}{I_0}}$
3[†]	C-F	$W_n(x) = \cosh\left[\frac{(a_2L)_n x}{L}\right] - \cos\left[\frac{(a_2L)_n x}{L}\right] - \frac{\cosh[(a_2L)_n] + \cos[(a_2L)_n]}{\sinh[(a_2L)_n] + \sin[(a_2L)_n]}\left\{\sinh\left[\frac{(a_2L)_n x}{L}\right] - \sin\left[\frac{(a_2L)_n x}{L}\right]\right\}$	$\omega_n = \left[\frac{(2n-1)\pi}{2L}\right]^2 \sqrt{\frac{D_{11}}{I_0}}$

Tab. 7.2 (continued)

No.	Name	Mode shape	Eigenvalues
4[†]	F-F	$W_n(x) = \cosh\left[\frac{(\alpha_2 L)_n x}{L}\right] - \cos\left[\frac{(\alpha_2 L)_n x}{L}\right] - \frac{\cosh[(\alpha_2 L)_n] - \cos[(\alpha_2 L)_n]}{\sinh[(\alpha_2 L)_n] - \sin[(\alpha_2 L)_n]}\left\{\sinh\left[\frac{(\alpha_2 L)_n x}{L}\right] - \sin\left[\frac{(\alpha_2 L)_n x}{L}\right]\right\}$	$\omega_n = \left[\frac{(2n+1)\pi}{2L}\right]^2\sqrt{\frac{D_{11}}{I_0}}$
5[†]	SS-C	$W_n(x) = \cosh\left[\frac{(\alpha_2 L)_n x}{L}\right] - \cos\left[\frac{(\alpha_2 L)_n x}{L}\right] - \frac{\cosh[(\alpha_2 L)_n] - \cos[(\alpha_2 L)_n]}{\sinh[(\alpha_2 L)_n] - \sin[(\alpha_2 L)_n]}\left\{\sinh\left[\frac{(\alpha_2 L)_n x}{L}\right] - \sin\left[\frac{(\alpha_2 L)_n x}{L}\right]\right\}$	$\omega_n = \left[\frac{(4n+1)\pi}{4L}\right]^2\sqrt{\frac{D_{11}}{I_0}}$
6[†]	SS-F	$W_n(x) = \cosh\left[\frac{(\alpha_2 L)_n x}{L}\right] + \cos\left[\frac{(\alpha_2 L)_n x}{L}\right] - \frac{\cosh[(\alpha_2 L)_n] + \cos[(\alpha_2 L)_n]}{\sinh[(\alpha_2 L)_n] + \sin[(\alpha_2 L)_n]}\left\{\sinh\left[\frac{(\alpha_2 L)_n x}{L}\right] + \sin\left[\frac{(\alpha_2 L)_n x}{L}\right]\right\}$	$\omega_n = \left[\frac{(4n+1)\pi}{4L}\right]^2\sqrt{\frac{D_{11}}{I_0}}$
7[†]	G-F	$W_n(x) = \cosh\left[\frac{(\alpha_2 L)_n x}{L}\right] - \cos\left[\frac{(\alpha_2 L)_n x}{L}\right] - \frac{\cosh[(\alpha_2 L)_n] - \cos[(\alpha_2 L)_n]}{\sinh[(\alpha_2 L)_n] - \sin[(\alpha_2 L)_n]}\left\{\sinh\left[\frac{(\alpha_2 L)_n x}{L}\right] - \sin\left[\frac{(\alpha_2 L)_n x}{L}\right]\right\}$	$\omega_n = \left[\frac{(4n-1)\pi}{4L}\right]^2\sqrt{\frac{D_{11}}{I_0}}$
8	G-SS	$W_n(x) = \sin\left[\frac{(2n-1)\pi x}{2L}\right]$	$\omega_n = \left[\frac{(2n-1)\pi}{2L}\right]\sqrt{\frac{D_{11}}{I_0}}$
9	G-G	$W_n(x) = \cos\left(\frac{n\pi x}{L}\right)$	$\omega_n = \left(\frac{n\pi}{L}\right)^2\sqrt{\frac{D_{11}}{I_0}}$
10[†]	G-C	$W_n(x) = \cosh\left[\frac{(\alpha_2 L)_n x}{L}\right] - \cos\left[\frac{(\alpha_2 L)_n x}{L}\right] - \frac{\sinh[(\alpha_2 L)_n] + \sin[(\alpha_2 L)_n]}{\cosh[(\alpha_2 L)_n] - \cos[(\alpha_2 L)_n]}\left\{\sinh\left[\frac{(\alpha_2 L)_n x}{L}\right] - \sin\left[\frac{(\alpha_2 L)_n x}{L}\right]\right\}$	$\omega_n = \left[\frac{(4n-1)\pi}{4L}\right]^2\sqrt{\frac{D_{11}}{I_0}}$

*SS, simply supported; **C, clamped; ***F, free; ****G, guided.
[†]The expression for the eigenvalues is given for $n \gg 1$. For the first modes, see Tab. 7.1.

7.2.2 Nonsymmetric laminate ($B_{11} \neq 0, I_1 \neq 0$)

To solve the case of a nonsymmetric laminate, it is easier to present eqs. (7.20) and (7.21) in a matrix form, where higher-order terms are neglected, leading to

$$\begin{bmatrix} A_{11}\frac{\partial^2}{\partial x^2} & -B_{11}\frac{\partial^3}{\partial x^3} \\ -B_{11}\frac{\partial^3}{\partial x^3} & D_{11}\frac{\partial^4}{\partial x^4} + \bar{N}_{xx}\frac{\partial^2}{\partial x^2} \end{bmatrix}\begin{Bmatrix} u_0(x,t) \\ w_0(x,t) \end{Bmatrix} + \begin{bmatrix} -I_0 & I_1\frac{\partial}{\partial x} \\ I_1\frac{\partial}{\partial x} & I_0 - I_2\frac{\partial^2}{\partial x^2} \end{bmatrix}\frac{\partial^2}{\partial t^2}\begin{Bmatrix} u_0(x,t) \\ w_0(x,t) \end{Bmatrix} = \begin{Bmatrix} 0 \\ 0 \end{Bmatrix} \tag{7.36}$$

The matrix presentation in eq. (7.36) includes axial compression $\bar{N}_{xx}$, rotary inertia I_2, and the coupling mass moment of inertia I_1, which couples both structurally and dynamically the two assumed deflections, u_0 and w_0. There is no general solution for eq. (7.36). Only for simply supported boundary conditions, we can present the natural frequencies of a nonsymmetric beam in a closed form solution. For this case, we assume that the deflections u_0 and w_0 have the following form:

$$\begin{Bmatrix} u_0(x,t) \\ w_0(x,t) \end{Bmatrix} = \begin{Bmatrix} \sum_{m=1}^{M} U_m \cos\left(\frac{m\pi x}{L}\right) e^{i\omega t} \\ \sum_{m=1}^{M} W_m \sin\left(\frac{m\pi x}{L}\right) e^{i\omega t} \end{Bmatrix} \equiv \begin{Bmatrix} \sum_{m=1}^{M} U_m \cos(\lambda x) e^{i\omega t} \\ \sum_{m=1}^{M} W_m \sin(\lambda x) e^{i\omega t} \end{Bmatrix} \tag{7.37}$$

where $i = \sqrt{-1}$, ω^2 is the circular natural frequency squared and $\lambda = m\pi/L$. Substituting eq. (7.37) into eq. (7.36) yields

$$\begin{bmatrix} A_{11}\lambda^2 & -B_{11}\lambda^3 \\ -B_{11}\lambda^3 & D_{11}\lambda^4 - \bar{N}_{xx}\lambda^2 \end{bmatrix} \begin{Bmatrix} U_m \\ W_m \end{Bmatrix} + \begin{bmatrix} -\omega^2 I_0 & \omega^2 I_1 \lambda \\ \omega^2 I_1 \lambda & -\omega^2\left(I_0 - I_2\lambda^2\right) \end{bmatrix} \begin{Bmatrix} U_m \\ W_m \end{Bmatrix} = \begin{Bmatrix} 0 \\ 0 \end{Bmatrix} \tag{7.38}$$

Then eq. (7.38) can be casted in the following form:

$$\begin{bmatrix} A_{11}\lambda^2 - \omega^2 I_0 & -B_{11}\lambda^3 + \omega^2 I_1\lambda \\ -B_{11}\lambda^3 + \omega^2 I_1 \lambda & D_{11}\lambda^4 - \bar{N}_{xx}\lambda^2 - \omega^2\left(I_0 - I_2\lambda^2\right) \end{bmatrix} \begin{Bmatrix} U_m \\ W_m \end{Bmatrix} = \begin{Bmatrix} 0 \\ 0 \end{Bmatrix} \tag{7.39}$$

To obtain a unique solution, the determinant of the matrix in eq. (7.39) must vanish, leading to the following characteristic equation:

$$A\left(\omega^2\right)^2 + B\left(\omega^2\right) + C = 0 \tag{7.40}$$

where

$$\begin{aligned} A &\equiv I_1^2\lambda^2 - I_0\left(I_0 - I_2\lambda^2\right) \\ B &\equiv A_{11}\lambda^2\left(I_0 - I_2\lambda^2\right) + I_0\left(D_{11}\lambda^2 - \bar{N}_{xx}\right)\lambda^2 - 2I_1B_{11}\lambda^4 \\ C &\equiv \left[A_{11}\bar{N}_{xx} - \left(A_{11}D_{11} - B_{11}^2\right)\lambda^2\right]\lambda^4 \end{aligned} \tag{7.41}$$

Solution of eq. (7.40) would provide the natural frequency for the simply supported nonsymmetric case. For a symmetric case ($B_{11} = I_1 = 0$), eq. (7.39) will not be coupled and the result will be the expression presented in eq. (7.31).

Appendix A presents a procedure to solve the case of a nonsymmetric beam having other boundary conditions.

7.3 Vibrations of columns – FOSDT approach

As derived in Chapter 6, the equations of motion for a general laminate using the FOSDT approach (see also [9–14]) can be written as (assuming $v_0 = \phi_y = 0$)

$$A_{11}\left(\frac{\partial^2 u_0}{\partial x^2}+\frac{\partial^3 w_0}{\partial x^3}\right)+B_{11}\frac{\partial^2 \phi_x}{\partial x^2}=I_0\frac{\partial^2 u_0}{\partial t^2}+I_1\frac{\partial^2 \phi_x}{\partial t^2} \tag{7.42}$$

$$KA_{55}\left(\frac{\partial^2 w_0}{\partial x^2}+\frac{\partial \phi_x}{\partial x}\right)-\bar{N}_{xx}\frac{\partial^2 w_0}{\partial x^2}=q+I_0\frac{\partial^2 w_0}{\partial t^2} \tag{7.43}$$

$$B_{11}\left(\frac{\partial^2 u_0}{\partial x^2}+\frac{\partial^3 w_0}{\partial x^3}\right)+D_{11}\frac{\partial^2 \phi_x}{\partial x^2}-KA_{55}\left(\frac{\partial w_0}{\partial x}+\phi_x\right)=I_2\frac{\partial^2 \phi_x}{\partial t^2}+I_1\frac{\partial^2 u_0}{\partial t^2} \tag{7.44}$$

One should note that eqs. (7.42)–(7.44) are for the case of uniform properties along the beam. For the particular case of properties varying along the x coordinate of the beam, the reader is referred to [12–14]. To solve the vibration problem, the lateral load, q, is set to zero leading to the following three coupled equations of motion:

$$A_{11}\left(\frac{\partial^2 u_0}{\partial x^2}+\frac{\partial^3 w_0}{\partial x^3}\right)+B_{11}\frac{\partial^2 \phi_x}{\partial x^2}=I_0\frac{\partial^2 u_0}{\partial t^2}+I_1\frac{\partial^2 \phi_x}{\partial t^2} \tag{7.45}$$

$$KA_{55}\left(\frac{\partial^2 w_0}{\partial x^2}+\frac{\partial \phi_x}{\partial x}\right)-\bar{N}_{xx}\frac{\partial^2 w_0}{\partial x^2}=I_0\frac{\partial^2 w_0}{\partial t^2} \tag{7.46}$$

$$B_{11}\left(\frac{\partial^2 u_0}{\partial x^2}+\frac{\partial^3 w_0}{\partial x^3}\right)+D_{11}\frac{\partial^2 \phi_x}{\partial x^2}-KA_{55}\left(\frac{\partial w_0}{\partial x}+\phi_x\right)=I_2\frac{\partial^2 \phi_x}{\partial t^2}+I_1\frac{\partial^2 u_0}{\partial t^2} \tag{7.47}$$

Assuming harmonic vibrations with a circular squared frequency, ω^2, and neglecting higher-order terms, we can rewrite eqs. (7.45)–(7.47) as

$$A_{11}\frac{d^2U}{dx^2}+B_{11}\frac{d^2\Phi}{dx^2}+\omega^2 I_0 U+\omega^2 I_1\Phi=0 \tag{7.48}$$

$$KA_{55}\left(\frac{d^2W}{dx^2}+\frac{d\Phi}{dx}\right)-\bar{N}_{xx}\frac{d^2W}{dx^2}+\omega^2 I_0 W=0 \tag{7.49}$$

$$B_{11}\frac{d^2U}{dx^2}+D_{11}\frac{d^2\Phi}{dx^2}-KA_{55}\left(\frac{dW}{dx}+\Phi\right)+\omega^2 I_2\Phi+\omega^2 I_1 U=0 \tag{7.50}$$

For symmetric laminates, we can write the following expressions:

$$A_{11}\frac{d^2U}{dx^2}+\omega^2 I_0 U=0 \tag{7.51}$$

$$KA_{55}\left(\frac{d^2W}{dx^2}+\frac{d\Phi}{dx}\right)-\bar{N}_{xx}\frac{d^2W}{dx^2}+\omega^2 I_0 W=0 \tag{7.52}$$

$$D_{11}\frac{d^2\Phi}{dx^2} - KA_{55}\left(\frac{dW}{dx} + \Phi\right) + \omega^2 I_2 \Phi = 0 \tag{7.53}$$

One should note that for the symmetric case, the first equation eq. (7.51) is not coupled with the other two equations, eqs. (7.52) and (7.53), which are coupled and have to be solved together.

7.3.1 Symmetric laminate ($B_{11} = 0$, $I_1 = 0$)

First, we shall present a general solution for the symmetric case presented by the two coupled equations (7.52) and (7.53). The two equations are decoupled to yield uncoupled equations

$$\left[D_{11} - \frac{\bar{N}_{xx}}{KA_{55}}\right]\frac{d^4W}{dx^4} + \left\{\omega^2\left[\frac{D_{11}I_0}{KA_{55}} + \left(1 - \frac{\bar{N}_{xx}}{KA_{55}}\right)I_2 + \bar{N}_{xx}\right]\right\}\frac{d^2W}{dx^2} + \omega^2 I_0\left[\frac{I_2\omega^2}{KA_{55}} - 1\right]W = 0 \tag{7.54}$$

$$\left[D_{11} - \frac{\bar{N}_{xx}}{KA_{55}}\right]\frac{d^4\Phi}{dx^4} + \left\{\omega^2\left[\frac{D_{11}I_0}{KA_{55}} + \left(1 - \frac{\bar{N}_{xx}}{KA_{55}}\right)I_2 + \bar{N}_{xx}\right]\right\}\frac{d^2\Phi}{dx^2} + \omega^2 I_0\left[\frac{I_2\omega^2}{KA_{55}} - 1\right]\Phi = 0 \tag{7.55}$$

One should note that without axial compression, $\bar{N}_{xx} = 0$ eqs. (7.54) and (7.55) will have the following form which is similar to what is given in [10]

$$D_{11}\frac{d^4W}{dx^4} + \omega^2\left[\frac{D_{11}I_0}{KA_{55}} + I_2\right]\frac{d^2W}{dx^2} + \omega^2 I_0\left[\frac{I_2\omega^2}{KA_{55}} - 1\right]W = 0 \tag{7.56}$$

$$D_{11}\frac{d^4\Phi}{dx^4} + \omega^2\left[\frac{D_{11}I_0}{KA_{55}} + I_2\right]\frac{d^2\Phi}{dx^2} + \omega^2 I_0\left[\frac{I_2\omega^2}{KA_{55}} - 1\right]\Phi = 0 \tag{7.57}$$

The general solutions for eqs. (7.54) and (7.55) (as well as for eqs. (7.56) and (7.57)) have the following form:

$$W(x) = A_1\cosh(s_1x) + A_2\sinh(s_1x) + A_3\cos(s_2x) + A_4\sin(s_2x) \tag{7.58}$$

$$\Phi(x) = B_1\cosh(s_1x) + B_2\sinh(s_1x) + B_3\cos(s_2x) + B_4\sin(s_2x) \tag{7.59}$$

while the constants A_1, A_2, A_3, A_4 and B_1, B_2, B_3, B_4 are interconnected by back-substituting eqs. (7.58) and (7.59) into the coupled equations (7.52) and (7.53) and have the following form:

$$B_1 = \frac{KA_{55}\cdot s_1}{D_{11}\cdot s_1^2 - KA_{55} + \omega^2 I_2}A_2 \equiv \alpha A_2 \tag{7.60}$$

$$B_2 = \frac{KA_{55} \cdot s_1}{D_{11} \cdot s_1^2 - KA_{55} + \omega^2 I_2} A_1 \equiv \alpha A_1 \tag{7.61}$$

$$B_3 = \frac{KA_{55} \cdot s_2}{\omega^2 I_2 - D_{11} \cdot s_1^2 - KA_{55}} A_4 \equiv \beta A_4 \tag{7.62}$$

$$B_4 = -\frac{KA_{55} \cdot s_2}{\omega^2 I_2 - D_{11} \cdot s_1^2 - KA_{55}} A_3 \equiv -\beta A_3 \tag{7.63}$$

and

$$s_1 = \sqrt{-\frac{b}{2a} + \frac{\sqrt{(b^2 - 4ac)}}{2a}} \tag{7.64}$$

$$s_2 = \sqrt{+\frac{b}{2a} + \frac{\sqrt{(b^2 - 4ac)}}{2a}} \tag{7.65}$$

where

$$a \equiv D_{11} - \frac{\bar{N}_{xx}}{KA_{55}}; \quad b \equiv \omega^2 \left[\frac{D_{11} I_0}{KA_{55}} + \left(1 - \frac{\bar{N}_{xx}}{KA_{55}}\right) I_2 + \bar{N}_{xx}\right]; \quad c \equiv \omega^2 I_0 \left[\frac{I_2 \omega^2}{KA_{55}} - 1\right] \tag{7.66}$$

The remaining constants A_1, A_2, A_3 and A_4 will be found after imposing the relevant boundary conditions.

The general boundary conditions for a nonsymmetric laminate are given in Tab. 7.3 and eq. (7.67).

Tab. 7.3: Out-of-plane boundary conditions for a nonsymmetric laminate column.

Name	Boundary conditions
Simply supported (or hinged) end	$W = 0$ and $B_{11} \frac{dU}{dx} + D_{11} \frac{d\Phi}{dx} = 0$
Clamped end	$W = 0$ and $\Phi = 0$
Free end	$B_{11} \frac{dU}{dx} + D_{11} \frac{d\Phi}{dx} = 0$ and $KA_{55}\left(\frac{dW}{dx} + \Phi\right) - \bar{N}_{xx} \frac{dW}{dx} = 0$
Guided end	$\Phi = 0$ and $KA_{55}\left(\frac{dW}{dx} + \Phi\right) - \bar{N}_{xx} \frac{dW}{dx} = 0$

The in-plane boundary condition for the nonsymmetric laminate is given as

$$A_{11} \frac{dU}{dx} + B_{11} \frac{d\Phi}{dx} = -\bar{N}_{xx} \quad \text{or} \quad U = 0 \tag{7.67}$$

One should note that for a symmetric case without axial compression load the expressions for the boundary condition simplify and are given in Tab. 7.4 and eq. (7.68).

Tab. 7.4: Out-of-plane boundary conditions for a symmetric laminate axial uncompressed column.

Name	Boundary conditions
Simply supported (or hinged) end	$W=0$ and $\frac{d\Phi}{dx}=0$
Clamped end	$W=0$ and $\Phi=0$
Free end	$\frac{d\Phi}{dx}=0$ and $\frac{dW}{dx}+\Phi=0$
Guided end	$\Phi=0$ and $\frac{dW}{dx}+\Phi=0$

The in-plane boundary condition for the symmetric laminate axial uncompressed column is given as follows:

$$\frac{dU}{dx}=0 \quad \text{or} \quad U=0 \tag{7.68}$$

Table 7.5 presents the characteristic equations for various boundary conditions for symmetric laminate axial uncompressed columns.

Tab. 7.5: Characteristic equations for natural vibrations of laminated composite columns using FOSDT approach.

No.	Name	Boundary conditions	Characteristic equation
1	SS-SS*	$W(0)=0 \quad d\Phi(0)/dx=0$ $W(L)=0 \quad d\Phi(L)/dx=0$	$\sin(s_2L)=0$ $(s_2L)_n=n\pi, \quad n=1,2,3,\ldots$
2	C-C**	$W(0)=0 \quad \Phi(0)=0$ $W(L)=0 \quad \Phi(L)=0$	$2\alpha-2\alpha\cos(s_2L)\cosh(s_1L)+$ $\left(\frac{\alpha^2-\beta^2}{\beta}\right)\sin(s_2L)\sinh(s_1L)=0$
3	C-F***	$W(0)=0 \quad \Phi(0)=0$ $d\Phi(L)/dx=0$ $dW(L)/dx+\Phi(L)=0$	$\alpha[s_1(s_1-\alpha)-s_2(s_2+\beta)]+$ $\left[s_1s_2\left(\frac{\beta-\alpha^2}{\beta}\right)+\alpha(\beta s_2-\alpha s_1)\right]\cosh(s_1L)\cos(s_2L)+$ $-\alpha(2s_1s_2+s_1\beta+s_2\alpha)\cosh(s_1L)\cos(s_2L)=0$
4	F-F	$d\Phi(0)/dx=0$ $dW(0)/dx+\Phi(0)=0$ $d\Phi(L)/dx=0$ $dW(L)/dx+\Phi(L)=0$	$2\alpha s_1(s_1+\alpha)-2\cos(s_2L)\cosh(s_1L)+$ $\frac{\beta^2s_2^2(s_1+\alpha)^2-\alpha^2s_1^2(s_2+\beta)^2}{\beta s_2(s_2+\beta)}\sin(s_2L)\sinh(s_1L)=0$

Tab. 7.5 (continued)

No.	Name	Boundary conditions	Characteristic equation
5	SS-C	$W(0)=0 \quad d\Phi(0)/dx=0$ $W(L)=0 \quad \Phi(L)=0$	$\beta\tanh(s_1L)=\alpha\tan(s_2L)$
6	SS-F	$W(0)=0 \quad d\Phi(0)/dx=0$ $d\Phi(L)/dx=0$ $dW(L)/dx+\Phi(L)=0$	$s_1\alpha(s_2+\beta)\tanh(s_1L)=-s_2\beta(s_1+\alpha)\tan(s_2L)$
7	G****-F	$\Phi(0)=0$ $dW(0)dx+\Phi(0)=0$ $d\Phi(L)/dx=0$ $dW(L)/dx+\Phi(L)=0$	$s_1\alpha(s_2+\beta)\tan(s_2L)=s_2\beta(s_1+\alpha)\tanh(s_1L)$
8	G-SS	$\Phi(0)=0$ $dW(0)dx+\Phi(0)=0$ $W(L)=0 \quad d\Phi(L)/dx=0$	$\cos(s_2L)=0(s_2L)_n=(2n-1)\frac{\pi}{2} \qquad n=1,2,3,\ldots$
9	G-G	$\Phi(0)=0$ $dW(0)dx+\Phi(0)=0$ $\Phi(L)=0$ $dW(L)dx+\Phi(L)=0$	$\sin(s_2L)=0$ $(s_2L)_n=n\pi, \quad n=1,2,3,\ldots$
10	G-C	$\Phi(0)=0$ $dW(0)dx+\Phi(0)=0$ $W(L)=0 \quad \Phi(L)=0$	$\alpha\tanh(s_1L)=-\beta\tan(s_2L)$

*SS, simply supported; **C, clamped; ***F, free; ****G, guided.

7.3.2 Nonsymmetric laminate ($B_{11} \neq 0, I_1 \neq 0$)

The way to solve a nonsymmetric laminate for the case of beams or columns, we shall follow the derivation performed by Abramovich and Livshits [17].

The equations of motion (7.45)–(7.47) will be presented in a matrix formulation (for $\bar{N}_{xx}=0$), namely

$$\begin{bmatrix} A_{11}\frac{\partial^2}{\partial x^2} & 0 & B_{11}\frac{\partial^2}{\partial x^2} \\ 0 & KA_{55}\frac{\partial^2}{\partial x^2} & KA_{55}\frac{\partial}{\partial x} \\ B_{11}\frac{\partial^2}{\partial x^2} & -KA_{55}\frac{\partial}{\partial x} & \left(D_{11}\frac{\partial^2}{\partial x^2} - KA_{55}\right) \end{bmatrix} \begin{Bmatrix} u_0(x,t) \\ w_0(x,t) \\ \phi_x(x,t) \end{Bmatrix} + \begin{bmatrix} -I_0 & 0 & -I_1 \\ 0 & -I_0 & 0 \\ -I_1 & 0 & -I_2 \end{bmatrix} \frac{\partial^2}{\partial t^2} \begin{Bmatrix} u_0(x,t) \\ w_0(x,t) \\ \phi_x(x,t) \end{Bmatrix} = \begin{Bmatrix} 0 \\ 0 \\ 0 \end{Bmatrix} \quad (7.69)$$

The associated boundary conditions are given by eq. (7.63) and Tab. 7.3. Following Ref. [17], nondimensional displacements and beam's length are defined as

$$\{\bar{q}\} = \{u_0/L, w_0/L, \phi_x\}^T, \qquad \xi \equiv x/L \quad (7.70)$$

yielding

$$\{q\} = \text{diagonal}\{L, L, 1\}\{\bar{q}\} \quad (7.71)$$

To obtain expressions for free vibrations, we assume that the solution for eq. (7.69) has the following form:

$$\{\bar{q}\} = \{U, W, \Phi\}^T e^{i\omega t} \equiv \{Q\} e^{i\omega t} \quad (7.72)$$

Substituting eq. (7.72) into eq. (7.69) leads to the following matrix equation:

$$\left[\begin{bmatrix} A_{11}\frac{\partial^2}{\partial x^2} & 0 & B_{11}\frac{\partial^2}{\partial x^2} \\ 0 & KA_{55}\frac{\partial^2}{\partial x^2} & KA_{55}\frac{\partial}{\partial x} \\ B_{11}\frac{\partial^2}{\partial x^2} & -KA_{55}\frac{\partial}{\partial x} & \left(D_{11}\frac{\partial^2}{\partial x^2} - KA_{55}\right) \end{bmatrix} + \omega^2 \begin{bmatrix} I_0 & 0 & I_1 \\ 0 & I_0 & 0 \\ I_1 & 0 & I_2 \end{bmatrix}\right] \begin{Bmatrix} L\cdot U \\ L\cdot W \\ \Phi \end{Bmatrix} = \begin{Bmatrix} 0 \\ 0 \\ 0 \end{Bmatrix} \quad (7.73)$$

Introducing nondimensional parameters and after some algebraic transformation, eq. (7.73) can be written as

$$\left[\begin{bmatrix} \zeta_1^2 & 0 & \zeta^2 \\ 0 & 1 & 0 \\ \zeta^2 b^2 & 0 & b^2 \end{bmatrix} \frac{\partial^2}{\partial \xi^2} + \begin{bmatrix} 0 & 0 & 0 \\ 0 & 0 & 1 \\ 0 & -1 & 0 \end{bmatrix} \frac{\partial}{\partial \xi} + \begin{bmatrix} p^2 & 0 & \eta^2 p^2 \\ 0 & b^2 p^2 & 0 \\ \eta^2 b^2 p^2 & 0 & (r^2 b^2 p^2 - 1) \end{bmatrix}\right] \begin{Bmatrix} U \\ W \\ \Phi \end{Bmatrix} = 0 \quad (7.74)$$

where the various parameters are defined as

$$p^2 \equiv \frac{\omega^2 I_0 L^4}{D_{11}}, \quad b^2 \equiv \frac{D_{11}}{KA_{55}L^2}, \quad \zeta^2 \equiv \frac{B_{11}L}{D_{11}}, \quad \zeta_1^2 \equiv \frac{A_{11}L^2}{D_{11}}, \quad r^2 \equiv \frac{I_2}{I_0 L^2}, \quad \eta^2 \equiv \frac{I_1}{I_0 L} \tag{7.75}$$

Assuming that the general solution for eq. (7.74) has the following form:

$$\{Q\} = \{\bar{Q}\}e^{im\xi} \tag{7.76}$$

with m being the eigenvalue and $\{\bar{Q}\}$ the eigenvector, substituting into eq. (7.74) leads to the following cubic algebraic equation:

$$As^3 + Bs^2 + Cs + D = 0, \quad s \equiv \frac{m^2}{p^2} \tag{7.77}$$

with

$$\begin{aligned} &A \equiv \zeta_1^2 - \zeta^4, \quad B \equiv 1 + (r^2 + b^2)\zeta_1^2 - \zeta^2\left(2\eta^2 + \zeta^2 b^2\right) \\ &C \equiv \zeta_1^2\left(r^2 b^2 - \frac{1}{p^2}\right) + (r^2 + b^2) - \eta^2\left(\eta^2 + 2\zeta^2 b^2\right) \\ &D \equiv r^2 b^2 - \frac{1}{p^2} - b^2\eta^4 \end{aligned} \tag{7.78}$$

Solving eq. (7.77), we can write the general solution for eq. (7.74) in the following form (see a detailed discussion regarding the format of the solution in [17]):

$$\begin{aligned} U &= A_1\gamma\mu \sinh(m_1\xi) + A_2\gamma\mu \cosh(m_1\xi) + A_3\lambda\delta \sin(m_2\xi) \\ &\quad - A_4\lambda\delta \cos(m_2\xi) + A_5\alpha\beta \sin(m_3\xi) - A_6\alpha\beta \cos(m_3\xi) \\ W &= A_1 \cosh(m_1\xi) + A_2 \sinh(m_1\xi) + A_3 \cos(m_2\xi) \\ &\quad + A_4 \sin(m_2\xi) + A_5 \cos(m_3\xi) + A_6 \sin(m_3\xi) \\ \Phi &= -A_1\mu \sinh(m_1\xi) - A_2\mu \cosh(m_1\xi) - A_3\lambda \sin(m_2\xi) \\ &\quad + A_4\lambda \cos(m_2\xi) - A_5\alpha \sin(m_3\xi) + A_6\alpha \cos(m_3\xi) \end{aligned} \tag{7.79}$$

where the various terms in eq. (7.79) are defined as follows:

$$m_1 \equiv \sqrt{s_1}p, \qquad m_2 \equiv \sqrt{-s_2}p, \qquad m_3 \equiv \sqrt{-s_3}p$$
$$\gamma \equiv \frac{\zeta^2 s_1 + \eta^2}{\zeta_1^2 s_1 + 1}, \qquad \delta \equiv \frac{\zeta^2 s_2 + \eta^2}{\zeta_1^2 s_2 + 1}, \qquad \beta \equiv \frac{\zeta^2 s_3 + \eta^2}{\zeta_1^2 s_3 + 1} \tag{7.80}$$
$$\mu \equiv \frac{s_1 + b^2}{\sqrt{s_1}}p, \qquad \lambda \equiv \frac{s_2 + b^2}{\sqrt{-s_2}}p, \qquad \alpha \equiv \frac{s_3 + b^2}{\sqrt{-s_3}}p$$

Imposing the adequate boundary conditions would lead to the finding of both the eigenvalues (the natural frequencies) and the eigenvectors (the modes of vibration). For this problem, six boundary conditions should be imposed as presented in Tab. 7.6.

Tab. 7.6: Boundary conditions for the nonsymmetrical laminated columns using the FOSDT approach.

Name	Boundary conditions
Clamped immovable end	$U = W = \Phi = 0$
Clamped movable end	$\zeta_1^2 \frac{dU}{d\xi} + \zeta^2 \frac{d\Phi}{d\xi} = W = \Phi = 0$
Simply supported immovable end	$\zeta^2 \frac{dU}{d\xi} + \frac{d\Phi}{d\xi} = W = U = 0$
Simply supported movable end	$\zeta^2 \frac{dU}{d\xi} + \frac{d\Phi}{d\xi} = W = \zeta_1^2 \frac{dU}{d\xi} + \zeta^2 \frac{d\Phi}{d\xi} = 0$
Free end	$\zeta^2 \frac{dU}{d\xi} + \frac{d\Phi}{d\xi} = \frac{1}{b^2}\left(\Phi + \frac{dW}{d\xi}\right) = \zeta_1^2 \frac{dU}{d\xi} + \zeta^2 \frac{d\Phi}{d\xi} = 0$

One should note (see also [17]) that only for a column or a beam with two simply supported (hinged) movable ends, an analytical solution exists. The characteristic equation obtained after demanding the determinant of the coefficients in eq. (7.79) has the following form:

$$\sin(m_3)\sin(m_2) = 0 \tag{7.81}$$

The solutions for eq. (7.81) have two series: the first one having bending dominated vibrations

$$m_3 = k\pi, \qquad k = 1, 2, 3, \ldots, n \tag{7.82}$$

while the second series has longitudinal dominated vibrations

$$m_2 = k\pi, \qquad k = 1, 2, 3, \ldots, n \tag{7.83}$$

7.4 Vibrations of plates – CLPT approach

7.4.1 Simply supported special orthotropic plates

The first case to be solved is sometimes called special orthotropic plates for which the bending–stretching coupling terms B_{ij} and the bending–twisting coefficients D_{16} and D_{26} are set to zero. Then taking eq. (6.77) from Chapter 6 assuming symmetry ($B_{ij} = D_{16} = D_{26} = I_1 = 0$) and zeroing the in-plane and out-of-plane loads, one obtains:

$$D_{11}\frac{\partial^4 w_0}{\partial x^4} + 2(D_{12} + 2D_{66})\frac{\partial^4 w_0}{\partial x^2 \partial y^2} + D_{22}\frac{\partial^4 w_0}{\partial y^4} + I_0\frac{\partial^2 w_0}{\partial t^2} - I_2\frac{\partial^2}{\partial t^2}\left(\frac{\partial^2 w_0}{\partial x^2} + \frac{\partial^2 w_0}{\partial y^2}\right) = 0 \quad (7.84)$$

where I_0 and I_2 are defined in eq. (7.2). Let us assume the following solution for the out-of-plane displacement w_0 (tacitly assuming harmonic vibrations with a circular frequency ω)

$$w_0(x, y, t) = W_{mn}\sin\left(\frac{m\pi x}{a}\right)\left(\frac{n\pi y}{b}\right)e^{i\omega t} \quad (7.85)$$

where a is the length and b is the width of the plate having a total thickness of t, and substituting in eq. (7.84) we get

$$D_{11}\left(\frac{m\pi}{a}\right)^4 + 2(D_{12} + 2D_{66})\left(\frac{m\pi}{a}\right)^2\left(\frac{n\pi}{b}\right)^2 + D_{22}\left(\frac{n\pi}{b}\right)^4 - \omega^2\left\{I_0 + I_2\left[\left(\frac{m\pi}{a}\right)^2 + \left(\frac{n\pi}{b}\right)^2\right]\right\} = 0 \quad (7.86)$$

The solution for eq. (7.86) has the following form, which presents the natural frequencies for a special orthotropic laminated plate

$$\omega_{mn}^2 = \left(\frac{\pi}{a}\right)^4 \frac{D_{11}m^4 + 2(D_{12} + 2D_{66})m^2n^2\left(\frac{a}{b}\right)^2 + D_{22}n^4\left(\frac{a}{b}\right)^4}{I_0 + I_2\left(\frac{\pi}{a}\right)^2\left[m^2 + n^2\left(\frac{a}{b}\right)^2\right]} \quad (7.87)$$

while the mode of vibration is given by

$$w_0(x, y) = W_{mn}\sin\left(\frac{m\pi x}{a}\right)\left(\frac{n\pi y}{b}\right) \quad (7.88)$$

One can see from eq. (7.87) that the inclusion of the rotary inertia I_2 tends to reduce the natural frequencies. For a square plate, $a = b$, when neglecting the rotary inertia, the general frequency would show as

$$\omega_{mn}^2 = \frac{\pi^4}{I_0 a^4}\left[D_{11}m^4 + 2(D_{12} + 2D_{66})m^2n^2 + D_{22}n^4\right] \quad (7.89)$$

The lowest frequency, sometimes called also the fundamental frequency, will occur at $m = n = 1$; namely, for a rectangular plate we will have

$$\omega_{11}^2 = \frac{\pi^4}{I_0 a^4}\left[D_{11} + 2(D_{12} + 2D_{66})\left(\frac{a}{b}\right)^2 + D_{22}\left(\frac{a}{b}\right)^4\right] \tag{7.90}$$

7.4.2 Simply supported on two opposite edges of special orthotropic plates

Using the Lévy method,[1] one can solve the vibrations of rectangular plates, where two opposite edges are on simply supported boundary conditions, while the other two sides can be either clamped, free or any other combination of boundaries. Therefore, the equilibrium equation, without in-plane loads, is given by eq. (7.84). The out-of-plane deflections can be written as

$$w_0(x, y, t) = W_m(x)\left(\frac{n\pi y}{b}\right)e^{i\omega t} \tag{7.91}$$

Substituting eq. (7.91) into (7.84) leads to the following differential equation:

$$D_{11}\frac{d^4 W_m}{dx^4} + \left[\omega^2 I_2 - 2(D_{12} + 2D_{66})\left(\frac{n\pi}{b}\right)^2\right]\frac{d^2 W_m}{dx^2} - \left\{\omega^2\left[I_0 + I_2\left(\frac{n\pi}{b}\right)^2\right] - D_{22}\left(\frac{n\pi}{b}\right)^4\right\}W_m = 0 \tag{7.92}$$

or

$$\hat{a}\frac{d^4 W_m}{dx^4} + \hat{b}\frac{d^2 W_m}{dx^2} - \hat{c}W_m = 0 \tag{7.93}$$

where

$$\hat{a} \equiv D_{11}, \quad b \hat{\equiv} \omega^2 I_2 - 2(D_{12} + 2D_{66})\left(\frac{n\pi}{b}\right)^2, \quad c \equiv \omega^2\left[I_0 + I_2\left(\frac{n\pi}{b}\right)^2\right] - D_{22}\left(\frac{n\pi}{b}\right)^4 \tag{7.94}$$

The general solution of eq. (7.92) has the following form:

$$W(x) = A_1 \sinh(\alpha_1 x) + A_2 \cosh(\alpha_1 x) + A_3 \sin(\alpha_2 x) + A_4 \cos(\alpha_2 x) \tag{7.95}$$

where

$$s_1 = \sqrt{\frac{-\hat{b} + \sqrt{\hat{b}^2 + 4\hat{a}\hat{c}}}{2\hat{a}}}, \quad s_2 = \sqrt{\frac{\hat{b} + \sqrt{\hat{b}^2 + 4\hat{a}\hat{c}}}{2\hat{a}}} \tag{7.96}$$

Application of the boundary conditions in the x direction will lead to the natural frequencies and their associated mode shapes in the x direction.

1 M. Lévy, Memoire sur la theorie des plaques elastiques planes, *J Math Pures Appl*, Vol. 3, p. 219, 1899.

7.5 Vibrations of cylindrical shells

The topic of shells is presented in detail in Chapter 6, which dealt with buckling of cylindrical shells under various loadings. This chapter presents ways of calculating natural frequencies and their associated mode shapes according to different theories available in the literature. Typical studies on this topic can be found in [18–27].

7.5.1 Isotropic shells

Figure 7.1 presents the dimensions of a typical isotropic circular cylindrical shell, for which the natural frequencies would be calculated.

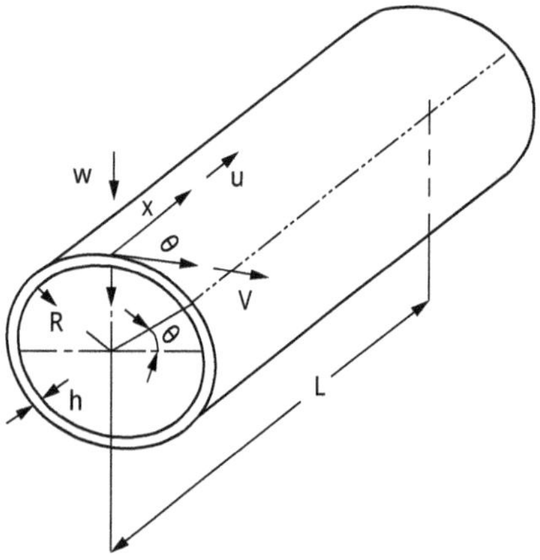

Fig. 7.1: A circular cylindrical shell: a schematic drawing.

Assuming the displacement field described in Fig. 7.1, and following the recommendations given in [25, 26], the governing equations of motion leading to the calculation of natural frequencies and their respective model shapes for a cylindrical shell can be presented in a matrix form:

$$\begin{bmatrix} L_{11} & L_{12} & L_{13} \\ L_{21} & L_{22} & L_{23} \\ L_{31} & L_{32} & L_{33} \end{bmatrix} \begin{Bmatrix} u(x,\theta,t) \\ v(x,\theta,t) \\ w(x,\theta,t) \end{Bmatrix} = \begin{Bmatrix} 0 \\ 0 \\ 0 \end{Bmatrix} \tag{7.97}$$

where

$$L_{31} = -L_{13}, \quad L_{32} = -L_{23}$$

with $L_{i,j}$, $i,j = 1,2,3$ being differential operators with respect to variables x, θ and t. The various differential operators would depend on the type of shell theory selected to solve the given problem.

To solve the problem, the first approximation for the axial, circumferential and radial displacements should be assumed in the following form (see also [23–25]):

$$\begin{aligned} u(x,\theta,t) &= \alpha e^{\lambda_n x}\sin(m\theta)e^{i\omega t} \\ v(x,\theta,t) &= \beta e^{\lambda_n x}\cos(m\theta)e^{i\omega t} \\ w(x,\theta,t) &= \gamma e^{\lambda_n x}\sin(m\theta)e^{i\omega t} \end{aligned} \tag{7.98}$$

where $\lambda_n = n\pi/L$ and n and m are half-wave numbers in the axial and circumferential directions, respectively, α, β and γ are constants to be determined, and ω is the circular frequency of the natural free vibration. Substituting eq. (7.98) into (7.97) leads to the following set of homogeneous equations, written in a matrix form:

$$\begin{bmatrix} \Gamma_{11} & \Gamma_{12} & \Gamma_{13} \\ \Gamma_{21} & \Gamma_{22} & \Gamma_{23} \\ \Gamma_{31} & \Gamma_{32} & \Gamma_{33} \end{bmatrix} \begin{Bmatrix} \alpha \\ \beta \\ \gamma \end{Bmatrix} = \begin{Bmatrix} 0 \\ 0 \\ 0 \end{Bmatrix} \quad \text{and} \quad \Gamma_{21} = -\Gamma_{12}, \quad \Gamma_{31} = -\Gamma_{13} \tag{7.99}$$

Note that the terms A_{ij}, $i,j = 1,2,3$ are functions of m, λ_n and Φ, with the last term being called the frequency parameter, defined as

$$\Phi^2 = \frac{(1-\upsilon^2)}{E}\rho R^2\omega^2 \tag{7.100}$$

The general matrix $[\Gamma_{ij}]$ is then presented for various available shell theories:

a. The form of the matrix for Donnell–Mushtari shell theory [25] is

$$[\Gamma_{ij}]^{D.-M.} = \begin{bmatrix} \Phi^2 + \lambda_n^2 - \frac{1-\upsilon}{2}m^2 & -\frac{1+\upsilon}{2}m\lambda_n & \upsilon\lambda_n \\ \frac{1+\upsilon}{2}m\lambda_n & \Phi^2 + \frac{1-\upsilon}{2}\lambda_n^2 - m^2 & m \\ -\upsilon\lambda_n & m & \Phi^2 - \left[1 + k(\lambda_n^2 - m^2)^2\right] \end{bmatrix} \tag{7.101}$$

where $k \equiv h^2/(12R^2)$ is the nondimensional thickness parameter. Note that this parameter is very small for small h/R ratios.

b. The form of the matrix for Love–Timoshenko shell theory [25] is

$$[\Gamma_{ij}]^{L.-T.} = \begin{bmatrix} \Phi^2+\lambda_n^2-\frac{1-\upsilon}{2}m^2 & -\frac{1+\upsilon}{2}m\lambda_n & \upsilon\lambda_n \\ \frac{1+\upsilon}{2}m\lambda_n & \Phi^2+(1+2k)\frac{1-\upsilon}{2}\lambda_n^2-(1+k)m^2 & m+mk\left(m^2-\lambda_n^2\right) \\ -\upsilon\lambda_n & m+mk\left(m^2-\lambda_n^2\right) & \Phi^2-\left[1+k(\lambda_n^2-m^2)^2\right] \end{bmatrix} \tag{7.102}$$

c. The form of the matrix for Arnold–Warburton shell theory [25] is

$$[\Gamma_{ij}]^{A.-W.} = \begin{bmatrix} \Phi^2+\lambda_n^2-\frac{1-\upsilon}{2}m^2 & -\frac{1+\upsilon}{2}m\lambda_n & \upsilon\lambda_n \\ \frac{1+\upsilon}{2}m\lambda_n & \Phi^2+(1+4k)\frac{1-\upsilon}{2}\lambda_n^2-(1+k)m^2 & m+mk\left[m^2-(2-\upsilon)\lambda_n^2\right] \\ -\upsilon\lambda_n & m+mk\left[m^2-(2-\upsilon)\lambda_n^2\right] & \Phi^2-\left[1+k\left(\lambda_n^2-m^2\right)^2\right] \end{bmatrix} \tag{7.103}$$

d. The form of the matrix for Houghton–Jones shell theory [25] is

$$[\Gamma_{ij}]^{H.-J.} = \begin{bmatrix} \Phi^2+\lambda_n^2-\frac{1-\upsilon}{2}m^2 & -\frac{1+\upsilon}{2}m\lambda_n & \upsilon\lambda_n \\ \frac{1+\upsilon}{2}m\lambda_n & \Phi^2+\frac{1-\upsilon}{2}\lambda_n^2-m^2 & m+mk\left[m^2-(2-\upsilon)\lambda_n^2\right] \\ -\upsilon\lambda_n & m+mk\left[m^2-(2-\upsilon)\lambda_n^2\right] & \Phi^2-\left[1+k(\lambda_n^2-m^2)^2\right] \end{bmatrix} \tag{7.104}$$

e. The form of the matrix for Flügge–Byrne–Lur'ye shell theory [25] is

$$[\Gamma_{ij}]^{F.-B.-L.} = \begin{bmatrix} \Phi^2+\lambda_n^2-(1+k)\frac{1-\upsilon}{2}m^2 & -\frac{1+\upsilon}{2}m\lambda_n & \upsilon\lambda_n-k\lambda_n\left[\lambda_n^2+(1-\upsilon)m^2\right] \\ \frac{1+\upsilon}{2}m\lambda_n & \Phi^2+(1+3k)\frac{1-\upsilon}{2}\lambda_n^2-m^2 & m\left[1-\frac{3-\upsilon}{2}k\lambda_n^2\right] \\ -\upsilon\lambda_n+k\lambda_n\left[\lambda_n^2+(1-\upsilon)m^2\right] & m\left[1-\frac{3-\upsilon}{2}k\lambda_n^2\right] & \Phi^2-k\left[\left(\lambda_n^2-m^2\right)^2-2m^2\right] \\ & & -(1+k) \end{bmatrix} \tag{7.105}$$

f. The form of the matrix for Reissner–Naghdi–Berry shell theory [25] is

$$\left[\Gamma_{ij}\right]^{\text{R.-N.-B.}}=\begin{bmatrix}\Phi^2+\lambda_n^2-\dfrac{1-\upsilon}{2}m^2 & -\dfrac{1+\upsilon}{2}m\lambda_n & \upsilon\lambda_n\\ \frac{1+\upsilon}{2}m\lambda_n & \Phi^2+(1+k)\left[\dfrac{1-\upsilon}{2}\lambda_n^2-m^2\right] & m\left[1+k\left(m^2-\lambda_n^2\right)\right]\\ -\upsilon\lambda_n & m\left[1+k\left(m^2-\lambda_n^2\right)\right] & \Phi^2-\left[1+k\left(\lambda_n^2-m^2\right)^2\right]\end{bmatrix} \tag{7.106}$$

g. The form of the matrix for Sanders shell theory [25] is

$$\left[\Gamma_{ij}\right]^{\text{S.}}=\begin{bmatrix}\Phi^2+\lambda_n^2-\left(1+\dfrac{k}{4}\right)\dfrac{1-\upsilon}{2}m^2 & -m\lambda_n\left[\dfrac{1+\upsilon}{2}-\dfrac{3k(1-\upsilon)}{8}\right] & \lambda_n\left(\upsilon-\dfrac{1-\upsilon}{2}km^2\right)\\ m\lambda_n\left[\dfrac{1+\upsilon}{2}-\dfrac{3k(1-\upsilon)}{8}\right] & \Phi^2+(1+k)m^2+\left(1+\dfrac{9k}{4}\right)\dfrac{1-\upsilon}{2}\lambda_n^2 & m\left[1+k\left(m^2-\dfrac{3-\upsilon}{2}\lambda_n^2\right)\right]\\ \lambda_n\left(\dfrac{1-\upsilon}{2}km^2-\upsilon\right) & m\left[1+k\left(m^2-\dfrac{3-\upsilon}{2}\lambda_n^2\right)\right] & \Phi^2-\left[1+k\left(\lambda_n^2-m^2\right)^2\right]\end{bmatrix} \tag{7.107}$$

h. The form of the matrix for Vlasov shell theory [25] is

$$\left[\Gamma_{ij}\right]^{\text{V.}}=\begin{bmatrix}\Phi^2+\lambda_n^2-\dfrac{1-\upsilon}{2}m^2 & -\dfrac{1+\upsilon}{2}m\lambda_n & \lambda_n\left[\upsilon-k\left(\dfrac{1-\upsilon}{2}m^2+\lambda_n\right)\right]\\ \dfrac{1+\upsilon}{2}m\lambda_n & \Phi^2-m^2+\dfrac{1-\upsilon}{2}\lambda_n^2 & m\left(1-\dfrac{3-\upsilon}{2}k\lambda_n^2\right)\\ \lambda_n\left[k\left(\dfrac{1-\upsilon}{2}m^2+\lambda_n\right)-\upsilon\right] & m\left(1-\dfrac{3-\upsilon}{2}k\lambda_n^2\right) & \Phi^2-(1+k)-k\left[\left(\lambda_n^2-m^2\right)^2-2m^2\right]\end{bmatrix} \tag{7.108}$$

i. The form of the matrix for Kennard simplified shell theory [25] is

$$\left[\Gamma_{ij}\right]^{\text{K.}}=\begin{bmatrix} \Phi^2+\lambda_n^2-\frac{1-\upsilon}{2}m^2 & -\frac{1+\upsilon}{2}m\lambda_n & \upsilon\lambda_n \\ \frac{1+\upsilon}{2}m\lambda_n & \Phi^2-m^2+\frac{1-\upsilon}{2}\lambda_n^2 & m\left[1+\frac{3k\upsilon}{2(1-\upsilon)}(1-m^2)\right] \\ -\upsilon\lambda_n & 0 & \Phi^2-\left[1+\frac{2+\upsilon}{2(1-\upsilon)}\right]-k\left[\left(\lambda_n^2-m^2\right)^2-\frac{4-\upsilon}{2(1-\upsilon)}m^2\right] \end{bmatrix} \tag{7.109}$$

To obtain the natural frequencies, the determinant of the matrix presented by eq. (7.99) should vanish. Solving the determinant leads to a cubic equation in terms of the frequency parameter, Φ^2. For a fixed value of m and λ_n, one obtains three positive and three negative roots, leading to the determination of Φ^2.

One should note that for circular cylindrical shell, there are four types of simply supported boundary conditions:

$$\begin{aligned} &\text{SS1:} \quad w=0, \quad M_{xx}=w_{,xx}=0, \quad N_{xx}=0, \quad N_{x\theta}=0 \\ &\text{SS2:} \quad w=0, \quad M_{xx}=w_{,xx}=0, \quad u=0, \quad N_{x\theta}=0 \\ &\text{SS3:} \quad w=0, \quad M_{xx}=w_{,xx}=0, \quad N_{xx}=0, \quad v=0 \\ &\text{SS4:} \quad w=0, \quad M_{xx}=w_{,xx}=0, \quad u=0, \quad v=0 \end{aligned} \tag{7.110}$$

The SS3 is known in the literature as the classical simply supported boundary conditions for circular cylindrical shells. The same is for the clamped boundary conditions:

$$\begin{aligned} &\text{C1:} \quad w=0, \quad w_{,x}=0, \quad N_{xx}=0, \quad N_{x\theta}=0 \\ &\text{C2:} \quad w=0, \quad w_{,x}=0, \quad u=0, \quad N_{x\theta}=0 \\ &\text{C3:} \quad w=0, \quad w_{,x}=0, \quad N_{xx}=0, \quad v=0 \\ &\text{C4:} \quad w=0, \quad w_{,x}=0, \quad u=0, \quad v=0 \end{aligned} \tag{7.111}$$

7.5.2 Laminated composite shells – a CLT approach

Leissa [25] presented in his report the Donnell–Mushtari governing equations for an orthotropic circular cylindrical shell having the following formulation:

$$\frac{\partial^2 u}{\partial s^2}+\frac{G_{x\theta}(1-\upsilon_x\upsilon_\theta)}{E_x}\frac{\partial^2 u}{\partial\theta^2}+\frac{G_{x\theta}(1-\upsilon_x\upsilon_\theta)+\upsilon_x E_\theta}{E_x}\frac{\partial^2 v}{\partial s\partial\theta}+\frac{\upsilon_x E_\theta}{E_x}\frac{\partial w}{\partial s}=\frac{\rho R^2(1-\upsilon_x\upsilon_\theta)}{E_x}\frac{\partial^2 u}{\partial t^2}$$

$$\frac{G_{x\theta}(1-\upsilon_x\upsilon_\theta)+\upsilon_x E_\theta}{E_x}\frac{\partial^2 u}{\partial s\partial\theta}+\frac{G_{x\theta}(1-\upsilon_x\upsilon_\theta)}{E_x}\frac{\partial^2 v}{\partial s^2}+\frac{E_\theta}{E_x}\frac{\partial^2 v}{\partial\theta^2}++\frac{E_\theta}{E_x}\frac{\partial w}{\partial\theta}=\frac{\rho R^2(1-\upsilon_x\upsilon_\theta)}{E_x}\frac{\partial^2 v}{\partial t^2}$$

$$\frac{\upsilon_x E_\theta}{E_x}\frac{\partial u}{\partial s}+\frac{E_\theta}{E_x}\frac{\partial v}{\partial\theta}+\frac{E_\theta}{E_x}w++k\left[\frac{\partial^4 w}{\partial s^4}+2\frac{\upsilon_x E_\theta+2G_{x\theta}(1-\upsilon_x\upsilon_\theta)}{E_x}\frac{\partial^4 w}{\partial s^2\partial\theta^2}+\frac{E_\theta}{E_x}\frac{\partial^4 w}{\partial\theta^4}\right]=$$

$$=-\frac{\rho R^2(1-\upsilon_x\upsilon_\theta)}{E_x}\frac{\partial^2 w}{\partial t^2}$$

(7.112)

where

$$s=x/R,k=h^2/\left(12R^2\right)$$

and E_x, υ_x and E_θ, υ_θ are the longitudinal and circumferential Young's modulus and Poisson's ratio, respectively.

References

[1] Murty, A.V.K. and Shimpi, R.P. Vibrations of laminated beams, Journal of Sound and Vibration, 36(2), 1974, 273–284.

[2] Teoh, L.S. and Huang, C.C. The vibration of beams of fiber-reinforced material, Journal of Sound and Vibration, 51(4), 1977, 467–473.

[3] Chandrashekhara, K., Krishnamurthy, K. and Roy, S. Free vibration of composite beams including rotary inertia and shear deformation, Composite Structures, 14, 1990, 269–279.

[4] Singh, G., Rao, G.V. and Iyengar, N.G.R. Analysis of the nonlinear vibrations of unsymmetrically laminated composite beams, AIAA Journal, 29(10), 1991, 1727–1735.

[5] Krishnaswamy, S., Chandrashekhara, K. and Wu, W.Z.B. Analytical solutions to vibration of generally layered composite beams, Journal of Sound and Vibration, 159(1), 1992, 85–99.

[6] Chandrashekhara, K. and Bangera, K.M. Free vibration of composite beams using a refined shear flexible beam element, Computers and Structures, 43(4), 1992, 719–727.

[7] Kant, T. and Swaminathan, K. Analytical solutions for free vibration of laminated composite and sandwich plates based on a higher-order refined theory, Composite Structures, 53, 2001, 73–85.

[8] Aagaah, M.R., Mahinfalah, M. and Jazar, G.N. Natural frequencies of laminated composite plates using third order shear deformation theory, Composite Structures, 72, 2006, 273–279.

[9] Abramovich, H. and Hamburger, O. Vibration of a uniform cantilever Timoshenko beam with translational and rotational springs and with a tip mass, Journal of Sound and Vibration, 154(1), 1992, 67–80.

[10] Abramovich, H. Shear deformation and rotary inertia effects of vibrating composite beams, Composite Structures, 20, 1992, 165–173.

[11] Abramovich, H. A note on experimental investigation on a vibrating Timoshenko cantilever beam, Journal of Sound and Vibration, 160(1), 1993, 167–171.

[12] Abramovich, H. Thermal buckling of cross-ply composite laminates using a first-order shear deformation theory, Composites Structures, 28, 1994, 201–213.

[13] Abramovich, H. Deflection control of laminated composite beams with piezoceramic layers- closed form solutions, Composite Structures, 43, 1998, 217–231.
[14] Abramovich, H., Eisenberger, M. and Shulepov, O. Vibrations and buckling of cross-ply non-symmetric laminated composite beams, AIAA Journal, 34(5), May 1996, 1064–1069.
[15] Abramovich, H. and Livshits, A. Free vibrations of non-symmetric cross-ply laminated composite beams, Journal of Sound and Vibration, 176(5), 1994, 597–612.
[16] Abramovich, H. and Livshits, A. Dynamic behavior of cross-ply laminated beams with piezoelectric layers, Composite Structures, 25(1–4), 1993, 371–379.
[17] Abramovich, H. and Livshits, A. Free vibrations of non-symmetric cross-ply laminated composite beams, Journal of Sound and Vibration, 176(5), 1994, 597–612.
[18] Bert, C.W. and Kumar, M. Vibration of cylindrical shells of bimodulus composite materials, Journal of Sound and Vibrations, 81(1), 1982, 107–121.
[19] Soedel, W. Simplified equations and solutions for the vibration of orthotropic cylindrical shells, Journal of Sound and Vibrations, 87(4), 1983, 555–566.
[20] Soldatos, K.P. A comparison of some shell theories used for the dynamic analysis of cross-ply laminated circular cylindrical panels, Journal of Sound and Vibrations, 97(2), 1984, 305–319.
[21] Jones, R.M. and Morgan, H.S. Buckling and vibration of cross-ply laminated circular cylindrical shells, AIAA Journal, 13(5), 1975, 664–671.
[22] Lim, C.W., Kitipornchai, S. and Liew, K.M. Comparative accuracy of shallow and deep shell theories for vibrations of cylindrical shells, Journal of Vibration and Control, 3, 1997, 119–143.
[23] Farshidianfar, A. and Oliazadeh, P. Free vibration analysis of circular cylindrical shell: comparison of different shell theories, International Journal of Mechanics and Applications, 2(5), 2012, 74–80.
[24] Singer, J. and Abramovich, H. Vibration techniques for definition of practical boundary conditions in stiffened shells, AIAA Journal, 17(7), July 1979, 762–769.
[25] Leissa, W. Vibration of shells, NASA SP-288, US Government Printing Office, Washington DC, 1973.
[26] Soedel, W. Vibrations of shells and plates, 3rd ed., Marcel Dekker, Inc., 2004, 553.
[27] DiGiovanni, P.R. and Dugundji, J. Vibrations of freely-supported orthotropic cylindrical shells under internal pressure, AFOSR Scientific Report, AFOSR 65-0640, ASRL TR 112–4 (AD 617 269), Feb. 1965.

Appendix A: General solution for a nonsymmetrical beam resting on any boundary conditions

Using eq. (7.36) and assuming no axial compression is applied ($\bar{N}_{xx} = 0$), while $I_1 = 0$ (by placing the beam's coordinate system in the middle plane of the beam) and the rotary moment of inertia, assumed to be being negligible ($I_2 = 0$), we have the following matrix notation:

$$\begin{bmatrix} A_{11}\dfrac{\partial^2}{\partial x^2} & -B_{11}\dfrac{\partial^3}{\partial x^3} \\ -B_{11}\dfrac{\partial^3}{\partial x^3} & D_{11}\dfrac{\partial^4}{\partial x^4} \end{bmatrix} \begin{Bmatrix} u_0(x,t) \\ w_0(x,t) \end{Bmatrix} + \begin{bmatrix} -I_0 & 0 \\ 0 & I_0 \end{bmatrix} \frac{\partial^2}{\partial t^2} \begin{Bmatrix} u_0(x,t) \\ w_0(x,t) \end{Bmatrix} = \begin{Bmatrix} 0 \\ 0 \end{Bmatrix} \tag{A.1}$$

Assuming that the nondimensional displacements have the following form:

$$\begin{Bmatrix} \dfrac{u_0(x,t)}{L} \\ \dfrac{w_0(x,t)}{L} \end{Bmatrix} = \begin{Bmatrix} U(x)e^{i\omega t} \\ W(x)e^{i\omega t} \end{Bmatrix} \tag{A.2}$$

while the axial nondimensional axis is $\xi = x/L$ (L is the length of the beam), and substituting the nondimensional expressions of the two beam's displacements into eq. (A.1), we get

$$\begin{bmatrix} \zeta_1^2 \dfrac{d^2}{d\xi^2} + p^2 & -\zeta^2 \dfrac{d^3}{d\xi^3} \\ -\zeta^2 \dfrac{d^3}{d\xi^3} & \dfrac{d^4}{d\xi^4} - p^2 \end{bmatrix} \begin{Bmatrix} U(\xi) \\ W(\xi) \end{Bmatrix} = \begin{Bmatrix} 0 \\ 0 \end{Bmatrix} \tag{A.3}$$

where

$$p^2 \equiv \frac{\omega^2 I_0 L^4}{D_{11}}, \quad b^2 \equiv \frac{D_{11}}{KA_{55}L^2}, \quad \zeta^2 \equiv \frac{B_{11}L}{D_{11}}, \quad \zeta_1^2 \equiv \frac{A_{11}L^2}{D_{11}} \tag{A.4}$$

The characteristic equation of eq. (A.3) has the following form:

$$\left(\zeta_1^2 - \zeta^4\right)s^3 + s^2 - \frac{\zeta_1^2}{p^2}s - \frac{1}{p^2} = 0, \qquad s \equiv \frac{m^2}{p^2} \tag{A.5}$$

Solving eq. (A.5), we can write the general solution for (A.3) in the following form (see a detailed discussion regarding the format of the solution in [17])

$$\begin{aligned} U &= A_1\gamma\mu \sinh(m_1\xi) + A_2\gamma\mu \cosh(m_1\xi) + A_3\lambda\delta \sin(m_2\xi) \\ &\quad - A_4\lambda\delta \cos(m_2\xi) + A_5\alpha\beta \sin(m_3\xi) - A_6\alpha\beta \cos(m_3\xi) \\ W &= A_1 \cosh(m_1\xi) + A_2 \sinh(m_1\xi) + A_3 \cos(m_2\xi) \\ &\quad + A_4 \sin(m_2\xi) + A_5 \cos(m_3\xi) + A_6 sin(m_3\xi) \end{aligned} \tag{A.6}$$

where the various terms in eq. (A.6) are defined as follows:

$$\begin{aligned} &m_1 \equiv \sqrt{s_1}p, \quad m_2 \equiv \sqrt{-s_2}p, \quad m_3 \equiv \sqrt{-s_3}p \\ &\gamma \equiv \frac{\zeta^2 s_1}{\zeta_1^2 s_1 + 1}, \quad \delta \equiv \frac{\zeta^2 s_2}{\zeta_1^2 s_2 + 1}, \quad \beta \equiv \frac{\zeta^2 s_3}{\zeta_1^2 s_3 + 1} \\ &\mu \equiv \sqrt{s_1}p, \quad \lambda \equiv \sqrt{-s_2}p, \quad \alpha \equiv \sqrt{-s_3}p \end{aligned} \tag{A.7}$$

Imposing the adequate boundary conditions would lead to the finding of both the eigenvalues (the natural frequencies) and the eigenvectors (the modes of vibration). For this problem, six boundary conditions should be imposed at each end as presented in Tab. A.1.

Tab. A.1: Boundary conditions for a nonsymmetric laminate beam (CLT approach).

Name	Boundary conditions
Clamped immovable end	$U = W = \frac{dW}{d\xi} = 0$
Clamped movable end	$\zeta_1^2 \frac{dU}{d\xi} - \zeta^2 \frac{d^2W}{d\xi^2} = W = \frac{dW}{d\xi} = 0$
Simply supported immovable end	$U = W = \zeta^2 \frac{dU}{d\xi} - \frac{d^2W}{d\xi^2} = 0$
Simply supported movable end	$\zeta^2 \frac{dU}{d\xi} - \frac{d^2W}{d\xi^2} = W = \zeta_1^2 \frac{dU}{d\xi} - \zeta^2 \frac{d^2W}{d\xi^2} = 0$
Free end	$\zeta^2 \frac{dU}{d\xi} - \frac{d^2W}{d\xi^2} = \zeta_1^2 \frac{d^2U}{d\xi^2} - \zeta^2 \frac{d^3W}{d\xi^3} = \zeta_1^2 \frac{dU}{d\xi} - \zeta^2 \frac{d^2W}{d\xi^2} = 0$

Appendix B: Matrix notation for the equilibrium equations using CLT approach

A convenient way of presenting the equilibrium equations at buckling and/or vibration is the following matrix form:

$$\begin{bmatrix} a_{11} & a_{12} & a_{13} \\ a_{21} & a_{22} & a_{23} \\ a_{31} & a_{32} & [a_{33} - N] \end{bmatrix} \begin{Bmatrix} u(x,y,t) \\ v(x,y,t) \\ w_0(x,y,t) \end{Bmatrix} + \frac{\partial^2}{\partial t^2} \begin{bmatrix} m_{11} & 0 & m_{13} \\ 0 & m_{22} & m_{23} \\ m_{13} & m_{23} & m_{33} \end{bmatrix} \begin{Bmatrix} u(x,y,t) \\ v(x,y,t) \\ w_0(x,y,t) \end{Bmatrix} = \begin{Bmatrix} 0 \\ 0 \\ q \end{Bmatrix} \tag{B.1}$$

where the various operators are given by

$$a_{11} \equiv A_{11}\frac{\partial^2}{\partial x^2} + 2A_{16}\frac{\partial^2}{\partial x \partial y} + A_{66}\frac{\partial^2}{\partial y^2}$$

$$a_{22} \equiv A_{22}\frac{\partial^2}{\partial y^2} + 2A_{26}\frac{\partial^2}{\partial x \partial y} + A_{66}\frac{\partial^2}{\partial x^2}$$

$$a_{33} \equiv D_{11}\frac{\partial^4}{\partial x^4} + 4D_{16}\frac{\partial^4}{\partial x^3 \partial y} + 2(D_{12} + 2D_{66})\frac{\partial^4}{\partial x^2 \partial y^2} + 4D_{26}\frac{\partial^4}{\partial x \partial y^3} + D_{22}\frac{\partial^4}{\partial y^4}$$

$$a_{12} = a_{21} \equiv A_{16}\frac{\partial^2}{\partial x^2} + (A_{12} + A_{66})\frac{\partial^2}{\partial x \partial y} + A_{26}\frac{\partial^2}{\partial y^2} \tag{B.2}$$

$$a_{13} = a_{31} \equiv -B_{11}\frac{\partial^3}{\partial x^3} - 3B_{16}\frac{\partial^3}{\partial x^2 \partial y} - (B_{12} + 2B_{66})\frac{\partial^3}{\partial x \partial y^2} - B_{26}\frac{\partial^3}{\partial y^3}$$

$$a_{23} = a_{32} \equiv -B_{16}\frac{\partial^3}{\partial x^3} - (B_{12} + 2B_{66})\frac{\partial^3}{\partial x^2 \partial y} - 3B_{26}\frac{\partial^3}{\partial x \partial y^2} - B_{22}\frac{\partial^3}{\partial y^3}$$

$$N \equiv \bar{N}_{xx}\frac{\partial^2}{\partial x^2} + 2\bar{N}_{xy}\frac{\partial^2}{\partial x \partial y} + \bar{N}_{yy}\frac{\partial^2}{\partial y^2}$$

and

$$m_{11} \equiv -I_0, \quad m_{22} \equiv -I_0, \quad m_{33} \equiv I_0 - I_2\left(\frac{\partial^2}{\partial x^2} + \frac{\partial^2}{\partial y^2}\right)$$
$$m_{13} \equiv I_1\frac{\partial}{\partial x}, \quad m_{23} \equiv I_1\frac{\partial}{\partial y}, \tag{B.3}$$

Appendix C: The terms of the matrix notation for the equilibrium equations using FSDPT approach

$$\begin{bmatrix} \hat{a}_{11} & \hat{a}_{12} & 0 & \hat{a}_{14} & \hat{a}_{15} \\ \hat{a}_{21} & \hat{a}_{22} & 0 & \hat{a}_{24} & \hat{a}_{25} \\ 0 & 0 & \hat{a}_{33} & \hat{a}_{34} & \hat{a}_{34} \\ \hat{a}_{41} & \hat{a}_{42} & \hat{a}_{43} & \hat{a}_{44} & \hat{a}_{45} \\ \hat{a}_{51} & \hat{a}_{52} & \hat{a}_{53} & \hat{a}_{54} & \hat{a}_{55} \end{bmatrix} \begin{Bmatrix} U_{mn} \\ V_{mn} \\ W_{mn} \\ \bar{\mathrm{E}}_{mn} \\ \mathrm{E}_{mn} \end{Bmatrix} = \begin{Bmatrix} 0 \\ 0 \\ 0 \\ 0 \\ 0 \end{Bmatrix} \tag{C.1}$$

where

$$
\begin{aligned}
&\hat{\alpha}_{11} \equiv A_{11}\left(\frac{m\pi}{a}\right)^2 + A_{66}\left(\frac{n\pi}{b}\right)^2 \hat{\alpha}_{12} = \hat{\alpha}_{21} \equiv (A_{12} + A_{66})\left(\frac{m\pi}{a}\right)\left(\frac{n\pi}{b}\right)\\
&\hat{\alpha}_{14} = \hat{\alpha}_{41} \equiv B_{11}\left(\frac{m\pi}{a}\right)^2 + B_{66}\left(\frac{n\pi}{b}\right)^2 \hat{\alpha}_{15} = \hat{\alpha}_{51} \equiv (B_{12} + B_{66})\left(\frac{m\pi}{a}\right)\left(\frac{n\pi}{b}\right)\\
&\hat{\alpha}_{22} \equiv A_{66}\left(\frac{m\pi}{a}\right)^2 + A_{22}\left(\frac{n\pi}{b}\right)^2 \hat{\alpha}_{24} = \hat{\alpha}_{42} \equiv (B_{12} + B_{66})\left(\frac{m\pi}{a}\right)\left(\frac{n\pi}{b}\right)\\
&\hat{\alpha}_{25} = \hat{\alpha}_{52} \equiv B_{66}\left(\frac{m\pi}{a}\right)^2 + B_{22}\left(\frac{n\pi}{b}\right)^2 \hat{\alpha}_{33} \equiv K\left[A_{44}\left(\frac{n\pi}{b}\right)^2 + A_{55}\left(\frac{m\pi}{a}\right)^2\right]\\
&\qquad\qquad\qquad\qquad\qquad\qquad - \bar{N}_{xx}\left(\frac{m\pi}{a}\right)^2 - \bar{N}_{yy}\left(\frac{n\pi}{b}\right)^2\\
&\hat{\alpha}_{34} = \hat{\alpha}_{43} \equiv KA_{55}\left(\frac{m\pi}{a}\right)\hat{\alpha}_{35} = \hat{\alpha}_{53} \equiv KA_{44}\left(\frac{n\pi}{b}\right)\hat{\alpha}_{44} \equiv D_{11}\left(\frac{m\pi}{a}\right)^2 + D_{66}\left(\frac{n\pi}{b}\right)^2 + KA_{55}\\
&\hat{\alpha}_{45} = \hat{\alpha}_{54} \equiv (D_{12} + D_{66})\left(\frac{m\pi}{a}\right)\left(\frac{n\pi}{b}\right)\hat{\alpha}_{55} \equiv D_{22}\left(\frac{n\pi}{b}\right)^2 + D_{66}\left(\frac{m\pi}{a}\right)^2 + KA_{44}
\end{aligned}
\tag{C.2}
$$

8 Dynamic buckling of thin-walled structures

8.1 Introduction

This chapter deals with what is called in the literature "dynamic" buckling of columns and plates (metal and composite materials). First, the term "dynamic" buckling will be explained and defined followed by examples from the literature. Then the equations of motions for columns and plates will be presented, and numerical and experimental results of tests performed on columns, plates and shells will be highlighted.

The topic of applying an axially time-dependent load onto a column was studied for many years, thus inducing lateral vibrations and eventually causing the buckling of the column. Sometimes this is called *vibration buckling*, as proposed by Lindberg [1]. As it is described in his fundamental report [1], the axial oscillating load might lead to unacceptable large vibrations amplitudes at a critical combination of the frequency and amplitude of the axial load and the inherent damping of the column. This behavior is presented in Fig. 8.1a (adapted from [1]), where an oscillating axial load induces bending moments that cause lateral vibrations of the column. As described in [1], the column will laterally vibrate at large amplitude when the loading frequency will be twice the natural lateral bending frequency of the column. The term, *vibration buckling*, used by Lindberg, presents some kind of similarity to vibration resonance. However, in the case of vibration resonance the applied load is in the same direction as the motion, namely in our case lateral to the column, and the resonance will occur when the loading frequency equals the natural frequency of the column. This type of vibration buckling was called by Lindberg as *dynamic stability of vibrations induced by oscillating parametric loading*. This type of resonance is also called in the literature *parametric resonance* (see an application of this type of dynamic stability in [2, 3]).

Another type of vibration type is sometimes also called pulse buckling, where the structure will be deformed to unacceptably large amplitudes as a result of a transient response of the structure to the dynamic axially applied load [1]. One should note that the sudden applied load might cause a permanent deformation due to plastic response of the column, a snap to a larger postbuckling deformation or simply a return to its undeformed state. This is pictured in Fig. 8.1b (adapted from [1]) where the response of the column to a sudden short-time axial load is shown.

One should note that buckling will occur when an unacceptably large deformation or stress is encountered by the column. The column can withstand a large axial load before reaching the buckling condition, provided the load duration is short enough. Under an intense, short duration axial load, the column would buckle into a very high-order mode as shown in Fig. 8.1b. Lindberg [1] claims that pulse buckling falls under the following mathematical definition: *dynamic response of structural systems induced by time-varying parametric loading*. Throughout this chapter, the pulse buckling will be equivalent to dynamic buckling.

https://doi.org/10.1515/9783111621104-008

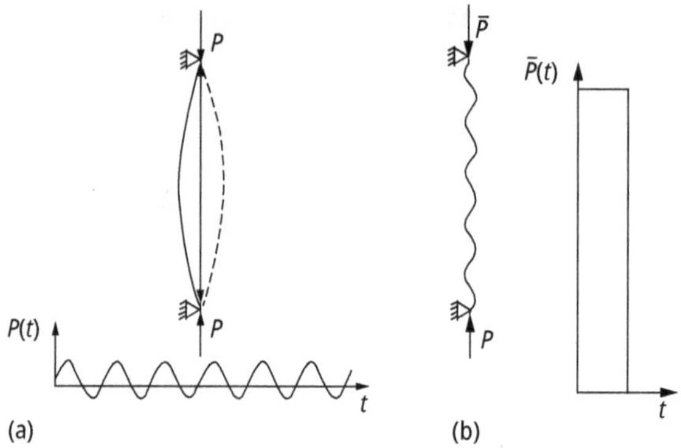

Fig. 8.1: (a) Buckling under parametric resonance and (b) pulse-type buckling.

The dynamic buckling of structures has been widely addressed in the literature. It started with the famous paper by Budiansky and Roth [4], through Hegglin's report on dynamic buckling of columns [5] and continued with Budiansky and Hutchinson [6] and Hutchinson and Budiansky [7] in the mid-sixties. Then more structures have been addressed as presented in typical references [1, 2, 8–43].

One of the most intriguing and challenging things is to define a criterion to clearly define the critical load causing the structure to buckle under the subjected pulse loading. As presented by Kubiak [32] and also by Ari Gur [13, 19, 22] and others in [18, 20, 24–26], a new quantity is introduced called DLF (dynamic load factor) to enable the use of the dynamic buckling criteria. It is defined as

$$\mathrm{DLF} \equiv \frac{\text{Pulse buckling quantity}}{\text{Static buckling quantity}} \equiv \frac{(P_{\mathrm{cr}})_{\mathrm{dyn.}}}{(P_{\mathrm{cr}})_{\mathrm{static}}} \tag{8.1}$$

According to Kubiak [32], the most popular criterion had been proposed by Volmir [10] for plates subjected to in-plane pulse loading. As quoted in [32], Volmir proposed the following criterion:

Dynamic critical load corresponds to the amplitude of pulse load (of constant duration) at which the maximum plate deflection is equal to some constant value k (k – half or one plate thickness).

Another very widely used criterion has been formulated and proposed by Budiansky and Hutchinson [6], based on an earlier work [4] and latter extended [7]. Originally, the criterion was formulated for shell-type structures but was used also for columns and plates. The criterion claims that "Dynamic stability loss occurs when the maximum deflection grows rapidly with the small variation of the load amplitude." This criterion is schematically presented in Fig. 8.2b, where $R\ (\lambda, t)$ is the response of the simply nonlinear model assumed in [6] and presented also in Fig. 8.2a, and λ is the nondimensional applied dynamic pulse-type compressive load. Figure 8.2c presents

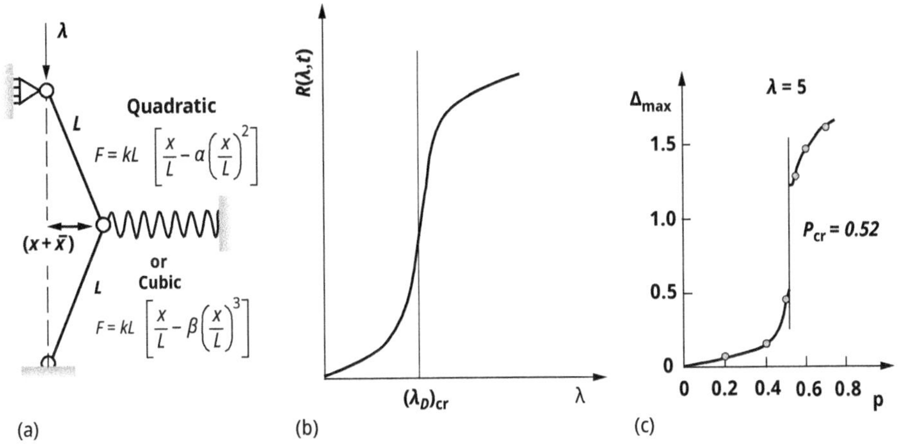

Fig. 8.2: (a) The nonlinear model (adapted from [6]), (b) the Budiansky and Hutchinson (B&H) schematic criterion (from [6]) and (c) the application of the B&H criterion to axisymmetric dynamic buckling of clamped shallow spherical shells (adapted from [4]).

the application of the criterion for the axisymmetric dynamic buckling of clamped shallow spherical shells as presented in [4].

Other dynamic buckling criteria were suggested and applied by Ari-Gur and Simonetta in their manuscript [22]. They formulated four criteria presented schematically in Fig. 8.3a–d. The first criterion (see Fig. 8.3a) correlates the maximum lateral deflection, W_m, to the pulse intensity defined as L_m. The first criterion is stated as "Buckling occurs when, for a given pulse shape and duration, a small increase in the pulse intensity causes a sharp increase in the rate of growth of the peak lateral deflection" [22]. The authors claim that this criterion can be used to both displacement and force loading types for a wide range of pulse frequencies; however, for very short pulse durations the results might be misleading. The reason for this is connected to the characteristics of the out-of-plane deflections which, for short pulse duration in the vicinity of buckling loads, turn out to have short wavelength patterns that are associated with smaller peak deflections. Therefore, the authors present the second criterion (Fig. 8.3b), which comes to answer the deficiency of the first criterion and is suitable to patterns of short-wavelength deflection shapes. It claims that "dynamic buckling occurs when a small increase in the pulse intensity causes a decrease in the peak lateral deflection" and is relevant to only impulsive loads and may be used in complimentary to the first criterion The last two criteria presented in [22] connect the intensity of the applied load versus the maximum response of the loaded edge, say at $x = 0$. According to the third buckling criterion (Fig. 8.3c), "buckling occurs when a small increase in the force intensity F_m causes a sharp increase in the peak longitudinal displacement U_m at $x = 0$." The buckling phenomenon under pulse loading is due to the diminishing of the structural resistance to an in-plane compressive load when the dynamic lateral deflections grow rapidly. The fourth buckling criterion associated with a displacement pulse (Fig. 8.3d) states that "buckling occurs when a small increase in the pulse displacement intensity U_m causes a transition of the peak reaction force F_m at $x = 0$ from compression to tension." This transition would come true, when the tensile force needed to keep the deforming structure at the prescribed U_m would be greater than the maximum compression at its loaded edge.

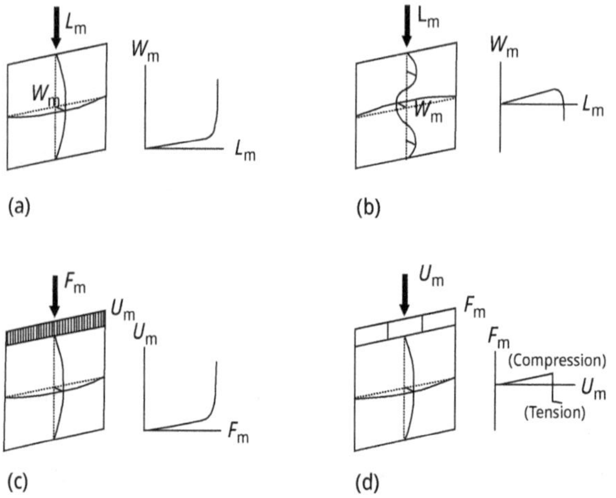

Fig. 8.3: Buckling criteria (adapted from [22]): (a) the first criterion, (b) the second criterion, (c) the third criterion and (d) the fourth criterion.

Another interesting criterion was suggested by Petry and Fahlbush [24], claiming that the Budiansky–Hutchinson criterion is very conservative for structures with stable postbuckling equilibrium path because it does not take into account load-carrying capacity of the structure. The criterion proposed is based on stress analysis and claims that "a stress failure occurs if the effective stress σ_E exceeds the limit stress of the material, σ_L; a dynamic response caused by an impact is defined to be dynamically stable if $\sigma_E \leq \sigma_L$ is fulfilled at every time everywhere in the structure." Using this criterion, which is claimed to be practical also for ductile materials (by using the yield stress σ_Y instead of σ_L), the DLF (see eq. (8.1)) is modified to

$$\mathrm{DLF} \equiv \frac{(N_F)_{\mathrm{dyn.}}}{(N_F)_{\mathrm{static}}} \tag{8.2}$$

where $(N_F)_{\mathrm{dyn}}$ and $(N_F)_{\mathrm{static}}$ are defined as the dynamic failure and static failure loads, respectively.

8.2 Dynamic buckling of columns

8.2.1 Dynamic buckling of columns using CLT

The column differential equations of motion under time-dependent axial compression, as presented in Fig. 8.4, are given by

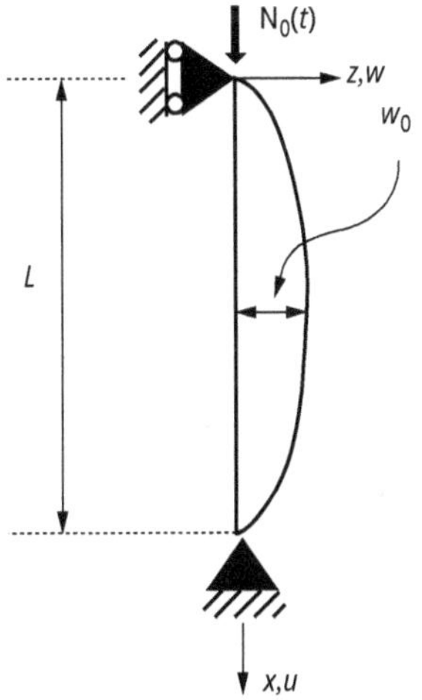

Fig. 8.4: A column with geometric initial imperfection, w_0, subjected to time-dependent axial compression ($N_0(t)$ is the axial compression at $x = 0$).

$$N_{x,x} = I_0\,\ddot{u} - I_1\,\ddot{w}_{,x} \tag{8.3}$$

$$\left[M_{x,x} + N_x \cdot w_{,x} + I_2 \cdot \ddot{w}_{,xx}\right] = I_0\,\ddot{w} + I_1\,\ddot{w}_{,x} \tag{8.4}$$

where

$$(I_0;I_1;I_2) = \rho \int_{-h/2}^{h/2} \left(1;\, z;\, z^2\right) dz \tag{8.5}$$

and $()_{,x}$ is the first partial differentiation with respect to x, $\ddot{u}$, $\ddot{w}$ are the second partial differentiation of the axial and lateral displacements with respect to time, t, h is the total thickness of the column and ρ is the relevant densities for each layer in the laminate.

Assuming that the force and moment resultants can be written as a function of the strains and curvatures, according to the CLT approach we obtain

$$\begin{Bmatrix} N_x \\ M_x \end{Bmatrix} = \begin{bmatrix} A_{11} & B_{11} \\ B_{11} & D_{11} \end{bmatrix} \begin{Bmatrix} \varepsilon_x \\ \kappa_x \end{Bmatrix} \tag{8.6}$$

where

$$(A_{11}; B_{11}; D_{11}) = \int_{-h/2}^{h/2} \bar{Q}_{11}(1; z; z^2)\,dz \tag{8.7}$$

and $\bar{Q}_{11}$ is the plane stress-reduced stiffness for each layer after transformation (for an exact expression, see Chapter 2). The strain displacements (u longitudinal displacement, w lateral displacement) relationships are

$$\begin{Bmatrix} \varepsilon_x \\ \kappa_x \end{Bmatrix} = \begin{Bmatrix} u_{,x} + \frac{1}{2}\left(w_{,x}^2 - w_{0,x}^2\right) \\ -(w - w_0)_{,xx} \end{Bmatrix} \tag{8.8}$$

Substitution of eqs. (8.6)–(8.8) into eqs. (8.3) and (8.4) yields

$$A_{11}(u_{,xx} + w_{,x} \cdot w_{,xx} - w_{0,x} \cdot w_{0,xx}) - B_{11}(w - w_0)_{,xxx} = I_0\,\ddot{u} - I_1\,\ddot{w}_{,x} \tag{8.9}$$

$$A_{11}\left[(u_{,x} \cdot w_{,x})_{,x} + \frac{3}{2}w_{,x}^2 \cdot w_{,xx} - w_{0,x}\left(w_{0,xx} \cdot w_{,x} - \frac{1}{2}w_{0,x} \cdot w_{,xx}\right)\right] +$$
$$B_{11}\left[u_{,xx} + w_{0,xx}(w - w_o)_{,x}\right]_{,x} - D_{11}[w - w_o]_{,xxxx} = I_0\,\ddot{w} + I_1\,\ddot{u}_{,x} \tag{8.10}$$

Assuming a symmetric laminate, $B_{11} = I_1 = 0$, then eqs. (8.9) and (8.10) are simplified to yield

$$A_{11}(u_{,xx} + w_{,x} \cdot w_{,xx} - w_{0,x} \cdot w_{0,xx}) = I_0\,\ddot{u} \tag{8.11}$$

$$A_{11}\left[(u_{,x} \cdot w_{,x})_{,x} + \frac{3}{2}w_{,x}^2 \cdot w_{,xx} - w_{0,x}\left(w_{0,xx} \cdot w_{,x} - \frac{1}{2}w_{0,x} \cdot w_{,xx}\right)\right] - D_{11}[w - w_o]_{,xxxx} = I_0\,\ddot{w} \tag{8.12}$$

One should note that all the terms in eqs. (8.11) and (8.12) are a function of both x and t.

The associated boundary conditions (pinned–pinned or simply supported) for the symmetric laminate are

$$\text{Lateral deflection:}\quad w(0, t) = w(L, t) = 0 \tag{8.13}$$

$$\text{Bending moment:}\quad D_{11} \cdot w_{,xx}(0, t) = D_{11} \cdot w_{,xx}(L, t) = 0 \tag{8.14}$$

$$\text{Axial force:.}\quad A_{11} \cdot u_{,x}(0, t) = N_x(0, t) = -N_0(t) = -N_0 \sin\frac{\pi t}{T} \tag{8.15}$$

$$\text{Axial displacement:}\quad u_{,x}(L, t) = 0 \tag{8.16}$$

while the initial conditions assume that $w(x, 0) = w_0 w_{,x}(x, 0) = u(0, 0) = \dot{u}\,(0, 0) = 0$.

Note that eqs. (8.11) and (8.12) can be used for an isotropic column, by letting $A_{11} = EA$, the axial stiffness (E = Young's modulus and A = cross section of the column) and the bending stiffness, $D_{11} = Eh^3/[12(1 - \upsilon^2)]$.

The solution for the equations of motion (8.11) and (8.12) can be obtained using energy methods, like the Galerkin method, using the boundary and initial conditions presented in eqs. (8.13)–(8.16). To apply the Galerkin approach, one should assume displacements that satisfy the boundary conditions of the problem, namely for the present case the following functions will be assumed:

$$w(x,t) = A(t)\sin\tfrac{\pi x}{L};\quad w_0(x) = A_0\sin\tfrac{\pi x}{L} \qquad u(x,0) = B(t)\cos\tfrac{\pi x}{L};\quad N_0(t) = N_0\sin\tfrac{\pi t}{T} \tag{8.17}$$

Multiplying eq. (8.11) by the expression for $u(x, 0)$, while eq. (8.12) is multiplied by $w(x, t)$ and integrating both from $x = 0$ till $x = L$, will exclude the x dependence, yielding two nonlinear time-dependent equations that have to be solved numerically.

For each value of N_0, the response of the beam w and u can be calculated for an assumed initial geometric imperfection, $w_0(x,t) = A_0 \sin\dfrac{\pi x}{L}$, where A_0 is a known amplitude (usually percentage of the columns thickness), for a constant time period T.

A typical drawing is presented in Fig. 8.5 (adapted from [18]). Calculating the dynamic buckling for each value of the $2T/T_b$ and dividing it by the static buckling load will yield the DLF for the tested or calculated case (see eq. (8.1)).

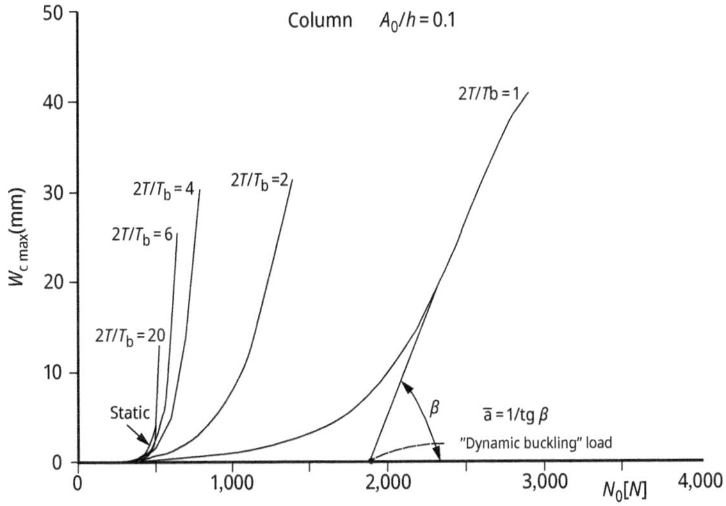

Fig. 8.5: The influence of the loading duration (T) on the column's response (adapted from [18]), T_b is the first natural period of the column, and A_0 is the amplitude of the initial geometric imperfection.

8.2.2 Dynamic buckling of columns using FOSDT

To solve the dynamic buckling of columns (see Fig. 8.4) using the FOSDT, we shall assume the following displacement field:

$$\begin{aligned} u_x(x,z,t) &= u(x,t) + z\phi_x(x,t) \\ u_z(x,z,t) &= w(x,t) - w_0(x) \end{aligned} \tag{8.18}$$

where the assumed variables u and w are displacements in the x and z directions, of a point on the mid-plane of the plate (namely at $z = 0$), respectively, and ϕ_x is a rotation about the x-axis, while w_0 is the initial geometric imperfection of the column. The equations of motion can be written as (see Chapters 6 and 7)

$$\begin{aligned} &N_{x,x} = I_0\,\ddot{u} + I_1\ddot{\phi}_x \\ &Q_{x,x} + \left[N_x \cdot (w_{,x} - w_{o,x})\right]_{,x} = I_0\,\ddot{w} \\ &M_{x,x} - Q_x = I_2\ddot{\phi}_x + I_1\,\ddot{u} \end{aligned} \tag{8.19}$$

In eq. (8.19), Q_x is the shear force, or the transverse force resultant defined as

$$Q_x = K\int_{-h/2}^{+h/2} \tau_{xy} \cdot dz \tag{8.20}$$

where K is the shear correction coefficient computed by equating the strain energy due to transverse shear stresses to the strain energy due to true transverse shear as calculated by a 3D elasticity theory while τ_{xy} is the transverse shear stress. For a homogenous beam having a rectangular cross section, $K = 5/6$ (see Chapter 6). The force and moment resultants as a function of displacements are given by

$$\begin{Bmatrix} N_x \\ M_x \\ Q_x \end{Bmatrix} = \begin{bmatrix} A_{11} & B_{11} & 0 \\ B_{11} & D_{11} & 0 \\ 0 & 0 & KA_{55} \end{bmatrix} \begin{Bmatrix} u_{,x} + \frac{1}{2}\left(w_{,x}^2 - w_{0,x}^2\right) \\ \phi_{x,x} \\ \phi_x + w_{,x} - w_{0,x} \end{Bmatrix} \tag{8.21}$$

As assumed in Section 8.2.1, the strains have the following form:

$$\begin{Bmatrix} \varepsilon_x \\ \gamma_{xz} \end{Bmatrix} = \begin{Bmatrix} u_{x,x} + \frac{1}{2}\left(w_{,x}^2 - w_{0,x}^2\right) \\ u_{x,z} + u_{z,x} \end{Bmatrix} = \begin{Bmatrix} u_{,x} + \frac{1}{2}\left(w_{,x}^2 - w_{0,x}^2\right) + z\phi_{x,x} \\ \phi_x + w_{,x} - w_{0,x} \end{Bmatrix} \tag{8.22}$$

Substituting eq. (8.21) into eq. (8.19) yields the following equations of motion expressed by the three assumed displacements, $u(x, t)$, $w(x, t)$, $\phi_x(x, t)$, and the known initial geometric imperfection, $w_0(x)$ (see also a similar derivation in [25]):

$$
\begin{aligned}
& A_{11}(u_{,xx}+w_{,x}\cdot w_{,xx}-w_{0,x}\cdot w_{0,xx})+B_{11}\phi_{x,xx}=I_0\ddot{u}+I_1\ddot{\phi}_x \\
& \left[A_{11}(u_{,xx}+w_{,x}\cdot w_{,xx}-w_{0,x}\cdot w_{0,xx})+B_{11}\phi_{x,xx}\right](w_{,x}-w_{0,x})+ \\
& \left[A_{11}\left(u_{,x}+\frac{1}{2}w_{,x}^2-\frac{1}{2}w_{0,x}^2\right)+B_{11}\phi_{x,x}\right](w_{,xx}-w_{0,xx})+KA_{55}(\phi_{x,x}+w_{,xx}-w_{0,xx})=I_0\ddot{w} \\
& B_{11}(u_{,xx}+w_{,x}\cdot w_{,xx}-w_{0,x}\cdot w_{0,xx})+D_{11}\phi_{x,xx}-KA_{55}(\phi_x+w_{,x}-w_{0,x})=I_2\ddot{\phi}_x+I_1\ddot{u}
\end{aligned}
\tag{8.23}
$$

For a symmetric layup, $B = I = 0$, then eq. (8.23) is simplified to yield

$$
\begin{aligned}
& A_{11}(u_{,xx}+w_{,x}\cdot w_{,xx}-w_{0,x}\cdot w_{0,xx})=I_0\ddot{u} \\
& [A_{11}(u_{,xx}+w_{,x}\cdot w_{,xx}-w_{0,x}\cdot w_{0,xx})](w_{,x}-w_{0,x})+ \\
& \left[A_{11}\left(u_{,x}+\frac{1}{2}w_{,x}^2-\frac{1}{2}w_{0,x}^2\right)\right](w_{,xx}-w_{0,xx})+KA_{55}(\phi_{x,x}+w_{,xx}-w_{0,xx})=I_0\ddot{w} \\
& D_{11}\phi_{x,xx}-KA_{55}(\phi_x+w_{,x}-w_{0,x})=I_2\ddot{\phi}_x
\end{aligned}
\tag{8.24}
$$

As in the previous section, one should note that all the terms in eq. (8.24) are a function of both x and t.

The associated boundary conditions (pinned–pinned or simply supported) for the symmetric laminate are

$$\text{Lateral deflection:}\quad w(0,t)=w(L,t)=0 \tag{8.25}$$

$$\text{Bending moment:}\quad D_{11}\cdot\phi_{x,x}(0,t)=D_{11}\cdot\phi_{x,x}(L,t)=0 \tag{8.26}$$

$$\text{Axial force:}\quad A_{11}\cdot u_{,x}(0,t)=N_x(0,t)=-N_0(t)=-N_0\sin\frac{\pi t}{T} \tag{8.27}$$

$$\text{Axial displacement:}\quad u_{,x}(L,t)=0 \tag{8.28}$$

while the initial conditions assume that $w(x,0) = w_0$, $w_{,x}(x,0)=u(0,0)=\dot{u}(0,0)=0$.

As before the analytic solution of the equations of motion cannot be obtained; therefore, energy methods like Galerkin method are suggested. Suitable assumed solutions for the case of simply supported boundary conditions are

$$
\begin{aligned}
& u(x,0)=A(t)\cos\frac{\pi x}{L},\qquad w(x,t)=B(t)\sin\frac{\pi x}{L},\qquad \phi_x(x,t)=C(t)\cos\frac{\pi x}{L} \\
& w_0(x)=A_0\sin\frac{\pi x}{L},\qquad N_0(t)=N_0\sin\frac{\pi t}{T}
\end{aligned}
\tag{8.29}
$$

The procedure described before in Section 8.2 is again employed to obtain the response of the column to a pulse-type loading. As shown in Fig. 8.6, based on the results presented in [25], the response of the column has a similar behavior to what had been presented in Fig. 8.5. Note that in Fig. 8.6b, the compressive strain is a linear function of the applied axial compression load; therefore, it can be compared with what it is presented in Fig. 8.5.

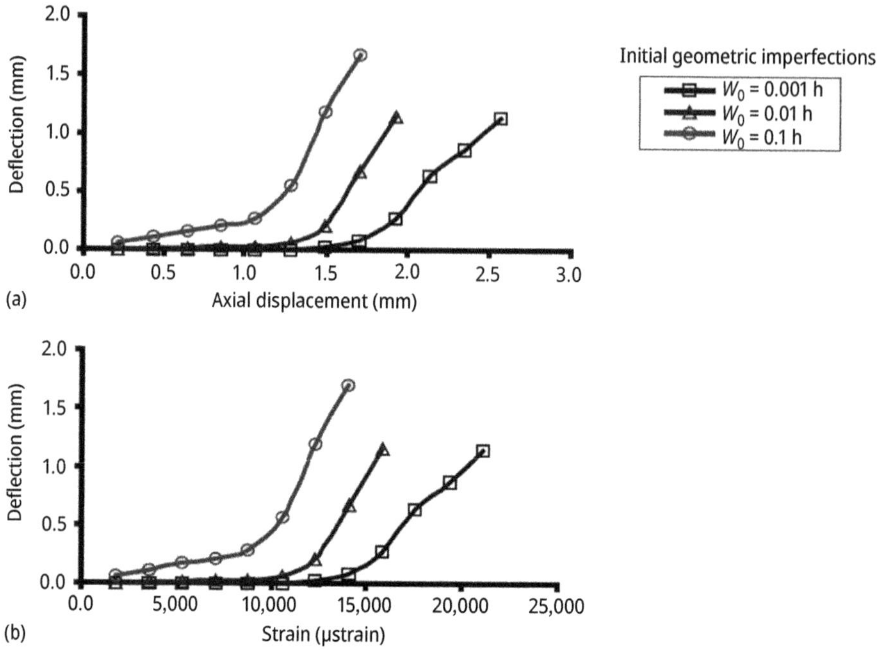

Fig. 8.6: The dynamic pulse buckling response of a column ($L = 150$ mm, width = 20 mm) for various initial geometric imperfections: (a) mid-span deflection versus maximal axial displacement at the impacted end; (b) mid-span deflection versus compressive strain at the neutral axis of the column mid-span (adapted from [25]).

8.3 Dynamic buckling of plates

The present derivation for the dynamic buckling of orthotropic rectangular plates under uniaxial loading is based on Ekstrom's fundamental study [11] presented already in 1973 using the models developed by Lekhnitskii for anisotropic plates [44]. Let us assume a simply supported rectangular orthotropic plate uniaxially loaded as presented in Fig. 8.7. We choose the natural axes of the material to coincide with the coordinate axes leading to the following in-plane stress–strain relations

$$\begin{Bmatrix} \sigma_x \\ \sigma_y \\ \tau_{xy} \end{Bmatrix} = \frac{1}{1-\upsilon_{xy}\cdot\upsilon_{yx}} \begin{bmatrix} E_x & \upsilon_{xy}E_y & 0 \\ \upsilon_{yx}E_x & E_y & 0 \\ 0 & 0 & (1-\upsilon_{xy}\cdot\upsilon_{yx})G_{xy} \end{bmatrix} \begin{Bmatrix} \varepsilon_x \\ \varepsilon_y \\ \gamma_{xy} \end{Bmatrix}$$
$$\equiv \begin{bmatrix} Q_{11} & Q_{12} & 0 \\ Q_{21} & Q_{22} & 0 \\ 0 & 0 & Q_{66} \end{bmatrix} \begin{Bmatrix} \varepsilon_x \\ \varepsilon_y \\ \gamma_{xy} \end{Bmatrix} \tag{8.30}$$

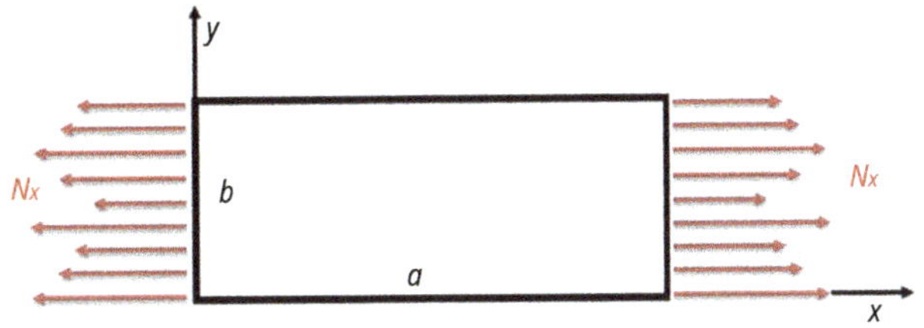

Fig. 8.7: A thin orthotropic plate uniaxial loaded.

$$\begin{Bmatrix} \varepsilon_x \\ \varepsilon_y \\ \gamma_{xy} \end{Bmatrix} = \begin{bmatrix} \frac{1}{E_x} & -\frac{\upsilon_{xy}}{E_x} & 0 \\ -\frac{\upsilon_{yx}}{E_y} & \frac{1}{E_y} & 0 \\ 0 & 0 & \frac{1}{G_{xy}} \end{bmatrix} \begin{Bmatrix} \sigma_x \\ \sigma_y \\ \tau_{xy} \end{Bmatrix} \tag{8.31}$$

and

$$Q_{12} = Q_{21} \quad \Rightarrow \quad \upsilon_{yx} = \frac{E_x}{E_y}\,\upsilon_{xy} \tag{8.32}$$

The dynamic buckling problem can be solved assuming out-of-plane initial geometric imperfections (w_0), and therefore, the strains will be assumed to have the following expressions:

$$\begin{Bmatrix} \varepsilon_x \\ \varepsilon_y \\ \gamma_{xy} \end{Bmatrix} = \begin{Bmatrix} \frac{\partial u}{\partial x} + \frac{1}{2}\left[\left(\frac{\partial w}{\partial x}\right)^2 - \left(\frac{\partial w_0}{\partial x}\right)^2\right] \\ \frac{\partial v}{\partial y} + \frac{1}{2}\left[\left(\frac{\partial w}{\partial y}\right)^2 - \left(\frac{\partial w_0}{\partial y}\right)^2\right] \\ \frac{\partial u}{\partial x} + \frac{\partial v}{\partial y} + \frac{\partial^2 w}{\partial x \partial y} - \frac{\partial^2 w_0}{\partial x \partial y} \end{Bmatrix} \tag{8.33}$$

where $w(x, y, t)$ is the total out of plane, $w_0\ (x, y)$ is the initial out-of-plane geometric displacement, and $u(x, y, t)$ and $v(x, y, t)$ are the in-plane displacements in the x and y directions, respectively.

The compatibility equation for the stress function $F\ (x, y, t)$ and the plate equations of motion (see [11, 44]) can be written as

$$\frac{1}{E_y}\cdot\frac{\partial^4 F}{\partial x^4} + \frac{1}{E_x}\cdot\frac{\partial^4 F}{\partial y^4} + 2\left(\frac{1}{2G_{xy}} - \frac{\upsilon_{xy}}{E_x}\right) = \left(\frac{\partial^2 w}{\partial x \partial y}\right)^2 - \left(\frac{\partial^2 w_0}{\partial x \partial y}\right)^2 + \frac{\partial^4 w_0}{\partial x^2 \partial y^2} - \frac{\partial^4 w}{\partial x^2 \partial y^2} \tag{8.34}$$

$$\begin{gathered} D_x\frac{\partial^4 (w - w_0)}{\partial x^4} + 2\left(D_1 + 2D_{xy}\right)\frac{\partial^4 (w - w_0)}{\partial x^2 \partial y^2} + D_y\frac{\partial^4 (w - w_0)}{\partial y^4} = \\ h\left[\frac{\partial^2 w}{\partial x^2}\cdot\frac{\partial^2 F}{\partial y^2} - 2\frac{\partial^2 w}{\partial x \partial y}\cdot\frac{\partial^2 F}{\partial x \partial y} + \frac{\partial^2 w}{\partial y^2}\cdot\frac{\partial^2 F}{\partial x^2} - \rho\frac{\partial^2 w}{\partial t^2}\right] \end{gathered} \tag{8.35}$$

where h is the plate thickness and ρ is the plate density. The flexural densities of the plate are defined as follows:

$$D_x \equiv \frac{E_x h^3}{12(1-\upsilon_{xy}\upsilon_{yx})}; \quad D_y \equiv \frac{E_y h^3}{12(1-\upsilon_{xy}\upsilon_{yx})}; \quad D_{xy} = \frac{G_{xy} h^3}{12}$$
$$D_1 \equiv \frac{E_x \upsilon_{yx} h^3}{12(1-\upsilon_{xy}\upsilon_{yx})} = \frac{E_y \upsilon_{xy} h^3}{12(1-\upsilon_{xy}\upsilon_{yx})} \tag{8.36}$$

The out-of-plane boundary conditions of the present problem, assuming that the initially straight edges of the plate remain straight after buckling, can be written as

$$@x=0,a \quad w=w_0=0 \quad \text{and} \quad \frac{\partial^2 w}{\partial x^2} = \frac{\partial^2 w_0}{\partial x^2} = 0$$
$$@y=0,b \quad w=w_0=0 \quad \text{and} \quad \frac{\partial^2 w}{\partial y^2} = \frac{\partial^2 w_0}{\partial y^2} = 0 \tag{8.37}$$

The in-plane boundary conditions have no restraint at $y = 0,b$ and in the x direction

$$@x=0,a \quad \frac{1}{b}\int_0^b \sigma_x dy = \frac{1}{b}\int_0^b \frac{\partial^2 F}{\partial y^2} dy = -P$$
$$@y=0,b \quad \frac{1}{a}\int_0^a \sigma_y dx = \frac{1}{a}\int_0^a \frac{\partial^2 F}{\partial x^2} dx = 0 \tag{8.38}$$

where P is the compression load acting on the plate.

The assumed solution for the w and the initial geometric imperfection, w_0, have the following forms:

$$w(x,y,t) = A(t)\sin\frac{m\pi x}{a}\sin\frac{n\pi y}{b} \tag{8.39}$$

$$w_o(x,y,t) = A_o \sin\frac{m\pi x}{a}\sin\frac{n\pi y}{b} \tag{8.40}$$

One should note that the expressions in eqs. (8.39) and (8.40) satisfy the boundary conditions assumed above (eq. (8.37)). Substituting those equations in the compatibility equation leads to the following relationship between the stress function F and the displacements

$$\frac{1}{E_y}\cdot\frac{\partial^4 F}{\partial x^4} + \frac{1}{E_x}\cdot\frac{\partial^4 F}{\partial y^4} + 2\left(\frac{1}{2G_{xy}} - \frac{\upsilon_{xy}}{E_x}\right) = \frac{m^2 n^2 \pi^4}{2a^2 b^2}\left[\cos\frac{2m\pi x}{a} + \cos\frac{2n\pi y}{b}\right]\left(A^2 - A_0^2\right) \tag{8.41}$$

The solution for eq. (8.41), which also satisfies the in-plane boundary conditions, eq. (8.38), can be written as

$$F(x,y,t) = \left(A^2 - A_0^2\right)\left[\frac{a^2 n^2 E_y}{32 b^2 m^2}\cos\frac{2m\pi x}{a} + \frac{b^2 m^2 E_x}{32 a^2 n^2}\cos\frac{2n\pi y}{b}\right] - \frac{P}{2}y^2 \tag{8.42}$$

Equation (8.42) is then inserted in eq. (8.34) yielding

$$\begin{aligned}&\frac{\pi^4}{h}\left[D_x\frac{m^4}{a^4} + 2\left(D_1 + 2D_{xy}\right)\frac{m^2 n^2}{a^2 b^2} + D_y\frac{n^4}{b^4}\right][A(t) - A_0] = \frac{m^2\pi^2}{a^2}P(t)\cdot A(t)\\ &-\rho\frac{d^2A(t)}{dt^2} + \frac{\pi^4}{8}\left[\frac{m^4}{a^4}E_x\cos\frac{2n\pi y}{b} + \frac{n^4}{b^4}E_y\cos\frac{2m\pi x}{a}\right]\left[A^2(t) - A_0^2\right]A(t)\end{aligned} \tag{8.43}$$

Note that eq. (8.43) is a nonlinear equation containing functions of the variables x, y and t. To eliminate the x, y dependency, the Galerkin method is applied which demands the multiplication of both sides of the equation by $\sin\frac{m\pi x}{a}\sin\frac{n\pi y}{b}\,dxdy$ and integrating over the middle plane of the plate yielding the following nonlinear time-dependent expression:

$$\begin{aligned}&\frac{d^2A(t)}{dt^2} + \frac{\pi^4}{\rho h}\left[D_x\frac{m^4}{a^4} + 2\left(D_1 + 2D_{xy}\right)\frac{m^2 n^2}{a^2 b^2} + D_y\frac{n^4}{b^4}\right][A(t) - A_0]\\ &-\frac{m^2\pi^2}{a^2\rho}P(t)\cdot A(t) + \frac{\pi^4}{24\rho}\left[\frac{m^4}{a^4}E_x + \frac{n^4}{b^4}E_y\right]\left[A^2(t) - A_0^2\right]A(t) = 0\end{aligned} \tag{8.44}$$

where m and n are both odd integers. The time dependence of the term $P(t)$ is chosen according to its time dependence, namely for a step function $P(t) = P_0$ (constant), for an impulse $P(t) = P_0\delta(t)$ (where $\delta(t)$ is the Kronecker delta[1]) or any other time function. Equation (8.44) can be nondimensionalized (assuming $n = 1$) to yield the following expression (see discussion in [11]):

$$\frac{d^2\xi}{d\tau^2} + S\left[\frac{m^4 + 2R_{12}m^2\beta^2 + R_{22}\beta^4}{4}(\xi - \xi_0) - m^2\tau\xi + \frac{(1 - \upsilon_{xy}\cdot\upsilon_{yx})}{8}\left(m^4 + R_{22}\beta^4\right)\left(\xi^2 - \xi_0^2\right)\xi\right] = 0 \tag{8.45}$$

where

$$\begin{aligned}&P_1 \equiv \frac{4D_x\pi^2}{a^2 h};\quad \xi \equiv \frac{A}{h};\quad \xi_0 \equiv \frac{A_0}{h};\quad \beta \equiv \frac{a}{b};\quad \tau \equiv \frac{P}{P_1} = \frac{c\cdot t}{P_1};\\ &R_{12} \equiv \frac{D_1 + 2D_{xy}}{D_x};\quad R_{22} \equiv \frac{D_y}{D_x} = \frac{E_y}{E_x};\quad S \equiv \frac{P_1^3\pi^2}{c^2 a^2\rho}.\end{aligned} \tag{8.46}$$

1 The function is 1 if the variables are equal, and 0 otherwise, namely:

$$\begin{aligned}\delta_{ij} &= 0 \quad \text{if} \quad i \neq j\\ \delta_{ij} &= 1 \quad \text{if} \quad i = j\end{aligned}$$

Equation (8.45) is solved using numerical methods for integration of nonlinear equations, like the famous Runge–Kutta methods.[2] Solutions for the response of the plate for various S values are presented in [11].

It is interesting to note that eq. (8.44) can be used to obtain the static plate buckling load and its natural frequencies, as well as the expression for large plate deflections.

Assuming the static case for a perfect plate,

$$\frac{d^2A(t)}{dt^2} = 0 \quad \text{and} \quad A_0 = 0 \tag{8.47}$$

and neglecting high-order terms like A^3, eq. (8.44) yields

$$\frac{\pi^2}{h}\left[D_x\frac{m^4}{a^4} + 2(D_1 + 2D_{xy})\frac{m^2n^2}{a^2b^2} + D_y\frac{n^4}{b^4}\right] - \frac{m^2}{a^2}P = 0 \tag{8.48}$$

The critical buckling load of the plate is obtained from eq. (8.48) by assuming $n = 1$, namely

$$P_{\text{cr}} = \frac{\pi^2 D_x}{b^2 h}\left[\left(\frac{mb}{a}\right)^2 + \frac{2(D_1 + 2D_{xy})}{D_x} + \frac{D_y}{D_x}\left(\frac{a}{mb}\right)^2\right]$$
$$\Rightarrow \quad (P_{\text{cr}})_{\text{min.}} \text{ occurs} \quad \text{when} \quad \frac{a}{b}\sqrt{\frac{D_y}{D_x}} = \text{integer} \tag{8.49}$$

Neglecting high-order terms like A^3 and assuming a perfect plate ($A_0 = 0$), eq. (8.44) is written as

$$\frac{d^2A(t)}{dt^2} + \frac{\pi^4}{\rho h}\left[D_x\frac{m^4}{a^4} + 2(D_1 + 2D_{xy})\frac{m^2n^2}{a^2b^2} + D_y\frac{n^4}{b^4}\right]A(t) - \frac{m^2\pi^2}{a^2\rho}P(t)\cdot A(t) = 0 \tag{8.50}$$

The second term in eq. (8.50) is the square of the natural frequency of the perfect plate, namely

$$\omega_{mn}^2 = \frac{\pi^4}{\rho h}\left[D_x\frac{m^4}{a^4} + 2(D_1 + 2D_{xy})\frac{m^2n^2}{a^2b^2} + D_y\frac{n^4}{b^4}\right] \tag{8.51}$$

Finally, assuming the static case, the expression for the postbuckling behavior for a perfect plate with large deflections is obtained from eq. (8.44), namely

2 Runge–Kutta methods are a family of implicit and explicit iterative methods (including the Euler methods routine), used in temporal discretization for the approximate solutions of ordinary differential equations developed approx. in 1900 by C. Runge and M. W. Kutta.

$$\left\{\frac{\pi^4}{h}\left[D_x\frac{m^4}{a^4}+2(D_1+2D_{xy})\frac{m^2n^2}{a^2b^2}+D_y\frac{n^4}{b^4}\right]-\frac{m^2\pi^2}{a^2}PA(t)\right\}+\frac{\pi^4}{24}\left[\frac{m^4}{a^4}E_x+\frac{n^4}{b^4}E_y\right]A^3=0 \tag{8.52}$$

Equation (8.52) shows that the lateral deflection would be zero until the critical buckling load given by eq. (8.49) is reached. After that point, nonzero deflections are possible while the load–deflection relation is cubic.

For further references dealing with the behavior of plates under dynamic buckling, the reader is referred to [16–20, 22, 24, 32, 36].

The definition of the buckling loads for plates for both static and dynamic cases is one of the often issues to be solved consistently. To assist the definition of the buckling load for a plate, a method is described in Appendix A, which can be easily applied for experimental and numerical data consisting of deflections versus applied load.

8.4 Dynamic buckling of thin-walled structures – numerical and experimental results

The dynamic buckling of thin-walled structures has been dealt in depth also from the experimental point of view, besides various calculations using different approaches. Figure 8.8a presents a shallow clamped spherical cap loaded by a sudden rectangular pressure q applied at time $t = 0$ and held constant for a given period of time and then suddenly removed (as described in [4]). Typical numerical results are then shown in Fig. 8.8b, showing the reduction of the critical nondimensional pressure by the increasing the duration of the applied pressure (see other results in [4]). It is interesting to note that the static critical load, from a certain nondimensional time duration (in this case from approximately $\bar{\tau} = 3$) and the static critical pressure, is above the dynamic critical pressure, whereas for short time duration the dynamic critical load is higher than the static one, as is normally known for short duration loads. The various nondimensional terms used in Fig. 8.8 were defined in [4] as

$$\bar{\tau}=\frac{ct}{R};\quad c=\sqrt{\frac{E}{\rho}};\quad p=\frac{q}{q_0};$$
$$\lambda=2\left[3\left(1-\upsilon^2\right)\right]^{1/4}\left(\frac{H}{h}\right)^{1/2};\quad q_0=\frac{2E}{\left[3\left(1-\upsilon^2\right)\right]^{1/2}}\left(\frac{h}{R}\right)^{1/2} \tag{8.53}$$

where E is the Young's modulus, υ is the Poisson's ratio and ρ is the density.

Similar results were reported in [24], where an isotropic plate (Fig. 8.9a) is suddenly loaded by a rectangular impulse and the response of the plate is presented using the term DLF as being defined in eq. (8.2) as a function of the nondimensional time duration, where T_s is the impact period and T_p is the natural period of the plate. As shown by other researchers, in the vicinity of $T_s/T_p = 1$, the term DLF is less than

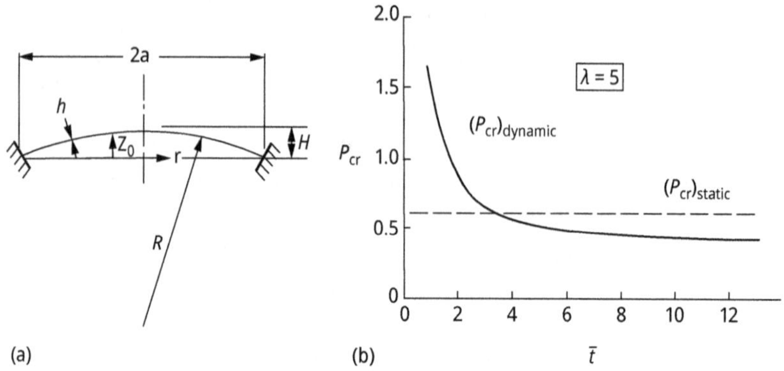

Fig. 8.8: (a) A clamped shallow spherical cap and (b) variation of the nondimensional critical pressure with nondimensional load duration (adapted from [4]).

unity, implying that the static load is higher than the dynamic one. Shifting this period ratio toward 0.5 shows a rapid increase in the value of DLF, reaching a maximal value of 3.6 in the vicinity of $T_s/T_p = 0.1$ (approx.); namely, as expected the plate can withstand high dynamic compressive stresses (3.6 the static buckling load of the plate) provided the time duration is very short. Impacting the plate with sinusoidal or triangular impulses displays the same behavior as the rectangular one (see Fig. 8.10 taken from [24]).

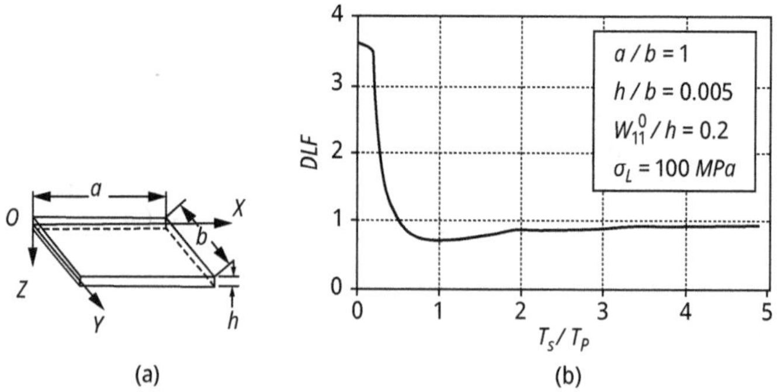

Fig. 8.9: (a) An isotropic flat plate and (b) variation of DLF with the nondimensional load duration (adapted from [24]).

Experimental results for laminated composite plates under impulse-type uniaxial loading obtained by dropping masses and measuring the response using a back-to-back strain gage bonded in the center of the plate are described in [20]. Figure 8.11 presents the experimental variation of the DLF term defined as

$$\mathrm{DLF} \equiv \frac{(\varepsilon_{\mathrm{cr}})_{\mathrm{dyn.}}}{(\varepsilon_{\mathrm{cr}})_{\mathrm{st.}}} \tag{8.54}$$

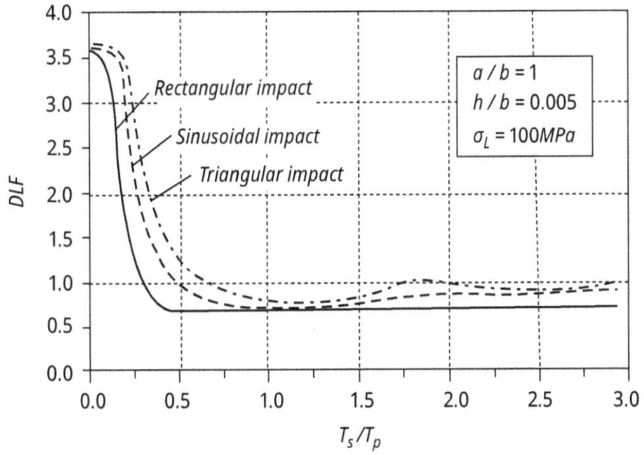

Fig. 8.10: Variation of DLF with the nondimensional load duration for three types of impulses (adapted from [24]).

as a function of the nondimensional period ratio $T_{\mathrm{imp}}/T_{\mathrm{b}}$, where T_{imp} is the period of the impact as measured by the bonded strain gages and T_{b} is the lowest natural frequency of the plate.

As shown before, DLF less than one was measured for some values of nondimensional period ratio, while for short periods, the DLF is higher than unity, reaching experimental values of up to 2. The properties of the specimens presented in Fig. 8.11 are summarized in Tab. 8.1, while the material properties are presented in Tab. 8.2 (using the data published in [20]). Table 8.3 presents additional data in the form of natural frequencies and static buckling strains as obtained during the tests described in [20]. One should note that except plate CM32, which was had clamped boundary conditions all around the perimeter of the plate, all the other plates had simply supported boundary conditions along the four edges of the plate.

As the measured response of the plates was in the form of strains, measured by two back-to-back bonded strain gages, the compression strain ε_{c} and the bending strain ε_{b} were calculated[3] and the axial compression load and lateral out-of-plane deflection were replaced by the compression and bending strains, respectively.

The definition of the static load was performed using the modified Donell's approach, which was found to be the most appropriate for plates with initial geometric imperfections, w_0 (see a discussion in [45]), having the following expression:

3 $\varepsilon_{\mathrm{c}} \equiv \frac{\varepsilon_1 + \varepsilon_2}{2}$; $\varepsilon_{\mathrm{b}} \equiv \frac{\varepsilon_1 - \varepsilon_2}{2}$ where ε_1 and ε_2 are the measured strains.

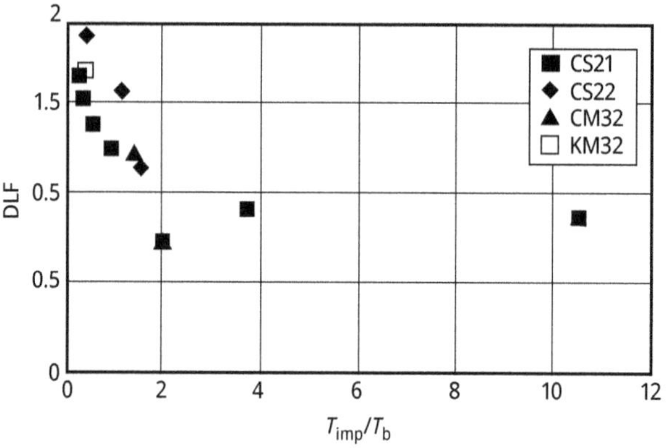

Fig. 8.11: Variation of DLF with the nondimensional load duration – experimental results (adapted from [20]).

Tab. 8.1: The properties of the tested plates (from [20]).

Specimen	Material	Lay-up	Total number of layers	Total thickness (measured) (mm)
CS21	Graphite-epoxy HT-T300 (Torey)	$(\pm 45°, \pm 45°, \pm 45°)_{sym}$	12	1.63
CS22	Graphite-epoxy HT-T300 (Torey)	$(\pm 45°, \pm 45°, \pm 45°)_{sym}$	12	1.63
CM32	Graphite-epoxy HT-T300 (Torey)	$(0°, \pm 45°, 90°, \pm 45°, 0°)$	9	1.125
KM32	Kevlar (Dupont)	$(0°, \pm 45°, 90°, \pm 45°, 0°)$	9	1.125

Tab. 8.2: The material properties of the tested plates (from [20]).

Material	E_{11} (MPa)	E_{22} (MPa)	G_{12} (MPa)	υ_{12}	υ_{21}
Graphite–epoxy HT-T300 (Torey)	122	8.55	3.88	0.32	0.022
Kevlar (Dupont)	70.8	5.5	2.05	0.34	0.026

Tab. 8.3: The experimental results – the tested plates (from [20]).

Specimen	AR $[a \times b]$ (mm^2)	f_{exp} (Hz)	$(\varepsilon_{cr})_{st.}$ (μs)
CS21	2 [150 × 300]	350	2,600
CS22	2 [150 × 300]	250	2,413
CM32*	1 [225 × 225]	393	90
KM32	1 [225 × 225]	186	400

*Nominal clamped boundary conditions.

$$\frac{\varepsilon_b}{\varepsilon_c} = \frac{\varepsilon_b + w_0}{(\varepsilon_{cr})_{st.} + a\left(\varepsilon_b^2 + 3\varepsilon_b w_0 + 2w_0^2\right)} \quad (8.55)$$

Using experimental data in the form of ε_b versus ε_c, one can curve fit it using eq. (8.55) to yield the static buckling strain $(\varepsilon_{cr})_{st.}$ together with the initial geometric imperfection w_0 and the constant a. Based on the work performed in [18], the dynamic buckling strain $(\varepsilon_{cr})_{dyn.}$ is determined by curve fitting the following equation to the experimental data (ε_b vs. ε_c):

$$\frac{\varepsilon_b}{\varepsilon_c} = \frac{\varepsilon_b + w_0}{(\varepsilon_{cr})_{dyn.} + a\varepsilon_b} \quad (8.56)$$

As for the static case, the two terms, the constant a and the initial geometric imperfection w_o, are also determined from the curve fitting process, yielding consistent results as presented in [20].

A dynamic buckling investigation was numerically performed in [39] on a curved laminated composite stringer stiffened panel. Details of the actual model, the dimensions of the stringer, the FE model and the mode shape at $f = 424$ (Hz) (the lowest frequency of the panel) are presented in Fig. 8.12.

The variation of the DLF (defined by eq. (8.1)) versus the nondimensional load duration is shown in Fig. 8.13, where T is the period of the applied load and $T_b = 1/f_b = 2{,}358.5$ μs is the lowest natural period of the stringered panel ($f_b = 424$ Hz). As shown before for other structural cases, also for this stringered panel, the DLF is lower than unity in the vicinity of the lowest natural frequency of the specimen, returning to values above unity for very short periods, while for long periods of time, the DLF tends to unity.

Investigations of shells under dynamic-type loading have been performed by many investigators, like in [21, 23, 26, 27, 30, 31, 38, 46, 47]. A comprehensive experimental and numerical investigation was performed in the Ph.D. thesis written by Eglis [38], in which composite cylinders were subjected to gradually and suddenly applied axial compression loads, including half-sine-shaped pulse loads leading to the eventually dynamic buckling of the tested specimens. Figure 8.14 shows typical numerical and experimental results for the variation of the DLF with the loading period. The general trend presented in Fig. 8.14 is that the DLF is above unity for the tests performed on type 1 shells [38]. The mode shapes for loading periods less, equal and above the lowest natural frequency (the largest period) are presented in Fig. 8.15. At $T < \tau/2$, the shell buckling modes are similar to the static buckling mode, but with a larger number of longitudinal waves, at load duration $T = \tau/2$ the buckling mode starts to transform to an axisymmetric mode while for $T > \tau/2$ leads to axisymmetric buckling mode together with a slight drop of DLF at a load duration of $T = 2a/c$ (a is the shell length and c the speed of sound in the shell). The natural periods for RTU no. 16, RTU no. 4 and RTU nos. 1–4 tested shells (see Fig. 8.14) were $\tau = 2.92$, 4.04 and 6.06 ms, respectively (see [38]).

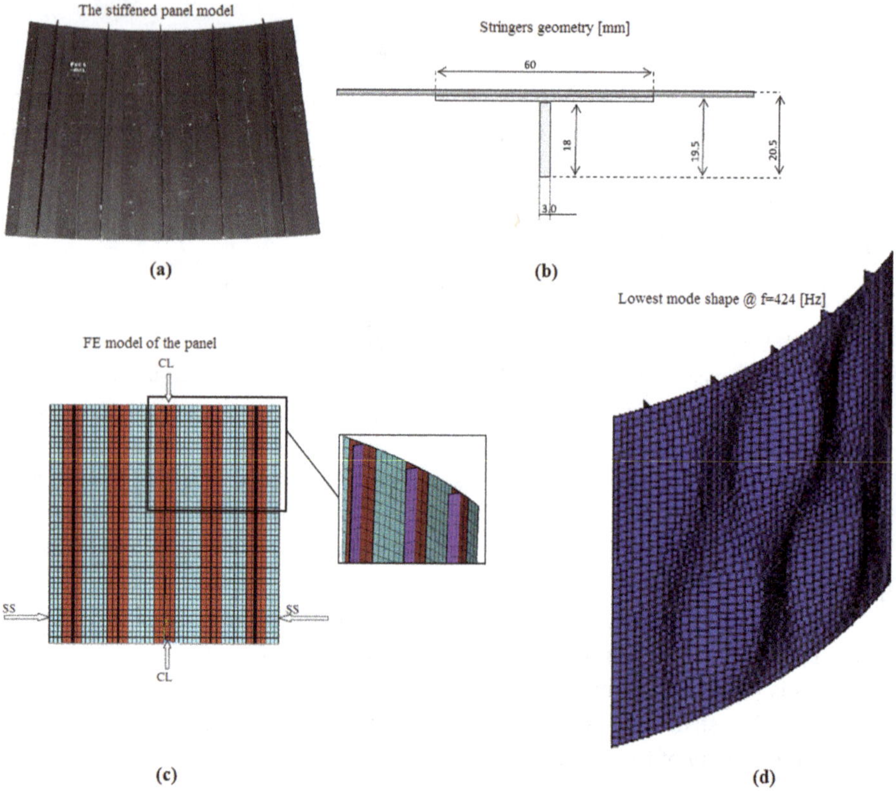

Fig. 8.12: Dynamic buckling investigation of a curved laminated composite stringer stiffened panel: (a) the stringered stiffened panel model, (b) geometric dimensions of the stringer stiffened panel, (c) the FE model and (d) the lowest mode shape of the stringered stiffened panel @ $f = 424$ Hz (adapted from [39]).

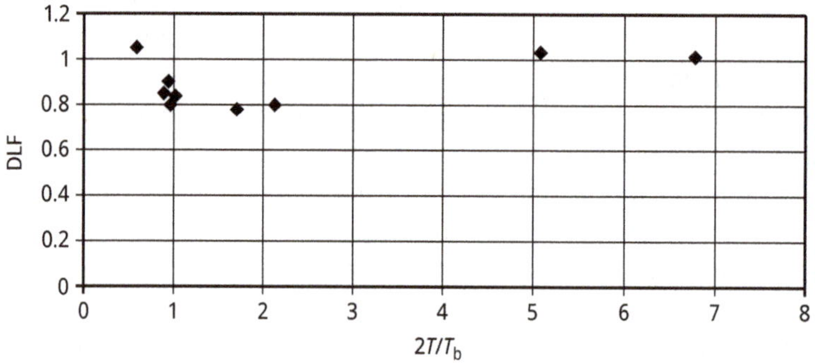

Fig. 8.13: Variation of DLF with the nondimensional load duration – experimental results (adapted from [39]).

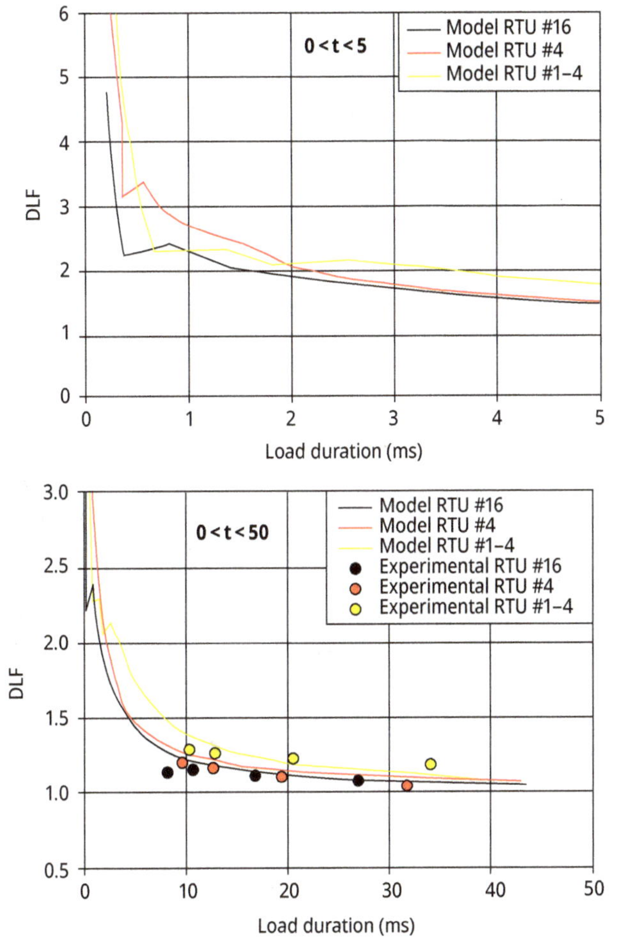

Fig. 8.14: Typical results from [38] – variation of DLF with load duration – experimental and numerical results of specimen type 1 (adapted from [38]).

Similar results are presented in [46], where numerical and experimental investigations of a composite laminated cylindrical shell under impulsive loading are presented. The tested shell (see Fig. 8.16a) had a laminate of [0°, – 45°, + 45°], which was nominally clamped at its edges, with a length of 230 mm (between the clamped edges), a diameter of 250 mm and manufactured from graphite–epoxy with the following properties: $E_{11} = 137.0$ GPa, $E_{22} = 9.81$ GPa, $G_{12} = 5.886$ GPa and $\nu_{12} = 0.34$ and 0.125 mm layer thickness. The dynamic loading was calculated using a dropping mass on the end plate with a mass of 32 kg. The calculated shell static mode shape is presented in Fig. 8.16b, while the dynamic responses for 746 and 2,170 μs impact duration are shown in Fig. 8.16c and d, respectively. As expected the static and the dynamic modes have different patterns. The calculated DLF for numerical and experimental

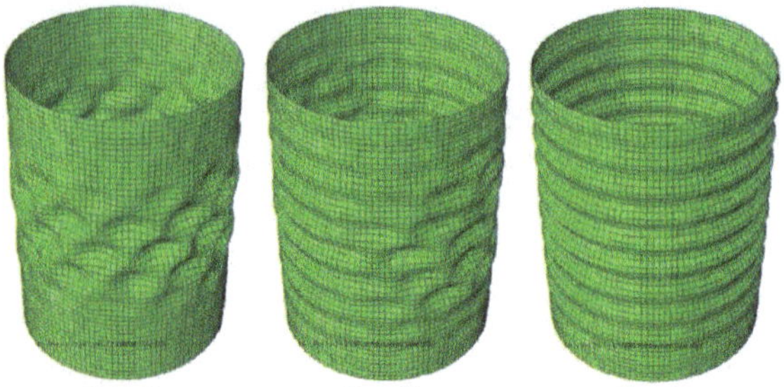

Fig. 8.15: Typical results from [38] – variation of dynamic buckling mode shapes (not to scale) with load duration (τ represents the longest natural bending period of the shell) – numerical results of specimen type 1 (adapted from [38]).

results is presented in Fig. 8.17 for various initial geometric imperfections. Although the experimental results do not go below DLF = 1, the numerical results show a DLF lower then unity in the vicinity of $T/T_b = 1$, for zero initial geometric imperfections, while for higher values of geometric imperfections, the DLF curve is higher the unity.

Another study [47] presents similar results to those described in [46], for the behavior of composite cylindrical shells having laminates of [0°/90°/0°/90°/90°/0°/90°/0°] and [0°/0°/60°/ – 60°/ – 60°/60°/0°/0°] under various durations of load impulse. A step pulse was used for the compressive loading, while calculations using the ABAQUS/Standard code yielded a lowest natural frequency of 427 Hz and a static linear buckling load of 97.19 kN for the [0°/90°/0°/90°/90°/0°/90°/0] laminate and 120.39 kN for the [0°/0°/60°/ – 60°/ – 60°/60°/0°/0°] laminate. ABAQUS/Explicit was used for the dynamic buckling analysis of the shells under the impulsive loading. Figure 8.18 presents the DLF, calculated using eq. (8.1), as a function of the period of the applied step impulse loading. For short periods the DLF is higher than unity, while from $T = 5$ ms it drops below unity for both laminates. As the largest natural period can be calculated to 2.342 ms, it again shows that in the vicinity of the lowest natural frequency the DLF might fall below unity also for laminated composite cylindrical shells.

In a relatively recent study [48], co-authored by the author of the present book, the dynamic instability of thin-walled laminated composite cylindrical shells was thoroughly investigated with interesting new results. The ABAQUS 2017 code [49] was used to model the cylindrical shell by 14852 S4R[4] shell elements. Static and dynamic calculations were numerically performed to investigate the response of the cylindrical shells. The instability of the cylindrical shells under pulse loading was investigated using the Dynamic Explicit code within the ABAQUS 2017 FE software [49], while varying the applied load duration, initial geometric imperfections and the shape of the

4 S4R is a 4-node, quadrilateral, stress/displacement shell element with reduced integration and large-strain formulation existing within the ABAQUS code.

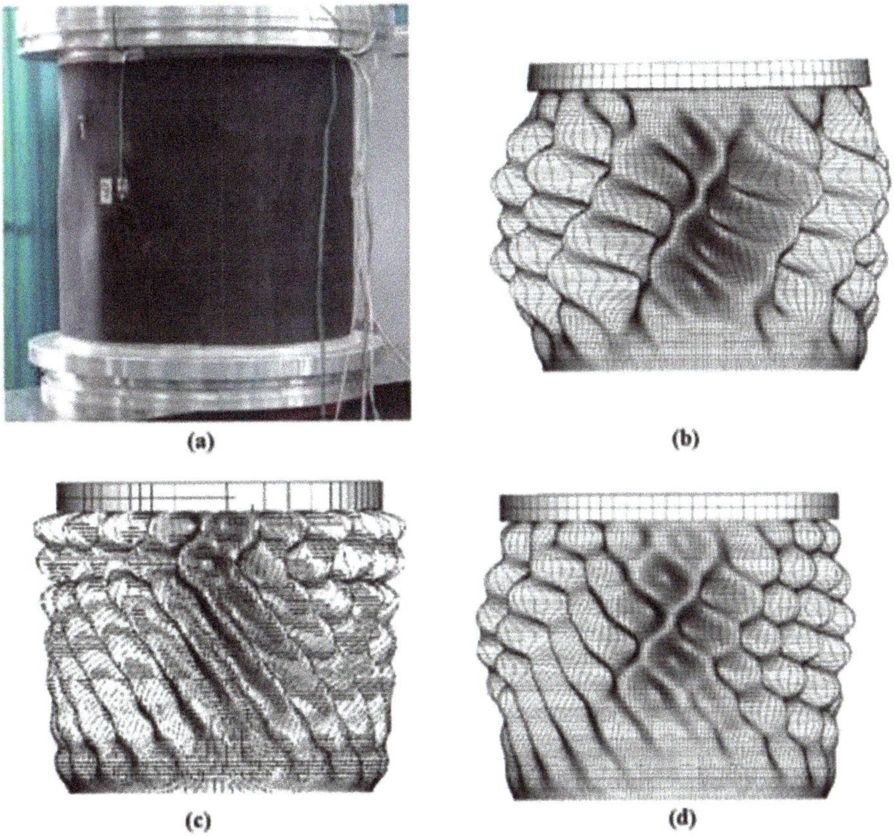

Fig. 8.16: Typical results from [46]: (a) the tested cylindrical shell, (b) the static mode shape, (c) dynamic mode shape @ 746 μs and (d) dynamic mode shape @ 2,170 μs (adapted from [46]).

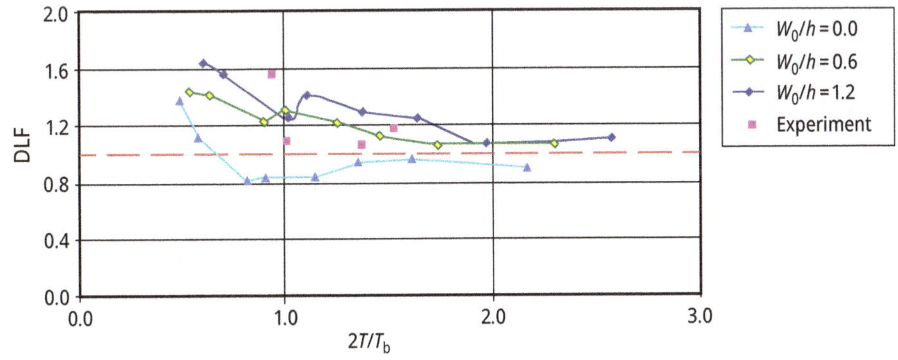

Fig. 8.17: DLF versus nondimensional impulse period ($T_b = 1/f_b$, where f_b is the lowest natural frequency of the shell) (adapted from [46]).

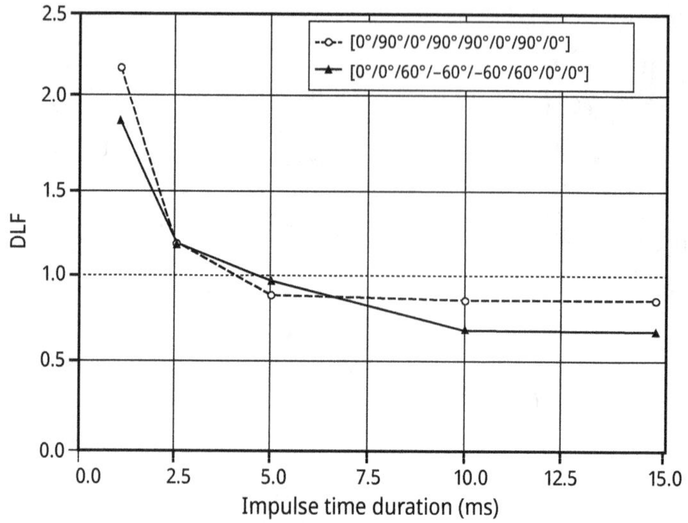

Fig. 8.18: DLF versus nondimensional impulse period ($T_b = 1/f_b$, where f_b is the lowest natural frequency of the shell) (adapted from [46]).

pulse load by using trapeze, triangular double triangular and half sine forms. Numerous graphs presenting the DLF for various parameters of the problem were generated. It is stated in the study [48] that the shape of the pulse loading has a great influence on the stability of the composite shell mainly for load duration smaller than its first natural bending period. A high value of DLF was obtained for loads having trapeze- and double-triangular-type shapes, which are characterized with the lowest value of the total pulse energy applied to the shell. A DLF lower than unity, namely the static buckling load is higher than the dynamic one, was observed only for the case in which the shell was loaded by a trapeze-type pulse having duration slightly higher than its first half natural period.

References

[1] Lindberg, H.E. Dynamic pulse buckling-theory and experiment, SRI International, DNA 6503H, Handbook, Vol. 333, Ravenswood Avenue, Menlo Park, California 94025, USA, 1983.

[2] Simitses, G.J. Dynamic stability of suddenly loaded structures, Springer Verlag, New York, USA, 1990, 290.

[3] Chung, M., Lee, H.J., Kang, Y.C., Lim, W.-B., Kim, J.H., Cho, J.Y., Byun, W., Kim, S.J. and Park, S.-H. Experimental study on dynamic buckling phenomena for supercavitating underwater vehicle, International Journal of Naval Architecture and Ocean Engineering, 4, 2012, 183–198. http://dx.doi.org/10.2478/IJNAOE-2013-0089.

[4] Budiansky, B. and Roth, R.S. Axisymmetric dynamic buckling of clamped shallow spherical shells, collected papers on instability of shell structures, NASA TN-D-1510, 761, 1962, 597–606.

[5] Hegglin, B. Dynamic buckling of columns, SUDAER No. 129, Department of Aeronautics & Astronautics, Stanford University, Stanford, California, USA, June 1962, 55.
[6] Budiansky, B. and Hutchinson, J.W. Dynamic buckling of imperfection sensitive structures, in Proceedings of the 11th international congress of applied mechanics, Götler, H. Ed., Springer-Verlag, Berlin, 1964, 636–651, 1966.
[7] Hutchinson, W.J. and Budiansky, B. Dynamic buckling estimates, AIAA Journal, 4(3), 1966, 525–530.
[8] Lock, M.H. A study of buckling and snapping under dynamic load, Air Force Report No. SAMSO-TR-68-100, Aerospace Report No. TR-0158(3240-30)-3 December 1967, 55.
[9] Burt, J.A. Dynamic buckling of shallow spherical caps subjected to a nearly axisymmetric step pressure load, Master Thesis, Naval Postgraduate School, Monterey, California 93940, USA, Sept 1971, 75.
[10] Volmir, S.A. Nonlinear dynamics of plates and shells, Science, 1972, 544, Moscow, USSR.
[11] Ekstrom, R.E. Dynamic buckling of a rectangular orthotropic plate, AIAA Journal, 11(12), 1973, 1655–1659.
[12] Lee, L.H.N. Dynamic buckling of an inelastic column, International Journal of Solids and Structures, 17(3), 1981, 271–279.
[13] Ari-Gur, J., Weller, T. and Singer, J. Experimental and theoretical studies of columns under axial impact, International Journal of Solids and Structures, 18(7), 1982, 619–641.
[14] Lee, L.H.N. and Ettestad, K.L. Dynamic buckling of an ice strip by axial impact, International Journal of Impact Engineering, 1(4), 1983, 343–356.
[15] Gary, G. Dynamic buckling of an elastoplastic column, International Journal of Impact Engineering, 1(4), 1983, 357–375.
[16] Dannawi, M. and Adly, M. Constitutive equation, quasi-static and dynamic buckling of 2024-t3 plates – experimental result and analytical modelling, Journal de Physique Colloques, 49(C3), 1988, C3-575–C3-588.
[17] Birman, V. Problems of dynamic buckling of antisymmetric rectangular laminates, Composite Structures, 12(1), 1989, 1–15.
[18] Weller, T., Abramovich, H. and Yaffe, R. Dynamic buckling of beams and plates subjected to axial impact, Computers and Structures, 32(3–4), 1989, 835–851.
[19] Ari-Gur, J. and Hunt, D.H. Effects of anisotropy on the pulse response of composite panels, Composite Engineering (Now Composites Part B), 1(5), 1991, 309–317.
[20] Abramovich, H. and Grunwald, A. Stability of axially impacted composite plates, Composite Structures, 32(1–4), 1995, 151–158.
[21] Schokker, A., Sridharan, S. and Kasagi, A. Dynamic buckling of composite shells, Computers and Structures, 59(1), 1996, 43–53.
[22] Ari-Gur, J. and Simonetta, R. Dynamic pulse buckling of rectangular composite plates, Composites Part B, 28, 1997, 301–308.
[23] Eslami, M.R., Shariyat, M. and Shakeri, M. Layerwise theory for dynamic buckling and postbuckling of laminated composite cylindrical shells, AIAA Journal, 36(10), 1998, 1874–1882.
[24] Petry, D. and Fahlbusch, G. Dynamic buckling of thin isotropic plates subjected to in-plane impact, Thin-Walled Structures, 38(3), 2000, 267–283.
[25] Zheng, Z. and Farid, T. Numerical studies on dynamic pulse buckling composite laminated beams subjected to an axial impact pulse, Composite Structure, 56(3), 2002, 269–277.
[26] Yaffe, R. and Abramovich, H. Dynamic buckling of cylindrical stringer stiffened shells, Computers & Structures, 81(9–11), 2003, 1031–1039.
[27] Abumeri, G.H. and Chamis, C.C., Probabilistic dynamic buckling of smart composite shells, NASA/TM-2003-212710, 2003, 15.

[28] Zhang, Z. Investigation on dynamic pulse buckling and damage behavior of composite laminated beams subject to axial pulse, Ph.D. thesis, Faculty of Engineering, Civil Engineering, Dalhousie University, Halifax, Nova Scotia, Canada, 2003, 228.
[29] Lindberg, H.E. Little book of dynamic buckling, LCE Science/Software, 18388 Chaparral Drive Penn Valley, CA 95946–9234, USA, September 2003, 39.
[30] Bisagni, C. Dynamic buckling tests of cylindrical shells in composite materials, 24th International Congress of the Aeronautical Sciences (ICAS2004), 29 August–3 September 2004, Yokohama, Japan.
[31] Bisagni, C. Dynamic buckling of fiber composite shells under impulsive axial compression, Thin-Walled Structures, 43(3), 2005, 499–514.
[32] Kubiak, T. Dynamic buckling of thin-walled composite plates with varying widthwise material properties, International Journal of Solids and Structures, 42, 2005, 5555–5567.
[33] Ji, W. and Waas, A.M. Dynamic bifurcation buckling of an impacted column, International Journal of Engineering Science, 46(10), 2008, 958–967.
[34] Kubiak, T. Dynamic buckling estimation for beam-columns with open cross-sections, Paper No. 13, in Proceedings of the twelfth international conference on civil, structural and environmental engineering computing, Topping, B.H.V., Costa Neves, L.F. and Barros, R.C. Eds., Civil-Comp Press, Stirlingshire, Scotland, 2009.
[35] Jabareen, M. and Sheinman, I. Dynamic buckling of a beam on a nonlinear elastic foundation under step loading, Journal of Mechanics of Materials and Structures, 4(7–8), 2009, 1365–1373.
[36] Michalska, K.K. About some important parameters in dynamic buckling analysis of plated structures subjected to pulse loading, Mechanics and Mechanical Engineering, 14(2), 2010, 269–279.
[37] Mania, R.J. Membrane-flexural coupling effect in dynamic buckling of laminated columns, Mechanics and Mechanical Engineering, 14(1), 2010, 137–150.
[38] Eglitis, E. Dynamic buckling of composite shells, Ph.D. thesis, Riga Technical University, Faculty of Civil Engineering, Institute of Materials and Structures, Riga, Latvia, 2011, 172.
[39] Abramovich, H. and Less, H. Dynamic buckling of a laminated composite stringer-stiffened cylindrical panel, Composite Part B, 43(5), July 2012, 2348–2358.
[40] Landa, A. The buckling resistance of structures subjected to impulsive type actions, Master Thesis, Norwegian University of Science and Technology, Department of Marine Technology NTNU Trondheim, Norway, Feb 2014, 104.
[41] Straume, J.G. Dynamic buckling of marine structures, Master Thesis, Norwegian University of Science and Technology, Department of Marine Technology NTNU Trondheim, Norway, June 2014, 148.
[42] Kuzkin, V.A. Structural model for the dynamic buckling of a column under constant rate compression, arXiv:1506.00427 [physics.class-ph], 2015.
[43] Mouhata, O. and Abdellatif, K. Dynamic buckling of stiffened panels, The 5th International Conference of Euro Asia Civil Engineering Forum (EACEF-5), Procedia Engineering, 125, 2015, 1001–1007.
[44] Lekhnitskii, S.G. Anisotropic plates, 2nd, translation from Russian, Gordon and Breach, New York, 1968, 534.
[45] Abramovich, H., Weller, T. and Yaffe, R. Application of a modified Donnell technique for the determination of critical loads of imperfect plates, Computers and Structures, 37, 1990, 463–469.
[46] Abramovich, H., Weller, T. and Pevzner, P. Dynamic buckling behavior of thin- walled composite circular cylindrical shells under axial impulsive loading, in Proceedings of AIAC-12, twelfth Australian international aerospace congress, Melbourne, 19–22 March 2007.
[47] Citra, V. and Priyadarsini, R.S. Dynamic buckling of composite cylindrical shells subjected to axial impulse, International Journal of Scientific & Engineering Research, 4(5), 2013, 162–165.

[48] Zaczynska, M., Abramovich, H. and Bisagni, C. Parametric studies on the dynamic buckling phenomenon of a composite cylindrical shell under impulsive axial compression, Journal of Sound and Vibration, 482, 2020, Paper Id: 115462, 17.
[49] ABAQUS 2017 code, https://www.3ds.com/products/simulia/abaqus, also ABAQUS Manuals, Version 5.7, Hibbitt: Karlsson and Sorensen, 1997.

Appendix A: Calculation of the critical buckling load of a uniaxial loaded plate from test results

One of the issues faced by a researcher is to correctly define the buckling load of a plate for the static and dynamic cases, based on experimental points obtained either by tests or numerically. The following method was already published and applied in [39] and is based on the work performed by Brown.[5] It was shown that for practical structural geometries and loading conditions, the following equation holds:

$$\delta_i^2 - \delta_0^2 = \alpha^2 h^2 \theta = \alpha^2 h^2 \left[\frac{P_i}{P_{cr}} - 1 + \frac{\delta_0}{\delta_i} \right] \tag{A.1}$$

where δ_i is the lateral plate deflection due to corresponding in-plane loading P_i; δ_0 is the initial lateral plate deflection; h is the thickness of the plate; and α is a constant accounting for the load configuration and boundary conditions. After some mathematical manipulations, eq. (A.1) can be rewritten as

$$P_i \delta_i = A_1 + A_2 \delta_i + A_3 \delta_i^3 \tag{A.2}$$

where

$$A_1 = -P_{\text{cr}} \delta_0; \quad A_2 = P_{\text{cr}} \left[1 - \frac{\delta_0^2}{\alpha^2 h^2} \right]; \quad A_3 = \frac{P_{\text{cr}}}{\alpha^2 h^2} \tag{A.3}$$

Equation (A.3) is presented in a suitable form for the application of the least square method to fit a curve to a series of given data points, P_i, δ_i. To do so, let us define the sum of the squares of the residuals for m data points as

$$\text{SUM} = \sum_{i=1}^{m} \left(P_i \delta_i - A_1 - A_2 \delta_i - A_3 \delta_i^3 \right)^2 \tag{A.4}$$

Taking partial derivatives of eq. (A.4) with respect to the coefficients A_1, A_2 and A_3 (see eq. (A.3)) and equating them to zero (thus minimizing the error) yields the following system of equations written in a matrix form:

5 Brown, V.L., Linearized least-squared technique for evaluating plate-buckling loads, Journal Engineering Mechanics, Vol. 116, No. 5, pp. 1050–1057, 1990.

$$\begin{bmatrix} m & \sum_{i=1}^{m}\delta_i & \sum_{i=1}^{m}\delta_i^3 \\ \sum_{i=1}^{m}\delta_i & \sum_{i=1}^{m}\delta_i^2 & \sum_{i=1}^{m}\delta_i^4 \\ \sum_{i=1}^{m}\delta_i^3 & \sum_{i=1}^{m}\delta_i^4 & \sum_{i=1}^{m}\delta_i^6 \end{bmatrix} \begin{Bmatrix} A_1 \\ A_2 \\ A_3 \end{Bmatrix} = \begin{Bmatrix} \sum_{i=1}^{m} P_i\delta_i \\ \sum_{i=1}^{m} P_i\delta_i^2 \\ \sum_{i=1}^{m} P_i\delta_i^4 \end{Bmatrix} \tag{A.5}$$

The three linear equations, presented in a matrix form in eq. (A.5), are then solved for the three unknowns A_1, A_2 and A_3. Substituting them into eq. (A.3) and the results in eq. (A.2) lead to the following cubic equation for the single unknown P_{cr}:

$$P_{cr}^3 - A_2 P_{cr}^2 - A_1^2 A_3 = 0 \tag{A.6}$$

Solving eq. (A.6) will yield the buckling load of the plate based on the displacement–force curve, generated either by experiments or by numerical calculations.

9 Optimization of thin-walled structures

9.1 Introduction

The concept of design optimization is based on using a mathematical method to arrive at the best possible design according to a selected number of desirable features. Since design is a decision-making process, there is a possibility of superior, alternative designs that exist. The use of mathematical expressions of the relevant natural laws, empirically obtained relationships, experience and geometry are combined to obtain an abstract description of the artifact designed [1]. These mathematical expressions can be used to analyze other possible designs and arrive at a more optimal solution [1]. The existence of multiple alternative designs obtainable from the mathematical model leads to the need to introduce criteria for comparing the alternative designs found from the mathematical model [1]. These criteria allow for the use of mathematical methods to find the optimal solution that satisfies all of the criteria. An important factor to consider during the optimization process is the prioritization of criteria, particularly for designs where multiple criteria are required to be optimized. In every design, the optimization of one criterion will have an impact on the other criteria. Prioritization of criteria allows the optimization to be applied in a way which ensures that the more important criteria are optimized while the less important ones remain within the desired design constraints. An example of this can be seen in the aerospace structure field, where weight reduction, cost reduction and structural rigidity are generally prioritized, even though these criteria generally have a negative impact on each other. Additionally, the optimization of a structural design could negatively impact the design from another point of view, such as aerodynamics, where size, shape and surface conditions have higher priority. This example illustrates not only the complexity of optimizing a single component of a design but ensuring that this optimization does not impact the overall design negatively. The example further illustrates the additional complexity in multidisciplinary optimization, particularly cases where optimization from the point of one design discipline negatively impacts that of another discipline requiring the introduction of trade-offs in the design. This high level of complexity has led to the logical exploitation of computers and software in order to allow for an increase in the speed, number of constraints and complexity of problems that can be simultaneously optimized. This has been exploited, particularly in recent years, where the drastic increase in computing power at a fraction of the historical cost has led to a plethora of optimization software [2]. This has led to a major challenge for researchers requiring them to design methods that can be implemented computationally by users with little knowledge of the source algorithms used by the software [2]. The results of an optimization method, whether manual, numerical or computational, are also highly dependent on the quality of the input model [1]. With the lack of obtainable knowledge on the intricate workings of modern software packages, the quality [2], the general concepts, methods and procedures for optimization will be discussed in the next section.

https://doi.org/10.1515/9783111621104-009

9.2 The optimization process

The following sections will provide a brief introduction of the necessary vocabulary, concepts and the design process with the inclusion of optimization within the process.

9.2.1 Vocabulary and concepts

The first concepts that will be discussed are those of the design variable, the design parameter and the design constant. The design variables are made up of quantities by which the model can be described, which can vary during the optimization process [1, 3]. During the design optimization, the design variable is considered an input of the model, which can be varied in order to obtain an optimum solution. The design variables can either be classified as continuous (having any value within a range of values) or discrete (having specific values from a list of permissible values) [2]. There also exist design parameters that are quantities by which the model can be described, which are set at a constant value during the optimization process. The choice of which quantities will be classified as variables or parameters is a subjective decision to be made by the designer [1]. The final type of quantity that can be used to describe a model is the design constants. These are constant values that are usually fixed by the underlying phenomenon instead of the model statement and are, therefore, uncontrollable by the designer, such as the gas constant [1]. These concepts can then be used to generate the objective or target function, which is discussed later.

The second concept that will be discussed is the objective function. The objective function is the specific criterion required to be optimized in order to characterize whether the design is the best one possible. It is a mathematical relation, which relates several model variables, parameters and constants as a function [4]. The purpose of any optimization problem is to find either a maximum or minimum value for the objective function, subject to the constraints of the problem that are further discussed below [4]. The formulation of the objective function is usually the most difficult part of the modeling process, with mathematical relations attempting to describe the function of the model within the conditions imposed by its environment [4]. It must be noted that for a standard optimization problem it is possible, if not highly likely, that the objective "function" can be a system of algebraic or differential equations or even a computerized procedure or subroutine [1]. The objective function can also be written in terms of multiple variables that would then classify the problem as multicriteria optimization [2].

The problem constraints are a set of mathematical relations that limit the possible optimized solutions based on the requirements of the design. The constraints can either be inequality constraints (formulated using > or < criteria) or equality constraints. The points of the objective function that satisfy the constraints of the system are known as feasible design points, with all the feasible design points making up the feasible region [3]. The feasible region will include a global minimum or maximum value of the objective function, which will be the optimized solution to the design problem. The feasible region can also include local minimum or maximum values, which give an optimized solution within a given range, however, will not be the most optimized solution within the entire feasible region. The minima or maxima that give a global minimum or maximum value of the objective function will be the best optimized solution of the design problem [3].

When the objective function and all the constraints of an optimization problem are linear with respect to the design variables, the optimization problem is said to be a linear problem [2]. These problems can be solved using a branch of mathematics known as linear programming [2]. If either the objective function or the constraints are nonlinear, the optimization problem is defined as a nonlinear problem.

The design problem can also be classified as constrained or unconstrained. Constrained design problems arise when explicit constraints are placed on the optimization variables. The constraints placed on the variables can be anything from simple bounds to a system of equalities and inequalities used to model highly complex relationships between variables in the mathematical model. These constraints can be linear, nonlinear or convex in nature and can be further defined by their smoothness, making them either differentiable or nondifferentiable.

The discussed vocabulary and concepts can now be used to describe the general procedure followed during design optimization and will be discussed in the following section.

9.2.1.1 The general design optimization process

The design process, while highly complex and variable in nature, will generally follow a specific set of steps, which can vary depending on the design itself. An example of a general procedure is shown in Fig. 9.1.

The optimization step, while always having been an important step in the past, has become the focus of much research in recent years, owing to the improvements in computing capabilities allowing for more complex optimization to be performed in a reduced amount of time. However, even with the increase in the number and complexity of methods for optimizing a design, the general procedure for optimization has remained similar.

The initial process of the optimization procedure is to set up the model, variables, parameters and constants. Once this has been set up, the objective function and constraints can be defined, usually in the following format [2]:

$$\text{Minimize: } f(x) \quad \text{such that: } g(x) \geq 0, h(x) = 0 \tag{9.1}$$

where x represents the optimization variables, $f(x)$ is the objective function, which can be made up of more than a single function, $g(x)$ represents the inequality constraints and $h(x)$ represents the equality constraints. Once the objective function and

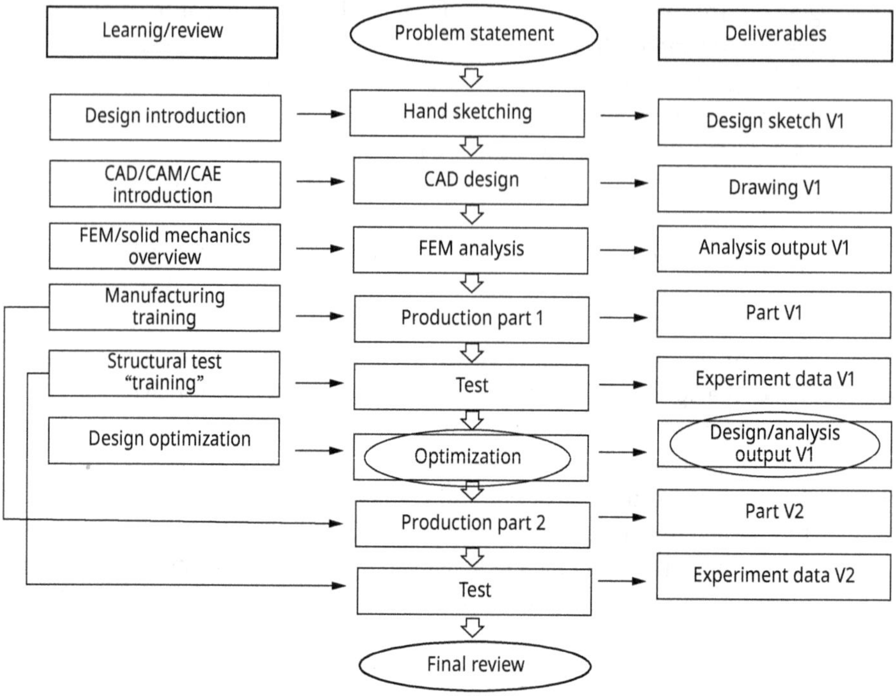

Fig. 9.1: Generalized design procedure.

constraints have been set up, the actual optimization process can begin. Using an initial design, an attempt is made to improve the design by adjusting the design variables using one of the methods described in the next section. By seeing if the objective func-

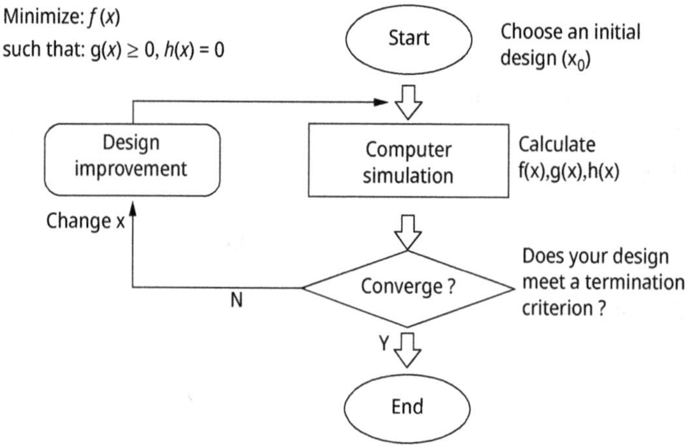

Fig. 9.2: Optimization process – a block diagram.

tion is larger/smaller, depending on if the objective function needs to be maximized/minimized, than the objective function of the previous design, the new design can be directly compared to the previous iteration to determine if it is a more optimized solution. This process is terminated when no improvement can be made in the design without violating any of the constraints [2] (see for example Fig. 9.2).

The procedure described earlier is a general approach to optimization, which is widely used in industry and academy. There exist many different methods of implementing this procedure, which are discussed later in the report.

9.3 Structural optimization

The principal aim of structural design is to create a structure that can carry the expected load in an efficient way [5]. This leads to a design that uses the least amount of material to carry the predicted loads in order to maximize its efficiency. The aim of using a minimum amount of material is usually synonymous by ensuring that the weight of the structure is minimized [5]. The optimization process is a useful tool for optimization of structures, particularly in the aerospace field where lightweight yet rigid structures are required.

Structural optimization is usually split into three main sections: size optimization, shape optimization and topology optimization. While the first two sections are easily explained in their naming, topology optimization attempts to distribute the least amount of material within a specified available area [6].

The main constraint for structural optimization is usually the required load, which the structure needs to carry; however, in real design problems, a multitude of other constraints, such as maximum deformation, fatigue requirements, damage resistance, buckling requirements and natural frequency of the structure, can be included in the optimization procedure [5].

9.4 Multidisciplinary and multiobjective design optimization

To obtain a design solution for many existing problems, it is usually necessary to solve the problem using a number of disciplines. It is very common for the optimization of one subsystem within a design, or even the optimization of a single subsystem from the point of view of a single discipline, to negatively impact the optimization of another subsystem or the subsystem being optimized from the point of view of another discipline. A classic example of this is the design of an aircraft wing, where from a structural point of view the designer would prefer a thicker wing cross section with internal structural strengthening, whereas from an aerodynamics point of view a very thin wing is more optimal. Both requirements are also not optimal for the design of the systems that are stored in the wing which require adequate space to be placed throughout the wing. If the wing was to be optimized by prioritizing a single discipline's requirements, it would negatively impact the design of the wing and the aircraft as a whole. It is for this reason that multidisciplinary design optimization exists, to ensure that the entire design is optimized incorporating all the relevant disciplines simultaneously. As expected, the inclusion of multiple disciplines within the simultaneous optimization problem significantly increases the complexity of the optimization problem.

The need for multidisciplinary optimization leads to the requirement to use a multiobjective design optimization method. These processes use multiple objective functions to find an optimum design [1]. It is generally impossible to arrive at an optimal design for every objective function, as objective functions are usually competing, which introduces the need for trade-offs within the optimization process [1]. The objective functions, relevant criteria, parameters and constants can be represented in a vector form and require specialized mathematics and optimization techniques to solve the optimization problem with the feasible solutions making up an attainable set instead of a region [1]. Another approach that is being taken to solve a multiobjective design optimization problem is to represent the objective function as a single scalar equation, which contains weighted contributions from all the objectives [3].

Most methods that use the vector form attempt to find the Pareto-optimal point. The Pareto efficiency, named after the Italian engineer and economist Vilfredo Pareto who used the concept for economic efficiency and income distribution, is a concept that looks for a Pareto-optimal point, where any change to one objective cannot be made without negatively affecting another objective [3]. A Pareto improvement exists if another optimal design point occurs, where an improvement can be made to one objective without negatively affecting, but not necessarily improving, any other objectives [3]. The set of all Pareto-efficient design points, usually depicted graphically, is called a Pareto set or frontier and is shown in Fig. 9.3.

The example in Fig. 9.3 shows a production-possibility frontier, where the frontier and the area left and below is a continuous room of choice. The brown points are examples of Pareto-optimal choices of production. Points off the frontier, such as K and L, are not Pareto efficient.

There are multiple methods for solving a single scalar setup, which heavily vary in their approach. While most of these methods are based on methods using calculus or numerical solving techniques [3], there is no simple procedure from which they all stem from; therefore, these methods, as well as other methods in general, will be discussed individually in the next section.

9.5 Methods of optimization

The previous section discussed the general approaches and definitions about the optimization of a design solution. This section will use these approaches and concepts to discuss a selection of methods that are used for optimal design analysis. While a mul-

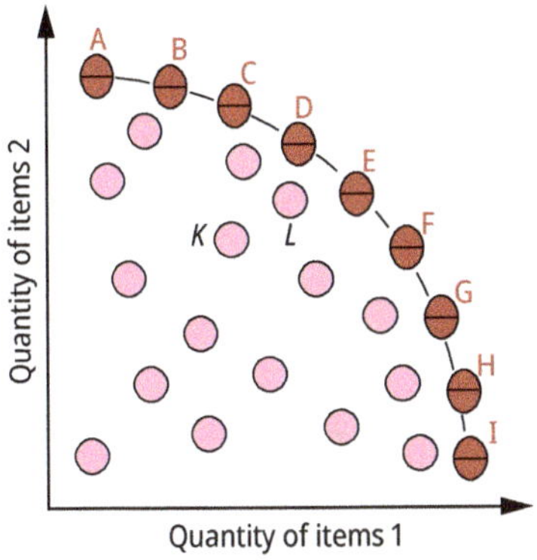

Fig. 9.3: An example of Pareto set.

titude of methods do exist, relying on calculus or numerical methods to obtain a solution, they all require some form of repetition or iteration. While defining a good mathematical model and objective functions from the start can greatly reduce the number of iterations required to arrive at a solution, the specific abilities of computers to perform billions of calculations in each second have made them a requirement for efficient implementation of any of the methods mentioned in what follows below [3, 5]. While simpler techniques can be done manually, it is vastly more inefficient than a computerized solution while limiting the complexity of the problem that can be solved.

The following sections are split into general categories of methods for solving optimization problems with specific examples of methods that fit into each category given.

9.6 Classical optimization techniques

The classical optimization techniques are a set of methods that have been developed from differential calculus and are, therefore, very useful for obtaining optimal solutions for functions that are continuous and differentiable [7]. The methods are analytic in nature and use calculus methods to find the optimal points for the objective function. Many practical problems involve objective functions that are not continuous or differentiable, limiting the number of scenarios to which classical optimization techniques can be applied [7]. The classical techniques, however, are also the basis for the numerical methods used to obtain solutions for practical problems. Classical optimization techniques can be used to find an optimal solution for single-variable functions, multivariable unconstrained functions and multivariable constrained functions with both equality- and inequality-type constraints [7]. Classical optimization methods usually lead to a set of nonlinear, simultaneous equations whose solution is required to obtain the optimum point, with the possibility of these equations being difficult to solve. The necessary and sufficient conditions required to use classical optimization techniques on each of the previously mentioned categories are presented below:

For single-variable optimization problems, the first necessary condition is that at a local minimum/maximum point within the range of the objective function with a dependence on only a single variable in which optimization is being performed, the first derivative of the function will be equal to zero [7]. This theorem does not state that all points that have a gradient equal to zero are minimum/maximum points as they can also be stationary points. A sufficient condition for classifying a minimum/maximum for this class of optimization problem states that if

$$f'(x^*) = f''(x^*) = \cdots = f^n(x^*) = 0 \text{ and } f^n(x^*) \neq 0 \tag{9.2}$$

Then the point is a minimum if the final derivative is greater than zero and n is even; it is a maximum point if the final derivative is negative and neither a maximum or minimum if n is odd [7]. These two theorems can be used to prove a local minimum/maximum point in a classical optimization technique.

The requirements for multivariable problems, both constrained and unconstrained, are immensely more complicated than those for single-variable problems and will therefore not be discussed in this chapter, particularly owing to the superiority of numerical methods when solving these problems. It is for this reason that numerical optimizations methods, particularly with the use of computers, are preferred for solving both complex single-variable problems and multivariable problems. Numerical methods, however, are usually based on classical methods but introduce an iterative estimation process to solve the problem instead of finding actual solutions for the problem.

9.7 Numerical methods of optimization

Numerical methods in mathematics are a highly useful tool for optimization problems. They use approximation methods to arrive at a solution to problems that are too complex for or cannot be solved using classical methods. A numerical method generally uses an initial value for the solution of a problem and, using a variation of developed techniques, iteratively changes the solution until it converges on a solution approximately equal to the actual solution. The iterative nature of these methods makes computers highly suitable tools for running these methods until a solution is reached. There is usually a trade-off between computational complexity and convergence speed when utilizing these methods. These methods are also useful, as a solution can be found without large amounts of analysis needing to be performed manually by the user. However, as with all numerical methods, the ability and speed of the method to converge to a realistic answer depend heavily on the model and the input by the user, as well as the starting feasible point used by the solver.

Many numerical methods and solvers exist for optimization problems. A few examples are next given and discussed in the following sections.

9.7.1 Linear programming

Linear programming is a method where the objective function and all of the equality and inequality constraints are linear with respect to the design variable. Owing to its early conception and relative simplicity, linear programming has been applied to an immeasurable number of optimization problems [7]. The relative simplicity of the objective function and constraints usually allows for a graphical solution to simpler real-world optimization problems. An example of the geometric solution is shown in Fig. 9.4.

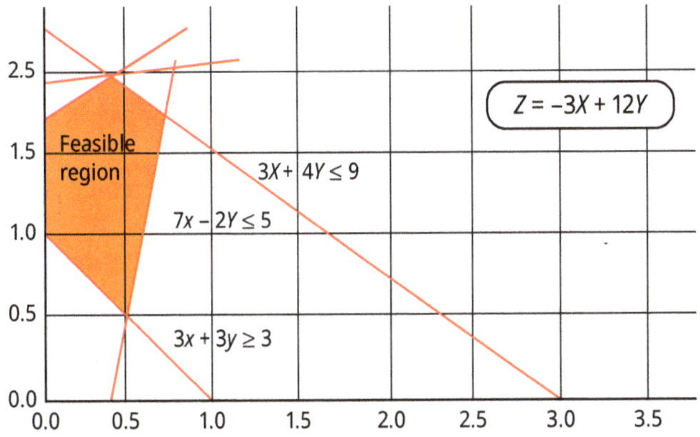

Fig. 9.4: Typical graphical solution to linear programming.

A method of solving a linear programming problem, known as the simplex method, utilizes the fact that if an optimum solution exists, it will be represented by a vertex of the feasible region. This allows for the selection of an optimal point from the set of all of the possible feasible points as the optimal point will be found on a vertex of the feasible region [7]. While examining each feasible point individually is inefficient, the simplex method improves the efficiency by only looking for a point that reduces the value of the objective function until a minimum is found [7].

The simplex method, while improving efficiency compared to checking each individual feasible point, still requires large amounts of computing power and memory to run; particularly for multivariable optimization processes. Other methods, such as the revised simplex method, dual simplex method and Karmarkar's interior method, can further improve efficiency of the linear programming solution method; however, they are still relatively computationally expensive and will not be discussed in this chapter [7].

9.7.2 Integer programming

Integer programming is similar to linear programming except some or all the variables are constrained to take on integer values. These possible variable values are not continuous, and while it is possible to use a linear programming approach and round off the answer, this approach can lead to violations of the constraints [7]. This is the only difference between linear programming and integer programming. Integer programming is useful for optimization problems, where the variables cannot physically take on a value that is not an integer. An example of that would be the number of products produced or number of people in a workforce. The integer variable can also represent a decision in which case it can only be represented by a 0 or a 1. This is known as binary integer programming. While there are some applications in engineering design, integer programming is more useful in fields such as production planning, scheduling and cellular networks.

9.7.3 Nonlinear programming

Nonlinear programming is used to solve general optimization problems, where either the objective function or the constraints are nonlinear. While it is possible to use classical optimization techniques to solve specific nonlinear problems, a large portion of them do not contain closed-form solutions to the problem requiring a numerical solving method to be implemented [7]. It is also possible, in some specific circumstances, to linearize any nonlinear components within the optimization model setup; however, this can result in a loss of generality of the model and the subsequent solution. Specific solution methods exist for different types of nonlinear programming problems, which usually depend on the type of nonlinear programming problem and the way in which the objective function or the constraints are characterized.

9.7.4 Quadratic programming

Quadratic programming is a type of nonlinear programming used for optimization problems, where the objective function contains quadratic terms, while the constraints are linear. Some examples of methods that are used to solve quadratic programming optimization problems are the interior point's method, active set method, augmented Lagrangian method, conjugate gradient method and an extension of the simplex method that is used to solve linear programming problems [7]. They have been successfully used in specific applications such as financial portfolio optimization, power generation optimization for electrical utilities and in engineering design optimization.

9.7.5 Stochastic programming

Stochastic programming is used to optimize cases in which some of the variables depend on random variables. These methods are usually based on statistical analysis with the randomness of the variables originating from several sources. Examples of real-world applications for optimization using stochastic programming include material properties (owing to variations in test results for the same material), dimensions of parts within a permissible range and flight loads placed on aircraft structures during operation [7]. Stochastic programming problems can be divided into stochastic linear, geometric dynamic or nonlinear programming problems [7]. Using classical optimization techniques, the approach usually involves converting stochastic programming problems into equivalent deterministic problems, which are then solved using the relevant numerical technique for the type of deterministic problem obtained [8]. These techniques use classical statistical techniques in place of calculus techniques to model the optimization problem.

9.7.6 Dynamic programming

Dynamic programming uses the approach of dividing the optimization problem into smaller optimization problems. This optimization method is particularly useful for engineering design of subsystems that all need to be optimized, where, owing to the need to make decisions sequentially at different stages of the design process (known as multistage decision problems), it is impossible to optimize the overall design as any changes would change the succeeding decisions that would be inefficient and impractical [7]. The dynamic programming optimization technique decomposes a multistage decision process into a sequence of single-stage decision problems, optimizing each stage individually [7]. The optimization technique used at each stage, whether classical, differential calculus, nonlinear programming or any other numerical method, is irrelevant to the overall process [7]. While it is possible to solve multistage decision problems using classical methods, it requires the number of variables to be small and the functions to be continuous and differentiable [7]. The use of nonlinear programming techniques can solve more difficult multistage decision problems; however, particularly with the introduction of randomness to the variables, most of the problems become unsolvable. Dynamic programming can handle all these complexities, owing to the procedure of breaking down the optimization problem into smaller optimization problems while being able to vary the optimization technique used at each stage according to which technique is most efficient and appropriate. Another advantage of dynamic programming is its ability to be used to optimize infinite or continuous problems as optimization occurs at each decision point whether there are a finite or infinite number of them. An example of this would be a missile system following a target where, even though the time frame is finite, the number of decisions made by the missile system is continuous as the target movement is measured and relayed to the missile system. A negative aspect of using dynamic programming is the "curse of dimensionality," a term that refers to the arising of various phenomena when organizing and analyzing data in high-dimensional spaces such as the ones needed for dynamic programming.

9.8 Advanced optimization techniques

The need for advanced optimization techniques arises out of the requirement to solve difficult optimization problems such as those that include multimodality, dimensionality and differentiability, which are usually associated with large-scale problems [9]. Traditional methods such as those previously discussed tend to fail for large-scale problems particularly if the objective function or the constraints are nonlinear. The previous methods also generally rely on obtaining information regarding the gradient of the function that is problematic if the function is not differentiable. Previously mentioned optimization techniques also tend to fail if there are many local maxima/minima [9]. It is for these reasons that more capable and robust optimization methods have become the focus of recent research. A large portion of these techniques are inspired by naturally occurring systems. A selection of techniques is discussed in the following sections.

9.8.1 Hill climbing

Hill climbing is a graph search algorithm, where a successor node is extended from the current path which is closer to the solution than the end of the current path. The hill climbing technique starts at a random solution in the feasible region and iteratively moves from a current solution to a better neighboring solution in the feasible space until a local optimum is reached [10]. The algorithm will only accept a downhill movement, where the quality of the neighboring solution is better than the current one, which can lead to the algorithm getting stuck on a local maxima/minimum instead of the global optimal point [10]. Attempts to improve the method have led to other methods such as simulated annealing, Tabu search (TS),[1] greedy randomize adaptive search procedure, variable neighborhood search, guided local search and iterated local search, all of which attempt to overcome the problem of stagnation on local optimum points instead of global ones [10]. Hill climbing is used in artificial intelligence in order to reach a goal state from a starting node.

9.8.2 Simulated annealing

The inspiration for the simulated annealing method comes from the metallurgical process of annealing, where the heating and cooling of a metal increase the crystal sizes and reduce defects. The heating allows the atoms within the crystal lattices to detach

1 TS is a global optimization algorithm and a metaheuristic or metastrategy for controlling an embedded heuristic technique. TS is a parent for a large family of derivative approaches that introduce memory structures in metaheuristics, such as reactive TS and parallel TS (from www.cleveralgorithms.com/nature-inspired/stochastic/tabu_search.html).

and wonder from their position, while the controlled cooling allows for the possibility of their finding of a configuration with a lower internal energy than the initial one.

The simulated annealing method compares each point within a feasible region to the state of a physical system where the function to be minimized is interpreted as the internal energy of the system in the current state. The method then attempts to bring the system from an arbitrary starting point to one with minimum internal energy, therefore, finding an optimal point in the feasible space. To simulate the annealing process, a temperature-like parameter is introduced and is controlled using the Boltzmann's probability distribution [7]. The accuracy of the solution obtained is unaffected by the chosen starting point with the only effect being on computational effort required to solve the problem [7]. The method can be used for mixed-integer, discrete and continuous problems.

9.8.3 Genetic algorithms

Genetic algorithms (GAs) are optimization methods inspired by evolutionary genetics, making use of concepts such as inheritance, mutation, selection and crossover. They are usually implemented as computer simulations in which a population of abstract representations ("chromosomes") of feasible solutions ("individuals") evolve toward a better solution. The initial population is completely random with improvements occurring in generations. The fitness of the whole population within each generation is evaluated by selected individuals chosen stochastically according to their fitness and modified by being mutated or recombined to form a new population, which is then used in the next iteration of the algorithm.

GAs are best suited to design problems with mixed continuous–discrete variables and discontinuous design spaces, which would be computationally expensive for standard nonlinear techniques [7]. Standard nonlinear techniques tend to find local optimum points, whereas GAs have a high probability of finding global optimal points. GAs do not rely on the gradient of the objective function but rather on the value of the objective function itself [7].

9.8.4 Ant colony optimization

The ant colony optimization method is inspired by the foraging process of ants. In reality, ants initially wonder randomly in search of food sources. When they find food, they return to the colony leaving a pheromone trail leading to the food source. Any other ants that come across the trail are likely to stop traveling randomly and follow the pheromones to the food source leaving their own pheromone trail thereby strengthening the trail. Owing to evaporation, the longer it takes for another ant to travel down the trail the less strong the pheromone trail will be causing a short path to a food source to have a stronger trail. The evaporation also stops convergence on a specific food source allowing for full exploration of the solution. When one ant finds a good and short path to a food source, the other ants are more likely to follow this path. If the path leads to a good food source, the number of ants on this path increases until all of the ants follow a single path. Simulating this search pattern within a feasible region allows for an optimum solution to be found. The simulation can be run continuously and react to real-time changes. This method of optimization is applicable to optimization in fields such as network routing and urban transport systems.

9.8.5 Neural network optimization

Neural network optimization methods are inspired by the large computational power of the nervous system and its ability for parallel processing [7]. A neural network is a massively parallel network of interconnected simple processors ("neurons") in which each neuron accepts the inputs from other neurons and computed an output that is sent to the output nodes. The network of processors maps an input vector from one space to another with the mapping being learnt instead of specified [7]. The network is described by individual processors, the network connectivity, the weighting of the connections between the processors and the activation function of each processor. The network can be trained by minimizing the mean squared error (MSE) between the actual output and the target output for all the input parameters by adjusting the weighting values in order to determine the optimal weighting values that lead to an optimum association of the input and output [7]. There are multiple methods and architectures of the neural network that can be used to solve optimization problems.

9.9 Gradient-based methods

Another way of classifying optimization methods is by separating the methods based on their approach to solving the optimization problem. Gradient-based methods use differential calculus to check the gradient of an objective function while moving from point to point in a feasible region until a gradient of zero (indicating a minimum/maximum point) is found. Gradient-based methods can be applied to both constrained and unconstrained problems. During each iteration of a gradient-based method, both the search direction and the step size between each iteration are required to be calculated. The different gradient-based methods, some of which are discussed below, are categorized by the method in which they compute the search direction. A problem with gradient-based methods arises from the need to determine if a point with a zero gradient is a global or local maximum/minimum point.

9.9.1 Unconstrained methods

These methods are applied to optimization problems without any constraints on the possible values of the optimization variables. The first gradient-based method for unconstrained problems, which will be discussed, is the steepest descent method.

The steepest descent method uses the theory that, for a differentiable function, the value of the function will decrease faster if the negative gradient of the function at a test point is used to determine the direction to the next test point until a minimum value is found [2]. To find the minimum point along this direction, a line search is then performed on the line in that direction [5]. By using the positive gradient for the method, a local maximum can be found instead of a minimum. A graphical representation of the method is presented in Fig. 9.5.

The steepest descent method is not very efficient for many problems and can suffer a condition known as zigzagging, which increases the time it takes for the method to converge on a solution. A method that attempts to correct for this error is the conjugate gradient method. The conjugate gradient method is an improvement of the steepest descent method, where the history of the gradients is considered when calculating the direction of the movement. This improves the efficiency of the method, which results in a decrease in convergence time [5].

Another related pair of gradient-based methods for unconstrained problems are the Newton method and the quasi-Newton method. The Newton method uses the value of the Jacobian or the Hessian matrix, a matrix of the second derivatives of the objective function at a specific point, at each iteration to numerically determine the minimum/maximum value of the problem [5]. However, calculating the Jacobian or Hessian can be computationally expensive at each iteration with the possibility of di-

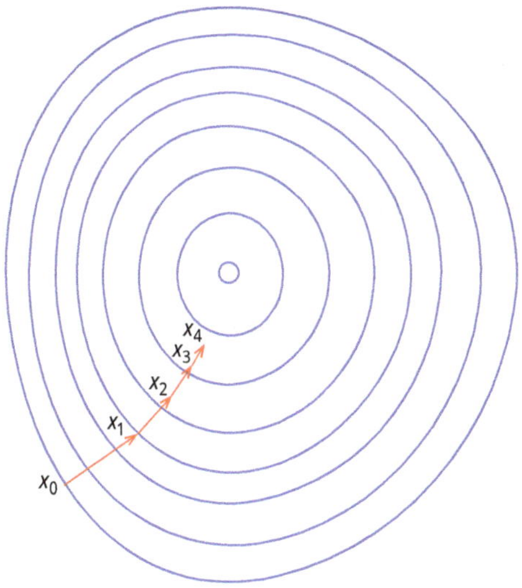

Fig. 9.5: Steepest descent method.

vergence being introduced to the problem. To overcome this, the quasi-Newton method uses estimates of the Hessian matrix, which are improved with the results from each iteration, to calculate a new search direction instead of the actual optimum point in each iteration [5].

9.9.2 Constrained methods

Some examples of gradient-based methods for constrained problems are the simplex method, sequential linear programming (SLP) method, sequential quadratic programming method, exterior penalty method, interior penalty method, generalized reduced gradient method, method of feasible directions and mixed integer programming which will be briefly discussed below.

Sequential or successive linear programming is a gradient-based method for approximately solving nonlinear optimization problems by reducing the problem using linearization to a series of linear approximations of the nonlinear optimization problem [2]. This method is particularly useful when the computation of the constraints, objective function and its derivatives are much larger than the calculation costs of the optimization operations, such as the search direction or step size [2]. The approximation involved in each cycle of the linear programming optimization generally results in a solution which is not the optimum solution; however, the solution will be more optimal than the initial design analyzed. The method is then repeated on the new solution, creating a sequence of linear programming problems giving this method its name [2].

Sequential or successive quadratic programming has the same concept of an SLP; however, it is applied to optimization problems with nonlinear quadratic approximations of the objective function instead of linearizing the objective function.

Penalty-based methods are used to convert a constrained problem into an effectively unconstrained problem by introducing a penalty term into the objective function, which ensures that the original constraints are either not violated or a solution that does violate them is penalized [5]. The method is repeated with decreasing penalty terms, producing a solution that approaches the optimal solution of the original constrained problem [5]. These methods are used for their simplicity and their ability to be applied to equality and inequality problems for both linear and nonlinear optimization problems [9]. The method does rely on iteration and is, therefore, more computationally expensive than the more sophisticated constraint following methods [5]. The exterior penalty method is a method that applies a penalty to a solution violating any of the problem constraints [2]. This method applies to penalties only if the point is found on the exterior of the feasible domain meaning that the method approaches the optimization problem from the unfeasible region and moves toward the feasible region. The interior penalty method uses a similar approach to the exterior penalty method; however, this method is restricted to solutions that are found in the feasible region only [5].

Generalized reduced gradient methods use the concept of projecting the search direction into the feasible region at a tangent to the constraints of the optimization problem [2]. They are suitable for large-scale, nonlinear structural optimization problems [11]. The method uses the addition of slack variables to inequality constraints to transform them to equality constraints while maintaining the total number of variables in the optimization problem, making the method completely general in terms of its application [11]. The method then uses the equalities to express the part of the variables, called basic variables, in terms of the remaining nonbasic variables in a manner similar to the simplex method [11]. This reduces the problem to a series of problems with only upper and lower bounds, which are then solved sequentially with all of the constraints being treated in the same manner [5, 11]. Using the derivatives of the constraints, a reduced gradient in the basic variables is found from which the search direction is calculated with nonbasic variables that need to remain positive or zero throughout the optimization procedure [5]. A line search method is then used in the direction calculated with a new reduced gradient when a nonbasic variable reaches zero [5].

The methods of feasible directions are specifically aimed at nonlinear optimization problems with inequality constraints [5]. These methods solve the problem of directions tangent to nonlinear constraints of an optimization problem causing a departure from the true constraints [5]. In these methods, the search direction is allowed to point at an angle into the feasible region [5]. The use of these methods ensures that the next solution considered during the optimization process is within the feasible region [1].

9.10 Heuristic methods

Heuristic methods are methods that rely on rules of thumb instead of established mathematical procedures to arrive at a near-optimum solution. These methods often use algorithms that were inspired by nature and are useful for solving large complex optimization problems. They are also able to circumvent the problem of getting stuck on local optimum points instead of global ones, usually by incorporating some form of randomness within the algorithm. Heuristic methods are generally faster and more efficient than traditional methods by sacrificing optimality, precision, accuracy or completeness for speed. A few of the heuristic methods have been discussed in a previous section of advanced optimization techniques. Another example of a heuristic method, the TS, is discussed below.

9.10.1 Tabu search

The TS method uses the premise of a local search technique, starting at a point and proceeding iteratively from one point to another local point until a termination criterion is met. Unlike the gradient descent method, the TS can move to a neighbor solution that is inferior to the current solution point while selecting the move from point to point using a modified neighborhood which helps in avoiding getting stuck at local minima/maxima. The method uses both short-term and long-term memory structures in order to construct the modified neighborhoods. This leads to the neighborhood being dynamic rather than static, such as in local search methods. The most common memory structure used in TSs is that of recency-based memory.[2] This memory structure records information about solution properties that change when moving from point to point. Recency-based memory structures keep track of solution properties that have changed in the recent past, labeling some solution properties as Tabu active. When a solution contains Tabu-active elements, these solutions become Tabu (or Taboo), removing them from the modified neighborhood and preventing them from being revisited. The use of adaptive memory creates a balance between search intensification and diversification. Intensification strategies modify search rules to encourage more move combinations and solution features which were seen to be historically good, while diversification strategies attempt to incorporate new attributes and attribute combinations that were not previously part of the previously generated solutions.

9.11 Optimization of topology of aerospace structures

The design of a structure usually requires optimization of three parameters of the structure: size, shape and topology. The optimization of the structure size attempts to find the optimum thickness distribution of components within the design which is based on the desire to minimize or maximize a physical quantity such as peak stress, deflection or a multitude of other parameters [6]. The main feature of size optimization is that the domain of the design model is fixed throughout the optimization procedure [6]. Shape optimization problems use the design domain as the optimization variable in an attempt to find the optimum shape of this domain [6]. Topology optimization of solid structures attempts to optimize the features of the structure such as the number and location and shape of holes and the connectivity of the design domain [6]. Figure 9.6 presents a schematic example of the above three types of structural optimizations.

Topology optimization assists with ensuring that a design has a minimum weight and that all components of the design are carrying maximum load so that no material is wasted in terms of the potential load-carrying capabilities [8]. This has led to vast improvements in the field since the early 2000s in both research and applications, as the optimization of the distribution of material and the load paths in the structure are of vast importance in all fields, particularly the aerospace field [8].

2 The first item in a list is initially distinguished from previous activities as important (primacy *effect*) and may be transferred to long-term *memory* by the time of recall. Items at the end of the list are still in short-term *memory* (*recency effect*) at the time of recall.

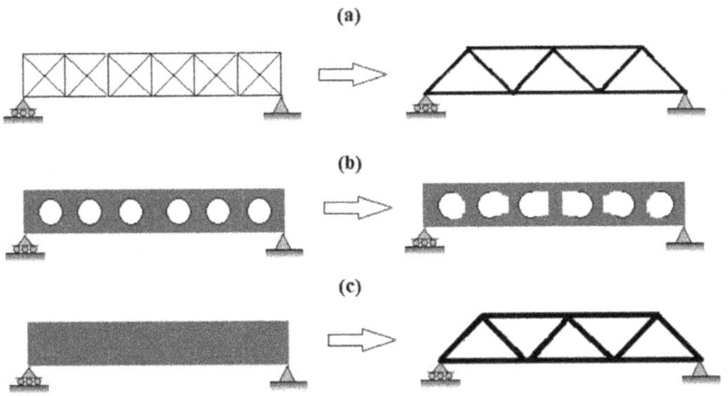

Fig. 9.6: Types of structural optimization problems: (a) truss structure sizing optimization, (b) shape optimization, (c) topology optimization (left: initial problem; right: optimized solution) (adapted from [10]).

A topology optimization problem is generally defined using the design loads, possible support conditions, the available domain size for the structure and the required holes or solid sections for other design considerations as constraints for the problem [6]. The constraints are used to optimize the distribution of the available material in the fixed domain size, usually by using a distribution function instead of a standard parametric function [6]. Topology optimization methods are focused on deciding whether a point in the design domain should contain material (a material point) or should not contain any material (a void) [6]. This led to the consideration of topology optimization as a binary optimization problem with a point in the design domain either containing material or not which would usually require an integer programming approach to solve [8], resulting in computationally expensive methods to solve such large-scale integer programming problems and, therefore, limited the success of applicable methods. This also led to the development of multiple methods that approached an optimization problem from different perspectives in order to avoid these problems [8]. The sections below discuss a selection of methods developed within the last 20 years to optimize topology problems in the structural field. The number of methods developed over the last 20 years is immense, and it is for that reason that this chapter will only focus on a select few methods that were considered to have made important advances in the field of topology optimization.

9.12 History of topology optimization

The first description of structural optimization was presented by Anthony G.M. Michell, an Australian mechanical engineer [12]. The field of structural optimization was only focused on and further developed nearly 50 years later with the invention of electronic computers, allowing for a great increase in efficiency of all procedures [13]. The success of topology optimization, even with the use of computer solutions, was limited until the late 1980s as the problem was approached as a binary optimization problem, which required large amounts of computing power and precluded the use of gradient free algorithms [8]. The first major improvement was made by Bendsöe and Kikuchi in 1988 when they proposed a homogenization-based method [14], which was the basis for the density-based methods discussed in the next section. The homogenization method uses density variables, which are linked to specific microstructure models, to optimize global structural performance. The density variable of each cell within the design domain is then modified iteratively to optimize the topology. Figure 9.7 displays schematically the different microstructure cells that were used for the optimization procedure.

The original homogenization method is extremely mathematically complicated, which prevents its general application. This led to the density-based methods to be presented and discussed in the next section.

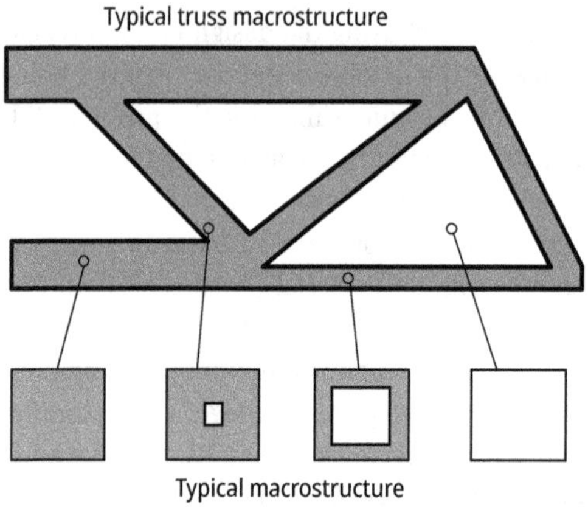

Fig. 9.7: The homogenization method cell structures (adapted from [8]).

9.13 Density-based methods

Density-based methods are the most widely used methods for topology optimization [15]. These methods operate by discretizing a fixed domain into finite elements and minimizing an objective function by deciding whether each element should contain materials or remain empty (void). The constraints are usually placed on the amount of material that is utilized. To avoid a challenging large-scale integer programming problem, discrete variables are replaced by continuous variables and iteratively driven to a discrete solution using an interpolation function, which interprets the continuous design variables as the material density of each element [15]. This is because the general discretized problems lack a solution [16] as for many problems the addition of more holes will decrease the objective function and, therefore, the optimum solution for the discretized design domain is dependent on the number of divisions of the domain (mesh dependence) [13]. The first density-based method introduced was the simplified isotropic with penalizations (SIMP) method [17] and will be discussed below.

The SIMP method was first introduced by Bendsøe [17] as an improvement of the previously devised homogenization method [14]. Its purpose was to artificially reduce the complexity of the homogenization method while improving the convergence of the binary topology optimization problem [13]. The SIMP method was physically justified by Bendsøe and Sigmund [18].

The SIMP method makes use of a power law relation to apply a penalization parameter onto the density design variable that is multiplied onto a material property (such as material stiffness, cost or conductivity) [13, 18]. The simplicity of the SIMP method has led to its widespread use both in industry and academia. The choice of the penalization parameter has a major impact on the quality of the solution obtained using the SIMP method, possibly causing either grayscale or convergence problems [13]. The experimentally verified value for the penalization parameter is a value of 3, which ensures physical realizability of elements with intermediate densities [18] and has been shown to make density gradients equal to topological derivatives for elasticity problems [19]. The penalization parameter, however, will only work correctly when the volume of the material is constrained, either as a direct constraint or indirectly through another constraint [13].

The SIMP method, while being simple and efficient to implement compared to the homogenization method, does have some shortfalls. When looking at the well-posedness of the SIMP method, when the penalization parameter is greater than 1, there is no guarantee that a solution to the optimization problem can be found [6]. Therefore, the use of numerical methods to solve the optimization problem, such as FEA (finite element analysis), introduces the possibility of mesh dependence of the solution [20]. Another problem is the possibility of the checkerboard phenomenon, where the optimum solution has adjacently varying material-filled elements and voids creating a checkerboard pattern. This is often seen in topology optimization problems particularly when lower-ordered finite elements are used to discretize the design domain [20]. An example of the checkerboard phenomenon is presented in Fig. 9.8.

Another problem is the possible existence of intermediate density values in the optimization solution, where the element has a density value that is neither 0 nor 1 and is therefore neither a pure solid-filled element nor a void [20].

The problems associated with the SIMP method have led to modified versions of the penalization method being developed, such as the rational approximation of material property (RAMP) method [19] and the SINH method [21, 22]. Unlike the SIMP method, the RAMP method has nonzero sensitivity (gradient) at zero density [23], rectifying numerical problem in regions with low-density values in the presence of design-dependent loading [15]. The RAMP method was introduced to alleviate nonconcavity of the original SIMP method [24] ensuring convergence to a fully binary solution. This feature does not play a major role in practical examples. The SINH

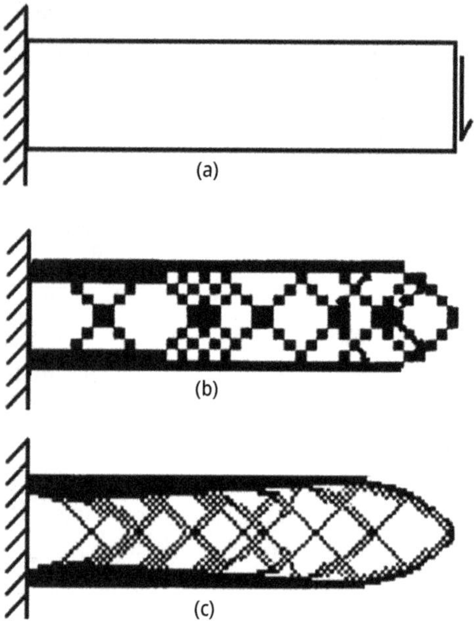

Fig. 9.8: Typical example of the checkerboard problem: (a) initial structure, (b) coarse optimization and (c) finer optimization (adapted from [6]).

method uses cost penalization, as cost can represent material weight, to penalize material volume instead of material parameters as seen in the previous two methods [15]. The various expressions for the three methods, the SIMP, the SINH and the RAMP, are described in the following equations, where p or q is the penalization parameter (with multiple values shown in Fig. 9.9) and ρ is the density variable:

$$
\begin{aligned}
\text{SIMP:}\quad & \zeta(p) = p^{p} \\
\text{SINH:}\quad & \zeta_1(p) = \frac{1 - \sin h\,[p(1-p)]}{\sin h\,(p)} \\
& \zeta_2(p) = p \\
\text{RAMP:}\quad & \zeta(p) = \frac{p}{1+q(1-p)}
\end{aligned}
\tag{9.3}
$$

This results in intermediate density cells carrying more volume with respect to loading than either material filled cells or voids, reducing the number of intermediate cells in the optimization problem [15]. A comparison of the results for the three methods is shown in Fig. 9.9.

In general, density-based methods are represented by smooth, differentiable problems that are easily solvable by gradient-based methods [25, 26], the method of moving asymptotes [27] or other mathematical programming-based methods [13]. The use of these optimization methods also allows for the systematic and straight-forward inclusion of any additional global constraints; however, when including local constraints, parametrization problems, particularly in stress constrained problems, may cause great difficulty when solving real problems [28].

9.14 Hard-kill methods

Hard-kill methods optimize a topology problem by incrementally adding or removing a material to a design domain, with the choice of the location to add or remove material being dictated by the heuristic method [15]. The heuristic analyses may or may not be based on the sensitivity information in the design domain which, unlike the density-based methods, is never relaxed [15]. A problem with hard-kill methods is the possibility that the solution that is obtained may be nonoptimal when the methods are implemented and used inadequately, particularly when a prescribed boundary support is broken during optimization of a statically indeterminate structure resulting in a completely changed structure from the initially defined one [29]. By far, the most popular and widely implemented hard-kill method is the evolutionary structural optimization (ESO) method and its derivatives which are discussed in more detail in the following section.

9.14.1 Evolutionary structural optimization methods

The first developed evolutionary method was the ESO method. The ESO method is based solely on the heuristic-based removal of material from an optimization design by gradually removing redundant or inefficient material from the structure until an optimal solution was found by achieving the presumed volume constraint [30]. This is achieved by calculating a criteria function, also known as a sensitivity number, for each element in the model, with the removal of elements having low sensitivity numbers [15]. Early versions of ESO were devised for use with FEA analyses for stress-based, stiffness or displacement problems using an iteratively increasing rejection ratio to remove increasing amounts of inefficient material from the FEA model until an optimum solution is reached [31]. An example of an ESO stiffness topology optimization process for a Michell-type structure[3] with two simple supports is shown in Fig. 9.10.

A major disadvantage in the ESO method comes from the way in which it is implemented. The method removes material, or volume, iteratively until with the re-

3 A. G. M. Michell formulated the criteria to be satisfied by all least-volume trusses with equal tensile and compressive yield stresses (see Michell, A. G. M., The limits of economy of material in frame-structures, *Philosophical Magazine*, 8 (47), 1904, pp. 589–597).

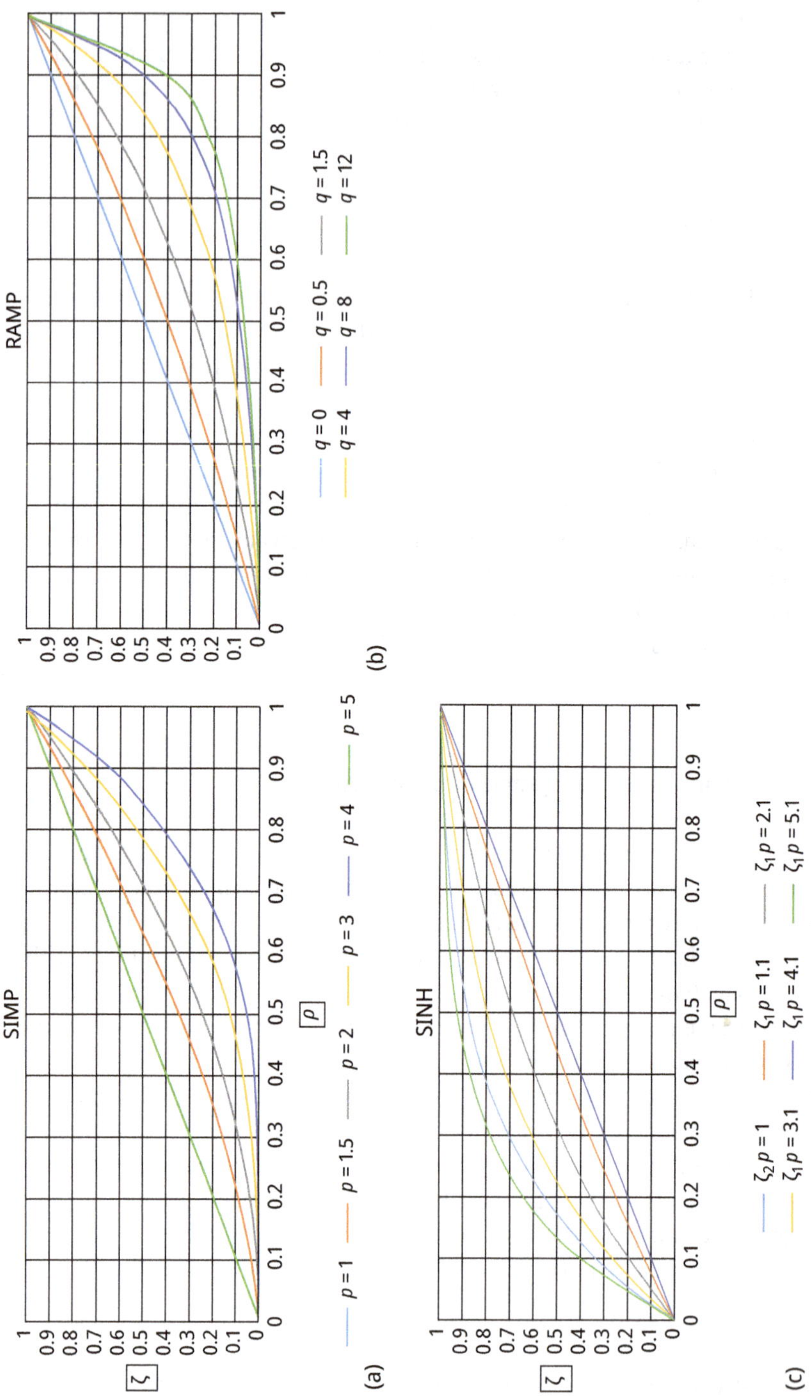

Fig. 9.9: Comparison between (a) SIMP, (b) RAMP and (c) SINH methods (adapted from [15]).

moval of any more material, a constraint is no longer satisfied. This, however, does not necessarily give an optimum solution as a material that has been removed in earlier iterations could be required for a more optimum design at a later stage in the process when other elements of a material have been removed [31]. This removed material cannot be recovered by using the ESO method. The early implementation of the ESO method also ignored many numerical problems that arose, such as checkerboarding, owing to its being a heuristic method. These deficiencies in the method led to the development of the bidirectional ESO (BESO) method.

The BESO method is an attempt to correct the disadvantage of not being able to reintroduce the removed material in the ESO method by simultaneously removing and adding material at each iteration of the optimization problem. A method for the removal and addition of elements was first designed in 1998 [32] and was applied for a stress-based optimization problem [33] and for a stiffness optimization problem [34] using FEA model bases. The method developed in the stiffness optimization problem [34] uses a linear extrapolation of the displacement field in the FEA model to estimate the sensitivity numbers of void elements, allowing the removal of elements with the lowest sensitivity numbers and the replacement of void elements with the highest sensitivity numbers. The method that makes use of the rejection ratio for the removal of a material and an inclusion ratio for addition of a material, both are not in relation to each other [31]. The selection of these values directly impacts the ability of the method to arrive at an optimum solution. The BESO method can be applied to initial full designs and an initial guess design and can even be used for 3D optimization.

The BESO method is a widely used optimization method mainly owing to its ability to obtain converged, checkerboard free, mesh-independent solutions for a multitude of designs with comparable optimal solutions to the SIMP method, as shown in Fig. 9.11, with the ability to start from a smaller than the domain guessed initial design, saving the computational time.

The BESO method, however, has been shown to be incorrect for specific optimization problems bringing the validity of the method into question [35]. While overcoming some shortfalls of the ESO method, the BESO method still has some common problems. The methods are fully heuristic; therefore, there is no mathematical proof that the obtained solution is the optimal one. Both the ESO and BESO methods rely on selecting the best solution by comparing many intuitively generated solutions, reducing the efficiency of the methods heavily. Both methods have many cases where the solution obtained is not the optimal solution or, cases, where they are completely broken down [35]. These problems have led to many debates in the academic world on the validity of ESO/BESO methods, particularly when compared to SIMP methods [35].

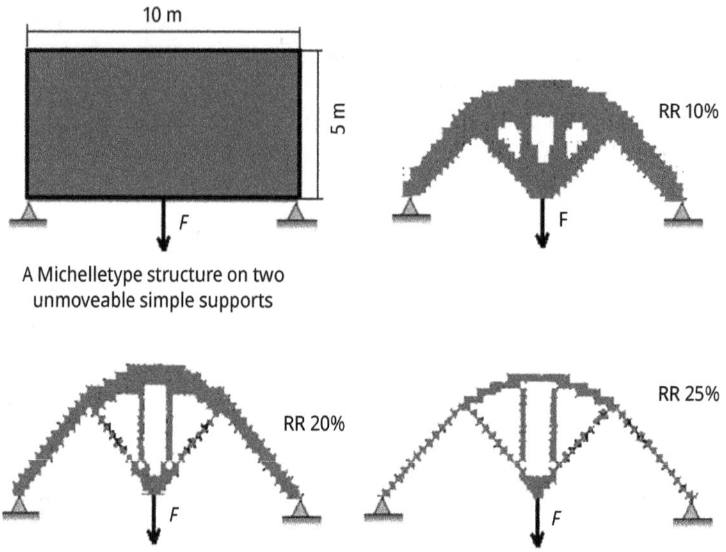

Fig. 9.10: ESO topologies with varying rejection rates (RR) (adapted from [30]).

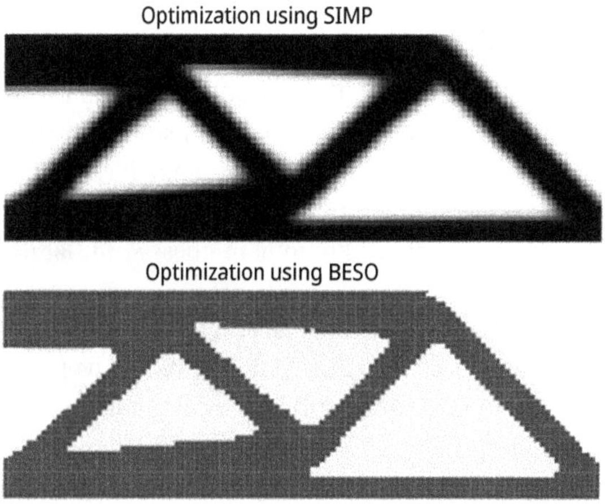

Fig. 9.11: Comparison of beam optimization using BESO and SIMP methods (adapted from [31]).

9.15 Boundary variation methods

Boundary variation methods are a relatively recent addition to the topology optimization field, having been adapted from shape optimization techniques. Compared to density-based methods, they use implicit functions to define structural boundaries instead of explicit parameterizations of the design domain. Boundary variation techniques generally have the advantage of producing very well-defined results, having little need for post-optimization processing or interpretation of results [15]. These methods are also distinguished from the shape optimization methods on which they are based as they allow for the movement of structural boundaries as well as the addition, removal and merging of voids within the model, making these methods as topological optimization methods [15]. There are two prominent boundary variation methods being researched, which will be discussed in the following sections.

9.15.1 Level-set methods

In level-set methods the boundary of the design is defined by the zero-level contour of the level-set function, while the structure is defined by the domain where the level-set function takes positive values. The shape of the geometric boundary is modified by controlling the motion of the level set according to optimization conditions and the physical problem by using the shape sensitivities of the objective function until a convergence criterion is met.

The conventional level-set approach, developed from the work of Wang et al. [36], uses the concept of the link between the velocity of the points on the structural boundary, the design sensitivity, the structural optimization process and the level-set method for boundary definition. Concurrently, a mathematical framework linking the velocity of the level-set boundaries to an adjoint shape sensitivity analysis for a stiffness optimization problem was developed [37–39]. These two direct methods are limited by their inability to create new holes in the level-set function away from the free boundary, which typically represents the outside of the design domain, and that they both are extremely dependent on the initial state of the design problem. This led to the inclusion of topological derivatives, representing the change in objective function with the introduction of infinitesimally small holes, in subsequent methods to allow for the nucleation of holes at any point in the design domain [39–41]. More recently, Dunning and Kim [42] make use of a secondary level-set function instead of topological derivatives to introduce holes. James and Martins [43] proposed using an extension of the conventional level-set method on body-fitted finite mesh models, allowing the design domain to be irregularly shaped.

The conventional level-set formulations make use of an explicit numerical method for solving the Hamilton–Jacobi partial differential equation (PDE) that controls the structural boundary. A problem associated with explicit numerical methods is that their convergence is dependent on the time step size chosen for the method. Additionally, the explicit method often causes the need to reinitialize the level-set function when they become too steep or too flat. This reduces the computational efficiency of the method. This has led to the development of several alternative level-set function methods that use alternative strategies to solve the Hamilton–Jacobi PDE. These methods will not be discussed in this chapter as they are too numerous both in types and methods of solving the PDE.

Level-set methods have some advantages when compared with other topology optimization methods. Checkerboarding can be significantly reduced, the optimal shape and topology can be obtained simultaneously, and the handling of any topology-dependent loads is more easily implemented when using level-set methods while the method is more versatile compared to the density-based methods [20, 44]. However, the level-set methods have their own distinct disadvantages. They have a slow convergence rate, particularly when using a Hamilton–Jacobi-based level-set method, and the required topology changes can only be obtained by pinching or merging of the boundaries in the original form with complicated workarounds required adding voids in the problem [20]. In general, there is large potential for the use of level-set methods; however, currently, most uses are in research and not commercial applications.

9.15.2 Phase field method

The phase field method for topology optimization is based on theories that were originally developed as a method to represent the surface dynamics of phase-transition phenomena such as solid–liquid transitions [45, 46]. While this method has been utilized in many fields, such as diffusion, crack propagation and multiphase flow, the method can also be applied to topology optimization problems. The method begins by specifying a phase field function over a design domain that is composed of two phases with a continuously varying boundary region between the phases having a thin finite thickness. This setup is shown in Fig. 9.12.

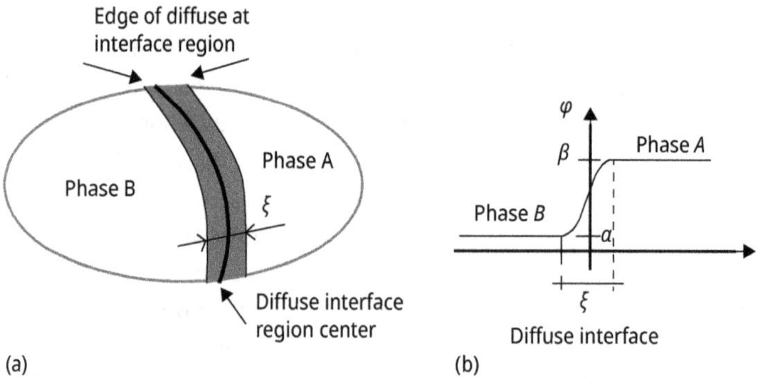

Fig. 9.12: (a) A 2D domain represented by a phase field function and (b) a 1D representation of the phase field function (adapted from [46]).

In phase field topology optimization this region defines structural boundaries and is modified by dynamic evolution equations of the phase field function. The main difference between the level-set and phase field methods lies in the fact that in the phase field method the boundary interface between phases is not tracked throughout optimization as it is when using level-set methods. The governing equations of phase transition are solved over the complete design domain without previous information about the location of the phase interface. In addition, phase field methods do not require the re-initialization step of level-set functions [15], which reduces the complexity of the problem [47]. Phase field methods, like level-set methods, have the advantage of being able to simultaneously solve shape and topology optimization problems [47]. However, it has been observed that phase field methods are dependent on the initial shape of the design model as well as the mesh size and number of iterations of the method on the optimization problem [47]. An example of these effects is shown in Fig. 9.13.

Initial work began on the phase field method in 2003 by Bourdin and Chambolle [48] with further developments being made from 2004 onward. However, because of the later start date as well as relatively less research being performed on phase field methods, the field has remained in the academic sphere and has not been widely applied in real-world applications. Additionally, the computational costs of the phase field method are relatively high, reducing the current applications of the method [49, 50]. The flexibility of the approach has the potential to allow for modeling coupled systems, not only focusing on a pure structural optimization problem, giving the method a large potential for future research [49].

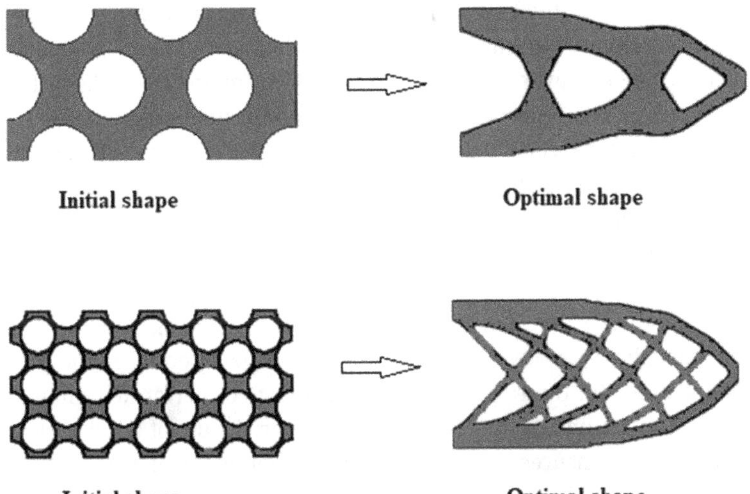

Fig. 9.13: Effect of initial shape to obtain the optimal shape using the phase field optimization (adapted from [47]).

9.16 Recently developed methods

The field of topology optimization is a vast topic. While shape and size optimization problems have been developed further than topology optimization, mainly due to the complexity of topology optimization comparatively, the possible benefits of a topologically optimized design outweigh those from size or shape optimization. Additionally, with the increased computing power available today, computationally expensive methods are becoming easier, quicker and cheaper to run while yielding better results. It is for these reasons that research into new, more efficient methods that give improved optimization results is a focus of the field of topology optimization. While many different new methods of topology optimization have been proposed, one of the most promising and easy-to-implement methods is a bio-inspired cellular division-based method which is discussed as follows.

9.16.1 Bio-inspired cellular division-based method

In 2010, Kobayashi proposed a topology optimization method inspired by the cellular division process in living organisms that is capable of generating discrete and continuum-like structures [49]. The method uses a developmental program to implicitly govern topology layout development in stages. The method, when driven by a GA, assigns a set of rules called a Lindenmayer or map-L system[4] which define the tasks of the developmental program as the design variables for the optimization problem [15]. By controlling these developmental rules, a diverse set of topological designs may be generated with few design variables [15].

An advantage of the method is its ability to couple easily onto the existing finite element tools to develop a topology optimization capability. Similar to ESO methods, this is done by using simple pre-processing and post-processing operations especially if the design is a multiphysics one. The method also produces topologically optimal solutions that are immediately ready to undergo shape and sizing optimization processes and has the potential for these optimization processes to be performed simultaneously with the topology optimization process [15]. However, the use of GA as a basis of the method immediately ensures that it is more computationally expensive than the most gradient-based methods.

The map-L systems are a type of grammar system capable of acting as rewriting systems that can generate developmental programs to describe the construction of a natural engineered system [51]. The maps are analogous to cellular layers with the

4 L-systems were introduced and developed in 1968 by Aristid Lindenmayer, a Hungarian theoretical biologist and botanist at the University of Utrecht. He used L-systems to describe the behavior of plant cells and to model the growth processes of plant development.

regions representing the cells and the edges in the walls of the cells. A series of production rules can then be applied in order to govern the processes that construct the map leading to complex topology being obtained. An example of the execution of a production rule is shown in Fig. 9.14.

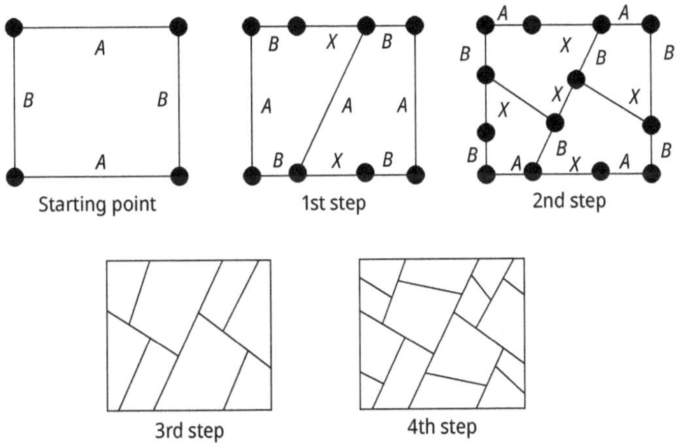

Fig. 9.14: The first four steps in a cellular division process (adapted from [52]).

More complex topologies can be obtained by utilizing additional rules and not just the division rule shown in Fig. 9.14. The geometry can also be stretched and superimposed onto nonrectangular domains that can also undergo shape changes. However, it must be noted that topologies generated by the map-L system have no physical or structural meaning and as such the geometry must be interpreted into a structural element [15]. Reference [53] presents the topology of a structure defined by the map-L system, projected onto a nonrectangular domain and interpreted into a 3D wing and spar layout for an aircraft component.

Topology optimization is performed by encoding the map-L system into a binary representation to use in GA. Several parameters that affect the topology are controlled by the optimizer, including those that control the growth and dynamics of the development, the developmental rules and the definition of the overall geometry or physical properties of the design model. The GA is, therefore, able to modify both the initial map and the rules that create the topology according to some fitness function.

The implicit representation of topology found in the GA of this method differs from the explicit representation, where a single design variable for each finite element corresponds to the genome that is utilized in other GA topology optimization methods. It is this representation that allows for fewer design variables and more design freedom while avoiding other issues such as maintaining domain connectivity throughout the evolution process [15].

9.17 Applications in the aerospace field

Having discussed a multitude of methods for topology optimization, the following section will look at some real-life applications of topology optimization in the aerospace field and, where possible, will give examples of the specific topology optimization methods discussed in the previous section.

9.17.1 General topology optimization applications

This section contains some general examples of topology optimization for structural components in the aerospace field. Most of the examples are presented as graphical results and have limited information on the method of optimization used, usually because the design process relied on optimization software where the specific method of optimization was not attainable.

When optimizing aircraft components, a vast majority of the optimization problems are either minimum compliance (stiffest design) problems or minimum weight designs. Optimizing topology has a great effect on both cases with some stating that out of the three structural optimization categories, this will have the greatest impact on the design.

9.17.2 Applications of density-based methods

A good example of topology optimization using a density-based method is observed in the optimization problem for an engine pylon of a large cargo aircraft [8]. The optimization was performed to comply with stiffness, strength and weight requirements of the design. The first consideration was to define the design and nondesign domains based on multiple restrictions imposed on the structure from sources such as aerodynamic restrictions and manufacturing restrictions.

The next consideration was the predicted flight loads. The entire envelope of predicted flight loads was analyzed with 24 possibly critical cases chosen for the topology optimization of the pylon, leading to the use of a polynomial interpolation model to incorporate all the loading cases and boundary conditions. The implemented topology optimization method was an extended formulation of the SIMP method.

Another use of density-based topology optimization is to solve aeroelastic problems. The first work to tackle this problem used a 3D Euler solver, linear finite element models and additional sensitivity analyses to design wing stiffeners for minimizing mass with constraints given by the values of the drag, lift and deflection [54]. While the chapter did reach a final design, the focus was on developing a methodology for similar problems and, as such, the pictures of the results are not presented here.

9.17.3 Applications of hard-kill methods

Hard-kill methods have been utilized far less frequently than density-based methods. This is partially owing to their fully heuristic nature and the possibility for presenting a nonoptimum result as an optimum one. However, there have still been attempts to use these methods in optimizing aircraft structures. An example of the optimization of an aircraft bulkhead component using a modified ESO method [55] is presented below.

The paper focuses on improving the optimization of the design of a bulkhead in an F/A 18 fighter aircraft. The goal was to produce a lighter bulkhead for the aircraft while satisfying the geometric, functional and strength requirements. A finite element analysis of the half symmetric 3D model of the current bulkhead was first performed. The von Mises stress was then taken as the ESO criterion. The ESO algorithm was used to reduce the weight of the structure by removing the inefficient material. At each optimization cycle, a finite element analysis was performed using NE-NASTRAN (a FE code). The optimization algorithm was then employed to modify the topology based on the von Mises stress field. At each iteration, an element of the structure was removed only if the calculated current von Mises stress was less than a reference value. The topology optimization presented in [55] succeeds in removing the material from regions not in the primary load path. This has led to a lower weight design with a more even distribution of the von Mises stresses.

9.17.4 Applications of boundary variation methods

The use of boundary variation methods in aerospace fields appears to be very limited. NASA has researched the optimization of a 3D wing for different purposes, with an example focusing on aeroelastic constraints [56] and another focusing on a common research model wing [57]. The example described later is focused on the topology optimization of a reinforced wing box for enhanced roll maneuvers [58].

The chapter looks for a solution to maximize aileron reversal speed of a wing torsion box by reinforcing the upper skin of the wing. A spectral level-set topology optimization method was used to solve the design problem. The results showed that similar topology optimum results were obtained for different initial conditions. Therefore, for any of the four initial setups, the final optimal solution is similar for all cases. This was considered a successful implementation of the optimization method; showing some form of independence between the initial setup and the solution.

9.17.5 Applications of bio-inspired cellular division-based method

While the bio-inspired division-based method is relatively new, there have been a few attempts to use the method for topology optimization in aircraft structures, such as the optimization of aircraft lifting surface [57] and the aeroelastic optimization of flapping wing venation [58]. The example below is for the topology optimization of a component in a jet engine [59]. The topology optimization of a bracket for drawing bars of a reverse device was performed using a software-based implementation of the bio-inspired cellular division-based method. Because the method was implemented using a software-based optimizing tool, the details of the method are scarce in the chapter.

The optimization procedure resulted in a weight saving of nearly 20% of the part being built out of steel powder by a selective laser melting method in order to be tested mechanically.

9.18 Conclusions

The field of optimization in general and particularly for structures is a vast one, which is continuously growing. Having been developed from mathematical first principles, most methods became practical and feasible after the introduction of the computer, which allowed for iterative procedures to be performed economically and quickly while increasing the complexity of the calculations that could be handled. With the continued increase in computational power and the focused research on both improving current methods and developing new methods, the field of optimization is continuously expanding and improving. The obvious benefits of design optimization are numerous. Not only can a design save on material and cost while ensuring that any part of the structure that is inefficient is removed, but the ability to run optimization simulations before having to manufacture test items increases the overall efficiency of the design process continuously.

The chapter covers some background to design optimization and some general optimization techniques while focusing on topology optimization methods and their applicability in the aerospace field. However, the amount of information contained within this chapter is just a small part of the immense field that is design optimization.

To enhance the optimization method applicability, Appendix A presents a method to design optimized structures using the response surface methodology (RSM) approach, as authored by Dr Kaspars Kalnins, from Riga Technical University, Latvia.

References

[1] Papalambros, P.Y. and Wilde, D.J. Principles of optimal design, modeling and computation, 2nd, The Press Syndicate of the University of Cambridge, Cambridge, 2000, 416.

[2] Haftka, R.T., Gürdal, Z. and Kamat, M.P. Elements of structural optimization, 2nd revised, Springer-Science+Business Media, B.V., 1992, 410.

[3] Adeli, H. Ed. Advances in design optimization, Chapman and Hall, London, UK, 1994, 590.

[4] Majid, K.I. Optimum design of structures, Newnes-Butterworths, London, 1974, 264.

[5] Rothwell, A. Optimization methods in structural design, ©, Springer International Publishing AG 2017, Gewerbestrasse 11, 6330 Cham, Switzerland, 2017, 314.
[6] Sigmund, O. and Bendsøe, M.P.. Topology optimization: Theory, methods and applications, Springer-Verlag, Berlin and Heidelberg, 392.
[7] Rao, S.S. Engineering optimization: Theory and practice, John Wiley & Sons, Inc., s.l., 2009, 814.
[8] Zhu, J.-H., Zhang, W.-H. and Xia, L. Topology optimization in aircraft and aerospace structures design, Archives of Computational Methods in Engineering, 23, 2015, 595–622.
[9] Rao, R.V. and Savsani, V.J. Mechanical design optimization using advanced optimization techniques, Springer, s.l., 2012, 233.
[10] Al-Betar, M.A. β -Hill climbing: An exploratory local search, Neural Computing and Applications, 28(1), 2017, 153–168.
[11] Lasdon, L.S., Fox, R.L. and Ratner, M.W. Nonlinear optimization using the generalized reduced gradient method, RAIRO – Operations Research – Recherche Opérationnelle, 8(V3), 1974, 73–103.
[12] Michell, A.G.M. The limits of economy of material in frame-structures. Philosophical Magazine, 8, 2010, 589–597. doi: 10.1080/14786440409463229.
[13] Sigmund, O. and Maute, K. Topology optimization approaches, Structural and Multidisciplinary Optimization, 48, 2013, 1031–1055.
[14] Bendsøe, M.P. and Kikuchi, N. Generating optimal topologies, Computer Methods in Applied Mechanics and Engineering, 71, 1988, 197–224.
[15] Deaton, J.D. and Grandhi, R.V. A survey of structural and multidisciplinary continuum topology optimization: Post 2000, structural and multidisciplinary optimization, Vol. 49, Springer–Verlag New York, Inc., Secaucus, NJ, USA, 2014, 1–38.
[16] Sigmund, O. and Petersson, J. Numerical instabilities in topology optimization: A survey on procedures dealing with checkerboards,mesh-dependencies and local minima, Structural Optimization, 16(1), 1998, 68–75.
[17] Bendsøe, M.P. Optimal shape design as a material distribution problem, Structural Optimization, 1, 1989, 193–202.
[18] Bendsøe, M.P. and Sigmund, O. Material interpolation schemes in topology optimization, Archive of Applied Mechanics, 69, 1999, 635–654.
[19] Stolpe, M. and Svanberg, K. An alternative interpolation scheme for minimum compliance optimization, Structural and Multidisciplinary Optimization, 22, 2001, 116–124.
[20] Guo, X. and Cheng, G.-D. Recent development in structural design and optimization, Acta Mechanica Sinica, 26(2010), 2010, 807–823.
[21] Bruns, T.E. A reevaluation of the SIMP method with filtering and an alternative formulation for solid–void topology optimization, Structural and Multidisciplinary Optimization, 30, 2005, 428–436.
[22] Zhou, M. and Rozvany, G.I.N. The COC algorithm, part II: Topological, geometrical and generalized shape optimization, Computer Methods in Applied Mechanics and Engineering, 89, 1991, 309–336.
[23] Pedersen, N.L. Maximization of eigenvalues using topology optimization, Structural and Multidisciplinary Optimization, 20, 2000, 2–11.
[24] Stolpe, M. and Svanberg, K. On the trajectories of penalization methods for topology optimization, Structural and Multidisciplinary Optimization, 21, 2001, 128–139.
[25] Sigmund, O. A 99 line topology optimization code written in Matlab. 2, s.l.: Springer, 2001, Structural and Multidisciplinary Optimization, 21, 2001, 120–127.
[26] Andreassen, E., Clausen, A., Schevenels, M., Lazarov, B.S. and Sigmund, O. Efficient topology optimization in MATLAB using 88 lines of code, Structural and Multidisciplinary Optimization, 43(1), 2011, 1–16. doi: 100.1007/s00158-010-0594-7.
[27] Krister, S. The method of moving asymptotes – A new method for structural optimization. International Journal for Numerical Methods in Engineering, 24, 1987, 358–373. doi: 10.1002/nme.1620240207.

[28] Duysinx, P. and Bendsoe, M.P. Topology optimization of continuum structures with local stress constraints, Numerical Methods in Engineering, 43, 1998, 1453–1478.
[29] Huang, X. and Xie, Y.M. A new look at ESO and BESO optimization methods, Structural and Multidisciplinary Optimization, 35, 2008, 89–92.
[30] Xie, Y.M. and Steven, G.P. A simple evolutionary procedure for structural optimization, Computers & Structures, 49, 1993, 885–896.
[31] Huang, X. and Xie, M. Evolutionary topology optimization of continuum structures: Methods and applications, Wiley, Chichester, United Kingdom, 2010, 238.
[32] Querin, O.M., Steven, G.P. and Xie, Y.M. Evolutionary structural optimisation (ESO) using a bidirectional algorithm, Engineering Computations, 15(8), 1998, 1031–1048, doi: 10.1108/02644409810244129.
[33] Querin, O.M., Young, V., Steven, G.P. and Xie, Y.M. Computational efficiency and validation of bi-directional evolutionary structural optimisation, Computer Methods in Applied Mechanics and Engineering, 189(2), 2000, 559–573.
[34] Yang, X.Y., Xei, Y.M., Steven, G.P. and Querin, O.M. Bidirectional evolutionary method for stiffness optimization, AIAA Journal, 37(11), 1999, 1483–1488, doi: 10.2514/2.626.
[35] Rozvany, G.I.N. A critical review of established methods of structural topology optimization. Structural and Multidisciplinary Optimization, 37, 2008, 217–237. doi: 10.1007/s00158-007-0217-0.
[36] Wang, M.Y., Wang, X. and Guo, D. A level set method for structural topology optimization, Computer Methods in Applied Mechanics and Engineering, 192, 2003, 227–246.
[37] Allaire, G., Jouve, F. and Toader, A.-M. A level-set method for shape optimization, Comptes Rendus Mathematique, 334(12), 2002, 1125–1130.
[38] Allaire, G., Jouve, F. and Toader, A.-M. Structural optimization using sensitivity analysis and a level-set method, Journal of Computational Physics, 194, 2004, 363–393.
[39] Blank, L., Garcke, H., Sarbu, L., Srisupattarawanit, T., Styles, V. and Voigt, A. Phase-field approaches to structural topology optimization, in Leugering, G., Engell, S., Griewank, A., Rannacher, R., Schulz, V., Ulbrich, M. and Ulbrich, S. Eds, Constrained optimization and optimal control for partial differential equations, Birkhäuser, Basel, Switzerland, 2012, 245–256.
[40] Burger, M., Hackl, B. and Ring, W. Incorporating topological derivatives into level set methods, Journal of Computational Physics, 194, 2004, 344–362.
[41] He, L., Kao, C.-Y. and Osher, S. Incorporating topological derivatives into shape derivatives based level set methods. Journal of Computational Physics, 225, 2007, 891–909. doi: 10.1016/j.jcp.2007.01.003.
[42] Dunning, P.D. and Kim, H.A. A new hole insertion method for level set based structural topology optimization. Numerical Methods in Engineering, 93, 2012, 118–134. doi: 10.1002/nme.4384.
[43] James, K. and Martins, J.R.R.A. An isoparametric approach to level set topology optimization using a body-fitted finite-element mesh, Computers & Structures, 90–91, 2012, 97–106.
[44] Chen, L.Q. Phase-field models for microstructure evolution, Annual Review of Materials Research, 32, 2002, 113–140.
[45] McFadden, G.B. Phase-field models of solidification, Contemporary Mathematics, 306, 2002, 107–146.
[46] Takezawa, A., Nishiwaki, S. and Kitamura, M. Shape and topology optimization based on the phase field method and sensitivity analysis, Journal of Computational Physics, 229, 2010, 2697–2718.
[47] Blaise, B. and Antonin, C. Design-dependent loads in topology optimization, ESAIM: Control, Optimisation and Calculus of Variations, 9, 2003, 19–48.
[48] Bourdin, B. and Chambolle, A. Design-dependent loads in topology optimization, ESAIM: Control, Optimisation and Calculus of Variations, 9, 2003, 19–48.
[49] Kobayashi, M.H. On a biologically inspired topology optimization method, Communications in Nonlinear Science and Numerical Simulation, 15, 2010, 787–802.

[50] Nakamura, A., Lindenmayer, A. and Aizawa, K. Some systems for map generation, The Book of L., Springer, Berlin, Heidelberg, 1986, 323–332. doi: 10.1007/978-3-642-95486-3_26.
[51] Kolonay, R.M. and Kobayashi, M.H. Optimization of aircraft lifting surfaces using a cellular division method, Journal of Aircraft, 52(2), 2015, 2051–2063.
[52] Krog, L., Tucker, A. and Rollema, G. Application of topology, sizing and shape optimisation methods to optimal design of aircraft components, in Proceedings of the 3rd Altair UK HyperWorks users Conference, Airbus UK Ltd., Copyright Altair Engineering, Inc., 2002.
[53] Maute, K. and Allen, M. Conceptual design of aeroelastic structures by topology optimization, Structural and Multidisciplinary Optimization, 27(1–2), 2004, 27–42, doi: 10.1007/s00158-003-0362-z.
[54] Dunning, P.D., Stanford, B.K. and Kim, A.K. Level-set topology optimization with aeroelastic constraints. 56th AIAA/ASCE/AHS/ASC Structures, Structural Dynamics, and Materials Conference, AIAA SciTech Forum, AIAA 2015-1128, 2015, doi:10.2514/6.2015-1128.
[55] Dunning, P.D., Stanford, B.K. and Kim, A.K. Aerostructural level set topology optimization for a common research model wing, 10th AIAA Multidisciplinary Design Optimization Conference, AIAA SciTech Forum, AIAA 2014–0634, 2014, doi:10.2514/6.2014-0634.
[56] Gomes, A.A. and Suleman, A. Topology optimization of a reinforced wing box for enhanced roll maneuvers, AIAA Journal, 46(3), 2008, 548–556.
[57] Kolonay, M.R. and Kobayashi, M.H. Optimization of aircraft lifting surfaces using a cellular division method, Journal of Aircraft, 52(6), 2015, 2051–2063, doi: 10.2514/1.C033138.
[58] Stanford, B., Beran, P. and Kobayashy, M.H. Aeroelastic optimization of flapping wing venation: A cellular division approach, AIAA Journal, 50(4), 2012, 938–951, doi: 10.2514/1.J051443.
[59] Faskhutdinov, R.N., Dubrovskaya, A.S., Dongauzer, K.A., Maksimov, P.V. and Trufanov, N.A. Topology optimization of a gas-turbine engine part. IOP Conference Series: Materials Science and Engineering, 177, 2017, 012077. doi: 10.1088/1757-899X/177/1/012077.

Appendix A: Response surface methodology

K. Kalinin
Riga Technical University, Latvia

RSM can be summarized as a collection of statistical tools and techniques for constructing and exploring an approximate functional relationship between a response variable and a set of design variables [1]. RSM has been widely used to simulate and analyze complex engineering problems in different industrial applications (see [2, 3]). The strategy used in RSM is to utilize approximation models, which are often referred to as metamodels or surrogate models, as they provide a model of the model, replacing the expensive simulation analyses during the process. This approximate functional relationship is typically constructed in the form of a low-order polynomial, referred to as a response surface (RS) approximation. RSM was originally developed for constructing and exploring approximate response functions based on physical experiments and results in smooth approximate response functions, thus effectively filtering out any noise.

RSM includes methods for selecting data points where the experiments should be evaluated, a process known as statistical design of experiments (DoEs) (as in [4, 5]); methods for solving the unknown coefficients of an RS approximation; and methods for evaluating the accuracy of the resulting RS approximation. The unknown coefficients of RS approximation are estimated from experimental data points by means of a process known as regression. These coefficients are estimated in such a way as to minimize the sum of the squares of the error terms, the so-called least squares criterion (e.g. [1, 6–8] with more than 300 year long history).

An alternative to least squares is maximum likelihood estimation method requiring an assumption about the probability distribution of the disturbance term. It is commonly assumed that the disturbances are identically and independently distributed normal variables with zero mean and constant variance.

An important research issue related to metamodeling is how to achieve good accuracy in a metamodel with a reasonable number of sample points and the distribution of the sample points in the domain of interest. This sampling technique is often referred to as a design of computer experiments (DoCE) and is implemented into numerical analysis to reduce the number of simulation runs without decreasing the accuracy of the metamodel.

The stages of metamodeling (see Fig. A1) can be considered as follows [9–11]:
- Development and experimental verification of a representative finite element model
- Formulation of the domain of interest and selection of variables to be used in approximation of sampled data responses
- Elaboration of space-filled experiment designs to determine sample points for which deterministic computer simulation is performed
- Building the approximating function (parametric or nonparametric approximations) employing the dataset of deterministic computer experiment (or alternatively employing a combined dataset of numerical and physical experimental data)
- Screening, parametric sensitivity analysis and validation of metamodel
- Design optimization and derivation of design guidelines

It should be noted that all these metamodeling stages can easily be extended by additive user-defined functions or methods, though the so-called curse of dimensionality is the limiting factor for application of the methodology to engineering processes and problems.

A1 Design of experiments and design of computer experiments

An experiment can be defined as a test or series of tests on a process or system that is performed to study the relationship between the input variables and the output responses of the process. In any experiment, the results and conclusions that can be drawn depend largely on the manner in which the data are collected [12]. In this context, the objectives of any experiment may be outlined as follows:
- determining which of the input variables, x, is most influential on the response, y.

- determining where to set the influential x's so that y is almost always near the desired nominal value.
- determining where to set the influential x's so that variability in y is small.

In attempting to fulfill these objectives, engineers and scientists frequently use a best-guess approach; however, this is limited by the technical or theoretical knowledge of the experimenter. The one-factor-at-a-time approach is an alternative approach that is also extensively used in practice, though this is limited to problem dimensions and fails to consider any possible interaction between input factors. So, there is an obvious need for an experimentation strategy, or an approach for DoE.

There is also an important distinction between physical experiments and computer experiments, so that separate experimentation strategies have been developed for DoCE. While physical experiments have statistical experimental error, numerical analysis is deterministic and results are obtained with 100% repetition and no statistical variance of model parameters (see [13, 14]), though there is numerical noise due to calculation or discretization errors (see, e.g. [15],). Many researchers (see [16–18]) argue that classical experimental designs are not well suited for sampling deterministic computer experiments, and Sacks et al. [17] state that the "classical notions of experimental blocking, replication and randomisation are irrelevant" when it comes to deterministic computer experiments that have no random error. This means that designs for deterministic computer experiments should "fill the space" of experiment design as opposed to possessing properties for estimating the variability in the data. Currently, there is a wide range of literature concerning different methods for DoCE (as in [4, 5]), which include many approaches for space-filling designs. It should be noted that the first space-filling design criterion [13, 19] for numerical experiments was proposed at Riga Technical University.

In this section the key concepts of DoE and DoCE are summarized, which includes the classical DoE and Latin hypercube (LH) design. This is followed by discussion on the techniques of space-filling and sequential design, which were developed for DoCE with LH designs.

A1.1 Classical design of experiments

Classical experimental designs are so named because they have been developed for what are considered to be the more "classical" applications of RS metamodeling: physical experiments, which are plagued by variability and random error (see [6–8, 2, 21]). Among these designs, the full-factorial, central composite and Box–Behnken designs (BBD) are well known and are easily generated, and a more detailed description of these types of designs can be found in the literature [21, 22].

In box-like design spaces, the range of a design variable between the upper and lower limits is typically divided into levels, and experiments are run only at the levels of the design variables. A full-factorial design contains all the combinations of the dif-

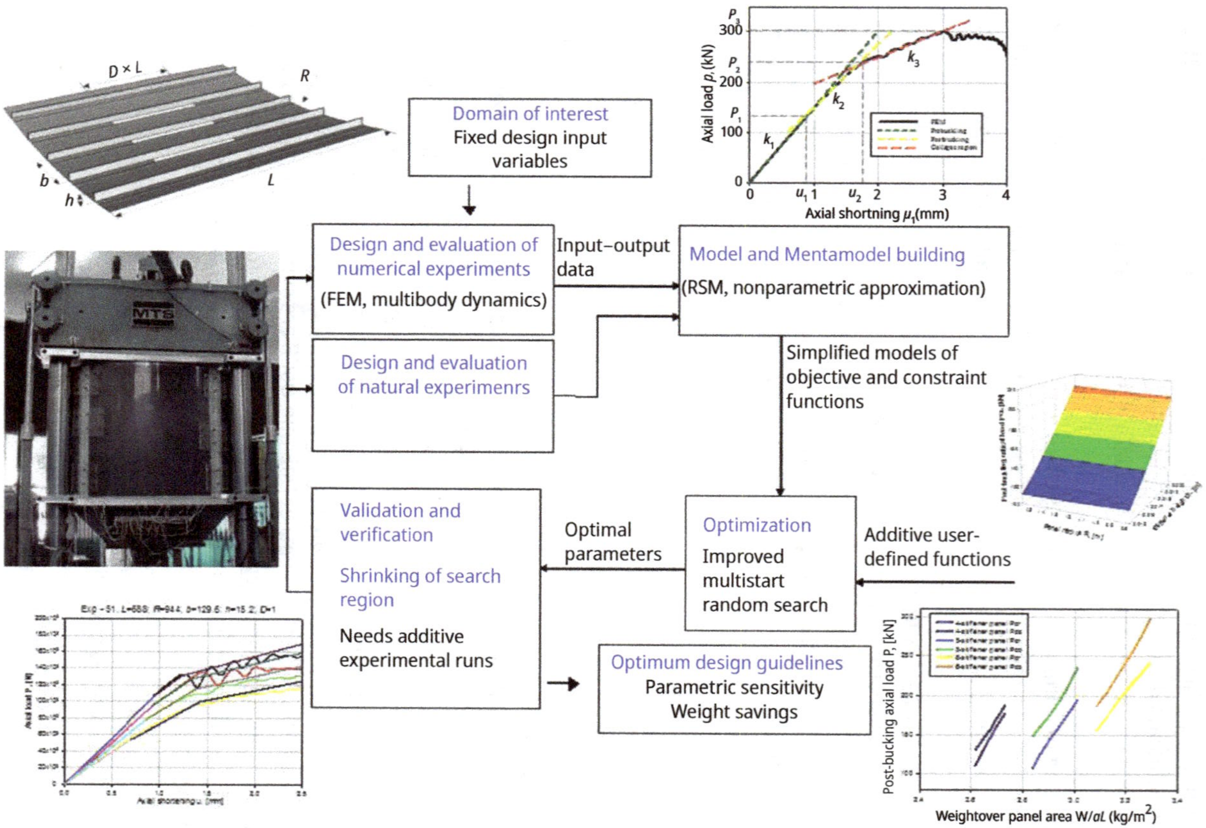

Fig. A1: Response surface flowchart example, to be used for optimization of a stiffened shell design.

ferent levels of all the design variables that are also commonly called factors. It is easy to understand that even with a modest number of factors, the full-factorial design approach becomes unreasonable in solving practical problems, as the number of experimental runs becomes very large and can be an order of magnitude greater than the number of parameters to be estimated. The number of experiments required by a factorial design is $l^{n_{dv}}$, where n_{dv} is the number of design variables and l is the number of levels per design variable. There are several special cases of the factorial design that are important to discuss because they are widely used and because they form the basis of other designs of considerable practical value. The most important of these is perhaps the $2^{n_{dv}}$ factorial design. Figure A2 shows a two-level factorial design for three design variables, that is, a 2^3-factorial design. Note that in the figure the location of experiments is denoted by black dots.

This design is particularly useful in the early stages of experimental work when only the different impact of the design variables on the results is of interest. Because there are only two levels per design variable, only a linear RS or at most the mixed terms of a quadratic RS approximation (first-order interactions) can be obtained from the $2^{n_{dv}}$ factorial design. The limitation of linear approximation inhibits the use of $2^{n_{dv}}$ factorial designs for many problems. When the true function that is being approximated by the RS approximation is expected to have considerable curvature, designers often apply $3^{n_{dv}}$ factorial design. The $3^{n_{dv}}$ factorial designs allow an estimation of all the terms of quadratic RS approximation. Also, RS approximations obtained from $3^{n_{dv}}$ factorial designs are generally accurate because they avoid extrapolation. An example of a $3^{n_{dv}}$ factorial design for a three-variable case with a total of 27 experiments is shown in Fig. A3.

A central composite design (CCD) is a combination of $2k$ factorial points, $2k$ star points and a center point for k factors as shown in Fig. A4. CCDs are the most widely used experimental design for fitting second-order RSs [1]. Different CCDs are defined by varying the distance from the center of the design space to the star points and include:

- ordinary central composite design – star points are placed at $\pm a$ ($a > 1$) from the center with the cube points placed at ±1 from the center;
- face-centered central composite design – star points are positioned on the faces of the cube.
- inscribed central composite design – star points are positioned $\frac{1}{a}$ from the center with the cube points placed at ±1.

It should be noted that experiments in center point are replicated at least three to five times to estimate the variance of output y.

A BBD is formed by combining $2k$ factorial and incomplete block designs and is an important alternative to CCDs because they require only three levels of each factor [8]. The BBD are found by perturbing combinations of two variables in turn, and adding to this n_c replicates of the nominal design. When the experiments are numerical in nature, instead of repeating the nominal design experiment slight perturbations

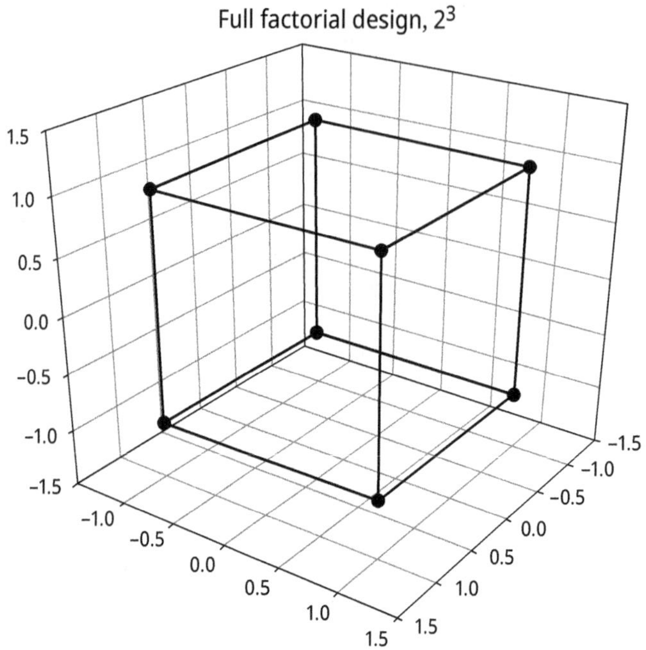

Fig. A2: Full-factorial design, $2^n = 2^3$.

are needed. The two variables vary from their nominal value to a low and high level, respectively, in all four possible ways. As there are different ways to select two variables out of n_{dv}, a particular BBD consists of totally $2n_{dv}(n_{dv} - 1) + n_c + 1$ points. This arrangement of the BBDs allows the number of design points to increase at the same rate as the number of polynomial coefficients. For many variables, the amount of points asymptotically approaches four times the number of terms required for a quadratic polynomial fit. As an example, a 7-variable BBD with no center-point replication holds 85 designs, nearly half of the 143 runs required by the CCD and only a fraction of the 2,187 needed for a full $3^{n_{dv}}$-level factorial design. The inevitable drawback of a BBD is that corner regions of the domain are poorly represented; therefore, corner approximations are based on extrapolation rather that interpolation. As a result, Myers et al. [1] warn that these designs should not be used when accurate predictions at the extremes (i.e., the corners) are important. An example of 13-point BBD for three factors is shown in Fig. A5.

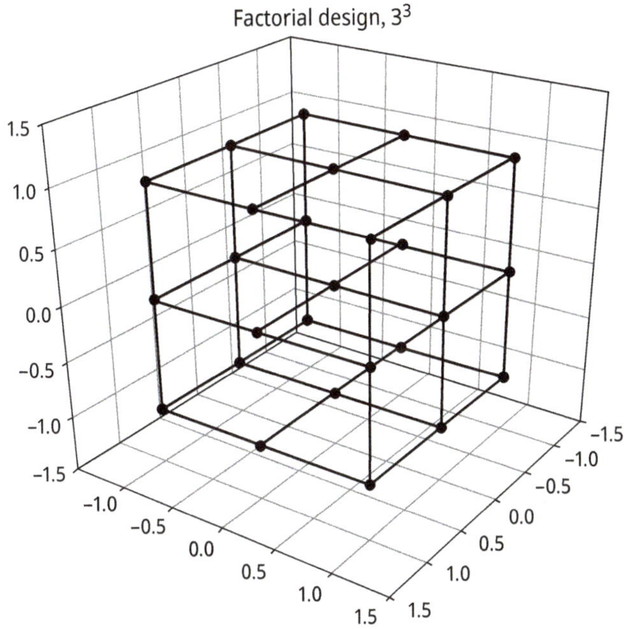

Fig. A3: Full-factorial design, $3^n = 3^3$.

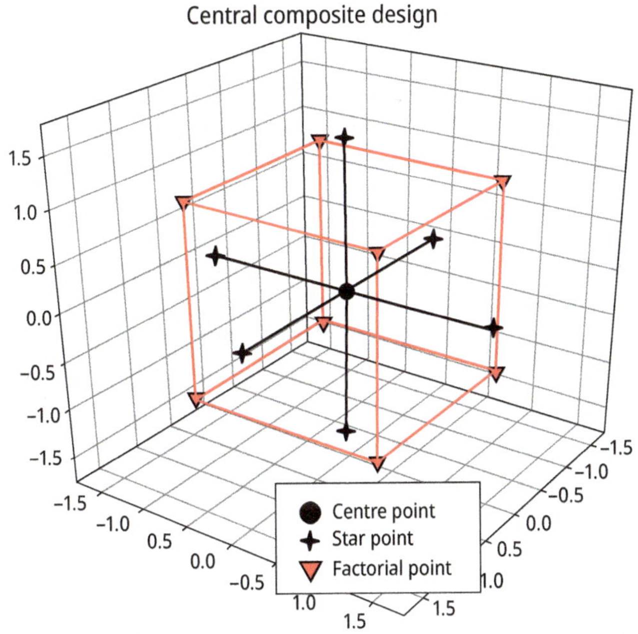

Fig. A4: Central composite design.

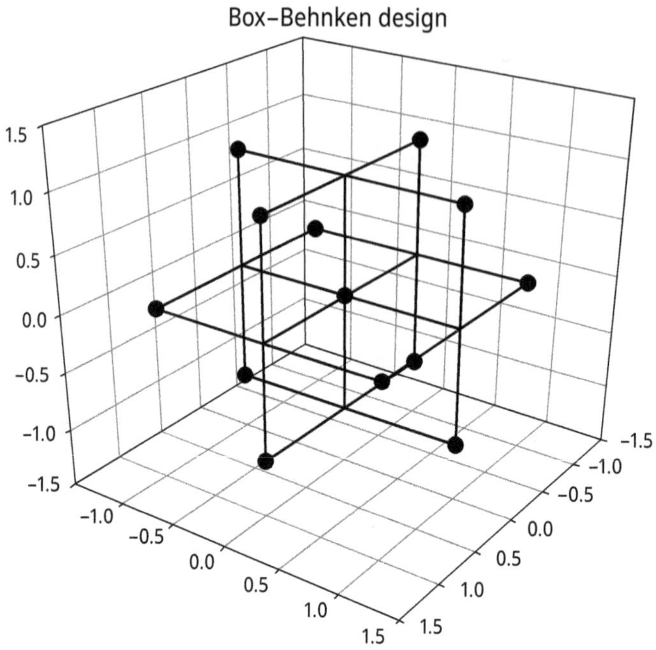

Fig. A5: Box–Behnken design.

A1.2 Latin hypercube designs

LH design was first proposed by Audze and Eglājs [13] and later McKay et al. [23] formulated the term "Latin hypercube." In the context of statistical sampling, a square grid containing sample positions is a Latin square only if there is only one sample in each row and each column. An LH is the generalization of this concept to an arbitrary number of dimensions, whereby each sample is the only one in each axis-aligned hyperplane containing it (see Fig. A6). When sampling a function of m variables, the range of each variable is divided into n equally probable intervals, where n-sample points are then placed to satisfy the LH requirements. Note that this forces the number of divisions, n, to be equal for each variable. Also note that this sampling scheme does not require more samples for more dimensions, and this independence is one of the main advantages of LH sampling.

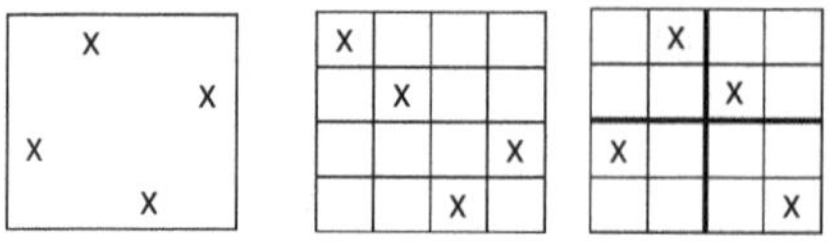

Fig. A6: A comparison between random sampling, LH sampling and orthogonal LH sampling.

Another advantage of LH designs is that random samples can be taken one at a time, regarding the samples that have already been taken. Orthogonal sampling adds the requirement that the entire sample space must be sampled evenly. Although more efficient, orthogonal sampling strategy is more difficult to implement since all random samples must be generated simultaneously. A 2D comparison in Fig. A6 demonstrates the difference between random sampling, LH sampling and orthogonal LH sampling and is explained as follows:

a) In random sampling new sample points are generated without considering the previously generated sample points. Thus, one does not necessarily need to know beforehand how many sample points are needed.
b) In LH sampling one must first decide how many sample points to use and for each sample point remember in which row and column the sample point was taken.
c) In orthogonal LH sampling, the sample space is divided into equally probable subspaces, where Fig. A6 shows four subspaces. All sample points are then chosen simultaneously making sure that the total ensemble of sample points is an LH sample and that each subspace is sampled with the same density.

In other words, LH sampling ensures that the ensemble of random numbers is a good representative of the real variability, whereas traditional random sampling is just an ensemble of random numbers without any guarantees. However, there is a general belief that the distribution of points in the design space should be regulated by a certain criterion, detailed in the following section, and therefore, the distribution will be regular, and experiments can be used not only for parametric approximations but also for nonparametric approximations.

A1.3 Space-filling criteria

Numerous space-filling experimental designs have been developed in an effort to provide more efficient and effective means for sampling deterministic computer experiments based on LH, as discussed previously. Different space-filling criteria for LH designs were proposed by many authors: maximin LH [24], minimal integrated MSE (IMSE) designs [17], orthogonal array-based LH designs [25], orthogonal LH [26], and IMSE and optimal LH [27].

Historically, the first space-filling experimental design criterion for numerical experiments was proposed by Riga Technical University researcher Vilnis Eglājs and published in 1977 [13], which is referred to as Eglajs or Audze–Eglais uniformity criterion [28]. For the analysis of deterministic computer experiments, Eglājs proposed the principle that the number of levels for each factor is equal to the number of runs and each level is used only once. In the situation where the response depends mainly on one factor (the number of which is unknown before experimentation) this principle provides the maximum amount of information about this dependency. Later, McKay et al. [23] introduced the name “Latin Hypercube” for DoEs of this type and showed that random LHs give better accuracy for the Monte Carlo integration than pseudo-random samples. The second principle proposed in Audze and Eglājs [13, 19] was that the experiments must fill the area of interest as uniformly as possible. For the measure of the uniformity, Audze and Eglājs [13] introduced the first space-filling criterion – the potential energy criterion

$$\Phi_{\text{Eglājs}} = \sum_{i=1}^{m} \sum_{j=i+1}^{m} \frac{1}{d_{ij}^{2}} \tag{A.1}$$

where m is the number of experimental points (runs) and d_{ij} is the Euclidean distance between points i and j. The LH designs with minimal value of the criterion (A.1) have good space-filling properties; however, the experimental points tend to spread out to the corners of the unit cube. Audze and Eglājs [13] also proposed the first coordinate exchange algorithm for construction of LHs with minimal value of the potential energy criterion. Morris and Mitchell [29] introduced a generalization of Eglājs criterion:

$$\Phi_{\text{Morris – Mitchell}} = \left[\sum_{i=1}^{n-1} \sum_{j=i+1}^{n} \frac{1}{d_{ij}^{p}} \right]^{1/p} \tag{A.2}$$

where

$$d(x_i, x_j) = d_{ij} = \left[\sum_{k=1}^{m} |x_{ik} - x_{jk}|^{t} \right]^{1/t}, \quad t = 1 \text{ or } 2 \tag{A.3}$$

The other category of space-filling designs are constructed by algorithmic approaches under certain optimality criteria, such as minimax and maximin designs [30]. The initial results on minimax and maximin distance designs were developed by Johnson et al. [24], and as they are based on the same principle, they are referred to as the MinDist criterion.

The MinDist criterion [24] seeks to maximize the minimum distance between any pair of points in the design:

$$\Phi_{\text{MinDist}} = \min_{u, v=1,\ldots,n} \sum_{i=1}^{m} (x_{ui} - x_{vi})^{2} \tag{A.4}$$

The entropy criterion was first proposed by Shewry and Wynn [31] and then used by Currin et al. [32]. The application of the entropy criterion for designs in unit cube [0,1] m is equivalent to the minimization of $E = -\log|C|$, where C is the $n \times n$ covariance matrix of the design with elements:

$$\Phi_{\text{Entropy}} = \exp\left\{-\Theta\sum_{u=1}^{m}\left|x_{iu}-x_{ju}\right|^{q}\right\},\quad 0<q\leq 2 \tag{A.5}$$

where $i,j = 1, \ldots, n$. The mostly used value of the parameter q is 2 [33], so the correlation between two points is a function of their Euclidean distance L_2, and Θ is set equal to 2.

Another measure used in constructing designs is the discrepancy $(D_c)^2$ [34–37]. The discrepancy measures how much the empirical distribution of the design points departs from the uniform distribution; in other words, the criterion averages the squared difference in the cumulative density function [38], and so minimum discrepancy designs are often called uniform designs. The discrepancy has been used to construct space-filling designs in computer experiments [28] and to construct designs for evaluating multiple integrals. Fang and Wang [34] gave the details of uniform designs and described several examples of discrepancy measurement. Uniform designs are a class of designs based on statistical applications of several theoretical methods [34]:

$$\Phi_{(D_C)^2} = \left(\frac{13}{12}\right)^{m} - \frac{2}{n}\sum_{u=1}^{n}\prod_{i=1}^{m}\left[1+\frac{1}{2}\left|x_{ui}-0.5\right|-\frac{1}{2}\left|x_{ui}-0.5\right|^{2}\right] + \\ +\frac{1}{n^{2}}\sum_{u=1}^{n}\sum_{v=1}^{n}\prod_{i=1}^{m}\left[1+\frac{1}{2}\left(\left|x_{ui}-0.5\right|+\left|x_{vi}-0.5\right|-\left|x_{ui}-x_{vi}\right|\right)\right] \tag{A.6}$$

The MSE criterion, as suggested by Fang and Wang [34] and Auzins [34], would give the root mean squared distance between the mesh points in design space R^m and the nearest point from experimental design D:

$$\Phi_{\text{MSE}} = \sqrt{\left(\frac{1}{N}\right)\sum_{v=1}^{N}\min_{u=1,\ldots,n}\left[\sum_{i=1}^{m}\left(w_{vi}-x_{ui}\right)^{2}\right]} \tag{A.7}$$

where w_v are points from a large sample in design space R^m, $v = 1, \ldots, m$, m is the number of points of the experimental design and n is the number of mesh points. Designs optimized according to the MSE criterion give points uniformly distributed in the design space and tend to minimize the expected MSE of the nonparametric approximations [14, 34, 39]). According to any criterion, to search for the optimal LH DoE is a very difficult task, though this can be achieved with methods of discrete optimization – coordinate exchange, multistart, threshold or simulated annealing methods [1, 33, 37, 40]. Comparisons of the different types of space-filling experimental designs are few, and often the novel space-filling design being described is compared against LH designs and random sampling [27], though rarely are space-filling designs compared with each other. Simpson [14, 18] gives a comparison of wide variety of space-filling designs against themselves and classical experimental designs. In a paper by Auzins [33], a comparison was given of many metamodeling test problems where experimental designs are sampled according to an optimality criterion and space-filling criteria such as MSE, Eglājs criterion, entropy criterion and discrepancy criterion. It is the decision of the author to state that in the case of second-order local polynomial approximation the use of the MSE criterion would be preferable, as it gives good accuracy of metamodels and the finding of optimal designs using the proposed algorithm is less difficult than optimization according to other criteria.

Further comparison of the various approaches is given below, where Fig. A7 shows several 3D 20-factor plans of experiments with different space-filling criteria, and Tab. A1 presents a comparison of different space-filling criteria and designs.

A1.4 Sequential design

Sequential designs can be obtained in a traditional way by adding new experimental points to the existing design created with space-filling criteria. These designs are called "adaptive," when the information about responses in previously created experimental points is used. Such designs are commonly needed when the initial amount of the sampling points is insufficient to derive a required level of approximation error. To obtain a sequential design from a fixed-size design space, Auzins et al. [33, 41–43] used the point arranging method. According to this method, first, an optimal LH design with large number of runs must be built. Then the point that gives the minimal worsening of optimality criterion by its elimination is moved out from the n-run design to build an n–1-run design. In this work and in [37], also the cross-validation approach was used where a point with the largest prediction error is selected as the new sample point. An accurate and efficient assessment of the prediction error must be made since in usual engineering optimization tasks several responses must be simultaneously approximated. A validation test was performed by Janushevskis et al. [39] and Auzins et al. [41–43], where the point in which the largest prediction error occurs for some test functions (six-hump camel, Branin) was directly added, and the authors found that maximal prediction error measures decreased when using fixed-size experimental designs or sequential designs without adaptation.

Sequential design of sample point addition is required to manage the fact that at the start of an experiment the optimal number of runs for metamodel building is unknown, but adding additional runs to an existing data set damages the uniformity of LHs. The classical hierarchical quasirandom sequences, which allow adding new sample points (experimental runs, trials), like Halton [44] and Sobol [45] sequences, give highly correlated components for four, five or more dimensions. Good results are given by adaptive experimental designs, in which the approximation or optimization information obtained from previous experimental runs is used for the choice of additional runs. However, this approach loses effectiveness in the practical engineering case when several responses must be simultaneously observed and used in optimization for quality criterion and constraint functions. In this work a recently developed approach for the creation of hierarchical degrees of freedom (DoEs) and the modification of a local polynomial approximation method is applied for the first time to engineering applications, and this approach has been shown to give almost the same or

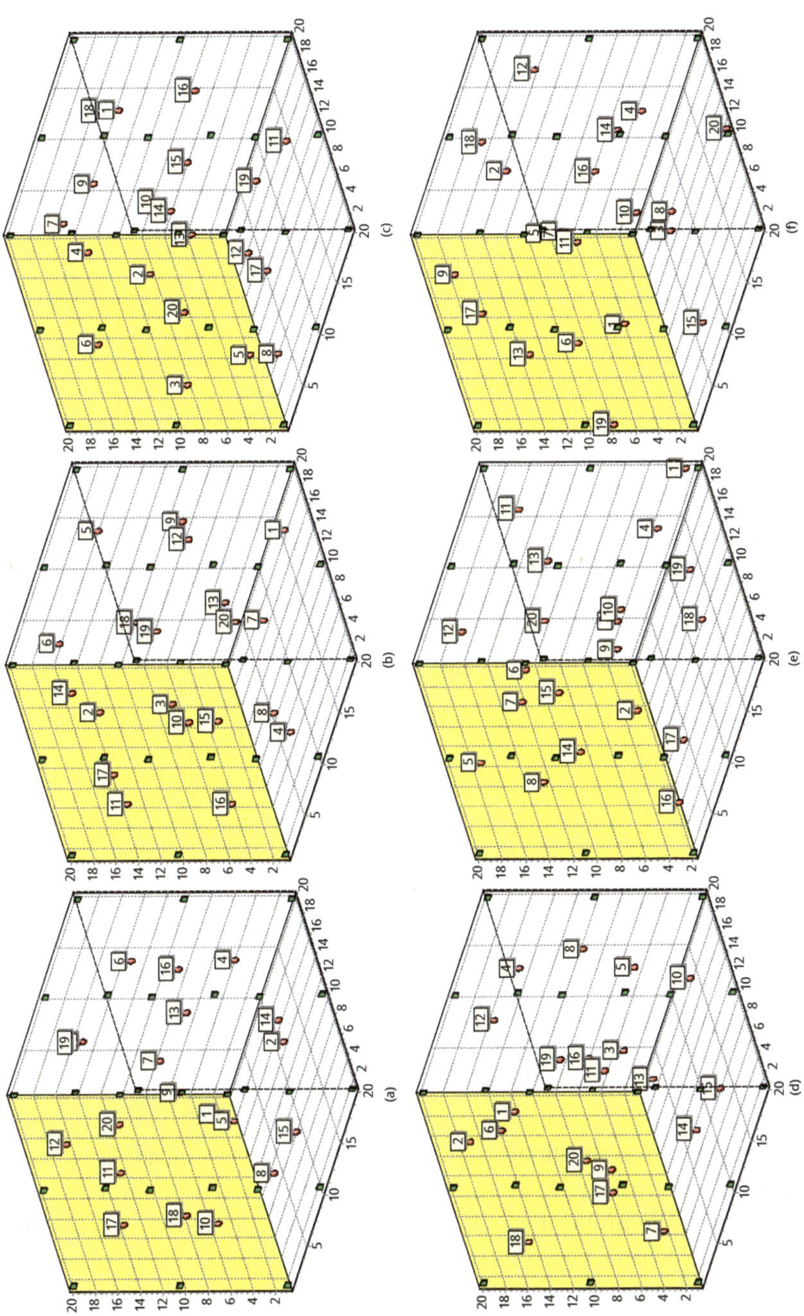

Fig. A7: Three-dimensional 20-variable designs of experiments using space-filling criteria: Eglājs (a), Morris–Mitchell (b), MinDist (c), entropy (d), discrepancy (e) and MSE (f).

Tab. A1: Comparison of different space-filling criterions and design.

	Eglājs	Morris–Mitchell	MinDist	Entropy	Discrepancy	MSE
Eglājs Φ_{Eglajs}	10.906*	11.043	11.083	11.004	11.236	11.202
MinDist Φ_{MinDist}	0.394	0.372	0.424*	0.337	0.302	0.307
MinDist between points	5–7	12–19	10–15	13–15	1–16	4–16
Entropy Φ_{Entropy}	11.606	11.350	11.538	10.983*	12.486	11.799
Discrepancy $\Phi_{(D_c)^2}$	0.00342	0.00363	0.00424	0.00366	0.00301*	0.00453
MSE Φ_{MSE}	0.4135	0.4154	0.4113	0.4133	0.4293	0.4084*
D^2	194,194	555,769	344,326	830,173	579,277	375,989
D^3	0	8.28	0.01	134.19	0.02	0.06

better accuracy of prediction than metamodels built with adaptive sampling methods and other nonparametric approximations.

The sequential design approach applied in this work is based on a compromise between space uniformity and subspace uniformity, including the uniformity of 1D projections, being used to build a combined criterion for any known criterion as the weighted sum of criteria for entire space and subspaces. For example, when we are interested in the uniformity of space filling of the entire space of all 1D projections and all 2D subspaces, we can build the combined criterion

$$\Phi_{\text{comb}} = \Phi + \frac{a}{m}\sum_{k=1}^{m}\Phi_k + \frac{2b}{m(m-1)}\sum_{k=1}^{m-1}\sum_{j=k+1}^{m}\Phi_{kj} \tag{A.8}$$

where Φ_k, Φ_{kj} are filling quality criterions for 1D and 2D subspaces, and a, b are weighting coefficients for 1D and 2D uniformity, respectively. For example, using criteria (A.7) and (A.8), we can build the complex criterion

$$\Phi_{\text{complex}} = \left[\sum_{i=1}^{N-1}\sum_{j=i+1}^{N}\left[\frac{1}{\sum_{k=1}^{m}\left|x_{ik}-x_{jk}\right|}\right]^{p} + \frac{a}{m}\sum_{i=1}^{N-1}\sum_{j=i+1}^{N}\sum_{k=1}^{m}\left[\frac{1}{\left|x_{ik}-x_{jk}\right|}\right]^{p}\right. \tag{A.9}$$
$$\left. + \frac{2b}{m(m-1)}\sum_{i=1}^{N-1}\sum_{j=i+1}^{N}\sum_{k=1}^{m-1}\sum_{l=k+1}^{m}\left[\frac{1}{\left|x_{ik}-x_{j}\right| + \left|x_{il}-x_{jl}\right|}\right]^{p}\right]^{1/-p}$$

To apply this equation, the weighting coefficients need to be determined. Using large values of weighting coefficients, a, we can obtain LHs with a nearly uniform distribution. The analysis of the problem can be facilitated by an approximate expression, which describes the dependency of the optimal criteria (A.7) and (A.8) values on the number of runs N and the number of factors m. Figure A8 shows this dependence graphically for $p = 50$, $t = 1$, $20 \le N \le 200$ and $1 \le m \le 6$.

Good approximation for the dependency of the optimal criterion values on the number of runs N and number of factors (variables) m gives the expression

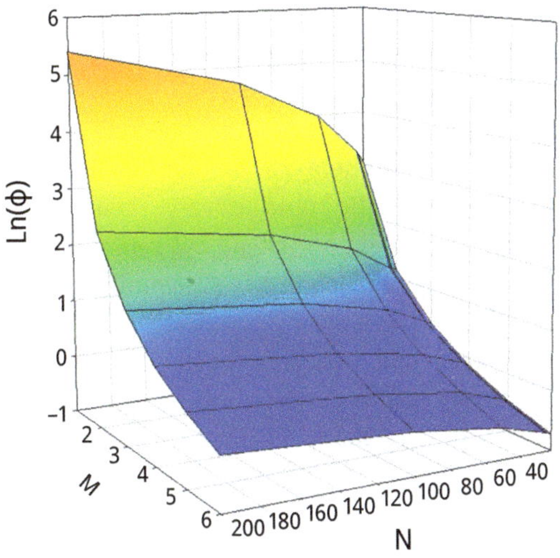

Fig. A8: Optimal values of criterion presented in eqs. (A.7) and (A.8).

$$\Phi_p^{\text{Optimal}} = \left(1 + \sqrt{(m-1)(0.04778 + 0.000422N)}\right)(Nm)^{1/p}\left(N^{1/m} - 1\right) \qquad \text{(A.10)}$$

This equation can be used for the control of the choice of the weighting coefficients a, b in the complex criterion (A.9).

The nonadaptive sequential design was implemented into the program code Relax [33]. Figure A9 gives an example of the use of this code for the optimization of a 2-factor 41-run design, obtained by adding one point at a time according to criterion (A.5), beginning from a 13-run optimal LH design. As further example of the use of the program in three dimensions, Fig. A10 shows the sequential design for a 3-factor 125-run experiment.

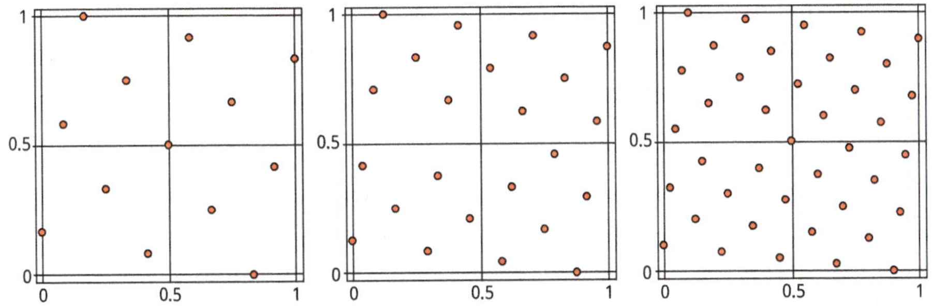

Fig. A9: Fixed-size Latin hypercubes optimized according to eqs. (A.9) and (A.10), $p = 50$.

A1.5 Fitting an approximation to given data

In this section, methods for determining an accurate RS for a given experimental data set are discussed. It is assumed that a relationship describing the performance measure of a phenomenon that it is under investigation exists, and that this relationship is a function of the design variables. It is important to note that this relationship does not have to be known explicitly or even be possible to be known exactly. The response that is to be approximated is denoted as y or $\hat{F}(x)$ and in principle can represent any measurable quantity such as stresses, critical force, deflection and structural eigenfrequencies.

In statistics, regression analysis is used to model relationships between random variables, determine the magnitude of the relationships between variables and can be used to make predictions based on the approximated models. Least square regression is one of the simplest, the most applicable and widely used methodologies for derivation of functional dependencies.

A1.5.1 Polynomial approximations

Low-order polynomial approximations are the most widely used and are described in every statistics book, and implemented in every statistics program. First-, second- and third-order polynomials can be expressed as follows:

$$\hat{F}(x) = b_0 + \sum_{i=1}^{m} b_i x_i \tag{A.11}$$

$$\hat{F}(x) = b_0 + \sum_{i=1}^{m} b_i x_i + \sum_{i=1}^{m} \sum_{j=i}^{m} b_{ij} x_i x_j \tag{A.12}$$

$$\hat{F}(x) = b_0 + \sum_{i=1}^{m} b_i x_i + \sum_{i=1}^{m} \sum_{j=i}^{m} b_{ij} x_i x_j + \sum_{i=1}^{m} \sum_{j=i}^{m} \sum_{k=j}^{m} b_{ijk} x_i x_j x_k \tag{A.13}$$

where m is total amount of variables; b_0 b_i b_{ij} b_{ijk} are the unknown coefficients of the regression functions.

As stated by Myers and Montgomery [46], there is a considerable practical experience indicating that second-order models work well in solving real RS problems. In general, it is thought that third- and higher-order polynomials can over-fit data, consequently avoiding construction of global behavior of the parameters. On the contrary, first-order polynomials are too simple and give prediction errors too high for use in science and engineering. In this chapter, a linear model will be used to derive parametric sensitivities and weight-saving parametric analysis, and third-order approximations will be compared with second-order and nonparametric approximations for RSs.

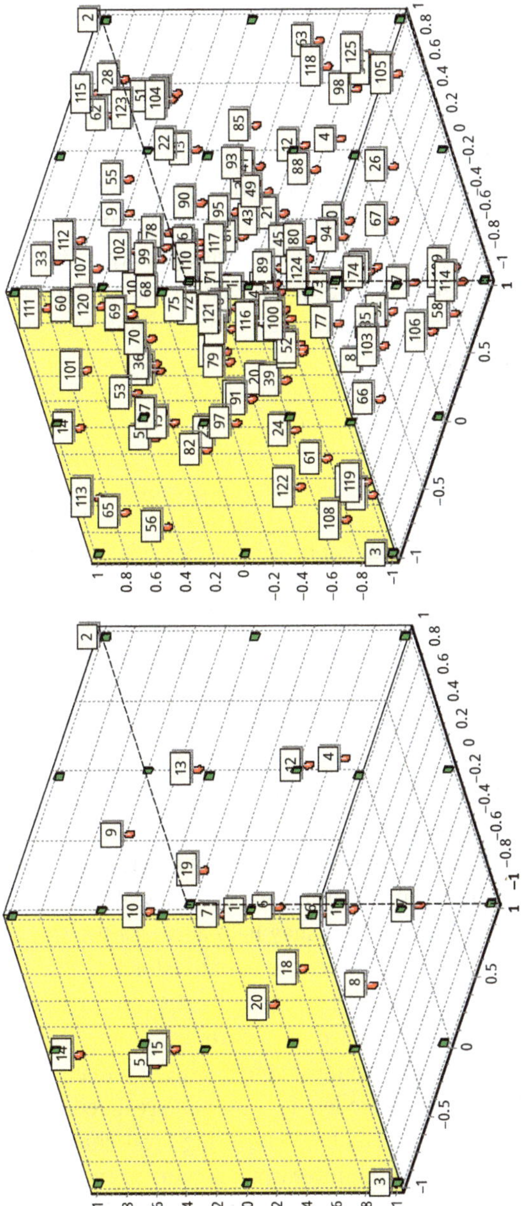

Fig. A10: Comparison of three-dimensional sequential design of 20 (left figure) and 125 factors (right figure).

A1.5.2 Partial polynomial approximations

In general, design variables in engineering problems are not equally distributed. For example, in the analysis of simple beam bending, beam length, height and thickness have different effects on the response. Moreover, common response parameters are affected by only some of the input variables, and the irrelevant parameters become approximation noise that decreases prediction accuracy. Considering this, it seems reasonable to use partial polynomials instead of full-order polynomial approximations. In such an assumption, polynomial parameters are used adequately with relevance to the input variables [47, 48]. However, one of the drawbacks for partial polynomials is the difficulty in selecting the essential parameters for an approximation function in the case of using nonorthogonal terms.

By generalizing the idea of polynomials, the nonlinear regression function is the linear combination of any kind of function. Such a model is written as follows:

$$\hat{F}(x) = b_0F_0(x_1, x_2, \ldots, x_D) + b_1F_1(x_1, x_2, \ldots, x_D) + \cdots + b_{m-1}F_{m-1}(x_1, x_2, \ldots, x_D) \quad \text{(A.14)}$$

where $F = \{F_i\}$ is a set of linearly independent functions consisting exclusively of m functions.

An exhaustive search of space is impractical, as there exist 2^M possible subsets of functions. The problem of function selection is to take a set of candidate functions and select a subset with the best performance. Eglājs [19] proposed the algorithm of selection of the terms for nonpolynomial regression function. In this work also the "correlation" approach for the optimal choice of the number of terms was proposed. This approach needs no additional experimental runs for the estimation of best choice of regression function but may give wrong results in some cases.

The selection procedure can provide better regression accuracy due to finite sample size effects – irrelevant functions may negatively affect the accuracy of regression [49–51]. In addition, reducing the number of functions may help decrease the cost of acquiring data and might make the regression models easier to understand.

One strategy is to normalize variables in scaled $[-1, 1]^m$ space and eliminate the regression parameters that have the least importance, therefore decreasing the approximation prediction error. This approach is the most simple and practical; however, variations of different parametrical interdependencies cannot be realized.

A more efficient approach conducts the search by considering local changes to the current set of attributes, selecting one and then iterating. For instance, the hill climbing approach considers both adding and removing functions at each decision point, which lets one retract an earlier decision without keeping explicit track of the search path. Within these options, one can consider all states generated by the operators and then select the best, or one can simply choose the first state that improves accuracy over the current set. A convenient paradigm for viewing the function selection approach is that of heuristic search, with each state in the search space specifying a subset of the possible functions. Two of the most promising sequential search algorithms are those proposed by Russel and Norvig [52], namely the sequential forward floating selection (SFFS) algorithm and the sequential backward floating selection (SBFS) algorithm. They improve the standard SFS (sequential floating selection) and SBS (sequential backward selection) techniques by dynamically changing the number of features included (SFFS) or removed (SBFS) at each step and by allowing the reconsideration of the features included or removed at the previous steps.

Another effective algorithm is random mutation hill climbing (RMHC) [53]. RMHC is a stochastic meta-algorithm. It simply runs an outer loop over hill climbing that stochastically iterates in any direction as long as it is possible to increase the value of the criterion function. Each step of the outer loop chooses a random initial condition x_0 to start hill climbing. The best x_m is kept: if a new run of hill climbing produces a better x_m than the stored state, it replaces the stored state.

A1.5.3 Nonparametric approximations

Nonparametric approximations such as multivariate adaptive regression splines (MARS), radial basis functions (RBFs) and krigings are recognized as the most precise for different orders of nonlinearity and problem scale response approximations, though by definition nonparametric approximations do not generate any tangible function. In design optimization, all nonparametric approximations are supreme over polynomial functions, though on the other hand implementation of response functions in different tools requires more advanced skills in statistics. Comparative studies of metamodeling techniques under multiple modeling criteria were performed by Jin et al. [37] providing insightful observations into performance of various metamodeling techniques under different testing problems. The comparative study presented three common advantages for polynomial regression models and MARS: they both have good transparency, which means it is easy to obtain the contributions of each input factor and the interaction among them; both methods take the least amount of time for model building; and both methods are resistant to numerical noise from the input data.

A methodology comparable to MARS is that of locally weighted polynomial regression (LOESS). In LOESS, originally proposed by Cleveland [53] and further developed by Cleveland and Devlin [54] and Cleveland and Grosse [55], at each point in the data set a low-degree polynomial is fit to a subset of the data, with explanatory variable values near the point whose response is being estimated. The polynomial is fit using the least weighted squares, giving more weight to points near the point whose response is being estimated and less weight to points further away. The value of the regression function for the point is then obtained by evaluating the local polynomial using the explanatory variable values for that data point. The LOESS fit is complete after regression function values have been computed for each of the N data points. Many of the details of this method, such as the degree of the polynomial model and the weights, are flexible. The range of choices for each part of the method and typical default values are briefly discussed in the following paragraph. LOESS is an efficient method that makes use of small data sets to share the ability to provide different types of easily interpretable statistical intervals for estimation, prediction, calibration and optimization.

To predict the value of the response function at point x, we use a second-order locally weighted polynomial approximation:

$$\hat{F}(x) = \beta_0 + \sum_{j=1}^{m} \beta_j x_j + \sum_{i=1}^{m} \sum_{k=j}^{m} \beta_{jk} x_j x_k \tag{A.15}$$

Unlike global (parametric) approximation, the coefficients β depend on x and are calculated by minimizing the weighted least squares

$$\beta = \arg\min_{\beta} \sum_{j \in N_X} w(x - x^j) \times \left(F^j - \beta_0 - \sum_{i=1}^{m} \beta_i x_{ji} - \sum_{i=1}^{m} \sum_{k=1}^{m} \beta_{ik} x_i^j x_k^j \right)^2 \tag{A.16}$$

where $\beta_0, \beta_i, \beta_{ik}$ are coefficients of the local quadratic approximation, N_x is the set of the nearest neighbors of the point x. Here we use a constant number of neighbors – bandwidth N_t. The weight function w depends on the Euclidean distance (in scaled $[-1, 1]^m$ space) between the points of interest x and the points of observation x^j. Let u be

$$u = (x, x^j) = \left\| \frac{x - x^j}{x - x^q} \right\| \tag{A.17}$$

where x^q is the furthest point in the neighborhood of point x. An often used weight function is the tricube function $(1 - u^3)^3$ [34, 40]; however, investigations in this work showed that a more accurate local approximation model can be obtained using a weight function of $(1 - u)^4$. Also, in this work, the Gaussian weight function

$$w(x, x^j) = \exp(-\alpha u^2) \tag{A.18}$$

is used to achieve a very accurate and smooth fit. In general, the Gaussian weight function with constant value $\alpha = 1/2$ is used in local approximations varying only the optimum bandwidth value. In this work, the Gaussian weight function was used with a particular weight function of fixed bandwidth value $N_t = N$, controlling the selection of the optimal fitting coefficient α. If the coefficient value is set to zero, then the global instead of the local parametric approximation is evaluated.

A main advantage of weighted least squares in comparison with other methods is the ability to handle regression situations in which the data points are of varying quality [56]. If the standard deviation of the random errors in the data is not constant across all levels of the explanatory variables, using weighted least squares with weights that are inversely proportional to the variance at each level of the explanatory variables yields the most precise parameter estimates possible. However, the biggest and largely unknown disadvantage of weighted least squares is the fact that when the weights are estimated from small numbers of replicated observations, the results of an analysis can be very bad and unpredictably affected. This is especially likely to be the case when the weights for extreme values of the predictor or explanatory variables are estimated using only a few observations. Therefore, it is recommended to remain aware of this potential problem and to only use weighted least squares when the weights can be estimated precisely relative to one another [57, 58].

A1.6 Evaluation of metamodel prediction accuracy

The accuracy of an RS approximation is expressed in terms of various error terms and statistical parameters that are representative of the predictive capabilities of the approximation [1]. Traditionally, the statistical DoE is performed using a minimum-variance criterion that assumes all errors to be random errors of variance, as shown in Tab. A2.

Tab. A2: Error measures for accuracy assessment.

Name	Equation	
Max. absolute error	$\max\lvert F_i - \hat{F}_i \rvert$	(A.19)
Average absolute error	$\frac{1}{n}\sum_{i=1}^{n}\lvert F_i - \hat{F}_i \rvert$	(A.20)
Mean square error	$\frac{1}{n}\sum_{i=1}^{n}\left(F_i - \hat{F}_i\right)^2$	(A.21)
Root mean square error	$\sqrt{\frac{\sum_{i=1}^{n}\left(F_i - \hat{F}_i\right)^2}{n}} = \sqrt{\text{MSE}}$	(A.22)
Relative root mean square error	$100\%\sqrt{\frac{\text{MSE}}{\text{Variance}}} = 100\%\sqrt{\frac{\frac{1}{n}\sum_{i=1}^{n}\left(F_i - \hat{F}_i\right)^2}{\frac{1}{n}\sum_{i=1}^{n}\left(F_i - \bar{F}\right)^2}}$	(A.23)

When finite element codes are used, the random errors associated with uncertainties of testing and measuring can be disregarded; therefore, for evaluation of approximation prediction accuracy different error estimation criteria should be used. A bootstrap or jackknife [59] error estimation methodology is one of the most promising methods; however, for effective application of these error estimation methodologies the amount of testing samples is not reasonable in engineering problems. The most effective approach for prediction accuracy can be a leave-one-out cross-validation error approach [52], in which each single experiment in turn removed from the data set, and the result for this removed experiment is estimated using the approximation built from the remaining data. Using this approach, the choice between polynomial approximations of full, partial and locally weighted different order polynomial functions can be evaluated. The standard deviation of approximation can be written in the form

$$\sigma = \sqrt{\frac{\sum_{i=1}^{N} \left(F(x^i) - \hat{F}(x^i)\right)^2}{N}} \tag{A.24}$$

The relative (percentage) standard deviation is calculated as

$$\sigma_{\%} = 100\% \frac{\sigma}{\sqrt{\frac{\sum_{i=1}^{N} \left(F(x^i) - \bar{F}\right)^2}{N}}} = 100\% \frac{\sigma}{\text{STD}} \tag{A.25}$$

where $\bar{F}$ is the mean of observed response values and STD stands for standard deviation from the mean value of experimental responses. This measure shows the degree to which the standard deviation of approximation is smaller in comparison with the approximation using a constant value.

The leave-one-out cross-validation error is calculated as

$$\sigma_{\text{cr}} = \sqrt{\frac{\sum_{i=1}^{N} \left(\hat{F}_{-i} - F_i\right)^2}{N}} \tag{A.26}$$

where $\hat{F}_{-i}$ denotes the prediction of the response of x^i using the metamodel created with $N - 1$ sample points and point i removed ($i = 1, 2, \ldots, N$).

Similarly, the cross-validation leave-one-out percentage error [33, 39] is calculated as

$$\sigma_{\text{cr/STD\%}} = 100\% \cdot \frac{\sigma_{\text{cr}}}{\text{STD}} \tag{A.27}$$

or alternatively the standard deviation mean value can be used

$$\sigma_{\text{cr/MEAN\%}} = 100\% \cdot \sqrt{\left(\frac{(1/n)\sum_{i=1}^{n} \left(\hat{F}_{-i} - F_i\right)^2}{(1/n)\sum_{i=1}^{n} F_i}\right)} \tag{A.28}$$

Usually leave-one-out cross-validation error gives an overvaluation of the prediction error, though it has the singular benefit of being able to estimate the prediction error without additional experimental runs. Various statistical tools are available in order to evaluate the predictive capabilities of an RS approximation from the data used to generate them (MiniTab, Design-Expert and modeFRONTIER). In this dissertation the calculation of cross-validation percentage error was performed using EDAOpt, a program developed at Riga Technical University, Machine and Mechanism Dynamics Research Laboratory [39, 41–43], and FUNSEL [50, 60].

A1.7 Sensitivity analysis of metamodels

Sensitivity analysis methods can be classified in a variety of ways. In a report by Frey and Patil [61], they are classified as (1) mathematical, (2) statistical and (3) graphical. In this classification, the focus is on sensitivity analysis techniques applied in addition to the fundamental modeling technique. For example, an analyst may be required to perform a deterministic analysis, in which case a mathematical method, such as nominal range sensitivity, can be employed to evaluate sensitivity. Alternatively, an analyst may perform a probabilistic analysis, using either frequentist or Bayesian frameworks, in which case statistical-based sensitivity analysis methods can be used [6, 62–65].

Mathematical methods assess the sensitivity of a model output to the range of variation of an input. These methods typically involve calculating the output for a few values of an input that represent the possible input range [66]. These methods do not address the variance in the output due to the variance in the inputs, but they can assess the impact of range of variation in the input values on the output [67]. In some cases, mathematical methods can be helpful in screening the most important inputs [68], which can also be used for verification and validation [69] and to identify inputs that require further data acquisition or research [70].

Statistical methods involve running simulations in which inputs are assigned probability distributions and assessing the effect of input variance on the output distribution [71, 72]. Depending on the method, one or more inputs are varied at a time. Statistical methods allow one to identify the effect of interactions among multiple inputs. The range and relative likelihood of inputs can be propagated using a variety of techniques such as Monte Carlo simulation, LH sampling and other methods. The sensitivity of the model results to individual inputs or groups of inputs can be evaluated by a variety of techniques, with Cullen and Frey [62], Fontaine and Jacomino [73] and Andersson et al. [71], all giving examples of the application of statistical methods. The statistical methods evaluated include regression analysis, analysis of variance, RS methods, Fourier amplitude sensitivity test and mutual information index.

Graphical methods give a representation of sensitivity in the form of graphs, charts or surfaces. Generally, graphical methods are used to give visual indication of how an output is affected by variation in inputs [74]. Graphical methods can be used as a screening method before further analysis of a model or to represent complex dependencies between inputs and outputs [75]. Graphical methods can be used to complement the results of mathematical and statistical methods for better representation [76, 77].

A1.8 Metamodels in optimization

RS approximations are often used to replace constraints or objective functions in optimization problems. Optimization problems, however, have the general tendency of exploiting weaknesses in the formulation of RS. In terms of RS approximation, the optimizer tends to drive the optimum design toward regions where the approximations are inaccurate. Due to the similarities between detailed numerical simulations and physical experiments, there has recently been growing interest in using RS approximations within structural optimization. When using polynomial RS approximations, in addition to providing the designer with an overall perspective of the response, spurious local minimum resulting from human and numerical errors are eliminated by using the least squares method to formulate smooth low-order polynomials.

In recent years, many researchers have exploited the advantages of using conventional (second-order polynomial) RS approximations that are more global in nature to integrate numerical analyses in an optimization environment. For example, in a monograph by Schmidt and Launsby [3] a manufacturing process of forging hammer and molding casting was optimized using a response approach. Similarly, using GAs a design and optimization procedure of laminated composite materials was compiled by Gürdal et al. [78]. Additionally, Thacker and Wu [79] used RS approximations to reduce the computational cost associated with reliability-based optimization problems using probabilistic methods, while Harrison et al. [80] and Sellar and Batill [81] reduced the cost of designing a stiffened composite panel using a GA. RBF formulated by GA in combination with neural network approximations was used in post-buckling optimization of composite stiffened panels by Bisagni and Lanzi [82]. Giunta et al. [15], Kaufman et al. [83] and Venter and Haftka [84] showed that RS approximations are valuable in solving optimization problems with nonsmooth functional behavior, as numerical noise is filtered out from the resulting approximate response function. Additionally, Mistree et al. [85] made extensive use of the global perspective of the response over the design space provided by global RS approximations, while Barthelemy and Haftka [86] used RS approximations to meet the organizational challenge posed by multidisciplinary optimization problems. More specifically, Ragon et al. [87] used RS approximations as a simple yet flexible interface between the global and local design codes in the design of a large airplane wing structure. Rodríguez et al. [88] applied the approach to multidisciplinary design optimization, constructing local RS approximations in a sequential manner based on design points evaluated at the discipline level of the current design cycle. With few exceptions [89], almost all the standard applications of RS approximations in structural optimization employ quadratic approximations, which still do not provide fully global approximations. For further details, Haftka and Gürdal [90] gave a wide overview of engineering problems optimized using RS methodology and also provided some of the benchmark cases frequently validated in structural design optimization.

References

[1] Meyer, R.H. and Nachtsheim, C.J. The coordinate-exchange algorithm for constructing exact optimal experimental designs, Technometrics, 37, 1995, 60–69.

[2] Evans, M. Optimisation of manufacturing processes: A response surface approach, Maney Publishing, 2003, 320.

[3] Schmidt, S.R. and Launsby, R.G. Understanding industrial design experiments, Air Academy Press, 4th, 1994, 768.

[4] Wu, C.F. and Hamada, M. Experiments – Planning, analysis, and parameter design optimization, John Wiley & Sons, Inc, 2000, 624.

[5] Santner, T.J., Williams, B.J. and Notz, W.I. The design and analysis of computer experiments, Springer Verlag, New York, 2003, 203.

[6] Box, G.E.P. and Tiao, G.C. Bayesian inference in statistical analysis, John Wiley and Sons Inc, New York, 1992, 608.

[7] Box, G.E.P. and Draper, N.R. On minimum-point second-order designs, Technometrics, 16, 1974, 613–616.

[8] Box, G.E.P. and Benhken, D.W. Some new three level designs for the study of quantitative variables, Technometrics, 2, 1960, 455–476.

[9] Rikards, R., Abramovich, H., Auzins, J., Korjakins, A., Ozolinsh, O., Kalnins, K. and Green, T. Surrogate models for optimum design of stiffened composite shells, composite structures, Vol. 63, Elsevier, 2004, 243–251.

[10] Rikards, R., Kalnins, K. and Ozolinsh, O. Delamination and skin-stringer separation analysis in composite stiffened shells, Topping, B.H.V. and Mota Soares, C.A. Eds, Computational structures technology, Proc. of 7th Int. Conf., Lisbon, 7–9 September 2004 Stirling, Civil-Comp Press, Vol. 47, 2004.

[11] Rikards, R., Auzins, J. and Kalnins, K. Surrogate modeling in design optimization of stiffened composite shells, in Neittaanmäki, P., Rossi, T., Majava, K. and Pironneau, O. Eds, Proc. European congress on computational methods in applied sciences and engineering, ECCOMAS-04, Jyväskylä, 24–28 July, CD-ed, Vol. 14, pages 2004.

[12] Montgomery, D.C. Design and analysis of experiments, Wiley and Sons Ltd, New York, 1997, 478.

[13] Audze, P. and Eglājs, V. New approach for planning out of experiments, problems of dynamics and strength, Vol. 35, Zinatne Riga (in Russian), 1977, 104–107.

[14] Simpson, T.W. A concept exploration method for product family design, in Ph.D. thesis, Georgia Institute of Technology, Georgia, USA, 1998.

[15] Giunta, A.A., Balabanov, V., Haim, D., Grossman, B., Mason, W.H., Watson, L.T. and Haftka, R.T. Wing design for a high-speed civil transport using a design of experiments methodology, in Proc. of the 6th AIAA/NASA/ISSMO symposium on multidisciplinary analysis and optimization, Bellevue, Washington, Paper No.96–4001, Part 1, 4–6 September, 1996, 168–183.

[16] Sacks, J., Schiller, S.B. and Welch, W.J. Designs for computer experiments, Technometrics, 34, 1989, 15–25.

[17] Sacks, J., Welch, W.J., Mitchell, T.J. and Wynn, H.P. Design and analysis of computer experiments, Statistical Science, 4(4), 1989, 409–435.

[18] Simpson, T.W., Peplinski, J., Koch, P.N. and Allen, J.K. On the use of statistics in design and the implications for deterministic computer experiments, design theory and methodology – DTM'97, in ASME paper, No.DETC97/DTM-3881, Sacramento, CA, 1997.Eglājs, 1981.

[19] Eglājs, V. Approximation of data by multi-dimensional equation of regression, problems of dynamics and strength, Vol. 39, Zinatne, Riga ((in Russian)), 1981, 120–125.

[20] Draper, N.R. and Lin, D.K.J. Small response-surface designs, Technometrics, 32, 1990, 187–194.

[21] Myers, R.H., Montgomery, D.C. and Anderson-Cook, C.M. Response surface methodology: Process, and product optimization using designed experiments, 4th, Wiley, 2016, 856.

[22] Vitali, R., Haftka, R.T. and Sankar, B.V. Correction response surface design of stiffened composite panel with a crack, in AIAA/ASME/ASCE/ AHS/ASC structures, structural dynamics and material conference paper AIAA No.99–1313, proceedings, St Louis, Missouri, 1999.

[23] McKay, M.D., Beckman, R.J. and Conover, W.J. A comparison of three methods for selecting values of input variables in the analysis of output from a computer code, Technometrics, 21, 1979, 239–245.

[24] Johnson, M.E., Moore, L.M. and Ylvisaker, D. Minimax and maximin distance designs, Journal of Statistical Planning and Inference, 26, 1990, 131–148.

[25] Tang, B. Orthogonal array-based Latin hypercubes, Journal of the American Statistical Association, 88, 1993, 1392–1397.

[26] Ye, K.Q. Column orthogonal Latin hypercubes and their application in computer experiments, Journal of American Statistical Association, 93, 1998, 1430–1439.

[27] Park, J.S. Optimal Latin-hypercube designs for computer experiments, Journal of Statistical Planning and Inference, 39(1), 1994, 95–111.

[28] Bates, S.J., Sienz, J. and Langley, D.S. Formulation of the Audze–Eglais uniform Latin hypercube design of experiments, Journal of Advances in Engineering Software, 34(8), 2003, 493–506.

[29] Morris, M.D. and Mitchell, T.J. Exploratory designs for computational experiments, Journal of Statistical Planning and Inference, 43(3), 1995, 381–402.

[30] Hardin, R.H. and Sloane, N.J.A. A new approach to the construction of optimal designs, Journal of Statistical Planning and Inference, 37, 1993, 339–369.

[31] Shewry, M. and Wynn, H.P. Maximum entropy sampling, Journal of Applied Statistics, 14, 1987, 165–170.

[32] Currin, C., Mitchell, T.J., Morris, M. and Ylvisaker, D. Bayesian prediction of deterministic functions with applications to the design and analysis of computer experiments, Journal of American Statistics Association, 86, 1991, 953–963.

[33] Auzins, J. Direct optimization of experimental designs, in 10th AIAA/ISSMO multidisciplinary analysis and optimization conference, AIAA paper, No.2004–4578, Albany, NY, 28 Aug.-2 Sep., 2004.

[34] Fang, K.T. and Wang, Y. Number-theoretic methods in statistics, Chapman & Hall, London, 1994, 344.

[35] Fang, K., Ma, X. and Winker, P. Centered L2-discrepancy of random sampling and Latin hypercube design and construction of uniform designs, Mathematics of Computation, 71, 2002, 275–296.

[36] Fang, K., Lu, X. and Winker, P. Lower bounds for centered and wrap-around L2-discrepancies and construction of uniform design by threshold accepting, Journal of Complexity, 19, 2003, 692–711.

[37] Jin, R., Chen, W. and Simpson, T. Comparative studies of metamodeling techniques under multiple modelling criteria, in 8th AIAA/NASA/USAF/ISSMO symposium on multidisciplinary analysis and optimization, Long Beach, CA, USA, AIAA Paper AIAA-2000–4801, 2000.

[38] Hickernell, F.J. A generalized discrepancy and quadrature error bound, Mathematics of Computation, 67, 1998, 299–322.

[39] Janushevskis, A., Akinfiev, T., Auzins, J. and Boyko, A. A comparative analysis of global search procedures, Proceedings of the Estonian Academy of Sciences, 10(4), 2004, 236–250.

[40] Ma, C.X. and Fang, K.T. A new approach to construction of nearly uniform designs, International Journal of Materials and Product Technology, 20, 2004, 115–126.

[41] Auzins, J., Janushevskis, A. and Rikards, R. Software tools for experimental design, metamodeling and optimization, book of abstracts, in Aifantis, E.C. Ed, 5th euromech solid mechanics Conf. ESMC, 5 August, 2003, Giapoulis, Thessaloniki, Greece, 2003, Vol. 182.

[42] Auzins, J., Kalnins, K. and Rikards, R. Sequential design of experiments for metamodeling and optimization, in Proc. of 6th world congress of structural and multidisciplinary optimization, 30 May – 03 June 2005, Rio de Janeiro, Brazil, 2005.

[43] Auzins, J., Janushevskis, J., Kalnins, K. and Rikards, R. Sequential metamodeling techniques for structural optimization, in Proc. of Intern. Conf. TCN CAE -05, 5–8 October 2005, Lecce, Italy, CD-ed., 20, pages, 2005 b.

[44] Halton, J.H. On the efficiency of certain quasi-random sequences of points in evaluating multi-dimensional integrals, Numerische Mathematik, 2, 1960, 84–90.

[45] Sobol, I.M. The distribution of points in a cube and approximate evaluation of integrals. Zhurnal Vychislitel'noi Matematiki I Matematicheskoi Fiziki, 7, 1967, 784–802. in Russian.

[46] Myers, R.H. and Montgomery, D.C. Response surface methodology: Process and product optimization using designed experiments, John Wiley and Sons, New York, 1995, 690.

[47] Antony, J. Design of experiments for engineers and scientists, Butterworth-Heinemann, 2003, 190.

[48] Minitab – User Manual Version 8.30, http://www.mscsoftware.com, 2006.

[49] Vapnik, N.V. Statistical learning theory, John Wiley, 1998, 768.

[50] Jekabsons, G. Heuristic approach for nonlinear multiple regression analysis, Master thesis, RTU, Riga, Latvia, 2005.

[51] Russell, S.J. and Norvig, P. Artificial intelligence: A modern approach, 3rd, Prentice-Hall, Upper Saddle River, New Jersey, 1995, 1132.

[52] Kohavi, R. A study of cross-validation and bootstrap for accuracy estimation and model selection, in Mellish, C.S. Ed, Proceedings of IJCAI-95, Montreal, Canada, Los Altos, CA, Morgan Kaufmann, 1995, 1137–1143.

[53] Cleveland, W.S. Robust locally weighted regression and smoothing scatterplots, Journal of the American Statistical Association, 74, 1979, 829–836.

[54] Cleveland, W.S. and Devlin, S.J. Locally weighted regression: An approach to regression analysis by local fitting, Journal of the American Statistical Association, 83, 1988, 596–610.

[55] Cleveland, W.S. and Grosse, E. Computational methods for local regression, Statistics and Computing, 1, 1991, 47–62.

[56] Croarkin, C. NIST/SEMATECH, e-handbook of statistical methods, http://www.itl.nist.gov/handbook/ , 2006.

[57] Carroll, R.J. and Ruppert, D. Transformation and weighting in regression, Chapman and Hall, New York, 1988, 264.

[58] Ryan, T.P. Modern regression methods, Wiley, New York, 1997, 515.

[59] Shao, J. and Tu, D. The Jackknife and bootstrap, Springer Series in Statistics, 1995, 446.

[60] Jēkabsons, G. and Lavendels, J. Labākā aproksimanta izvēle iespējamo variantu kopā, computer science, Vol. 5, 18 Scientific Proceedings of Riga Technical University, Riga, 2004, 24–31.

[61] Frey, H.C. and Patil, S. Identification and review of sensitivity analysis methods, Proc. Of NCSU/USDA Workshop on Sensitivity Analysis Method, 2001.

[62] Cullen, A.C. and Frey, H.C. Probabilistic techniques in exposure assessment, Plenum Press, New York, 1999, 336.

[63] Saltelli, A., Tarantola, S. and Chan, K. A quantitative, model independent method for global sensitivity analysis of model output, Technometrics, 41(1), 1999, 39–56.

[64] Saltelli, A., Chan, K. and Scott, E.M. Sensitivity analysis, John Wiley and Sons, Ltd., West Sussex, England, 2009, 494.

[65] Weiss, R. An approach to Bayesian sensitivity analysis, Journal of the Royal Statistical Society Series B-Methodological, 58(4), 1996, 739–750.

[66] Salehi, F., Prasher, S.O., Amin, S., Madani, A., Jebelli, S.J., Ramaswamy, H.S. and Drury, C.T. Prediction of annual Nitrate-N losses in drain outflows with artificial neural networks, Transactions of the ASAE, 43(5), 2000, 1137–1143.

[67] Morgan, M.G. and Henrion, M. Uncertainty: A guide to dealing with uncertainty in quantitative risk and policy analysis, Cambridge University Press, Cambridge, NY, 1990, 332.

[68] Brun, R., Reichert, P. and Kunsch, H.R. Practical identifiability analysis of large environmental simulation models, Water Resources Research, 37(4), 2001, 1015–1030.

[69] Wotawa, G., Stohl, A. and Kromp, K.H. Estimating the uncertainty of a Lagrangian photochemical air quality simulation model caused by inexact meteorological input data, Reliability Engineering & System Safety, 57(1), 1997, 31–40.

[70] Ariens, G.A., Van Mechelen, W., Bongers, P.M., Bouter, L.M. and van der Wal, G. Physical risk factors for neck pain, Scandinavian Journal of Work Environment & Health, 26(1), 2000, 7–19.

[71] Andersson, F.O., Aberg, M. and Jacobsson, S.P. Algorithmic approaches for studies of variable influence, contribution and selection in neural networks, Chemometrics and Intelligent Laboratory Systems, 51(1), 2000, 61–72.

[72] Neter, J., Kutner, M.H., Nachtsheim, C.J. and Wasserman, W. Applied linear statistical models, 4th, WCB McGraw-Hill, Chicago, IL, 1996, 1408.

[73] Fontaine, T.A. and Jacomino, V.M.F. Sensitivity analysis of simulated contaminated sediment transport, Journal of the American Water Resources Association, 33(2), 1997, 313–326.

[74] Geldermann, J. and Rentz, O. Integrated technique assessment with imprecise information as a support for the identification of best available techniques (BAT, OR Spektrum, 23(1), 2001, 137–157.

[75] McCamley, F. and Rudel, R.K. Graphical sensitivity analysis for generalized stochastic dominance, Journal of Agricultural and Resource Economics, 20(2), 1995, 403–403.

[76] Stiber, N.A., Pantazidou, M. and Small, M.J. Expert system methodology for evaluating reductive dechlorination at TCE sites, Environmental Science and Technology, 33(17), 1999, 3012–3020.

[77] Critchfield, G.C. and Willard, K.E. Probabilistic analysis of decision trees using Mote Carlo simulation, Medical Decision Making, 6(1), 1986, 85–92.

[78] Gürdal, Z., Haftka, R. and Hajela, P. Design and optimization: Laminated composite materials, Wiley and Sons Ltd, New York, 1999, 352.

[79] Thacker, B.H. and Wu, Y.T. A new response surface approach for structural reliability analysis, in Proc. of the 33rd AIAA/ASME/ASCE/AHS/ASC structures, structural dynamics and materials conference, Dallas, Texas, Paper No.92–2408, Part 2, 13–15 April, 1992, 586–593.

[80] Harrison, P.N., Le Riche, R. and Haftka, R.T. Design of stiffened composite panels by genetic algorithm and response surface approximations, in Proc. of the 36th AIAA/ASME/ASCE/AHS/ASC structures, structural dynamics, and materials conference, New Orleans, Louisiana, Paper No.95–1163, Part 1, 10–13 April 1995, 58–65.

[81] Sellar, R.S. and Batill, S.M. Concurrent subspace optimization using gradient-enhanced neural network approximations, in Proc. of the 6th AIAA/NASA/ISSMO symposium on multidisciplinary analysis and optimization, Bellevue, Washington, Paper No.96–4019, Part 1, 4–6 September 1996, 319–330.

[82] Bisangi, C. and Lanzi, L. Post-buckling optimization of composite stiffened panels using neural networks, Composites Structures, 58, 2002, 237–247.

[83] Kaufman, M., Balabanov, V., Burgee, S.L., Giunta, A.A., Grossman, B., Mason, W.H., Watson, L.T. and Haftka, R.T. Variable-complexity response surface approximations for wing structural weight in HSCT design, Journal of Computational Mechanics, 18(2), 1996, 112–126.

[84] Venter, G. and Haftka, R.T. Minimum-bias based experimental design for constructing response surfaces in structural optimization, in Proc. of the 38th AIAA/ASME/ASCE/AHS/ASC structures, structural dynamics, and materials conference, Kissimmee, Florida, Paper No.97–1053, Part 2, 7–10 April 1997, 1225–1238.

[85] Mistree, F., Patel, B. and Vadde, S. On modeling objectives and multilevel decisions in concurrent design, advances in design automation, Gilmore, B.J., Hoeltzel, D., Dutta, D. and Eschenauer, H. Eds., ASME, New York, 1994, 151–161. ASME DE-Vol. 69–2.

[86] Barthelemy, J.F.M. and Haftka, R.T. Approximation concepts for optimum structural design-A review, Structural Optimization, 5(3), 1993, 129–144.

[87] Ragon, S., Gürdal, Z., Haftka, R.T. and Tzong, T.J. Global/local structural wing design using response surface techniques, in Proc. of the 38th AIAA/ASME/ASCE/AHS/ASC structures, structural dynamics, and materials conference, Kissimmee, Florida, Paper No.97–1051, Part 2, 7–10 April 1997, 1204–1214.

[88] Rodríguez, J.F., Renaud, J.E. and Watson, L.T. Trust region augmented Lagrangian methods for sequential response surface approximation and optimization, in Proc. of the 1997 ASME design engineering technical conferences, Sacramento, California, Paper No.DETC97DAC3773, 14–17 September 1997.

[89] Houten, M.H., Schoofs, A.J.G. and Van Campen, D.H. Response surface techniques in structural optimization, 1st world congress of structural and multidisciplinary optimization, Gosler, Germany, 28 May–2 June 1995, 89–94.

[90] Haftka, R. and Gürdal, Z. Elements of structural optimization, 3rd, Kluwer Academic Publishers, 1992, 468.

10 Structural health monitoring (SHM)

10.1 Introduction

Structural health monitoring (SHM) is an emerging technique developed from nondestructive testing (NDT),[1] which combines advanced sensing technology with intelligent algorithms to interrogate the structural "health" condition of a given structure. Unlike conventional maintenance procedures based on NDTs, the SHM system, also known as health and usage monitoring system, makes the embedded/bonded sensors an integral part of the structure and operates with a minimal manual intervention. The potential benefits of using an SHM system might include improved reliability and safety and reduced operating costs [1]. Furthermore, when used to monitor new manufacturing processes, SHM-based procedures might also improve the design and ensure the quality of the products [2]. The field of SHM is considered to emerge from another relatively new field of smart structures and includes various disciplines such as materials, fatigue, NDT, sensors, signal processing, communication systems and decision-making processes. It is an interdisciplinary field, with a very active research community (a dedicated journal on this field being well evaluated by the scientists[2]), with industrial deployments being demonstrated as very effective.

A typical SHM system consists of three key subsystems: diagnosis, damage prognosis (DP) and predictive maintenance (PM) (see Fig. 10.1). Diagnosis deals with the sensing mechanism, which usually relies on active or passive in situ sensors. The sensors can be either wired or wireless, covering a large area of inspection on the monitored structure. Usually, the measurements are recorded and logged, in real time or periodically as part of the maintenance policy, using a centralized analysis station.

DP involves taking the data acquired from the diagnosis subsystem and performing analysis that combines damage evolution models with probability of detection models. The purpose of DP is to assess the structural integrity of the structure and its remaining lifetime under assumed operational and service conditions. Relying on the DP, the prevailing scheduled maintenance policy of the structure can be replaced by PM, also known as condition-based maintenance (CBM), which allows scheduling and optimizing the corrective maintenance actions required in order to ensure the structural integrity of an aircraft or any other system.

---

1 NDT is a wide group of analysis techniques used in science and technology industry to evaluate the properties of a material, component or system without causing damage. The terms "nondestructive examination" (NDE), "nondestructive inspection" (NDI) and "nondestructive evaluation" (NDE) are alternative terms.

2 Structural Health Monitoring, https://journals.sagepub.com/home/shm.

https://doi.org/10.1515/9783111621104-010

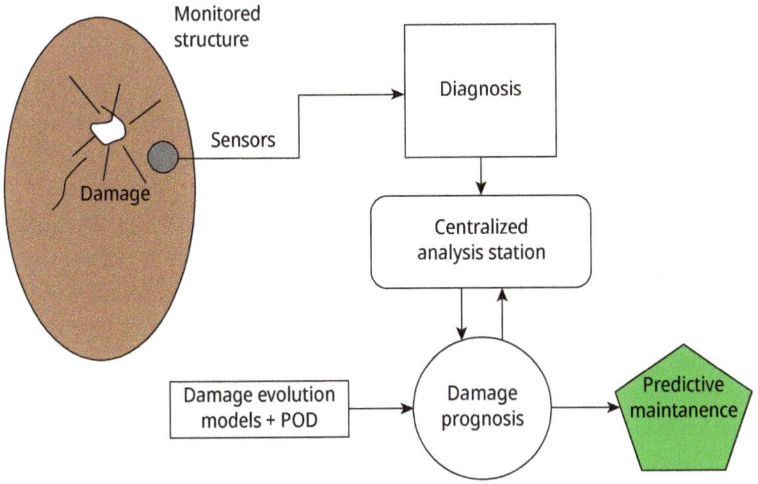

Fig. 10.1: A schematic typical SHM system.

10.1.1 Diagnosis

As stated earlier, the diagnosis subsystem refers to the sensing mechanism existing in every SHM system. The sensors constantly record at least one parameter related to the monitored structure, which is expected to obtain irregular values in the case of structural integrity deterioration. Electrical conductivity, mechanical strain, mechanical stiffness and wave propagation characteristics are few examples of parameters that can be monitored over time, each by its own sensing mechanism.

The literature divides the sensor mechanism into two categories: passive and active sampling.

Passive sampling-based sensors operate by detecting responses due to perturbations of ambient conditions in the monitored structure, without using any external source of energy. This category includes pressure-sensitive polymers, shape memory alloys (SMAs) and comparative vacuum monitoring (CVM) sensors, which can passively detect crack propagation. Another type of passive sensing, considered to be much more advanced than the previous examples, is the acoustic emission (AE) measurement using piezoelectric wafers.

Active sampling-based sensors require externally supplied energy in order to function. This energy can be in the form of mechanical stress or electromagnetic waves. This category includes the electrical and magnetic impedance sensors, piezoelectric wafer-active sensors (PWASs) and optical fiber sensors (OFSs), which require an external laser for taking measurements.

Appendix A presents the various sensors and the way they operate.

10.1.2 Damage prognosis

DP attempts to forecast the structural integrity of an aircraft structural component by assessing its current damage state, estimating the future loading environments, and then performing a prediction, through simulations and past experience, of the remaining applicable lifetime of the component. There are two main approaches for conducting DP as part of SHM. The first is to monitor load sequences/cycles, which are then used to estimate the assumed accumulated fatigue damages. The other way is to use the diagnosis system to directly determine the occurrence, size and location of the damages in the structural component, and then to constantly estimate its remaining lifetime. The former approach of DP is based on operating load monitoring (OLM), and the latter one is based on a process called pattern recognition (PR). This section briefly introduces both of the DP approaches and the relevant terminology associated with them. Even though the latter approach, which deals with the identification and classification of the actual damage, is considered more advanced and attractive, these two approaches are presented for a better understanding of the full picture.

10.1.2.1 Operating load monitoring

Loads can be monitored only through the parameters describing them. Here (again) there are two approaches. Loads monitoring can be performed globally by analyzing the flight parameters recorded in the flight data recorder, which is found in almost every aircraft. Flight parameters such as height, speed, acceleration, position of control surfaces and flap positions determine the aerodynamic forces acting on the aircraft structure. Using the manufacturer-supplied strength reports, numeric simulations and several assumptions, the loads acting on the aircraft can be indirectly deduced and monitored during the aircraft lifetime.

The second approach, which is much more "straightforward," is based on taking local measurements of the mechanical strain at several discrete points. The data are analyzed and directly converted, usually through a preliminary calibration test, to the aerodynamic loads acting on the aircraft. Both ways have their advantages and disadvantages. The former approach involves much less data processing issues, but demands full information regarding the aircraft structure, which is (intentionally) not always provided by the aircraft manufacturer. On the other hand, the latter approach only requires a fundamental knowledge regarding the structural loads acting on the monitored structure in order to adequately place the sensors on the aircraft components. However, it demands a large amount of data processing, in direct relationship to the number of the sensors. An OLM system allows the identification of unusual flight events that could lead to a static failure and the estimation of the number of load cycles leading to a fatigue failure. Therefore, loads monitoring can assist the operator to indirectly forecast damages before occurrence and thus to define a more effective maintenance policy.

10.1.2.2 Pattern recognition (PR)

A number of definitions can be proposed to define damage. Intuitively, damage is a material, structural or functional failure. Therefore, one of the fundamental assumptions taken while designing SHM system is that material or structural failure in a structure will lead to a change in its static and/or dynamic behavior. In other words, one can think of the data recorded, containing the static and/or dynamic information of a structure, as describing its "healthy" pattern, and a change in this recorded pattern may be associated with the existence of a new damage. This approach is known in the professional jargon as "statistical pattern recognition."[3]

PR, originally a term borrowed from "machine learning," describes the ability to (automatically) recognize patterns in multivariable datasets [3]. The notion of PR appears in many literature sources in the context of damage identification and classification. As explained earlier, after having the ability to identify and classify all types of damages in a structure, the step toward having DP abilities is quite small. Here, three types of the most common applications of PR are introduced and briefly discussed.

10.1.2.3 A strain measurement under constant static loading

Consider a "healthy" structural component, that is, without any significant damage, which is subjected to a constant static loading. As a result, a mechanical strain field develops in the structural component, obeying Hooke's law. As long as the component remains "healthy," and after the cancellation of all environmental effects, the same external static loading will lead to the same corresponding mechanical strain field. Monitoring the mechanical strain field, either using local strain sensors (strain gauges, fiber Bragg gratings (FBGs), Lead Zirconate Titanate (PZTs), etc.) or a distributed sensing (optical fiber backscattering), allows the location of areas in a component that might have been damaged.

10.1.2.4 Modal analysis

Modal analysis relates to the dynamic properties of a structure under vibrational excitation. The dynamic properties of a structure are determined by how it responds to forced and free vibrations. Forced vibrations occur when the structure is forced to vibrate at a particular frequency by a periodic input of force. Free vibrations are caused by a one-time excitation. In this case, the structure vibrates in one or more particular frequencies that are called natural frequencies/resonant frequencies. Since the resonant frequencies do not depend on the external one-time excitation, they can be thought of a unique pattern of the structure, which can be monitored.

3 From the writer's point of view, the word "statistical" can be omitted since there are several SHM applications that do not necessarily involve statistics.

Williams and Messina [4] formulated a correlation coefficient that compares the resonant frequencies of a structure with a finite element model (FEM)-based prediction for damage identification and localization purposes. For that, they defined a damage index, δD_i, which represents a linear stiffness reduction in the i element of the model. Thus, the fractional change in the k resonant frequency, δf_k, can be described as follows:

$$\delta f_k = \sum_{i=1}^{m} \frac{\delta f_k}{\delta D_i} \delta D_i \tag{10.1}$$

where i and m represent the element index and the total number of elements, respectively, and the expression

$$\frac{\delta f_k}{\delta D_i} = \frac{\{\phi_k\}^T [K_i] \{\phi_k\}}{8\pi^2 \{\phi_k\}^T [M_i] \{\phi_k\}} \tag{10.2}$$

is the sensitivity of the k frequency to the damage index at the i element, $\{\phi_k\}$ being the k mode shapes, and $[M_i]$ and $[K_i]$, respectively, the mass and stiffness matrices of the element i.

Williams and Messina conducted several experiments on aluminum rods and compared the predicted changes in the resonant frequencies, δf, to the actual changes measured in the experiments, Δf, by using the cross-correlation technique:

$$C = \frac{\left| \{\Delta f\}^T \{\Delta f\} \right|^2}{\left[\{\Delta f\}^T \{\Delta f\} \right] \cdot \left[\{\Delta f\}^T \{\Delta f\} \right]} \tag{10.3}$$

where C represents the cross-correlation coefficient that takes the value of 1.0 in the case of a perfect match between the model and experiment, and 0.0 in the case of a mismatch. They found that 10–15 resonant frequencies are needed to provide sufficient information about a damage. Furthermore, they also suggested a couple of techniques to improve the accuracy of the damage estimation. Additional information can be found in their paper [4].

Many other works based on a modal analysis have been conducted in the past two decades, demonstrating creative ideas to identify, classify and even predict damages. However, the major challenge in this kind of PR technique is related to damage sensitivity. Modal analysis is in fact a global method and is therefore limited to the identification of local damage, which affects the dynamic behavior of the entire monitored structure [5]. In other words, damage must have become more significant before being identifiable. This is an issue especially in composite-based components. For example, a minor delamination, which does not necessarily affect the global dynamic response of the structural component, may lead to a structural failure. This issue requires further investigation.

10.1.2.5 Principal component analysis

Principal component analysis (PCA) is another familiar PR-based technique. This technique involves multivariable analysis, which reduces complex datasets to a lower dimension, thereby revealing simplified patterns, which are often hidden in the complexity of the raw data. Generally, PCA enables to determine the more important patterns in the measured data, which provides more information regarding the structural integrity of the component under test.

A fundamental PCA of a multiple sensor-based SHM system can be described in three main stages: First, it is necessary to normalize and arrange the collected data in a $[m \times n]$ matrix A, which represents the measurements acquired from n sensors in m experimental trials. Experimental trial is defined as a measurement taken from each sensor at a specific moment. The next stage is the calculation of the covariance matrix, symbolized by C_A, which measures the degree of linear relationship within the dataset among all possible pairs of the sensors. C_A is calculated as follows (T stands for "transpose"):

$$C_A = \frac{1}{m-1} A^T A \tag{10.4}$$

In the last stage, the eigenvectors and eigenvalues of the covariance matrix are derived using the following relation:

$$C_A \cdot V = V \cdot \Lambda \tag{10.5}$$

where the eigenvectors of C_A are the columns of V, and the eigenvalues are the diagonal elements of Λ (the off-diagonal elements are zero). The columns of the matrix V are arranged according to the corresponding eigenvalues in descending order. The principal components are defined as the eigenvectors of the covariance matrix scaled by their corresponding eigenvalues.

10.1.3 Predictive maintenance

As stated at the beginning of the present chapter, once the DP stage was executed, the prevailing scheduled maintenance policy of the monitored structure or the monitored system can be replaced by a PM, also known as CBM. This will allow scheduling and optimizing the corrective maintenance actions required in order to ensure the structural integrity of the monitored system or structure. The benefits from applying the PM are:

a. Reduced maintenance time – automatic reports for strategic maintenance scheduling and pro-active repairs can lead to a reduction of up to 20–50% in maintenance time, yielding a decrease in overall maintenance costs.
b. Better efficiency – insights based on analytical procedures would improve the overall equipment effectivity, by reducing unnecessary maintenance procedures, leading to extended system life and enabling the application of root cause system analysis to pinpoint issues ahead of failure.

c. Improved customer satisfaction – providing the customer with automatic alerts for parts to be replaced in parallel with timely maintenance services would lead to a better predictability of the product, thus boosting the satisfaction of the customers.
d. Competitive advantage – a company applying the PM would be seem as a strong one with costumers continuously appreciating it due to its reliable and better products.
e. Additional revenues – a company can earn additional revenues by offering its PM digital services to their customers, thus generating a new growth engine for the company.

10.2 Applications

The use of SHM in industry is somehow restricted to two main sectors: the aerospace and the civil engineering, although other sectors also use the method (naval engineering and car industry). In the present chapter we shall address the aerospace and the civil engineering sectors.

Beside industry, the academy is very active in the SHM area, delivering many new manuscripts on advanced topics of this research area. To familiarize with the topic applications, the reader is referred to the various reviews [5-8].

10.2.1 Aerospace applications

The aerospace sector demands high reliability for its vehicles, therefore the SHM was first introduced in this industry. References [9-28] are only typical manuscripts, written on SHM issues. The increasing use of laminated composite materials in the various passenger aircraft leads to numerous papers on how to increase the reliability of those structures. Scala et al. [9] present a report on the use of AE method to monitor and thus ensure the structural integrity of an aircraft. Staszewski et al. [10] address the issue of impact being inflicted to a composite made structure. The research presents the use of an active approach, 3D laser vibrometry; to scan the structure and by revealing the change in the Lamb wave response amplitudes, the severity of the local delamination can be estimated. An additional approach, the authors called "passive means" involves the use of an array of piezoelectric patches, serving as sensors, to detect the waves being caused by an applied impact on the composite structure. Using genetic algorithms,[4] the authors succeeded in locating the impact. Giurgiutiu et al. [11] use PWAS to monitor the onset and the progress of fatigue cracks and corrosion in aging aircraft

4 Genetic algorithm is a method that is used to optimize a given problem and is based on natural selection, the same process that drives biological evolution. The genetic algorithm repeatedly modifies a population of individual solutions.

structure. The electromechanical (EM) impedance technique is used to detect near field damage, while the wave propagation approach is applied for far-field damage detection. Heida and Platenkamp [12] derived in-service NDT guidelines for aerospace structures made of composite materials, mainly carbon fiber-reinforced structures. They recommend using tap test to detect and find the size and depth of the damage due to impact, delaminations and debonds. The application of the shearography[5] and thermography[6] methods seemed to be less applicable due to its poor-to-moderate capability to characterize a defect, when compared to ultrasonic inspection. However, the authors recommend the use of noncontact techniques, like thermography, for special shape configurations like curved panels and/or repaired structures, and for detection of water entry in honeycomb structures.

Giurgiutiu and Soutis [13] recommend the use of enhanced SHM to increase the reliability and integrity of composite structures, while Assler and Telgkamp [14] from Airbus advocate the use of smart structures approach, in the design phase of an aircraft, thus enabling SHM during the manufacturing and operation phases of the vehicle. Finlayson [15] present the use of AE and acousto-ultrasonic NDT methods to monitor the health of aerospace structures. Pieczonka et al. [16] present a theoretical and experimental study for detection of damage. They used guided ultrasonic waves together with 3d laser Doppler[7] vibrometry. The Lamb waves are excited by surface bonding piezoelectric actuators. Comparing between in-plane and out-of-plane wave vector components leads to the reliable detection of the structural damage. D'Angelo and Rampone [17] address the diagnosis issue within the SHM approach. They present a high-performance computing parallel implementation of a novel learning algorithm, U-BRAIN,[8] allowing them to process multiparameter data involved in composite specimen testing. They claim that their approach yielded a "defect classifier in aerospace structures." A different approach is presented by Oliver [18], in his Ph.D. thesis for the diagnosis issue. He developed and validated a damage identification approach based on a statistical least-squares damage identification algorithm based on concepts of parameter estimation and model update. The data used by the algorithm are the frequency response function-based residual force vectors coming from the distributed vibration measurements, which is updating an FEM through statistically weighted least-squares minimization leading to the location and the size of the damage, its uncertainty and a new updated model.

---

5 Shearography, sometimes called speckle pattern shearing interferometry, is an NDT approach like holographic interferometry. It uses coherent light or coherent soundwaves yielding strain measurement output of the measured structure.

6 Thermography is an NDT approach that uses infrared cameras usually to detect radiation in the long-infrared range of the electromagnetic spectrum due to changes in the measured specimen.

7 Doppler vibrometry is a scientific instrument used to make noncontact vibration measurements of a surface using a laser.

8 U-BRAIN (*U*ncertainty-managing *B*atch *R*elevance-based *A*rtificial *IN*telligence) is designed for learning DNF Boolean formulas from partial truth tables, possibly with uncertain values or missing bits.

Boukabache et al. [19] designed an advanced sensor, based on a piezoelectric transducer, capable of generating Lamb waves (shear-type waves) and measuring those waves to detect the damage and its severity. In addition, the manuscript presents an analog interface procedure for the measurements and an experimental validation of the new sensor.

Wandowski et al. [20] checked three NDT methods to assess the integrity of composite aerospace structures. They used laboratory samples made of CFRP (carbon fiber-reinforced polymer) and GFRP (glass fiber-reinforced polymer). The application of the EM impedance method on CFRP specimens revealed the delamination by measuring the real part of the impedance, namely the resistance. The delamination caused the shifting of some resonance frequencies, which could be detected by the resistance characteristics. The second method applied was laser vibrometry, a noncontact NDT technique. They had used two types of waves, standing (vibration-based application) and propagating (guided wave-based application) ones. The third method was Terahertz spectroscopy, using electromagnet radiation in the range of 01–3 THz. The GFRP specimens were scanned yielding B scans and C scans, from which the delamination could be located.

Diamanti and Soutis [21] used the Lamb waves NDT technique to monitor the composite layers of an aerospace structure, to ensure no delamination was developed, thus increasing the reliability of structures. Baker et al. [22] address another important issue, namely the patched cracks in an aircraft structure. Monitoring the health of the patch was done in their study using strain gages, although they had evaluated also the use of Bragg grating optical fibers. Maier et al. [23] address the same topic also in their report on SHM of repairs. The state of the art to detect cracks, delamination, impacts and corrosion of aircraft structures using AE, acoustic ultrasonic, phase array ultrasonic FBGs and CVM is outlined and discussed. Schnars and Henrich [24] give a review of the application of various NDT methods in the composite aerospace structures. Manual and automatic ultrasonic measurement using either single element transducers or phase array detectors applying pulse echo or transmission modes is highlighted in their report. Other NDT methods, like resonance methods, shearography and thermography, mentioned earlier in this chapter are also described. Laser ultrasonic method is recommended for contactless measurements of composite structures. Hsu [25] addresses also the application of NDT methods, for composite structures. The focus is on the practice applications of these methods.

Chang and Lan [26] address the issue of SHM being applied on aging transport aircraft, which experienced severe atmospheric turbulence. Due to the lack of knowledge on the structural flexibility of these old structural parts, the structural response is achieved using the fuzzy logic modeling of the aerodynamic loads.

Finally, Terroba et al. [27] present a research study on the measurement of the structural integrity of a target drone equipped with four FBG sensors. The system was found to be reliable and can detect barely visible structural damage in the high-loaded front fuselage of the drone.

10.2.2 Civil applications

Spencer et al. [28] present a short review of the use of smart sensors technology for SHM of civil engineering structures like bridges, highways and buildings. They define a smart sensor as a MEMS[9] sensor with wireless capabilities. Various aspects of the use and the respective hardware are outlined. Swartz et al. [29] address the same topic of the use of smart sensors. Wireless sensors serve as the basic means to monitor the health of civil structures. Their sensors would include accelerometers, strain gages, geophones, whose data have to be transmitted to a central computer for further analysis. Typical results for wind turbines structures and bridges are presented.

Sun et al. [30] review the smart sensing technologies available for SHM of civil engineering structures. Typical sensors as fiber optic, piezoelectric transducers, magnetostrictive transducers and self-diagnostic fiber-reinforced composite are cited as good candidates to measure the health of the structure monitored. Typical laboratory and on-site applications of these sensors are evaluated and compared.

Yi et al. [31] apply a hybrid method named optimal sensor placement strategy to optimize the number of sensors and their locations on a monitored structure. The on-site use of the method on a high TV tower (Guangzhou New TV Tower, China) proved to be effective with a friendly software interface. Yi et al. [32] also studied the use of global positioning system to monitor the displacement of high-rise structures due to wind, thermal variation and earthquake-induced responses.

Dhakal et al. [33] review wired and wireless technologies, with subtechniques like impedance-based, nondestructive evaluation using vibration signature, limit strain measurement, data fusion method and inverse method and their applications in civil infrastructures like buildings, bridges and towers.

Kim et al. [34] present an interesting technique to digital processing of images containing actual cracks and differ them from crack-like noise pattern like dark shadows, stains, lumps and holes, characteristic of concrete structures. The technique is based on machine learning process.

Li et al. [35] also use the learning machine process to accurately monitor the displacement of a deep foundation pit at early construction stage and predict its behavior with time.

Qu et al. [36] present a new approach of using an amplitude-adaptive wavelet transform, to detect deep or shallow delaminations in a concrete slab, using the impact echo experimentally recorded. The authors report good results for the civil structures they had tested.

9 MEMS – microelectrical mechanical system.

10.2.3 General applications

Various NDT methods are found in the literature. Park et al. [37] review the piezoelectric impedance-based health monitoring method, while Oromiehie et al. [38] use the Bragg grating sensors to characterize process-induced defects in automated fiber placement manufacturing of composites. Cha and Wang [39] improved the learning machine method by adding an unsupervised mode, as it requires only enough intact states at the training of the machine. A novel detection-based structural damage localization using a density peak-based fast clustering algorithm is described in detail. The same topic, the use of machine learning, is addressed by Demarie and Sabia [40] for long-term SHM and Jin and Jung [41] in their study.

Other topics like guided wave excitation in composite plates [42], the use of fuzzy c-means clustering algorithm to detect double defects in plates [43], damage-scattered wave extraction in an integral stiffened isotropic plate [44] and the use of changes in the curvature mode shapes of the monitored structure to detect damage [45] are worth to be mentioned to show the diversibility of the SHM method.

10.3 Monitoring natural frequencies of composite beams for detection of damage

This section is aimed at providing the insight of a method to monitor the natural frequencies to detect the damage in a given structure as elaborated by Penias and Abramovich [46], in the first author master thesis.

The numerical study was performed on a rectangular beam, with a high length-to-width ratio (1:20) (see Fig. 10.2 for beam's geometry). The beam was constructed of 16 graphite-epoxy As4–12k/E7K8 layers (see Fig. 10.3 for the beams' cross section).

The cantilever beam's first 10 natural frequencies and their associated mode shapes were calculated numerically, using the ANSYS FE code. Delamination was created by changing the boundary condition of certain elements (by choice) to "unbounded," thus creating a perfect delamination. A typical delamination, which was implemented in the beam between various layers, is displayed in Fig. 10.4.

Numerical results show that the natural frequencies around the *Z*-axis (sideway bending) remain the same (as expected); therefore, these modes were disregarded. The natural modes around the *X*-axis (torque) and the *Y*-axis (bending) show decrease in the natural frequencies as the delamination is closer to the mid-layer of the beam (Figs. 10.5 and 10.6).

Delamination of different sizes was considered, all symmetric with respect to the *X*-axis. The numeric results show that, as expected, the natural frequencies in the bending and torque modes decrease with the increase in the delamination size. Figure 10.7 shows the results for delamination of different sizes and different layers, for the fourth

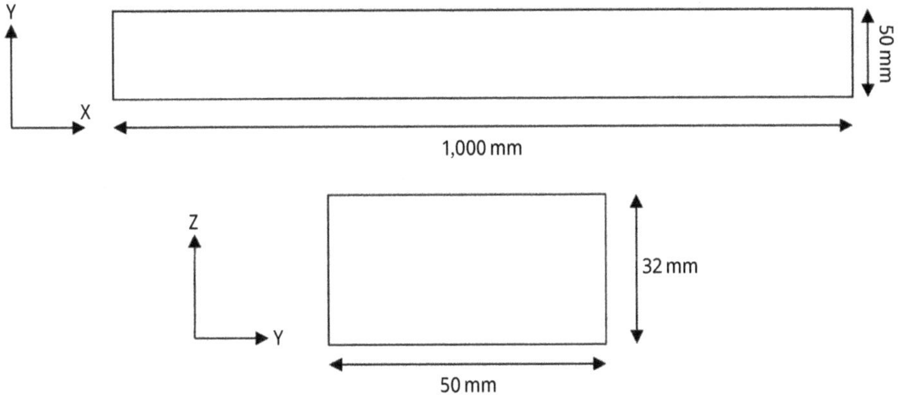

Fig. 10.2: The geometry of the beam's model.

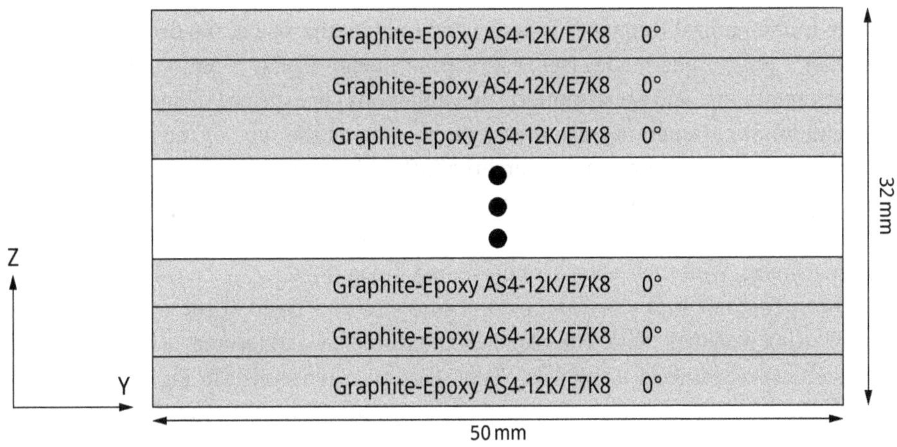

Fig. 10.3: The model's cross section.

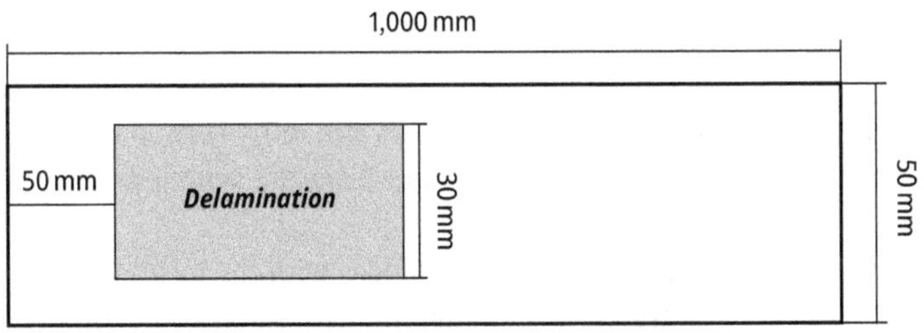

Fig. 10.4: The delamination model (not to scale).

bending mode and second torque mode as typical examples (all modes showed the same behavior).

To determine the effect of delamination's location, a delamination, with fixed sizes, was implemented in different locations (along the *X*-axis) in the mid-layer of the beam (see Fig. 10.8). The results for the first four bending modes, presented in Fig. 10.9, display various behaviors of the natural frequency as the delamination change its location (Fig. 10.9).

The results, presented in Fig. 10.9, show that the change in the natural frequencies for every bending mode resembles its mode shape, which indicated that the change in the natural frequency is affected by the movement of the delaminated area with respect to the mode shape.

A similar investigation was performed for the *Y*-axis, for which different delaminations were inspected in various ways (Fig. 10.10) – symmetric and nonsymmetric with respect to the *Y*-axis. Results presented in Fig. 10.11 show that the bending modes (e.g., fourth mode) are not influenced by the delamination location along the *Y*-axis, and the change in the natural frequency was dictated only by the size of the delamination. On the other hand, torque modes (see, e.g., second mode) display a noticeable difference between symmetric and nonsymmetric delaminations (as expected) along the *Y*-axis.

Additional parameters were investigated, such as the type of the material used, the influence of the width of the beam, the influence of the boundary conditions and various other laminates. The results are found in [46].

To validate the numerical model, an experimental campaign was performed on four specimens, with their geometry presented in Fig. 10.12.

The specimens that were clamped at one end and freed at the other end were excited using a shaker table. The response measured by accelerometer bonded on the specimens was recorded using a PC. Typical results are shown in Fig. 10.13 [46]. Although the four tested specimens were not 100% identical, the results present a detectable reduction in the natural frequencies.

The conclusions from the study were:

a. The effect of delamination on the bending and torque modes of natural frequencies is larger, as the delamination is closer to the midlayer of the beam.
b. Natural bending and torque frequencies are reduced, as the delamination is larger.
c. Bending frequencies are affected by the delamination's size only. Torque frequencies are affected by the deviation of the delamination from the symmetry line of the specimen.
d. The variation of the natural bending frequencies changes with the location of the delamination on the *X*-axis. The frequency change plot resembles the relevant mode shape.
e. Delamination in laminated composite beams has a significantly larger effect on their natural frequencies when compared with delaminated isotropic materials.

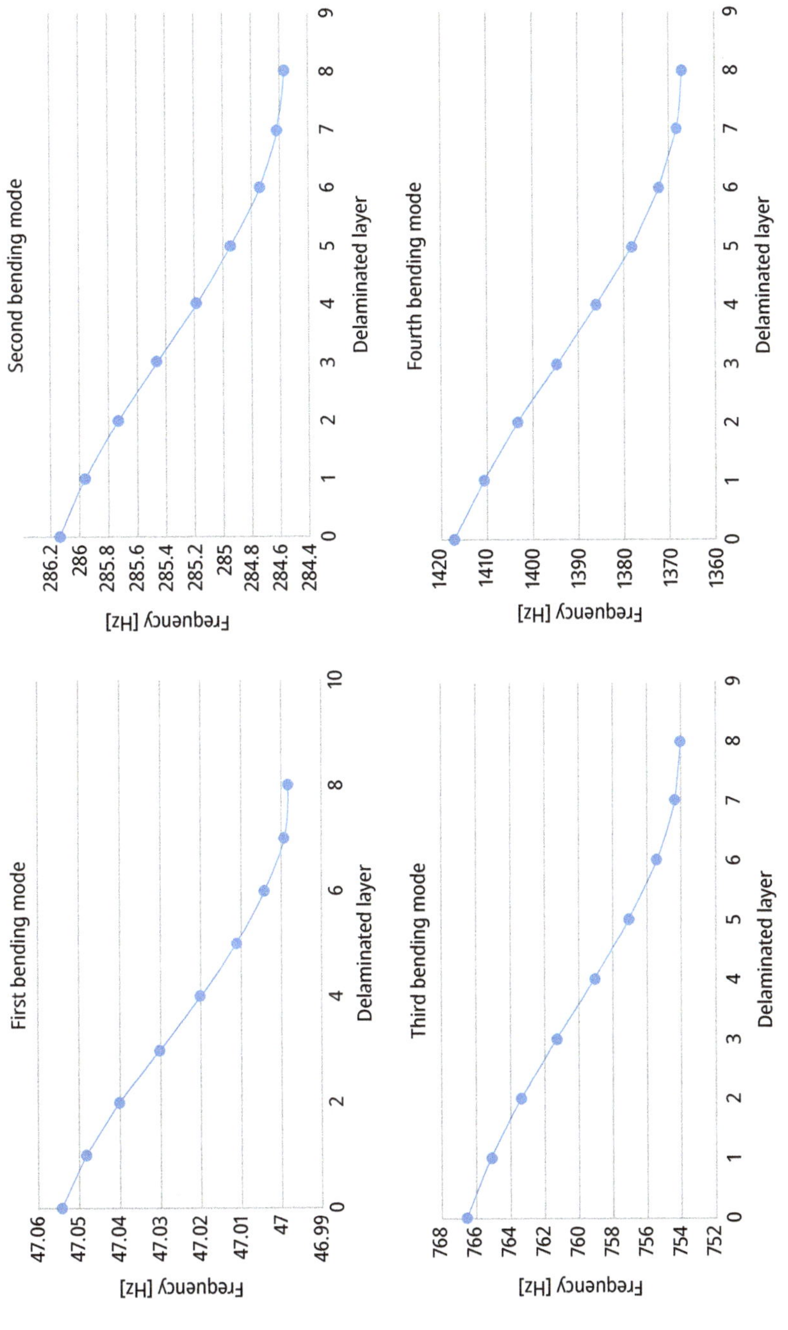

Fig. 10.5: The first four bending modes – natural frequencies for different delamination layers.

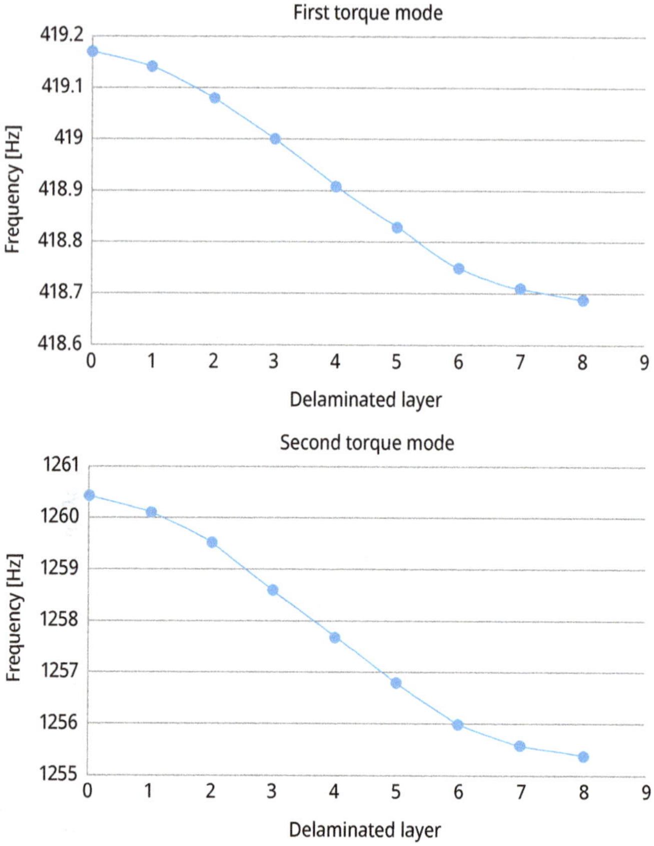

Fig. 10.6: The first two torque modes – natural frequencies for different delamination layers.

f. Clamped–clamped beams tend to show a more significant change in their natural frequencies compared with cantilever and simply supported beams.
g. Tests performed on four specimens, with and without damage, showed the reduction in the natural frequencies due to the presence of delaminations.

All the above conclusions show that a simple measurement of the first few bending and torque natural frequencies of a beam might provide a lot of information regarding the presence of a delamination:
- Delaminated layer – by the change in bending mode frequencies
- Delamination size – by change in both torque and bending modes
- Delamination location:
  - Along the *X*-axis – by difference between different bending modes
  - Along the *Y*-axis – by change in the torque modes

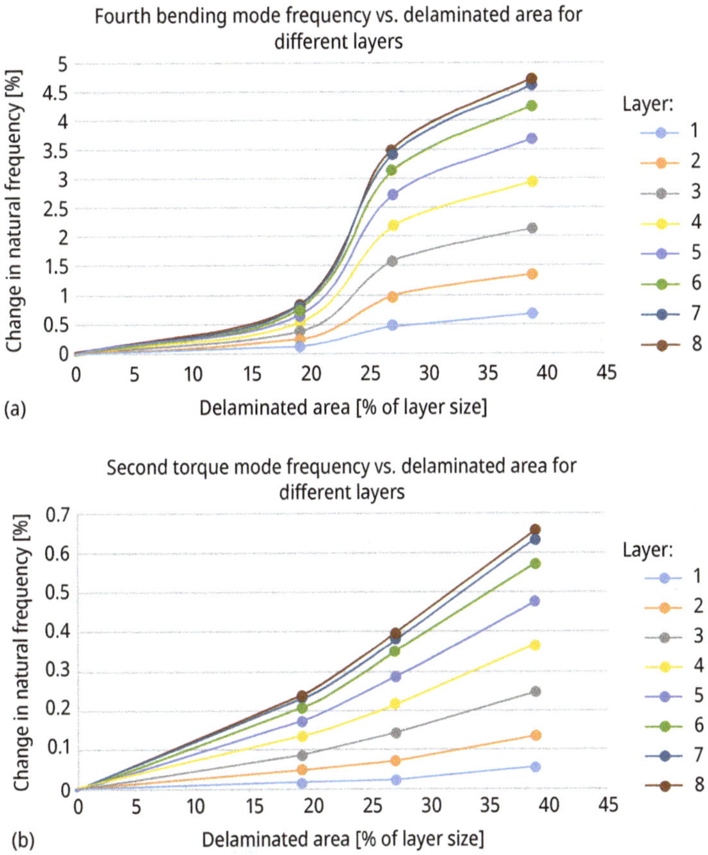

Fig. 10.7: Change in frequency with change in delamination area and delamination layer: (a) fourth bending mode and (b) second torque mode.

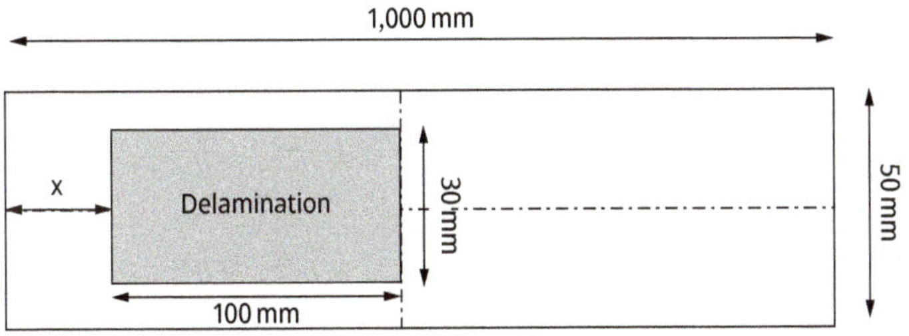

Fig. 10.8: Delamination location along the *X*-axis (*X* symbolizes the distance from the beam's edge).

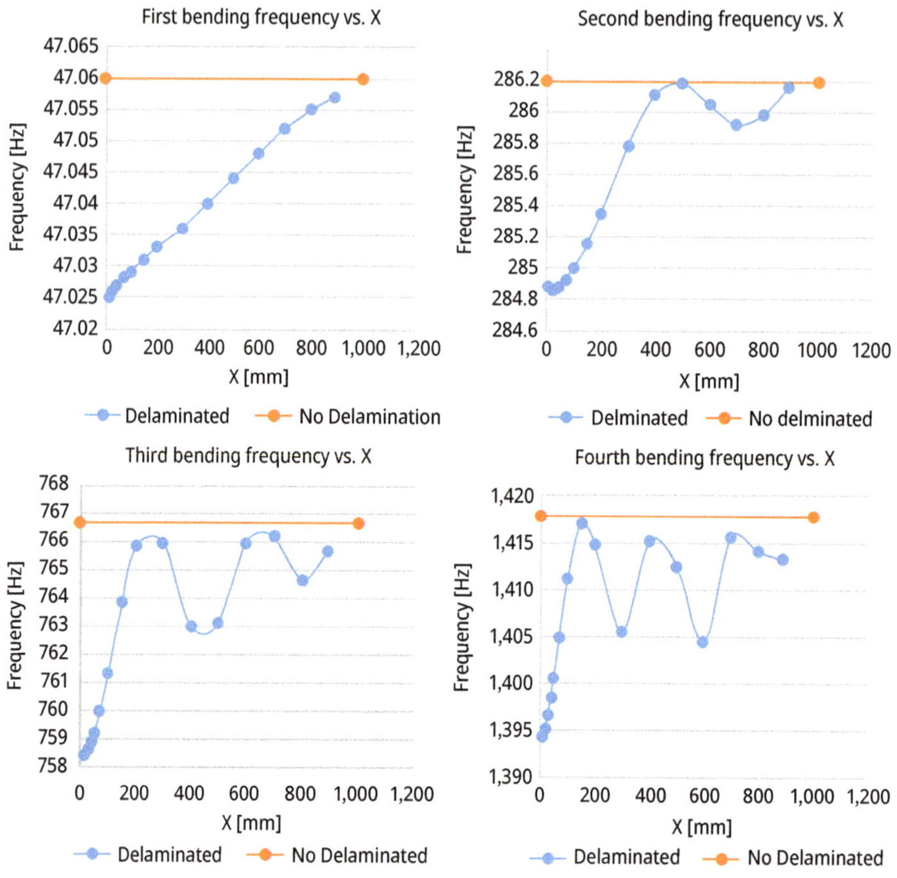

Fig. 10.9: Natural frequencies for different values of *X* – the first four bending modes.

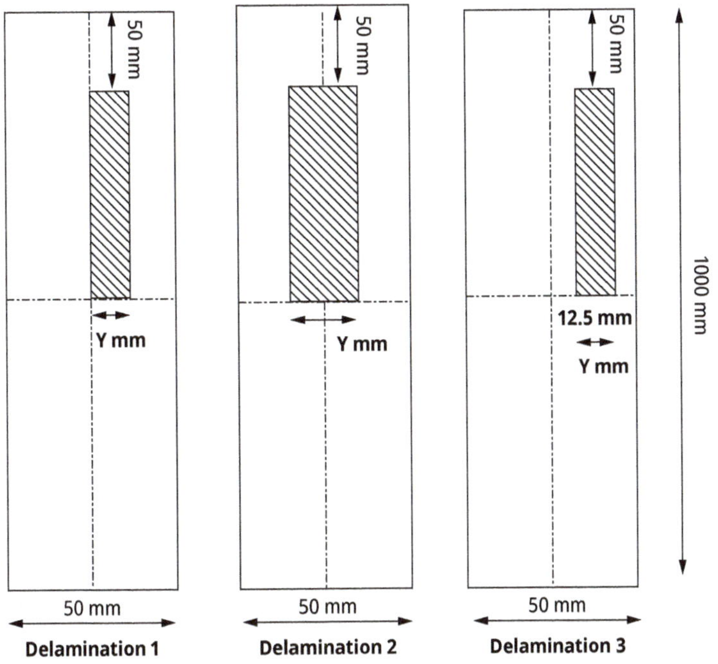

Fig. 10.10: The geometry for the various delaminations used in the *Y*-axis direction.

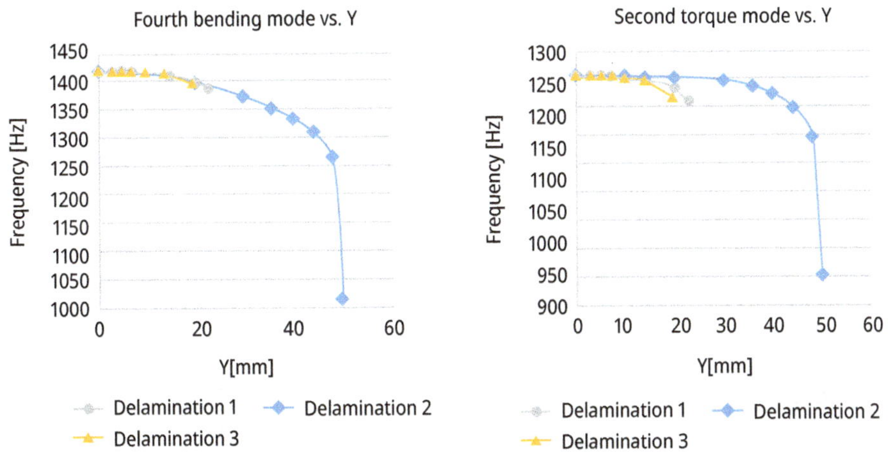

Fig. 10.11: The change in the fourth bending and second torque modes for different *Y* values.

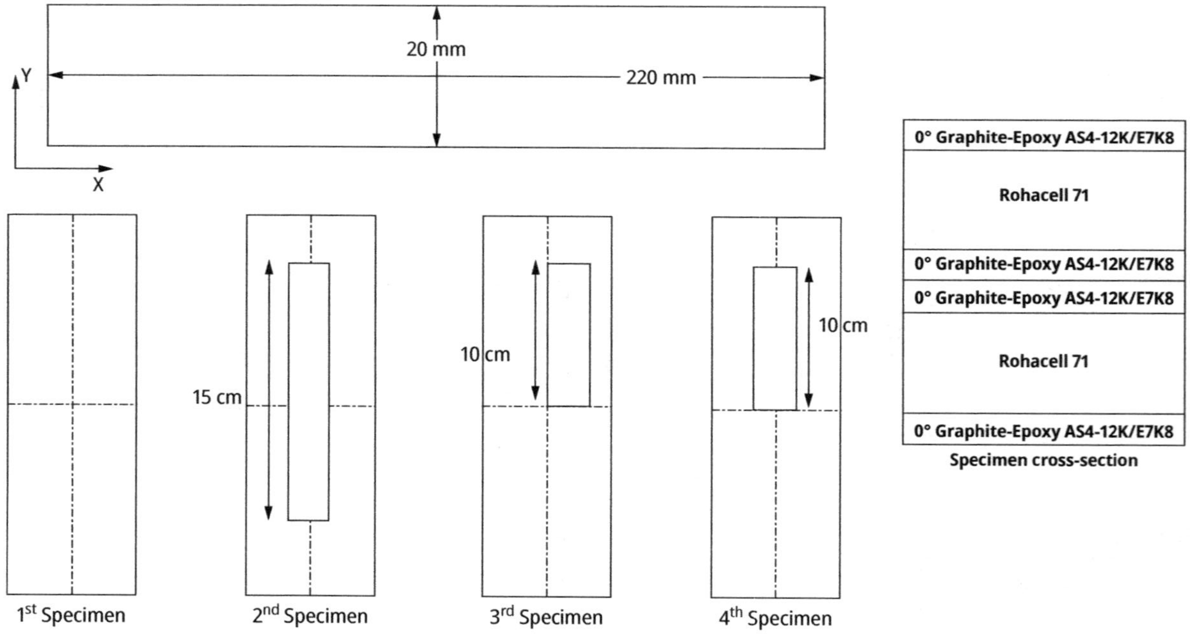

Fig. 10.12: The geometry of the experimental specimens.

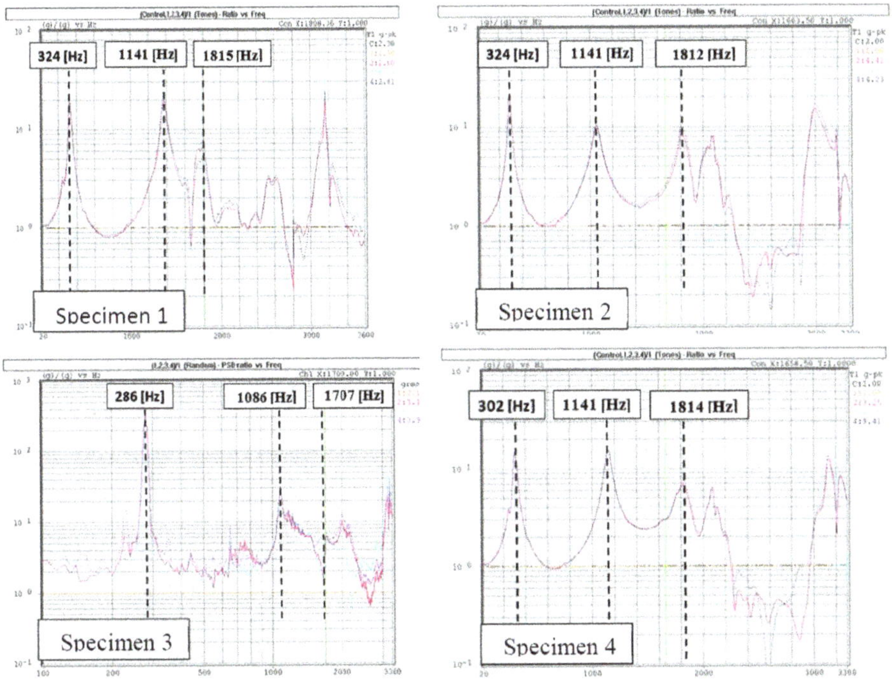

Fig. 10.13: The first experimental natural frequencies for the four tested specimens.

To increase the chances of detecting delaminations in laminated composite beams, it is recommended to clamp the beam at its both sides.

References

[1] Cai, J., Qiu, L., Yuan, S., Shi, L., Liu, P.P. and Liang, D. Structural health monitoring for composite materials, Ch. 3, in Composites and their applications, Hu, N. Ed. INTECHOPEN, 2012, 25. doi: 10.5772/3353.

[2] De Baere, D., Strantza, M., Hinderdael, M., Devesse, W. and Guillame, P. Effective structural health monitoring with additive manufacturing, in EWSHM – 7th European Workshop on Structural Health Monitoring, La Cité, Nantes, France, July 8–11, 2014, 10.

[3] Jain, A.K., Duin, R.P.W. and Mao, J. Statistical pattern recognition, a review, IEEE Transactions on Pattern Analysis and Machine Intelligence, 22(1), 2000, 4–37.

[4] Messina, A.E., Williams, J. and Contursi, T. Structural damage detection by a sensitivity and statistical-based method, Journal of Sound and Vibration, 216(5), 1998, 791–808.

[5] Sohn, H., Farrar, C.R., Hemez, F., Shunk, D.D., Stinemates, D.W., Nadler, B.R. and Czarnecki, J.J. A review of structural health monitoring literature 1996–2001, Los Alamos National Laboratory, LA-13976-MS Feb, 2004, 311.

[6] Farrar, C.R. and Worden, K. An introduction to structural health monitoring, Philosophical Transactions of the Royal Society A, 365, 2007, 303–315. doi: 10.1098/rsta.2006.1928.

[7] Del Grosso, A.E. Structural health monitoring: research and practice, in Ilki, A. Ed Proc. of the 2nd Conference on Smart Monitoring, Assessment and Rehabilitation of Civil Structures (SMAR2013), Istanbul Turkey, 9–11 Sept 2013.
[8] Kim, J.-T., Sim, S.-H., Cho, S., Yun, C.-B. and Min, J. Recent R&D activities on structural health monitoring in Korea, Structural Monitoring and Maintenance, 3(1), 2016, 91–114. doi: 10.12989/smm.2016.3.1.091.
[9] Scala, C.M., Bowles, S.J. and Scott, I.G. The development of acoustic emission for structural integrity monitoring of aircraft (U), Department of Defense, Defense, Science and Technology Organization, Aeronautical Research Laboratory, Melbourne, Victoria, Australia, Aircraft materials Report, Vol. 120, 1989, 34.
[10] Staszewski, W.J., Mahzan, S. and Traynor, R. Health monitoring of aerospace composite structures-Active and passive approach, Composite Science and Technology, 69, 2009, 1678–1685.
[11] Giurgiutiu, V., Zagrai, A. and Bao, J.J. Piezoelectric wafer embedded active sensors for aging aircraft structural health monitoring, Structural Health Monitoring, 1(1), 2002, 41–61.
[12] Heida, J.H. and Platenkamp, D.J. In-service inspection guidelines for composite aerospace structures, in 18th World Conference on Nondestructive Testing, Durban, South- Africa, 16–20 April 2012, 14.
[13] Giurgiutiu, V. and Soutis, C. Enhanced composites integrity through structural health monitoring, Applied Composite Materials, 19(5), 2011, 813–829. doi: 10.1007/s10443-011-9247-2.
[14] Assler, H. and Telgkamp, J. Design of aircraft structures under special consideration of NDT, in 9th European Conference on NDT (ECNDT 2006), Berlin, Germany, 25–29 Sept. 2006.
[15] Finlayson, R.D., Friesel, M., Carlos, M., Cole, P. and Lenain, J.C. Health monitoring of aerospace structures with acoustic emission and acousto-ultrasonics, Insight, 43(3), 2001, 4.
[16] Pieczonka, Ł., Ambroziński, Ł., Staszewski, W.J., Barnoncel, D. and Pérès, P. Damage detection in composite panels based on mode-converted Lamb waves sensed using 3D laser scanning vibrometer, Optics and Lasers in Engineering, 99, 2017, 80–87. doi: 10.1016/j.optlaseng.2016.12.017.
[17] D'Angelo, G. and Rampone, S. Diagnosis of aerospace structure defects by a HPC implemented soft computing algorithm, in 2014 IEEE Metrology for Aerospace (MetroAeroSpace 2014), Vol. 5, Benevento, Italy, 29–30 June 2014, doi: 10.1109/MetroAeroSpace.2014.6865959.
[18] Oliver, J.A. Frequency response function based damage identification for aerospace structures, Ph.D. thesis, University of California, San Diego, CA, 2015, 583.
[19] Boukabache, H., Escriba, C. and Fourniols, J.-Y. Toward smart aerospace structures: design of a piezoelectric sensor and its analog interface for flaw detection, Sensors, 14, 2014, 20543–20561. doi: 10.3390/s141120543.
[20] Wandowski, T., Malinowski, P., Radzienski, M., Opoka, S., Ostachowicz, W. Assessment methods for composite aerospace structures, in Araújo, A.L., Mota Soares, C.A., et al Ed Proc. of 7th ECCOMAS Thematic Conference on Smart Structures and Materials (SMART 2015), @IDMEC, 2015, 15.
[21] Diamanti, K. and Soutis, C. Structural health monitoring techniques for aircraft composite structures, Progress in Aerospace Sciences, 46, 2010, 342–352.
[22] Baker, A., Rajic, N. and Davis, C. Towards a practical structural health monitoring technology for patched cracks in aircraft structure, Composite: Part A, 40, 2009, 1340–1352.
[23] Maier, A., Benassi, L. and Stolz, C. Structural health monitoring of repairs, NATO Research and Technology Organization, RTO-EN-AVT-156, Battle Damage Repair Techniques and Procedures on Air Vehicles-Lessons Learned and Prospects, May 2010, URL: http://www.rta.nato.int/pubs/rdp.asp?RDP=RTO-EN-AVT-156, 5 Apr. 2011.
[24] Schnars, U. and Henrich, R. Application of NDT methods on composite structures in aerospace industry, in CDCM 2006-Conference on Damage in Composite materials, Online-Proc. on ndt.net. 18./19.09.2006, Germany, 2006, 8.
[25] Hsu, D.K. Nondestructive inspection of composite structures: methods and practice, in 17th World Conference on Nondestructive Testing, Shanghai, China, 25–28 Oct. 2008, 14.

[26] Chang, R.C. and Lan, C.E. Structural health monitoring of transport aircraft with fuzzy logic modelling, Mathematical Problems in Engineering, 2013, 2013, 11. Article ID 640852, doi: 10.1155/2013/640852.
[27] Terroba, F., Frövel, M. and Atienza, R. Structural health and usage monitoring of an unmanned turbojet target drone, Structural Health Monitoring, 18(2), 2018, 635–650. doi: 10.1177/1475921718764082.
[28] Spencer, B.F., Jr., Ruiz-Sandoval, M. and Kurata, N. Smart sensing technology for structural health monitoring,in 13th World Conference on Earthquake Engineering (WCEE), Vancouver, B.C., Canada, 1–6 Aug. 2004, Paper # 1791, 13.
[29] Swartz, R.A., Zimmerman, A. and Lynch, J.P. Structural health monitoring system with the latest information technology, in Proc. of 5th Infrastructure & Environmental Management Symposium, Yamaguchi, Japan, 28 Sept. 2007, 28.
[30] Sun, M., Staszewski, W.J. and Swamy, R.N. Smart sensing technologies for structural health monitoring of civil engineering structures, Advances in Civil Engineering, 2010, 2010, 13. Article ID 724962, doi: 10.1155/2010/724962.
[31] Yi, T.-H., Li, H.-N. and Gu, M. Optimal sensor placement for structural health monitoring based on multiple optimization strategies, Structural Design of Tall and Special Buildings, 20(7), 2011, 881–900. doi: 10.1002/tal.712.
[32] Yi, T.-H., Li, H.-N. and Gu, M. Recent research and applications of GPS-based monitoring technology for high-rise structures, Structural Control and Health Monitoring, 20(5), 2013, 649–670.
[33] Dhakal, D.R., Neupane, K., Thapa, C. and Ramanjaneyulu, G.V. Different techniques of structural health monitoring, International Journal of Civil, Structural, Environmental and Infrastructure Engineering Research and Development (IJCSEIERD), 3(2), 2013, 55–66.
[34] Kim, H., Ahn, E., Shin, M. and Sim, S.-H. Crack and non-crack classification from concrete surface images using machine learning, Structural Health Monitoring, 2018, 1–14. doi: 10.1177/1475921718768747.
[35] Li, X., Liu, X., Li, C.Z., Hu, Z., Shen, G.Q. and Huang, Z. Foundation pit displacement monitoring and prediction using least squares support vector machines based on multi-point measurement, Structural Health Monitoring, 2018, 1–10. doi: 10.1177/1475921718767935.
[36] Qu, H., Li, T. and Chen, G. Adaptive wavelet transform: definition, parameter optimization algorithms, and application for concrete delamination detection from impact echo responses, Structural Health Monitoring, 2018, 1–10. doi: 10.1177/1475921718776200.
[37] Park, G., Sohn, H., Farrar, C.R. and Inman, D.J. Overview of piezoelectric impedance-based health monitoring and path forward, The Shock and Vibration Dogest, 55(6), 2003, 451–463.
[38] Oromiehie, E., Prusty, B.G., Compston, P. and Rajan, G. Characterization of process-induced defects in automated fiber placement manufacturing of composite using Bragg grating sensors, Structural Health Monitoring, 15(6), 2016, 706–714. doi: 10.1177/1475921716685935.
[39] Cha, Y.-J. and Wang, Z. Unsupervised novelty detection-based structural damage localization using a density peaks-based fast clustering algorithm, Structural Health Monitoring, 17(2), 2017, 313–324. doi: 10.1177/1475921717691260.
[40] Demarie, G.V. and Sabia, D. A machine learning approach for the automatic long-term structural health monitoring, Structural Health Monitoring, 2018, 1–19. doi: 10.1177/1475921718779193.
[41] Jin, -S.-S. and Jung, H.-J. Vibration-based damage detection using online learning algorithm for output-only structural health monitoring, Structural Health Monitoring, 17(4), 2018, 727–746. doi: 10.1177/1475921717717310.
[42] Mei, H. and Giurgiutiu, V. Guided wave excitation and propagation in damped composite plates, Structural Health Monitoring, 2018, 1–25. doi: 10.1177/1475921718765955.

[43] Chen, S., Zhou, S., Chen, C., Li, Y. and Zhai, S. Detection of double defects for plate-like structures based on a fuzzy c-means clustering algorithm, Structural Health Monitoring, 2018, 1–10. doi: 10.1177/1475921718772042.
[44] He, J., Leser, P.E., Leser, W.P. and Yuan, F.-G. IWSHM 2017: damage-scattered wave extraction in an integral stiffened isotropic plate: a baseline-subtraction-free approach, Structural Health Monitoring, 17(6), 2018, 1365–1376. doi: 10.1177/1475921718769232.
[45] Pandey, A.K., Biswas, M. and Samman, M.M. Damage detection from changes in curvature mode shapes, Journal of Sound and Vibration, 145(2), 1991, 321–332.
[46] Penias, S. and Abramovich, H. Monitoring natural frequencies of composite structures as a means to detect defects, M.Sc. Thesis (in Hebrew), Faculty of Aerospace Engineering, Technion, I.I.T, Vol. 32000, Haifa, Israel, 2017, 110.

Appendix A: Passive and active sensors

A1 Passive sensors

A1.1 Pressure-sensitive polymers

Pressure-sensitive polymers are materials that will change their properties due to applied stress. Two types that are discussed here are pressure-sensitive paint (PSP) and pressure-sensitive film (PSF).

A1.1.1 Pressure-sensitive paint

PSP has been around for some time (see [1–11]. PSP was developed at NASA to enable better measurements of the drag loadings on airfoils. The technology behind PSP is based on the sensitivity of certain luminescent molecules to the presence of oxygen. Using a process called oxygen quenching, the fluorescence of the molecules changes as a function of the pressure in oxygen molecules. The measurement system, depicted in Fig. A1, consists of a light source to excite the oxygen and the luminescence molecules contained in the paint layer, and a photodetector to detect the long-wavelength emission to be translated into pressure applied on the paint.

A1.1.2 Pressure-sensitive film

PSF is based on a simple but reliable technology. The film consists of a microencapsulated color layer and a color-developing layer sandwiched between two polyester layers as presented in Fig. A2. When a defined pressure is exceeded, the microcapsules would break and would react with the adjacent color-forming layer leading to a change in the film color. The color presented is monochromatic, but the shading of the color will indicate pressure level. A calibration table is supplied by the manufacturing of the film (see, e.g., Fuji company[10]). One should note that the measurement of the pressure is not

10 http://www.fujifilm.com/products/prescale/prescalefilm/#See_All.

continuous, and the change in color is at discrete levels of pressure, causing the user to use various PSFs for different load levels, and also the films come in a variety of pressure ranges (0.05–300 MPa).

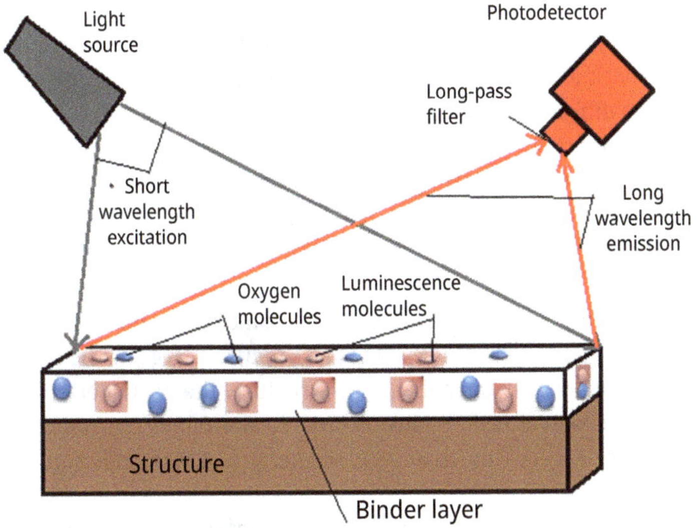

Fig. A1: A schematic PSF test setup.

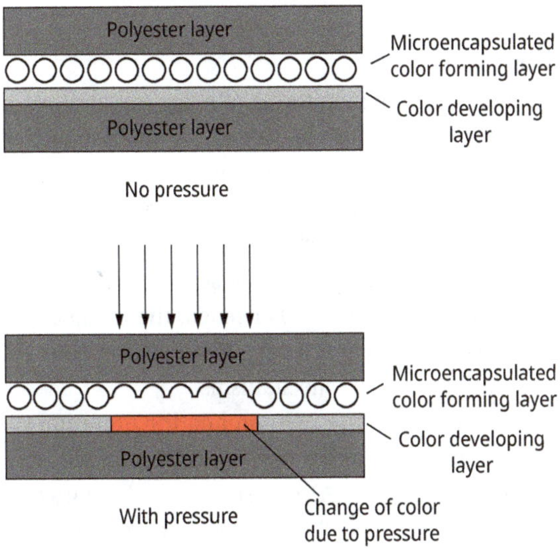

Fig. A2: A schematic PSF way of activating due to the application of pressure on the film.

A1.2 Shape memory alloys

SMAs are materials that “would remember” their original shape and go back to their original shape after deformation under a stimulus in the form of heating/cooling and/or application of mechanical stresses. Probably, the most known SMA is the one abbreviated as NITINOL (Nickel Titanium Naval Ordnance Laboratory) developed by William Buehler and Frederick Wang of the US Naval Ordnance Laboratory in 1962 [12]. Some SMAs can go back to a shape different from their original shape under a stimulus, thus holding two different shapes and are called two-way SMAs.

SMAs have two different phases with different crystal structures leading to large different properties: martensite, a low-temperature phase; and austenite, a high-temperature phase. Contrary to a typical transformation that involves the diffusion of atoms, the phase change in SMAs occurs by a shear lattice distortion, which makes the change reversible.

SMAs also show a property called pseudoelasticity sometimes, whereby they show almost rubber-like mechanical behavior. SMAs can produce a large deformation compared to most other metals. Due to their lattice structure, the pseudoelasticity property enables the SMA to recover from relatively large strains (up to 7%), although some hysteresis can be experienced. This dissipation of energy by SMAs finds applications in vibration dampers.

Their property of changing structural phases from martensite to austenite and vice versa, applying mechanical stress, allows the use of SMA as sensors to mechanical loads.

A1.3 Comparative vacuum monitoring

CVM offers a novel application for in situ, real-time monitoring of crack initiation and/or propagation. CVM measures the differential pressure between very small diameter tubes containing a low vacuum alternating with similar tubes at atmosphere in a simple manifold-type structure. A schematic picture of the sensor, manufactured by the SMS Company,[11] mounted on a structural part containing a crack is shown in Fig. A3.

The CVM system consists of three primary components: a sensor, a fluid flow meter and a stable source of low vacuum. As no adhesive is needed, the method utilizes a direct measurement of the test surface making it a part of the sensor, thus resulting in a high reliability [13, 14] and built-in fail-safe mechanisms.

Note that an increase in ΔP pressure (between the vacuum and the atmospheric pressures) would be measured when a crack would cross two adjacent channels, one at vacuum (V) and the second one at atmospheric (A) pressure. As schematically shown in Fig. A3, any shape development of a flaw will cause a flow of air from the

11 www.smsystems.com.au/.

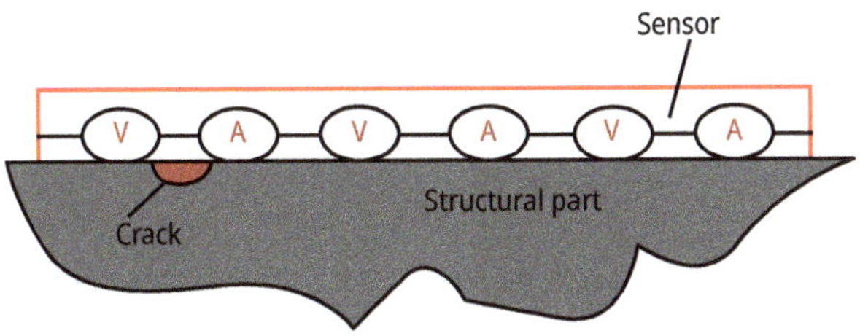

Fig. A3: A schematic drawing of the sensing of the crack using CVM (V, vacuum pressure; A, atmospheric pressure).

atmosphere to the vacuum channel, through the passage between them. A transducer would be used to measure the fluid flow between the small tubes; the rate of fluid flow would be an indication of the size of the flaw.

A1.4 Acoustic emission

AE is the phenomenon of sound and ultrasound stress wave radiation in structures, which undergo deformation or fracture processes. When a structure is subjected to an external stimulus (change in pressure, load or temperature), localized AE sources trigger the release of stress waves that propagate to the surface. These stress waves are recorded by several sensors and then are analyzed in order to locate and classify these AE sources (see Fig. A4).

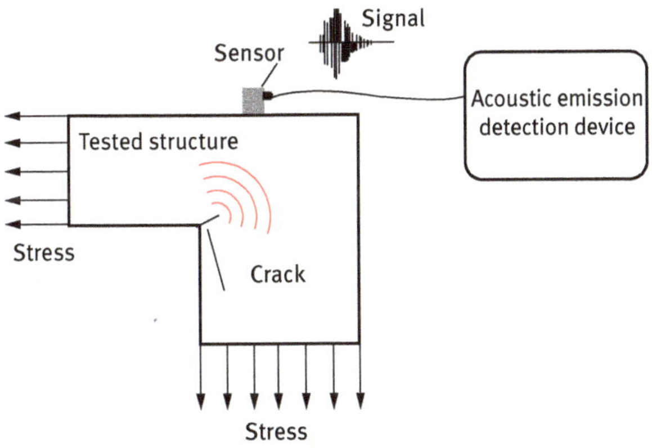

Fig. A4: Schematic of AE source, a crack, triggered due to external stress, detected and transmitted to a centralized device prior to further analysis.

Common examples of AE sources can be growth of cracks, slip and dislocation movements, and melting and phase transformations in metals. In composites, matrix cracking, fiber breakage and delamination are typical modes of failure and each contributes its own unique AE (see [15]).

AE sensing is considered to be *passive* since the only energy required for activating the sensors is supplied by the wave signal, which is spontaneously released in the case of existence of damage/discontinuity in the *loaded* structure. AE usually provides an immediate indication relating to the risk of failure of a component. In addition, using multiple AE sensors, a large area of inspection can be monitored. These sensors can even be permanently mounted for applications demanding periodic inspections, saving the need for preliminary preparations.

However, AE-based systems can only qualitatively gauge how much damage is contained in a structure. In order to obtain quantitative results about size, depth and overall acceptability of a part, other sensing techniques should be involved.

Another aspect that should be mentioned is how to define a loading procedure, which ensures optimal damage identification. Currently, there is no standard, which gives certain guidelines regarding the loading procedures, and therefore most of the applications today are based on the accumulated experience of the designers. The loading procedure becomes a challenge mostly in the case of large aircraft components, which are assembled on aircrafts. It will be interesting and feasible to investigate new techniques for implementing *local* loading, that is, loading only area of interests instead of all the aircraft. Such techniques can be based on either dynamic or thermal loading. Usually, the sensors utilized for AE are piezoelectric (see Fig. A5). See also Chapter 14 of the present book.

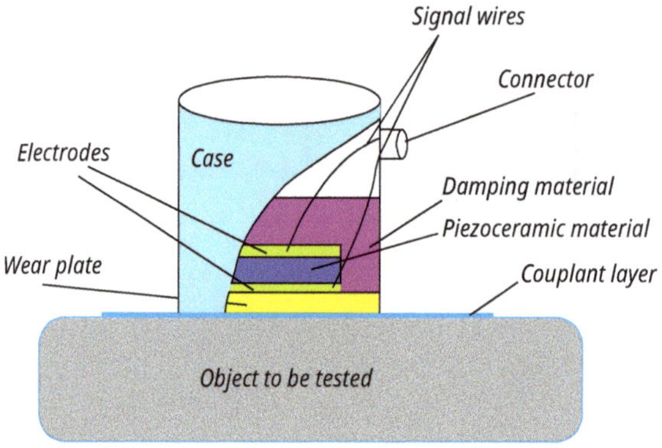

Fig. A5: Schematic drawing (not to scale) of an acoustic emission (AE) piezoelectric sensor.

A2 Active sensors

A2.1 Electrical impedance

The sensor is capable of measuring the changes in the impedance of the tested specimen and thus provides adequate output to correlate the measured change in the impedance to the size of the flaw. Depending on the application, a change in the resistivity, capacitance or inductance (the structure's impedance, *Z*) of the tested structure in the presence of a flaw can be detected, providing the intact structure was previously measured.

A2.2 Magnetic impedance

The magnetoimpedance sensor is based on the change in the complex impedance (*Z*) that ferromagnetic conductors experience (like amorphous microwires) when an external magnetic field (*H*) is applied and a high frequency (ω) alternating current flows through them. The sensor is applied on the tested structure, and due to external disturbances and/or flaws, it will change its magnetic impedance, thus measuring the reason for its change. Note that the method would require a reference data, for the undisturbed, perfect structure.

A2.3 Piezoelectric wafer active

The brothers Jacque and Pierre Curie first discovered piezoelectricity at the end of the nineteenth century, who demonstrated it on crystals of tourmaline, quartz, topaz, cane sugar and Rochelle salt [12]. The word *Piezoelectricity* comes from the Greek word *piesi*, which means *pressure*, and *electricity*, which implies the production of electricity resulting from the pressure. Piezoelectric materials, whether inherently piezoelectric or artificially poled, can be used to convert mechanical energy to electrical energy and vice versa. The constitutive equations describing these materials are written as follows:

$$\begin{aligned} S &= s^E T + d^t E \\ D &= dT + \varepsilon^T E \end{aligned} \tag{A.1}$$

where S is the mechanical strain, E is the electric field, T is the mechanical stress, D is the electrical displacement (all these variables are second-order tensors in the general case); d^t, s^E and ε^T, respectively, represent the matrix of the transpose piezoelectric effect, the mechanical compliance tensor (obtained at $E = 0$) and the permittivity (obtained at $T = 0$). This EM coupling property enables the utilization of these unique materials both as strain sensors and mechanical actuators. The voltage–strain curve does not exhibit linear behavior in its entire range. However, there is always a linear part in the curve used in practice for the sensing.

Piezoelectric ceramics, also called piezoceramics, is a subgroup of piezoelectric materials, which exhibit simultaneous actuator and sensor behavior, and therefore, are particularly attractive for diagnosis systems especially in the form of PWAS, which can be permanently attached to the structure [12]. The main advantage of PWAS over conventional ultrasonic probes is in their small size, lightweight, low profile and inexpensive cost. In addition, their frequency bandwidth is several orders of magnitude larger than that of conventional modal analysis equipment, permitting effective modal diagnosis in a relatively wide frequency band. The diagnosis process can be described as follows: The PWAS transmitter generates Lamb waves in the structure [16]. The generated Lamb waves travel through the structure and are reflected or diffracted by the structural boundaries, discontinuities and damage. The waves arrive at the PWAS receiver, where they are transformed into electrical signals. Using at least two PWAS, one can detect structural anomalies, that is, cracks, corrosions, delaminations and other damage [17]. PWAS can be used in several sensing applications as shown in Fig. A6.

Since the PWAS sensing technology has been found to be useful for detecting damages in composites, much effort has been recently invested in the qualification of this technology. This effort includes the investigation of sensor transduction under mechanical static and cyclic loading [18] and its reliability over time [19].

A2.4 Optical fiber

Among the available sensing technologies, the use of OFS to implement SHM appears quite attractive. Optical fibers are flexible, passive, tolerant to environmental conditions and insensitive to electromagnetic disturbances. In addition, due to their small diameter, it can be easily embedded within composite materials. The most common implementation of a fiber optic-based SHM system involves one of the two sensing principles: site-specific sensing, performed by using FBG sensors, or distributed sensing, performed by using Rayleigh backscattering.

A2.4.1 Fiber Bragg grating

Site-specific sensing using FBG sensors allows the measurement of mechanical strain and temperature changes in several locations along an optical fiber. An FBG sensor is a type of *distributed Bragg reflector*[12] constructed in a short segment of optical fiber, which reflects a *particular* wavelength of light and transmits all others (see Fig. A7). This particular wavelength reflected, called the Bragg wavelength, is given as follows:

$$\lambda_B = 2n_e\Lambda \tag{A.2}$$

12 A structure formed from multiple layers of alternating materials with varying refractive indices.

where n_e is the effective refractive index characterizing the FBG, and Λ represents the grating period. The linear dependence of the reflected wavelength λ_B on the grating period Λ enables to easily deduce changes in the grating period Λ by measuring the reflected wavelength λ_B.

An external mechanical load or temperature change causes a uniform change in the grating period Λ, resulting in a shift in the reflected wavelength λ_B. By measuring the difference in the reflected wavelength λ_B, the strain and the temperature of the optical fiber at the FBG location can be derived by using the following relation:

$$\frac{\Delta\lambda_B}{\lambda_B} = C_S\varepsilon + C_T\Delta T = (1 - P_e)\varepsilon + (\alpha_\Lambda + \alpha_n)\Delta T \tag{A.3}$$

where

$$(1 - P_e) = C_S \qquad (\alpha_\Lambda + \alpha_n) = C_T$$

where C_s is the strain coefficient that is related to the strain optic coefficient P_e and C_T is the coefficient of temperature, obtained by the summation of the optical fiber core thermal expansion coefficient, α_Λ, and the thermo-optic coefficient α_n.

A2.4.2 Rayleigh backscattering distributed strain sensing

Rayleigh backscattering in an optical fiber is caused by random fluctuations in the refractive index profile along the length of the fiber. For a given fiber, the scatter amplitude as a function of distance is a random but fixed property. Similar to FBG sensors, external stimulus such as mechanical strain or change in the temperature causes a shift in the local reflected spectrum [20]. In data processing, the complex data acquired from the optical fiber are divided into interval length, also called "sensor length," which are separated in a constant predefined gap. The shift in the reflected spectrum is calculated upon each "sensor length" separately and independently. Thus, the optical fiber can be modeled as a series of weak FBG sensors, which are placed side by side in a constant predefined gap (see Fig. A8). The predefined length of the sensors and the gap between the sensors, respectively, affect the spectral resolution and the signal-to-noise ratio. Therefore, these two parameters should be optimized to obtain the desired spatial resolution with maximum measurement accuracy.

Additional information associated with Rayleigh-distributed strain and temperature sensing can be found in Ref. [21].

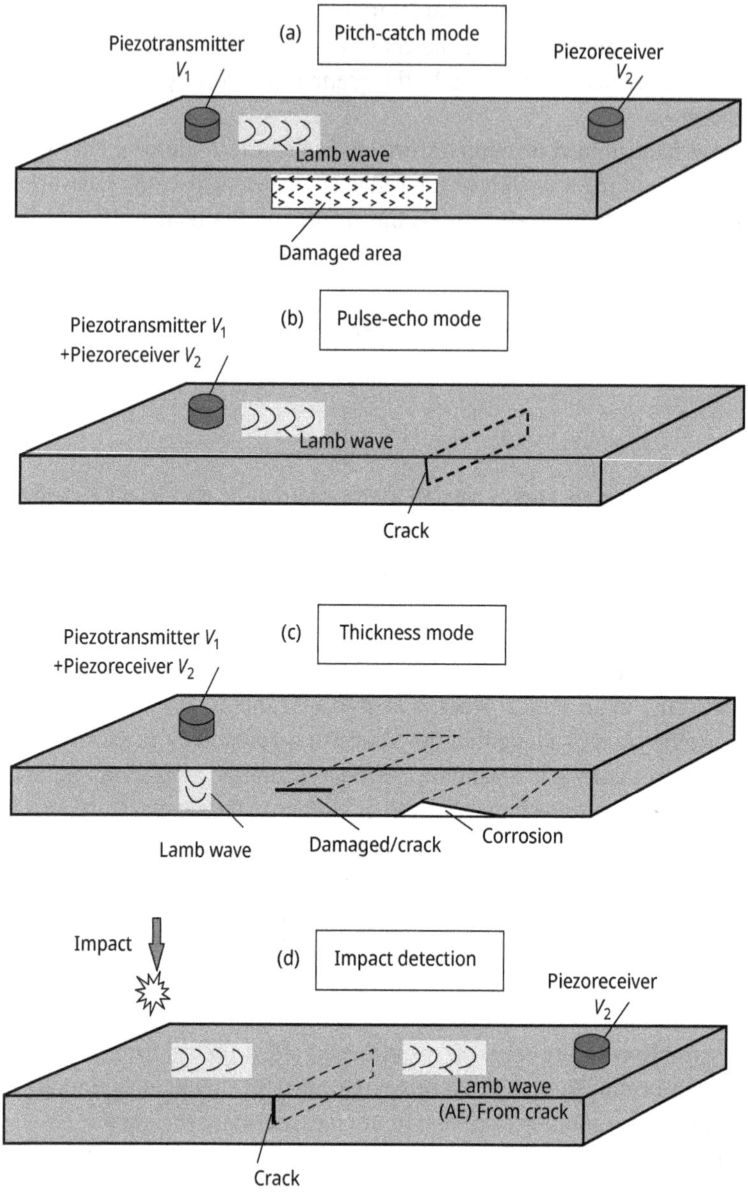

Fig. A6: Implementations of PWAS as traveling wave transducers for damage detection: (a) pitch-catch mode, (b) pulse-echo mode, (c) thickness mode and (d) detection of impacts and acoustic emission (AE) (adapted from [11]).

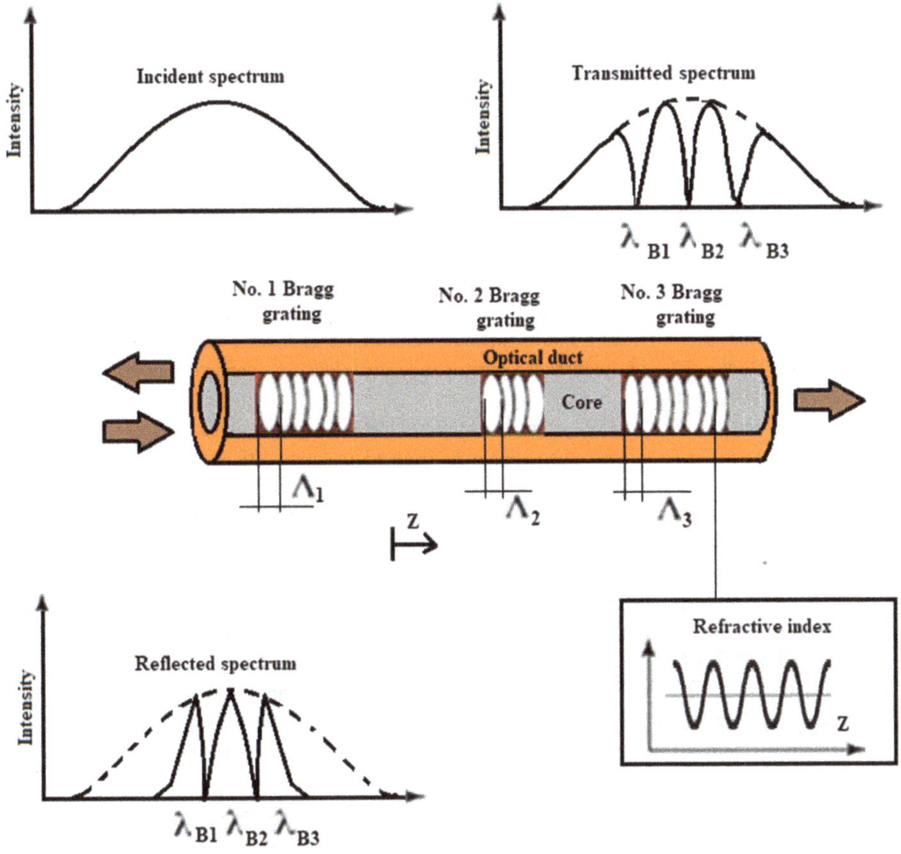

Fig. A7: Optical fiber containing several FBG sensors (each sensor reflects a particular wavelength of light only), depending on its grating period (adapted from [14]).

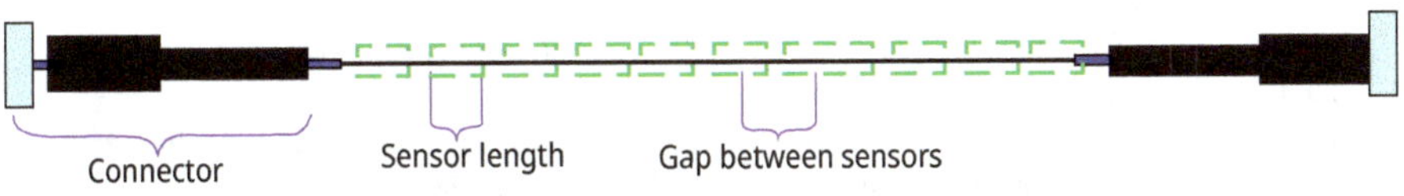

Fig. A8: A schematic drawing of an optical fiber, virtually divided into intervals, representing the sensors along the optical fiber that are separated by a constant gap.

References

[1] Kavandi, J., Callis, J., Gouterman, M., Khalil, G., Wright, D. and Green, E. Luminescence barometry in wind tunnels, Review of Scientific Instruments, 61(11), 1990, 3340–3347.

[2] Morris, M.J., Benne, M.E., Crites, R.C. and Donovan, J.F. Aerodynamic measurements based in photoluminescence, presented at the 31st Aerospace Sciences Meeting and Exhibit, Reno, NV, 11–14 Jan. 1993, Paper 93–0175.

[3] McLachlan, B. and Bell, J. Pressure-sensitive in aerodynamic testing, Experimental Thermal and Fluid Science, 10(4), 1995, 470–485.

[4] Liu, T., Campbell, B., Burns, S. and Sullivan, J. Temperature and pressure-sensitive luminescent paints in aerodynamics, Applied Mechanical Reviews, 50(4), 1997, 227–246.

[5] Liu, T. and Sullivan, J.P. Pressure and Temperature Sensitive Paints, Springer-Verlag, Berlin, 2005, 292.

[6] Lakowicz, J.R. Principles of Fluorescence Spectroscopy, 3rd, Kluwer Academic/Plenum Publishers, New York, 1999, 795636.

[7] Engler, R. and Klein, C. DLR PSP System: Intensity and lifetime measurements, in Proc. 17th International Congress on Instrumentation in Aerospace Simulation Facilities, Pacific Grove, CA, USA, 1997, 46–56.

[8] Holmes, J. Analysis of radiometric, lifetime and fluorescent imaging for pressure sensitive paint, Aeronautical Journal, 102(1014), 1998,, 189–194.

[9] Bell, J.H., Schairer, T.E., Hand, L.A. and Mehta, R.D. Surface pressure measurements using luminescent coatings, Annul Review of Fluid Mechanics, 33, 2001, 155–206.

[10] Mitsuo, K., Egami, Y., Asai, K., Suzuki, H. and Mizushima, H. Development of lifetime imaging system for pressure-sensitive paint, presented at the 22nd AIAA Aerodynamic Measurement Technology and Ground Testing Conference, St. Louis, MO, June 24–26, 2002, Paper 2002–2909.

[11] Watkins, A.N., Jordan, J.D., Leighty, B.D., Ingram, J.L. and Oglesby, D.M. Development of next generation lifetime PSP imaging systems, Proc. 20th Int. Congr. Instrumentation in Aerospace Facilities, Gottingen, Germany, 2003, 372–382.

[12] Abramovich, H. Intelligent Materials and Structures, ©, Walter de Gruyter GmbH, Berlin/Boston, 2016, 386.

[13] Stehmeier, H. and Speckmann, H. Comparative Vacuum Monitoring (CVM)-Monitoring of fatigue cracking in aircraft structures, 2nd European Workshop on Structural Health Monitoring, Munich, Germany, 7–9 July 2004, 9.

[14] Roach, D. Real time crack detection using mountable comparative vacuum monitoring sensors, Smart Structures and Systems, 5(4), 2009, 317–328.

[15] Professionals & Educators, Introduction to Acoustic Emission testing, Education Resource Center, 2014. [Online]. Available: https://www.nde-ed.org/EducationResources.

[16] Kessler, S.S., Spearing, S.M. and Soutis, C. Damage detection in composite materials using Lamb wave methods, Smart Materials and Structures, 11(2), 2002, 269–278.

[17] Giurgiutiu, V. Structural damage detection with piezoelectric wafer active sensors, Journal of Physics Conference Series, 305, 2011, 012123, 10.

[18] Abramovich, H., Tsikchotsky, E. and Klein, G. An experimental investigation on PZT behavior under mechanical and cycling loading, Journal of the Mechanical Behavior of Materials, 22(3–4), 2013, 129–136.

[19] Bach, M., Dobmann, N., Eckstein, B., Moix-Bonet, M. and Stolz, C. Reliability of co-bonded piezoelectric sensors on CFRP structures, 9th International Workshop on Structural Health Monitoring, Sept 10–12, Stanford University, CA, USA, 2013, 9.

[20] Kreger, S.T., Gifford, D.K., Froggatt, M.E., Soller, B.J. and Wolfe, M.S. High resolution distributed strain or temperature measurements in single- and multi-mode fiber using swept-wavelength interferometry, Optical Fiber Sensors, Technical Digest (CD) (optical Society of America), 2006, paper ThE42.

[21] Samiec, D. Distributed fiber-optic temperature and strain measurement with extremely high spatial resolution, Photonik International, 2012, 10–13.

11 Vibration correlation technique

11.1 Introduction

Thin-walled structures, like columns, plates and shells, are liable to buckling when subjected to axial compressive loads, torsional loads, external or internal pressure or a combination of these loads. Although the buckling loads of these basic structures can be analytically calculated, their experimental values are found to be less than the numerical predictions. This is due to the real boundary conditions of the tested specimens, the initial imperfections of the structure and the load eccentricity induced during the tests. Not all these factors can be a priori considered, leading to overprediction of the buckling loads using analytical formulas or by modeling the thin structure with a finite element code. The discrepancy between experimental and numerical/analytical predictions leads to the including of a knockdown factor (less that unity). This factor multiplies the numerical/analytical buckling load value to yield a design buckling value for the thin-walled structures. As the knockdown factor is based on the lower value for all the experimental results included in the database for a certain thin-walled structure, the resulting structure will have higher thickness, thus reducing its stiffness/mass advantage. The striving of an engineer is to be able to nondestructively predict the actual in situ buckling loads of a thin-walled structure and thus save weight of the structure. One of these methods is the vibration correlation technique (VCT).

11.2 Application of VCT to thin-walled structures under axial compression

The VCT consists of measuring the natural frequencies of a loaded structure and monitoring their change, while increasing the applied load. Assuming that the vibrational modes are like the buckling ones, one can draw a curve, displaying the natural frequencies squared versus the applied load and extrapolating the curve to zero frequency would yield the predicted buckling load of the tested structure. Abramovich [1] dedicated a whole chapter in his book to review the VCT approach and its applications. The subject was also presented in detail in [2].

Besides its capability to nondestructively predict the buckling load of thin-walled structures, the approach can also determine the actual in situ boundary condition of the structures, and therefore, the VCT is usually classified into two main groups according to their approach: (1) determination of in situ boundary conditions and (2) direct prediction of buckling loads.

The VCT method has been successfully applied to beams and columns axially loaded (see, e.g., [2–17]), yielding a straight line between the frequency squared and

https://doi.org/10.1515/9783111621104-011

the compressive load for both theoretical and experimental cases. Taking into account the differential equation for an isotropic column with constant properties EI along its length L, compressed by an axial loading P and undergoing small vibrations at a circular frequency ω, which can be presented as

$$EI\frac{\partial^4 W(x)}{\partial x^4}+P\frac{\partial^2 W(x)}{\partial x^2}-\rho A\omega^2 W(x)=0 \tag{11.1}$$

one can easily find the buckling load P_{cr} and its fundamental frequency f_1 for a column on simply supported boundary conditions. Their relevant expressions are

$$P_{cr}=\frac{\pi^2 EI}{L^2} \quad \omega_1^2=\frac{\pi^4 EI}{\rho AL^4} \quad \text{but} \quad \omega_1=2\pi f_1 \Rightarrow \quad f_1^2=\frac{\pi^2 EI}{4\rho AL^4} \tag{11.2}$$

where ρA is the mass per unit length. Using the definitions in eq. (11.2), it is easy to find the following relationship between the compressive load and the natural basic frequency of the column:

$$\left(\frac{f}{f_1}\right)^2+\left(\frac{P}{P_{cr}}\right)=1 \tag{11.3}$$

where f and f_1 are the measured frequency and its value at zero compressive load, respectively, and p and P_{cr} are the experimental applied compressive load and its buckling value at zero frequency, respectively. Although the expression presented in eq. (11.3) was obtained using simply supported boundary conditions at both ends of the column, other boundary conditions, like the clamped–clamped ones, would also comply with the linear relation (see eq. (11.3)), as displayed in Tab. 11.1. As shown in Tab. 11.1, the maximal deviation from a straight line would be at $(f/f_1)^2 = 0.5$ with a value of 0.008312.

Tab. 11.1: Comparison between simply supported (SS–SS) and clamped (C–C) at both ends of the compressed column.

$\left(\frac{P}{P_{cr}}\right)$	$(f/f_1)^2$	
	SS–SS	C–C
0.0	1.0	1.000000
0.1	0.9	0.902735
0.2	0.8	0.804965
0.3	0.7	0.706670
0.4	0.6	0.607795
0.5	0.5	0.508312
0.6	0.4	0.408180
0.7	0.3	0.307342
0.8	0.2	0.205744
0.9	0.1	0.103297
1.0	0.0	0.00000

More on this issue can be found in [1, 2].

One should note that other cases of columns, like a compressed column on Winkler-type foundation and laminated symmetric and nonsymmetric compressed columns using either classical lamination theory (CLT)) or first-order shear deformation theory (FOSDT) [1], would also comply with the expression presented by eq. (11.3).

These cases are summarized in Tabs. 11.2 and 11.3.

Tab. 11.2: Columns' schematic drawings.

Case	**Schematic drawing of the case**
An isotropic column, axially compressed	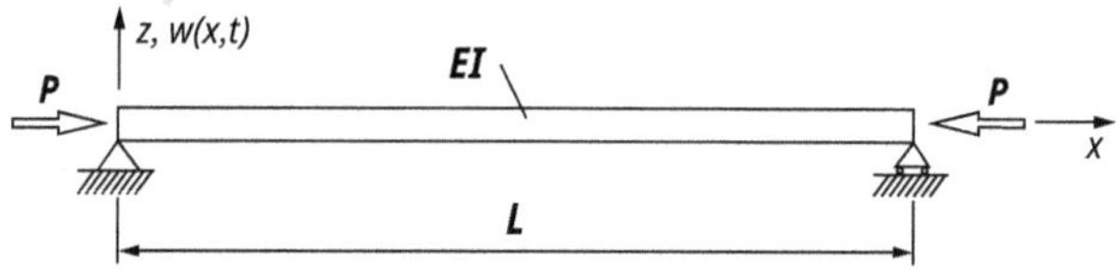
An isotropic column, axially tensioned	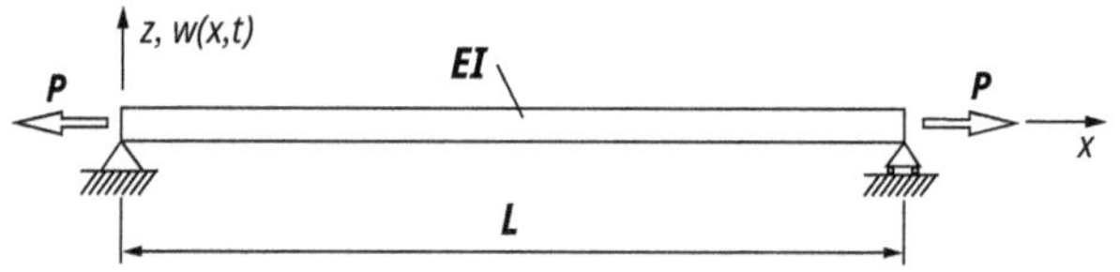
An isotropic column axially compressed on a Winkler-type foundation	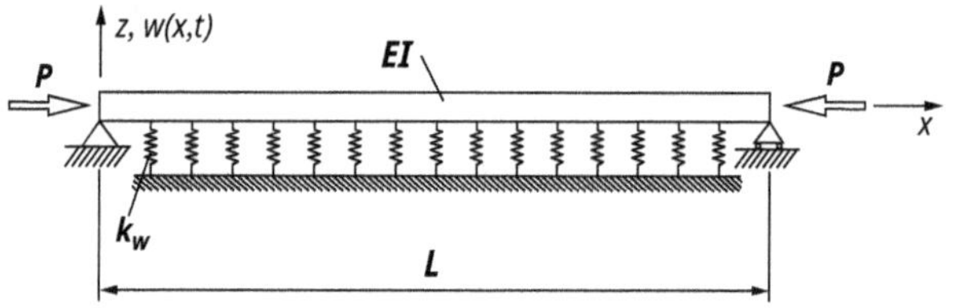
A symmetric laminated composite column under compression using CLT	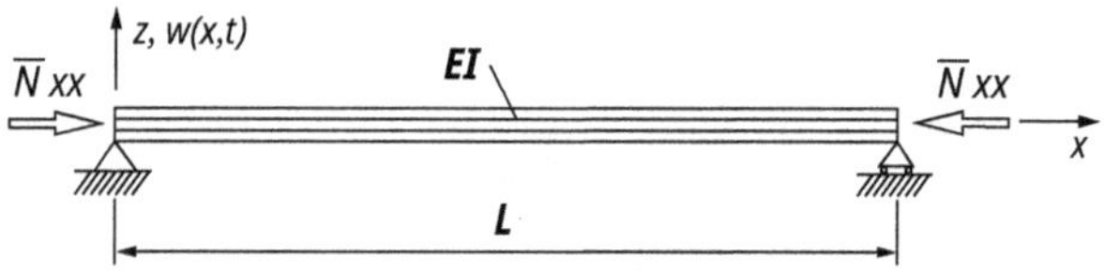
A nonsymmetric laminated composite column under compression using CLT	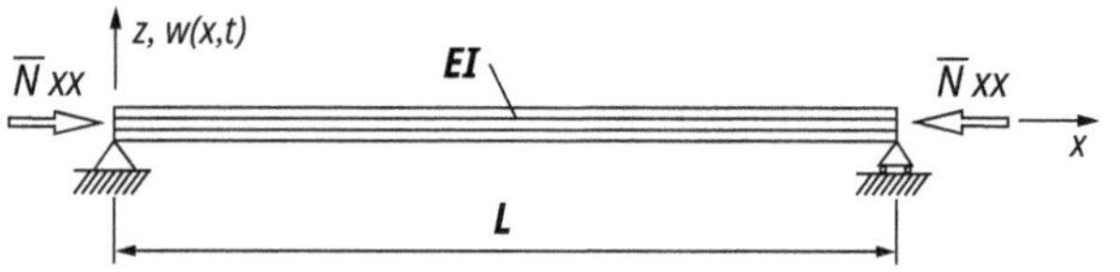
A symmetric laminated composite column under compression using FOSDT	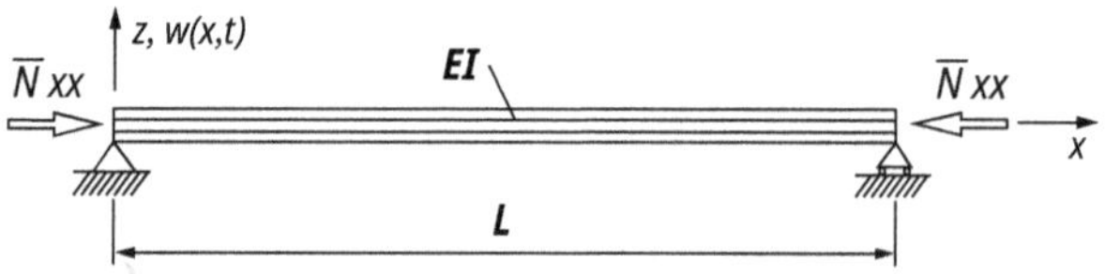

Tab. 11.2 (continued)

Case	Schematic drawing of the case
A nonsymmetric laminated composite column under compression using FOSDT	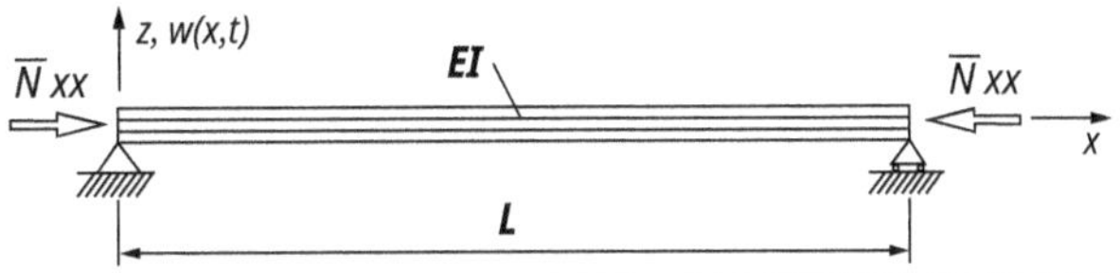
An isotropic column, axially compressed due to thermal loading	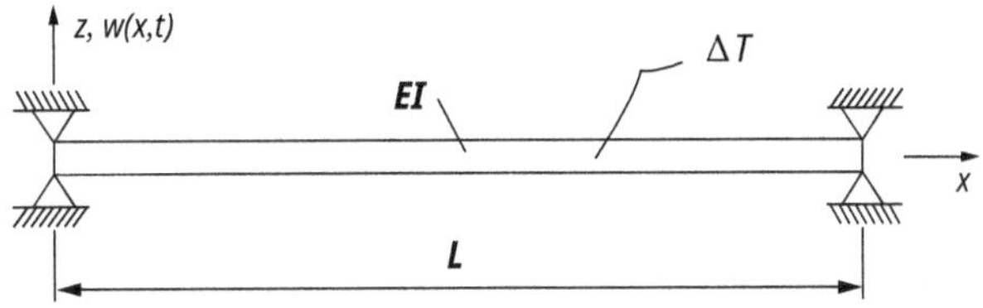

Tab. 11.3: Governing equations for columns and load versus natural frequency relationships.

Case	Governing equation	Load versus natural frequency relationship
An isotropic column, axially compressed	$EI\frac{\partial^4 w(x,t)}{\partial x^4}+P\frac{\partial^2 w(x,t)}{\partial x^2}+\rho A\frac{\partial^2 w(x,t)}{\partial t^2}=0$	$\left(\frac{f}{f_1}\right)2+\frac{P}{P_{cr}}=1$
An isotropic column, axially tensioned	$EI\frac{\partial^4 w(x,t)}{\partial x^4}-P\frac{\partial^2 w(x,t)}{\partial x^2}+\rho A\frac{\partial^2 w(x,t)}{\partial t^2}=0$	$1+\frac{P}{P_{cr}}=\left(\frac{f}{f_1}\right)^2$
An isotropic column axially compressed on a Winkler-type foundation	$EI\frac{\partial^4 w(x,t)}{\partial x^4}+P\frac{\partial^2 w(x,t)}{\partial x^2}+$ $+k_W\cdot w(x,t)+\rho A\frac{\partial^2 w(x,t)}{\partial t^2}=0$	$\left(\frac{f}{f_1}\right)^2+\frac{P}{P_{cr}}=1$
A symmetric laminated composite column under compression using CLT	$D_{11}\frac{\partial^4 w(x,t)}{\partial x^4}+\bar{N}_{xx}\frac{\partial^2 w(x,t)}{\partial x^2}+$ $+I_2\frac{\partial^4 w(x,t)}{\partial t^2\partial x^2}-I_0\frac{\partial^2 w(x,t)}{\partial t^2}=0$	$\left(\frac{f}{f_1}\right)^2+\frac{\bar{N}_{xx}}{(\bar{N}_{xx})_{cr}}=1$
A nonsymmetric laminated composite column under compression using CLT	$D_{11}\frac{\partial^4 w(x,t)}{\partial x^4}+\bar{N}_{xx}\frac{\partial^2 w(x,t)}{\partial x^2}+$ $+I_2\frac{\partial^4 w(x,t)}{\partial t^2\partial x^2}-I_0\frac{\partial^2 w(x,t)}{\partial t^2}=0$	$\left(\frac{f}{f_1}\right)^2+\frac{\bar{N}_{xx}}{(\bar{N}_{xx})_{cr}}=1$
A symmetric laminated composite column under compression using FOSDT*	$A_{11}\left(\frac{\partial^2 u}{\partial x^2}+\frac{\partial^3 w}{\partial x^3}\right)=I_0\frac{\partial^2 u}{\partial t^2}$ $KA_{55}\left(\frac{\partial^2 w}{\partial x^2}+\frac{\partial \phi}{\partial x}\right)-\bar{N}_{xx}\frac{\partial^2 w}{\partial x^2}=I_0\frac{\partial^2 w}{\partial t^2}$ $D_{11}\frac{\partial^2 \phi}{\partial x^2}-KA_{55}\left(\frac{\partial w}{\partial x}+\phi\right)=I_2\frac{\partial^2 \phi}{\partial t^2}$	$\left(\frac{f}{f_1}\right)^2+\frac{\bar{N}_{xx}}{(\bar{N}_{xx})_{cr}}=1$

Tab. 11.3 (continued)

Case	Governing equation	Load versus natural frequency relationship
A nonsymmetric laminated composite column under compression using FOSDT	$A_{11}\left(\frac{\partial^2 u}{\partial x^2}+\frac{\partial^3 w}{\partial x^3}\right)+B_{11}\frac{\partial^2 \phi}{\partial x^2}=I_0\frac{\partial^2 u}{\partial t^2}+I_1\frac{\partial^2 \phi}{\partial t^2}$ $KA_{55}\left(\frac{\partial^2 w}{\partial x^2}+\frac{\partial \phi}{\partial x}\right)-\bar{N}_{xx}\frac{\partial^2 w}{\partial x^2}=I_0\frac{\partial^2 w}{\partial t^2}$ $B_{11}\left(\frac{\partial^2 u}{\partial x^2}+\frac{\partial^3 w}{\partial x^3}\right)+D_{11}\frac{\partial^2 \phi}{\partial x^2}-KA_{55}\left(\frac{\partial w}{\partial x}+\phi\right)$ $=I_2\frac{\partial^2 \phi}{\partial t^2}+I_1\frac{\partial^2 u}{\partial t^2}$	$\left(\frac{f}{f_1}\right)^2+\frac{\bar{N}_{xx}}{(\bar{N}_{xx})_{cr}}=1$
An isotropic column, axially compressed due to thermal loading	$EI\frac{\partial^4 w(x,t)}{\partial x^4}+(A\cdot E\cdot \alpha\cdot \Delta T)\frac{\partial^2 w(x,t)}{\partial x^2}+$ $+\rho A\frac{\partial^2 w(x,t)}{\partial t^2}=0$	$\left(\frac{f}{f_1}\right)^2+\frac{\Delta T}{(\Delta T)_{cr}}=1$

* $K = 5/6$ for a rectangular cross section. The accurate value of the shear coefficient is $\kappa = 10(1+\upsilon)/(12+11\upsilon)$ for a rectangular cross section and $\kappa = 6(1+\upsilon)/(7+6\upsilon)$ for a solid circular cross section, where υ is the Poisson's ratio.

The application of VCT on perfect plates and shells, for compressive loading, is also being dealt in the literature as presented in [17–44], respectively. For perfect rectangular isotropic plates (see Fig. 11.1), loaded only by compressive loads, the differential equation of motion can be written as [1]

$$D\left[\frac{\partial^4 W(x,y)}{\partial x^4}+2\frac{\partial^4 W(x,y)}{\partial x^2 \partial y^2}+\frac{\partial^4 W(x,y)}{\partial y^4}\right]-\rho h\omega^2 W(x,y)$$
$$=N_x\frac{\partial^2 W(x,y)}{\partial x^2}+N_y\frac{\partial^2 W(x,y)}{\partial y^2} \tag{11.4}$$

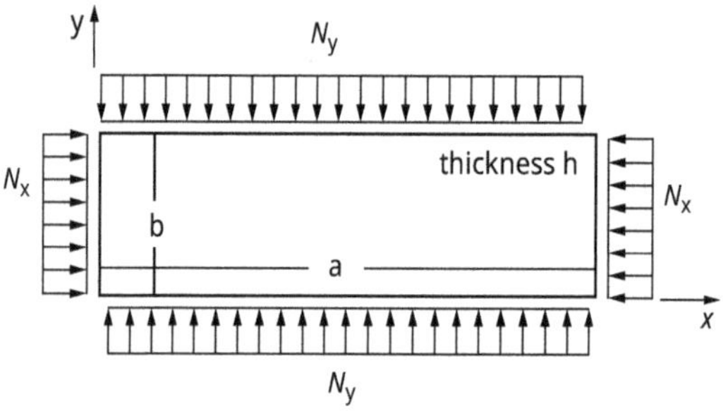

Fig. 11.1: A rectangular flat plate compressed in its two perpendicular directions.

where

$$D = \frac{Eh^3}{12(1-\upsilon^2)}$$

where E is the Young' modulus, h is the plate thickness and ρh is the mass per unit area of the plate.

N_x and N_y are the in-plane loads per unit length in the x and y directions, respectively.

For a simply supported all-around case, one obtains the following expressions for the natural frequencies and the buckling loads for compression in the x-only or y-only directions:

$$\begin{aligned} \hat{\omega}_{mn}^2 &= D\frac{\pi^4}{\rho h}\left[\left(\frac{m}{a}\right)^2 + \left(\frac{n}{b}\right)^2\right]^2 \\ N_{xcr}^{mn} &= -D\pi^2\frac{a^2}{m^2}\left[\left(\frac{m}{a}\right)^2 + \left(\frac{n}{b}\right)^2\right]^2 \\ N_{ycr}^{mn} &= -D\pi^2\frac{b^2}{n^2}\left[\left(\frac{m}{a}\right)^2 + \left(\frac{n}{b}\right)^2\right]^2 \end{aligned} \tag{11.5}$$

Keeping in mind the expressions in eq. (11.5) and using eq. (11.4), one obtains the following equation:

$$\frac{N}{N_{xcr}^{mn}} + \frac{N}{N_{ycr}^{mn}} + \left(\frac{\omega}{\hat{\omega}_{mn}}\right)^2 = 1 \Rightarrow \frac{N}{N_{xcr}^{mn}} + \frac{N}{N_{ycr}^{mn}} + \left(\frac{f}{\hat{f}_{mn}}\right)^2 = 1 \tag{11.6}$$

which is similar to eq. (11.3), namely, the frequency squared is linearly dependent on the compressive loads, as shown above for columns. This linear equation was also shown to be true in experiments; see, for example [18],. However, as in real life plates have some initial imperfections, the theoretical linear relationship ceased to be true as the compressive load is approaching the buckling load [21]. At the buckling load, it does not occur at zero frequency, and afterwards, the curve starts to rise again with an increase in the compressive load. Experimentally, the point where the curve changes its tendency would be the VCT-predicted buckling load.

As pointed out in [1, 2], also for cylindrical shells compressed axially a linear theoretical curve exists for the square of the lowest natural frequency and the applied compressed load. However, when trying to apply the VCT to a cylindrical compressed shell, the linear relationship predicts higher and wrong buckling loads. The literature presents two applications to correctly predict the buckling loads of cylinders. The first one is attributed to Souza et al. [30–32] that suggested using the following relationship:

$$(1-p)^2+\left(1-\xi^2\right)\left(1-f^4\right)=1 \tag{11.7}$$

where ξ is the “experimental” knock-down factor based on the results of the test, at relatively low loads. The procedure starts with the acquisition of the natural frequencies at zero axial load. Then the load is increased, and the nondimensional frequency f is calculated for each load step by normalizing the measured frequency at a compression load P by the frequency at zero compression. Then the load P is also normalized by the numerical buckling load P_{cr} to yield the variable p. At about 60% of the calculated buckling load the test is stopped, a straight line is drawn, starting at point $[(1-f^4)=0, (1-p)^2=1]$ till point $[\xi^2, (1-f^4)=1]$. The value of ξ^2 is determined as the cross-point between the oblique line and the vertical line from $(1-f^4)=1$. The application of eq. (11.7) was shown to provide good results for stringer-stiffened circular isotropic shells [30]. Another application is the empiric relationship suggested by Arbelo et al. [34, 35]. The empirical approach is based on the modification of eq. (11.7) yielding the following relationship:

$$\begin{aligned} &1-p=f^2=1-\left(1-f^2\right) \\ &\Rightarrow (1-p)^2=\left[1-\left(1-f^2\right)\right]^2 \end{aligned} \tag{11.8}$$

Then a graph of $(1-p)^2$ versus $\left(1-f^2\right)$ is constructed based on a typical test. Following the definitions of p and f, presented in eq. (11.7), a best-fit second-order equation is approximated based on the experimental points. The fitted second-order polynomic curve of the measured experimental natural frequencies would have the following expression:

$$(1-p)^2=\alpha\left(1-f^2\right)^2+\beta\left(1-f^2\right)+\chi \tag{11.9}$$

with the values of the constants α, β and χ being determined by the best-fit process. Finding the minimal point of the second-order equation presented in eq. (11.9) yields

$$\left(1-f^2\right)_{\min}=-\frac{\beta}{2\alpha}\rightarrow(1-p)^2_{\min}=\chi-\frac{\beta^2}{4\alpha}\equiv\xi^2 \tag{11.10}$$

with ξ being the “knock-down factor” which represents the drop in the shell load carrying capacity, like Souza et al. [30] method, presented earlier, leading to the prediction of the in situ tested specimen using the following form:

$$P_{\text{pred.}}=(1-\xi)P_{cr} \tag{11.11}$$

The use of Arbelo’s empirical method provided very good results, as shown in [33–44].

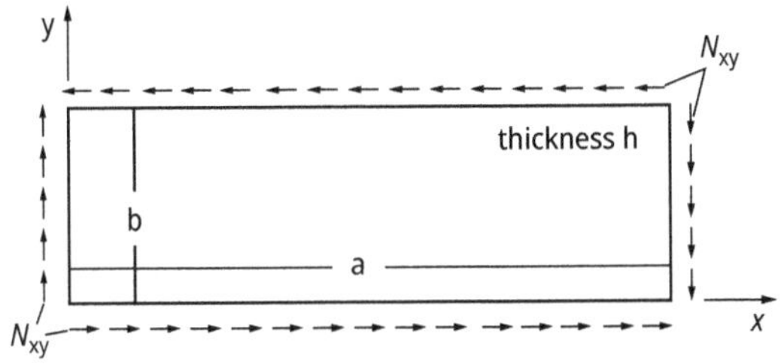

Fig. 11.2: A flat rectangular plate under pure shear, N_{xy}.

11.3 Application of the VCT to thin-walled structures under pure shear

When trying to apply the VCT to flat rectangular plates loaded in pure shear, due to the complexity of the problem, the linear relationship presented before for the case of compression does not hold anymore, and a new procedure should be used. The influence of shear on the natural frequencies of a flat plate was rarely treated in the literature (see [45–60]) with no decisive formulation. Therefore, a new application aimed to allow predicting the buckling of a flat plate under pure shear, by monitoring its natural frequencies, is presented further.

Figure 11.2 presents a rectangular isotropic flat plate, having the thickness h, length a and width b under pure shear loads N_{xy} per unit length.

The differential equation for this case can be written as

$$D\left[\frac{\partial^4 W(x,y)}{\partial x^4}+2\frac{\partial^4 W(x,y)}{\partial x^2\partial y^2}+\frac{\partial^4 W(x,y)}{\partial y^4}\right]-\rho h\omega^2 W(x,y)=2N_{xy}\frac{\partial^2 W(x,y)}{\partial x\partial y} \tag{11.12}$$

where

$$D=\frac{Eh^3}{12(1-\upsilon^2)}$$

Due to its problematic mathematical appearance[1] of eq. (11.12), there is no closed-form expression for the critical shear (or shear stress) loading. Only approximate solutions are available in the literature. One of these solutions is the one presented by Timoshenko and Gere [61] and can be presented as

1 While the left-hand side of the equation has an odd differentiation of the unknown $W(x,y)$, the left-hand side of the equation presents even differentiation of the unknown. Assuming a double sine (or cosine) for the unknown $W(x,y)$, as in the case of compressive loaded plate, will not eliminate the sine (or cosine) terms from both sides of the equation, thus preventing the solution of the equation in a closed form.

$$\tau_{cr} = k\frac{\pi^2 D}{b^2 h} \Rightarrow \left(N_{xy}\right)_{cr} = k\frac{\pi^2 D}{b^2} \tag{11.13}$$

The factor k depends on the aspect ratio (a/b) of the flat rectangular plate as presented in Tab. 11.4 (from [61]).

Tab. 11.4: Variation of the *k* factor with plate aspect ratio, a/b.

a/b	1.0	1.2	1.4	1.5	1.6	1.8	2.0	2.5	3.0	4.0
k	9.34	8.0	7.3	7.1	7.0	6.8	6.6	6.1	5.9	5.7

For larger aspect ratios, the following approximation should be used (see [61]):

$$k = 5.35 + 4\left(\frac{b}{a}\right)^2 \tag{11.14}$$

The natural frequency $\hat{\omega}_{mn}^2$ expression for the unloaded case keeps its value as it was shown in eq. (11.5).

To investigate the relationship between the shear load and the natural frequency of a flat rectangular plate, a finite element model using ANSYS Workbench 2020 (Student Version) [62] was constructed and its results are presented further.

The finite element model consists of a 100×100 mm^2 plate with a thickness of 1 mm. The chosen material has a Young's modulus of 200 GPa, a Poisson's ratio of 0.3 and a density of 7,850 kg/m^3. The basic square model had 50×50 Quad elements. To enable simulations for various aspect ratios (AR = a/b), the basic model was enlarged in the x direction, yielding 50*AR elements. Figure 11.3 presents the application of the shear forces and the applied boundary conditions to simulate the demanded simply supported all-around plate circumference. Note that for all-around clamped boundary conditions, the rotations around x and y should be zero.

The buckling load calculated by the finite element code is displayed in Tab. 11.5 for various aspect ratios of the plate compared to the theoretical values obtained through eqs. (11.13) and (11.14).

As one can see, the results from the finite element code are in very good agreement with the theoretical ones, the largest deviation being 1% for AR = 3, for plates on simply supported all-around plates. The limited results for the all-around clamped rectangular plates are presented in Tab. 11.6, where the theoretical results are calculated using the equations developed in [53].

In contrast to what have been found for the all-around simply supported plates (see Tab. 11.5), the results of the all-around clamped plates, as presented in Tab. 11.6, show a good match between the numerical and theoretical calculations for low aspect ratios, while for higher a/b ratios the matching is deteriorating. However, one has to note that the theoretical results of [53] are only approximated one, and therefore, the numerical FE calculations are acceptable.

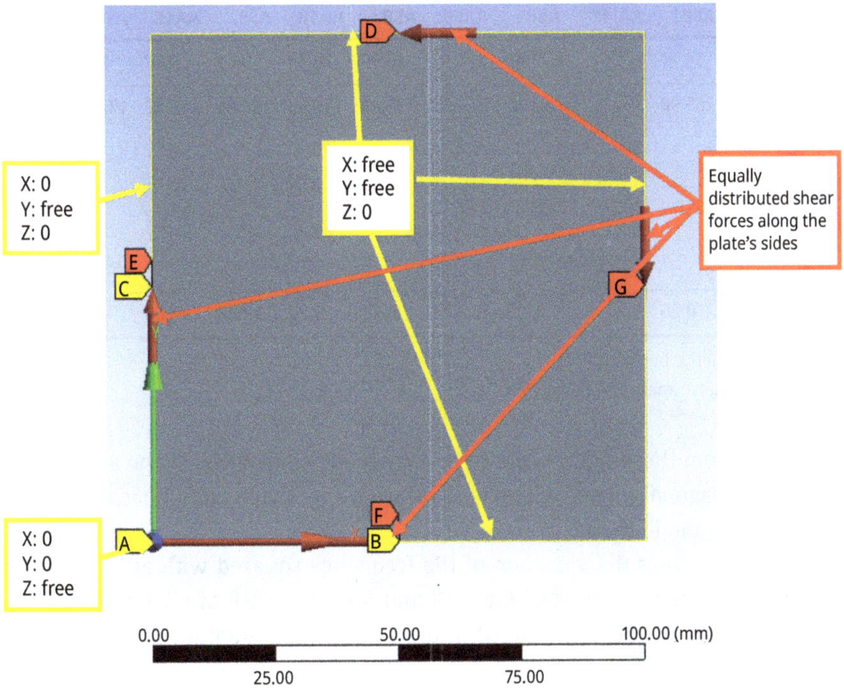

Fig. 11.3: Boundary conditions and shear load application – the simply supported all-around case.

Tab. 11.5: Theoretical and numerical buckling loads for various aspect ratios – simply supported all-around plates.

a/b	**1.0**	**1.5**	**2.0**	**3.0**	**5.0**	**7.0**	**10.0**
k	9.34	7.10	6.60	5.90	5.55	5.43	5.41
Theoretical buckling force (N/mm)	168.83	128.34	119.30	106.65	100.32	98.40	97.79
Numerical (FE) buckling force (N/mm)	168.20	127.70	118.40	105.60	100.00	98.15	97.40
% deviation	0.37	0.50	0.75	0.99	0.32	0.25	0.34

Typical buckling mode shapes for both types of boundary conditions, simply supported and clamped all-around plates, are presented in Fig. 11.4.

It turned out that the buckling mode shape and the frequency mode shape at zero loading are not identical for pure shear loading in contrast to axial compression case, for which these modes are the same. The influence of the shear loading on the shape of the frequency modes is next presented for various modes, aspect ratio and boundary conditions, as depicted in Figs. 11.5–11.13.

Tab. 11.6: Theoretical and numerical buckling loads for various aspect ratios – clamped all-around plates.

a/b	1.0		1.5		2.0		3.0		∞	
	ASB*	**AASB****	**ASB**	**AASB**	**ASB**	**AASB**	**ASB**	**AASB**	**ASB**	**AASB**
k	14.71	–	11.50	11.93	10.90	10.34	9.62	9.67	8.98	8.98
Theoretical buckling force (N/mm)	265.90	–	207.87	215.65	197.03	186.90	194.86	185.82	172.56	172.56
Numerical (FE) buckling force (N/mm)	265.20	207.50	185.60	172.70	–					
% deviation	0.26	–	0.18	3.93	6.16	0.70	12.83	7.60	–	–

*ASB, averaged symmetric buckling [53].
**AASB, averaged antisymmetric buckling [53].

As can be seen from these figures, the frequency mode changes its shape and rotates along the main diagonal line of the plate's surface till its shape would become identical to the buckling mode shape.

Figure 11.14 displays the variation of the frequency squared with an increase in the shear load, for aspect ratios of AR = 1, 2 and 3 for the case of all-around simply supported rectangular plate. The fourth-order polynomial equation fitted to the FE calculated values is presented for each graph and includes its relevant R^2 value (how good is the fitted polynomial equation for the given data, with $R^2 = 1$ been a 100% fitting). Similar graphs are presented in Fig. 11.15 for the case of all-around clamped rectangular plates.

The calculated buckling loads, using the various fitted polynomial equations, are presented in Tab. 11.7, together with the reference buckling loads, as calculated by the FE code.

Table 11.8 presents the various deviations of the predicted buckling loads from its relevant numerical buckling loads for various aspect ratios for the case of all-around simply supported and all-around clamped boundary conditions. One can see that the predicted buckling loads when considered 100% $N_{xy\ \mathrm{cr}}$ of the points are in a very close proximity with the numerical buckling loads (deviations between 0.33% and −1.83%). However, the prediction procedure should use much less points to be used as a VCT approach. As depicted in Tab. 11.8, considering 70% $N_{xy\ \mathrm{cr}}$ points also leads to a good proximity to the numerical buckling loads (deviations between −1.56% and 4.32%). This proximity deteriorates for points up to 60% $N_{xy\ \mathrm{cr}}$ (deviations between −4.98% and 16.44%) and up to 50% $N_{xy\ \mathrm{cr}}$ (deviations between −7.10% and 40.53%). Similar behavior is expected also for all-around clamped boundary conditions.

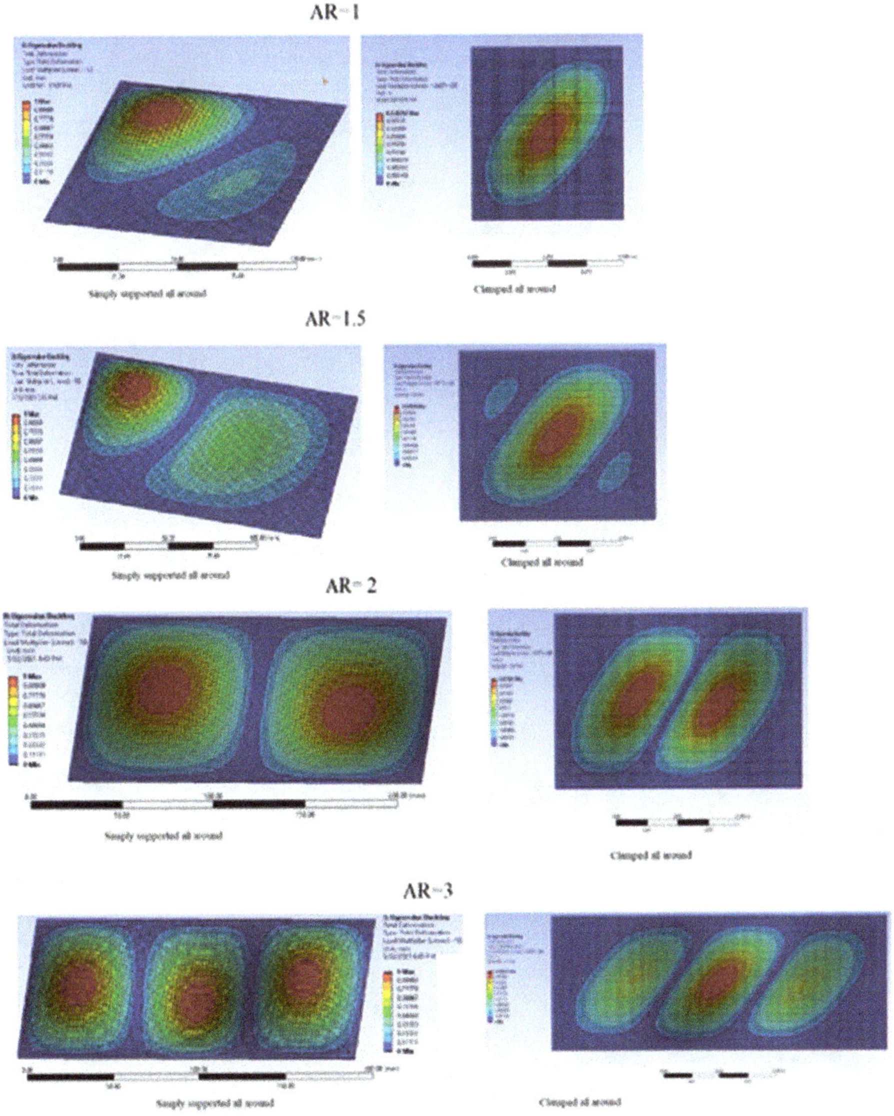

Fig. 11.4: Buckling mode shapes for various aspect ratios (AR) and boundary conditions.

Based on the earlier results, to use the present method as a VCT approach, it is recommended to use points up to 60% or 70% of the calculated buckling loads.

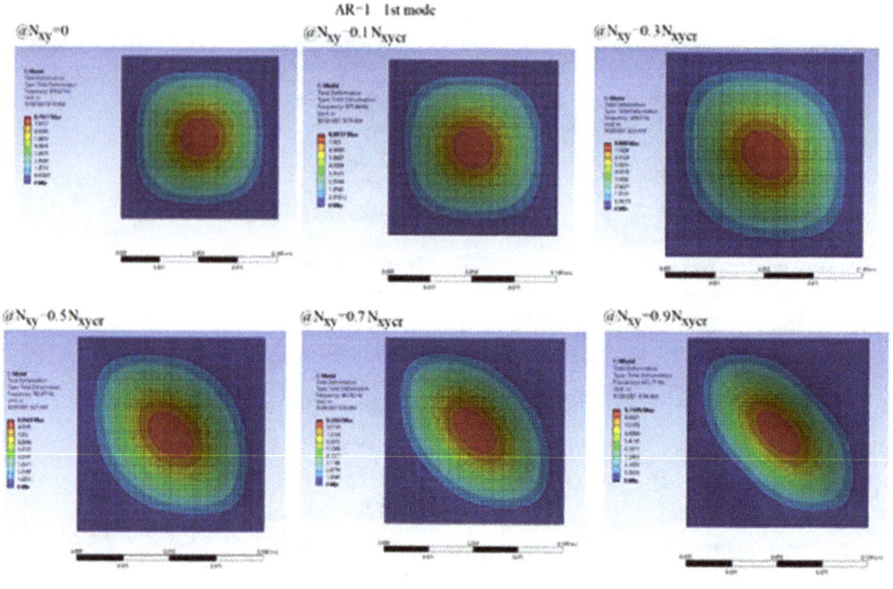

Fig. 11.5: The evolution of the first frequency mode with an increase in the shear load for AR = 1 for an all-around clamped rectangular plate.

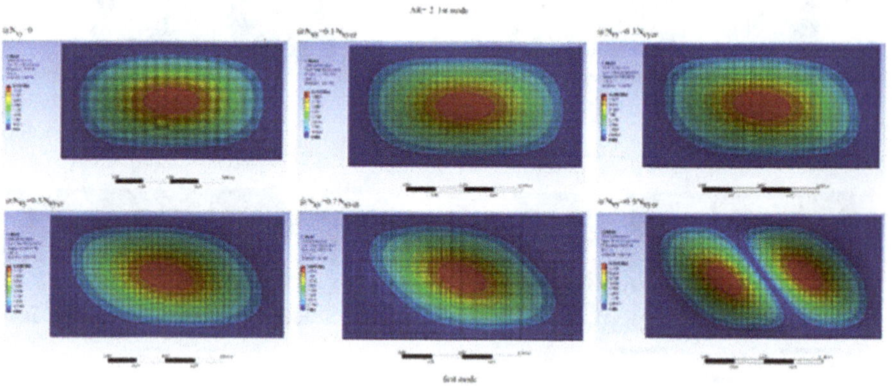

Fig. 11.6: The evolution of the first frequency mode with an increase in the shear load for AR = 2 for an all-around clamped rectangular plate.

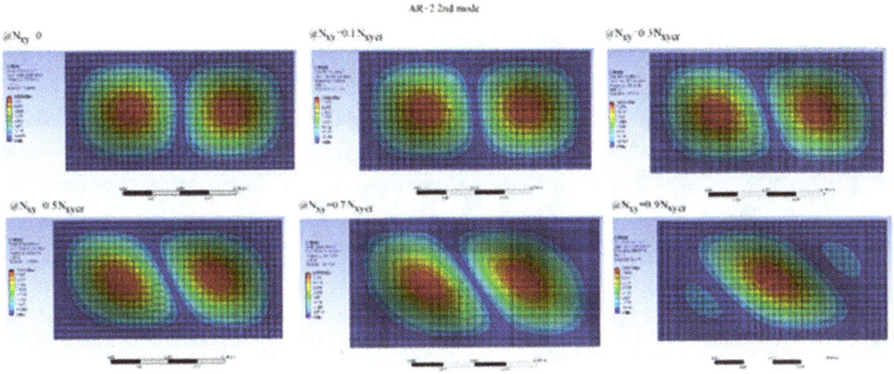

Fig. 11.7: The evolution of the second frequency mode with an increase in the shear load for AR = 2 for an all-around clamped rectangular plate.

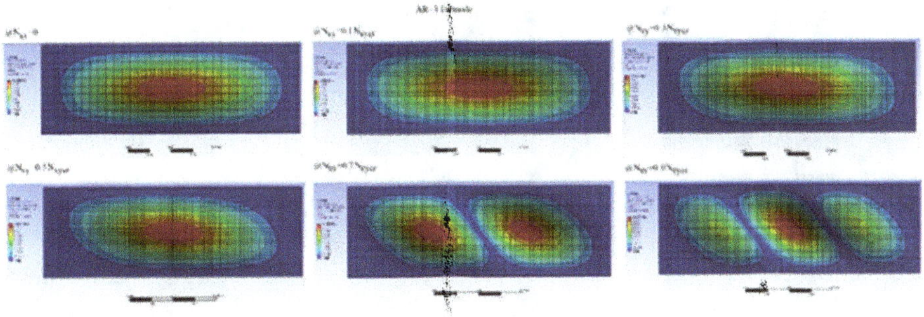

Fig. 11.8: The evolution of the first frequency mode with an increase in the shear load for AR = 3 for an all-around clamped rectangular plate.

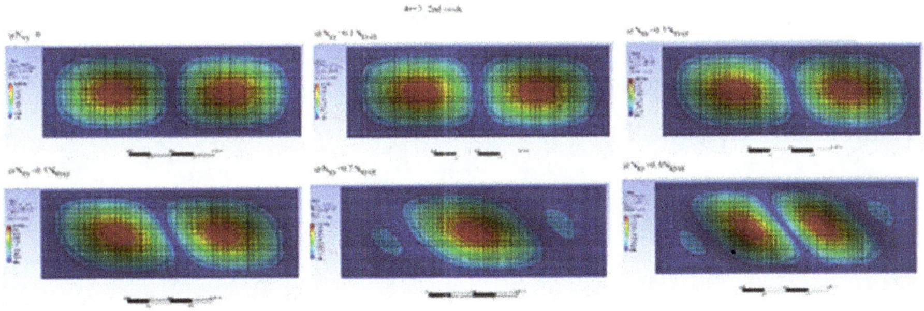

Fig. 11.9: The evolution of the second frequency mode with an increase in the shear load for AR = 3 for an all-around clamped rectangular plate.

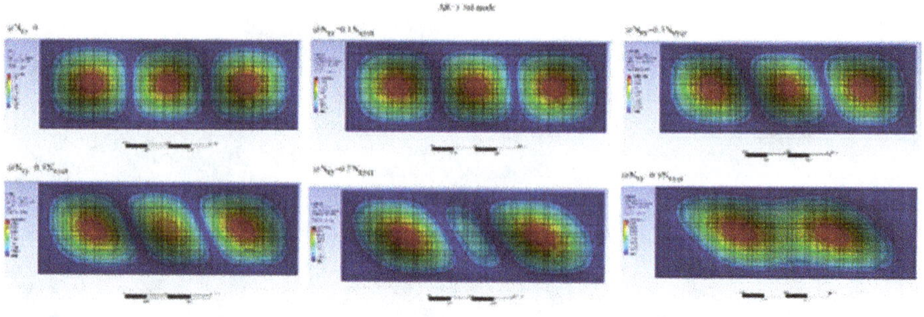

Fig. 11.10: The evolution of the third frequency mode with an increase in shear load for AR = 3 for an all-around clamped rectangular plate.

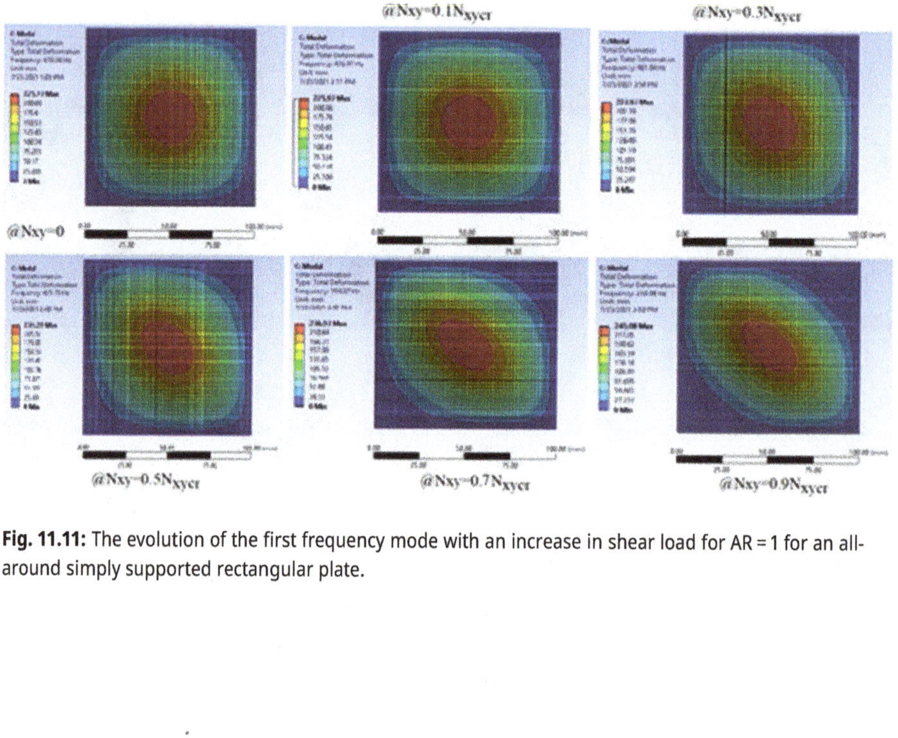

Fig. 11.11: The evolution of the first frequency mode with an increase in shear load for AR = 1 for an all-around simply supported rectangular plate.

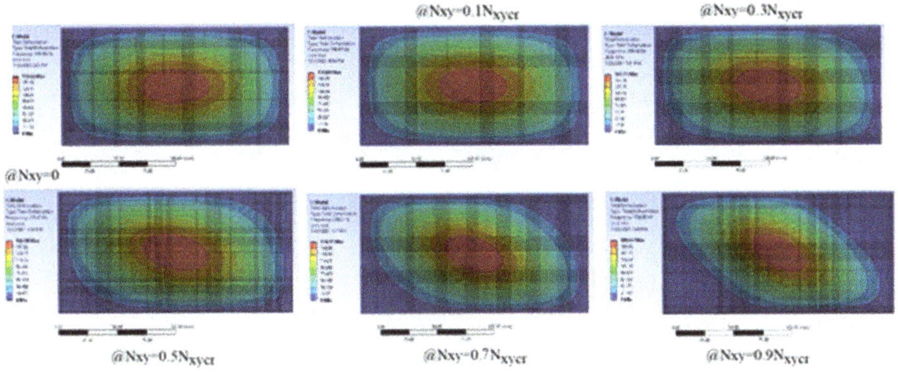

Fig. 11.12: The evolution of the first frequency mode with an increase in shear load for AR = 2 for an all-around simply supported rectangular plate.

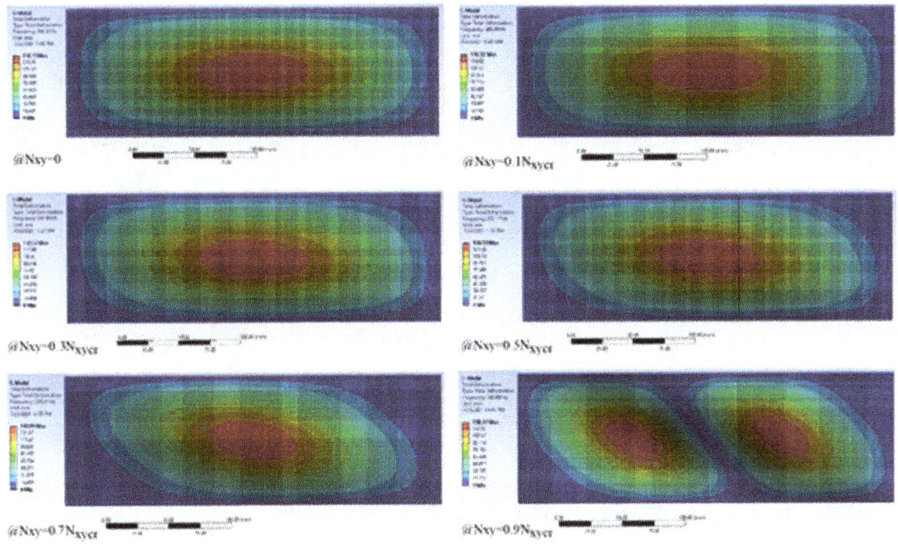

Fig. 11.13: The evolution of the first frequency mode with an increase in shear load for AR = 3 for an all-around simply supported rectangular plate.

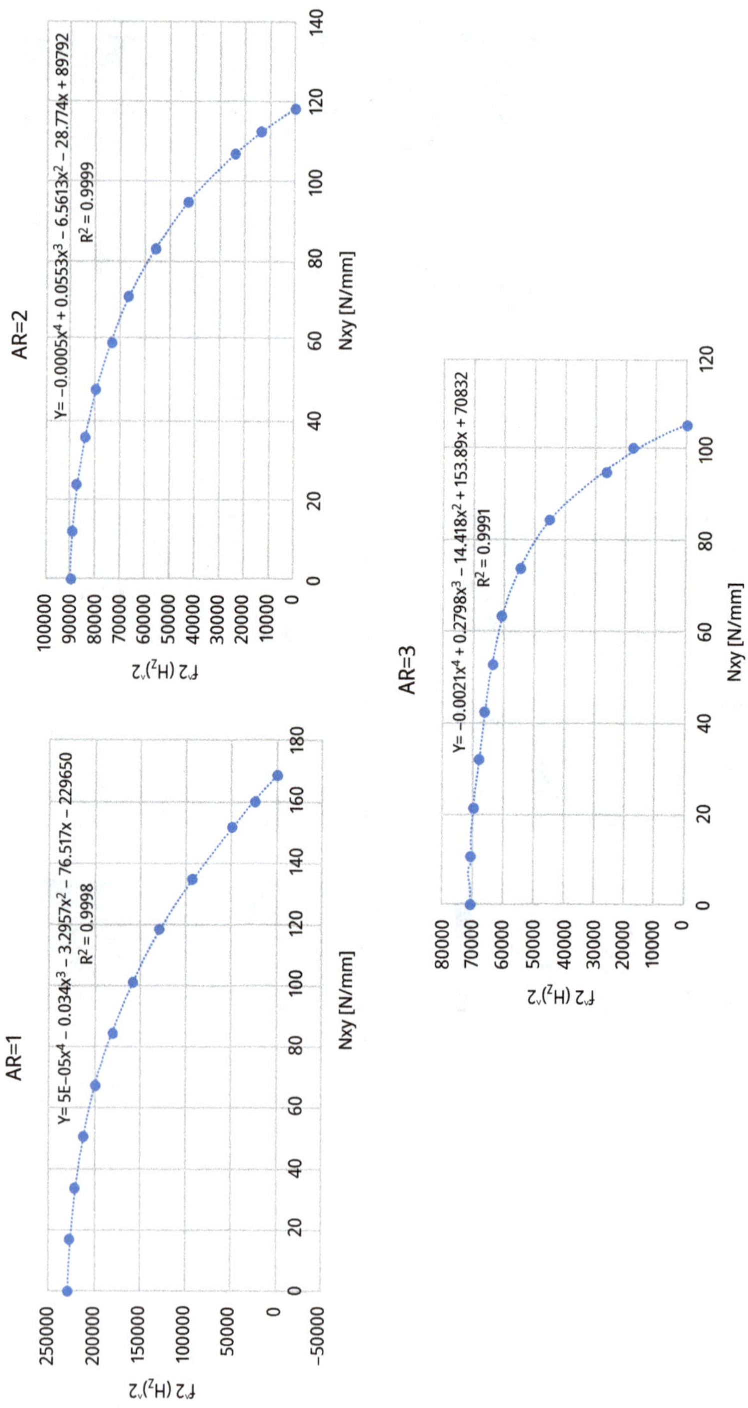

Fig. 11.14: Frequency squared versus applied shear load for various aspect ratios of the plate all-around simply supported boundary conditions – first mode.

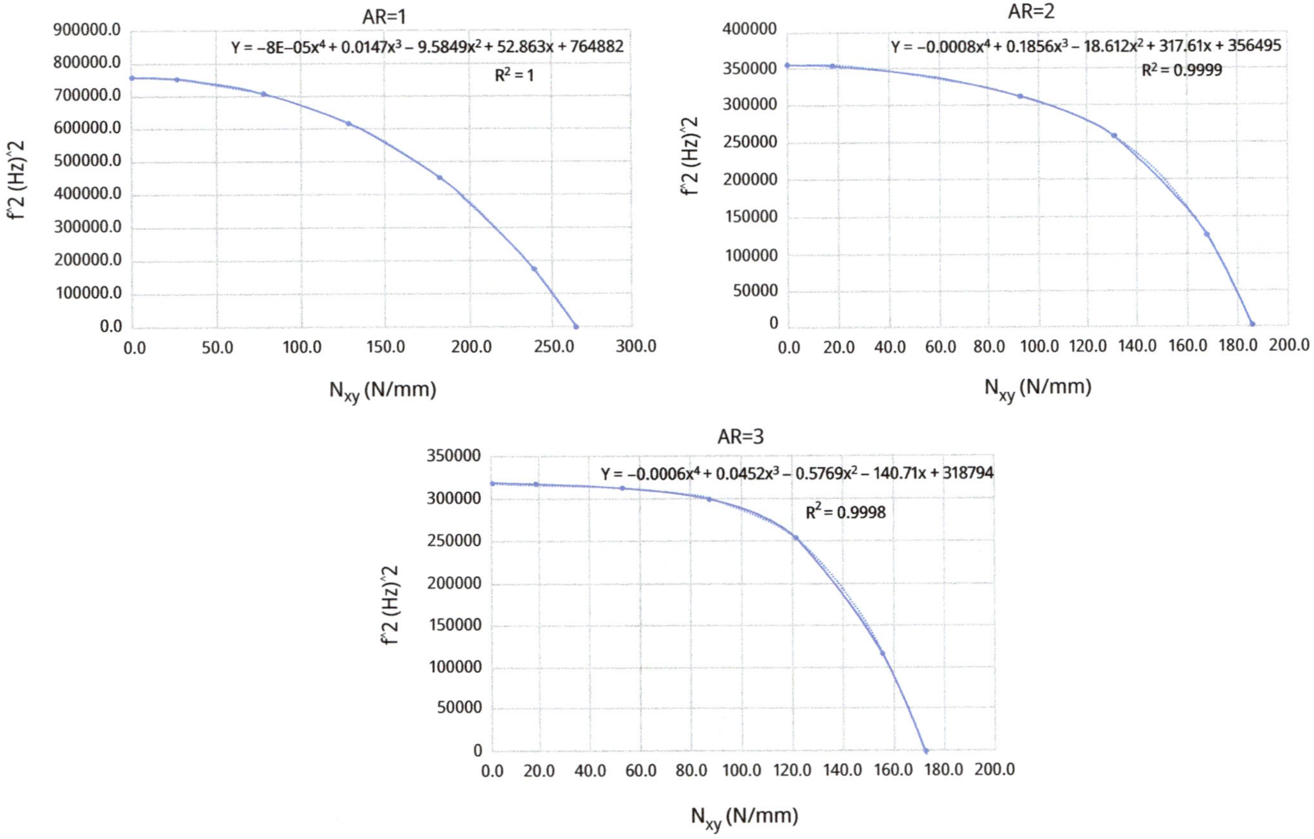

Fig. 11.15: Frequency squared versus applied shear load for various aspect ratios of the plate-all-around clamped boundary conditions – first mode.

Tab. 11.7: Numerical and predicted buckling loads for various aspect ratios and amount of data.

a/b	**1.0**	**2.0**	**3.0**
All-around simply supported conditions			
Numerical (FE) buckling force (N/mm)	168.20	118.40	105.60
Calculated buckling force using the polynomial equation (N/mm): $y = 5 \times 10^{-05}x^4 - 0.034x^3 - 3.2957x^2 - 76.517x + 229,630$ $R^2 = 0.9998$ (for all frequencies till 100% $N_{xy\ cr}$)	168.76		
Calculated buckling force using the polynomial equation (N/mm): $y = 6 \times 10^{-05}x^4 - 0.0253x^3 - 4.9142x^2 - 13.339x + 229,347$ $R^2 = 0.9999$ (for all frequencies till 70%*$N_{xy\ cr}$)	173.99		
Calculated buckling force using the polynomial equation (N/mm): $y = -7 \times 10^{-05}x^4 - 0.0015x^3 - 6.2917x^2 + 10.482x + 229,319$ $R^2 = 0.9998$ (for all frequencies till 60%*$N_{xy\ cr}$)	165.33		
Calculated buckling force using the polynomial equation (N/mm): $y = -0.0005x^4 + 0.0711x^3 - 9.8269x^2 + 61.38x + 229,284$ $R^2 = 0.9995$ (for all frequencies till 50%*$N_{xy\ cr}$)	120.69		
Calculated buckling force using the polynomial equation (N/mm): $y = -0.0004x^4 + 0.063x^3 - 9.2619x^2 + 68.721x + 119,650$ $R^2 = 0.9996$ (for all frequencies till 100% $N_{xy\ cr}$)			
Calculated buckling force using the polynomial equation (N/mm): $y = -0.0006x^4 + 0.0762x^3 - 9.1338x^2 + 51.072x + 119,732$ $R^2 = 0.9998$ (for all frequencies till 70%*$N_{xy\ cr}$)			
Calculated buckling force using the polynomial equation (N/mm): $y = 0.00018x^4 - 0.0496x^3 - 3.5888x^2 - 21.933x + 119,799$ $R^2 = 1$ (for all frequencies till 60%*$N_{xy\ cr}$)			
Calculated buckling force using the polynomial equation (N/mm): $y = -0.0004x^4 + 0.0338x^3 - 6.668x^2 + 11.72x + 119,782$ $R^2 = 1$ (for all frequencies till 50%*$N_{xy\ cr}$)			

Tab. 11.7 (continued)

a/b	1.0	2.0	3.0
Calculated buckling force using the polynomial equation (N/mm): $y = -0.0005x^4 + 0.0553x^3 - 6.5613x^2 + 28.774x + 89,792$ $R^2 = 0.9999$ (for all frequencies till 100% $N_{xy\ cr}$)		116.23	
Calculated buckling force using the polynomial equation (N/mm): $y = -0.0008x^4 + 0.1032x^3 - 8.6856x^2 + 56.624x + 89,765$ $R^2 = 0.9996$ (for all frequencies till 70%*$N_{xy\ cr}$)		114.42	
Calculated buckling force using the polynomial equation (N/mm): $y = -0.00006x^4 - 0.0712x^3 - 1.5764x^2 - 29.913x + 89,838$ $R^2 = 0.9999$ (for all frequencies till 60%*$N_{xy\ cr}$)		97.71	
Calculated buckling force using the polynomial equation (N/mm): $y = -0.0006x^4 + 0.0512x^3 - 5.7754x^2 + 12.644x + 89,818$ $R^2 = 1$ (for all frequencies till 50%*$N_{xy\ cr}$)		110.62	
Calculated buckling force using the polynomial equation (N/mm): $y = -0.0021x^4 + 0.2798x^3 - 14.418x^2 + 153.89x + 70,832$ $R^2 = 0.9991$ (for all frequencies till 100% $N_{xy\ cr}$)			103.99
Calculated buckling force using the polynomial equation (N/mm): $y = 6 \times 10^{-05}x^4 - 0.0253x^3 - 4.9142x^2 - 13.339x + 229,347$ $R^2 = 0.9999$ (for all frequencies till 70%*$N_{xy\ cr}$)			108.48
Calculated buckling force using the polynomial equation (N/mm): $y = -0.00009x^4 - 0.0866x^3 + 0.25x^2 - 27.883x + 71,040$ $R^2 = 0.9997$ (for all frequencies till 60%*$N_{xy\ cr}$)			90.645
Calculated buckling force using the polynomial equation (N/mm): $y = -0.0005x^4 + 0.0711x^3 - 9.8269x^2 + 61.38x + 229284$ $R^2 = 0.9995$ (for all frequencies till 50%*$N_{xy\ cr}$)			127.385
All-around clamped boundary conditions			
Numerical (FE) buckling force (N/mm)	265.2	185.6	172.7
Calculated buckling force using the polynomial equation (N/mm): $y = -8 \times 10^{-05}x^4 + 0.0147x^3 - 9.5849x^2 + 52.863x + 764,882$ $R^2 = 1.0000$ (for all frequencies till 100% $N_{xy\ cr}$)	263.1		

Tab. 11.7 (continued)

a/b	**1.0**	**2.0**	**3.0**
Calculated buckling force using the polynomial equation (N/mm): $y = -0.0008x^4 + 0.1856x^3 - 18.612x^2 + 317.61x + 356,495$ $R^2 = 0.9999$ (for all frequencies till 100% $N_{xy\ cr}$)		187.1	
Calculated buckling force using the polynomial equation (N/mm): $y = -0.0006x^4 + 0.0452x^3 - 0.5769x^2 - 140.71x + 318,794$ $R^2 = 0.9998$ (for all frequencies till 100% $N_{xy\ cr}$)			169.9

Tab. 11.8: Predicted buckling loads for various aspect ratios and their deviation from the numerical buckling load.

a/b	**Num. buckling load (N/mm)**	**Predicted buckling load (N/mm) (till 100% $N_{xy\ cr}$)**	**Predicted buckling load (N/mm) (till 70% $N_{xy\ cr}$)**	**Predicted buckling load (N/mm) (till 60% $N_{xy\ cr}$)**	**Predicted buckling load (N/mm) (till 50% $N_{xy\ cr}$)**
All-around simply supported boundary conditions					
1.0	168.20	168.76 (dev. = 0.33%)*	173.99 (dev. = 3.09%)	165.33 (dev. = −4.98%)	120.60 (dev. = −27.06%)
2.0	118.40	116.23 (dev. = −1.83%)	114.42 (dev. = −1.56%)	97.71 (dev. = 14.6%)	110.62 (dev. = 13.21%)
3.0	105.60	103.99 (dev. = −1.52%)	108.48 (dev. = 4.32%)	90.645 (dev. = 16.44%)	127.384 (dev. = 40.53%)
All-around clamped boundary conditions					
1.0	265.2	263.1 (dev. = −0.79%)	–	–	–
2.0	185.6	187.1 (dev. = 0.80%)	–	–	–
3.0	172.7	169.9 (dev. = −1.62%)	–	–	–

*The deviation (dev.) was calculated according to the following formula:

(Num. buckling load predicted buckling load/Num.buckling load)

A positive value of the deviation would mean that the predicted buckling load is higher than the numerical one, while negative values would mean that the predicted buckling load is lower than the numerical buckling load.

11.4 Conclusions

A research was initiated to investigate the changes in the natural frequencies of flat isotropic rectangular on all-around simply supported boundary conditions under increasing pure shear, till buckling. This case differs completely from the cases of plates under compressive loads for which the square of the natural frequency is inversely linear to the applied compressive loads. The frequency squared for a plate under increasing shear shows a non-linear behavior, which should be defined. The results of the investigation for flat plates with aspect ratios of AR = 1, 1.5, 2 and 3 show that:

- The relationship between the applied shear load and the frequency squared can be approximated using a fitted fourth-order polynomial equation, when using frequencies from zero till buckling. The deviation between the predicted buckling load and the calculated FE buckling load is very low and ranges from 0.33% to –1.83%.
- Using only part of the points, from zero to 70% $N_{xy\ \mathrm{cr}}$, and fitting a fourth-order polynomial equation, yields also good results with deviations ranging between –1.56% and 4.32%, which, from the engineering point of view, can be considered a good tool to predict buckling under pure shear.
- Trying to reduce the points used to fit a fourth-order polynomial equation up to 60% $N_{xy\ \mathrm{cr}}$ or 50% $N_{xy\ \mathrm{cr}}$ would predict buckling loads with large deviations (larger than 5%) from the numerical FE buckling load.
- It is recommended that the user would use points up to 70% $N_{xy\ \mathrm{cr}}$ to predict the pure shear buckling load of a flat isotropic rectangular plate on simply supported boundary conditions with a high level of confidence.
- It was observed that the first mode changes its shape during the increase in the shear load and rotates to yield the inclined buckling mode of the plate under shear.
- One has to remember that once geometric imperfections are taken into account, and the plate is no more flat, the equations of motion would change and the frequency squared versus the applied shear load relationship might be different. For plates under compressive loads, experiments show that the frequency squared does not diminish to zero at the buckling load. Instead, the reduction of the frequency squared changes its tendency at the buckling load and starts increasing its value. The load, at which the curve changes its tendency, would be defined as experimental buckling load. This behavior is expected to appear also in experiments on plates under pure shear loads as depicted in Fig. 11.16. The figure presents a square plate rotated by 45° under tensile load, *P*. This tensile load *P* transforms into shear forces *S*, which act on the specimen. Once the *P* force is known, the *S* shear force is easily

calculated to be equal to $0.707P$. The graph is taken from the ongoing study performed by the author of this book. Additional details can be found in [1].

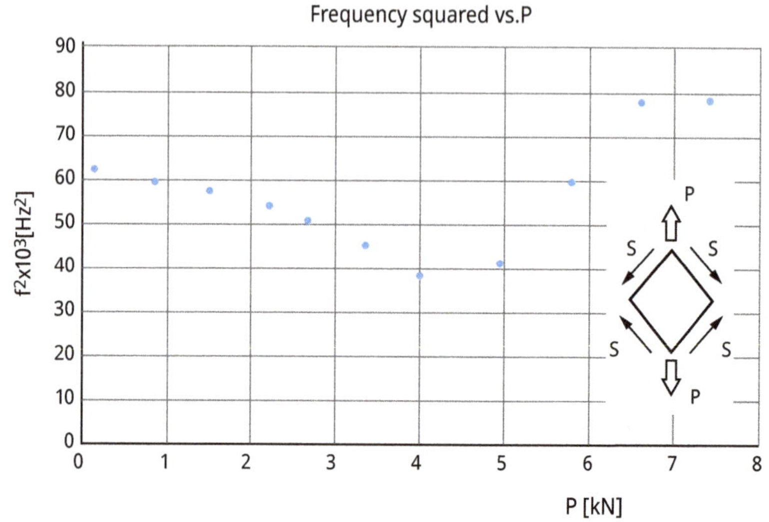

Fig. 11.16: Experimental first natural squared frequency versus applied tensile loads which transform into pure shear acting on the specimen.

References

[1] Abramovich, H. Ch. 5, in Advanced topics of thin walled structures, World Scientific Publishing Co. Pte. Ltd., 5 Toh Tuck Link, Singapore 596224, 2021, 404.

[2] Abramovich, H. The vibration correlation technique – a reliable nondestructive method to predict buckling loads of thin walled structures, Thin Walled Structures, 159, Feb. 2021, Paper Id: 107308, 17.

[3] Lurie, H. Effective end restraint of columns by frequency measurements, Journal of Aeronautical Science, 18, 1951, 556–557.

[4] Lurie, H. Lateral vibrations as related to structural stability, Journal of Applied Mechanics, 19 June 1952, 195–204.

[5] Bokaian, A. Natural frequencies of beams under tensile axial loads, Journal of Sound and Vibration, 142(3), 1990, 481–498.

[6] Bokaian, A. Natural frequencies of beams under compressive axial loads, Journal of Sound and Vibration, 126(1), 1988, 49–65.

[7] Dhanaraj and Palaninathan, R. Free vibration of initially stressed composite laminates, Journal of Sound and Vibration, 142(3), 1990, 365–378.

[8] Abramovich, H. Natural frequencies of timoshenko beams under compressive axial loads, Journal of Sound and Vibration, 157(1), 1992, 183–189.

[9] Abramovich, H. A new insight on vibrations and buckling of a cantilevered beam under a constant piezoelectric actuation, Composite Structure, 93, 2011, 1054–1057.

[10] Abramovich, H. Axial stiffness variation of thin walled laminated composite beams using piezoelectric Patches- a new experimental insight, Journal of Aeronautics and Aerospace Research, 3(2), 2016, 97–105.

[11] Johnson, E.E. and Goldhammer, B.F. The determination of the critical load of a column or stiffened panel in compression by the vibration Method, Proc. Society for Experimental Stress Analysis, 11(1), 1953, 221–232.

[12] Jacobson, M.J. and Wenner, M.L. Predicting buckling loads from vibration data, Experimental Mechanics, 8(10), Oct. 1968, 35N–38N.

[13] Coulter, B.A. and Miller, R.E. Vibration and buckling of beam-columns subjected to non-uniform axial loads, International Journal for Numerical Methods in Engineering, 23, 1986, 1739–1755.

[14] Lake, M.S. and Mikulas, M.M. Buckling and Vibration Analysis of a Simply Supported Column with a Piecewise Constant Cross, Section, NASA TP 3090, Mar. 1991, 18.

[15] Plaut, R.H. and Virgin, L.N. Use of frequency data to predict buckling, Journal of Engineering Mechanics, 116(10), 1990, 2330–2335.

[16] Virgin, L.N. and Plaut, R.H. Effect of axial load on forced vibrations of beams, Journal of Sound and Vibration, 168(3), 1993, 395–405.

[17] Virgin, L.N. Vibration of axially loaded structures, Cambridge University Press, Cambridge CB2 8RU, UK, 2007, 351, The Edinburgh Building.

[18] Chailleux, A., Hans, Y. and Verchery, G. Experimental study of the buckling of laminated composite columns and plates, International Journal of Mechanical Sciences, 17, 1975, 489–498.

[19] Carpinteri, A. and Paggi, M. A theoretical approach to the interaction between buckling and resonance instabilities, Journal of Engineering Mathematics, 78, 2013, 19–35, https://doi.org/10.1007/s10665-011-9478-0.

[20] Abramovich, H. Stability and vibrations of thin walled composite structures, © 2017 Elsevier Ltd, Woodhead Publishing Limited, the Officers' Mess Business Centre, Royston Road, Duxford, CB22 4QH, United Kingdom; 50 Hampshire Street, 5th Floor, Cambridge, MA 02139, United States; The Boulevard, Langford Lane, Kidlington, OX5 1GB, United Kingdom, 778.

[21] Massonnet, C. Le Voilement Des Plaques Planes Sollicitees Dans Leur Plan (Buckling of Plates), Final report of the 3rd Congress of the International Association for Bridge and Structural Engineering, Liege, Belgium, Sept. 1948, 291–300.

[22] Abramovich, H., Govich, D. and Grunwald, A. Buckling prediction of panels using the vibration correlation technique, Progress in Aerospace Sciences, 78, 2015, 62–73.

[23] Mandal, P. Prediction of buckling load from vibration measurements, in Drew, H.R. and Pellegrino, S. Eds., New approaches to structural mechanics, shells and biological structures. Solid mechanics and its applications, vol. 104, Springer, Dordrecht, 2002, https://doi.org/10.1007/978-94-015-9930-6_15.

[25] Singhatanadgid, P. and Sukajit, P. Experimental determination of the buckling load of rectangular plates using vibration correlation technique, Structural and Mechanical Engineering, 7(3), 2011, 331–349, https://doi.org/10.12989/sem.2011.37.3.331.

[26] Shahgholian-Ghahfarokhi, D., Aghaei-Ruzbahani, M. and Rahimi, G. Vibration correlation technique for the buckling load prediction of composite sandwich plates with Iso-grid cores, Thin-Walled Structures, 142, 2019, 392–404.

[27] Radhakrishnan, R. Prediction of buckling strengths of cylindrical shells from their natural frequencies, Earthquake, Engineering and Structural Dynamics, 2, 1973, 107–115.

[28] Abramovich, H. and Singer, J. Correlation between vibrations and buckling of cylindrical shells under external pressure and combined loading, Israel Journal of Technology, 16(1–2), 1978, 34–44.

[29] Singer, J. and Abramovich, H. Vibration correlation techniques for definition of practical boundary conditions in stiffened shells, AIAA Journal, 17(7), 1979, 762–769.

[30] Souza, M.A., Fok, W.C. and Walker, A.C. Review of experimental techniques for thin- walled structures liable to buckling. Part I- neutral and unstable buckling, Expedition Technology, 7(9), 1983, 21–25.
[31] Souza, M.A., Fok, W.C. and Walker, A.C. Review of experimental techniques for thin- walled structures liable to buckling, Part II- stable buckling, Expedition Technology, 7(10), 1983, 36–39.
[32] Souza, M.A. and Assaid, L.M.B. A new technique for the prediction of buckling loads from nondestructive vibration tests, Experimental Mechanics, 31(2), 1991, 93–97.
[33] Jansen, E.L., Abramovich, H. and Rolfes, R. The direct prediction of buckling loads of shells under axial compression using VCT – A revisited and upgraded approach, in: ICAS 2014, 29th Congress of the International Council of the Aeronautical Sciences, St. Petersburg, Russia, Sept 2014, 7–12.
[34] Arbelo, M.A., De Almeida, S.F.M., Donadon, M.V., Rett, S.R., Degenhardt, R., Castro, S.G.P. et al. Vibration correlation technique for the estimation of real boundary conditions and buckling load of unstiffened plates and cylindrical shells, Thin-Walled Structures, 79, 2014, 119–128.
[35] Arbelo, M.A., Kalnins, K., Ozolins, O., Skukis, E., Castro, S.G.P. and Degenhardt, R. Experimental and numerical estimation of buckling load on unstiffened cylindrical shells using a vibration correlation technique, Thin-Walled Structures, 94, 2015, 273–279.
[36] Kalnins, K., Arbelo, M.A., Ozolins, O., Skukis, E., Castro, S.G.P. and Degenhardt, R. Experimental nondestructive test for estimation of buckling load on unstiffened cylindrical shells using vibration correlation technique, Shock and Vibration, 2015, 2015, 8. Article Id. 729684.
[37] Skukis, E., Ozolins, O., Kalnins, K. and Arbelo, M.A. Experimental test for estimation of buckling load on unstiffened cylindrical shells by vibration correlation technique, Procedia Engineer, 172, 2017, 1023–1030.
[38] Skukis, E., Ozolins, O., Andersons, J., Kalnings, K. and Arbelo, M.A. Applicability of the vibration correlation technique for estimation of the buckling load in axial compression of cylindrical isotropic shells with and without circular cutouts, Shock and Vibration, 2017, Article Id. 2983747.
[39] Shahgholian-Ghahfarokhi, D. and Rahimi, G. Buckling load prediction of grid- stiffened composite cylindrical shells using the vibration correlation technique, compos, Science and Technology, 167, 2018, 470–481.
[40] Labans, E., Abramovich, H. and Bisagni, C. An experimental vibration-buckling investigation on classical and variable angle tow composite shells under axial compression, Journal of Sound and Vibration, 449, 2019, 315–329.
[41] Franzoni, F., Degenhardt, R., Albus, J. and Arbelo, M.A. Vibration correlation technique for predicting the buckling load of imperfection-sensitive isotropic cylindrical shells: An analytical and numerical verification, Thin-Walled Structures, 140, 2019, 236–247.
[42] Franzoni, F., Odermann, F., Labans, E., Bisagni, C., Arbelo, M.A. and Degenhardt, R. Experimental validation of the vibration correlation technique robustness to predict buckling of unstiffened composite cylindrical shells, Composite Structure, 2019, Article Id. 111107.
[43] Franzoni, F., Odermann, F., Wilckens, D., Kalnins, K., Arbelo, M.A. and Degenhardt, R. Assessing the axial buckling load of a pressurized orthotropic cylindrical shell through vibration correlation technique, Thin-Walled Structures, 137, 2019, 353–366.
[44] Shahgholian-Ghahfarokhi, D. and Rahimi, G.H. Prediction of the critical buckling load of stiffened composite cylindrical shells with lozenge grid based on the nonlinear vibration Analysis, Modares Mechanical Engineering, 18(4), 2018, 135–143.
[45] Bassily, S.F. and Dickinson, S.M. Bucking and vibration of in-plane loaded plates by a unified ritz approach, Journal of Sound and Vibration, 59, 1978, 1–14.
[46] Kennedy, D. and Lo, K.I. Critical buckling predictions for plates and stiffened panels from natural frequency measurements, Journal of Physics: Conference Series, 1106, 2018, 8, Article Id 012018.
[47] Bambill, D.V. and Rossit, C.A. Coupling between transverse vibrations and instability phenomena of plates subjected to in-plane loading, 2013, Journal of Engineering, 2013, 7. Article ID 937596.

[48] Srivastava, A.K.L. and Pandey, S.R. Effect of in-plane forces on frequency parameters, International Journal of Scientific and Research Publications, 2(6), 2012, 1–18.

[49] Carrera, E., Nali, P., Lecca, S. and Suave, M. Effects of in-plane loading on vibration of composite plates, Shock & Vibration, 19, 2012, 619–664.

[50] Stowell, E.Z. Critical shear stress of an infinite long flat plate with equal elastic restraints against rotation along the parallel edges, NACA Wartime Report ARR-3K12, Nov. 1943, Washington, USA, 32.

[51] Stein, M. and Neff, J. Buckling Stresses of Simply Supported Rectangular Flat Plates in Shear, NACA TN 1222, March 1947, Washington, USA, 13.

[52] Budiansky, B., Connor, R.W. and Stein, M. Buckling in Shear of Continuous Flat Plates, NACA TN 1565, April 1948, Washington, USA, 24.

[53] Budiansky, B. and Connor, R.W. Buckling Stresses of Clamped Rectangular Flat Plates in Shear, NACA TN 1559, May 1948, Washington, USA, 11.

[54] Peters, R.W. Buckling tests of flat rectangular plates under combined shear and longitudinal compression, NACA TN 1750, Nov. 1948, Washington, USA, 14.

[55] Buchert, K.P. Stability of ALCLAD Plates, NACA TN 1986, Dec. 1949, Washington, USA, 33.

[56] Warburton, G.B. The vibration of rectangular plates, Proceedings of the Institution of Mechanical Engineers, Series E, 168, 1954, 371–384.

[57] Gerard, G. and Becker, H. Handbook of Flat Plates- Part I – Buckling of Flat Plates, NACA TN 3781, Jul. 1957, Washington, USA, 102.

[58] Piscopo, V. Buckling analysis of rectangular plates under combined action of shear and uniaxial stresses, International Journal of Mechanical, Aerospace, Industrial, Mechatronic and Manufacturing Engineering, 4(10), 2010, 1010–1017.

[59] Zhu, Q. and Wang, X. Free vibration of thin isotropic and anisotropic rectangular plates by the discrete singular convolution algorithm, International Journal for Numerical Methods in Engineering, 86, 2011, 782–800.

[60] Weaver, P. and Nemeth, M.P. Improved design formulae for buckling of orthotropic plates under combined loading, AIAA Journal, 46(9), 2008, 2391–2396.

[61] Timoshenko, S.P. and Gere, J.M. Theory of elastic stability, Ch. 9, 2nd, International student edition, McGraw-Hill International Book Company, Singapore, 1985, 541.

[62] ANSYS Workbench 2020 R2 (Student version), 2020, www.ansys.com.

12 Morphing flying vehicles

12.1 Introduction

The topic of morphing flying vehicles has grown exponentially in the past two decades (2000–2020) yielding a big number of articles in the literature with typical references being listed in [1–25]. The word "morph" is said to come from the word "metamorphosis," which means in Greek "form" or "a transformation." Weisshaar [6] in his article claims that "morphing aircraft are also known as variable geometry or polymorphous aircraft."

When talking about aircraft, the various researchers were inspired by insects' wings structures (see as an example [21, 22]). Their main output was in the form of changing the leading and trailing edges of wing profiles, schematically presented in Fig. 12.1. The morphing process demands large structural deformations leading to stretching of membranes, flexural deformation in thin plates or using auxetic materials having positive Poisson's ratio [9]. Actuation for morphing is realized through an external force field or through embedded active materials that respond to an external stimulus like piezoelectric and/or shape memory alloys (SMAs) [1–7].

One has to remember that these morphing advance wings are still not implemented on real flying vehicles although many lab and flying of models were performed and published. As presented in various studies like [1, 4–6], the aim of continuously changing the wing shape was to optimize its performance at various stages of flight such as take-off, cruising, loitering and landing.

One has to remember that morphing wings is not a novel concept per se, as in fact as far back as the Wright brothers' glider; in 1903, a mechanical actuation was used to twist the wing of the aircraft, and thus, control was achieved during flight [26].

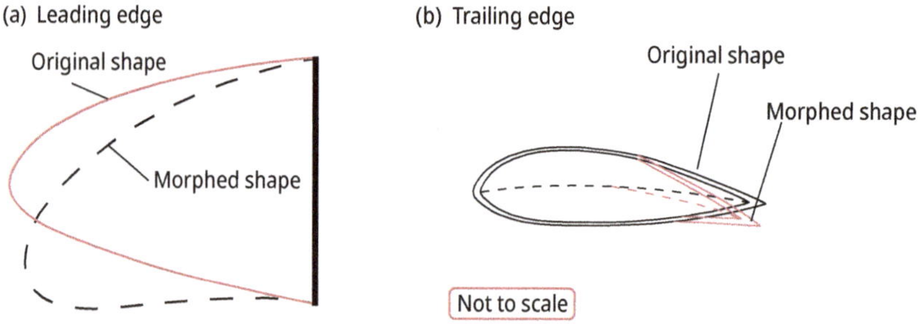

Fig. 12.1: Schematic drawings for morphing of leading and trailing edges of an airfoil.

Browsing the recent history reveals that morphing wings were primarily using the variable sweep concept as schematically presented in Fig. 12.2. The swept wing was proven beneficial at supersonic speeds; however, at subsonic flight it caused the ap-

https://doi.org/10.1515/9783111621104-012

pearance of span-wise flow along the wing, which reduced performance. The variable sweep type of morphing reduces the adverse effect of the swept wing configuration while flying in the less favorable state of subsonic flight by decreasing (or eliminating all together) the sweep angle. The outcome is an aircraft that has an optimized wing surface for several working points.

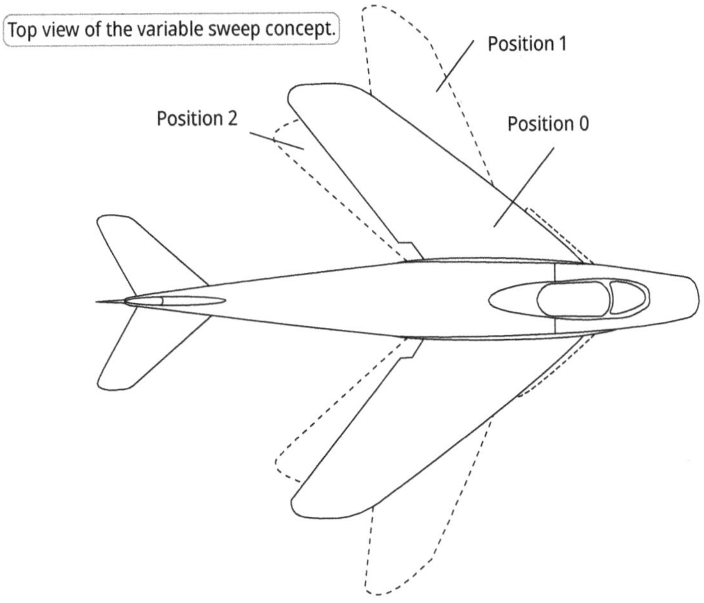

Fig. 12.2: Schematic drawing of the variable swept wing.

12.2 A new morphing concept

The morphing concept presented herein is composed of a hinged double-segmented half span, which can be mechanically folded in such a manner that the outbound wing is tucked underneath the inbound wing (as described in Fig. 12.3).[1] The resulting airfoil of the folded plane is the combination of the inbound airfoil and the inverse outbound airfoil (see Fig. 12.4). Thus, an alternating transition occurs from a high aspect ratio aircraft with a fitting airfoil (which will be referred to as "glider" throughout this chapter) into a high-velocity, improved maneuverability aircraft with a significantly shorter wing and a second, more favorable to high Reynolds airfoil (referred to as "aerobatic" throughout this chapter). One should remember that both aircrafts are motorized and the notation "glider" refers to the glider-like geometry of the wing.

1 The morphing concept is based on the study presented in [27].

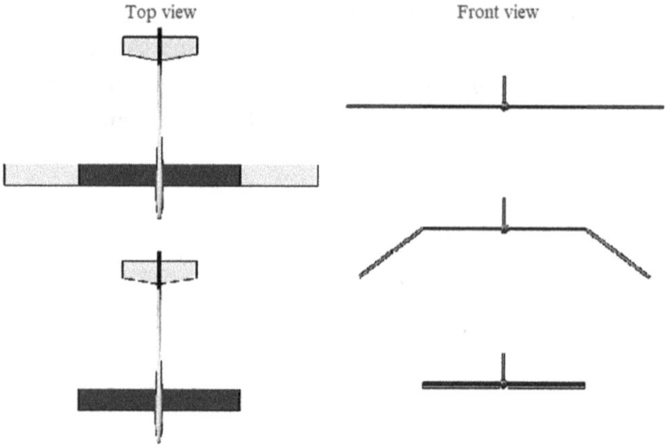

Fig. 12.3: Schematic drawing of the morphing concept: top view (left) and front view (right).

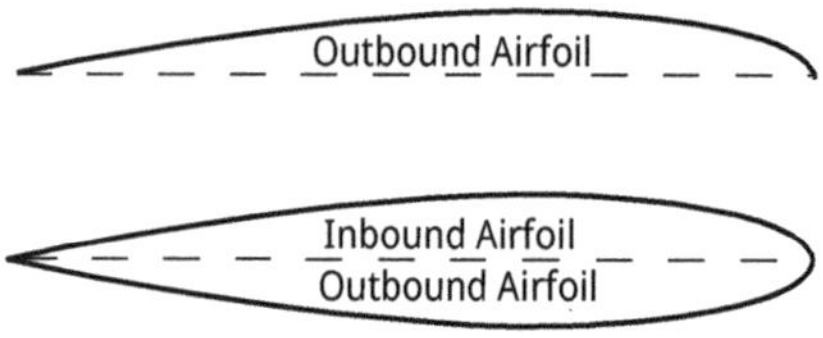

Fig. 12.4: Schematic drawing of the airfoil configuration of the glider (top) and the aerobatic phases (bottom).

The morphing concept enables the transition from the geometry of the "glider" to the "aerobatic" type, and their respective data are presented in Tab. 12.1.

Tab. 12.1: Aircraft main data.

	Glider (G)	**Aerobatic (A)**
Wingspan	b	$b/2$
Aspect ratio	Æ	Æ/2
Wing area	S	$S/2$
Wing loading	w/S	2 w/S
Airfoil thickness	t	(</ =) 2 t

The folding process influences the design of the control surfaces located on the wing (like flaps and ailerons). Thus, a simplification is assumed using elevons (which enables both elevator and aileron control) on the horizontal stabilizer.

The motivation of unfolding from the aerobatic configuration into the glider configuration is primarily rooted in endurance enhancement as will be described in the following section.

The motivation of folding from the glider configuration into the aerobatic configuration is due to maneuverability improvement in (and not limited to) both pitch and roll maneuver allowing for the ability to perform a swift nose-down or tight roll.

First, an analysis is performed to assess the additional endurance gained by the morphing capabilities compared to standard aircraft.

A suggested folding algorithm is then outlined, and certain aspects of the morphing process are highlighted such as the roll stability during the morphing process and the required elevon necessary to maintain such stability.

Finally, an airfoil selection process is detailed and wind tunnel test phase results will be presented.

12.2.1 Endurance improvement and morphing efficiency

In general, the process of designing an aircraft for two distinct working points is quite complex due to the fact that every parameter chosen for one working point greatly influences or even dictates the parameters of the second working point. The present proposed aircraft concept consists of solely two singular points, folded and retracted, that might not be airworthy for the intermediate angles beside the two extremes. This sub-chapter describes the parametric study aimed at pinpointing a viable and economically desirable solution according to specific flight mission, which may serve as a design methodology for future conceptual designs.

12.2.1.1 Mission's scenario

The design process will start with the examination of the designated flight mission, which is a given input parameter. We will define the term χ[2] as the overall part of the flight mission that the aircraft is required to perform at low velocity (or at glider configuration). Every flight mission is contained within $0 \le \chi \le 1$, where $\chi = 1$ represents a mission that relies solely on a low-velocity and $\chi = 0$ represents a mission that relies solely on a high-velocity working point (or as the aerobatic configuration). Thus, the portion of the flight mission that is performed at high velocity is $1 - \chi$.

In the present study, it was assumed that the required portions of the flight which are at low and high velocity (essentially defining χ) are either given to the designer or a direct result of the flight mission requirements. However, it is possible that the designer will extract that data from standard flight mission profiles as shown in Figs. 12.5 and 12.6. These figures represent flight mission profiles and their corresponding χ. For simplification, the path segments, which are required to perform at low velocity and would later be suited for the glider configuration (e.g., due to high aspect ratio or increased

2 The list of abbreviations is presented in Appendix A.

wing area), are marked in blue, while the equivalent high-velocity segments which would later be suited for the aerobatic configuration are marked in red.

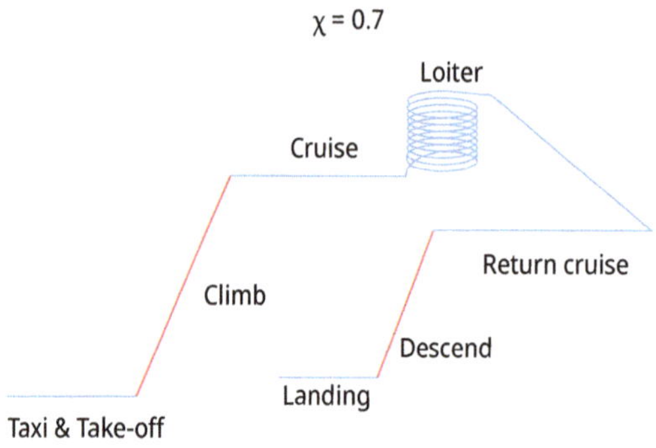

Fig. 12.5: Schematic mission profile of a surveillance UAV.

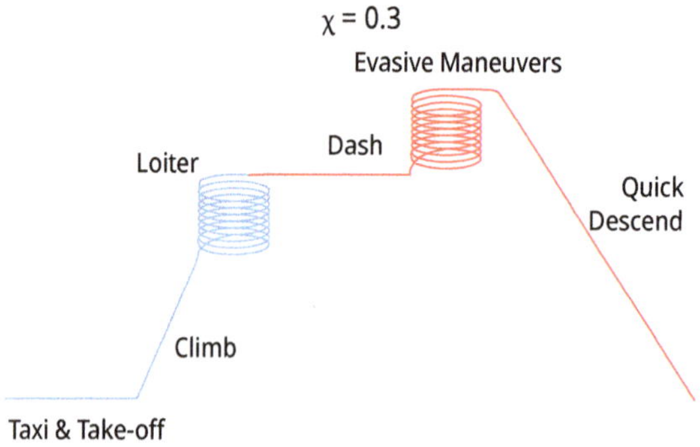

Fig. 12.6: Schematic mission profile of an attack plan.

12.2.1.2 Working points and morphing efficiency

For the two aircraft versions, presented in Tab. 12.1 and Figs. 12.3 and 12.4, the assumption is that during the low-velocity portion of the flight (χ phases) the aircraft is maintaining a leveled flight, while during the high-velocity and maneuvering portion of the flight ($1 - \chi$ phases), the aircraft is performing a maneuver, assumed to be equivalent to that of a steady climb. One should note that the maneuvering phase may also be assumed to be a steady vertical or horizontal maneuver as well as any number of plausible representations. However, as a preliminary study that aims at asserting the viability of the morphing concept, this is a reasonable assumption.

Therefore, under the above assumptions, the power required from the engine in the case of the first working point (low-velocity, leveled flight) is given by

$$P_{\text{leveled}} = D \cdot V \tag{12.1}$$

while the power required from the engine in the case of the second working point (high-velocity, steady climb) is given by

$$P_{\text{maneuver}} = D \cdot V + W \cdot V \cdot \sin \gamma \tag{12.2}$$

Then the overall power required to maintain a mission by any of the two aircraft types is given by

$$P_{*} = \chi \cdot P_{\text{leveled}} + (1-\chi) \cdot P_{\text{maneuver}} \tag{12.3}$$

The drag force and its associated drag coefficient will then be calculated to be

$$D = \frac{1}{2} \cdot C_D \cdot \rho \cdot V^2 \frac{b^2}{Æ}, \tag{12.4}$$

where

$$C_D = C_{\text{fe}} \frac{S_{\text{wet}} \cdot Æ}{b^2} + \frac{C_L^2}{\pi \cdot Æ} = C_{\text{fe}} \frac{S_{\text{wet}} \cdot Æ}{b^2} + \left[\frac{2\frac{W}{S}}{\rho \cdot V^2}\right]^2 \cdot \frac{1}{\pi \cdot Æ}$$

Using eqs. (12.1) and (12.4), the expected power versus velocity of the glider can be calculated for any combination of wing loading and aspect ratio for given values for the ρ, e[3] and C_{fe} terms. Since the parameters for the aerobatic configuration are defined using that of the glider, a second calculation was made for that aircraft as well. The results are presented in Fig. 12.7 for a specific wing loading of 420 N/m^2 and an aspect ratio of 32.2 for the glider configuration ($Æ = 16.1$ for the aerobatic configuration). The other required parameters were: $\rho = 1.225$ kg/m^3, $e = 0.8$, and the skin friction drag for the glider and aerobatic plane was taken as $C_{\text{fe}} = 0.005$ and $C_{\text{fe}} = 0.0055$, respectively.

The minimum point on each line represents the minimum required power, which, in theory, is the desired design point when striving for optimal endurance noted as "V_G" and "V_A."

As pointed out in [28], the conditions for maximum endurance can be written as

$$C_D^+ = 4 \cdot C_{D_0} \quad C_L^+ = \sqrt{\frac{3 \cdot C_{D_0}}{K}} \quad V^+ = \sqrt{\frac{2\frac{W}{S}}{3\frac{C_{D_0}}{K}}} \tag{12.5}$$

3 The Oswald efficiency is a correction factor that represents the change in drag with lift of a three-dimensional wing or airplane as compared with an ideal wing having the same aspect ratio and an elliptical lift distribution.

where

$$K = \frac{C_D - C_{D_0}}{C_L^2}$$

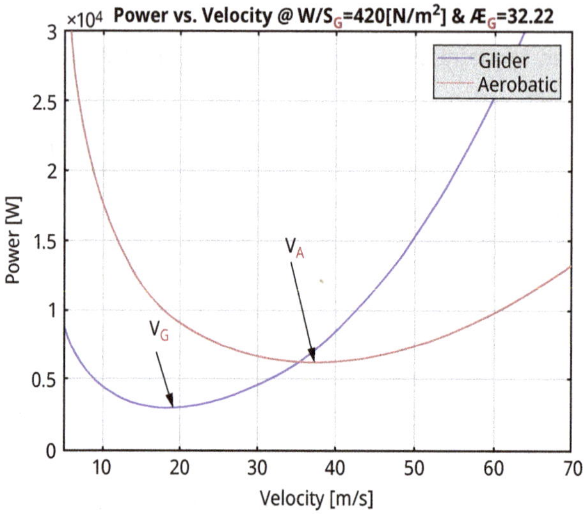

Fig. 12.7: Required power versus velocity for the glider and aerobatic configurations at a specific wing load of 420 N/m^2 and an aspect ratio of 32.2.

Notice that the endurance improvement is rooted in the ability of the morphing aircraft to "cherry pick" between the two configurations, which yields lower power requirement at each velocity. However, since the morphing mechanism is bound to correspond with additional weight, it is crucial that we quantify the advantages of such morphing aircraft over its fixed counterparts so that we could compare the two against each other and make an educated decision.

The added efficiency of the morphing concept over a standard glider-like aircraft is defined by eq. (12.6a), while the added efficiency of the morphing concept over the standard aerobatic-like aircraft is defined by eq. (12.6b). Note that the added efficiency is defined as a percentage of improvement in the power requirement:

$$\Psi_G = \frac{P_G - P_{\text{morphing}}}{P_G} \tag{12.6a}$$

$$\Psi_A = \frac{P_A - P_{\text{morphing}}}{P_A} \tag{12.6b}$$

One should note, as presented earlier, that the morphing concept has the potential of contributing toward minimizing the flight power requirements. In order to realize this potential, we must quantify this contribution and compare it against the weight increase due to morphing mechanisms. The algorithm of the analysis used to determine the added efficiency of the morphing aircraft (Ψ_G, Ψ_A) is schematically depicted in Fig. 12.8.

A few remarks regarding the block diagram are presented in Fig. 12.8:

- **Aspect ratio selection**: The initial input parameter for the analysis was chosen to be the aspect ratio. However, it is possible to start the analysis with any number of valid parameters, like wing area, χ and wing loading.
- **Drag assumptions**: Skin friction coefficient (C_{fe}) and Oswald efficiency number (e) were chosen from the experience of the authors. Wetted area was also assumed to be a function of wing area and airfoil thickness so that $S_{wet} = f(S, t)$; this assumption serves to simplify the drag calculation.
- **Drag calculation**: A wing-loading selection was made and using eq. (12.4) with the above input and assumptions yields drag data at specific wing loading and aspect ratio.
- **Minimum power velocity calculation**: The lift coefficient and drag force are estimated. The velocity at which the power requirement is minimal is calculated for the glider configuration, and accordingly, the minimum power requirement velocity of the aerobatic configuration is also computed.
- **Flight scenario impact**: A flight scenario is introduced into the analysis using χ and the morphing efficiency is calculated using eqs. (12.6a) and (12.6b). To obtain results at various wing loadings, an iteration loop is set from the drag calculation phase recalculating the efficiency at different wing loading values.

A typical graph of the morphing efficiency for a glider's aspect ratio of $Æ_G = 8$ and a flight scenario of $\chi = 0.6$ is presented in Fig. 12.9. Browsing Fig. 12.9, one can see that the morphing efficiency is quite constant above a minimal wing loading. Thus, for most applications, an average can be made, and a single value of efficiency per aspect ratio and flight scenario is obtained. The stagnation in efficiency (see Fig. 12.9) is because of the wing loading on the minimum power velocity, which serves as a control loop that maintains a steady value throughout the examined wing loading spectrum.

- **Final morphing efficiency calculation**: Given a single value for the efficiency at every χ, an iteration loop is set from the aspect ratio selection (initial phase) calculating of the morphing efficiency at different flight scenarios for a given spectrum of aspect ratios. Figures 12.10 and 12.11 depict the results of the analysis for the morphing aircraft efficiency over the standard glider and aerobatic aircraft, respectively. Each line represents a specific χ, and it is apparent that the flight scenario and aspect ratio are the main factors in determining the benefit of the present morphing concept. As expected, for given flight scenarios that are defined with a larger value of χ (flight scenarios with increased phases of steady flight such as loiter and others) the morphing efficiency or the "gain" from morphing over a standard glider is less prominent (see Fig. 12.10). In a similar fashion, for flight scenarios that are defined with the same larger value of χ the morphing efficiency or the "gain" from morphing over aerobatic aircraft is increased (see Fig. 12.11).

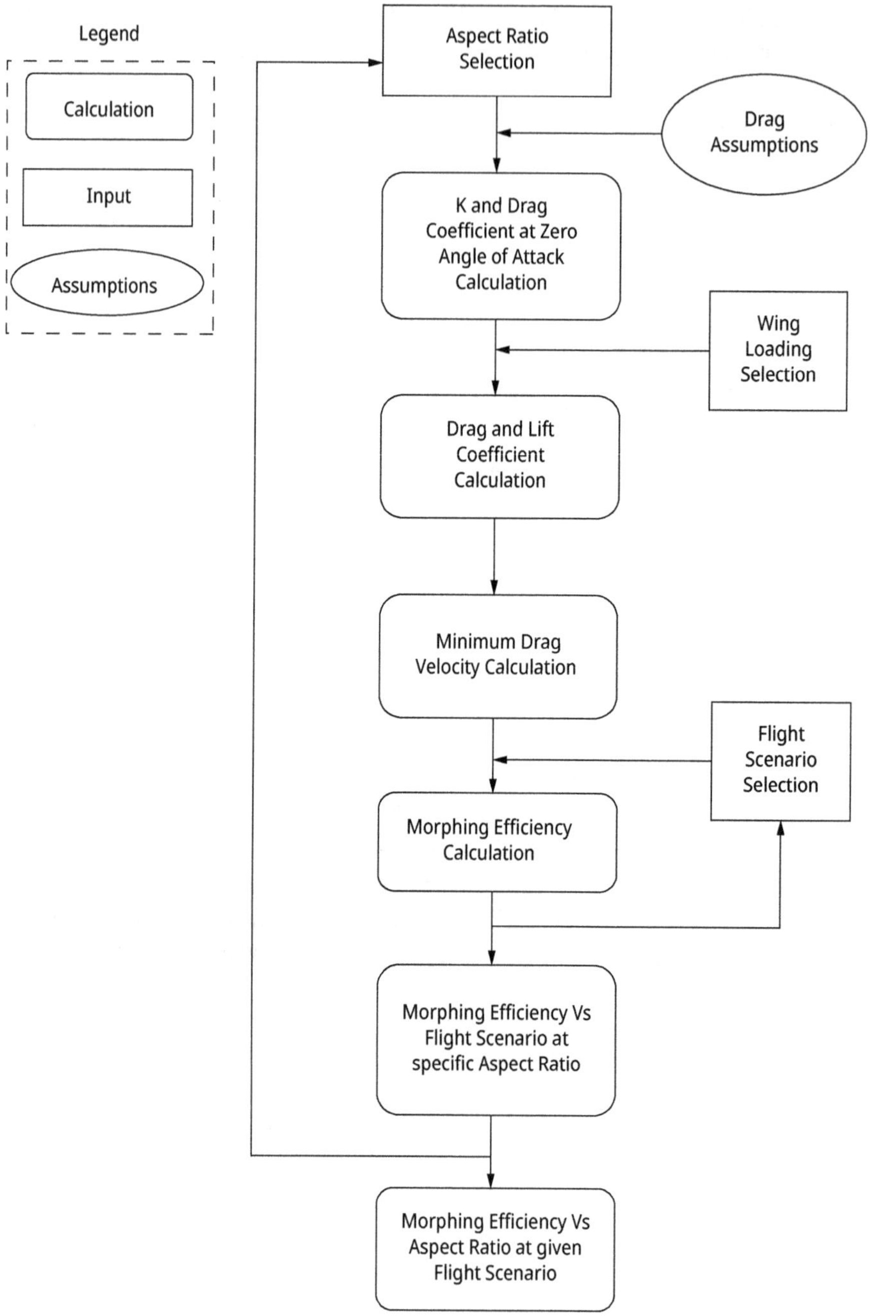

Fig. 12.8: Block diagram for the various morphing efficiency analysis phases.

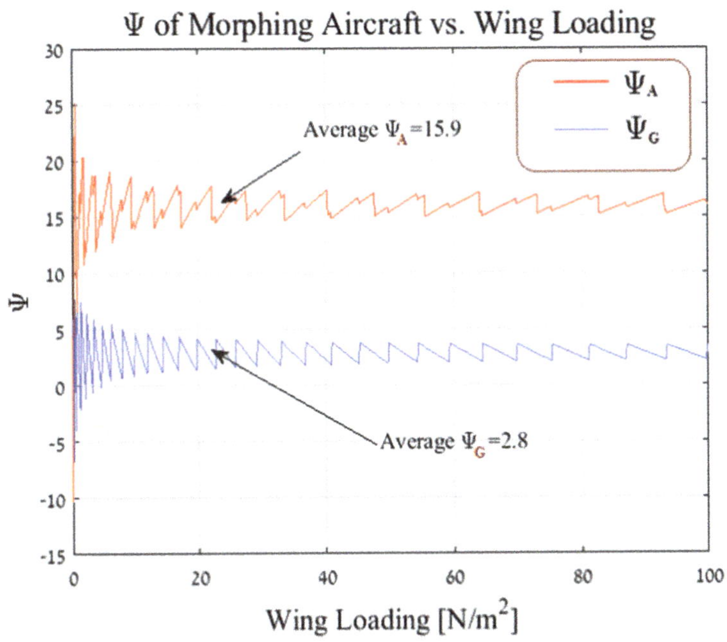

Fig. 12.9: Morphing efficiency Ψ versus wing loading at $Æ_G = 8$ and $\chi = 0.6$.

Furthermore, it is visible that for this typical example, morphing efficiency Ψ_A is significantly higher than Ψ_G, meaning that the power requirements are substantially lower for the morphing aircraft over its standard aerobatic counterpart. This is due to the behavior of the power versus velocity as is presented in Fig. 12.7. During flight phases where the standard aerobatic aircraft is required to maintain the glider's velocity (V_G), the power is significantly increased; subsequently, the overall efficiency is increased. However, during flight phases where the standard glider is required to maintain the aerobatic aircraft's velocity (V_A), the power is indeed increased however not by the same magnitude, leading to less "profit" for the morphing concept and a lower morphing efficiency.

Note that this conclusion is not an absolute one and is bound to the assumptions made during the analysis pertaining to the drag and wetted surface functions.

12.2.1.3 Weight considerations

The additional weight due to the morphing mechanism has an adverse effect on the morphing efficiency. Since the morphing aircraft is complicated and costly to design compared with a standard aircraft, a designer may view a severe negative effect on the morphing efficiency as a deterrent to the specific design. Thus, an analysis was performed to quantify this degradation in the morphing efficiency.

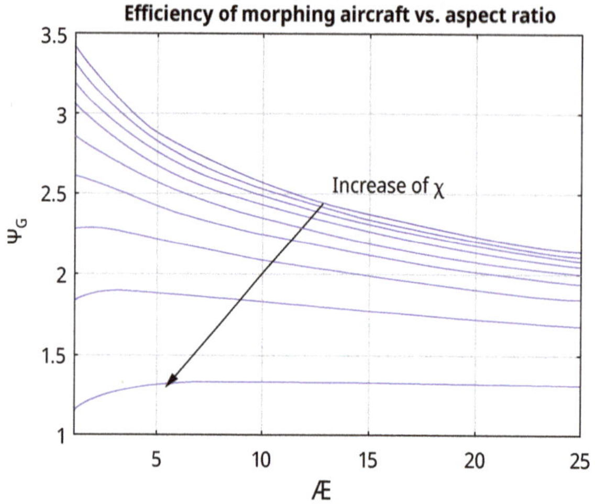

Fig. 12.10: Morphing efficiency, Ψ, over standard glider versus aspect ratio, Æ.

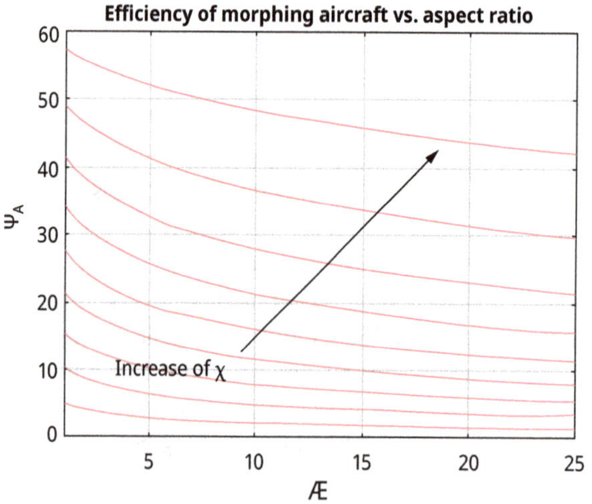

Fig. 12.11: Morphing efficiency, Ψ, over standard aerobatic aircraft versus aspect ratio, Æ.

The analysis presented so far took into consideration the aircraft's weight within the selection of the wing loading. The aircraft weight did not contain added weight due to morphing, meaning that the efficiencies calculated represented an ideal case. Furthermore, it was shown that for a given aspect ratio the change in wing loading did not alter the morphing efficiency, and the latter was subsequently averaged throughout the analysis. The weight of the morphing mechanism was calculated as a percentage of the standard glider's weight and noted dW_{mm}. This term includes all mechanisms and structural parts that were added to gain the ability to morph either directly or indirectly. Thus, the weight of the morphing aircraft can be represented as a function of the standard aircraft weight and the added mass as follows:

$$W_M = W \cdot (1 + dM_{\text{mm}}) \tag{12.7}$$

Similarly, the wing loading of the morphing aircraft can be written as

$$\frac{W_M}{S} = \frac{W}{S} \cdot (1 + dM_{\text{mm}}) \tag{12.8}$$

For each flight scenario χ, a morphing efficiency calculation was performed for a spectrum of aspect ratios in the same manner that was previously presented. The weight of the morphing mechanism was initially set as zero, and the minimum drag velocity for the ideal case was calculated. The analysis was then repeated for a variety of morphing mechanism weights; however, the velocity at which the original aircraft is most favorable was unaltered to compare between the configurations. Figures 12.12 and 12.13 present the result of the analysis with the same assumptions used in the analysis presented in Section 12.2. As expected, the weight of the mechanism has a significant effect on morphing efficiency; a 5% increase is translated into a reduction of approximately 20% in the aerobatic morphing efficiency (see Fig. 12.12). The decrease in efficiency progresses with additional mechanism weight and at a certain value of dW_{mm} will present a negative efficiency as shown in Fig. 12.13, resulting in a specific design that would have shorter endurance compared with a glider that performs the original flight scenario without morphing.

The task of defining the morphing mechanism weight is not a simple one, albeit certain mechanisms such as motors or wiring can be pinpointed to the morphing concept; however, structural components such as flexible skin, spars or folding ribs perform tasks, which are mutual to both the morphing and standard structural design. In these cases, the prospective designer will need to assess the weight of a standard structure as well as that of the morphing one in order to identify the increment. This means that the prospective designer may need to perform a conceptual design for the glider and aerobatic aircraft prior to designing the morphing one. However, this is a somewhat intuitive preluding phase of the morphing conceptual design.

In the presented results, the morphing efficiency's decrease is significant while still maintaining a sizeable improvement compared to the aerobatic standard aircraft (see Fig. 12.12). Therefore, the designer would need to choose an efficiency above which the morphing is still favorable over the standard counterparts. Furthermore, Figs. 12.12 and 12.13 depict maximum additional weight of 5% of the overall aircraft weight. Therefore, additional studies are needed to ratify the validity of this assumption.

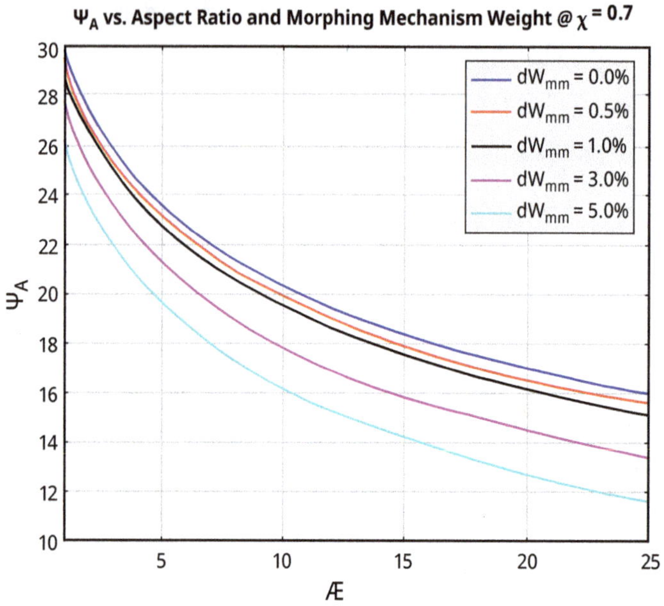

Fig. 12.12: Ψ_A versus aspect ratio, Æ, for several morphing mechanisms weighed at a flight scenario $\chi = 0.7$.

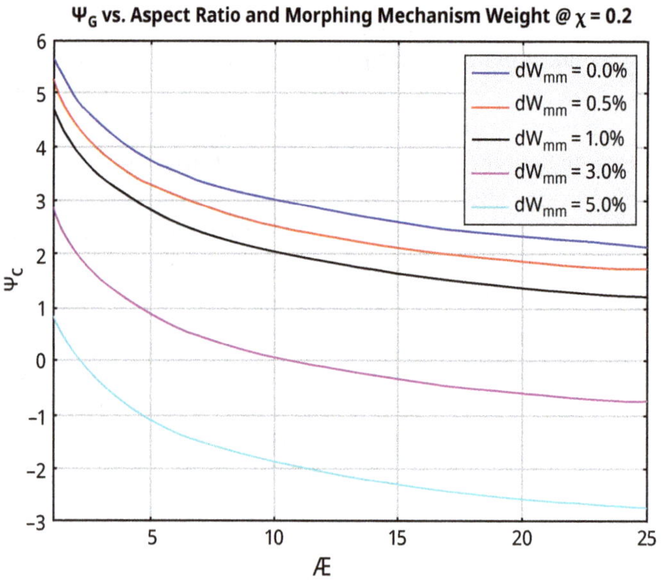

Fig. 12.13: Ψ_G versus aspect ratio, Æ, for several morphing mechanisms weighed at a flight scenario $\chi = 0.2$.

12.2.2 Morphing algorithm

During the transition from one working point to the other, the wingspan, aspect ratio and lift surfaces are substantially altered. Additionally, the direction of the lift vector in the outbound surfaces undergoes a 180 °change during the folding or unfolding process.

Provided a purely symmetrical folding phase is performed, the horizontal component of the lift vectors from the two outbound wing sections works to negate each other and we are left to deal with the change in the vertical components. Throughout the folding process, the overall lift of the aircraft is reduced with correlation to the magnitude of the folding. Theoretically, given an angle of attack of $\alpha = 0°$ and equal outbound and inbound lift surfaces, the lift at the point of completed folding phase is $L = 0$ due to the nature of symmetric airfoils at $\alpha = 0°$.

Thus, if lift is desired at the end of the folding process, a positive angle of attack must be maintained during the process. Alternatively, a nonsymmetric airfoil may be chosen for the folding configuration. However, this option carries a complication penalty to be described in detail in Section 12.2.3. A third option may include the usage of a smaller outbound wing surface compared to the inbound surface, in a manner that would serve to maintain some effective lift surface post-folding. However, this approach reduces the overall level of morphing and thus can be considered as counter-efficient. Furthermore, reality dictates that the transition will not be completely symmetric as presumed earlier. Hence, a roll moment (as well as a yaw moment) may be introduced due to differences in the horizontal component of the lift vector in both outbound sections. Consequently, a correction is required using the elevons at the tail. It is possible to achieve the same roll control using ailerons/airbrakes at the inbound (static) portion of the wing as well as to use ailerons at the outbound section of the wing. However, the latter necessitates a complicated control algorithm, and both of the aileron-based solutions would present a mechanical complexity to the already challenging wing design.

The roll moment due to asymmetry, the ability to control said roll and the lift change from one configuration to the next are all affected by the velocity of flight. While it is intuitive that the folding (or unfolding) will take place at either V_G or V_A (the velocity of the previous state or the velocity of the following state), it is possible that prior to transition, the velocity will be set to a separate value that would allow for efficient transition accompanied by an increase or decrease to either V_G or V_A.

One plausible solution may be to reduce the velocity below V_{stall} and thus reduce the effect of asymmetry and lift change. Theoretically speaking, at $V = 0$ m/s the issue is eliminated altogether with the drawback of losing the use of the control surfaces and overall lift during that period. This may be acceptable for UAV, given a relatively swift transition and sufficient altitude. Given the complications listed above, it is imperative that the transition phase is done swiftly and with a high level of symmetry. Figure 12.14 depicts the proposed overall transition algorithm.

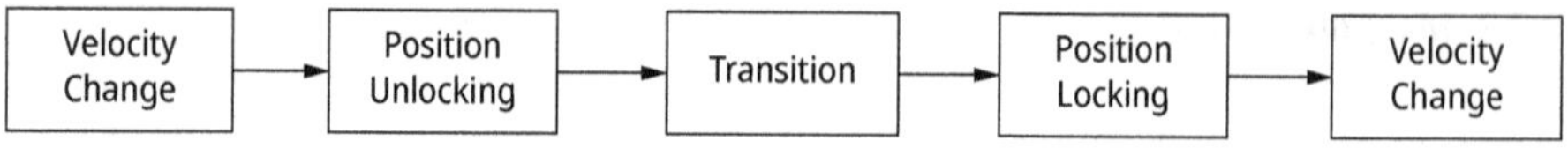

Fig. 12.14: The proposed morphing algorithm.

It is visible from Fig. 12.14 that the folding and unfolding are preceded and then succeeded by an unlocking and a locking phase. A locking mechanism is necessary to maintain the two end geometries, regardless of the aircraft's position and forces that act upon it. This locking mechanism can be implemented within the transition mechanism as is presented in most control surfaces, such a constant hydraulic pressure or continuous power supply to a motor as in the case of electric servo-controlled surfaces. However, the high level of modification introduced to the structure in this case means that vast forces are required to maintain the end positions (as high as one half of the overall lift of the wing) as opposed to most moving mechanisms in the aircraft.

The transition (or morphing) phase is dissimilar when discussing folding compared to unfolding. A mechanism designed to fold must counter the lift of the outbound surface in order to create movement, the same forces during the opposite process will assist the mechanism, and the lift may cause the surfaces to unfold on their own. In this case, it is possible that the mechanism will only have to control the timing of the transition so that the symmetry is maintained. A possible solution is to perform the folding phase at low velocity, such that the counting forces on the mechanism are negligible, while the unfolding phase is done at high velocity so that the wing attempts to unfold itself. This would allow for a relatively swift morphing for a low-powered morphing mechanism.

12.2.2.1 Roll stability during the morphing process

As stated in the previous paragraph, it is quite probable that the folding process will not be completely symmetrical; hence, a roll moment is expected. In order to keep the proposed morphing concept feasible, an analysis was performed to ensure that the aircraft can perform a roll-wise stable folding maneuver.

First, an extreme case scenario is defined. The folding process can be discussed as the forming of an anhedral[4] at the half span of each wing, noted by the angle Γ. As the folding process proceeds, the anhedral is increased until a complete folded state is achieved; thus, the anhedral angle is of the range of $0° < \Gamma < 180°$.

The roll moment around the center of gravity of a single half wing (inbound and outbound section) is given in eq. (12.9). For the sake of simplicity, the center of gravity is assumed to be in mid-span of the wing. Furthermore, the roll moment due to the lift generated by the static inbound wing section was omitted since it would be canceled when the second half of the wing is factored in

4 Anhedral means a downward inclination of an aircraft's wing.

$$\mathcal{L}_{\text{Anhedral}} = L \cdot \cos(\Gamma) \cdot \left[\frac{b}{4} + \frac{b}{8} \cdot \cos(\Gamma)\right] + L \cdot \sin(\Gamma) \cdot \left[\frac{b}{8} \cdot \sin(\Gamma)\right] \tag{12.9}$$

The roll moment was calculated for a complete folding process of one-half wing. An assumption is made as to a difference in the anhedral angle of the second folding wing, such as that it is lagging by a $\Delta\Gamma$ after the first anhedral. Figure 12.15 depicts the roll moment at the center of gravity for both anhedrals, as calculated for a UAV with a wingspan of $b = 2$ m, a weight of $W = 8$ kg and a lift coefficient of $C_L = 0.7$.

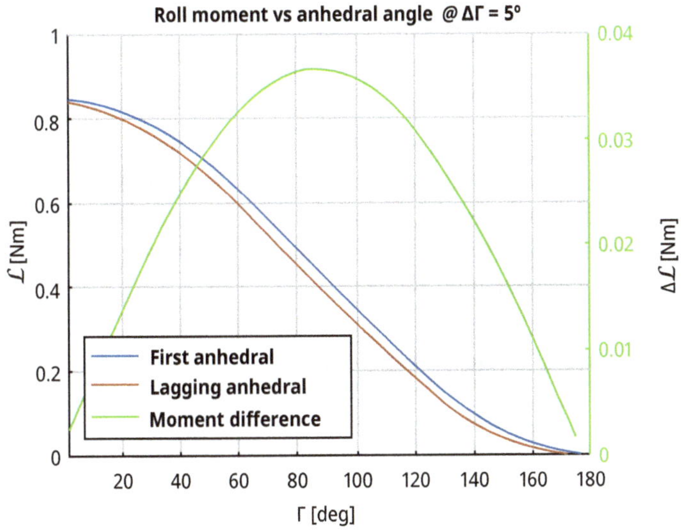

Fig. 12.15: Roll moment of a folding wing and a second lagging wing versus angle of anhedral.

It can be inferred from Fig. 12.15 that the maximum difference in the roll moment is given for $\Gamma \approx 90°$, meaning that the most extreme scenario is when the first folding segment reaches the point of a half fold (outbound wing segment is pointing downward). However, this result is plotted for the specific case of $\Delta\Gamma = 5°$, the conclusion as to the position of the wing segments at extreme roll moment would differ when the lag is increased. The anhedral angle of maximum roll is decreased with the increase of the lag between the two wings. In the present study, the actual moment difference between the two wings is used for further calculations. Nonetheless, for an anhedral lag of $\Delta\Gamma < 15°$, it is a fair assumption to define the maximum roll moment position as the downward facing position of the outbound wing section ($\Gamma = 90°$).

To sustain a steady flight, the roll moment caused by the lag in the folding wings needs to be reconciled by the elevons, the equilibrium being expressed as follows:

$$\mathcal{L}_{\text{Anhedral}_{\text{1st}}} - \mathcal{L}_{\text{Anhedral}_{\text{lag}}} = \frac{1}{2} \cdot \rho \cdot V^2 \cdot S_H \cdot b_H \cdot C_{\mathcal{L}_{\delta a}} \cdot \delta a \tag{12.10}$$

Substituting eq. (12.9) into eq. (12.10) yields

$$\frac{b}{8} \cdot L \cdot [2 \cdot \cos(\Gamma_{\text{1st}}) + 1] - \frac{b}{8} \cdot L \cdot \left[2 \cdot \cos\left(\Gamma_{\text{lag}}\right) + 1\right] = \frac{1}{2} \cdot \rho \cdot V^2 \cdot S_H \cdot b_H \cdot C_{\mathcal{L}_{\delta a}} \cdot \delta a \tag{12.11}$$

The last equation can be further simplified, and the required deflection angle can be calculated as a function of the aircraft geometry and the wings anhedral angles as follows:

$$\delta a = \frac{1}{4} \cdot \frac{b}{b_H} \cdot \frac{S}{S_H} \cdot \frac{C_L}{C_{\mathcal{L}_{\delta a}}} \left[\cos(\Gamma_{\text{1st}}) - \cos\left(\Gamma_{\text{lag}}\right)\right] \tag{12.12}$$

An assumption was made as to the wing to horizontal tail span ratio and area ratio as well as an assumption regarding the airfoil selection for the two surfaces. Figure 12.16 depicts the required elevon[5] deflection angle versus a range of anhedral lags, assuming that the first wing segment is positioned at the maximum roll angle. The calculation was repeated for several horizontal tail volumes.

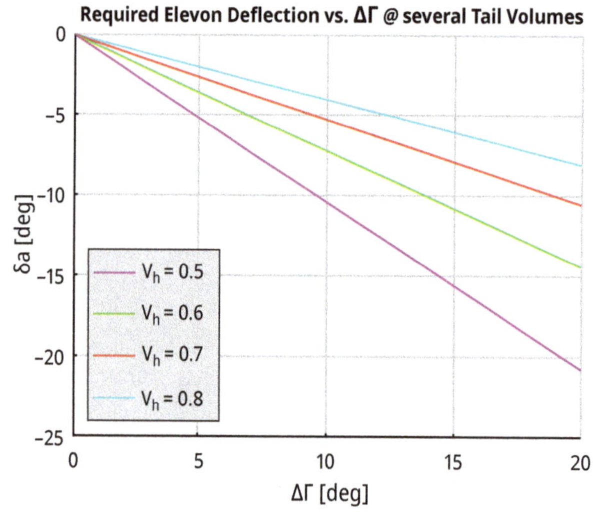

Fig. 12.16: Elevon deflection angle versus ΔΓ for several horizontal tail volumes.

5 Elevons are aircraft control surfaces that combine the functions of the elevator (used for pitch control) and the aileron (used for roll control), hence the name.

It can be deduced from Fig. 12.16 that for a conservative horizontal tail volume of a glider $V_H = 0.5$, the required elevon angle is quite reasonable for a wide range of $\Delta\Gamma$. This means that a roll-wise stable flight is possible. Moreover, the trade-off between the quality of the mechanical solution which limits the lag and the horizontal tail volume can be achieved and factored in for future design.

12.2.3 Airfoil selection

12.2.3.1 Dominant aerobatic aircraft working point

If we were to designate the aerobatic aircraft as the dominant or the "original," upon retraction, a knifed edge of some sort would be generated in the leading edge of the glider as portrayed in Fig. 12.17. The figure presents several airfoils that were considered for high-velocity working point and the result of unfolding them.

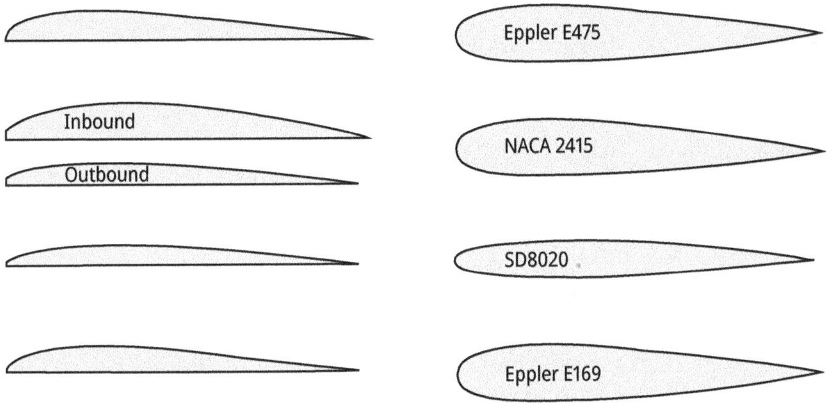

Fig. 12.17: Airfoils samples, originally selected for high velocity (right) and their resulting airfoil upon unfolding (left).

For purely symmetric airfoils (i.e., Eppler 475), the resulting unfolding airfoil is identical along the entire unfolded surface of the wing and is an exact geometrical half of the retracted airfoil. However, for the other cases, the result would be a dissimilar airfoil for the inbound (or "static") wing surface and the outbound (or "retractable") wing surface (i.e., NACA 2,415). This attribute might be used toward an improvement of the aerodynamic properties of the wing; however, this would not be easy to design.

The sharp-edged airfoils stand in contrast to "traditional" smooth and round leading-edge concepts and as such are quite unintuitive.

12.2.3.2 Dominant glider aircraft working point

If a glider configuration is defined as the dominant working point, upon folding, a gap would appear at the leading edge of the aerobatic configuration. Figure 12.18 illustrates several of the ideas considered. One should also note that concepts where the fixed and folding surfaces contain dissimilar airfoils are not presented. However, the following discussion can shed light on these concepts as well.

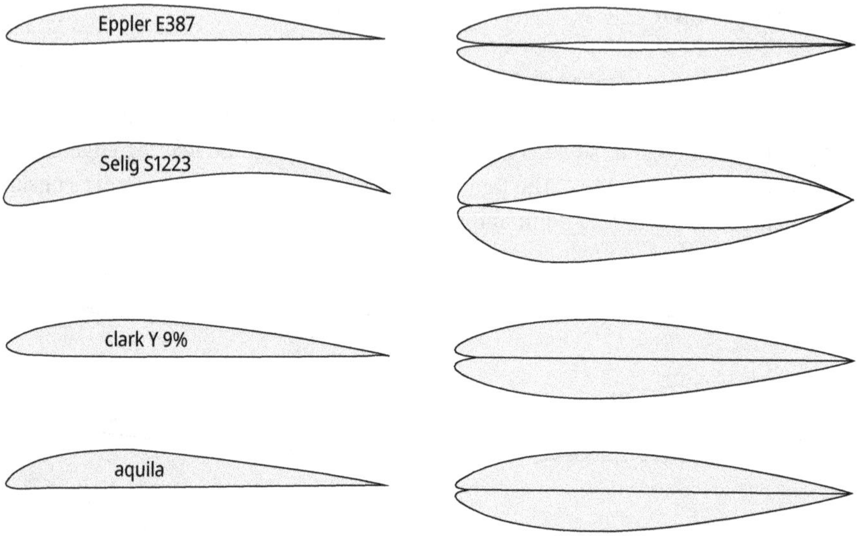

Fig. 12.18: Airfoils samples, originally selected for glider aircraft (left) and their resulting airfoil upon unfolding (right).

The chord-wise length of the gap, which is a defining character of the original airfoil, was denoted by the term Δ and presented in percentage of the chord. For example, the value $\Delta = 10.8$ implies a gap that is protruding into the front 10.8% of the aerobatic airfoil.

Initial attempts to analyze these airfoils using standard airfoil analysis codes were deemed to be futile, as the gap produced either nonconvergence or outrageous aerodynamic characteristics.

Common sense would dictate that the gap would have some adverse effect on the aerodynamic qualities of the airfoil in comparison to that of a standard leading edge. Two methods of improving the gaped airfoil were suggested:

1. Surface panel

According to this concept, a single panel on the lower or upper surface of the wing would retract and be pushed forward in a manner that would eliminate the gap. This concept can be realized by either two methods; either two panels from both the static and folding wings are used to close the gap or a single panel originating from the static wing would solely perform the gap closing action. The latter method serves to reduce mechanism weight at the wing tips. The surface concept, if implemented, would result in a new airfoil, in which the leading-edge gap is sealed from the airflow and can be visualized (in the most general way) as an airfoil with a single panel that connects the forward most points of the upper and lower surface (see Fig. 12.19). Several airfoils were compared to an equivalent NACA 4-digit approximation airfoil. This was done in order to evaluate the adverse effect of the leading-edge gap. This comparison might be considered fair since the NACA 4 series airfoils represent improvements to the shape of the leading edge with minimal effect to the remainder of the chordwise geometry of the airfoil and thus can alleviate some analysis complications for early-stage design. Note that the closing of the leading edge via a NACA 4-digit approximation is valid for preliminary design and is likely to give a good indication for the mechanically closed gap as well as the original opened gap airfoil. As for the deeper gaped airfoils, the approximation is not as productive. It poorly predicts the mechanically closed gap and is further intuitive that these approximations will not correlate with the opened gap airfoil.

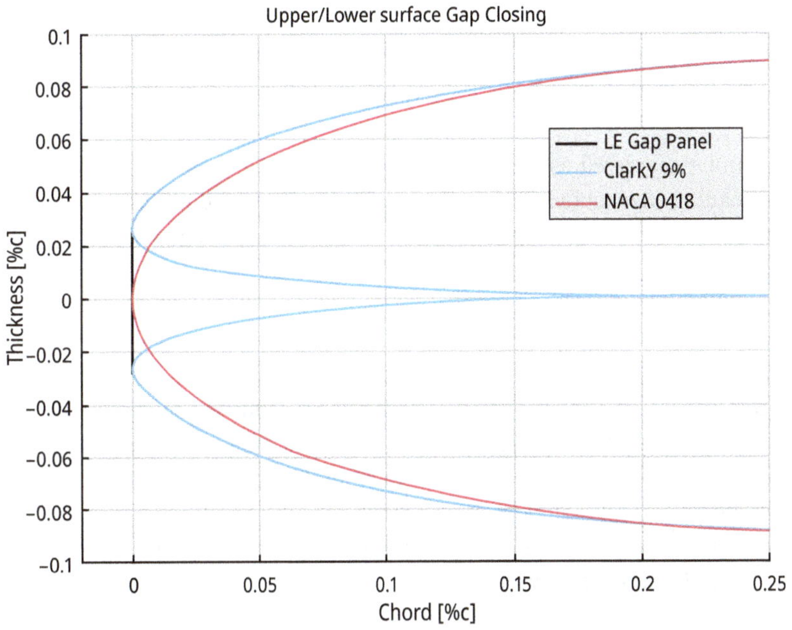

Fig. 12.19: Closing of a ClarkY 9% airfoil using an upper or lower section and a NACA 4-digit approximation.

2. Thickness reduction

Thickness reduction is an amending concept, which eliminates the gap entirely by means of allowing the two wing sections to contract along their lower surface toward one another. This would be achieved by a flexible skin at the front lower surface, which allows that two wing sections to merge into a slicker gap-less airfoil. The contracting form used for this procedure is based on two contact points between the two lower surfaces of the mirrored airfoil: an aft contact point at the trailing edge and a forward contact point at the origin of the gap (see Fig. 12.20).

Let us denote the angle of retraction as the angle between the chord line and the symmetry plane of the two airfoils with the symbol β (see Fig. 12.20) so that the angle between the two chords lines of the airfoils is 2β.

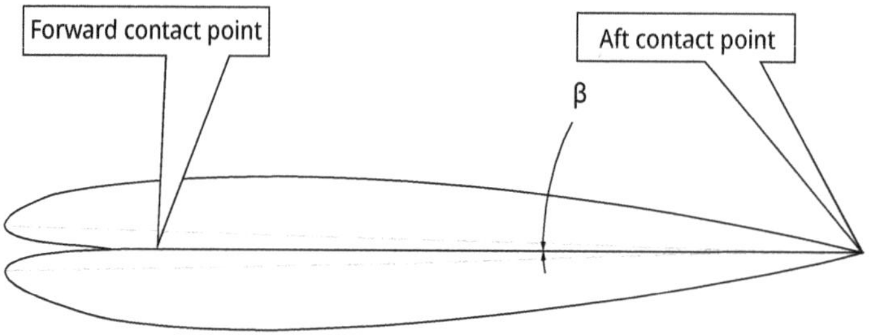

Fig. 12.20: Definition of parameters for thickness reduction method.

The gap also can be expressed as a function of the distance between the symmetrical plane and the leading edge, noted by the term y_{LE}. Reduction in β diminishes the leading-edge distance to the symmetry plane $y_{LE} = 0$, meaning that the gap is eliminated. The correlation between the two terms is represented as follows:

$$\beta = \tan^{-1}\left(\frac{y_{LE}}{C}\right) \tag{12.13}$$

An investigation of the resulting parameters of thickness reduction method was performed; the airfoil used was an initially long-gaped airfoil, the outcome of a ClarkY 9% folding, possessing a 30.7%c gap prior to reduction. Figure 12.21 provides a visual representation of the assimilation of the two airfoils.

Note that at $\beta = 0$, the gap does not exist, and the "new" airfoil can be assessed via conventional analysis tools that were unavailable up to this point.

12.2.4 Wind tunnel models and testing

The addition of a leading-edge gap to the examined airfoils resulted in non-convergence of traditional airfoil analysis tools (this is true for even a small gap). Thus, a more empirical method was used to determine the aerodynamic properties of the unique airfoil.

Four wing sections were built to provide adequate answers for various queries raised during the present research. The set of sections assumed that the dominant working point is that of a glider transforming at via folding to an aerobatic aircraft. The airfoil used for the low velocity is that of ClarkY 9% due to a substantial leading-edge gap that is produced after folding (see Fig. 12.18), which might aggravate any phenomena related to the gap. The first section was that of an unaltered ClarkY 9% which serves as a control point for further reference and comparison (see Fig. 12.22).

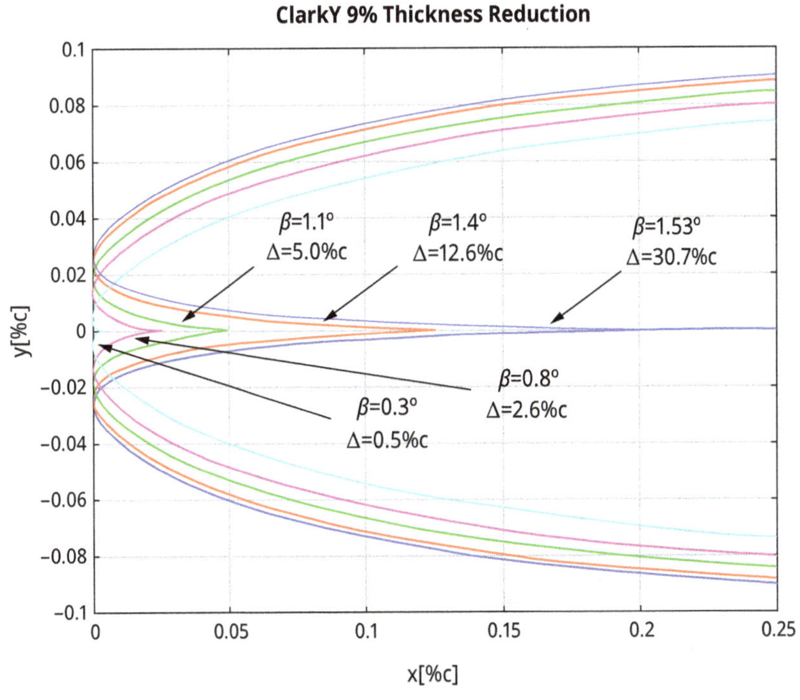

Fig. 12.21: Visualization of thickness reduction as a means to reduce leading-edge gap.

The second section was that of a completely folded form, meaning a ClarkY 9% on top of the other forming a 30.7% leading gap (see Fig. 12.22). This section was noted as "G2A30.7" for being the result of a glider-like aircraft to aerobatic-like aircraft transformation, resulting in a 30.7% chord leading-edge gap. To accurately mimic the geometry of a folded airfoil, this section was actually built as two separate ClarkY 9% sections that were combined together at later phases of the construction process.

While the first two sections served as the fixed folded or unfolded airfoils, the two remaining sections were used to ascertain airfoil improving techniques discussed earlier.

Section 3 was used to examine the attribute of thickness reduction method. The β angle between the two airfoils of a folded configuration decreased until the leading-edge gap was zero ($\Delta = 0$). This airfoil was denoted as "G2A0.0."

The fourth and the last section was used to examine the attribute of surface panel method. This was achieved by using the second section (G2A30.7) for a second test while closing the leading-edge gap in a mechanical manner using a smooth rigid tape.

The test models were constructed as a "realistic" wing section with a wingspan of 0.6 m and a chord of 0.25 m, using a D-box leading edge and spars configuration as well as balsa ribs enclosed in a polyester film. For visual purposes the upper surface of the ClarkY 9% was colored blue, while the lower surface was colored yellow. The sections were fixed during the wind tunnel tests using a steel fixture place at the aft lower surface of each section (see Fig. 12.23).

Fig. 12.22: Models for wind tunnel testing: ClarkY 9% (top), G2A30.7 (middle) and G2A0.0 (bottom).

The four models described above were tested in the subsonic wind tunnel at the Faculty of Aerospace Engineering, Technion, IIT. Each model was tested at velocities ranging from 10 to 30 m/s, corresponding to Reynolds numbers of 1.6×10^5 to 5.0×10^5 and at angle of attack ranging from $-25°$ to $+28°$ (see Fig. 12.24).

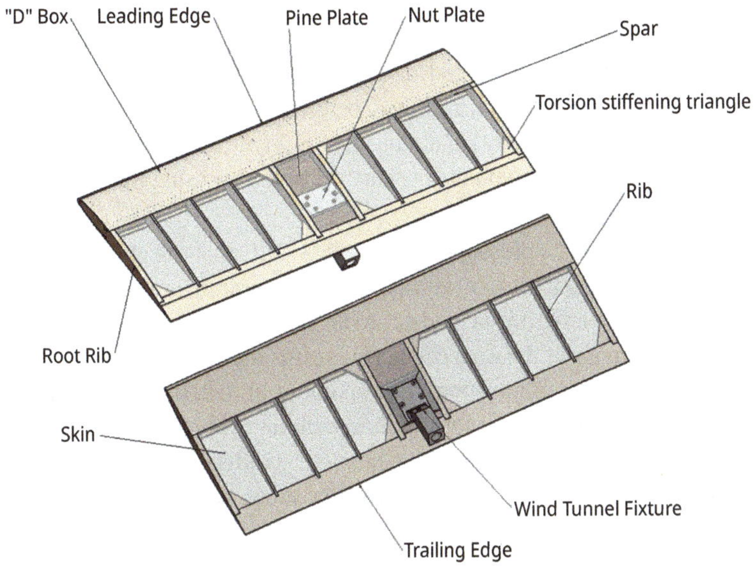

Fig. 12.23: Test model wing section view.

Fig. 12.24: G2A30.7 wing segment fixed to the test apparatus during wind tunnel test.

12.3 Results and discussion

Figures 12.25–12.27 present the lift, drag and pitch moment (respectively) coefficient versus the angle of attack for the four airfoil models tested at $V = 20$ m/s. The figures also depict results of the ClarkY9% as calculated using the Xfoil code [29]. In addition, Tab. 12.2 summarizes the main aerodynamic parameters obtained using the wind tunnel experiments.

It can be seen from the graphs presented in Figs. 12.25–12.27 that the glider's airfoil (ClarkY 9% Model) has a superior lift to drag ratio and is overall favorable compared to the other three airfoils. The other three models are all symmetric, yielding zero lift at $\alpha = 0°$ and a symmetric behavior of the drag coefficient versus the angle of attack.

Upon folding into G2A30.7 configuration, the maximum lift coefficient is reduced as well as the maximum angle of attack (at which maximum lift occurs). The drag is adversely affected as expected while the pitching moment coefficient is reduced; thus, the folded configuration is less stable and more readily allows the "nose-down" maneuver as described earlier.

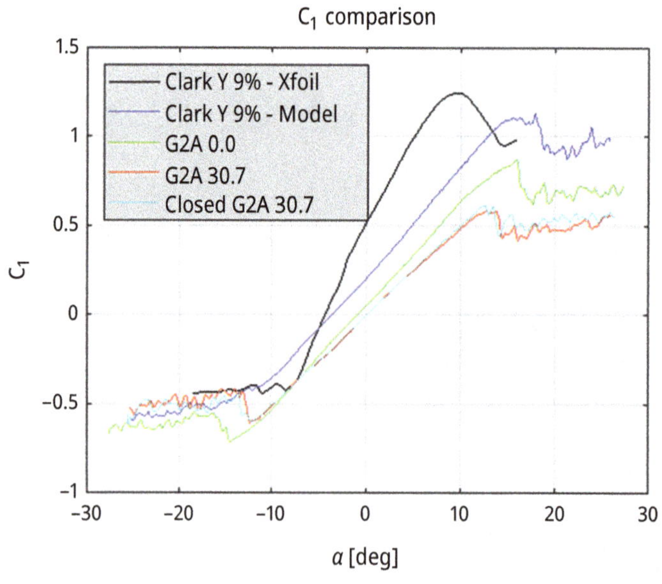

Fig. 12.25: Lift coefficient versus angle of attack comparison of the four tested wing segments.

The difference in the results calculated by the analysis and those received by the wind tunnel tests can be explained by the effect of a finite wing in the case of the wind tunnel test model as well as manufacturing inaccuracies of the handmade test models. Attempting to improve the folded G2A30.7 airfoil by closing it with a front panel had a minimal effect on the aerodynamic coefficients. It was suggested that the

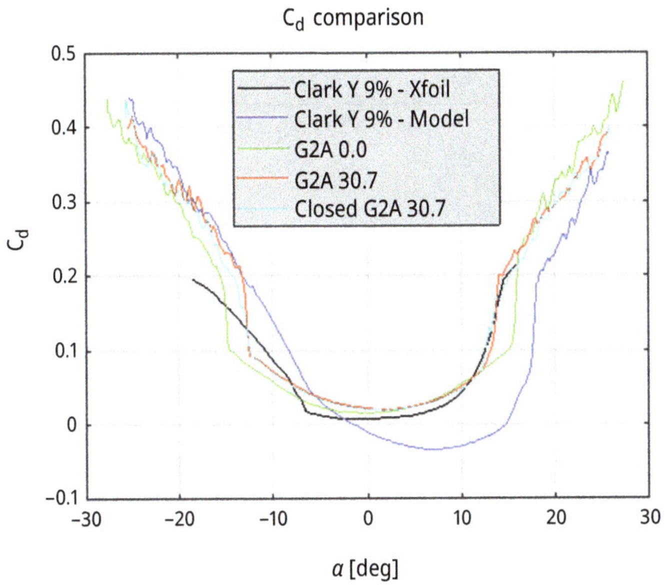

Fig. 12.26: Drag coefficient versus angle of attack comparison of the four tested wing segments.

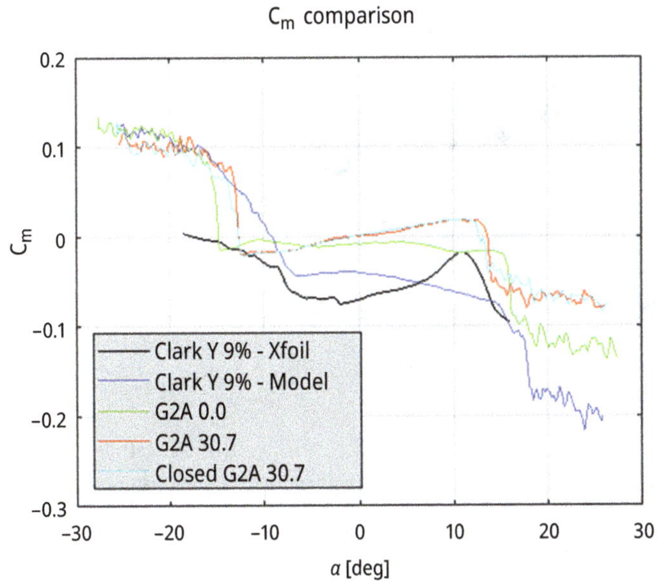

Fig. 12.27: Pitch moment coefficient versus angle of attack comparison of the four tested wing segments.

Tab. 12.2: Experimental main aerodynamic parameters.

Parameters	ClarkY9%	G2A0.0	G2A30.7	Closed G2A30.7
C_{l_α} (1/rad)	3.51	3.35	2.81	2.87
C_{d_0}	0.05	0.01	0.02	0.02
C_{M_0}	−0.04	−0.009	9.4×10^{-5}	0.001
$C_{l\max}$	1.10	0.87	0.58	0.61
α_{stall} (°)	16.4	16.0	13.7	12.5

lack of improvement in the aerodynamic coefficients is possibly due to the airflow creating a stagnation region at the gap's vicinity and thus effectively "closing" the gap in G2A30.7 model, in much the same manner that was used in the wind tunnel test using a closed model via a straight vertical panel.

A second set of wind tunnel tests was held using smoke as a visual aid to assess the validity of the above assumption. Figure 12.28 depicts the smoke test results on the G2A30.7 model. It was discernible that the airflow indeed tended to bypass the leading-edge gap and thus effectively "closing" it without the use of an actual leading-edge panel.

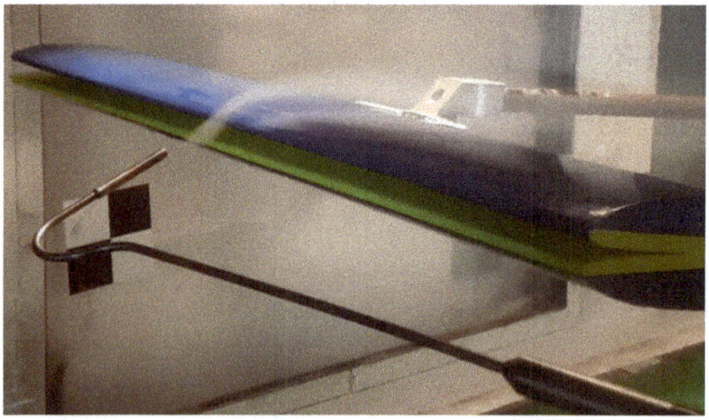

Fig. 12.28: Smoke-aided wind tunnel test photo on a G2A30.7 model.

Reducing the gap using thickness reduction, which results in the G2A0.0 configuration, had shown to be more promising in comparison with the previous technique since it produced visible lift, drag and pitching moment coefficient improvement at a given angle of attack. The maximum angle of attack is restored to that experienced by the original glider airfoil. The pitching moment is the optimal among the four for acute diving maneuver, and the drag is lowered compared to the remaining symmetric airfoils.

12.4 Conclusions

A novel concept for a morphing aircraft capable of altering between two distinct working points was designed, calculated and presented. It was shown based on a detailed analysis that for every aspect ratio, wing loading and flight scenario defined by a designer, an efficiency factor can be assigned to the morphing aircraft.

This efficiency represents the endurance improvement of the specific configuration compared with standard fixed wing aircraft with a typical low-velocity geometry and a typical high-velocity geometry. In general, the overall morphing efficiency over aerobatic aircraft was one scale larger than that of the efficiency over the glider aircraft.

Airfoil selection is a crucial issue, and two main options for selecting an airfoil were presented. It was shown by an analysis and wind tunnel tests that a design process which assigns an adequate airfoil for the unfolded phases and a resulting leading-edge-gaped airfoil for the folded phases is the preferable alternative.

Two correction methods were used to reduce the effect of the leading-edge gap on the aerodynamic forces. Among the two, the thickness reduction method proved to be more promising, as it reduced the adverse effect of the gap by up to 50%.

References

[1] Barbarino, S., Bilgen, O., Ajaj, R.M., Friswell, M.I. and Inman, D.J. A review of morphing aircraft, Journal of Intelligent Materials Systems and Structures, 22(9), 2011, 823–877.

[2] Barbarino, S., Pecora, R., Lecce, L., Concilio, A., Ameduri, S. and De Rosa, L. Airfoil structural morphing based on SMA actuator series: Numerical and experimental studies, Journal of Intelligent Material Systems & Structures, 22, 2011, 987–1004.

[3] Chillara, V.S.C. and Dapino, M.J. Shape memory alloy-actuated bistable composites for morphing structures, in Proc. SPIE 10596, Behavior and Mechanics of Multifunctional Materials and Composites XII, 22 March 2018, 1059609, 8.

[4] Chillara, V.S.C. and Dapino, M.J. Review of morphing laminated composites, Applied Mechanics Reviews, ASME, 72, 2020, Paper Id.: 010801, 16.

[5] Apuleo, G. Chapter 2 – aircraft morphing: An industry vision, in Concilio, A., Dimino, I., Lecce, L. and Pecora, R. Eds. Morphing wing technologies, Elsevier, Oxford, UK, 2018, 85–101, 970.

[6] Weisshaar, T.A. Morphing aircraft technology – new shapes for aircraft design, multifunctional structures / integration of sensors and antennas (pp. 01-1-01-20), in Meeting Proceedings RTO-MP-AVT-141, Overview 1, Neuilly-sur-Seine, France, RTO, 2006, 20.

[7] Weisshaar, T.A. Morphing aircraft systems: Historical perspectives and future challenges, Journal of Aircraft, 50(2), 2013, 337–353.

[8] Bowman, J., Sanders, B. and Weisshaar, T., Evaluating the Impact of Morphing Technologies on Aircraft Performance, 2002, AIAA Paper No. 2002-1631.

[9] Thill, C., Etches, J., Bond, I., Potter, K. and Weaver, P.M.S. Aeronautical Journal, 112(1129), 2008, 117–139.

[10] Sofla, A.Y.N., Meguid, S.A., Tan, K.T. and Yeo, W.K. Shape Morphing of Aircraft Wing: status and Challenges, Materials and Design, 31(3), 2010, 1284–1292.

[11] Lachenal, X., Daynes, S. and Weaver, P.M. Review of morphing concepts and materials for wind turbine blade applications, Wind Energy, 16, 2013, 283–307.
[12] Karakals, A., Machairas, T., Solomou, A., Riziotis, V. and Saravanos, D. Design and simulation of morphing airfoil sections with SMA actuators for wind turbine rotors, in ICAST2014, 25th Inter. Conf. on Adaptive Structures and Technologies, October 6–8th, The Hague, The Netherlands, 12.
[13] Baghdadi, M., Elkoush, S., Akle, B. and Elkhoury, M. Dynamic shape optimization of a vertical-Axis wind turbine via blade morphing technique, Renewable Energy, 154, 2020, 239–251.
[14] Wlezien, R.W., Horner, G.C., McGowan, A.R., Padula, S.L., Scott, M.A., Silcox, R.J., et al. The aircraft morphing program, in AIAA Paper No. 98-1927, 39th Structures, Structural Dynamics, and Materials Conference and Exibit, April 20–23, 1998, Long Beach, CA, USA, 14.
[15] McGowan, A.-M.R., Horta, L.G., Harrison, J.S. and Raney, D.L. Research activities within NASA's morphing program, NASA, Langley Research Center, Hampton, VA 23681-2199, USA, 2000, Accession No. ADP010487, 11.
[16] McGowan, A.-M.R., Vicroy, D.D., Busan, R.C. and Hahn, A.S. Perspectives on highly adaptive of morphing aircraft, in Proc. of the NATO Research Technology Organization Applied Vehicle Technology Panel Symposium, RTO-MP-AVT-168, 2009, Evora, Portugal, 20–24 April 2009, 14.
[17] Daynes, S. and Weaver, P.M. Review of shape-morphing automobile structures: Concepts and outlook, Proceedings of the Institution of Mechanical Engineers, Part D, 227(11), 2013, 1603–1622.
[18] Chu, W.-S., Lee, K.-T., Song, S.-H., Han, M.-W., Lee, J.-Y., Kim, H.-S., et al. Review of biomimetic underwater robots using smart actuators, International Journal of Precision Engineering and Manufacturing, 13(7), 2012, 1281–1292.
[19] Wang, Z. and Yang, Y. Design of a variable-stiffness compliant skin for a morphing leading edge, Applied Sciences, 11(3165), 2021, 18.
[20] Jensen, P.D.L., Wang, F., Dimono, I. and Sigmund, O. Topology optimization of large-scale 3D morphing wing structures, Actuators, 10(217), 2021, 16.
[21] Lentink, D., Müller, U.K., Stamhuis, E.J., De Kat, R., Van Gestel, W., Veldhuis, L.L.M., et al. How swifts control their glide performance with morphing wings, Nature, 446(7139), 2007, 1082–1085.
[22] Forterre, Y., Skotheim, J.M., Dumais, J. and Mahadevan, L. How the venus flytrap snaps, Nature, 433(7024), 2005, 421–425.
[23] Rhodes, O. Optimal design of morphing structures, Ph.D. thesis at Department of Aeronautics, Imperial College London, U.K., 2012, 219.
[24] Zanjani, J.S.M., Louyeh, P.Y., Tabrizi, I.E., Al-Nadhari, A.S. and Yidiz, M. Thermo-responsive and Shape-morphing CF/GF composite skin: Full field experimental measurement, theoretical predictions and finite element analysis, Thin-Walled Structures, 160, 2021, Paper Id.:106874, 16.
[25] Bruyneel, M., Mawet, A., Marinone, E. and Carossa, G. DEMMOW-detailed model of a morphing wing: Development of the full wing model, in 9th European Conference for Aeronautics and Space Science (EUCASS), June 27–July 1 2022, Lille, France, 10.
[26] Culick, F.E.C. The wright brothers: First Aeronautical Engineers and test pilots, AIAA Journal, 41, 2003, 985–1006.
[27] Geva, A., Abramovich, H. and Arieli, R. Investigation of a morphing wing capable of airfoil and span adjustment using a retractable folding mechanism, Aerospace, 6(8), August 2019, 85, doi: 10.3390/aerospace6080085. 24.
[28] Shevell, R.S. Chapter 15, in Fundamentals of Flight, Prentice Hall, Englewood Cliffs, NJ, USA, 1989, 405.
[29] Drela, M. XFOIL: An analysis and design system for low reynolds number airfoils, in Conference on Low Reynolds Number Airfoil Aerodynamics, Springer, Berlin/Heidelberg, Germany, 1989, 1–12.

Appendix A

Æ	Aspect ratio
b (m)	Wingspan
b_H (m)	Horizontal stabilizer span
C_{D_0}	Zero-lift drag coefficient
C_D^+	Drag coefficient at maximum endurance condition
C_d	Airfoil drag coefficient
C_{fe}	Skin friction drag
C_L^+	Lift coefficient at maximum endurance condition
$C_{l\alpha}$	Lift coefficient derivative of angle of attack
C_l	Airfoil lift coefficient
C_M	Pitch coefficient
c (m)	Airfoil chord
dW_{mm} (%W)	Added weight due to morphing mechanism
e	Oswald efficiency number[6]
$\mathscr{L}$ (Nm)	Roll moment
m (kg)	Mass
P_A (W)	Total mission required power for an aerobatic aircraft
P_G(W)	Total mission required power for a glider
$P_{leveled}$ (W)	Power required for leveled flight
$P_{maneuver}$ (W)	Power required for a maneuver or climb
$P_{morphing}$ (W)	Total mission required power for a morphing aircraft
P^* (W)	Total mission required power
S (m^2)	Wing area
S_H (m^2)	Horizontal stabilizer area
S_{wet} (m^2)	Wetted area
t (%c)	Percentage thickness of airfoil
UAV	**U**nmanned **a**erial **v**ehicle
V^+ (m/s)	Velocity at maximum endurance condition
V_a (m/s)	Ideal aerobatic plane velocity for endurance
V_H (m^3)	Horizontal stabilizer volume
V_g (m/s)	Ideal glider velocity for endurance
V_{stall} (m/s)	Stall velocity
W (N)	Gross weight of the aircraft
W_M (N)	Weight of morphing aircraft
W_M/S (N/m^2)	Wing loading of morphing aircraft
W/S (N/m^2)	Wing loading
y_{LE} (%c)	Percentage leading-edge distance normal to folded airfoils symmetry plane
Γ (°)	Anhedral angle
Ψ_A	Morphing added efficiency over aerobatic plane
Ψ_G	Morphing added efficiency over glider

6 The Oswald efficiency is a correction factor that represents the change in drag with lift of a three-dimensional wing or airplane as compared with an ideal wing having the same aspect ratio and an elliptical lift distribution.

α (rad)	Angle of attack
β (°)	Angle of airfoil retraction
γ (rad)	Climbing angle
ρ (kg/m^3)	Air density
χ	Part of flight scenario governed by gliding conditions

13 Large deflections of columns and plates

13.1 Introduction

Thin-walled structures, like columns/beams and plates, are liable to large deflections either in their postbuckling states or due to lateral loads applied along the surface of the specimens. Whereas the cases of small deflections show a linear relationship between the applied loading and the resulting deflection, the large deflection cases display a nonlinear relationship between the two, leading to very complicated solutions.

The large deflections of beams had been investigated in-depth starting with Euler [1], Lagrange [2] and many other researchers presented in references [3–27].

The behavior of plates, leading to large lateral displacements, had also been investigated (see typical references in [28–64]); however, no closed-form analytical solutions are available due to the complexity of the problem.

13.2 Large deflections of columns

13.2.1 Large deflections of columns – literature review

The topic of large deflections of columns (or beams) started with Euler's study in 1759 [1] and continues up to nowadays. Many efforts had been directed to solve the postbuckling behavior of a cantilever column to find its shape in the large deflection zone (see typical examples in [1–5]). Other investigators, like in [6–9], solved the problem of a cantilever beam loaded laterally by a concentrated tip load. Rohde [10] presents a solution for a cantilever beam loaded with uniformly distributed load for large deflections, while Conway [11] addressed the issue of large deflections of a simply supported beam carrying a central concentrated load, which was later revisited (in 1961) by Wang et al. [12], and displaying new numerical methods, while assuming an inextensible beam. In a more recent paper, Bélendez et al. [13] presented interesting results for large and small deflections for a cantilever beam loaded laterally at its free end, while Lubis and Tanti [14] presented in their study the ways to solve the well-known problem of large deflection of thin cantilever beams having rectangular cross sections and loaded by a lateral tip force, with interesting numerical integration backed by experimental results. Shvartsman [15] studied the large deflection problem of a nonuniform spring-hinged cantilever beam under a tip-concentrated follower force, while in [16–18] the behavior of a cantilever beam made from a nonlinear material type named Ludwick was determined in the large deflection zone.

Banerjee et al. [19] addressed the problem of cantilever beams having geometric nonlinearities and presenting new analytical and numerical solutions for the large deflections issue, while Tolou and Herder [20] presented a semi-analytical approach to

https://doi.org/10.1515/9783111621104-013

solve the case of large displacements of compliant beams under a point load. Kocatürk et al. [21] investigated the same topic, applying it to a cantilever beam.

An interesting thesis is presented by Nishawala [22] summing up the issue of beams and plates under large lateral displacements.

Khavvaji et al. [23] presented in their paper a new approach to analyze and solve the large displacement behavior of prismatic and nonprismatic beams made of non-homogeneous material under combined loading and various boundary conditions.

Sitar et al [24] claimed in their paper to have a simple method to determine the large displacement states for arbitrary curved planar elastic, and Khosravi and Jani [25] presented new methods like Bernstein spectral and convolution quadrature for solving the case of large deflections of cantilever beams.

The last references [26, 27] deal with large deflections of inextensible cantilevers, presenting their theories, modeling and simulations. Both are applicable for aeroelastic problems, where the wings underwent large displacements.

To introduce the reader to the methods for solving the elastica problems, a few cases are next presented.

13.2.2 Large deflections of columns – the Euler elastica case

Figure 13.1 presents a column in its postbuckling state, after buckling at load of

$$P_{\text{cr}} = \frac{\pi^2 EI}{4L^2} \tag{13.1}$$

which is known to be the smallest load for the case of a cantilever beam, axially loaded [3]. Note that the origin of the axes is at point O, located at the free tip of the cantilever; EI is the bending stiffness of the column, while L is its length before bending. According to the drawing in Fig. 13.1, the exact curvature of the column can be written as [3] $d\theta/ds$ and multiplying it by the column's bending stiffness would yield the following expression (while the change in the column's length due to the application of the compression load is neglected)

$$EI\frac{d\theta}{ds} = -P \cdot y \tag{13.2}$$

Equation (13.2) presents two terms that are connected by the following equation: $dy/ds = \sin\theta$. Inserting this equation in eq. (13.2), after once being differentiated with respect with θ, yields a second-order differential equation for the deflection curve of the column

$$EI\frac{d^2\theta}{ds^2} = -P \cdot \sin\theta \tag{13.3}$$

Note that eq. (13.3) is a nonlinear second-order differential equation. The way to obtain a solution is described in detail in Ref. [3]. In what follows, only the important steps are presented.

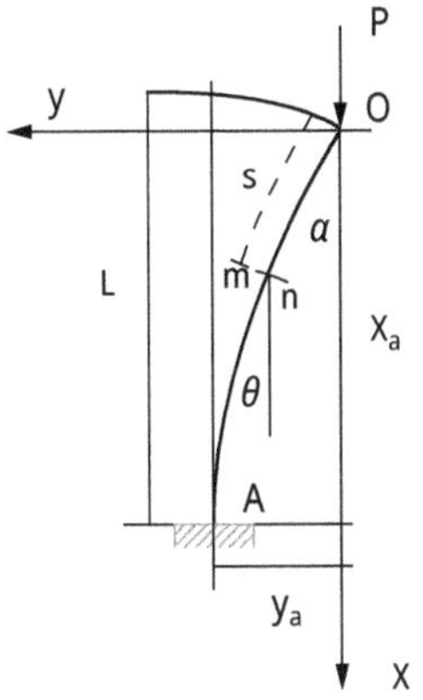

Fig. 13.1: A cantilever beam after its axial buckling.

Multiplying both sides of eq. (13.3) by $d\theta$ and integrating yields the solution to the problem:

$$\frac{1}{2}\int \frac{d}{ds}\left[\frac{d\theta}{ds^2}\right]^2 ds = -\lambda^2 \cdot \int \sin\theta \cdot d\theta \Rightarrow \frac{1}{2}\left[\frac{d\theta}{ds^2}\right]^2 = \lambda^2 \cos\theta + A \tag{13.4}$$

where

$$\lambda^2 = \frac{P}{EI} \quad \text{and} \quad A = \text{const}$$

Constant A in eq. (13.4) is found by applying the following boundary conditions:

$$\text{At point } O{:}\, \theta = \alpha \quad \text{and} \quad \frac{d\theta}{ds} = 0 \text{ [due to M(0,0) = 0]} \tag{13.5}$$

yielding

$$A = -\lambda^2 \cos\alpha \tag{13.6}$$

Substituting the value of the constant A into eq. (13.4) yields the expression for $d\theta/ds$, namely

$$\left[\frac{d\theta}{ds}\right]^2 = 2\lambda^2(\cos\theta - \cos\alpha) \Rightarrow \frac{d\theta}{ds} = \pm\lambda\sqrt{2(\cos\theta - \cos\alpha)} \tag{13.7}$$

Consulting Fig. 13.1 reveals that the expression $d\theta/ds$ is negative; therefore, we can write

$$ds = -\frac{d\theta}{\lambda\sqrt{2(\cos\theta - \cos\alpha)}} \Rightarrow L = \int_{\alpha}^{0} ds = -\int_{0}^{\alpha} ds = \int_{0}^{\alpha} \frac{d\theta}{\lambda\sqrt{2(\cos\theta - \cos\alpha)}} \tag{13.8}$$

or

$$L = \int_{0}^{\alpha} \frac{d\theta}{2\lambda\sqrt{\left(\sin^2\frac{\theta}{2} - \sin^2\frac{\alpha}{2}\right)}}$$

To obtain the length L, the integral in eq. (13.8) must be evaluated. As suggested in [3], the following notations are used to solve the problem:

$$\begin{aligned}
&\left[\frac{d\theta}{ds}\right]^2 = 2\lambda^2(\cos\theta - \cos\alpha) \Rightarrow \frac{d\theta}{ds} = \pm\lambda\sqrt{2(\cos\theta - \cos\alpha)}\\
&\eta \equiv \sin\left(\frac{\alpha}{2}\right);\ \sin\varsigma \equiv \frac{1}{\eta}\sin\left(\frac{\theta}{2}\right) \to \sin\left(\frac{\theta}{2}\right) = \sin\varsigma\cdot\sin\left(\frac{\alpha}{2}\right)\\
&0 \le \theta \le \alpha \Rightarrow 0 \le \sin\varsigma \le 1 \quad \text{or} \quad 0 \le \varsigma \le \frac{\pi}{2}\\
&\text{moreover: } \frac{d\theta}{2}\cos\left(\frac{\theta}{2}\right) = \eta\cdot\cos\varsigma\cdot d\varsigma\\
&\Rightarrow d\theta = \frac{2\eta\cdot\cos\varsigma\cdot d\varsigma}{\cos\left(\frac{\theta}{2}\right)} = \frac{2\eta\cdot\cos\varsigma\cdot d\varsigma}{\sqrt{1-\sin^2\left(\frac{\theta}{2}\right)}} = \frac{2\eta\cdot\cos\varsigma\cdot d\varsigma}{\sqrt{1-\eta^2\sin^2\varsigma}}
\end{aligned} \tag{13.9}$$

and

$$\sqrt{\sin^2\left(\frac{\alpha}{2}\right) - \sin^2\left(\frac{\theta}{2}\right)} \equiv \eta\cos\varsigma$$

Substituting the various notations presented in eq. (13.9) yields

$$L = \int_{0}^{\alpha} \frac{d\theta}{2\lambda\sqrt{\left(\sin^2\frac{\theta}{2} - \sin^2\frac{\alpha}{2}\right)}} = \frac{1}{\lambda}\int_{0}^{\pi/2} \frac{d\varsigma}{\sqrt{(1-\eta^2\sin^2\varsigma)}} \equiv \frac{1}{\lambda}\kappa(\eta) \tag{13.10}$$

where $\kappa(\eta)$ is known to be *the complete elliptical of the first kind* to be found in the literature.[1]

For small values of the angle α, $\eta \equiv \sin(\alpha/2) \to 0$, eq. (13.10) can be written as

---

1 A complete elliptical of the first kind can be found for example on the web: complete elliptic integral $K(k)$, $E(k)$ (chart) calculator – high-accuracy calculation (https://keisan.casio.com/exec/system/1180573454) or in Hammersley, J. M., Tables of complete elliptic integrals, *Journal of Research of the National Bureau of Standards*, Vol. 50, No. 1, January 1953, Research Paper 2386, pp. 43–43.

$$L = \frac{1}{\lambda}\int_0^{\pi/2} = \frac{\pi}{2\lambda} = \frac{\pi}{2}\sqrt{\frac{EI}{P}}$$
$$\Rightarrow P = \frac{\pi^2 EI}{4L^2} = P_{\text{cr-cantilever}} \tag{13.11}$$

For other larger values for α, η, the κ integral as well as the force P will also increase as can be seen in Tab. 13.1.

$$L = \int_0^{\alpha} \frac{d\theta}{2\lambda\sqrt{\left(\sin^2\frac{\theta}{2} - \sin^2\frac{\alpha}{2}\right)}} = \frac{1}{\lambda}\int_0^{\pi/2} \frac{d\varsigma}{\sqrt{\left(1 - \eta^2 \sin^2\varsigma\right)}} \equiv \frac{1}{\lambda}\kappa(\eta) \tag{13.12}$$

To calculate the deflections for the free end, the horizontal one Y_o and the vertical one X_o point O, we use the formulation detailed in [3] to yield

$$Y_o = \frac{2\eta}{\lambda}\int_0^{\pi/2} \sin\varsigma \cdot d\varsigma = \frac{2\eta}{\lambda}$$
$$X_o = \frac{2}{\lambda}\int_0^{\pi/2} \sqrt{1 - \eta^2 \sin^2\varsigma} \cdot d\varsigma - L = \frac{2E(\eta)}{\lambda} - L \tag{13.13}$$

The expression $E(\eta)$ is named *the complete elliptic integral of the second type* to be found in the same reference as the $\kappa(\eta)$. Typical results for various values of the variable $\eta \equiv \sin(\alpha/2)$ are presented in Tab. 13.1.

Tab. 13.1: Length, load, horizontal and vertical displacement values as a function of the tip inclination angle α.

α	$\eta = \sin(\alpha/2)$	$\kappa(\eta)$	L	P/P_{cr}*	$E(\eta)$	X_o#	Y_o##
0°	0	1.57080 = $\pi/2$	L	1.00000	1.57080 = $\pi/2$	1	0
20°	0.173648	1.58284	1.58284/λ	1.01500	1.55889	0.96973L	0.21941L
40°	0.342020	1.62003	1.62003/λ	1.06366	1.52380	0.88120L	0.42224L
60°	0.500000	1.68575	1.68575/λ	1.15172	1.46746	0.74102L	0.59321L
80°	0.642788	1.78677	1.78677/λ	1.29389	1.39314	0.55940L	0.71950L
100°	0.766044	1.93558	1.93558/λ	1.51839	1.30554	0.34899L	0.79154L
120°	0.866025	2.15652	2.15652/λ	1.88480	1.21106	0.12316L	0.80317L
140°	0.939693	2.50455	2.50455/λ	2.54226	1.11838	−0.10692L	0.75039L
160°	0.984808	3.15339	3.15339/λ	4.03009	1.04011	−0.34032L	0.62460L
170°	0.996195	3.83174	3.83174/λ	5.95049	1.01266	−0.47143L	0.51997L
176°	0.999391	4.74272	4.74272/λ	9.11622	1.00258	−0.57721L	0.42144L
179°	0.999962	6.12778	6.12778/λ	15.21831	1.00021	−0.67355L	0.32637L

$$^{*}P_{cr}=\pi^2 EI/(4L^2),\quad P/P_{cr}=4\cdot\kappa^2(\eta)/\pi^2;\ ^{\#}X_0=\frac{[2\cdot E(\eta)-\kappa(\eta)]}{\kappa(\eta)}L;\ ^{\#\#}Y_o=\frac{2\cdot\eta}{\kappa(\eta)}L.$$

13.2.3 Large deflections of columns – a cantilever under lateral tip force

As written and described in [6, 7] and presented in Fig. 13.2, a cantilever beam is loaded by a tip load P. A point on the beam will be identified by four symbols (x, y, s, θ) and the original length of the beam is L. Note that the subscript L is used to identify the value of the four symbols at the free end of the beam and the location of the load P is $(L, 0, L, 0)$. The bending stiffness of the beam is EI, and the point O is the origin of the coordinates.

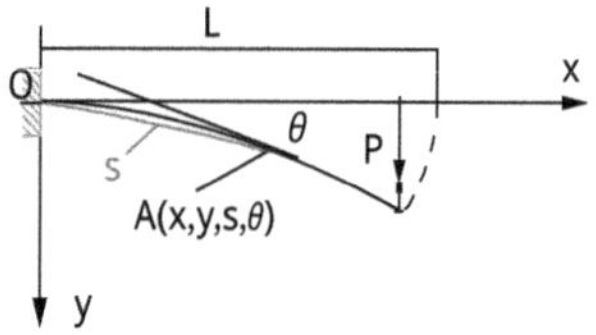

Fig. 13.2: A cantilever beam under tip load P.

Then the bending moment M and the change in the angle θ with the arc length s are written as

$$M=P(x_L-x)\Rightarrow\frac{d\theta}{ds}=\lambda^2(x_L-x)\tag{13.14}$$

where

$$\lambda^2=\frac{P}{EI}$$

Integrating eq. (13.14) while adding a new parameter yields

$$\frac{d\theta}{ds}=\frac{d\theta}{dx}\cdot\frac{dx}{ds}=\frac{d\theta}{dx}\cdot\cos\theta$$

therefore

$$\int\lambda^2(x_L-x)dx=\int\cos\theta d\theta\tag{13.15}$$

or

$$\lambda^2\left(x_L\cdot x-\frac{x^2}{2}\right)+A=\sin\theta$$

Equation (13.15) must satisfy the boundary conditions at $x=0$, namely the vanishing of the angle θ, yields

$$\lambda^2\left(x_L\cdot x-\frac{x^2}{2}\right)+A=\sin\theta$$

$$\text{but at}\quad x=0\quad \sin\theta=0$$

$$\Rightarrow A=0 \tag{13.16}$$

$$\Rightarrow \sin\theta=\lambda^2\left(x_L\cdot x-\frac{x^2}{2}\right);\qquad \sin\theta_L=\lambda^2\cdot\frac{x_L^2}{2}$$

and

$$\sin\theta_L-\sin\theta=\lambda^2\cdot\frac{x_L^2}{2}-\lambda^2\left(x_L\cdot x-\frac{x^2}{2}\right)=\frac{\lambda^2}{2}\left(x_L^2-2\cdot x_L\cdot x-x^2\right)=\frac{\lambda^2}{2}\left(x_L-x\right)^2$$

$$\Rightarrow (x_L-x)=\sqrt{\left[\frac{2}{\lambda^2}(\sin\theta_L-\sin\theta)\right]}$$

Substitution of eq. (13.16) into (13.14) yields

$$\frac{d\theta}{ds}=\frac{d\theta}{dy}\cdot\frac{dy}{ds}=\frac{d\theta}{dy}\cdot\sin\theta$$

but

$$\frac{d\theta}{ds}=\lambda^2\cdot(x_L-x)=\lambda^2\sqrt{\left[\frac{2}{\lambda^2}(\sin\theta_L-\sin\theta)\right]}=\sqrt{\left[2\cdot\lambda^2(\sin\theta_L-\sin\theta)\right]}$$

$$\Rightarrow \sqrt{\left[2\cdot\lambda^2(\sin\theta_L-\sin\theta)\right]}=\frac{d\theta}{dy}\cdot\sin\theta$$

$$\rightarrow y=\int_0^{\theta}\frac{\sin\theta\cdot d\theta}{\sqrt{\left[2\cdot\lambda^2(\sin\theta_L-\sin\theta)\right]}} \tag{13.17}$$

and

$$y_L=\int_0^{\theta_L}\frac{\sin\theta\cdot d\theta}{\sqrt{\left[2\cdot\lambda^2(\sin\theta_L-\sin\theta)\right]}}$$

As suggested in [6], a transformation of variables is proposed to be

$$\cos\left(\frac{\pi}{4}-\frac{\theta}{2}\right)=\cos\left(\frac{\pi}{4}-\frac{\theta_L}{2}\right)\sin\varsigma=\eta\cdot\sin\varsigma \tag{13.18}$$

Using eqs. (13.18) and (13.17) would change into

$$y_L = \int_{\theta_L}^{0} \frac{\sin\theta \cdot d\theta}{\sqrt{\left[2\cdot\lambda^2(\sin\theta_L - \sin\theta)\right]}} \frac{1}{\lambda}\int_{\pi/2}^{\delta} \frac{(2\cdot\eta^2\cdot\sin^2\varsigma - 1)\cdot d\varsigma}{\sqrt{(1-\eta^2\cdot\sin^2\varsigma)}}$$

with

$$\eta = \cos\left(\frac{\pi}{4} - \frac{\theta_L}{2}\right); \quad \sin\delta = \frac{\cos\left(\frac{\pi}{4}\right)}{\eta} \tag{13.19}$$

$$\Rightarrow y_L = \frac{1}{\lambda}\left[\kappa(\eta) - 2E(\eta) - \kappa'(\eta,\delta) + 2E'(\eta,\delta)\right]$$

where $\kappa(\eta), E(\eta)$[2] are the first and second complete elliptic integrals, respectively, and $\kappa'(\eta,\delta)$ and $E'(\eta,\delta)$[3] are the first and second incomplete integrals, respectively. As described in [7], the following expression holds:

$$\lambda \cdot L = \kappa'\left(\eta, \frac{\pi}{2}\right) - E'(\eta,\delta) \tag{13.20}$$

Using the variable δ as independent variables, corresponding values of η and $\lambda^2 L^2$ can be evaluated leading to the two nondimensional terms, F_x and F_y. Their variation is presented in Fig. 13.3 (adapted from [7]). The first expression is given by $F_x = x_L/L$, while the second one is $y_L/(\lambda^2 \cdot L^3/3) = 3 \cdot y_L/(\lambda^2 \cdot L^3)$.

13.2.4 Large deflections of columns – a cantilever uniform distributed load

Another approach to calculate the behavior of a cantilever beam loaded by a uniform load q is presented in this section (see Fig. 13.4). The method assumes a beam with uniform cross sections; a bending stiffness of EI, a length L and elastic in extensional deformations (see [10]). The Bernoulli–Euler equations can be written as

$$\frac{d\theta}{ds} = \frac{M}{EI} = -\frac{q\cdot s\cdot\frac{s}{2}\cdot\cos\theta}{EI}$$
$$\Rightarrow \frac{dM}{ds} = -q\cdot s\cdot\cos\theta = EI\frac{d^2\theta}{ds^2} \tag{13.21}$$

2 The first and second complete elliptic integrals can be calculated from the following address: complete elliptic integral $K(k)$,$E(k)$ (chart) calculator – high-accuracy calculation (https://keisan.casio.com/exec/system/1180573454).

3 First incomplete elliptic integrals can be calculated from the following address: incomplete elliptic integral of the first kind $F(\varphi,k)$ calculator – high-accuracy calculation (https://keisan.casio.com/exec/system/1124498950), while the second incomplete integral from https://keisan.casio.com/exec/system/180573455.

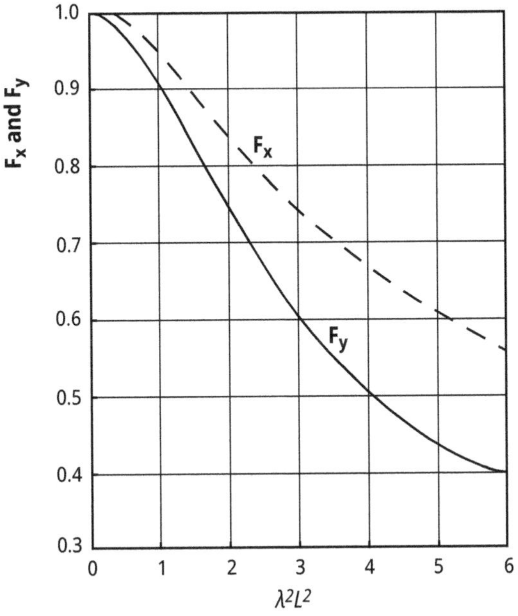

Fig. 13.3: The nondimensional lateral deflection and the nondimensional horizontal contraction as a function of parameter $\lambda^2 L^2$.

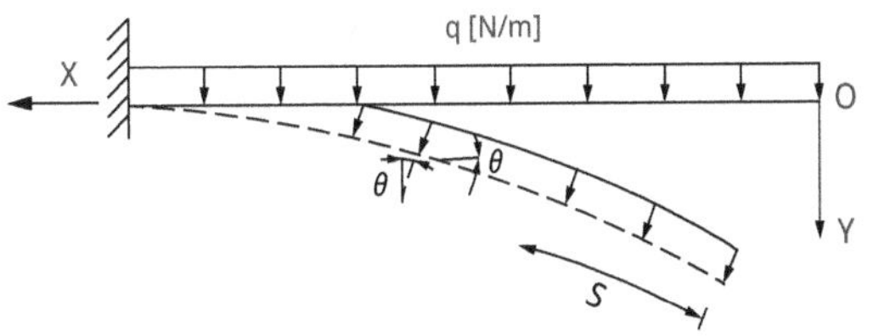

Fig. 13.4: A cantilever beam under uniform distributed load.

Equation (13.21) is a nonlinear equation and to obtain a solution some series are assumed [10] as

$$\theta = \sum_{n=0}^{\infty} A_n \cdot s^n \rightarrow \frac{d\theta}{ds} = \sum_{n=0}^{\infty} n \cdot A_n \cdot s^{n-1} \rightarrow \frac{d^2\theta}{ds^2} = \sum_{n=0}^{\infty} n \cdot (n-1) \cdot A_n \cdot s^{n-1} \tag{13.22}$$

Applying the boundary conditions of the present problem yields

Boundary conditions:

$$@s=0 \quad \theta=\upsilon \quad \text{and} \quad \frac{d\theta}{ds}=0 \tag{13.23}$$

$$\Rightarrow A_0=\upsilon;\ A_1=0$$

where υ is an arbitrary constant. Then eq. (13.21) can be written as

$$-q\cdot s\cdot \cos\theta = EI\frac{d^2\theta}{ds^2}$$

$$\rightarrow \frac{d^2\theta}{ds^2}=-\frac{q\cdot s}{EI}\cos\theta$$

but

$$\cos\theta = \cos(\upsilon-\alpha) = \cos\upsilon\cos\alpha - \sin\upsilon\sin\alpha \tag{13.24}$$

and

$$\upsilon=\sum_{n=2}^{\infty}A_n\cdot s^n$$

$$\Rightarrow \frac{d^2\theta}{ds^2}=-\frac{q\cdot s}{EI}(\cos\upsilon\cos\alpha-\sin\upsilon\sin\alpha)$$

where α is a new variable to denote the angle of the beam. Expanding the terms sin α and cos α in eq. (13.24) in series of s and equating coefficients yield:

$$\begin{aligned}
&A_2=0\\
&A_3=-\frac{q\cdot\cos\upsilon}{6EI}\\
&A_6=A_3\frac{q\cdot\sin\upsilon}{30EI}\\
&A_{3n+4}=A_{3n+5}=\cdots=0, \qquad n=0,1,2,\ldots,\infty\\
&\Rightarrow \theta=\upsilon+\sum_{m=1}^{\infty}A_{3m}\cdot s^{3m}
\end{aligned} \tag{13.25}$$

Since $dy/ds=\sin\theta$, the following expressions [10] can be written as

$$x = \int_0^s \cos\theta \cdot ds = \alpha_1 \cos\upsilon - \alpha_2 \sin\upsilon$$

$$y = \int_0^s \sin\theta \, ds = \alpha_1 \sin\upsilon + \alpha_2 \cos\upsilon \tag{13.26}$$

where

$$\alpha_1 = s - \frac{A_3 \cdot s^7}{14} - \frac{A_3 \cdot A_6 \cdot s^{10}}{10} - \cdots$$

$$\alpha_2 = \frac{A_3 \cdot s^4}{4} + \frac{A_6 \cdot s^7}{7} + \frac{\left(A_9 - \frac{A_3^3}{6}\right) \cdot s^{10}}{10} + \cdots$$

Using the expression in eq. (13.26), the equation of the moment would be

$$M = EI \sum_{m=1}^{\infty} 3 \cdot m \cdot A_{3m} \cdot s^{3m-1}; \quad \upsilon = -\sum_{m=1}^{\infty} A_{3m} \cdot L^{3m}. \tag{13.27}$$

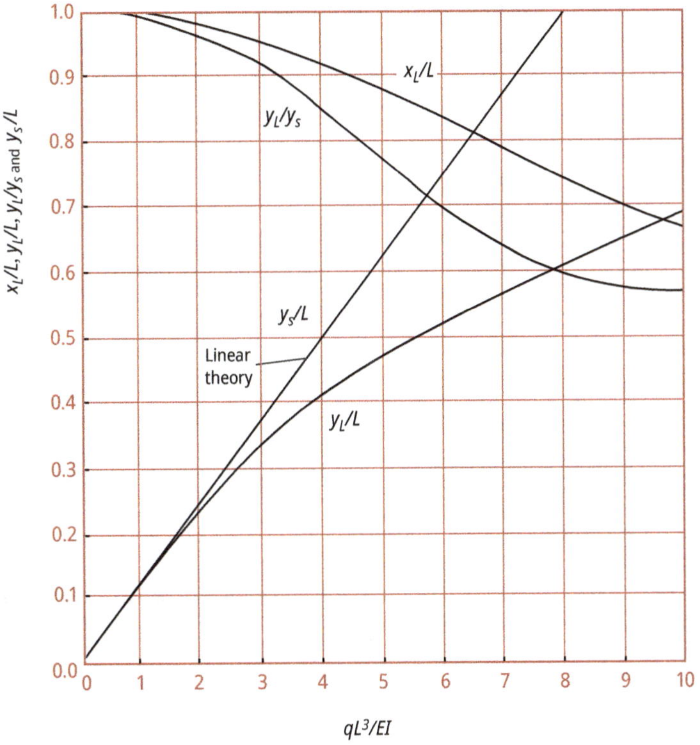

Fig. 13.5: The variation of x_L/L, y_L/L, y_L/y_s and y_s/L with the parameter qL^3/EI (adapted from [10]).

Substituting $s = L$ and $\upsilon = (q \cdot L^3)/(6 \cdot EI)$, as pointed out in [10], leads to the first approximation for the maximal deflection of the beam:

$$y \simeq s \cdot \sin \upsilon + \frac{A_3 \cdot s^4}{4} \cos \upsilon \simeq \frac{q \cdot L^4}{8 \cdot EI} \tag{13.28}$$

The expression in eq. (13.28) matches the small deflections theory.

Figure 13.5 displays the variation of the nondimensional horizontal contraction, x_L/L, the maximal deflection of the total length y_L/L and the y_L/y_s the maximal deflection to maximal elementary theory deflection and the small deflection theory y_s/L as a function of parameter qL^3/EI.

13.3 Large deflections of rectangular plates

13.3.1 Large deflections of plates – literature review

The behavior of flat plates[4] subjected to various in-plane and out-of-plane loading has attracted much attention due to its technological importance. This vast problem can be divided into many subproblems depending on the parameters to be investigated such as plate thickness, perimeter shape, material properties, small versus large deflections and shear deformability. A thin-walled plate made of isotropic material, loaded by transversal pressure is one of the classical problems in elasticity, with an enormous number of studies being written on the subject (see, e.g., [28–30]).

Plates undergoing small deflections that do not the exceed fraction of the plate thickness (usually $< 0.2 \cdot t$, t = thickness) have shown linear behavior with good satisfactory analytical solutions as described in [28–30]. However, for larger deflections of a few times the plate thickness and higher, a nonlinear behavior exists caused by stretching of the midsurface of the plate, thus increasing the transverse plate stiffness. The load–deflection graph line ceases to be linear as a function of the transverse pressure, and the whole stress distribution is a function of the out-of-plane and the in-plane deformations. In 1910, Theodore von Kármán [31] made a major breakthrough for the plate large deflections problem. He published a set of nonlinear differential equations that describes the large deflection behavior, considering the in-plane deformations and stresses. They are known in the literature as Föppl–von Kármán equations, named after August Föppl [32] and Theodore von Kármán [31], or in short, von Kármán equations, which have the following form:

4 This chapter is based on the manuscript [64]: Hakim, G. and Abramovich, H., Large deflection of thin-walled plates under transverse loading – investigation of the generated in-plane stresses, Materials 2022, 15(4), Paper Id: 1577, 28p.

$$\frac{\partial^4 F}{\partial x^4} + 2\frac{\partial^4 F}{\partial x^2 \partial y^2} + \frac{\partial^4 F}{\partial y^4} = E\left[\left(\frac{\partial^2 w}{\partial x \partial y}\right)^2 - \frac{\partial^4 w}{\partial x^2 \partial y^2}\right]$$
$$\frac{\partial^4 w}{\partial x^4} + 2\frac{\partial^4 w}{\partial x^2 \partial y^2} + \frac{\partial^4 w}{\partial y^4} = \frac{q}{D} + \frac{t}{D}\left[\frac{\partial^2 F}{\partial y^2}\frac{\partial^2 w}{\partial x^2} - 2\frac{\partial^2 F}{\partial x \partial y}\frac{\partial^2 w}{\partial x \partial y} + \frac{\partial^2 F}{\partial x^2}\frac{\partial^2 w}{\partial y^2}\right] \tag{13.29}$$

where E is the plate's Young's modulus, w is the out-of-plane deflection, t is the thickness of the plate, q is the transverse uniform pressure and D is the flexural bending presented as

$$D = \frac{Et^3}{12(1-\upsilon^2)} \tag{13.30}$$

with ν being the Poisson's ratio of the plates material and F is the airy function [30] defined as

$$\sigma_{xx} = \frac{\partial^2 F}{\partial y^2}; \ \sigma_{yy} = \frac{\partial^2 F}{\partial x^2}; \ \tau_{xy} = -\frac{\partial^2 F}{\partial x \partial y} \tag{13.31}$$

It is interesting to note, as pointed out by Bakker et al. [33], that eqs. (13.29) and (13.30) are a simplification of Marguerre [34] equations for plates having initial imperfections and subjected to in-plane and transverse loads (the initial imperfection is taken as zero in eqs. (13.29) and (13.30)).

Unfortunately, von Kármán's equations set turned out to be very difficult and cumbersome to solve. To date, there are still no closed-form analytic solutions for rectangular flat plate that satisfy both the differential equations and the boundary conditions. Nevertheless, many approximate and numerical solutions were published (see typical examples in [33, 35–50]). The approximated methods presented in the literature would suggest solutions, usually having severe limitations. Most of them are not easy to use, and the nonlinear nature of the plate hardening effect in large deflection mode is not evident.

The following literature review tries to establish the state of the art for rectangular and circular plates undergoing large deflections due to lateral loading.

Browsing the literature reveals that the nonlinear behavior of flat plates using von Kármán's two equations was mainly investigated for the transverse deflections of the plate, with the in-plane (the membrane) generated stresses being less discussed. NACA had allocated a lot of efforts to investigate the issue by publishing technical reports in the years 1941–1951 (see typical reports in [35–41]). Various methods were employed such as multiplying double Fourier series results in quadruple series with a large number of terms leading to numerical tables and graphs with the membrane stresses in the x and y directions being calculated at the midpoint of a square plate and midpoint edges and the shear stresses set to zero [36]. Another interesting study is presented in [37] for a clamped plate having an aspect ratio of 1.5 and undergoing large deflections. Their results differ only by 3% from an infinitely long plate, thus implying that long plates should be treated as infinitely long and the in-plane stress distribution along lines parallel to the edges going through the plate center not changing significantly.

The studies presented by Levy [36, 37] and Wang [39, 40] solved the large deflection problem by employing two finite differences schemes, the successive approximations and relaxation method to yield a good comparison with Levy's results. He considered an all-around immovable clamped plate and all-around simply supported movable boundary conditions. Both square and rectangular plates are presented with the in-plane stresses being calculated at three points, center of the plate, long-edge midpoint and short-edge midpoint, without indicating the presence of compression-type stress. One should also note that the study presented in [41] for sandwich-type plates, for which numerical and experimental results are presented and well-compared.

In 1954, Berger [42] assumed that the strain energy due to the second invariant of the middle surface strain can be neglected leading to a solution of the von Kármán's equations set. This neglection would mean that for the large deflection case, the plate's bending resistance is low, and the plate would behave like a pure membrane. The study presents results for circular and rectangular flat plates for both clamped and simply supported boundary conditions, with deflections and stresses being presented graphically and numerically at certain points on the plate.

In 1969–1970, Scholes and Bernstein [43] and Scholes [44] presented approximate large deflections solutions using energy methods for all-around simply supported rectangular plates [43] and all-around clamped plates [44]. A good comparison with experimental results is reported in [43]. They employed Timoshenko and Woinowsky-Krieger's [30] mentioned idea to divide the loading path into a first part which would cause bending, and a secondary part leading to membrane stretching, and the load–deflection curve calculated by a finite differences scheme. The clamped case being dealt in [44] presents stresses and deflection calculations and comparison with measured results. Maximal values are presented for a pressure-loaded plate to enable efficient design.

Li-Zhou and Shu [45] use the perturbation variational method to solve the large deflections problem of rectangular plates under transverse pressure leading to analytical expression for displacements and stresses. They report a good comparison with available experiments.

Bert et al. [46] also address von Kármán's equations for orthotropic rectangular plates. The solution is obtained and presented using the differential quadrature method. The boundary condition used in their study was all-around simply supported and all-around clamped, both immovable. They report that deflections, membrane and bending stress are in good agreement to known solutions. Values of the stresses are calculated at the plate's midpoint as a function of the applied transverse load. Yeh and Liu [47] also addressed the issue of approximated analytic solution for the orthotropic von Kármán's equations. The presented solution leads to an expression for self-mode frequency. Numerical and graphical solutions are presented only for deflections while the stress distribution being not dealt in the study.

More recent studies, such as by Wang and El-Sheikh [48], present results for the von Kármán's equations by multiplying Fourier series, getting quadruple sums and equating similar terms in the results series. The output is a nonlinear algebraic equations system with one, three, four, six or nine equations and unknowns according to the number of terms taken for the series. This system is solved for every desired point on the plate. For that, the authors use numerical tools based on generalized reduced gradient method. They also present a closed-form solution for the midpoint deflection using the first term only having the form of $q = \alpha \cdot w + \gamma \cdot w^3$ (q = transverse load, w = lateral deflection and α, γ = fitted constants). However, there are indications from other sources that the use of only one term is not accurate enough and has serious deviations from reality. An interesting solution for the Föppl–von Kármán equations set is presented by Bakker et al. [33] using an approximated analytic solution. Thanks to the simplicity of the trial function, the bending and the membrane loading influences are separated, easing the solution process. The results are compared to ANSYS FEA (Engineering Simulation Software, Canonsburg, PA, USA) results with less than 10% difference. Six combinations of boundary conditions and four loading cases make the results presentation rather complex. Stresses are presented using various formulas without any graphical outcome.

Ugural [49] in his book (Chapter 10) presents approximate solutions for circular thin plate S–I (simply supported, immovable edge). However, for the solution for a thin rectangular plate, he assumes membrane-only stresses (no bending resistance) at the midpoint, with SSSS–I boundary conditions.

Razdolsky [50] also presents approximate solutions for rectangular SSSS-I rectangular plates with deflections and stresses calculated for several aspect ratios. He converts the stress expressions of Levy [36–39] through minimum potential energy method to computer executable algorithms. His square plate deflection curve is found to be between Timoshenko and Woinowsky-Krieger [30] and Levy [36, 37] curves, while for the stresses no direct comparison is presented.

Turvey and Osman [51] performed numerical analysis with finite differences dynamic relaxation of square isotropic Mindlin (shear deformable) plates.

Paik et al. [52] have developed complex expressions for thin plate large deflection using the Galerkin method. Their example includes both transverse and axial edge compression load. Since the results show only in-plane loads on the edges, it is difficult to compare it to a case without these loads.

Nishawala [53] handles both nonlinear beams and plates. For plates, both movable and immovable edges are displayed. Several other sources are compared for deflections, but without stress. He suggests a third-degree polynomial load–deflection expression for the plate midpoint.

Jianqiao [54] uses both boundary elements and finite element to calculate deflection and midpoint stress for both simply supported and clamped immovable edges.

Abayakoon [55] studied beams as the main subject, while presenting also plates using a third-degree polynomial midpoint deflection expression. A deflection comparison is made to Timoshenko and Woinowsky-Krieger [30] and others. Stresses are simulated for ribs stiffened plates only, which cannot be compared with thin plates.

Seide [56] presents an expression for deflection, but for stresses it is limited to infinitely long plate, which cannot be used for a square plate.

Parker [57] solved the plate problem with finite differences. He presents membrane stress results with good agreement with Levy [36], but less good with Wang [40].

Belardi et al. [58] analyzed a circular plate made of shear deformable orthotropic composite materials. While the material has Cartesian XY orthotropy, the other variables in the analysis are polar. The shear deformations are calculated using the first-order shear deformation theory. Deflections and rotations are presented. Stresses are considered, but without presented results.

Plaut [59] in a recent study uses Reissner theory for plates that allows large strains (not to be confused with shear deformation) for circular and annular thin plates with both movable and immovable BCs. Results for various loading cases are presented, but for deflection only. No in-plane stresses are considered.

Finally, Shufrin et al. [60] solved the problem of laminated rectangular plates under large deflections with a semi-analytic method considering the coupling coefficients' tension-bending and bending-twisting. The nonlinear partial differential equations are converted to an iterative process of ordinary nonlinear differential equation according to Kantorovich method. The result is large mathematical expressions, calculated and compared to ANSYS FEA (Engineering Simulation Software, Canonsburg, PA, USA) with a good agreement. Several cases of local loads (patch-type load) are also demonstrated. In-plane stresses including shear stress are partially given along certain lines, loading arrangements.

Based on the above references, it seems that the membrane stress distribution along a plate under large deflections has been neglected in the literature. Despite the extensive research conducted in this field, there is still a lack of graphic pictures that describe the in-plane tensile and shear stresses generated in the plate as a function of the transverse loading. One should remember that under a transverse distributed load, the plate would deform. The general shape of the deformed plate is rather intuitive and easy to predict. However, unlike the deflection shape, the membrane stress fields created within the loaded plate are beyond our natural perception and are generally unknown. This is a real problem when trying to find areas with high tensile or compressive stress.

13.3.2 Large deflections of plates – calculations

The membrane stresses generated by the large deflection's regime are investigated, and their distribution and critical values are calculated and accompanied by graphical figures. The behavior of thin rectangular plates having various aspect ratios under transverse constant pressure was calculated for increasing loading parameters. Four types of boundary conditions were applied: all-around simply supported, with the in-plane movement being allowed (movable) or restricted (immovable), and all-around clamped edges with movable or immovable edges. The calculations were performed using Siemens Simcenter Femap with Nastran Ver. 2021.1 code (finite element, Siemens, Germany) [61].

The first model consisted of a thin square flat plate made of isotropic material with linear elastic response. The plate's dimensions were 6.28×6.28 m^2 (length × width). The plate had a thickness t of 12 mm and was made of isotropic plastics polycarbonate. The material polycarbonate was a transparent tough elastic polymer that is used in many engineering applications such as aircraft cockpit canopy, safety goggles, compact disks and greenhouse glazing. The type used in the present study was isotropic and has a Young's modulus of 2.4 GPa, Poisson's ratio 0.38, density 1,200 kg/m^3, tensile strength 63 MPa, high impact strength and price of about 2.7 €/kg. The xyz axes origin was at the plate's midpoint with z being normal to the surface. The plate was loaded in the z direction with an evenly distributed transverse load of $q = 800$ Pa or $q = 75$ Pa (see Fig. 13.6).

Two out-of-plane general boundary conditions were applied: the first one being all-around simply supported (transverse deflection and moment being zero and designated as SSSS) and the second one all-around clamped (transverse deflection and rotation being zero and short named CCCC). In addition, in-plane boundary conditions were used: allowing free movement of the plate's edges in the x and y directions nicknamed movable (M) or preventing this movement in both x and y directions leading to the case of immovable (I) (see Fig. 13.7).

As a result, four combinations of boundary conditions: SSSS-M, SSSS-I, CCCC-M and CCCC-I were used throughout the present study.

To calculate the nonlinear response of the transverse loaded plate, the Siemens Simcenter Femap with Nastran Ver. 2021.1 code [61] was used, enabling a high-resolution visualization of both the deflections and the generated stresses on the plate. The FEA process has three main steps: model preparation, running the solver and postprocessing. Model preparation steps included various topics: definition of the geometry of the plate including its thickness, definition of its material, meshing the surface using 2D CQUAD4 element (capable of modeling in-plane, bending and transverse shear behavior), application of the distributed load on the plate surface and defining the boundary conditions. Note that the CQUAD4 element was a quadrilateral one, with bending and membrane stiffness, and had four grid points and five Gauss points. The four types of BC were alternatively set on the surface perimeter as

- SSSS-M: TZ, designated 3.
- CCCC-M: TZ, RX, RY, designated 3, 4, 5.
- SSSS-I: TX, TY, TZ, designated T.
- CCCC-I: TX, TY, TZ, RX, RY, RZ, designated F.

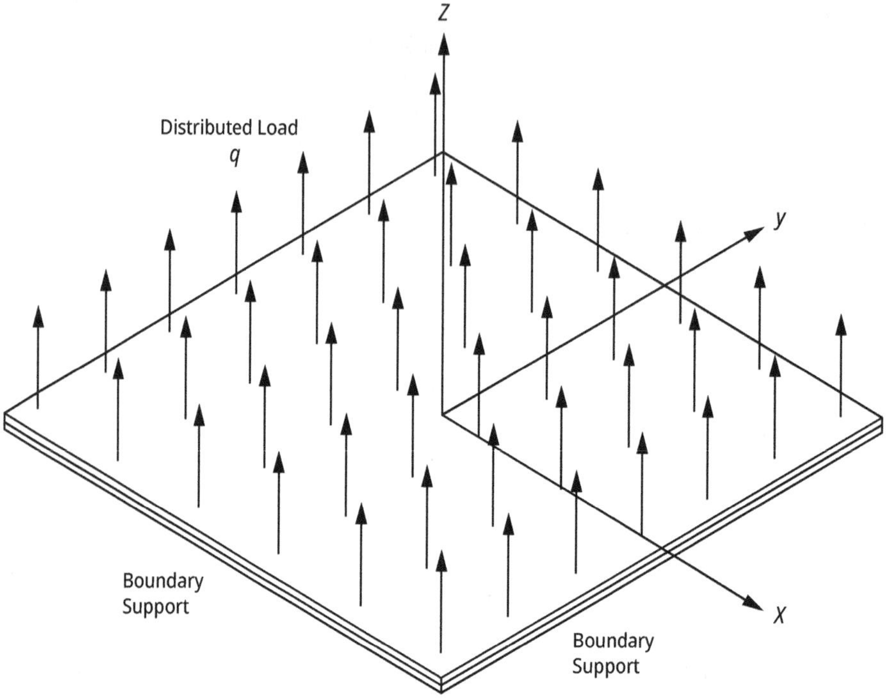

Fig. 13.6: The applied load and the relevant axes.

Additionally, the FEA required the elimination of all free body degrees of freedom (DOF). Therefore, virtual BC was added, where necessary: plate midpoint may have had TX, TY, designated as 1, 2, and one edge may have had additional TX or TY. Finally, the type of analysis was defined, for our case SOL106, which included nonlinear large deflections. The second step was the analyzing of the model constructed in the previous step. The third step, the postprocessing, contained a very rich set of tools allowing observation and reporting any requested feature of the plate performance. For the present application, four output vectors were chosen to display the deformation and contour (color) styles of the results: total translation (deflection), plate X membrane force, plate Y membrane force and plate XY membrane force. The resulting pictures (2D and 3D) were then saved to be used in the report.

The mesh density had an influence on both the accuracy and the calculation time. In the following convergence study, several models with various mesh densities were tested for deflection and force versus calculation duration. The highest density 500 × 500 was taken as the reference – 100%. The results are shown in Tab. 13.2 and Fig. 13.8.

Using the above results, the 100 × 100 mesh density was selected for the entire work, having a good balance between accuracy and calculation duration, and no significant changes occurred in higher densities.

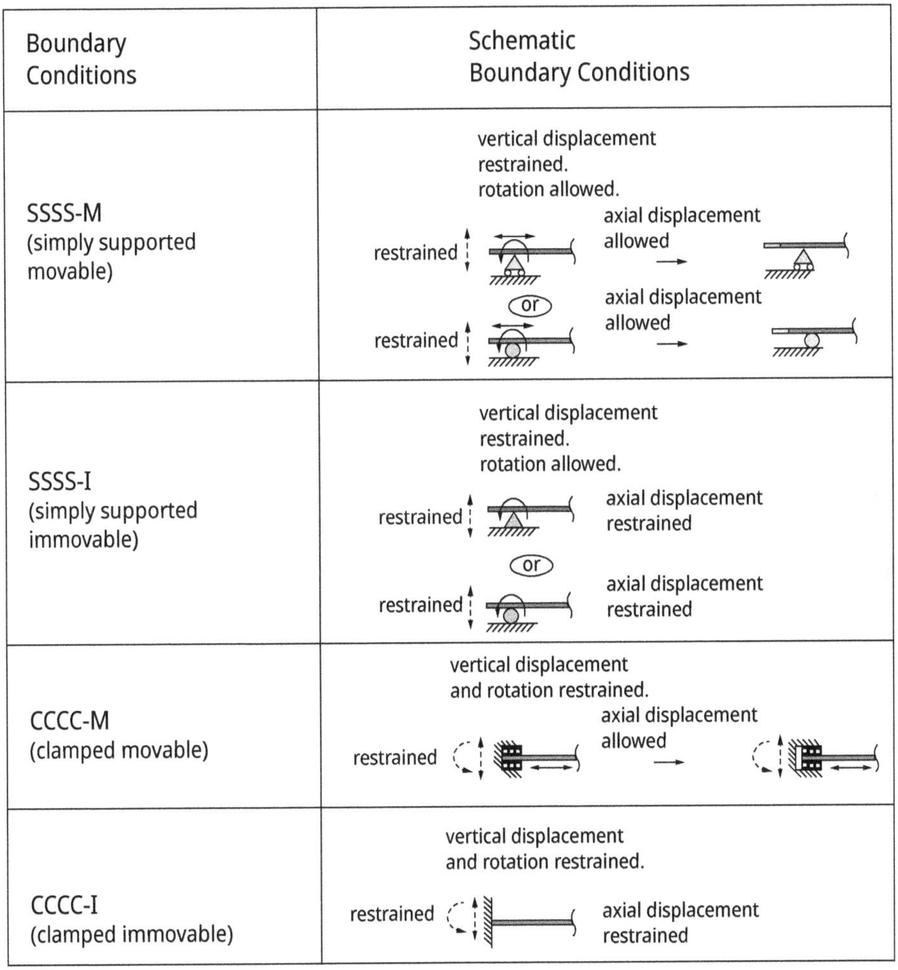
Boundary Conditions
Schematic Boundary Conditions
SSSS-M
(simply supported movable)
vertical displacement restrained.
rotation allowed.
restrained
axial displacement allowed
or
restrained
axial displacement allowed
SSSS-I
(simply supported immovable)
vertical displacement restrained.
rotation allowed.
restrained
axial displacement restrained
or
restrained
axial displacement restrained
CCCC-M
(clamped movable)
vertical displacement and rotation restrained.
restrained
axial displacement allowed
CCCC-I
(clamped immovable)
vertical displacement and rotation restrained.
restrained
axial displacement restrained

Fig. 13.7: The schematic boundary conditions used in the present chapter (adapted from [62]).

Tab. 13.2: Convergence study.

Mesh size	Number of elements in the mesh	Calculation duration (min)	Midpoint deflection (%)	Max *X* force (%)	Midpoint *X* force (%)
10 × 10	100	0.183	83.468	78.516	98.617
20 × 20	400	0.267	93.709	83.567	99.647
50 × 50	2,500	0.533	99.030	94.438	99.988
100 × 100	10,000	1.97	99.766	98.249	99.998
200 × 200	40,000	8.92	99.942	99.568	99.999
500 × 500	250,000	162	100	100	100

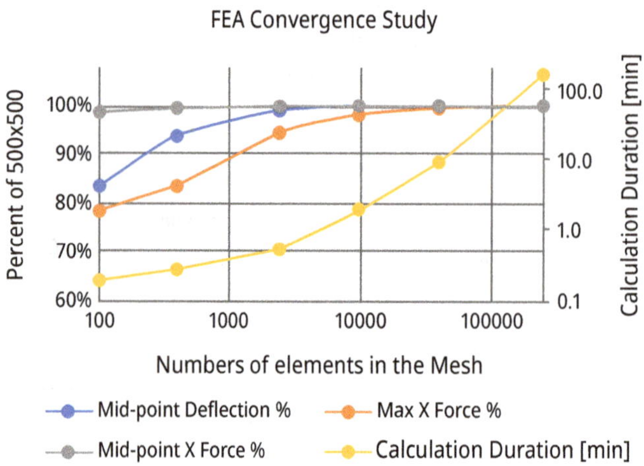

Fig. 13.8: Finite element convergence analysis versus calculation convergence and mesh density.

A nonlinear static analysis with 20 steps was applied, where the transverse load gradually increased, while deflections and stresses were recalculated at each step. This analysis was repeated for each of the four boundary conditions, described above.

In Femap, the analysis program 10. Nonlinear static uses SOL 106 [63] which can handle many nonlinear situations such as

- large deflection with small strains;
- nonlinear stress–strain material response;
- material plastic yield;
- geometric nonlinearities;
- creep behavior;
- snap-in mechanism;
- physical contact between objects; and
- thin-shell buckling.

For the present case, however, the only nonlinear parameter was large deflection that caused system stiffening due to in-plane stresses that accumulated additional elastic energy, resulting in a nonlinear load–deflection response. The way to handle this nonlinearity was to increase the load in 20 steps, where in each step the program used a linear formulation but performed several iterations, until convergence of the energy and the load was obtained. The program may have added additional steps when the convergence was slow. A complete description of SOL 106 and nonlinear static analysis is available (see the link in the reference list [63]).

To validate the FEA used here, a comparison of results presented by Timoshenko and Woinowsky-Krieger [30, p. 427] was carried out. A Femap model with the same plate configuration as in [30] was used.

The properties of the thin plate model are presented in Tab. 13.3, and the results are displayed in Fig. 13.9.

Tab. 13.3: The data used for the validation step.

Size	Length × width $a \times a$ (m^2) = 1 × 1
Thickness	h (m) = 0.006
Young's modulus	E (GPa) = 2.4 (isotropic material)
Poisson's ratio	ν = 0.316
Distributed load (uniform)	q (Pa) = 871
BCs	SSSS-I (four-edged simply supported, in-plane immovable)
Mesh	100 × 100 plate, Cquad4 elements (plane stress)
Analysis	Nonlinear static, 15 steps
Model run time	t (s) = 35

Note that points A, B, C in Fig. 13.8 are located five elements away from the plate edge.

As shown in Fig. 13.9, there is an excellent agreement of the present FEA results with Timoshenkoand Woinowsky-Krieger's results for points A, C and D, while for point B a maximal deviation up of 16.5% was detected. In view of the excellent agreement for points A, C and D and the deviation for point B, it is suggested that an error might occurred on the drawing of the original graph (Timoshenko and Woinowsky-Krieger's book [30]), and it might have a higher inclination, eventually coinciding with the present FEA results.

Figure 13.10a, b presents the 2D and 3D views of X direction membrane forces, as calculated by the present FEA code.

Therefore, the present results presented above using the Siemens FEA code are viable and correctly represent the stresses of the plate. Typical results are next displayed, with more being found in [64].

Figure 13.11a–d presents the deformed shape of a square plate at a transverse load of 800 Pa for the four boundary conditions used in the present study. As can also be seen from Tab. 13.4, the in-plane boundary conditions (movable vs. immovable) play a major role in stiffening the lateral deflection of the plate. Restricting the in-plane boundary conditions reduces the transverse deflection by a factor of 3.3 for the all-around simply supported (SSSS) boundary conditions or by a factor of 2.7 for the all-around clamped case (CCCC). Also note the large, normalized transverse deflections (see Tab. 13.2) experienced by the plate for movable in-plane boundary conditions (26.83 for SSSS and 21.58 for CCCC) in comparison with immovable in-plane boundary conditions (8.13–8.0 for SSSS and CCCC).

The calculated membrane forces are presented next. Note that as the cross section of the plate is constant, the distribution of the membrane forces also depicts exactly the stresses on the plate (a division of the membrane force by a factor 12,000 would yield the stress in at the same point).

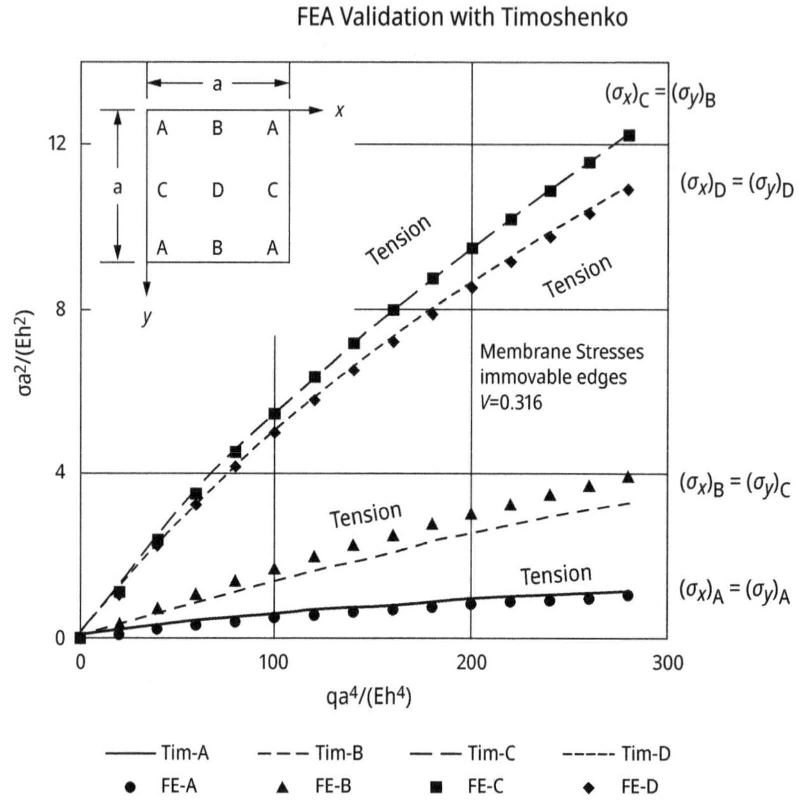

Fig. 13.9: Nondimensional stress versus applied pressure, *q*: the validation case.

Figure 13.12a–d presents 2D and 3D views of the *x* and *y* membrane forces' distribution generated on the plate on SSSS-M boundary conditions due to large transverse deflections. As the investigated plate is a square, it is obvious that the *y* membrane forces map (or the *y* stresses map) (see Fig. 13.11c) has an identical appearance to the *x* membrane forces map (or the *x* stresses map) after 90° rotation in the *xy* plane (see Fig. 13.11a). Therefore, when the *y* membrane stress is sought, one would be referred to the *x* stress map rotated by 90°.

The associated membrane stresses maps for a CCCC-M plate are given in Fig. 13.13a, b, displaying interesting fluctuations along the edges of the plate.

Based on Fig. 13.12a–d and 13.13a,b, some preliminary observations can be put forward:

- The maps for the distributions of σ_{xx} and σ_{yy} membrane stresses are symmetrical relative to both *x*- and *y*-axes.
- The middle plate area encounters tension stresses in both *x* and *y* directions, with moderate changes of the amplitude.

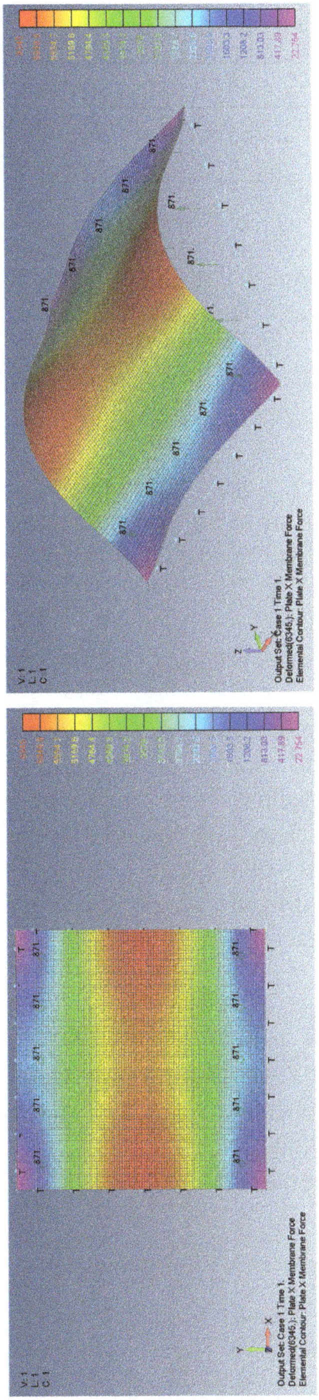

Fig. 13.10: The FEA model 2D (a) and 3D (b) views of *X* direction membrane forces.

Tab. 13.4: Midpoint deflections for a square plate.

Boundary conditions	**SSSS-M**	**SSSS-I**	**CCCC-M**	**CCCC-I**
Midpoint deflection w_0 (mm)	322.0	97.6	259.0	96.0
w_0/t	26.83	8.13	21.58	8.0

- At the plate edges, namely at $y = \pm 3.14$ m high σ_{xx} compression stress is visible, with a similar behavior at $x = \pm 3.14$ m high σ_{yy} compression stress. The presence of compressive stress on the plate's edges might lead to local buckling at those areas.
- At the plate's corners, very sharp changes in the stress amplitudes are encountered possibly due to the relatively coarse mesh at those locations, as the mesh distribution was kept constant across the plate.
- The maximal midpoint tensile stress (in both x and y directions) for the SSSS-M case is 1.684 MPa, while the maximal compression stress on the plate's edges reaches the value of − 10.767 MPa. For the CCCC-M case, these stresses reach the values of 1.518 and − 6.674 MPa, accordingly.

The behavior of the square plate, while applying immovable boundary conditions, is presented by Fig. 13.14a–d.

Unlike the movable boundary conditions, the immovable SSSS-I and CCCC-I present only tensile forces. The σ_{xx} and σ_{yy} membrane stresses maintain their symmetry relative to both x and y axes; tension stresses in both x and y directions exist on the plate middle area, with moderate changes, while in the vicinity of the plate edges at $x \approx 0, y = \pm 3.14$ m, high tension values of σ_{yy} are present as well as at $x = \pm 3.14$ m, $y \approx 0$, where high σ_{xx} is generated by the large deflections of the plate. Note that, approximately at the plate corners, one can find low stress values. The maximal midpoint tensile stress (in both x and y directions) for the SSSS-I case is 2.095 MPa, while the maximal tension stress on the plate's edges reaches the value of 2.404 MPa. For the CCCC-I case, these stresses reach the values of 1.993 and 2.037 MPa, accordingly.

The distribution of the shear forces and accordingly the τ_{xy} stresses is presented in Fig. 13.15a–d for SSSS-M and CCCC-M, Fig. 13.16 (a detail of Fig. 14.15b) and in Fig. 13.17a–d for and SSSS-I and CCCC-I boundary conditions. Note that for a better visualization of the shear stresses at the plate corner, the mesh was increased to 500 × 500, and the result is presented in Fig. 13.16.

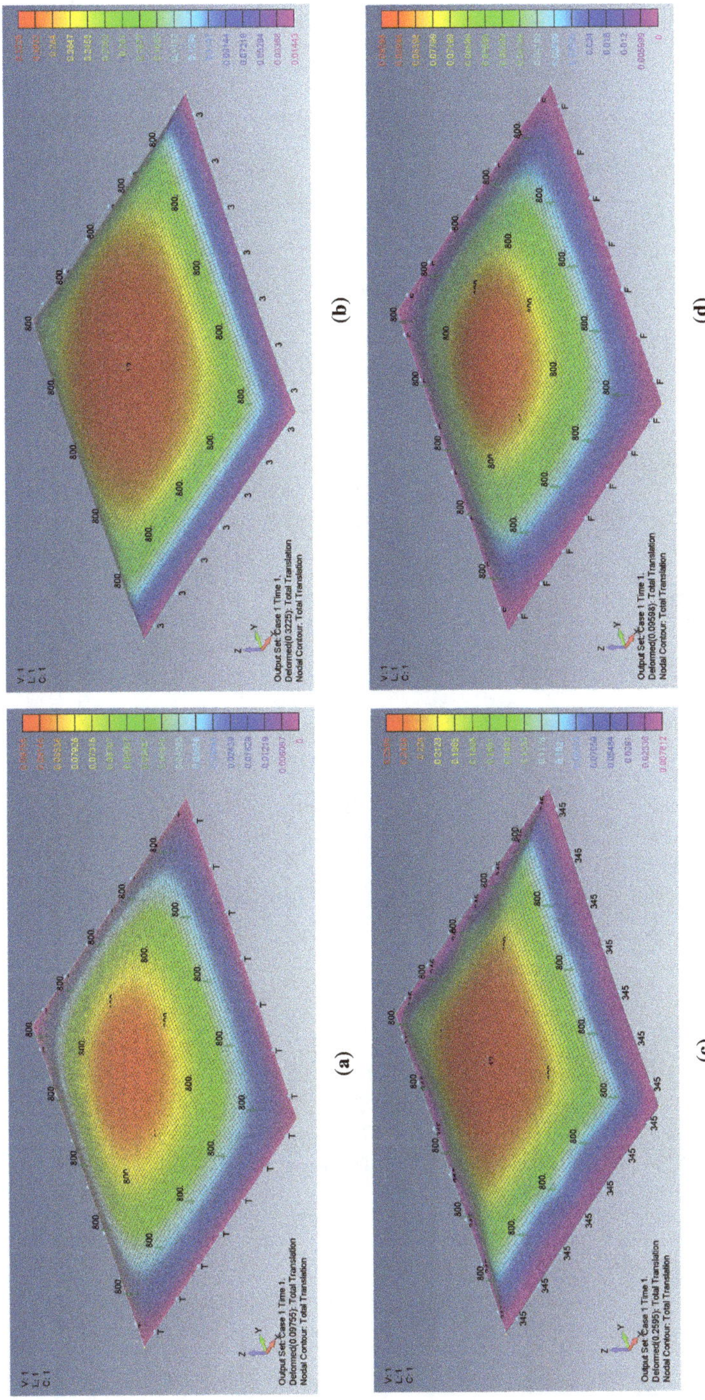

Fig. 13.11: The shape of the transverse deflection for a square plate: (a) SSSS-M, (b) SSSS-I, (c) CCCC-M and (d) CCCC-I.

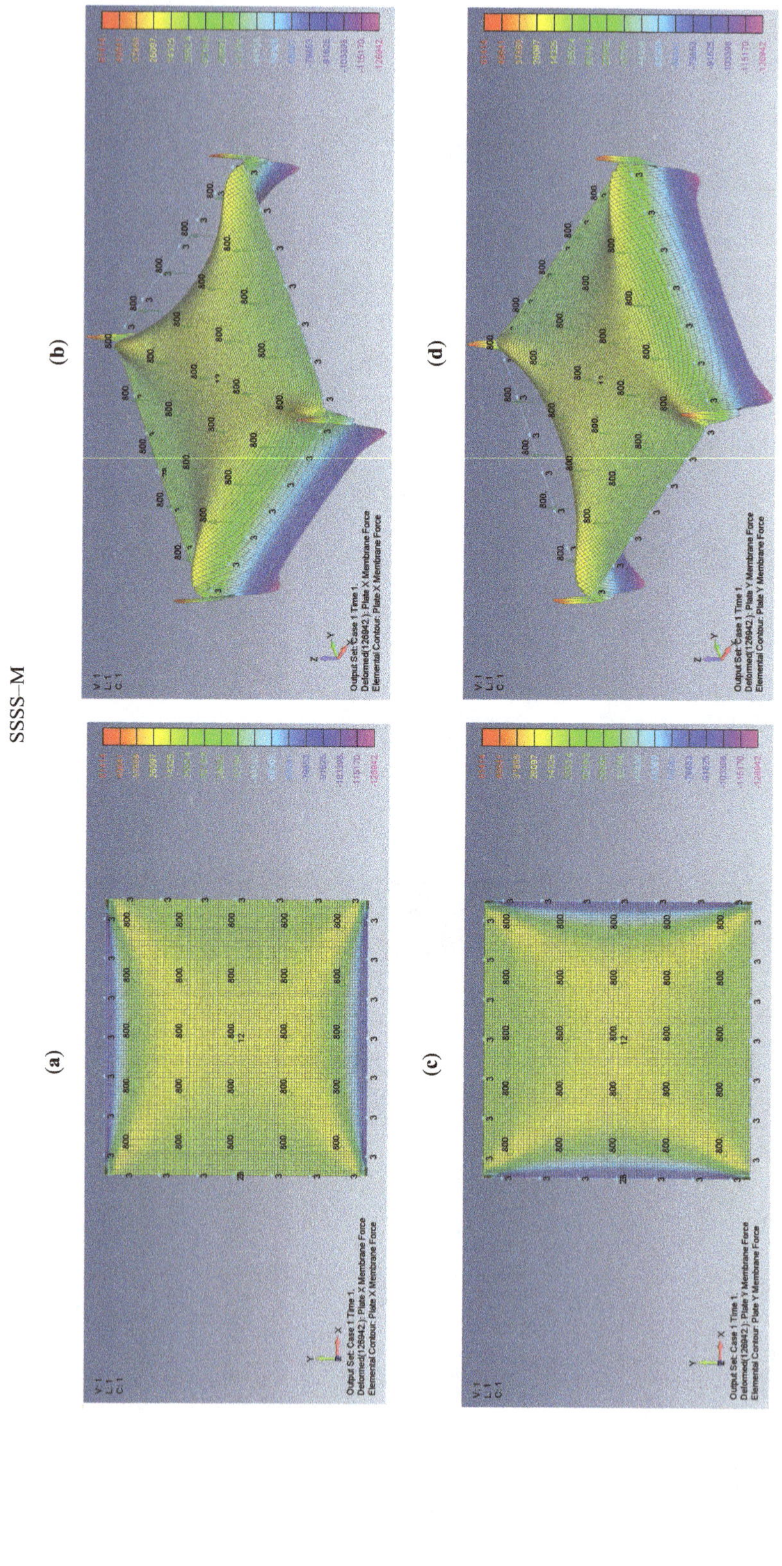

Fig. 13.12: The shapes of the membrane forces for a square plate on SSSS-M boundary conditions: (a) top view of the *x* membrane force (2D view), (b) a 3D view of the *x* membrane force, (c) top view of the *y* membrane force (2D view) and (d) a 3D view of the *y* membrane force.

CCCC–M

(a) (b)

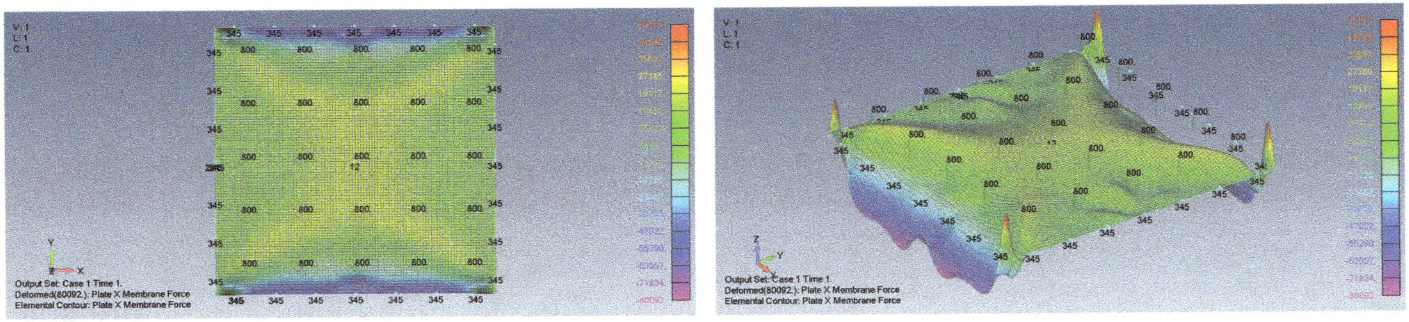

Fig. 13.13: The shapes of the membrane forces for a square plate on CCCC-M boundary conditions: (a) top view of the *x* membrane force (2D view) and (b) a 3D view of the *x* membrane force.

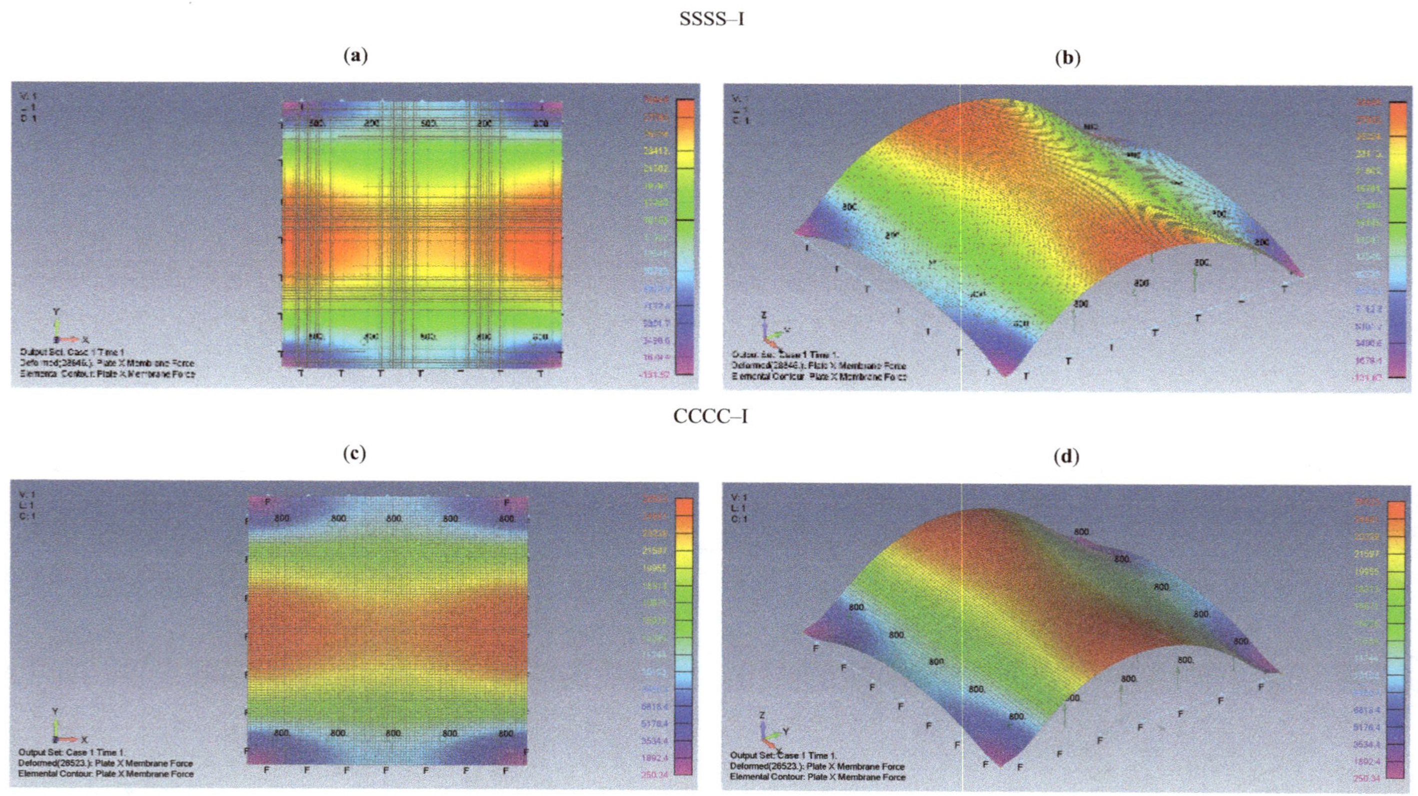

Fig. 13.14: The shapes of the *x* and *y* membrane forces for a square plate on SSSS-I boundary conditions: (a) top view of the x membrane force (2D view), (b) a 3D view of the *x* membrane force and on CCCC-I boundary conditions, (c) top view of the *x* membrane force (2D view) and (d) a 3D view of the *x* membrane force.

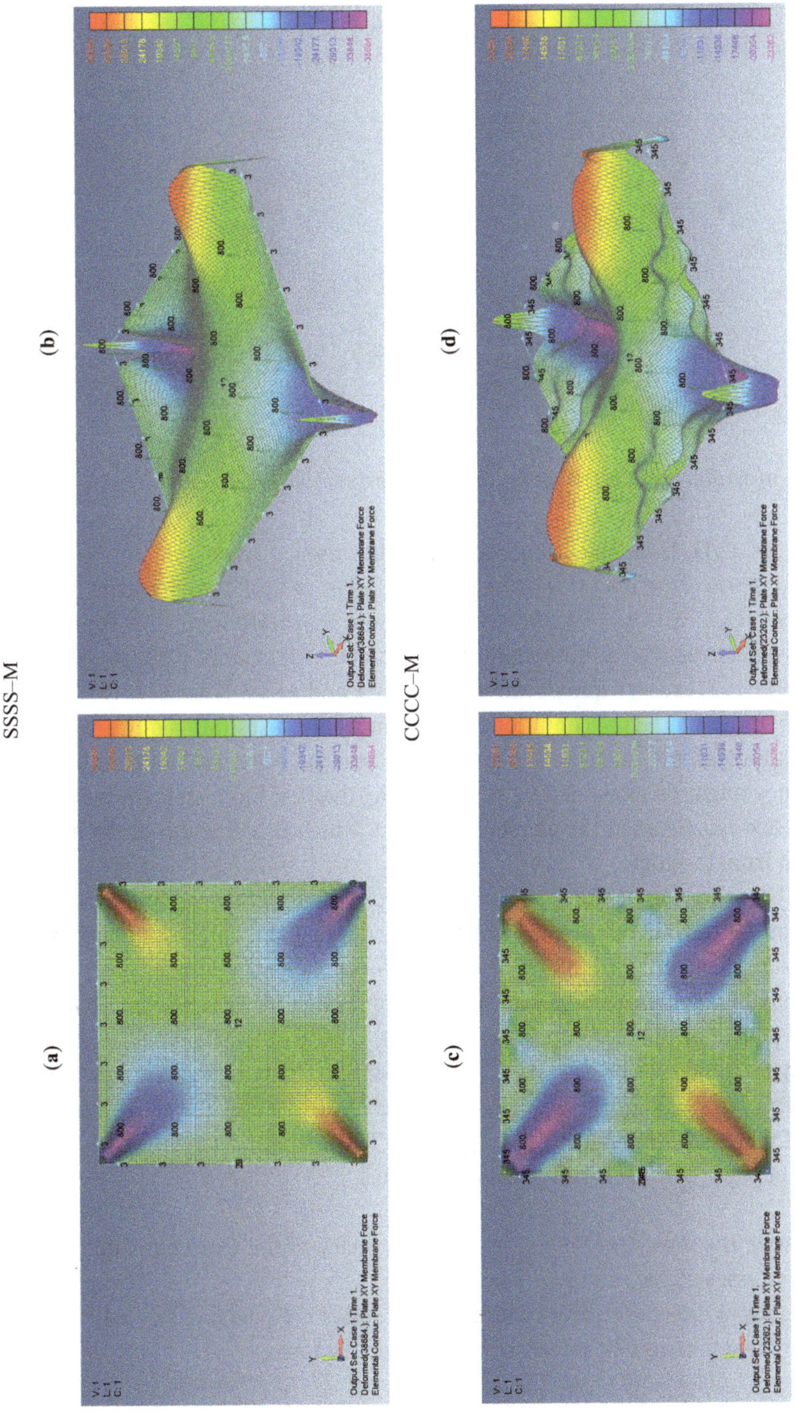

Fig. 13.15: The shapes of the shear *xy* membrane forces for a square plate with movable BCs: (a) SSSS-M top view of the *xy* membrane force (2D view), (b) SSSS-M 3D view of the *xy* membrane force, (c) CCCC-M top view of the *xy* membrane force (2D view) and (d) CCCC-M 3D view of the *xy* membrane force.

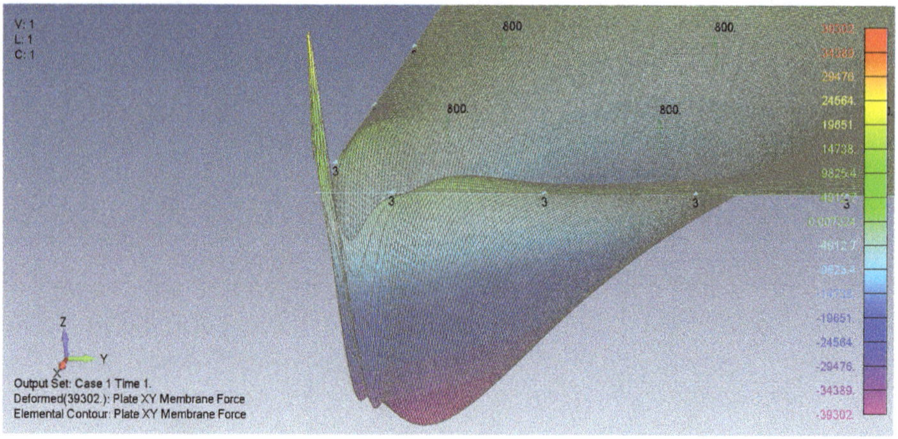

Fig. 13.16: A 3D detail of the *xy* membrane forces at the plate corner for SSSS-M.

Based on Figs. 13.15–13.16, one can observe the following:
- The τ_{xy} stress function is antisymmetric relative to both x and y axes, namely $\tau_{xy}(x,y) = \tau_{xy}(-x,-y) = -\tau_{xy}(x,-y) = -\tau_{xy}(-x,y)$. In addition, the function is symmetrical relative to both square main diagonals, namely $\tau_{xy}(x,y) = \tau_{xy}(y,x)$, $\tau_{xy}(x,y) = \tau_{xy}(-x,-y)$, and its value along the x and y axes is zero.
- Although the boundary conditions for the movable cases (M) require $\tau_{xy} = 0$ on the plate edges, the calculated shear stresses near the edges are not zero. This discrepancy might be explained by remembering that the finite element membrane forces are calculated at the element midpoint, which is half width of the element distant from the edge.
- Very sharp changes of the shear stresses values are encountered at the plate corners for the movable (M) cases, which is like the movable (M) cases tensile stress variations depicted in Fig. 13.12b, d and 13.13b.
- The maximal values of the shear stresses are located on the two main diagonals of the plate. For the SSSS-M case we get $(\tau_{xy})_{max} = \pm 3.224\text{MPa}$, while for CCCC-M boundary conditions we get $(\tau_{xy})_{max} = \pm 1.938\text{MPa}$. Changing the boundary conditions from movable (M) to immovable (I) drastically reduces the shear stresses to yield for the SSSS-I a value of $(\tau_{xy})_{max} = \pm 0.278\text{MPa}$, while for the CCCC-I case a value of $(\tau_{xy})_{max} = \pm 0.174\text{MPa}$ was calculated.

To summarize the topic, Tab. 13.5 presents a summary of the extremal values of the generated stresses σ_{xx}, σ_{yy} and τ_{xy}.

As already written, the cases of large aspect ratios are presented in [64], and the reader is invited to read that reference to complete the topic.

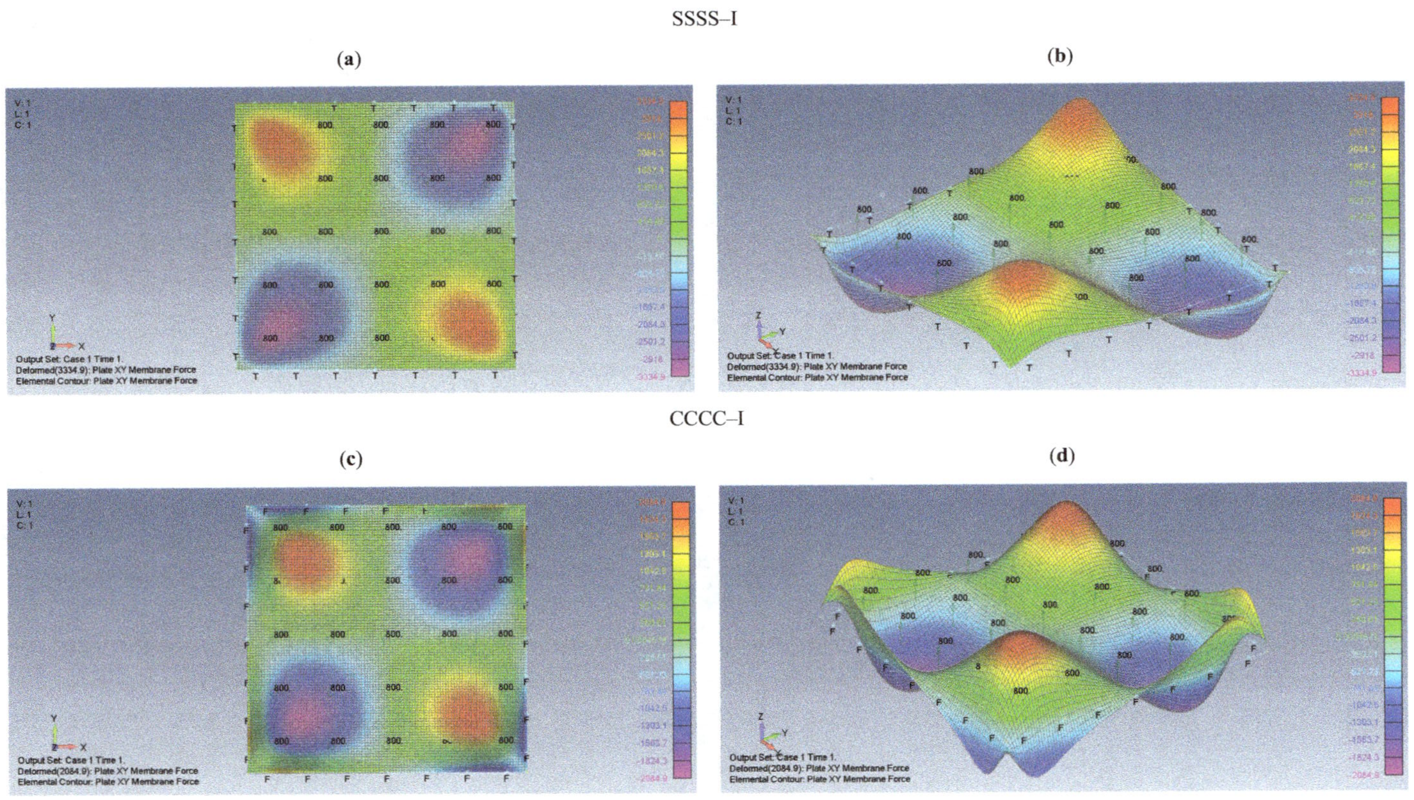

Fig. 13.17: The shear *xy* membrane forces for a square plate with immovable BCs: (a) SSSS-I top view of the *xy* membrane force (2D view), (b) SSSS-I 3D view of the *xy* membrane force, (c) CCCC-I top view of the *xy* membrane force (2D view) and (d) CCCC-I 3D view of the *xy* membrane force.

Tab. 13.5: Maximal values of the membrane stresses for rectangular flat plates.

Aspect ratio (AR)	1	2	5
Distributed load q (Pa)	800	75	75
Boundary conditions	SSSS-M		
σ_{xx} (MPa) tensile stress @ plate midpoint	1.684	0.906	1.880
σ_{yy} (MPa) tensile stress @ plate midpoint	1.684	0.0684	0.0126
σ_{xx} (MPa) compression stress @ plate edges	−10.767	−2.077	−2.713
σ_{yy} (MPa) compression stress @ plate edges	−10.767	−2.087	−2.653
τ_{xy} (MPa) shear stress @ 45° line from the corner	±3.224	±0.607	±0.738
Boundary conditions	SSSS-I		
σ_{xx} (MPa) tensile stress @ plate midpoint	2.095	0.295	0.217
σ_{yy} (MPa) tensile stress @ plate midpoint	2.095	0.553	0.558
σ_{xx} (MPa) tensile stress @ plate edges	2.404	0.465	0.458
σ_{yy} (MPa) tensile stress @ plate edges	2.404	0.583	0.572
τ_{xy} (MPa) shear stress @ 45° line from corner	±0.278	±0.064	±0.068
Boundary conditions	CCCC-M		
σ_{xx} (MPa) tensile stress @ plate midpoint	1.584	0.780	1.111
σ_{yy} (MPa) tensile stress @ plate midpoint	1.584	0.0629	0.00760
σ_{xx} (MPa) compression/tensile stress @ plate edges	−6.674	−1.227	−1.576/ + 1.533
σ_{yy} (MPa) compression/tensile stress @ plate edges	−6.674	−1.047	−1.176/ + 1.506
τ_{xy} (MPa) @ 45° line from corner into plate	±1.938	±0.299	±0.371
Boundary conditions	CCCC-I		
σ_{xx} (MPa) tensile stress @ plate midpoint	1.993	0.261	0.192
σ_{yy} (MPa) tensile stress @ plate midpoint	1.993	0.490	0.494
σ_{xx} (MPa) compression stress @ plate edges	2.037	0.377	0.370
σ_{yy} (MPa) compression stress @ plate edges	2.037	0.515	0.505
τ_{xy} (MPa) shear stress @ 45° line from corner	±0.174	±0.0352	±0.0388

Based on the investigation presented above, the following conclusions were drawn:
- The general shape of the plate deflection is rather similar for all the investigated cases. The various four applied boundary conditions do not significantly change the appearance of the deformed plate.
- Enabling the in-plane movement of the plate would generate higher membrane stresses for both SSSS and CCCC boundary conditions in comparison with restricting this movement yielding an immovable boundary condition.
- The stresses generated on the plate due to its large transverse deflections for the clamped cases CCCC-M and CCCC-I are consistently lower than those on simply supported cases SSSS-M and SSSS-I (see Tab. 13.5).
- Compression stresses would appear on the plate edges for both movable SSSS and CCCC boundary conditions. This should be considered during the design of the plate to prevent local buckling of the plate.

- To prevent local buckling of the plate, it is recommended to assure CCCC or SSSS immovable boundary conditions, leading to only tensile stresses.
- The existence of tensile stresses in the movable cases at relatively high aspect ratio AR = 5 suggests checking how these stresses asymptotically approach zero for the case of infinitely long plate, where the membrane tensile stress must be zero.

References

[1] Euler, L. Sur la force des colonnes (in English: concerning the Strength of Columns), Mem Ad Berlin, 13, 1759, 252–282.

[2] Lagrange, J.L. Sur la figure des colonnes (in English: concerning the Figure of Columns), Miscellanea Taurinensia, 5, 1770–1773, Also in: Œuvres de Lagrange, Vol. 2, Gauthier-Villars, Paris, France, 1868, 125–170.

[3] Timoshenko, S.P. and Gere, J.M. Theory of elastic stability, 2nd Edition, McGraw-Hill International Book Company, Inc, 1961, 280, Ch. 2.7, 76–82.

[4] Timoshenko, S.P. History of strength of materials, Dover Publications, Inc, New York, 425.

[5] Levien, R. The Elastic: A Mathematical History, Technical Report No. UCB/EECS-2008-103, http://www.eecs.berkeley.edu/Pubs/TechRpts/2008/EECS-2008-103.html August 2008, 27.

[6] Barten, H.J. On the deflection of a cantilever beam, Quarterly of Applied Mathematics, 2(2), July 1944, 168–171.

[7] Barten, H.J. Corrections to My Paper: on the deflection of a cantilever beam, Quarterly of Applied Mathematics, 3(3), October 1945, 275–276.

[8] Bisshop, K.E. and Drucker, D.C. Large deflection of cantilever beams, Quarterly of Applied Mathematics, 3(3), October 1945, 272–275.

[9] Conway, H.D. The nonlinear bending of thin circular rods, the American society of mechanical engineers, Applied Mechanics Division presented at the Diamond Jubilee Annual Meeting, Vol. 111, Chicago, November13 - 18 1955, 4.

[10] Rohde, F.V. Large deflections of a cantilever beam with uniformly distributed load, Quarterly of Applied Mathematics, 11(3), 1953, 337–338.

[11] Conway, H.D. The large deflection of simply supported beams, Philosophical Magazine, 38, 1947, 905–911.

[12] Wang, T.M., Lee, S.L. and Zienkiewicz, O.C. A numerical analysis of large deflections of beams, International Journal of Mechanical Sciences, 3, 1961, 219–228.

[13] Bélendez, T., Neipp, C. and Bélendez, A. Large and small deflections of a cantilever beam, European Journal of Physics, 23(3), May 2002, 371–379.

[14] Lubis, A. and Tanti, N. Large deflections analysis of thin cantilever beams using numerical integration and experimental procedures, Seminar Nasional Tahuna teknik Mesin (SNTTM) V, Universitas Indonesia, 21–23, November 2006, 9.

[15] Shvartsman, B.S. Large deflections of a cantilever beam subjected to a follower force, Journal of Sound and Vibration, 304, 2007, 969–973.

[16] Eren, I. Determining large deflections in rectangular combined loaded cantilever beams made of non-linear ludwick type material by means of different arc length assumptions, Sādhanā, 33(Part, 1), Feb 2008, 45–55.

[17] Borboni, A. and De Santis, D. Large deflection of a non-linear, elastic, asymmetric ludwick cantilever beam subjected to horizontal force, vertical force and bending torque at the free end, Meccanica, 49, 2014, 1327–1336.

[18] Borboni, A., De Santis, D., Solazzi, L., Villafane, J.H. and Faglia, R. Ludwick cantilever beam in large deflection under vertical constant load, The Open Mechanical Engineering Journal, 10, 2016, 23–37.
[19] Banerjee, A., Bhattacharya, B. and Mallik, A.K. Large deflection of cantilever beams with geometric non-linearity: analytical and numerical approaches, International Journal of Non-Linear Mechanics, 43(5), June 2008, 366–376.
[20] Tolou, N. and Herder, J.L. A semi-analytical approach to large deflections in compliant beams under point loading, Mathematical Problems in Engineering, 2009, Article Id. 910896, 13.
[21] Kocatürk, T., Akbas, Ş.D. and Şimsek, M. Large deflection static analysis of a cantilever beam subjected to a point load, International Journal of Scientific Engineering and Applied Science, 2(4), 2010, 1–13.
[22] Nishawala, V. A study of large deflection of beams and plates, M.Sc. thesis submitted to the Graduate School –New Brunswick Rutgers, The State University of New, Jersey, USA, Jan. 2011, 118.
[23] Khavvaji, A., Pashaei, M.H., Dardel, M. and Alashti, R.A. Large deflection analysis of compliant beams of variable thickness and non-homogenous material under combined load and multiple boundary conditions, International Transaction C: Aspects, 25(4), December 2012, 353–362.
[24] Sitar, M., Kosel, F. and Brojan, M. A simple method for determining large deflection states of arbitrarily curved planar elastica, Arch, Applied Mechanics, 84, 2014, 263–275.
[25] Khosravi, M. and Jani, M. Numerical resolution of large deflections in cantilever beams by bernstein spectral method and convolution quadrature, International Journal of Nonlinear Analysis and Applications, 9(1), 2018, 117–127.
[26] McHugh, K.A. Large deflection inextensible beams and plates and their responses to non-conservative forces: theory and computations, Ph.D. thesis submitted and accepted at the Department of Mechanical Engineering and Materials Science in the Graduate School of Duke University, USA, 2020, 225.
[27] Deliyianni, M., Gudibanda, V., Howell, J. and Webster, J.T. Large deflections of inextensible cantilevers: Modeling, theory and simulations, Mathematical Modelling of Natural Phenomena, 15, 2020, Article Id. 44, 34.
[28] Kirchhoff, G. Vorlesungen über Mathematische Physic, Druck und Verlag von B.G. Teubner, Leipzig, Germany, 1876, 481.
[29] Love, A.E.H. A treatise on the mathematical theory of elasticity, Dover Publications, Nassau County, NY, USA, 1944, 571.
[30] Timoshenko, S. and Woinowsky-Krieger, S. Theory of plates and shells, 2nd Edition, McGraw-Hill Book Company, New York, NY, USA, 1987, 611.
[31] von Kármán, T. Festigkeitsprobleme im maschinenbau (strength problems in mechanical engineering), Encyclopädie der Mathematischen Wissenschaften, B.G. Teubner Verlag, Germany, 1910, 311–385.
[32] Föppl, A. Vorlesungen über technische mechanik (lectures on technical mechanics), Bd., vol. 5, B.G. Teubner., Leipzig, Germany, 1907, 132.
[33] Bakker, M., Rosmanit, M. and Hofmeyer, H. Approximate large-deflection analysis of simply supported rectangular plates under transverse loading using plate post-buckling solutions, Thin-Walled Structures, 46, 2008, 1224–1235.
[34] Marguerre, K. Zur theorie der gekrümmter platte grosser formänderung (on the theory of curved plate with large displacements), in Proceedings of the 5th international congress for applied mechanics, Cambridge, MA, USA, 12–26 September 1938, 93–101.
[35] Ramberg, W., McPherson, E.A. and Levy, S. Normal-pressure tests of rectangular plates; NACA Report 748, NACA-National Bureau of Standards, Washington, D.C., USA, 1941, 24.
[36] Levy, S. Bending of rectangular plates with large deflections; NACA Report 737, NACA- National Bureau of Standards, Washington, DC, USA, 1942, 19.

[37] Levy, S. and Greenman, S. Bending of large deflection of clamped rectangular plate with length-width ratio of 1.5 under normal pressure; NACA TN-853, NACA- National Bureau of Standards, Washington, DC, USA, 1942, 47.

[38] Levy, S. Large deflection theory of curved sheet; NACA TN-895, NACA: National Bureau of Standards, Washington, DC, USA, 1943, 30.

[39] Wang, C.T. Nonlinear large-deflection boundary-value problems of rectangular plates, NACA TN-1425, NACA- National Bureau of Standards, Washington, DC, USA, 1948, 113.

[40] Wang, C.T. Bending of rectangular plates with large deflections, NACA TN-1462, NACA-National Bureau of Standards, Washington, DC, USA, 1948, 34.

[41] Yen, K.T., Gunturkum, S. and Pohle, V.F. Deflections of simply supported rectangular sandwich plate subjected to transverse loads, NACA TN-2581, NACA-National Bureau of Standards, Washington, DC, USA, 1951, 39.

[42] Berger, M.H. A new approach to the analysis of large deflections of plates, Ph.D. Thesis, California Institute of Technology, Pasadena, CA, USA, 1954, 68.

[43] Scholes, A. and Bernstein, E.L. Bending of normally loaded simply supported rectangular plates in the large-deflection range, Journal of Strain Analysis, 4, 1969, 190–198.

[44] Scholes, A. Application of large-deflection theory to normally loaded rectangular plates with clamped edges, Journal of Strain Analysis, 5, 1970, 140–144.

[45] Li-Zhou, P. and Shu, W. A perturbation-variational solution of the large deflection of rectangular plates under uniform load, Applied Mathematics and Mechanics, 7, 1986, 727–740.

[46] Bert, C.W., Jang, S.K. and Striz, A.G. Nonlinear bending analysis of orthotropic rectangular plates by the method of differential quadrature, Computational Mechanics, 5, 1989, 217–226.

[47] Yeh, F. and Liu, W. Nonlinear analysis of rectangular orthotropic plates, International Journal of Mechanical Sciences, 33, 1991, 563–578.

[48] Wang, D. and El-Sheikh, A.I. Large-deflection mathematical analysis of rectangular plates, Journal of Engineering Mechanics, 131, 2005, 809–821.

[49] Ugural, A.C. Plates and shells theory and analysis, CRC Press, Taylor & Francis Group, Boca Raton, FL, USA, 2018, 618.

[50] Razdolsky, A.G. Large deflections of elastic rectangular plates, International Journal for Computational Methods in Engineering Science, 16, 2015, 354–361.

[51] Turvey, G. and Osman, M. Elastic large deflection analysis of isotropic rectangular mindlin plates, International Journal of Mechanical Sciences, 32, 1990, 315–328.

[52] Paik, J.K., Park, J.H. and Kim, B.J. Analysis of the elastic large deflection behavior for metal plates under non-uniformly distributed lateral pressure with in-plane loads, Journal of Applied Mathematics, 2012, 2012, Article Id.:734521.

[53] Nishawala, V.V. A study of large deflection of beams and plates, MSc Thesis, New Brunswick, Rutgers, The State University of New Jersey, New Brunswick, NJ, USA, 2011.

[54] Ye, J. Large deflection of imperfect plates by iterative Be-Fe method, Journal of Engineering Mechanics, 120, 1994, 431–444.

[55] Abayakoon, S.B.S. Large deflection elastic-plastic analysis of plate structures by the finite strip method, Ph.D. Thesis, Department of Civil Engineering, The University of British Columbia, Vancouver, BC, Canada, 1987.

[56] Seide, P. Large deflections of prestressed simply supported rectangular plates under uniform pressure, International Journal of Non-Linear Mechanics, 13, 1978, 145–156.

[57] Parker, H.P. A numerical solution to the general large deflection plate equations, SAE Transmission, 74, 1966, 236–243.

[58] Belardi, V.G., Fanelli, P. and Vivio, F. On the radial bending of shear-deformable composite circular plates with rectilinear orthotropy, European Journal of Mechanics – A/Solids, 86, 2021, Article Id.: 104157.

[59] Plaut, R.H. Generalized Reissner analysis of large axisymmetric deflections of thin circular and annular plates, International Journal of Solids and Structures, 203, 2020, 131–137.

[60] Shufrin, I., Rabinovitch, O. and Eisenberger, M. A semi-analytical approach for the non-linear large deflection analysis of laminated rectangular plates under general out-of-plane loading, International Journal of Non-Linear Mechanics, 43, 2008, 328–340.

[61] Siemens. Simcenter Femap with Nastran Ver. 2021 1., Siemens Digital Industries Software, siemens.com/software, 2021. Available online: https://www.plm.automation.siemens.com/global/en/products/simcenter/femap.html .

[62] Erochko, J. An introduction to structural analysis, 1st Edition, Carleton University, Ottawa, ON, 2020, 384.

[63] Siemens. Simcenter Nastran Basic Nonlinear Analysis user's Guide, Siemens Industry Software: Plano, TX, USA, 2020, Available online https://docs.plm.automation.siemens.com/data_services/resources/scnastran/2020_1/help/tdoc/en_US/pdf/basic_nonlinear.pdf.

[64] Hakim, G. and Abramovich, H. Large deflection of thin-walled plates under transverse loading-investigation of the generated in-plane stresses, Materials, 15(4), 2022, Paper Id: 1577, 28.

14 Typical NDT methods – laser shearography and acoustic emission

14.1 Introduction and literature review

Nondestructive testing (NDT) is a wide group of analysis techniques used in science and industry to evaluate the properties of a material, component or system without causing damage [1]. The terms "nondestructive examination" (NDE), "nondestructive inspection" (NDI) and "nondestructive evaluation" (NDE) are also commonly used to describe this technology [2, 3]. The NDT does not permanently alter the inspected specimen, leading to a highly valuable technique that can save both money and time in product evaluation, troubleshooting and research.

Today, modern NDTs are used in manufacturing, fabrication and in-service inspection to ensure product integrity and reliability, to control manufacturing processes, to lower production costs and to maintain a uniform quality level.

In aviation, the NDT field is of crucial importance in order to maintain the integrity of the aircraft structure, as well as missiles and spacecraft structures, and prevent aviation disasters arising from flaws created during production or developed during service. Therefore, aircraft manufacturers and companies that operate them are required to meet a strict standard of NDT in production and in-service. During service, the aircraft experiences a complex and variable load regime in various areas of the fuselage and the wings. These loads are often not adequately represented by strength and fatigue analyses performed during the design stages. Design gaps and structural findings that develop during service are treated with careful performance of NDTs as defined in the aviation standard.

14.1.1 NDT for metallic structures

Figure 14.1 presents the six mostly used NDT methods for application on metallic structures. A brief description of these methods is presented as follows:

Note: The material presented in this chapter is based on my two master students' theses:

1. Elbaz, Y., Hybrid approach to predict and optimize flaw detection capabilities using laser shearography, Master Thesis, Faculty of Aerospace Engineering, Technion, I.I.T., 32000 Haifa, Israel, January 2022, 142p.
2. Seri. Y., Assessment of load bearing capabilities of CFRP filament wound structures, Master Thesis, Faculty of Aerospace Engineering, Technion, I.I.T., 32000 Haifa, Israel, September 2020, 97p.

https://doi.org/10.1515/9783111621104-014

14.1.1.1 Eddy current (EC)

This method uses the fact that when an alternating current coil induces an electromagnetic field into a conductive test piece, a small current is created around the magnetic flux field, much like a magnetic field is generated around an electric current. The flow pattern of this secondary current, called an "eddy" current, will be affected when it encounters a discontinuity in the test piece, and the change in the EC density can be detected and used to characterize the discontinuity causing that change. The coil must be placed within the proximity of a test surface where the changing magnetic field permeates the conductive material (see a schematic drawing in Fig. 14.2). The advantages of the method are:

- Very good sensitivity to surface defects
- Capability to detect flaws in multistructures (up to about 14 layers)
- Can detect flaws through surface coatings
- Little precleaning required
- Portability

The disadvantages of EC are:

- Very susceptible to magnetic permeability changes
- Applicable only on conductive materials (therefore excludes testing of fiber-reinforced materials)
- Will not detect defects parallel to the surface
- Not applicable for large areas and/or complex geometries

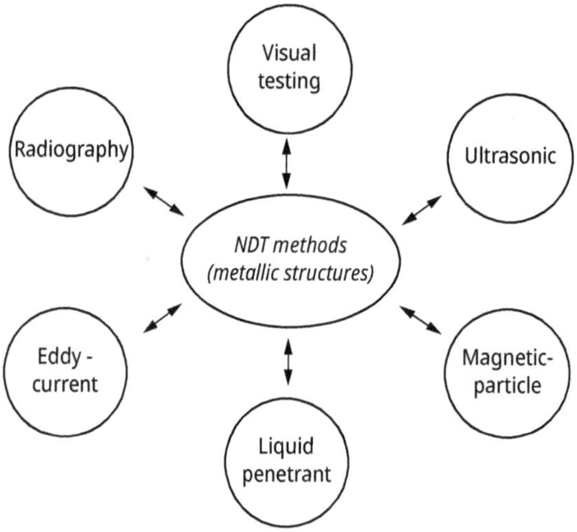

Fig. 14.1: Common NDT methods for metallic structures.

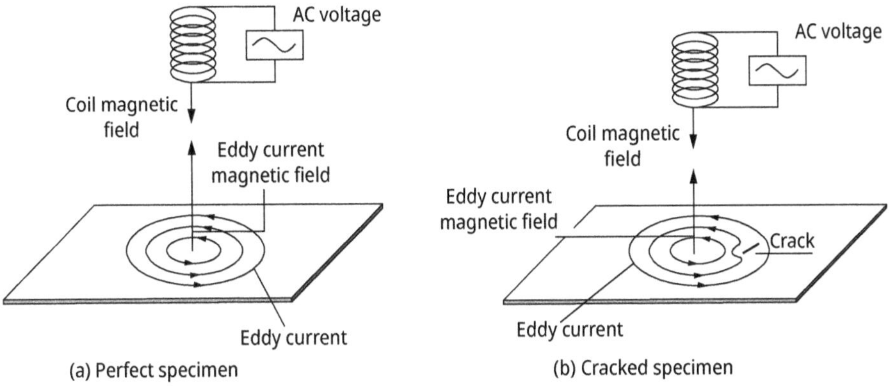

Fig. 14.2: A schematic drawing for the eddy current method.

14.1.1.2 Magnetic particle testing (MT)

MT uses magnetic fields to locate surface and near-surface discontinuities in ferromagnetic materials. When the magnetic field encounters a discontinuity transverse to the direction of the magnetic field, the flux lines produce a magnetic flux leakage field of their own. Because magnetic flux lines do not travel well in air, when very fine-colored ferromagnetic particles ("magnetic particles") are applied to the surface of the part, the particles will be drawn into the discontinuity, reducing the air gap and producing a visible indication on the surface of the part. The magnetic particles may be a dry powder or suspended in a liquid solution, and they may be colored with a visible dye or a fluorescent dye that fluoresces under an ultraviolet ("black") light (see a schematic drawing in Fig. 14.3).

The advantages of the method are:
- Simple to use
- Inexpensive
- Rapid results
- Higher sensitivity than visual inspection
- Requires a small area preparation

The disadvantages of the method are:
- Detection only on the surface or subsurface of the specimen
- No indication of flaw depths
- Applicable only on magnetic materials
- The detection is required in two directions

14.1.1.3 Liquid penetrant testing (PT)

The principle of liquid PT or dye PT (see Fig. 14.4) is based on the following phenomenon: when a very low-viscosity liquid is applied to the surface of a part, it will penetrate into fissures and voids open to the surface. Once the excess penetrant is removed, the penetrant trapped in these voids will flow back out, creating an indication. PT can be performed on magnetic and nonmagnetic materials. Penetrants may be "visible," meaning they can be seen in ambient light, or fluorescent, requiring the use of a "black" ambiance.

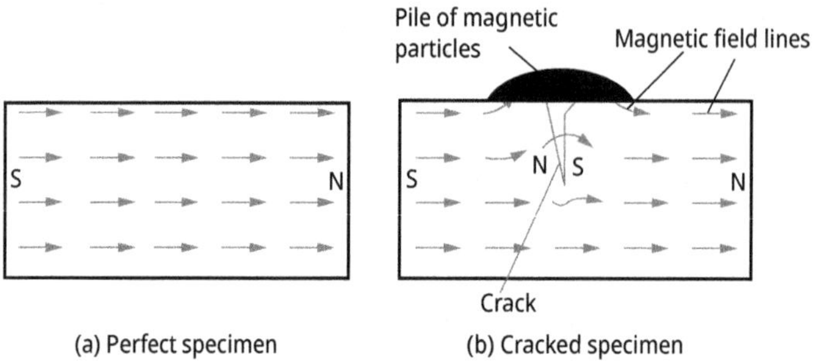

Fig. 14.3: A schematic drawing for the magnetic particle method.

The advantages of the method are:
- Simple and inexpensive
- Versatile and portable
- Application to ferrous, nonferrous, nonmagnetic and complex-shaped materials, nonporous and of any dimension.
- Can detect cracks, seams, lack of bonding and more
- Works through thin coating

The disadvantages are:
- Can detect only surface flaws
- Needs surface cleaning before and after the inspection
- To be used only on nonporous materials
- Deformed surfaces might prevent detection of the flaws
- Penetrants might be toxic or hazardous

14.1.1.4 Radiography testing (RT)

Industrial RT involves exposing a test object to penetrating radiation so that the radiation passes through the inspected specimen, and a recording medium is placed against the opposite side of that specimen. The recording media can be industrial X-ray film or one of several types of digital radiation detectors (see schematic description in Fig. 14.5). With both, the radiation passing through the test object exposes the media, causing an end effect of having darker areas where more radiation has passed

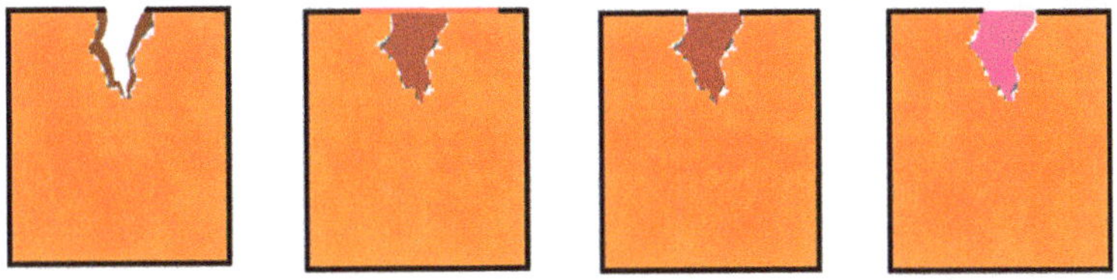

Fig. 14.4: Liquid penetrant testing: a schematic view of the four main steps.

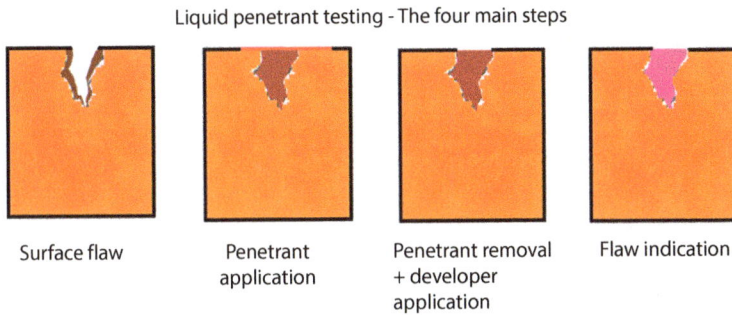

through the part and lighter areas where less radiation has penetrated. If there is a void or defect in the part, more radiation passes through, causing a darker image on the film or detector.

The advantages of this method are:
- Can inspect all types of materials
- Capable of detecting surface and subsurface flaws
- Capable of inspecting complex shapes, hidden surfaces and multilayered structures without disassembly
- Minimal preparation time
- Standardized with many reference standards available

The disadvantages are:
- Requires extensive operator training and skill
- The output does not indicate the depth of the flaw
- Relatively expensive equipment and investment is required
- Possible radiation hazard to the personnel
- Both sides of the specimen should be accessible

14.1.1.5 Ultrasonic testing (UT)

Ultra-high-frequency sound is introduced into the part being inspected, and if the wave sound (the pulse ray) hits a material with a different acoustic impedance (density and acoustic velocity), some of the wave will reflect back the echo ray to the sending unit and can be presented on a screen or can be saved on the PC hard disk for further processing (see Fig. 14.6). By knowing the speed of the sound through the part and the time required for the sound to return to the sending unit, the distance to the reflector can be determined. Sound is introduced into the part using an ultrasonic transducer ("probe") that converts electrical impulses from the UT machine into sound waves and then converts the returning sound back into electric impulses that can be displayed as a visual representation on a digital or LCD screen. Because ultra-

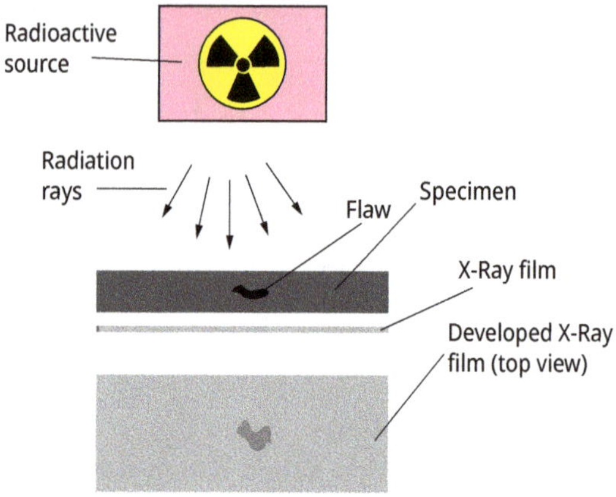

Fig. 14.5: Radioactive testing: a schematic view.

sound will not travel through air, a liquid or gel called "coupling agent" is used between the face of the transducer and the surface of the part to allow the sound to be transmitted into the part.

The advantages of this method are:
- One side inspection
- Complete nondestructive technique
- Highly reliable
- Wide measurement and application range
- Accurate and repeatable
- Flaw size and shape can be directly seen and evaluated
- Provides the depth of the flaw

The disadvantages are:
- Requires experienced technicians
- Test objects should be water resistant
- Difficulty on detecting volumetric defects
- Limitation for thick-walled components
- Demands direct touching of the specimen

14.1.1.6 Visual testing (VT)

Visual testing (VT) is the most commonly used test method in industry. Because most test methods require that the operator looks at the surface of the part being inspected, visual inspection is inherent in most of the other test methods (see typical inspection in Fig. 14.7). As the name implies, VT involves the visual observation of the surface of a test object to evaluate the presence of surface discontinuities. VT inspections may be by direct viewing, using line-of-sight vision, or may be enhanced with the use of optical instruments such as magnifying glasses, mirrors, borescopes, charge-coupled devices (CCD) and computer-assisted viewing systems. Corrosion, misalignment of parts, physical damage and cracks are just some of the discontinuities that may be detected by visual examinations.

The inherent advantages of this method are:
- Inexpensive
- Easy to train
- Portable
- Minimal specimen preparation
- No special equipment needed

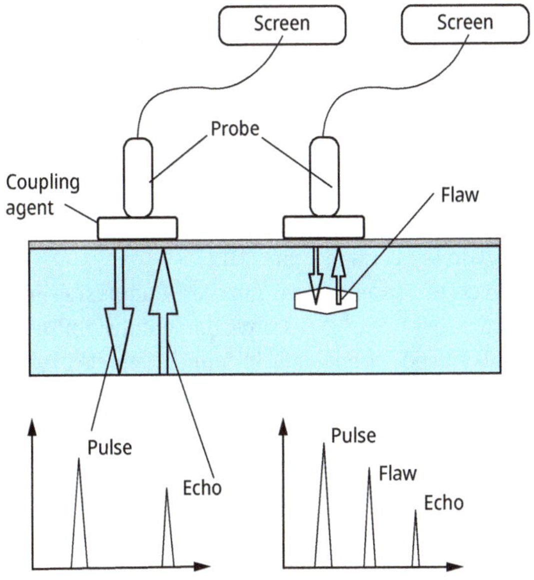

Fig. 14.6: Ultrasonic testing: a schematic view.

The disadvantages are:
- It displays only surface indications
- Capable of detecting only visible large flaws
- Possible misinterpretation of the found flaws

14.1.2 NDT for composite structures

In recent years, the aviation world has experienced significant changes in the field of aircraft structure. One of the important ones is the significant integration of elements made of composite materials due to their unique benefits [4–7]. Today, these materi-

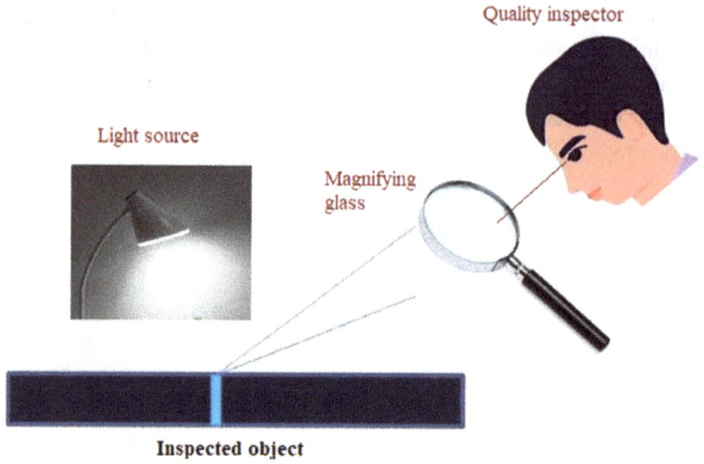

Fig. 14.7: Visual testing: a schematic view of one possible scenario.

als are integrated both as secondary structural parts, as has been done for many years, and as primary structural parts, carrying a significant load on the aircraft [8–10]. This trend is expected to intensify in the coming years [11].

This trend is also relevant for spacecraft, satellite and missiles. Composites are used in spacecraft and satellites for which lightweight and environmental stability is critical to mission success. They are also used extensively in launch vehicles for a growing number of applications. Solid rocket motors and pressure vessels for fuel and gas storage are typically reinforced by composites [12].

There are many forms of composite structures. In general, the difference between the structures depends on the type of material (polymer/metal), types of fibers and matrix and the order of the layers of materials. The two main composite structure types in the aviation world are laminated composites (monolithic composites) and sandwich structure. Laminated composites are made from completely bonded thin basic layers. These layers can be composites themselves: for instance, fibrous composite layer [13]. Sandwich structures can be classified as composite materials consisting of two or more individual components of differing properties which when combined result in high-performance material [14].

However, composite materials also have disadvantages. The main ones are the sensitivity of the final product to the manufacturing process (the way the composite structures are fabricated) [15] and the limitations of NDT methods.

Flaws in composite materials may be formed either during the manufacturing process or in the course of the normal service life of the component [16]. While it can be assumed that strict quality assurance will lead to a reduction in manufacturing flaws, service flaws may arise as a result of poor design, a change in the operating spectrum (fatigue) or an external impact [17].

There are different types of flaws, which can be formed in production and service [18]: porosity, inclusion of foreign bodies, fiber misalignment, waviness, debonding, matrix cracking, fiber–matrix debonding, delaminations, fiber breakage, and so on (see, e.g., Fig. 14.8).

The common and significant flaw from the strength and fatigue aspects is the delamination between the various elements in composite structures. These flaws may develop into different areas of the aircraft (wing skin, fuselage, etc.) and cause structural failure and even a flight safety incident.

These flaws pose a serious interference toward the effective performance of the structure [19, 20]. The combination of the sensitivity of the composite materials to the manufacturing process and the gaps in proving them over a lifetime [21] raises the vital need to achieve the testing ability to detect flaws in a nondestructive manner – both after (or in situ) the manufacturing process and during service. Effective and reliable NDT capability is the guarantee for early detection of structural flaws and monitoring of their formation and progress to maintain the integrity of the aircraft structure and ensure flight safety.

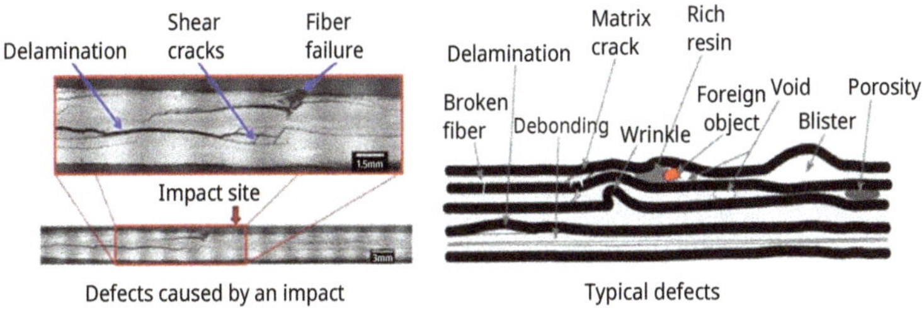

Fig. 14.8: Defects due to impact and a bunch of typical defects in a composite material.

Unfortunately, flaw detection is a challenging task because the flaws are normally hidden in between the layers and are invisible [22]. Testing of composite structures is a global challenge due to the faster progress of production capabilities in relation to the testing ability, as has happened in the past in the world of metals.

In a composite structure, there is an inherent variability of its medium. This variability is manifested both at the level of the individual layer (micro), which is made of fiber and matrix and has a fiber arrangement in varying directions, and at the macrolevel. There are different structural elements such as layers and cores. This medium variability constitutes a challenge in transferring excitation energy (thermal, mechanical, acoustic, electrical, etc.) to the structure medium and its absorption back in a satisfactory SNR (signal-to-noise ratio), in order to analyze the integrity of the structure.

Various factors that might affect the NDT performance applied on a composite structure are thickness of layers, orientation of plies, arranging the elements and other things. In addition, composite structures may contain elements that scatter or absorb the excitation energy needed to detect the flaws. These elements (such as

Rohacell[1] foam and honeycomb core) limit the ability to penetrate to the depth of the medium tested and might reduce the ability to detect flaws.

A number of nondestructive techniques have been proposed to provide answers to the above-presented problems: tapping UT (already presented in Section 14.1.1), laser shearography testing (LST), infrared thermography testing (IRT) and acoustic emission (AE) and their derivatives. These methods use an energy source (thermal, acoustic or mechanical) to inject energy into the structure medium, absorb its reflection on the surface with a high SNR and thus detect flaws in the depth of the medium. This principle meets the challenges of the composite materials, which is a variable medium that is not continuous and sometimes has elements that dissipate or absorb energy and therefore make it difficult to detect flaws in the depth of the tested medium.

Figure 14.9 presents the five most used NDT methods for application on composite structures. A brief description of the tapping and IRT methods is then presented, while the UT had already been presented in Section 14.1.1 and the LST and AE will be discussed in Sections 14.2 and 14.3, respectively:

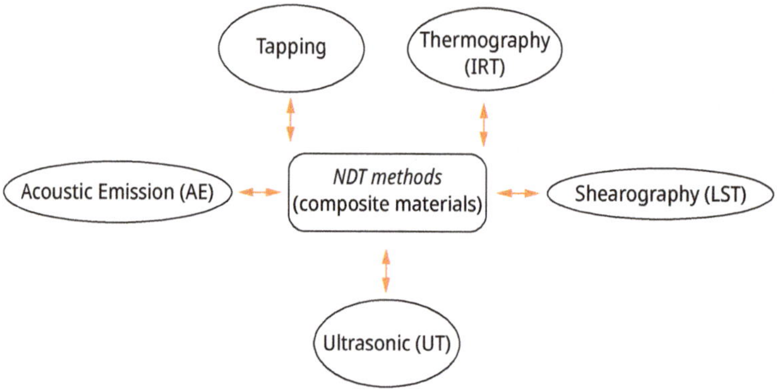

Fig. 14.9: Common NDT methods for composite-type structures.

14.1.2.1 Tapping testing

This method is one of the oldest approaches of NDT and is used nowadays to test laminated composite structures (see a schematic drawing in Fig. 14.10).

When a hammer strikes a structure, the characteristics of the impact are dependent on the local impedance of the structure and on the hammer used. Damage such as an adhesive debonding and fatigue damage results in a decrease in structural stiffness and hence a change in the nature of impact. Note that in-plane forces should be

1 Rohacell foam = a closed-cell rigid foam based on polymethacrylimide.

considered when the tapping test is performed. The literature reports two kinds of tap test methods:

The first one is based on the impact force produced when the structure is struck by a hammer and requires the time histories of the force input on the test structure. This is done using a transducer in the tap head so that no transducer needs to be attached to the structure.

The second type is based on the sound pressure radiating from the structure. This approach needs the time histories of the sound pressure. The sound-based tap test requires a Fourier transform of the sound pressure histories of the damaged structure to be computed and compared with an undamaged structure. The tap test will find defects only in the region of tap, so it is necessary to tap each part of the structure under investigation. The tap test uses the difference between the measured impact force histories or the sound pressure histories of a healthy structure and a locally damaged structure. For structurally radiated sound, the sound field is directly coupled to the structural motion.

The advantages of the method are:
- Simple detection of debonding and delamination
- Audible sonic testing (coin tapping) at 10–20 Hz
- Easy to be applied for daily uses
- Repeatable
- Sophisticated automated tap testers, not relying on a practiced hearing are available

The disadvantages are:
- Even the simple version relies on the well-practiced hearing of the inspection person
- The advanced method requires dedicated tools
- A better performance is obtained on thinner sheet thicknesses

Fig. 14.10: Common NDT tap testing methods for composite-type structures.

14.1.2.2 IRT

This method is defined as a noncontact mapping and analysis of thermal patterns emitted from the surface of the specimen. The method is suitable to test laminated composite structures (see a schematic drawing in Fig. 14.11). Passive thermography and active thermography are the two types available, with the active one being mostly used for composite NDT. The passive thermography uses infrared thermal cameras to detect damage formation as a function of the applied loading.

The "conventional" active thermography uses a heat input in the form of air blowers, or heat removal by keeping the specimen in a refrigerator. This application is most suitable for cases with high-temperature differences between perfect sectors and sectors with flaws. The "advanced" active thermography uses a controlled heat input by flash lamps, leading to a "heat wave" progression through the specimen's material. The fast camera images are synchronized with the heat input and then analyzed. This leads to better detection of the flaws, like delaminations, providing their depth information.

The advantages of the method are:
- Can be used to test large sectors for delamination, water inclusions and debonding.
- Noncontact method
- Relatively short measurement time
- Suitable to be applied on a large number of materials such as aluminum, steel, plastics, CFRP (carbon fiber-reinforced plastics), glass-reinforced plastic and honeycomb
- Provides an economical solution for 100% testing

The disadvantages are:
- Requires a well-trained inspection person
- Requires dedicated expensive equipment
- The postprocessing time might be long

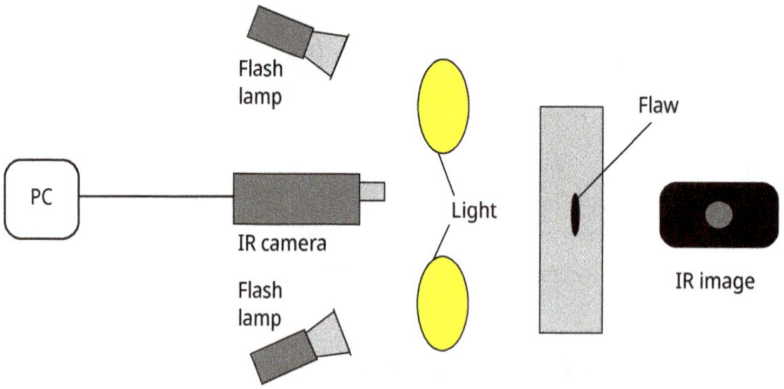

Fig. 14.11: Common NDT thermography testing using IR camera and flash lamps: a schematic view.

14.2 Laser shearography

Hung and Taylor [23] were the first to introduce the laser shearography testing (LST, named also speckle pattern shearing interferometry – SPSI) as a full-field strain analysis, noncontacting and noncontaminating technique. Since this initiation, many studies were performed and published in the literature (see typical examples in [24–54]).

Notice that the principles of the LST method are based on physical principles (electromagnetic waves, light and matter interaction, wave interference, laser, interferometry, etc.), which were introduced many years ago, but due to technological limitations, system development, which will be a practical solution, has been delayed until the recent years.

LST is an NDT technology based on the optical derivative analysis of the displacement perpendicular to the surface being examined under excitation. This method uses laser and interferometric imaging, which can be used to detect subsurface flaws in the depth of the medium [24–28] (see Fig. 14.12).

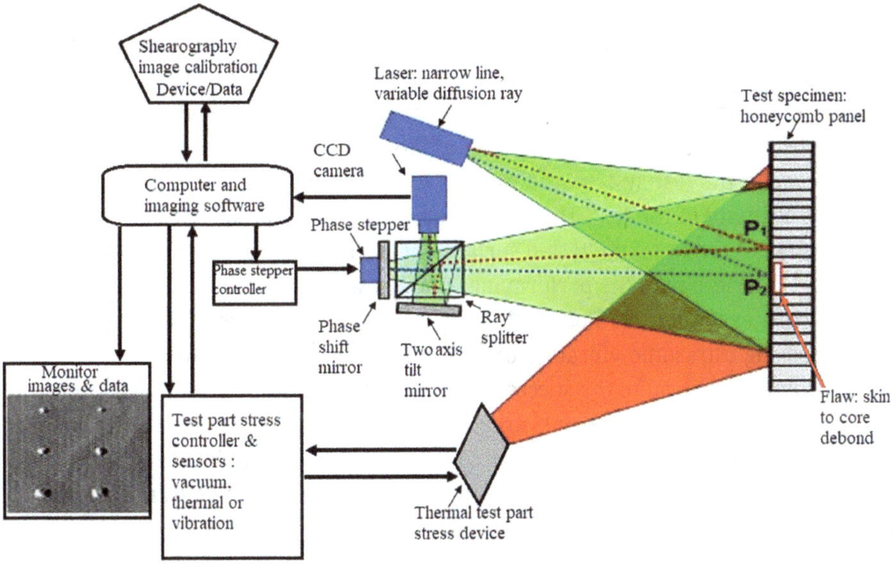

Fig. 14.12: A schematic description of the LST system (adapted from [71]).

The method offers optimal conditions for direct measurement of strain information on the test specimen. Because flaws usually generate strain concentrations, LST reveals a flaw within the tested part by identifying flaw-induced strain anomalies. Flaws in the part will give rise to strain concentrations when the part is loaded, and the strain concentrations form a fringe pattern to be used to detect and analyze the flaws [23]. LST applies laser light to the surface of the tested part, while the part is nonstressed and the resulting image is picked up by a CCD and stored on a computer.

The specimen's surface is then stressed, and a new image is generated, recorded and stored. The computer then superimposes the two patterns, and if flaws such as voids or debonding are present, the flaws can be revealed by the developed patterns. Discontinuities and flaws as small as a few micrometers in size can be detected in this manner.

Unlike other methods of speckle interferometry, for example, ESPI (electronic speckle pattern interferometry), the interference in LST is generated by two identical, laterally sheared object beams. This effect can be realized by a shear element, which is located between the object and camera. Since no additional reference beam is needed, the setup for LST is simple and with no need of vibration isolation compared to other coherent optical techniques. Thus, compared with ESPI, LST is relatively insensitive to disturbances from the environment and is a robust method to be applicable in industrial environments.

LST has a high sensitivity to local deformation over the surface of the tested part. This deformation is due to internal and/or external discontinuity in the subject being excited. In practice, using the LST system, very small perpendicular deformations (*Z*-axis) can be detected (2–20 nm), but there is a dependence on the environmental conditions that might affect the test results.

The derivative of the surface route is the optical shear of the laser speckle pattern. A cube beam splitter in a Michelson interferometer (see Fig. 14.13) performs this mentioned shear. Controlling the size and direction of the shear vector (by moving the interferometer mirrors) provides the ability to reduce the relative displacement effect between the test system and the test structure and adjust the directional and general sensitivity of the test. These allow performing an optimization for identification of minor changes on the tested surface and at the same time reducing the effects that might interfere with the performance of the test.

Common excitation methods in LST are heating, pressure and vibration (mainly acoustic, using ultrasonic vibrators or transducers). Choosing an appropriate excitation technique is critical to the success of the test. Appropriate excitation (e.g., heating of a surface in a metal honeycomb structure) will result in a different displacement of the surface in normal area (without discontinuities) relative to the surface above the discontinuity. Optical shear of the surface will lead to a more accurate detection of the discontinuity from a top view (C-scan display).

Thus, although the laser beam does not penetrate below the tested surface, the LST method can be used to detect discontinuities such as impact damage, nonadhesion, delaminations, inclusions, porosity near the surface, wrinkles, fiber bridging and cracks.

The output of LST is called a "shearogram." The shearogram is created by processing the interference pattern derivative of the tested surface in both strain states of the surface being tested – before and after the excitation (sometimes even during it). Due to the use of the optical derivative of the test surface, a typical indication form of discontinuity, as obtained by LST, is a "double lobe" as displayed in Fig. 14.14a.

In this way, the shearogram is a phase change map, which represents at each point the phase change of the laser beam (which returns from the surface being tested

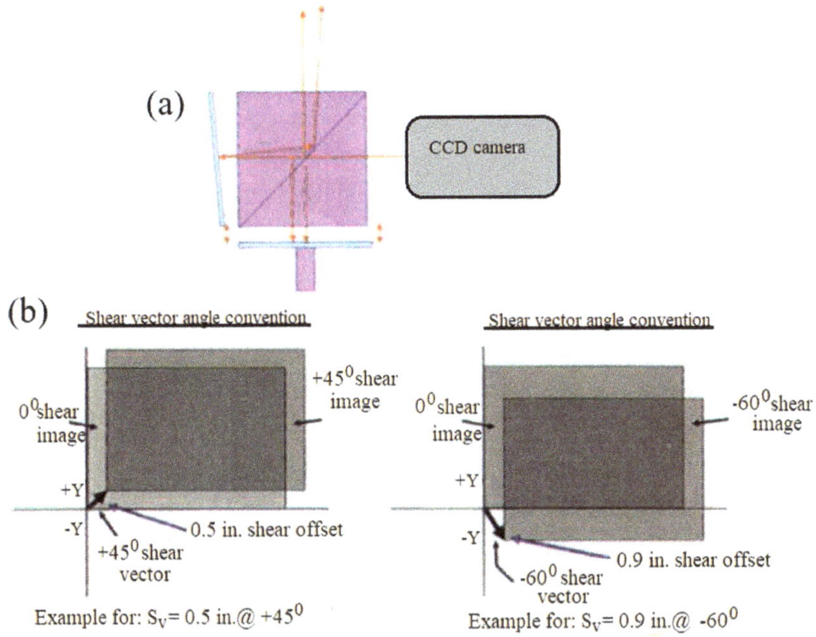

Fig. 14.13: An optical shear scheme: (a) Michelson's interferometer and (b) offset of the area photographed according to the shear vector (adapted from [71]).

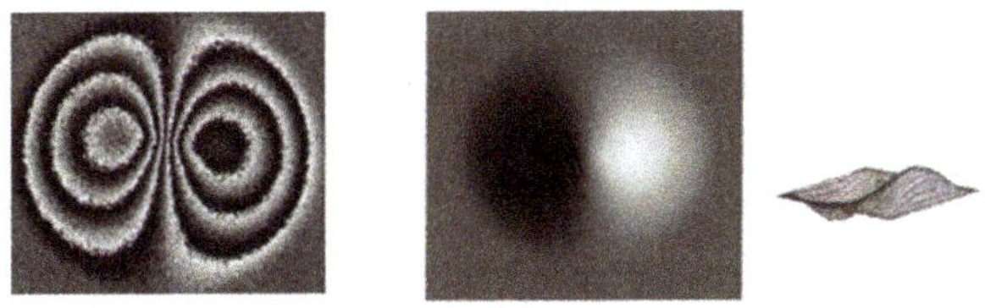

(a) wrapped phase map (b) unwrapped phase map

Fig. 14.14: A "double lobe": a typical indication of discontinuity obtained by LST (from [71]).

and is captured by the camera detector). These phase changes are created by the vertical displacement difference of the surface being tested – before and after excitation.

There are two ways to display the shearogram:

A "wrapped phase map" display, where the fringes can be seen, with each fringe representing an "equal-strain" line.

An "unwrapped phase map" display is the product of an integration process of all the fringes in a certain direction (defined by the shear vector), where each fringe represents a phase cycle from 0 to 2π. Thus, a normalized image is obtained, which is a general average of the derivative – white (positive derivative) and black (negative derivative), as shown in Fig. 14.14b.

Remember that the LST method is a comparative nondestruction test that requires the use of a calibration standard.

The LST method appears in the international standard for the training of NDT technicians (NAS410, EN4179 and SNT-TC-1A), and in practical standards (ASTM E 2581-07).

The LST governing equations (the mathematical model) are presented in Appendix A.

The advantages of the method are:
- Receiving the test result in near real time
- Inspection of large areas in a relatively short time (due to a large FOV – field of view), especially by automating the test system
- The method does not require contact with the tested part, does not contaminate and does not wet the surface.
- The method can be used to detect discontinuities immediately at the end of the production process.
- There is no need for access to either side. Items can be inspected on one side only (depending on the defects required for detection).
- The equipment can be used on a wide range of materials and surface geometries.
- LST results are easy to interpret and analyze relative to other NDT results.

The disadvantages of the method are:
- It is sometimes difficult to obtain a uniform return from the surface in dark bodies.
- In items with very high curvature, there might be areas where the return of the laser will cause the camera to saturate.
- The camera and specimen are required to remain stable during the test. Relative displacement might impair the test results.
- The flaw depth cannot be extracted directly from a shearogram.

14.3 Acoustic emission

The application of AE testing for aerospace structures started in the 1960s after failure of motor cases that passed the shop's proof-pressure hydro-tests but failed during fire tests. Since then, AE technology is widely applied for inspection of composite aerospace structures over the world and especially for composite structures of planes, rocket bodies, radomes, motor cases and composite-overlapped pressure vessels. It is also used for other industries like maritime, civil and mechanical engineering.

AE is a phenomenon of sound and ultrasound (stress) wave radiation in materials that undergo deformation or fracture processes, as presented in Fig. 14.15. It is an inspection method that uses the release of ultrasonic stress waves to locate and identify defects in materials. Note that these ultrasonic stress waves are not introduced from external sources, and they originate from the inspected material. In the literature, the AE testing is also named acoustic emission, acoustic testing or acoustic NDT.

Crack propagation in loaded solid materials such as metals and composites results in a fast release of potential energy in the form of stress waves with frequencies typically between 20 kHz and 1 MHz. These waves propagate along the structure for distances of several meters and are detected by piezoelectric sensors. As an acoustic wave travels on or through the surface of specimen, any defect it encounters can change that wave, both in terms of its speed and in terms of its amplitude. Special analysis of the detected AE waves is then performed to locate AE flaw sources, identify flaw type and evaluate the rate of flaw propagation and its sensitivity to load/stress/operational changes.

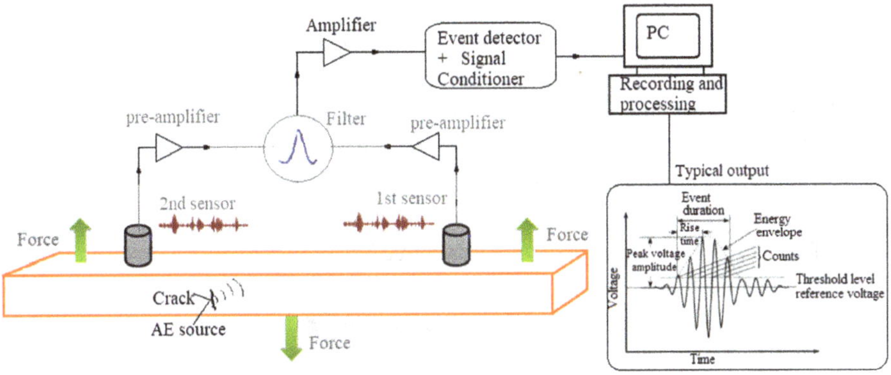

Fig. 14.15: A schematic test setup for acoustic emission testing (from [72]).

Main sources of AE in composite structures are matrix cracking, delamination, fiber cracking and fiber pullout. In addition, other sources of AE due to impact and leaks can be readily detected and assessed by the AE technology.

For other types of materials, AEs would happen when a material is under stress, either from holding a heavy load or from extremes of temperature. These emissions typically correspond with some kind of defect or damage being done to the structure emitting them, and this damage is what inspectors are looking for when they do an AE test. Typical sources of AE could be phase transformation, thermal stresses, cool-down cracking and melting.

In general, the inspectors would apply this method to find and monitor corrosion, coating removal, flaws/defects of welding, leaks in storage tanks and/or pipe systems and partial discharges from components subjected to high voltage.

Note that as displayed in Fig. 14.15, the most common set of transducers for AE testing consists of two sets of interdigital transducers, which is a device made of two interlocking comb-shaped arrays of metallic electrodes arranged like a zipper. One of the transducers converts electric field energy into mechanical wave energy, and the other transducer converts the mechanical wave energy back into an electric field.

The method is highly applied for inspection in pressure vessels, wind turbines, structural integrity, aircraft longevity estimation, bridges, concrete corrosion and mine wall stability.

The topic of AE was largely covered in the literature, including books and numerous manuscripts.

A list of typical sources is presented in [55–70].

The AE NDT method has several dozens of standards, procedures and test methods issued by various international organizations such as ASTM and ASME. The following standards were applicable partially or at whole for examination of the wing structure:
1. ASTM E 569 Standard Practice for Acoustic Emission Monitoring of Structures during Controlled Stimulation.
2. ASTM E 2076 Standard Test Method for Examination of Fiberglass Reinforced Plastic Fan Blades Using Acoustic Emission.
3. ASTM E 650 Guide for Mounting Piezoelectric Acoustic Emission Sensors.
4. ASTM E 750 Standard Practice for Characterizing AE Instrumentation.
5. ASTM E 976 Guide for Determining the Reproducibility of Acoustic Emission Sensor Response.
6. ASTM E 1316 Terminology for Nondestructive Examinations.
7. ASTM E 2374 Guide for Acoustic Emission System Performance Verification.
8. ASTM E 2533 Standard Guide for Nondestructive Testing of Polymer Matrix Composites Used in Aerospace Applications.
9. ASTM E 2661 / E 2661M-10 Standard Practice for Acoustic Emission Examination of Plate-like and Flat Panel Composite Structures Used in Aerospace Applications.
10. ASME Standard: Section V, Article 13, Boiler & Pressure Vessel Code, Continuous Acoustic Emission Monitoring.

For other applications, the following standards should be followed:

11. ASME Boiler and Pressure Vessel Code: Section XI, Division 1, Article IWA-2000, Examination and Inspection (IWA-2234), Acoustic Emission Examination
12. ASME Boiler and Pressure Vessel Code: Section XI, Division 1, Code Case N-471, Acoustic Emission for Successive Inspections
13. ASME Boiler and Pressure Vessel Code: Section XI, Division 1, Code Case No. N-471, Acoustic Emission for successive inspections – Supplement 1 Guidance information for acoustic emission monitoring of pressure boundaries during operation
14. ASME Boiler and Pressure Vessel Code: Section XI, Appendix, Acoustic Emission Monitoring of Nuclear Reactor Pressure Boundaries during Operation
15. ASME RTP-1-1995: Standard Guide to Test Methods and Standards for Nondestructive Testing of Advanced Ceramics
16. ASTM C 1175: Standard Guide to Test Methods and Standards for Nondestructive Testing of Advanced Ceramics

17. ASTM E 543: Standard Specification for Agencies Performing Nondestructive Testing
18. ASTM E 569: Standard Practice for Acoustic Emission Monitoring of Structures During Controlled Stimulation
19. ASTM E 650: Standard Guide for Mounting Piezoelectric Acoustic Emission Sensors
20. ASTM E 749: Standard Practice for Acoustic Emission Monitoring During Continuous Welding
21. ASTM E 750: Standard Practice for Characterizing Acoustic Emission Instrumentation
22. ASTM E 751: Standard Practice for Acoustic Emission Monitoring During Resistance Spot-Welding
23. ASTM E 976: Standard Guide for Determining the Reproducibility of Acoustic Emission Sensor Response
24. ASTM E 1065: Standard Guide for Evaluating Characteristics of Ultrasonic Search Units
25. ASTM E 1067: Standard Practice for Acoustic Emission Examination of Fiberglass Reinforced Plastic Resin (FRP) Tanks/Vessels
26. ASTM E 1106: Standard Test Method for Primary Calibration of Acoustic Emission Sensors
27. ASTM E 1118: Standard Practice for Acoustic Emission Examination of Reinforced Thermosetting Resin Pipe (RTRP)
28. ASTM E 1139: Standard Practice for Continuous Monitoring of Acoustic Emission from Metal Pressure Boundaries
29. ASTM E 1211: Standard Practice for Leak Detection and Location Using Surface-Mounted Acoustic Emission Sensors
30. ASTM E 1212: Standard Practice for Quality Management Systems for Nondestructive Testing Agencies
31. ASTM E 1316: Standard Terminology for Nondestructive Examination
32. ASTM E 1359: Standard Guide for Evaluating Capabilities of Nondestructive Testing Agencies
33. ASTM E 1419: Standard Practice for Examination of Seamless, Gas-Filled, Pressure Vessels Using Acoustic Emission
34. ASTM E 1495: Standard Guide for Acousto-Ultrasonic Assessment of Composites, Laminates, and Bonded Joints
35. ASTM E 1544: Standard Practice for Construction of a Stepped Block and Its Use to Estimate Errors Produced by Speed-of-Sound Measurement Systems for Use on Solids
36. ASTM E 1736: Standard Practice for Acousto-Ultrasonic Assessment of Filament-Wound Pressure Vessels
37. ASTM E 1781: Standard Practice for Secondary Calibration of Acoustic Emission Sensors

38. ASTM E 1888/E 1888M: Standard Practice for Acoustic Emission Examination of Pressurized Containers Made of Fiberglass Reinforced Plastic with Balsa Wood Cores
39. ASTM E 1930: Standard Practice for Examination of Liquid-Filled Atmospheric and Low-Pressure Metal Storage Tanks Using Acoustic Emission
40. ASTM E 1932: Standard Guide for Acoustic Emission Examination of Small Parts
41. ASTM E 2075/E 2075M: Standard Practice for Verifying the Consistency of AE Sensor Response Using an Acrylic Rod
42. ASTM E 2076/E 2076M: Standard Practice for Examination of Fiberglass Reinforced Plastic Fan Blades Using Acoustic Emission
43. ASTM E 2191/E 2191M: Standard Practice Method for Examination of Gas-Filled Filament-Wound Composite Pressure Vessels Using Acoustic Emission
44. ASTM E 2374: Standard Guide for Acoustic Emission System Performance Verification
45. ASTM E 2478: Standard Practice for Determining Damage-Based Design Stress for Fiberglass Reinforced Plastic (FRP) Materials Using Acoustic Emission
46. ASTM E 2533: Standard Guide for Nondestructive Testing of Polymer Matrix Composites Used in Aerospace Applications
47. ASTM E 2598: Standard Practice for Acoustic Emission Examination of Cast Iron Yankee and Steam Heated Paper Dryers
48. ASTM E 2661/E 2661M: Standard Practice for Acoustic Emission Examination of Plate-like and Flat Panel Composite Structures Used in Aerospace Applications
49. ASTM E 2863/E 2863M: Standard Practice for Acoustic Emission Examination of Welded Steel Sphere Pressure Vessels Using Thermal Pressurization
50. ASTM E 2907: Standard Practice for Examination of Paper Machine Rolls Using Acoustic Emission from Crack Face Rubbing
51. ASTM F 914/F 914M: Standard Test Method for Acoustic Emission for Aerial Personnel Devices Without Supplemental Load Handling Attachments
52. ASTM F 1430/F 1430M: Standard Test Method for Acoustic Emission Testing of Insulated and Non-Insulated Aerial Personnel Devices with Supplemental Load Handling Attachments
53. ASTM F 1797: Standard Test Method for Acoustic Emission Testing of Insulated and Non-Insulated Digger Derricks
54. ASTM F 2174: Standard Practice for Verifying Acoustic Emission Sensor Response
55. ASTM E 2374: Standard Guide for Acoustic Emission System Performance Verification
56. CEN EN 1071-3 2005: Advanced technical ceramics – Methods of test for ceramic coatings – Part 3: Determination of adhesion and other mechanical failure modes by a scratch test
57. CEN EN 1330-1 1998: Non-destructive testing – Terminology – Part 1: List of general terms

58. CEN EN 1330-2 1998: Non-destructive testing – Terminology – Part 2: Terms common to the non-destructive testing methods
59. CEN EN 1330-9 2009: Non-destructive testing – Terminology – Part 9: Terms used in acoustic emission testing
60. CEN EN 12817: 2010: LPG Equipment and accessories – Inspection and requalification of LPG tanks up to and including 13 m^3
61. CEN EN 12819 2009: LPG equipment and accessories – Inspection and requalification of LPG tanks greater than 13 m^3
62. CEN ISO/TR 13115 2011: Non-destructive testing – Methods for absolute calibration of acoustic emission transducers by the reciprocity technique (ISO/TR 13115:2011)
63. CEN EN 13445-5 2009: Unfired pressure vessels – Part 5: Inspection and testing (Annex E)
64. CEN EN 13477-1 2001: Non-destructive testing – Acoustic emission – Equipment characterization – Part 1: Equipment description
65. CEN EN 13477-2 2010: Non-destructive testing – Acoustic emission – Equipment characterization – Part 2: Verification of operating characteristic
66. CEN EN 13554 2011: Non-destructive testing – Acoustic emission – General principles
67. CEN EN 13480-5 2012: Metallic industrial piping – Part 5: Inspection and testing
68. CEN EN 14584 2013: Non-destructive testing – Acoustic emission – Examination of metallic pressure equipment during proof testing – Planar location of AE sources
69. CEN EN 15495 2007: Non-destructive testing – Acoustic emission – Examination of metallic pressure equipment during proof testing – Zone location of AE sources
70. CEN EN 15856 2010: Non-destructive testing – Acoustic emission – General principles of AE testing for the detection of corrosion within metallic surroundings filled with liquid
71. CEN EN 15857 2010: Non-destructive testing – Acoustic emission – Testing of fibre-reinforced polymers – Specific methodology and general evaluation criteria
72. CEN EN ISO 16148 2006: Gas cylinders – Refillable seamless steel gas cylinders – Acoustic emission testing (AT) for periodic inspection (ISO 16148:2006)
73. CEN ISO/TR 25107 2006: Non-destructive testing – Guidelines for NDT training syllabuses (ISO/TR 25107:2006)
74. CEN CR 13935 2000: Non-destructive testing – Generic NDE data format model

The advantages of the method are:
- It provides a direct measure of failure mechanisms.
- A highly sensitive method
- The output is within seconds
- The method is nondestructive without influencing the properties of the specimen's material
- The method provides a global monitoring of a given structure

- Applicable in hazardous environments, high pressure, high temperature or radiation
- The method can be applied remotely

The disadvantages of the method are:
- It might be slow to be applied
- Cannot detect static defects (defects that do not move, grow or change with time)
- It might be difficult to be used due to weak AE signals and increased noise
- The output of the method is the location of the defect, namely a qualitative estimation of the defect

References

[1] Cartz, L. Nondestructive testing, ASM International, 1995, 212.
[2] Hellier, C. Handbook of nondestructive evaluation, McGraw-Hill, 2003, 1.1., 594.
[3] Introduction to Nondestructive Testing. at www.asnt.org.
[4] Dutton, S., Kelly, D. and Baker, A. Composite materials for aircraft structures – 2nd Edition, AIAA, 2004, 400.
[5] Mrazova, M. Advanced composite materials of the future in aerospace industry, INCAS Bulletin, 5(3), 2013, 139–150.
[6] Campbell, F.C. Structural composite materials, ASM International, 2010, 629.
[7] Thori, P., Sharma, P. and Bhargava, M. An approach of composite materials in industrial machinery: advantages, disadvantages and applications, International Journal of Research in Engineering and Technology, 2(12), 2013, 350–355.
[8] Hale, J. Boeing 787 from the ground up, Aero Magazine, 24, Quarter, 04, 2006, 17–23.
[9] Hiken, A. The evolution of the composite fuselage: a manufacturing perspective, in aerospace engineering, Dekoulis, G. Ed., IntechOpen, 2018, 1–30.
[10] Xu, B. and Li, H.Y. Advanced composite materials and manufacturing engineering, selected, peer reviewed papers from the 2012 international conference on advanced composite materials and manufacturing engineering (CMME2012) October 13–14, 2012, Beijing, China, Durnten-Zurich, Switzerland Enfield, NH: Trans Tech Publications, 2012, 414.
[11] Research and Markets, Aerospace composites materials market report: trends, forecast and competitive analysis, 2019, 304.
[12] Composites applications for space article, at https://nasampe.org/page/CompositesApplicationsforSpace.
[13] Different types of composites in construction and their uses, at https://theconstructor.org/composite/composites-construction-uses/1570/.
[14] What is a sandwich structure, at https://www.twi-global.com/technical-knowledge/faqs/faq-what-is-a-sandwich-structure.
[15] Liu, L., Zhang, B.M., Wang, D.F. and Wu, Z.J. Effects of cure cycles on void content and mechanical properties of composite laminates, Composite Structures, 73(3), 2006, 303–309.
[16] Ghobadi, A. Common type of damages in composites and their inspections, World Journal of Mechanics, 7, 2017, 24–33.
[17] Reifsnider, K.L. Damage in composite materials, Chapter 2, in fatigue of composite materials, Vols. 19991, 519, Reifsnider, K.L. Ed., 11–77.

[18] Jollivet, T., Peyrac, C. and Lefebvre, F. Damage of composite materials, Procedia Engineering, 66, 2013, 746–759.
[19] de Almeida, S.F.M. and Neto, Z.D.S.N. Effect of void content on the strength of composite laminates, Composite Structures, 28(2), 1994, 139–148.
[20] Costa, M.L., de Almeida, S.F.M. and Rezende, M.C. The influence of porosity on the interlaminar shear strength of carbon/epoxy and carbon/bismaleimide fabric laminates, Composites Science and Technology, 61(14), 2001, 2101–2108.
[21] Gholizadeh, S. A review of non-destructive testing methods of composite materials, Procedia Structural Integrity, 1, 2016, 50–57.
[22] Dhulkhed, S. ANSYS simulation as a feasibility study for high repetition laser ultrasonic non-destructive evaluation (NDE), Master thesis submitted at Concordia University, Montreal, Quebec, Canada, Dec. 2016, 125.
[23] Akbari, D., Soltani, N. and Farahani, M. Numerical and experimental investigation of defect detection in polymer materials by means of digital shearography with thermal loading, Part B: Journal of Engineering Manufacture, 227(3), 2013, 430–442.
[24] Sharpe, W.N., Jr. Ed. Springer handbook of experimental solid mechanics, 2008, 1095.
[25] Yang, L., Chen, F., Steincheng, W. and Hung, M.Y. Digital shearography for nondestructive testing: Potentials, limitations, and applications, Journal of Holography and Speckle, 1, 2005, 1–11.
[26] Hung, Y.Y. A speckle-shearing interferometer: a tool for measuring derivatives of surface displacement, 11(2), 1974, 132–135.
[27] Francis, D., Tatam, R.P. and Groves, R.M. Shearography technology and applications: a review, Measurement Science and Technology, 21(10), 2010, Paper Id: 102001, 29.
[28] Bossi, R.H. Technical editor. Aerospace NDT, ASNT industry handbook, Chapter 12: shearographic and holographic testing, Vol. 32, 2014, 464.
[29] Yang, F., Ye, X., Qiu, Z., Zhang, B., Zhong, P., Liang, Z.Y., et al. The effect of loading methods and parameters on defect detection in digital shearography, Results in Physics, 7, 2017, 3744–3755.
[30] Leendertz, J.A. and Butters, J.N. An image-shearing speckle-pattern interferometer for measuring bending moments, Journal of Physics E: Scientific Instruments, 6, 1973, 1107–1110.
[31] Hung, Y.Y. A speckle-shearing interferometer: a tool for measuring derivatives of surface-displacements, Optics Communications, 11(2), 1974, 132–135.
[32] Shang, H.M., Toh, S.L. and Chau, F.S. Locating and sizing disbonds in glass fibre-reinforced plastic plates using shearography, Journal of Engineering Materials and Technology, 113(1), 1991, 99–103.
[33] Nokes, J. and Cloud, G.L. The application of three interferometric techniques to the NDE of composite materials, in Proceedings of the SPIE 1993 international symposium on optics, imaging, and instrumentation, San Diego, CA, US, 1993, 18–26.
[34] Maji, A.K. and Satpathi, D. Assessment of electronic shearography for structural inspection, Experimental Mechanics, 37(2), 1997, 197–204.
[35] Hung, Y.Y. Applications of digital shearography for testing of composite structures, Composites Part B: Engineering, 30, 1999, 765–773.
[36] Sirohi, R.S., Tay, C.J., Shang, H.M. and O'Shea, D.C. Nondestructive assessment of thinning of plates using digital shearography, Optical Engineering, 38(9), 1999, 1582–1585.
[37] Zou, G.P., Lu, J. and Wang, W.W. Application of electronic shearography speckle pattern interferometry to nondestructive testing of wood material, Journal of Harbin Engineering University, 30, 2009, 357–361.
[38] Huang, Y.H., Ng, S.P. and Liu, L. NDT&E using shearography with impulsive thermal stressing and clustering phase extraction, Optics and Lasers in Engineering, 47(7–8), 2009, 774–781.
[39] DeAngelis, G., Meo, M., Almond, D.P., Pickering, S.G. and Angioni, S.L. A new technique to detect defect size and depth in composite structures using digital shearography and unconstrained optimization, NDT&E International, 45(1), 2012, 91–96.

[40] Zhanwei, L., Jianxin, G., Huimin, X. and Wallace, P. NDT capability of digital shearography for different, Materials Optics and Lasers in Engineering, 49(12), 2011, 1462–1469.
[41] Chen, X. Computational and experimental approach for non-destructive testing by laser shearography, Master thesis, submitted to Mechanical Engineering Department, Worcester Polytechnic Institute, Worchester, MA, USA, 2014, 113.
[42] Naumov, A., Sikorski, R., Sandomirsky, S., Naumov, M. and Buesking, K. Thermoshearographic methodology for strain mapping, 58th International Instrumentation Symposium, ISA, 2012, 9.
[43] Wang, X., Gao, Z., Yang, S., Gao, C., Sun, X., Wen, X., et al. Application of digital shearing speckle pattern interferometry for thermal stress, Measurement, 125, 2018, 11–18.
[44] Hou, R., Wu, A. and Li, Y. Numerical simulation study on thermal shearography testing, Proceedings of the First Symposium on Aviation Maintenance and Management-Volume I, Lecture Notes in Electrical Engineering, 296, (Wang, J. editor), 2014, 53–67, 654.
[45] Choi, I.-Y., Hong, K.-M. and Ko, K.-S. Measurement of aluminum liner internal defect deformation using shearography and FEM verification, Journal of Korean Society of Manufacturing Technology Engineers, 22(4), 2013, 686–692.
[46] Gryzagoridis, J., Oliver, G. and Findeis, D. Modal frequency versus shearography in detecting and locating voids/delaminations in sandwich composites, Insight-Non-Destructive Testing and Condition Monitoring, 55(5), 2013, 9.
[47] Liu, H., Guo, S., Chen, Y.F., Tan, C.Y. and Zhang, L. Acoustic shearography for crack detection in metallic plates, Smart Materials and Structures, 27, 2018, Paper Id.: 085018, 10.
[48] Mihaylovaa, E., Naydenovaa, I., Duignanb, B., Martina, S. and Toala, V. Photopolymer diffractive optical elements in electronic speckle pattern shearing interferometry, Optics and Lasers in Engineering, 44, 2006, 965–974.
[49] Feng, Z., Gao, Z., Zhang, X., Wang, S., Yang, D., Yuan, H., et al. A polarized digital shearing speckle pattern interferometry system based on temporal wavelet transformation, Review of Scientific Instruments, 86(9), 2015, Paper Id.: 093102, 7.
[50] Liu, B., Guo, X.M. and Qi, G.J. Quality evaluation of rubber-to-metal bonded structures based on shearography, science China: Physics, Mechanics and Astronomy, 58(7), 2015, Paper Id.: 0742024, 7.
[51] Findeis, D., Gryzagoridis, J. and Asur, E. Phase unwrapping applied to portable digital shearography, IV Conferencia Panamericana de END Buenos Aires, Argentina, 2007, 11.
[52] Waldner, S.P. Quantitative strain analysis with image shearing speckle pattern interferometry (shearography), Doctoral thesis submitted to the Swiss Federal Institute of Technology, Zurich Switzerland, 2000, 126.
[53] Asundi, A.K. MATLAB for photomechanics – a primer, Elsevier Science, 1, 2002, 198.
[54] Murukeshana, V.M., Feia, L.Y., Krishnakumarb, V., Onga, L.S. and Asundi, A. Development of matlab filtering techniques in digital speckle pattern interferometry, Optics and Lasers in Engineering, 39, 2003, 441–448.
[55] Matthews, J.R. Ed. Acoustic emission, vol. 2 nondestructive testing monographs and tracts, Gordon and Breach Science Publisher, 1983, 167.
[56] Grosse, C.U. and Ohtsu, M. Ed. Acoustic emission testing: basics for research-applications in civil engineering, Springer-Verlag, Berlin Heidelberg, Germany, 2008, 397.
[57] Grosse, C.U., Ohtsu, M., Aggelis, D.G., and Shiotani, T. Ed. Acoustic emission testing: basics for research-applications in engineering, 2nd, Springer-Verlag, Berlin Heidelberg, Germany, 2022, 746.
[58] Sikorski, W. Ed. Acoustic emission-research and applications, IntechOpen, 2013, 223.
[59] Shen, G., Wu, Z. and Zhang, J. Ed. Advances in acoustic emission technology, Proc. of the World Conference on Acoustic Emission-2015, Springer International Publishing, Switzerland, 2017, 418.
[60] Drouillard, T. Acoustic emission: a bibliography with abstracts, 2013, 808.
[61] Hamstad, M.A. A review: Acoustic emission, a tool for composite-materials studies, Experimental Mechanics, 26, 1986, 7–13.

[62] Holford, K.M. Acoustic emission-basic principles and future direction, Strain, 36(2), 51–54.
[63] Ono, K. Acoustic emission in materials research- A review, 29, 2011, 244–308.
[64] Scruby, C.B. An introduction to acoustic emission, Journal of Physics E: Scientific Instruments, 20(8), 1987, 946–953.
[65] Gao, L., Zai, F., Su, S., Wang, H., Chen, P. and Liu, L. Study and application of acoustic emission testing in fault diagnosis of low-speed heavy-duty gears, Sensors, 11, 2011, 599–611.
[66] Beattie, A.G. Acoustic emission non-destructive testing of structures using source location technique, Sandia Report, SAND2013-7779, 2013, 128.
[67] Gholizadeh, S., Leman, Z. and Baharudin, B.T.H.T. A review of the application of acoustic emission technique in engineering, Structural Engineering and Mechanics, 54(6), 2015, 1075–1095.
[68] Lainé, E., Grandidier, J.-C., Cruz, M., Gorge, A.-L., Bouvy, C. and Vaes, G. Acoustic emission description from a damage and failure scenario of rotomoulded polyolefin sandwich structure subjected to internal pressure for storage, Mechanics & Industry, 21, 2020, Paper Id.: 105, 18.
[69] Bu, F., Xue, L., Zhai, M., Huang, X., Dong, J., Liang, N., et al. Evaluation of the characterization of acoustic emission of brittle rocks from the experiment to numerical simulation, Scientific Reports, 12, 2022, Paper Id.: 498, 16.
[70] Barbosh, M., Dunphy, K. and Sadhu, A. Acoustic emission-based damage localization using wavelet-assisted deep learning, Journal of Infrastructure Preservation and Resilience, 3, 2022, Paper Id.: 6, 24.
[71] Elbaz, Y. Hybrid approach to predict and optimize flaw detection capabilities using laser shearography, Master Thesis, Faculty of Aerospace Engineering, Technion, I.I.T., 32000 Haifa, Israel, January 2022, 142.
[72] Seri, Y. Assessment of load bearing capabilities of CFRP filament wound structures, Master Thesis, Faculty of Aerospace Engineering, Technion, I.I.T., 32000 Haifa, Israel, September 2020, 97p.

Appendix A: the LST governing equations

The experimental setup of digital LST with single-beam illumination is presented in Fig. A1 [24]. As can be seen from the figure, the tested object is illuminated by an expanded laser beam. The light reflected from the object surface is focused on the image plane of an image CCD camera, where a modified Michelson interferometer is implemented in front of its lens instead of the optical shearing element.

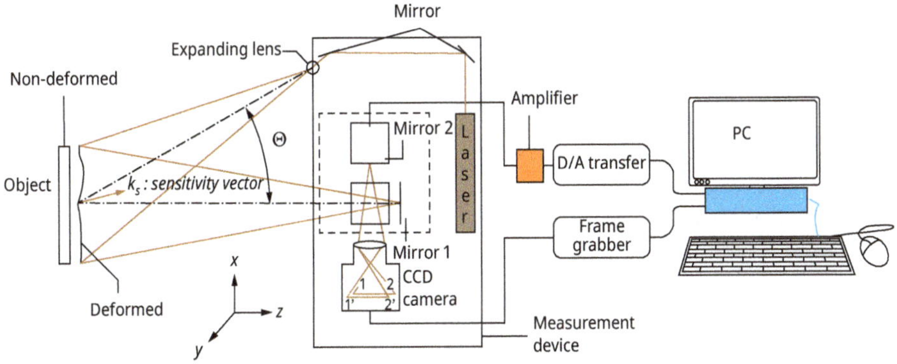

Fig. A1: The experimental setup of LST using a single-beam illumination (from [71]).

A pair of laterally sheared images of the tested object is generated on the image plane of the CCD camera by turning mirror 1 of the Michelson interferometer to a very small angle from the normal position. Two images with a certain shearing δ interfere with each other and generate a shearing speckle pattern interferogram.

The rays from point $A_1 = A_1(x, y)$ on the object are mapped into two points $A'_1 = P1(x, y)$ and $A''_1 = A''_1(x + \delta x', y)$ in the image plane (see Fig. A2). Similarly, the point $A_2 = A_2(x + \delta x', y)$ on the object surface is mapped to two points $A'_2 = A'_2(x + \delta x', y)$ and $A''_2 = A''_2(x + 2\delta x', y)$. At the point $(x + \delta x', y)$ in the shearography interferogram, points A'_2 and A''_1 are superposed.

The light intensity at this point can be expressed as

$$I = (a_1 e^{i\theta_1} + a_2 e^{i\theta_2})(a_1 e^{i\theta_1} + a_2 e^{i\theta_2}) = a_1^2 + a_2^2 + 2a_1 a_2 \cos(\theta_1 - \theta_2) = a_1^2 + a_2^2 + 2a_1 a_2 \cos(\phi_{12}) \tag{A.1}$$

where $\phi_{12} = \theta_1 - \theta_2$ is the random phase and $a_1 e^{i\theta_1}$ and $a_2 e^{i\theta_2}$ are temporally constant object beams from A_1 and A_2, respectively.

When the object or the specimen is deformed, an optical path change occurs due to the surface displacement of the object. This optical path change induces a relative phase change between the two interfering points. Thus, the intensity distribution of the speckle pattern is slightly altered and is mathematically represented by

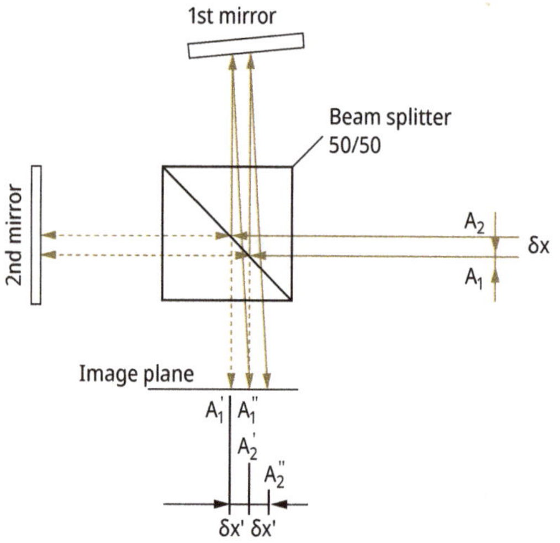

Fig. A2: The schematic shearing interferometry (from [71]).

$$I' = (a_1 e^{i(\theta_1+\Delta_1)} + a_2 e^{i(\theta_2+\Delta_2)})(a_1 e^{i(\theta_1+\Delta_1)} + a_2 e^{i(\theta_2+\Delta_2)})$$
$$\Rightarrow I' = {a_1}^2 + {a_2}^2 + 2a_1 a_2 \cos(\theta_1 - \theta_2 + \Delta_1 - \Delta_2) \tag{A.2}$$
$$\Rightarrow I' = {a_1}^2 + {a_2}^2 + 2a_1 a_2 \cos(\phi_{12} + \Delta_{12})$$

where $\Delta_{12} = \Delta_1 - \Delta_2$ is the relative phase difference between the two-phase changes at points A_1 and A_2.

Pixel-by-pixel digital subtraction of the two intensity distributions yields a macroscopic fringe pattern (i.e., the shearogram or the shearography correlogram):

$$I = |I - I'| = 2a_1 a_2 \cos|\cos(\phi_{12}) - \cos(\phi_{12} + \Delta_{12})| =$$
$$= 4a_1 a_2 |\sin\left(\phi_{12} + \frac{\Delta_{12}}{2}\right)\sin\left(\frac{\Delta_{12}}{2}\right)| \tag{A.3}$$

Equation (A.3) represents a high-frequency carrier $\sin(\phi_{12} + 0.5 \cdot \Delta_{12})$ modulated by a low-frequency factor $\sin(\Delta_{12}/2)$, which depends on small local deformations of the object due to loading.

If each speckle covers one pixel, the pixels will be black ($Is = 0$) when the intensity of both images is identical, namely $\Delta = 2\pi\ k$ ($k = 0, 1, 2, 3, \ldots$). The brightness of the pixel in the resulting image, Is, will increase as the difference between the intensities I and I' increases. Pixels with the same brightness generate macroscopic lines (fringes) in the resulting image, Is.

The visible fringes of an interferogram describe the distribution of relative phase changes. Figure A.3 shows the geometric relation between the illumination vector k_1, the observation vector k_2 and the deformation vector of a point A on an object surface d.

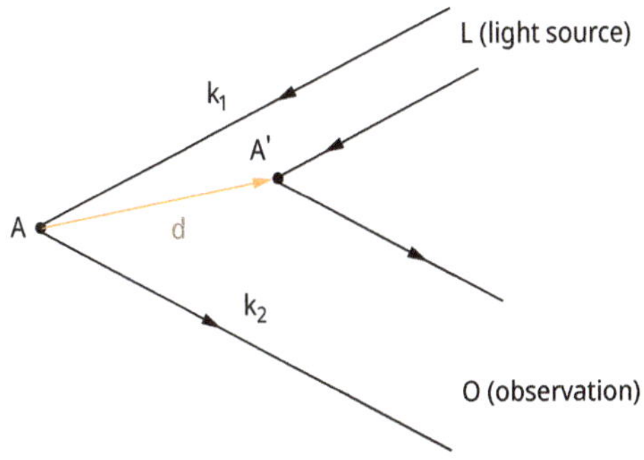

Fig. A3: The schematic appearance of the various vectors (from [71]).

The relative phase difference can be shown to be

$$\Delta = k_2 d - k_1 d = k_s d \tag{A.4}$$

where

$$k_s = k_2 - k_1$$

Using the x, y, z coordinates, the term Δ can be expressed as

$$\Delta = k_s(ue_x + ve_y + we_z) = uk_s e_x + vk_s e_y + wk_s e_z) \tag{A.5}$$

where u, v and w are the components of the deformation vector. e_x, e_y and e_z are the unit vectors in the x-, y- and z-directions, respectively.

The sensitivity vector k_s lies along the bisector of the angle between the illumination and viewing directions and is theoretically not the same for each point of the object surface due to the changing illumination and observation directions. However, it can be assumed that the sensitivity vector k_s at each point of the investigated surface is equal when the dimensions of the object are small compared with the distances between the laser and the object or the camera and the object. To determine the sensitivity vector k_s, the center point of the object is chosen (see Fig. A4).

The magnitude of the sensitivity vector can be expressed as

$$|k_s| = \frac{4\pi}{\lambda}\cos\left[\frac{\theta_{xz}}{2}\right] \tag{A.6}$$

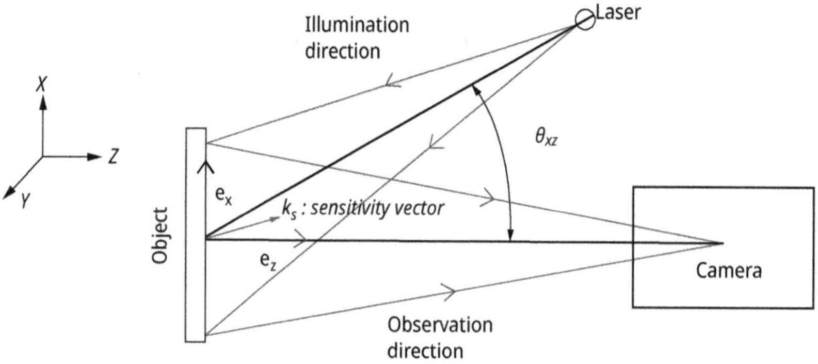

Fig. A4: The schematic geometric relation of the light source L, the observation position O and the object A (from [71]).

Note that in LST, the deformation derivatives can be measured directly. Using (A.5), the relative phase difference Δ_{12} can be written as

$$\begin{aligned} &\Delta_{12} = \Delta_1 - \Delta_2 = u_1 k_s e_x + v_1 k_s e_y + w_1 k_s e_z - u_2 k_s e_x + v_2 k_s e_y + w_2 k_s e_z \\ &\Rightarrow \Delta_{12} = \delta u k_s e_x + \delta v k_s e_y + \delta w k_s e_z \end{aligned} \tag{A.7}$$

When the images are sheared in the x-direction, (A.7) is represented by

$$\Delta_x = \delta x\left(\frac{\delta u}{\delta x}k_s e_x + \frac{\delta v}{\delta x}k_s e_y + \frac{\delta w}{\delta x}k_s e_z\right) \tag{A.8}$$

where δx is the amount of shear in the x-direction on the object surface. Similarly, for shearing in the y-direction, the relative phase change is

$$\Delta_y = \delta y\left(\frac{\delta u}{\delta y}k_s e_x + \frac{\delta v}{\delta y}k_s e_y + \frac{\delta w}{\delta y}k_s e_z\right) \tag{A.9}$$

where δy is the amount of shear in the y-direction on the object surface. Thus, the shear vector is defined as follows:

$$\vec{S} = \delta x \cdot \hat{e}_x + \delta y \cdot \hat{e}_y \tag{A.10}$$

Assuming very small shears, δx and δy, eqs. (A.8) and (A.9) can be rewritten as

$$\begin{aligned} \Delta_x &= \delta x \cdot \left[\frac{\partial u}{\partial x}k_s\hat{e}_x + \frac{\partial v}{\partial x}k_s\hat{e}_y + \frac{\partial w}{\partial x}k_s\hat{e}_z\right] \\ \Delta_y &= \delta y \cdot \left[\frac{\partial u}{\partial x}k_s\hat{e}_x + \frac{\partial v}{\partial x}k_s\hat{e}_y + \frac{\partial w}{\partial x}k_s\hat{e}_z\right] \end{aligned} \tag{A.11}$$

The expressions presented in eq. (A.11) are the fundamental equations for LST and describe the whole field correlation fringes as contours of constant first derivative of deformation. For this reason, the shearogram depicts the strain concentrations directly and it is therefore suitable for NDT.

A further simplification of eq. (A.11) can be done by adjusting the direction of illumination. When the direction of illumination is in the x-, z-plane, the sensitivity vector lies in the x-, z-plane as well. Using Fig. A3 and eq. (A.6), the following expression can be derived to be

$$\begin{aligned} k_s\hat{e}_x &= |k_s|\sin\left(\frac{\theta_{xz}}{2}\right) = \left(\frac{4\pi}{\lambda}\right)\cos\left(\frac{\theta_{xz}}{2}\right)\sin\left(\frac{\theta_{xz}}{2}\right) = \left(\frac{2\pi}{\lambda}\right)\sin\theta_{xz} \\ k_s\hat{e}_y &= 0 \\ k_s\hat{e}_z &= |k_s|\cos\left(\frac{\theta_{xz}}{2}\right) = \left(\frac{4\pi}{\lambda}\right)\cos\left(\frac{\theta_{xz}}{2}\right)\cos\left(\frac{\theta_{xz}}{2}\right) = \left(\frac{2\pi}{\lambda}\right)(1+\cos\theta_{xz}) \end{aligned} \tag{A.12}$$

Using eqs. (A.10) and (A.11), it follows that:

$$\begin{aligned} \Delta_x &= \delta x \cdot \left[\frac{\partial u}{\partial x}k_s\hat{e}_x + \frac{\partial v}{\partial x}k_s\hat{e}_y + \frac{\partial w}{\partial x}k_s\hat{e}_z\right] = \frac{2\cdot\pi\cdot\delta x}{\lambda}\left[\sin\theta_{xz}\frac{\partial u}{\partial x} + (1+\cos\theta_{xz})\frac{\partial w}{\partial x}\right] \\ \Delta_y &= \delta y \cdot \left[\frac{\partial u}{\partial x}k_s\hat{e}_x + \frac{\partial v}{\partial x}k_s\hat{e}_y + \frac{\partial w}{\partial x}k_s\hat{e}_z\right] = \frac{2\cdot\pi\cdot\delta y}{\lambda}\left[\sin\theta_{xz}\frac{\partial u}{\partial x} + (1+\cos\theta_{xz})\frac{\partial w}{\partial y}\right] \end{aligned} \tag{A.13}$$

The expressions presented in eq. (A.13) show that the illumination angles are characterized by θ_{xz} and the shearing displacements by Δ_x or Δ_y.

As shown in eq. (A.13), a shearogram usually contains the in-plane as well as the out-of-plane terms of the strain tensor. When the illumination direction is adjusted normal to the object surface, the angle of illumination θ_{xz} becomes zero; thus, $\sin\theta_{xz} = 0$ and $\cos\theta_{xz} = 1$. The out-of-plane shear components $\partial w/\partial x$ and $\partial w/\partial y$ can be obtained from the following equations:

$$\begin{aligned}\Delta_x &= \frac{4\cdot\pi\cdot\delta x}{\lambda}\cdot\frac{\partial w}{\partial x} \Rightarrow \frac{\partial w}{\partial x} = \frac{\Delta_x\cdot\lambda}{4\cdot\pi\cdot\delta x}\\ \Delta_y &= \frac{4\cdot\pi\cdot\delta y}{\lambda}\cdot\frac{\partial w}{\partial y} \Rightarrow \frac{\partial w}{\partial y} = \frac{\Delta_y\cdot\lambda}{4\cdot\pi\cdot\delta y}\end{aligned} \tag{A.14}$$

The expressions in eq. (A.14) present the relation between the phase map (shearogram) $\Delta_{(x,y)}$ and the derivative of the vertical surface displacement in the direction of the shear vector, $\partial w_{(x,y)}/\partial\vec{s}$, which is defined as follows:

$$\frac{\partial w_{(x,y)}}{\partial\vec{s}} = \frac{\Delta_{(x,y)}\cdot\lambda}{4\cdot\pi\cdot\delta s} \tag{A.15}$$

15 Monitoring natural frequencies to yield material properties

15.1 Introduction

Throughout history, a designer or a structural engineer would like to be sure about the capability of a structure to withstand the loads applied on it. To ascertain its design, the structural properties of the structure should be a priori of his design. The obtained properties like mass, gravitational acceleration, tensile force in a string and moment of inertia are suggested to monitor the natural frequencies of simple structures and calculate the needed parameter. The literature presents a vast number of manuscripts, dealing with this issue (see [1–20]). Good reviews can be found in [1, 3, 6], while the use of vibrational data for the medical sector is addressed in [7, 18]. All the other references address mechanical issues for various simple and advanced structures. In what follows, simple structures like pendulums, strings and beams are outlined both analytically and experimentally to measure various properties. One has to remember the vibration correlation technique (VCT) aimed at nondestructively predict the buckling loads of thin-walled structures like columns, plates and cylindrical shells presented in detail in Chapter 11 of this book.

15.2 Pendulums

15.2.1 Mathematical pendulum

Figure 15.1 presents a drawing of a mathematical pendulum with a mass M, a massless string with a length L swinging at point O at an angle θ. Applying the D'Alembert law, one obtains the following equation:

$$\begin{aligned}&Mg\ \sin\theta\cdot L+ML^2\frac{d^2\theta}{dt^2}=0\Rightarrow\frac{d^2\theta}{dt^2}+\frac{g}{L}\sin\theta=0\\&\text{for small}\ \theta:\ \frac{d^2\theta}{dt^2}+\frac{g}{L}\theta=0\end{aligned}\tag{15.1}$$

The solution of eq. (15.1) for small θ angles ($\sin\theta\approx\theta$) has the form:

$$\theta(t)=Ae^{i\omega t},\quad \text{where}\quad \omega^2=\tfrac{g}{L}\tag{15.2}$$

and ω is the natural frequency of the motion. The period can be written as

https://doi.org/10.1515/9783111621104-015

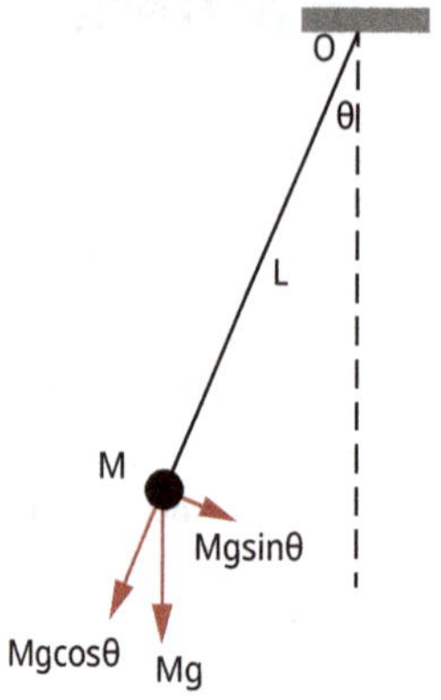

Fig. 15.1: A mathematical pendulum.

$$T = \frac{2\pi}{\omega} = 2\pi\sqrt{\frac{L}{g}} \tag{15.3}$$

To find the value of the gravitation acceleration g, one has to measure the period of the motion and the length of the string to yield

$$g = \frac{4\pi^2 L}{T^2} \tag{15.4}$$

To increase the accuracy of the calculated gravitation acceleration g, one would use a long string.

15.2.2 Physical pendulum

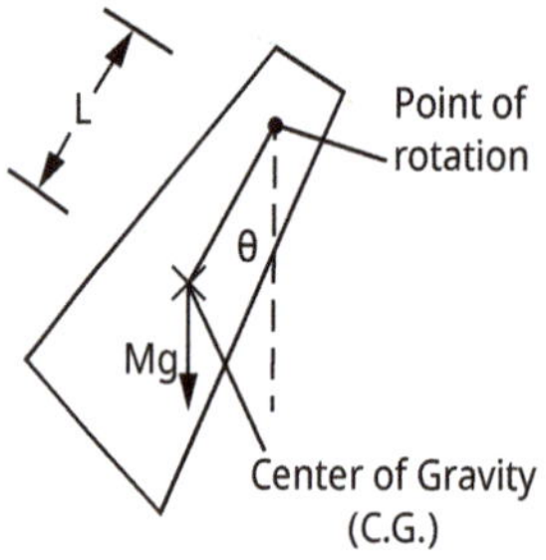

Fig. 15.2: A physical pendulum.

Figure 15.2 presents a drawing of a physical pendulum, with a mass M, and its center of gravity located at a distance L from the point of rotation. The pendulum is deflected at an angle θ. Applying the D'Alembert law, one obtains the following equation (similar to the process done before for a mathematical pendulum):

$$\begin{aligned} &Mg \sin\theta \cdot L + I\frac{d^2\theta}{dt^2} = 0 \Rightarrow \frac{d^2\theta}{dt^2} + \frac{Mg}{I}L\sin\theta = 0 \\ &\text{for small } \theta: \quad \frac{d^2\theta}{dt^2} + \frac{Mg}{I}L\cdot\theta = 0 \end{aligned} \tag{15.5}$$

The solution of eq. (15.5) for small θ angles ($\sin\theta \approx \theta$) has the form:

$$\theta(t) = Ae^{i\omega t}, \quad \text{where} \quad \omega^2 = \frac{Mg\cdot L}{I} \tag{15.6}$$

where I is the inertia moment of the pendulum relative to its point of rotation and ω is the natural frequency of the motion. The period can be written as

$$T = \frac{2\pi}{\omega} = 2\pi\sqrt{\frac{I}{Mg\cdot L}} \tag{15.7}$$

To find the value of the inertia moment, I, one has to measure the period of the motion and the mass of the pendulum to yield

$$I = Mg\cdot T^2 \cdot \frac{L}{4\pi^2} \tag{15.8}$$

To increase the accuracy of the calculated inertia moment, I, one would use a long string. In case this value, I, is known, while the mass is the unknown to be found, we can use the following equation:

$$M = 4\pi^2 \cdot \frac{I}{g\cdot L\cdot T^2} \tag{15.9}$$

with all the parameters on the right-hand side of eq. (15.9) being known a priori.

15.2.3 Torsional pendulum

Figure 15.3 presents a drawing of a torsional pendulum, with a mass having a moment of inertia I, hanged on a wire, spring or a thin column, having a torsional rigidity of κ. The pendulum is deflected at an angle θ and then the whole system is oscillating between $+\theta$ and $-\theta$, where $\theta = 0$ is the equilibrium point. Applying the D'Alembert law for small angles of rotation ($\sin\theta \approx \theta$), one obtains the following equation (similar to the process done before for a mathematical pendulum):

$$\kappa\theta + I\frac{d^2\theta}{dt^2} = 0 \Rightarrow \frac{d^2\theta}{dt^2} + \frac{\kappa}{I}\theta = 0 \tag{15.10}$$

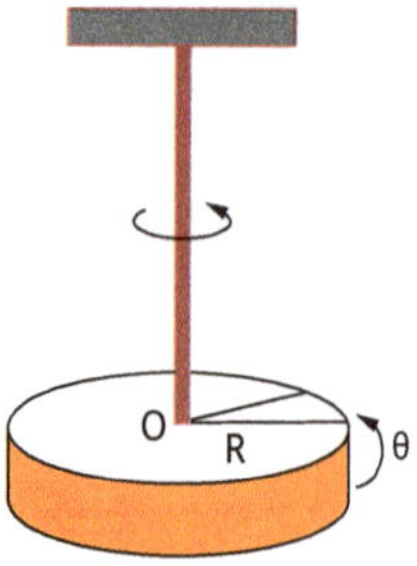

Fig. 15.3: A torsional pendulum.

As shown before, the solution of eq. (15.10) has the form:

$$\theta(t) = Ae^{i\omega t}, \quad \text{where} \quad \omega^2 = \frac{\kappa}{I} \tag{15.11}$$

where I is the inertia moment of the pendulum relative to its point of rotation and ω is the rotational natural frequency of the motion. The period can be written as

$$T = \frac{2\pi}{\omega} = 2\pi\sqrt{\frac{I}{\kappa}} \tag{15.12}$$

Note that the units of κ are N*m or kg*m^2/s^2 and the units of I are kg*m^2.

To find the value of the inertia moment, I, provided the value of κ is known, one has to measure the period of the motion to yield

$$I = \kappa \frac{T^2}{4\pi^2} \tag{15.13}$$

To increase the accuracy of the calculated inertia moment, I, one would use a long string. In case this value, I, is known, while the κ is the unknown to be found, we can use the following equation:

$$\kappa = 4\pi^2 \cdot \frac{I}{T^2} \tag{15.14}$$

with all the parameters on the right-hand side of eq. (15.14) being known a priori.

15.3 Strings

The string equation of motion, undergoing small lateral vibration (see Fig. 15.4), can be written as

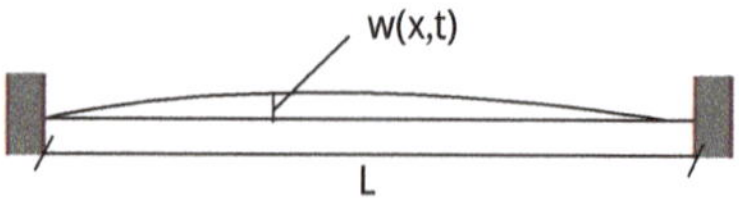

Fig. 15.4: A schematic string.

$$\rho(x)\frac{\partial^2 w(x,t)}{\partial t^2} - N(x)\frac{\partial^2 w(x,t)}{\partial x^2} = 0 \tag{15.15}$$

where $\rho(x)$ and $T(x)$ are the string mass density and tensile force, respectively. Assuming these two parameters are constant along the string, we obtain

$$\frac{\partial^2 w(x,t)}{\partial t^2} - \lambda^2\frac{\partial^2 w(x,t)}{\partial x^2} = 0 \tag{15.16}$$

where

$$\lambda^2 = N$$

The solution of eq. (15.16) can be written as

$$w(x,t) = \sum_{n=1}^{\infty}\left[A_n \sin\left(\frac{n\pi\lambda t}{L}\right) + B_n \cos\left(\frac{n\pi\lambda t}{L}\right)\right]\sin\left(\frac{n\pi x}{L}\right) \tag{15.17}$$

The eigenvalue and its associated eigenfunctions can be written as

$$\begin{aligned} &\omega_n^2 = \frac{n^2\pi^2}{L^2}\cdot\frac{N}{\rho} \qquad \phi_n(x) = \sin\left(\frac{n\pi x}{L}\right) \\ &n = 1, 2, 3, \ldots \end{aligned} \tag{15.18}$$

Therefore, to find the tension of the string, one must know its mass density and the tensile load N can be calculated after measuring its frequency. The easy way is to monitor the first frequency (at $n = 1$) to yield

$$\begin{aligned} &\omega_n^2 = \frac{n^2\pi^2}{L^2}\cdot\frac{N}{\rho} \Rightarrow N = \frac{\omega_n^2\cdot\rho\cdot L^2}{n^2\pi^2} = \frac{(2\pi f)^2\cdot\rho\cdot L^2}{n^2\pi^2} \\ &\Rightarrow \text{for} \quad n = 1 \quad N = 4f^2\cdot\rho\cdot L^2 \end{aligned} \tag{15.19}$$

Note the application of the method presented in this section is realized in Ref. [11].

15.4 Columns, beams or rods

15.4.1 Longitudinal vibrations of beams

Figure 15.5 represents schematically a rod having the length L and an axial stiffness of EA (E being its Young's modulus and A its cross section). The rod is assumed to be thin and uniform along its length.

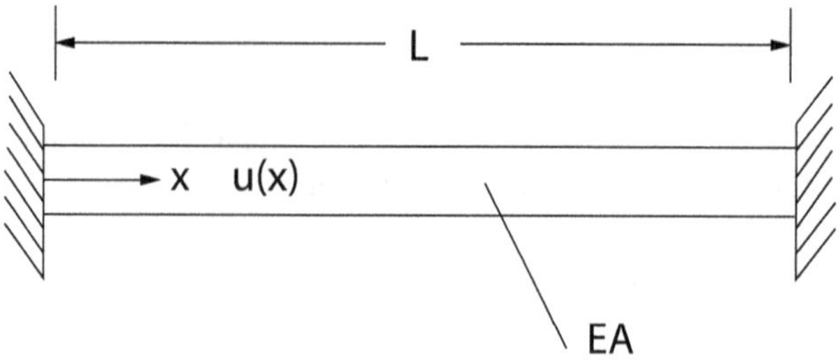

Fig. 15.5: A schematic beam.

The equation of motion for longitudinal vibrations has the form (see Ref. [21]):

$$\rho A \frac{\partial^2 u(x,t)}{\partial t^2} - EA \frac{\partial^2 u(x,t)}{\partial x^2} = 0$$

or

$$\frac{\partial^2 u(x,t)}{\partial t^2} - \kappa^2 \frac{\partial^2 u(x,t)}{\partial x^2} = 0 \tag{15.20}$$

where

$$\kappa^2 = \frac{E}{\rho}$$

Note that eq. (15.20) is like eq. (15.16), developed in the previous sub-chapter for a tensioned string. The solution for eq. (15.20) for a clamped–clamped rod is similar to that presented by eq. (15.17), namely

$$u(x,t) = \sum_{n=1}^{\infty} \left[A_n \sin\left(\frac{n\pi\kappa t}{L}\right) + B_n \cos\left(\frac{n\pi\kappa t}{L}\right) \right] \sin\left(\frac{n\pi x}{L}\right) \tag{15.21}$$

The eigenvalue and its associated eigenfunctions for clamped–clamped rod can be written as

$$\omega_n^2 = \frac{n^2\pi^2}{L^2} \cdot \frac{E}{\rho} \qquad \phi_n(x) = \sin\left(\frac{n\pi x}{L}\right) \qquad n = 1, 2, 3, \ldots \tag{15.22}$$

Therefore, to find the Young's modulus of the rod, one has to know its mass density ρ and to measure its frequency. The easy way is to monitor the first frequency (at $n = 1$) to yield

$$\omega_n^2 = \frac{n^2\pi^2}{L^2} \cdot \frac{E}{\rho} \Rightarrow E = \frac{\omega_n^2 \cdot \rho \cdot L^2}{n^2\pi^2} = \frac{(2\pi f)^2 \cdot \rho \cdot L^2}{n^2\pi^2}$$
$$\Rightarrow \text{for} \quad n = 1 \quad E = 4f^2 \cdot \rho \cdot L^2 \tag{15.23}$$

Alternatively, one can find the mass density, ρ, once the Young's modulus E is known to yield (for $n = 1$)

$$\omega_n^2 = \frac{n^2\pi^2}{L^2} \cdot \frac{E}{\rho} \Rightarrow \rho = \frac{n^2\pi^2}{L^2} \cdot \frac{E}{\omega_n^2} = \frac{n^2\pi^2}{L^2} \cdot \frac{E}{(2\pi f)^2}$$
$$\Rightarrow \text{for} \quad n = 1 \quad \rho = \frac{E}{4f^2L^2} \tag{15.24}$$

One should note that the boundary conditions at the ends of the rod have a great influence on its natural frequencies. For a free–free boundary conditions, the solution for eq. (15.20) should be

$$w(x,t) = \sum_{n=1}^{\infty} \left[A_n \sin\left(\frac{n\pi\kappa t}{L}\right) + B_n \cos\left(\frac{n\pi\kappa t}{L}\right)\right] \cos\left(\frac{n\pi x}{L}\right) \tag{15.25}$$

For other boundary conditions, please consult Appendix A at the end of this chapter.

15.4.2 Torsional vibrations of beams

Figure 15.6 represents schematically a rod having the length L and a torsional stiffness GI_p (G being its shear modulus and I_p is the polar moment of inertia). The rod is assumed circular thin and uniform along its length and the boundary conditions at its end can be either free–free, as depicted in the figure or other types of holding the rod. The cross section can be either circular or rectangular. The longitudinal axes would be x, while to single displacement would be θ, a rotation around the center of the cross section.

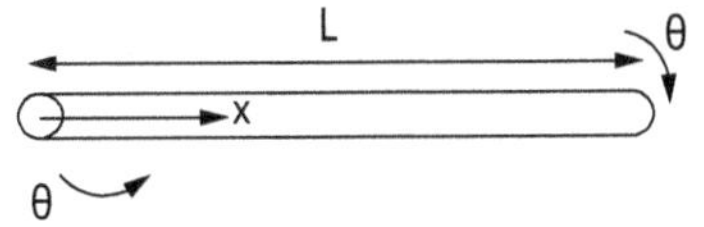

Fig. 15.6: A schematic free–free rod.

The equation of motion for torsional vibrations has the form (see [21]):

$$\rho I_p \frac{\partial^2 \theta(x,t)}{\partial t^2} - GI_p \frac{\partial^2 \theta(x,t)}{\partial x^2} = 0$$

or

$$\frac{\partial^2\theta(x,t)}{\partial t^2} - \lambda^2 \frac{\partial^2\theta(x,t)}{\partial x^2} = 0 \tag{15.26}$$

where

$$\lambda^2 = \frac{GI_p}{\rho I_p}$$

Note that I_p is the polar moment of inertia of the rod and ρ is its mass density. Also

$$G = \frac{E}{2(1+\upsilon)} \tag{15.27}$$

The general solution of eq. (15.26) can be written as

$$\theta(x,t) = e^{i\omega t}\left[A \sin\frac{\omega}{\lambda}x + B\cos\frac{\omega}{\lambda}x\right] \tag{15.28}$$

with the constants A and B to be determined by the boundary conditions of the problem.

For the case of clamped–free torsional vibrations, the boundary conditions would be

$$\theta(0,t) = 0 \quad \text{and} \quad \frac{\partial\theta(L,t)}{\partial x} = 0 \tag{15.29}$$

Enforcing the boundary conditions (eq. (15.29)) on the proposed solution (eq. (15.28)) yields

$$\begin{aligned} &\theta(0,t) = 0 && \Rightarrow B = 0 \\ &\frac{\partial\theta(L,t)}{\partial x} = 0 && \Rightarrow \cos\frac{\omega L}{\lambda} = 0 \quad \text{or} \quad \frac{\omega L}{\lambda} = \left(n + \frac{1}{2}\right)\pi, \quad n = 1,2,3,\ldots \end{aligned} \tag{15.30}$$

Therefore,

$$\omega = \left(n + \frac{1}{2}\right)\frac{\pi}{L}\lambda = \left(n + \frac{1}{2}\right)\frac{\pi}{L}\sqrt{\frac{G}{\rho}}$$

For the case presented in Fig. 15.3, namely a rod clamped at $x = 0$ and a disk having a moment of inertia J_0 attached at the other end (at $x = L$), the boundary conditions can be written as

$$\theta(0,t) = 0 \quad \text{and} \quad GI_p\frac{\partial\theta(L,t)}{\partial x} = J_0 \cdot \omega^2 \cdot \theta(L,t) \tag{15.31}$$

Using eq. (15.28) with the boundary conditions written in eq. (15.32) yields

$$
\begin{aligned}
&\theta(0,t)=0 \qquad \Rightarrow B=0\\
&GI_p\frac{\partial\theta(L,t)}{\partial x}=J_0\cdot\omega^2\cdot\theta(L,t)\\
&\text{or}\quad GI_p\frac{\omega}{\lambda}\cos\frac{\omega L}{\lambda}=J_0\cdot\omega^2\cdot\sin\frac{\omega L}{\lambda}\\
&\Rightarrow \tan\frac{\omega L}{\lambda}=\frac{GI_p}{J_0}\cdot\frac{1}{\lambda\omega}=\frac{GI_p}{J_0}\cdot\frac{L\cdot\lambda}{\lambda^2\omega L}=\frac{GI_p}{J_0}\cdot\frac{\lambda}{\omega L}\cdot\frac{L}{\lambda^2}\\
&\text{if}\quad \frac{\omega L}{\lambda}\equiv\delta \qquad \delta\cdot\tan\delta=\frac{GI_p}{J_0}\cdot\frac{L}{G}\rho=\frac{I_p}{J_0}\cdot L\cdot\rho=\frac{J_{rod}}{J_0}
\end{aligned}
\tag{15.32}
$$

where

$$J_{rod}=I_p\cdot L\cdot\rho$$

Another interesting configuration would be a rod on free–free boundary conditions undergoing torsional vibrations. The boundary conditions for this case are:

$$\frac{\partial\theta(0,t)}{\partial x}=0 \quad\text{and}\quad \frac{\partial\theta(L,t)}{\partial x}=0 \tag{15.33}$$

Enforcing the boundary conditions (eq. (15.32)) on the proposed solution (eq. (15.28)) yields

$$
\begin{aligned}
&\frac{\partial\theta(0,t)}{\partial x}=0 \quad \Rightarrow A=0\\
&\frac{\partial\theta(L,t)}{\partial x}=0 \quad \Rightarrow \sin\frac{\omega L}{\lambda}=0 \quad\text{or}\quad \frac{\omega L}{\lambda}=n\cdot\pi,\quad n=1,2,3,\ldots
\end{aligned}
\tag{15.34}
$$

Therefore,

$$\omega=n\frac{\pi}{L}\lambda=\frac{n\pi}{L}\sqrt{\frac{G}{\rho}}$$

Therefore, to find the shear modulus of the rod, G, one has to know its mass density ρ and to measure its frequency for a given configuration of boundary conditions. Taking the clamped–free boundary conditions, the easy way is to monitor the first frequency (at $n=1$) to yield

$$
\begin{aligned}
&\omega_n^2=\left[\left(n+\frac{1}{2}\right)\frac{\pi}{L}\right]^2\frac{G}{\rho}\Rightarrow G=\frac{\omega_n^2\cdot\rho\cdot L^2}{\left[n+\frac{1}{2}\right]^2\pi^2}=\frac{(2\pi f)^2\cdot\rho\cdot L^2}{\left[n+\frac{1}{2}\right]^2\pi^2}\\
&\Rightarrow \text{for}\quad n=1 \qquad G=\frac{4f^2\cdot\rho\cdot L^2}{2.25}
\end{aligned}
\tag{15.35}
$$

Alternatively, to measure the mass density ρ, while the shear modulus G is known, we shall use the following equation:

$$\omega_n^2 = \left[\left(n+\frac{1}{2}\right)\frac{\pi}{L}\right]^2 \frac{G}{\rho} \Rightarrow \rho = \left[\left(n+\frac{1}{2}\right)\frac{\pi}{L}\right]^2 \frac{G}{\omega^2} = \frac{\left[n+\frac{1}{2}\right]^2 \pi^2 \cdot G}{(2\pi f)^2 \cdot L^2}$$
$$\Rightarrow \text{for} \quad n=1 \qquad \rho = \frac{2.25 \cdot G}{4f^2 \cdot L^2} \tag{15.36}$$

15.4.3 Lateral vibrations of beams

The topic of lateral vibrations of beams has been already presented in Chapter 11 of this book. Only the important topics will be presented next. A typical beam resting on simply supported boundary conditions is presented in Fig. 15.7. The bending stiffness is EI, its mass per unit length is ρA (A being its cross section), and its length is L. The Euler–Bernoulli assumption is used to derive the equation of motion of the beam; namely, the shear deformations are neglected and small amplitude vibration is used.

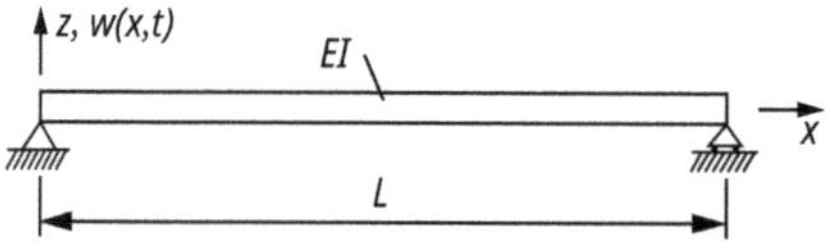

Fig. 15.7: A schematic simply supported–simply supported beam.

Based on the above assumptions, the equation of motion can be written as (see also [21])

$$EI\frac{\partial^4 w(x,t)}{\partial x^4} + \rho A\frac{\partial^2 w(x,t)}{\partial t^2} = 0 \tag{15.37}$$

where $w(x,t)$ is the out-of-plane displacement. Assuming harmonic vibrations, the solution of eq. (15.37) has the following form:

$$w(x,t) = W(x)e^{i\omega t} \tag{15.38}$$

where ω is the angular frequency of the column and $W(x)$ is a function to be next determined. Substituting eq. (15.38) into eq. (15.37) provides the following expression:

$$EIW_{xxxx} - \rho A\omega^2 W = 0$$
$$W_{xxxx} - \frac{\rho A\omega^2}{EI} W = 0 \qquad \Rightarrow W_{xxxx} - \lambda^4 W = 0 \tag{15.39}$$

where

$$\lambda^4 = \frac{\rho A \omega^2}{EI}$$

where $[]_{xxxx}$ stands for four times differentiation with respect to x.

The general solution for eq. (15.39) is

$$W(x) = C_1 \cosh \lambda x + C_2 \sinh \lambda x + C_3 \cos \lambda x + C_4 \sin \lambda x \tag{15.40}$$

where the constants C_1, C_2, C_3 and C_4 are to be determined by the boundary conditions of the beam at its both ends.

This topic was already presented in Chapter 7 of this book. Therefore, two tables from this chapter are again presented here for an isotropic beam (Tabs. 15.1 and 15.2).

Tab. 15.1: Characteristic equations and their relevant eigenvalues for natural vibrations of beams.

No.	Name	Characteristic equation	Eigenvalues
1	SS–SS*	$\sin(\lambda L) = 0$ $(\lambda L)_n = n\pi, \quad n = 1, 2, 3, \ldots$	$\omega_n = \left(\frac{n\pi}{L}\right)^2 \sqrt{\frac{EI}{\rho A}}$
2	C–C**	$\cos(\lambda L)\cosh(\lambda) = 1$ $(\lambda L)_n = 4.73004, 7.85321, 10.9956, \ldots, \frac{(2n+1)\pi}{2}$	$\omega_1 = \left(\frac{4.73004}{L}\right)^2 \sqrt{\frac{EI}{\rho A}}$
3	C–F***	$\cos(\lambda L)\cosh(\lambda L) = -1$ $(\lambda L)_n = 1.87510, 4.69409, 7.85340, \ldots, \frac{(2n-1)\pi}{2}$	$\omega_1 = \left(\frac{1.87351}{L}\right)^2 \sqrt{\frac{EI}{\rho A}}$
4	F–F	$\cos(\lambda L)\cosh(\lambda L) = 1$ $(\lambda L)_n = 4.73004, 7.85321, 10.9956, \ldots, \frac{(2n+1)\pi}{2}$	$\omega_1 = \left(\frac{4.73004}{L}\right)^2 \sqrt{\frac{EI}{\rho A}}$
5	SS–C	$\tan(\lambda L) = \tanh(\lambda L)$ $(\lambda L)_n = 3.9266, 7.0686, 10.2102, \ldots, \frac{(4n+1)\pi}{4}$	$\omega_1 = \left(\frac{3.9266}{L}\right)^2 \sqrt{\frac{EI}{\rho A}}$
6	SS–F	$\tan(\lambda L) = \tanh(\lambda L)$ $(\lambda L)_n = 3.9266, 7.0686, 10.2102, \ldots, \frac{(4n+1)\pi}{4}$	$\omega_1 = \left(\frac{3.9266}{L}\right)^2 \sqrt{\frac{EI}{\rho A}}$
7	G****–F	$\tan(\lambda L) = -\tanh(\lambda L)$ $(\lambda L)_n = 2.3650, 5.4978, 8.6394, \ldots, \frac{(4n-1)\pi}{4}$	$\omega_1 = \left(\frac{2.3650}{L}\right)^2 \sqrt{\frac{D_{11}}{I_0}}$
8	G–SS	$\cos(\lambda L) = 0$ $(\lambda L)_n = (2n-1)\frac{\pi}{2} \quad n = 1, 2, 3, \ldots$	$\omega_n = \left[\frac{(2n-1)\pi}{2L}\right] \sqrt{\frac{EI}{\rho A}}$

Tab. 15.1 (continued)

No.	Name	Characteristic equation	Eigenvalues
9	G–G	$\sin(\lambda L)=0$ $(\lambda L)_n=n\pi, \quad n=1,2,3,\ldots$	$\omega_n=\left(\frac{n\pi}{L}\right)^2\sqrt{\frac{EI}{\rho A}}$
10	G–C	$\tan(\lambda L)=-\tanh(\lambda L)$ $(\lambda L)_n=2.3650, 5.4978, 8.6394, \ldots, \frac{(4n-1)\pi}{4}$	$\omega_1=\left(\frac{2.3650}{L}\right)^2\sqrt{\frac{EI}{\rho A}}$

*SS, simply supported; **C, clamped; ***F, free; G****, guided.

To calculate the bending stiffness EI of a beam (while ρA is known) or alternatively its mass per unit length ρA (while EI is known), one needs to measure the natural frequency of a beam on a given boundary conditions. Provided the boundary conditions are the exact one presented in Tab. 15.1, one can use the following expressions:

Tab. 15.2: Mode shapes for natural vibrations of isotropic beams.

No.	Name	Mode shape
1	SS–SS	$W_n(x)=\sin\left(\frac{n\pi x}{L}\right)$
2	C–C	$W_n(x)=\cosh\left[\frac{(\alpha_2L)_nx}{L}\right]-\cos\left[\frac{(\alpha_2L)_nx}{L}\right]-\frac{\cosh[(\alpha_2L)_n]-\cos[(\alpha_2L)_n]}{\sinh[(\alpha_2L)_n]-\sin[(\alpha_2L)_n]}\left\{\sinh\left[\frac{(\alpha_2L)_nx}{L}\right]-\sin\left[\frac{(\alpha_2L)_nx}{L}\right]\right\}$
3	C–F	$W_n(x)=\cosh\left[\frac{(\alpha_2L)_nx}{L}\right]-\cos\left[\frac{(\alpha_2L)_nx}{L}\right]-\frac{\cosh[(\alpha_2L)_n]+\cos[(\alpha_2L)_n]}{\sinh[(\alpha_2L)_n]+\sin[(\alpha_2L)_n]}\left\{\sinh\left[\frac{(\alpha_2L)_nx}{L}\right]-\sin\left[\frac{(\alpha_2L)_nx}{L}\right]\right\}$
4	F–F	$W_n(x)=\cosh\left[\frac{(\alpha_2L)_nx}{L}\right]-\cos\left[\frac{(\alpha_2L)_nx}{L}\right]-\frac{\cosh[(\alpha_2L)_n]-\cos[(\alpha_2L)_n]}{\sinh[(\alpha_2L)_n]-\sin[(\alpha_2L)_n]}\left\{\sinh\left[\frac{(\alpha_2L)_nx}{L}\right]-\sin\left[\frac{(\alpha_2L)_nx}{L}\right]\right\}$
5	SS–C	$W_n(x)=\cosh\left[\frac{(\alpha_2L)_nx}{L}\right]-\cos\left[\frac{(\alpha_2L)_nx}{L}\right]-\frac{\cosh[(\alpha_2L)_n]-\cos[(\alpha_2L)_n]}{\sinh[(\alpha_2L)_n]-\sin[(\alpha_2L)_n]}\left\{\sinh\left[\frac{(\alpha_2L)_nx}{L}\right]-\sin\left[\frac{(\alpha_2L)_nx}{L}\right]\right\}$
6	SS–F	$W_n(x)=\cosh\left[\frac{(\alpha_2L)_nx}{L}\right]+\cos\left[\frac{(\alpha_2L)_nx}{L}\right]-\frac{\cosh[(\alpha_2L)_n]+\cos[(\alpha_2L)_n]}{\sinh[(\alpha_2L)_n]+\sin[(\alpha_2L)_n]}\left\{\sinh\left[\frac{(\alpha_2L)_nx}{L}\right]+\sin\left[\frac{(\alpha_2L)_nx}{L}\right]\right\}$

Tab. 15.2 (continued)

No.	Name	Mode shape
7[†]	G–F	$W_n(x)=\cosh\left[\frac{(\alpha_2L)_nx}{L}\right]-\cos\left[\frac{(\alpha_2L)_nx}{L}\right]-\frac{\cosh[(\alpha_2L)_n]-\cos[(\alpha_2L)_n]}{\sinh[(\alpha_2L)_n]-\sin[(\alpha_2L)_n]}\left\{\sinh\left[\frac{(\alpha_2L)_nx}{L}\right]-\sin\left[\frac{(\alpha_2L)_nx}{L}\right]\right\}$
8	G–SS	$W_n(x)=\sin\left[\frac{(2n-1)\pi x}{2L}\right]$
9	G–G	$W_n(x)=\cos\left(\frac{n\pi x}{L}\right)$
10	G–C	$W_n(x)=\cosh\left[\frac{(\alpha_2L)_nx}{L}\right]-\cos\left[\frac{(\alpha_2L)_nx}{L}\right]-\frac{\sinh[(\alpha_2L)_n]+\sin[(\alpha_2L)_n]}{\cosh[(\alpha_2L)_n]-\cos[(\alpha_2L)_n]}\left\{\sinh\left[\frac{(\alpha_2L)_nx}{L}\right]-\sin\left[\frac{(\alpha_2L)_nx}{L}\right]\right\}$

$$\omega_n^2=\left(\frac{n\pi}{L}\right)^4\frac{EI}{\rho A}\Rightarrow EI=\omega_n^2\cdot\rho A\cdot\left(\frac{L}{n\pi}\right)^4=(2\pi f)^2\cdot\rho A\cdot\left(\frac{L}{n\pi}\right)^4$$

or

$$\omega_n^2=\left(\frac{n\pi}{L}\right)^4\frac{EI}{\rho A}\Rightarrow\rho A=\left(\frac{n\pi}{L}\right)^4\frac{EI}{\omega^2}=\left(\frac{n\pi}{L}\right)^4\frac{EI}{(2\pi f)^2}$$

$$\Rightarrow\text{for}\quad n=1\qquad EI=\frac{4f^2\cdot\rho A}{\pi^2}L^4 \tag{15.41}$$

or

$$\Rightarrow\text{for}\quad n=1\qquad \rho A=\frac{\pi^2\cdot EI}{4f^2L^4}$$

As the actual boundary conditions in a given test are not exactly the ones it meant to be, some discrepancies could arise for the above measurements. To obtain exact results for the terms calculated, the following section will present an experimental approach.

15.4.4 Lateral vibrations of beams – an experimental approach

It is known that the analytical boundary conditions cannot be completely replicated in real-type experiments.

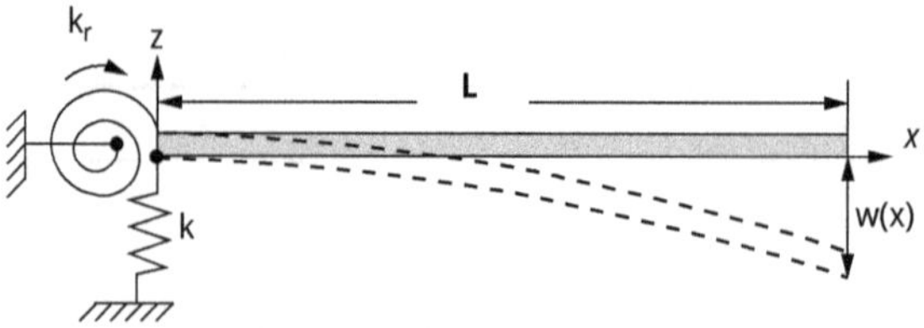

Fig. 15.8: A schematic cantilever-type beam restrained at $x = 0$.

Therefore, to understand the importance of the boundary conditions on the natural frequencies of a beam, a cantilever-type beam is considered, as presented in Fig. 15.8. To correctly simulate the experimental clamped boundary condition, the restrained is done using rotational k_r and translation k springs, as presented in Fig. 15.8. The equation of motion for this case is given by eq. (15.37). The boundary conditions of this problem can be written as

$$\begin{aligned} &@x=0 \qquad -kw(0,t)=EI\frac{\partial^3 w(0,t)}{\partial x^3}; \quad k_r\frac{\partial w(0,t)}{\partial x}=EI\frac{\partial^2 w(0,t)}{\partial x^2}; \\ &@x=L \qquad \frac{\partial^2 w(L,t)}{\partial x^2}=0; \qquad \frac{\partial^3 w(L,t)}{\partial x^3}=0 \end{aligned} \tag{15.42}$$

The solution of eq. (15.39) in the presence of the boundary conditions presented by eq. (15.42) can be written as (see also [10, 13])

$$W(x)=C_1\cosh\lambda x+C_2\sinh\lambda x+C_3\cos\lambda x+C_4\sin\lambda x \tag{15.43}$$

Applying the boundary conditions yields the following four equations:

$$\begin{aligned} &-k(C_1+C_3)-EI\lambda^3(C_2-C_4)=0 \\ &k_r\lambda(C_2+C_4)-EI\lambda^2(C_1-C_3)=0 \\ &\lambda^2(C_1\cosh\lambda L+C_2\sinh\lambda L-C_3\cos\lambda L-C_4\sin\lambda L)=0 \\ &\lambda^3(C_1\sinh\lambda L+C_2\cosh\lambda L+C_3\sin\lambda L-C_4\cos\lambda L)=0 \end{aligned} \tag{15.44}$$

Casting eq. (15.44) in matrix form yields

$$\begin{bmatrix} -k & -EI\lambda^3 & -k & EI\lambda^3 \\ -EI\lambda^2 & k_r\lambda & EI\lambda^2 & k_r\lambda \\ \lambda^2\cosh\lambda L & \lambda^2\sinh\lambda L & -\lambda^2\cos\lambda L & -\lambda^2\sin\lambda L \\ \lambda^3\sinh\lambda L & \lambda^3\cosh\lambda L & \lambda^3\sin\lambda L & -\lambda^3\cos\lambda L \end{bmatrix} \begin{Bmatrix} C_1 \\ C_2 \\ C_3 \\ C_4 \end{Bmatrix} = \begin{Bmatrix} 0 \\ 0 \\ 0 \\ 0 \end{Bmatrix} \tag{15.45}$$

To obtain an unique solution, the determinant of the matrix in eq. (15.45) should be zero. This would yield a characteristic equation for the present case in the form of a transcendental equation with infinite solutions to be obtained either numerically or graphically. We shall denote these solutions as α_n.

Note that for $k \to 0$ and $k_r \to 0$, the characteristic equation would be

$$1 - \cos\lambda_n L \cosh\lambda_n L = 0 \tag{15.46}$$

which is exactly the characteristic equation for a free–free beam (see Tab. 15.1).

For the case $k \to \infty$ and $k_r \to \infty$, the characteristic equation reduces to the clamped–free case, namely

$$1 + \cos\lambda_n L \cosh\lambda_n L = 0 \tag{15.47}$$

The test setup is presented in Fig. 15.9 and consists of a commercial Fourier Force Sensor DT 272,[1] with the spring constants being $k_r = 17.837$ Nm and $k = 6{,}400$ N/m, with an accuracy of ± 2% and resolution (12 bit) 0.005 N for a scale range of ± 10 N. The force sensor is mounted on a support through data acquisition system and data studio software to a PC. The mass of the beam specimen was roughly 10 g. A typical output is presented in Fig. 15.10 for an aluminum beam.

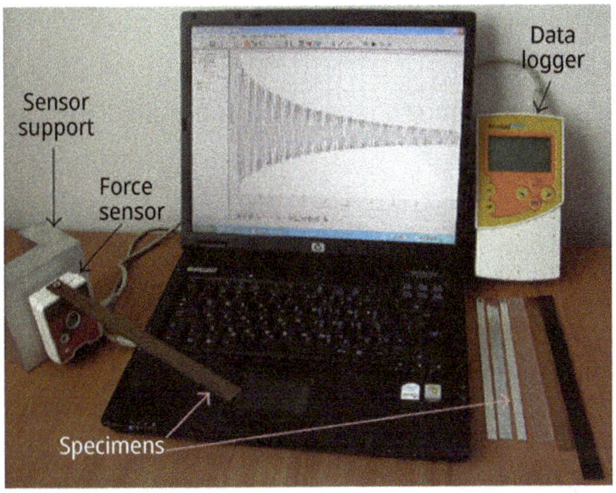

Fig. 15.9: The test setup.

1 https://lab.4lykzografou.gr/wp-content/uploads/2016/01/LA_Fourier-Sensor-Guide.pdf

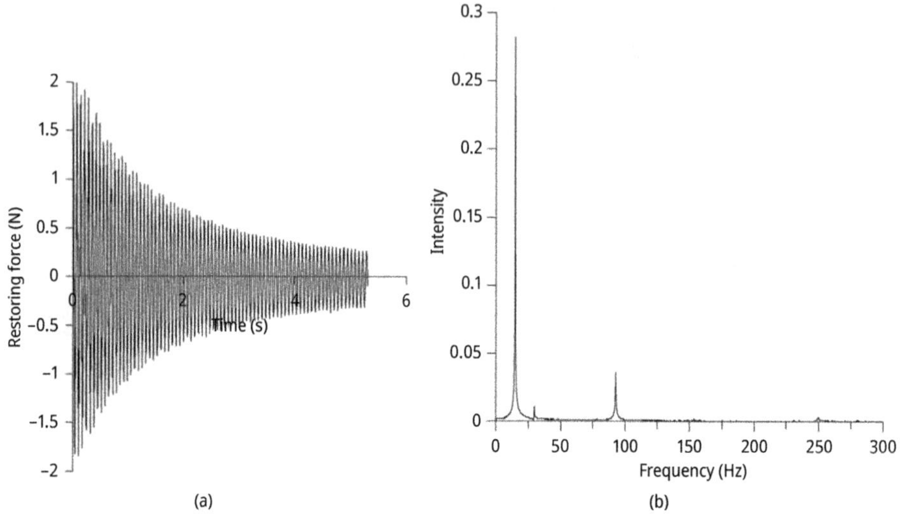

Fig. 15.10: Typical output from the tests: an aluminum beam with $L = 0.282$ m, width = 1.35 cm and thickness $h = 1.55$ mm: (a) dynamic response detected by the force sensor at a sample rate of 1 kHz and (b) identified natural frequencies: $f_1 = 14.7$ Hz, $f_2 = 92.8$ Hz and $f_3 = 249.5$ Hz.

Once the natural frequencies are known, the Young's modulus can be calculated according to the following expression:

$$(\lambda L)_n = \alpha_n$$

but

$$\lambda^4 = \frac{\rho A \omega^2}{EI}$$

therefore (15.48)

$$\lambda_n^4 L^4 = \frac{\rho A \omega^2}{EI} = \alpha_n^4$$

$$\Rightarrow E = \frac{\rho A \omega^2}{\alpha_n^4 \cdot I \cdot L^4} = 4\pi^2 f^2 \frac{\rho A}{\alpha_n^4 \cdot I \cdot L^4}$$

References

[1] Grady, J.E. and Meyn, E.H. Vibration Testing of Impact-Damaged Composite Laminates, NASA TM-4115, 1989, 7.

[2] Adams, D.E., Gothamy, J., Decker, P., Lamb, D. and Gorsich, D. Analysis of passive vibration measurement and data interrogation issues in health monitoring of a HMMWV using a dynamic simulation model, International Journal of Materials and Manufacturing, 1(1), 2009, 235–242.

[3] Doebling, S.W., Farrar, C.R., Prime, M.B. and Shevitz, D.W. Damage identification and health monitoring of structural and mechanical systems from changes in their vibration characteristics: a literature review, Los Alamos National Laboratory, LA-13070-MA report, UC 900, 1996, 127.

[4] Mantena, P.R. Frequency-domain vibration analysis for characterizing the dynamic mechanical properties of materials, in Proceedings of 1966 ASEE annual conference, Washington State University, WA, USA, Session 1626, 1996, 7.

[5] Dilena, M. and Morassi, A. Structural health monitoring of rods based on natural frequency and antiresonant frequency measurements, Structural Health Monitoring, 82(2), 2009, 149–173.

[6] Sinou, -J.-J. A review of damage detection and health monitoring of mechanical systems from changes in the measurement of linear and non-linear vibrations, in Sapri, R.C. Ed., Mechanical vibrations: Measurement, effects and control, Nova Science Publishers, Inc, 2009, 643–702.

[7] Bediz, B., Özgüven, H.N. and Korkusuz, F. Vibration measurements predict the mechanical properties of human Tibia, Clinical Biomechanics, 25, 2010, 365–371.

[8] Rafiee, M., Mehrabadi, S.J. and Saleh, N.R. Analytical solutions for the torsional vibrations of variables cross-sections rods, International Journal of Engineering and Applied Sciences (IJEAS), 2(4), 2010, 64–71.

[9] Al-Khazali, H.A.H. and Askari, M.R. The experimental analysis of vibration monitoring in system rotor dynamic with validate results using simulation data, International Scholarly Research Network (ISRN) Mechanical Engineering, 2012, Article ID: 981010, 2012, 17.

[10] Digilov, R.M. and Abramovich, H. Flexural vibration test of a beam elastically restrained at one end: A new approach for Young's modulus determination, Advance in Materials Science and Engineering, 2013, Article ID: 329530, 2013, 6.

[11] Rainieri, C., Gargaro, D. and Fabbrocino, G. Vibration-based monitoring of tensile loads: system development and application, EWSHM – 7th European Workshop on Structural Health Monitoring, IFFSTTAR, Inria, Université de Nantes, Jul 2014, Nantes, France, 1616–1623.

[12] Hamm, P., Schänzlin, J., Francke, W. and Scheuble, S. Monitoring of timber bridges by repeatedly measuring the natural frequency, World Conference on Timber Engineering, WCTE2016, Aug. 22–25, 2016, Vienna, Austria, 8.

[13] Digilov, R.M. and Abramovich, H. The impact of root flexibility on the fundamental frequency of a restrained cantilever beam, International Journal of Mechanical Engineering Education, 45(2), 2017, 184–193.

[14] Collini, L., Garziera, R. and Riabova, K. Vibration analysis for monitoring of ancient tie-rods, Hindawi Shock and Vibration, 2017, Article ID: 7591749, 2017, 11.

[15] Orak, M.S., Nasrollahi, A., Ozturk, T., Mas, D., Ferrer, B. and Rizzo, P. Non-contact smartphone-based monitoring of thermally stressed structures, Sensors, 18(1250), 2018, 15.

[16] Lei, X. and Wu, Y. Research on mechanical vibration monitoring based on wireless sensor network and sparse bayes, Journal of Wireless Communications and Networking, 2020, Article ID: 20202:225, 13.

[17] Hassannejad, R., Hosseini, S.A. and Hamidi, B.A. Influence of non-circular cross section shapes on torsional vibration of a micro-rod based on modified couple stress theory, Acta Astronautica, 178, 2021, 805–812.
[18] Rosenberg, N., Rosenberg, O., Politch, J.H. and Abramovich, H. Optimal parameters for the enhancement of human osteoblast-like cell proliferation *in vitro* via shear stress induced by high frequency mechanical vibration, IberoAmerican Journal of Medicine, 3(3), 2021, 204–211.
[19] Lima-Rodriguez, A., Garcia-Manrique, J., Dong, W. and Gonzalez-Herrera, A. A novel methodology to obtain the mechanical properties of membranes by means of dynamic tests, Membranes, 12(282), 2022, 17.
[20] Mirasoli, G., Brutti, C., Groth, C., Mancini, L., Porziani, S. and Biancolini, M.E. Structural health monitoring of civil structures through FEM high-fidelity modelling, IOP Conference Series: Materials Science and Engineering, 1214, 2022, Article ID: 012019, 16.
[21] Thomson, W.T. Vibration theory and applications, George Allen & Unwin ltd, London, Ruskin House, Museum Street, 1965, 384.

Appendix A: Longitudinal vibrations of rods with various boundary conditions

Note that $\omega = 2\pi f$ and $\kappa^2 = \frac{\rho}{E \cdot g}$

No.	Schematic drawing	Left B.C.	Right B.C.	Natural frequencies	Mode shapes	Characteristic equation
1	Clamped–clamped 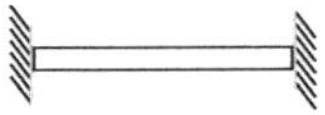	$u(0,t)=0$	$u(L,t)=0$	$\omega_n=\frac{n\pi\kappa}{L}$, $n=1,2,3,\ldots$	$u(x,t)=A_n\sin\frac{n\pi x}{L}$	$\sin\frac{\omega L}{\kappa}=0$
2	Clamped–free 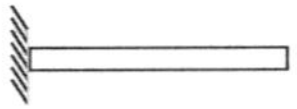	$u(0,t)=0$	$\frac{\partial u(L,t)}{\partial x}=0$	$\omega_n=\frac{(2n+1)\pi\kappa}{2L}$, $n=1,2,3,\ldots$	$u(x,t)=A_n\sin\frac{(2n+1)\pi x}{2L}$	$\cos\frac{\omega L}{\kappa}=0$
3	Free–free 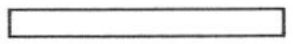	$\frac{\partial u(0,t)}{\partial x}=0$	$\frac{\partial u(L,t)}{\partial x}=0$	$\omega_n=\frac{n\pi\kappa}{L}$, $n=0,1,2,3,\ldots$	$u(x,t)=A_n\cos\frac{n\pi x}{L}$	$\sin\frac{\omega L}{\kappa}=0$
4	Clamped–attached mass M 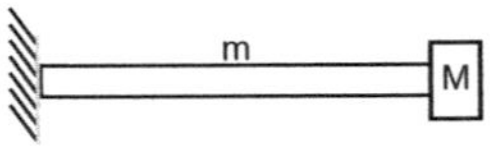	$u(0,t)=0$	$EA\frac{\partial u(L,t)}{\partial x}=-M\frac{\partial^2 u(L,t)}{\partial t^2}$	$\omega_n=\frac{\chi_n\kappa}{L},\chi=\frac{\omega L}{\kappa}$, $n=1,2,3,\ldots$	$u(x,t)=A_n\sin\frac{\omega_n x}{\kappa}$	$\frac{\omega L}{\kappa}\tan\frac{\omega L}{\kappa}=\frac{m}{M}$
5	Clamped–attached spring k 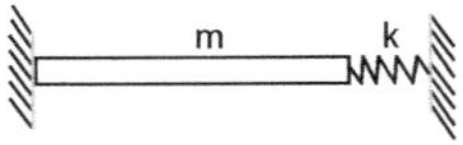	$\frac{\omega L}{\kappa}\tan\frac{\omega L}{\kappa}=\frac{m}{M}$	$EA\frac{\partial u(L,t)}{\partial x}=-k\cdot u(L,t)$	$\omega_n=\frac{\chi_n\kappa}{L},\chi=\frac{\omega L}{\kappa}$, $n=1,2,3,\ldots$	$u(x,t)=A_n\sin\frac{\omega_n x}{\kappa}$	$\frac{\omega L}{\kappa}\tan\frac{\omega L}{\kappa}=-\frac{m\omega^2}{k}$
6	Free–attached mass M 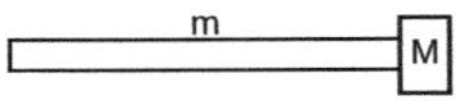	$\frac{\partial u(0,t)}{\partial x}=0$	$EA\frac{\partial u(L,t)}{\partial x}=-M\frac{\partial^2 u(L,t)}{\partial t^2}$	$\omega_n=\frac{\chi_n\kappa}{L},\chi=\frac{\omega L}{\kappa}$, $n=1,2,3,\ldots$	$u(x,t)=A_n\cos\frac{\omega_n x}{\kappa}$	$\tan\frac{\omega L}{\kappa}=-\frac{\omega L}{\kappa}\cdot\frac{m}{M}$
7	Free–attached spring k 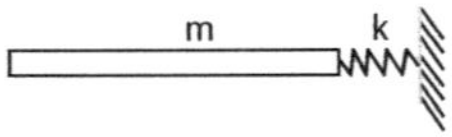	$\frac{\partial u(0,t)}{\partial x}=0$	$EA\frac{\partial u(L,t)}{\partial x}=-k\cdot u(L,t)$	$\omega_n=\frac{\chi_n\kappa}{L},\chi=\frac{\omega L}{\kappa}$, $n=1,2,3,\ldots$	$u(x,t)=A_n\cos\frac{\omega_n x}{\kappa}$	$\frac{\omega L}{\kappa}\cot\frac{\omega L}{\kappa}=\frac{AE}{Lk}$

(continued)

(continued)

No.	Schematic drawing	Left B.C.	Right B.C.	Natural frequencies	Mode shapes	Characteristic equation
8	Free–attached mass *M* and spring *k* 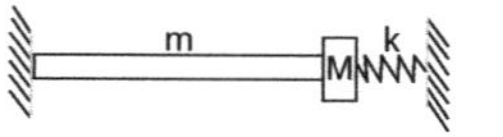	$u(0,t)=0$	$EA\frac{\partial u(L,t)}{\partial x}=-M\frac{\partial^2 u(L,t)}{\partial t^2}-ku(L,t)$	$\omega_n=\frac{\chi_n\kappa}{L},\chi=\frac{\omega L}{\kappa}$ $n=1,2,3,\ldots$	$u(x,t)=A_n\sin\frac{\omega_n x}{\kappa}$	$\frac{\omega L}{\kappa}\cot\frac{\omega L}{\kappa}=\frac{\omega^2L^2}{\kappa^2 m}M-\frac{k}{AE}L$

16 Buckling and vibrations of arches, trusses and frames

16.1 General terminology

16.1.1 Arches

An arch can be defined as a vertical structure spanning across an opening and capable of supporting the weight of a given load in the form of a roof, bridge or a wall located above them, as can be seen schematically in Fig. 16.1. This basic element is frequently used in civil and mechanical engineering structures. Under static compressive loading the arch is liable to buckle, while time-dependent forces might induce vibrations of the element.

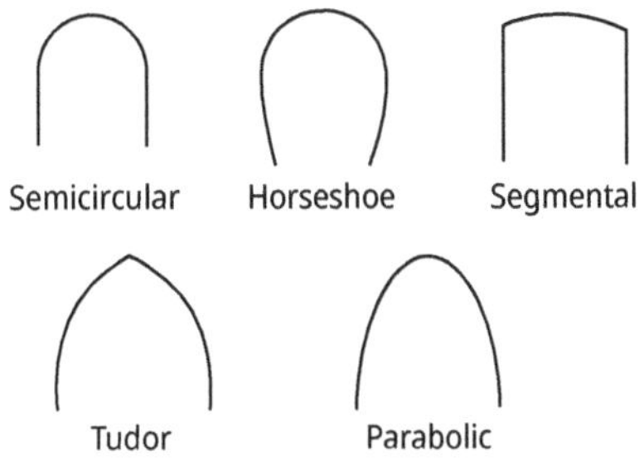

Fig. 16.1: Typical schematic arches.

16.1.2 Trusses

A truss is a structural element designed to carry various kinds of loads keeping its geometry intact under application of tractions. It consists of straight bars connected at their ends by nodes or joints, as it is schematically displayed in Fig. 16.2. The joints are assumed to be pin-connected; namely, no bending moments exist at the nodes and only forces can be transmitted along the bars connected at each joint. The forces can be applied only at the nodes, so transverse loads do not exist. The self weight of each bar is normally ignored relative to the various forces applied on the truss. Finding the internal forces acting in each bar due to the external loads can be calculated at each node. If a bar is under compressive loads, it might buckle making the whole truss unstable.

https://doi.org/10.1515/9783111621104-016

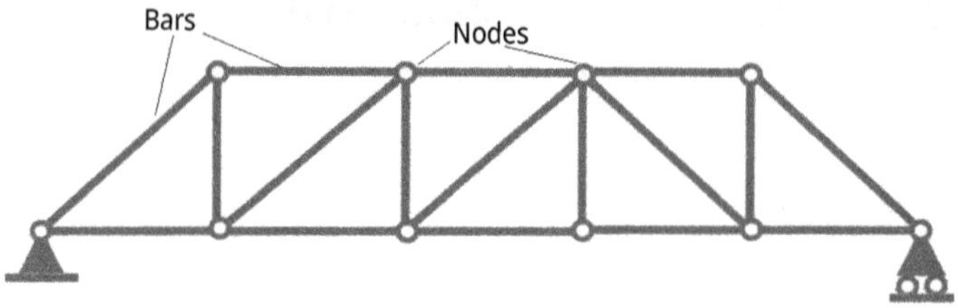

Fig. 16.2: A typical schematic truss.

16.1.3 Frames

A frame (see typical schematic forms in Fig. 16.3) is also a basic structure consisting of straight or curved members (beams) rigidly connected at its various joints. In contrast to trusses, each member of the frame can carry bending moments, axial, lateral and shear forces. The frame is calculated and designed to be either statically determined or indeterminate which means that the internal tractions might or might not be dependent on the deformation and structural stiffness.

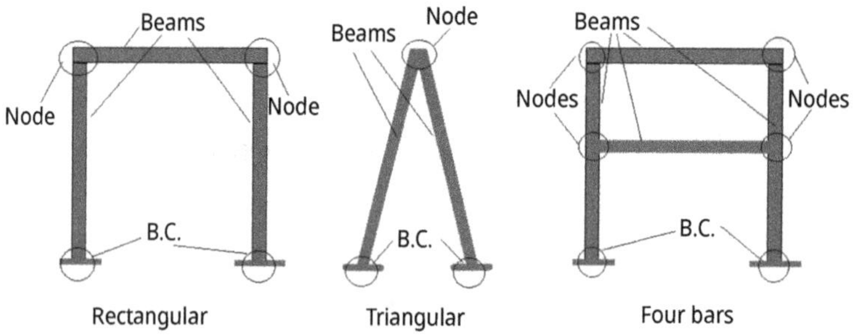

Fig. 16.3: Typical schematic frames.

Note that a truss or a frame can have also a tri-dimensional shape, thus increasing its solution complexity. Each member of the frame might individually buckle under compressive loads, or alternatively the whole frame might collapse as a single rigid structure. External loads can also induce vibrations of the frame and the natural frequencies of the whole frame, and each individual member must be evaluated to prevent undesirable effects.

In conclusion, frames offer robustness against lateral pressures and design versatility, while trusses are best when applying distributed loads over large spans with minimum amount of weight.

16.2 Buckling and vibration of arches

16.2.1 Buckling of arches – basic concepts

Arches belong to a special structural group named thin-walled structures which have an improved stiffness-to-weight ratio. However, when these structures are compressed, they are liable to buckling. Arches might have three types of buckling, snap-through (for shallow arches), in-plane buckling or out-of-plane buckling (for tall arches), as presented schematically in Fig. 16.4.

Buckling of arches has been widely addressed in the literature, with a typical list of references described in Refs. [1–25].

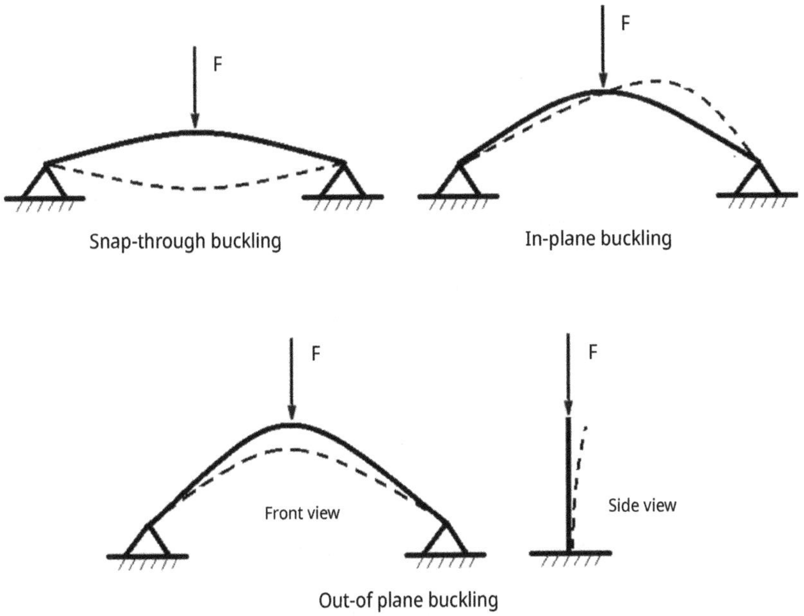

Fig. 16.4: Typical schematic buckling modes of arch.

The following chapter will deal with the basic equations of buckling, with more details to be found in typical books, like Refs. [2, 3], or other dedicated references.

The differential equation of motion for the deflection curve of any thin bar having a circular center line was established already in 1883 by Boussinesq [4] and latter on developed again by Timoshenko in 1935 [1] and included in his book [2] in 1963. It has the following form (see [2]):

$$\frac{d^2w}{ds^2} + \frac{w}{R^2} = -\frac{M}{EI} \tag{16.1}$$

or alternatively

$$\frac{d^2w}{d\theta^2}+w=-\frac{MR^2}{EI} \tag{16.2}$$

where s and θ are the coordinates (length and angle) along the curved bar, respectively, R is the initial radius of curvature of the center line of the bar, w is the in-plane radial displacement, EI is the flexural rigidity of the bar, and M denotes the bending moment acting on it.

16.2.2 Bending and buckling of rings

Consider a ring having a radius of R being compressed by two loads N_x along its diameter, as shown in Fig. 16.5a:

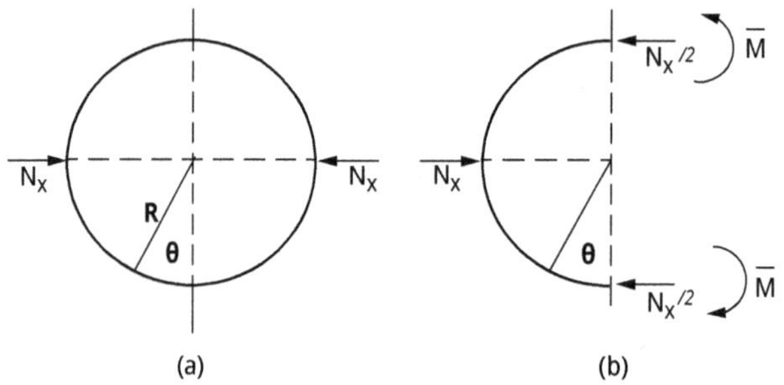

Fig. 16.5: (a) A ring under two compressive loads, N_x, and (b) equilibrium for half of the ring.

Using Fig. 16.5b, it is easy to write the distribution of the bending moment, as a function of the applied load N_x, and the bending moments $\bar{M}$ at its ends, to yield

$$M=\bar{M}+\frac{N_x}{2}R(1-\cos\theta) \tag{16.3}$$

Substitution of eq. (16.3) into eq. (16.2) provides the differential equation of the present problem, namely

$$\frac{d^2w}{d\theta^2}+w=-\frac{R^2}{EI}\left[\bar{M}+\frac{N_x}{2}R(1-\cos\theta)\right]=-\frac{\bar{M}R^2}{EI}-\frac{N_xR^3}{2EI}(1-\cos\theta) \tag{16.4}$$

The solution of eq. (16.4) has the following form (see [1]):

$$w(\theta)=A\sin\theta+B\cos\theta-\frac{\bar{M}R^2}{EI}-\frac{N_xR^3}{2EI}+\frac{N_xR^3}{4EI}\theta\sin\theta \tag{16.5}$$

The two constants A and B can be found from conditions of symmetry, namely

$$\frac{dw(0)}{d\theta}=0 \mapsto A=0 \text{ and } \frac{dw(\pi/2)}{d\theta}=0 \mapsto B=\frac{N_x R^3}{4EI} \tag{16.6}$$

The expression for the bending moment $\bar{M}$ can be found (see [1]) using Castigliano's theory to yield

$$\bar{M}=\frac{N_x R}{2}\left(\frac{2}{\pi}-1\right) \tag{16.7}$$

Substituting eqs. (16.6) and (16.7) into eq. (16.5) provides the expression for the radial deflection, w, having the following form

$$w(\theta)=\frac{N_x R^3}{4EI}\left(\cos\theta+\theta\sin\theta-\frac{4}{\pi}\right) \tag{16.8}$$

The next case to be presented is a ring under uniform pressure, q, as depicted in Fig. 16.6a and 16.6b.

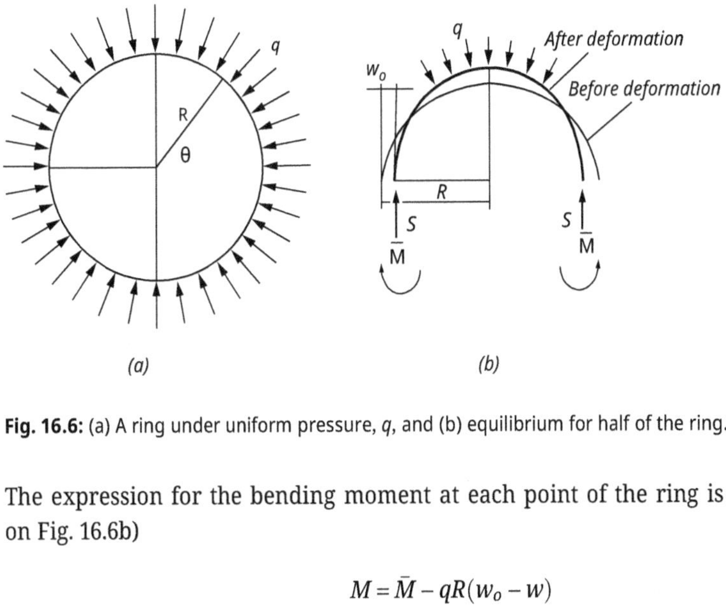

Fig. 16.6: (a) A ring under uniform pressure, q, and (b) equilibrium for half of the ring.

The expression for the bending moment at each point of the ring is given by (based on Fig. 16.6b)

$$M=\bar{M}-qR(w_o-w) \tag{16.9}$$

Substitution of eq. (16.9) into eq. (16.2) yields

$$\frac{d^2w}{d\theta^2}+w=-\frac{R^2}{EI}[\bar{M}-qR(w_o-w)] \tag{16.10}$$

Equation (16.10) can be rearranged leading to the following form:

$$\frac{d^2w}{d\theta^2} + w\left(1 + \frac{qR^3}{EI}\right) = -\frac{R^2}{EI}[\bar{M} - qRw_o] \tag{16.11}$$

Defining the term $\left(1 + \frac{qR^3}{EI}\right) = \tau^2$, one can reach the form for the solution of eq. (16.11) yielding

$$w(\theta) = A\sin\tau\theta + B\cos\tau\theta + \frac{qR^3w_o - \bar{M}R^2}{EI + qR^3} \tag{16.12}$$

The two constants A and B can be found from conditions of symmetry, namely

$$\frac{dw(0)}{d\theta} = 0 \mapsto A = 0 \text{ and } \frac{dw(\pi/2)}{d\theta} = 0 \mapsto B\sin\tau\frac{\pi}{2} = 0 \text{ or } \sin\tau\frac{\pi}{2} = 0 \tag{16.13}$$

The smallest root for eq. (16.13) would be for $\tau = 2$ leading to

$$\left(1 + \frac{qR^3}{EI}\right) = \tau^2 = 2^2 = 4 \mapsto q_{\text{cr}} = \frac{3EI}{R^3} \tag{16.14}$$

The result of eq. (16.14) is the buckling pressure of a ring, and the buckling mode shape would be an ellipse.

16.2.3 Buckling of arches

Considering the circular arch presented in Fig. 16.7, we would like to derive the expression for the critical of the pressure that will buckle the arch.

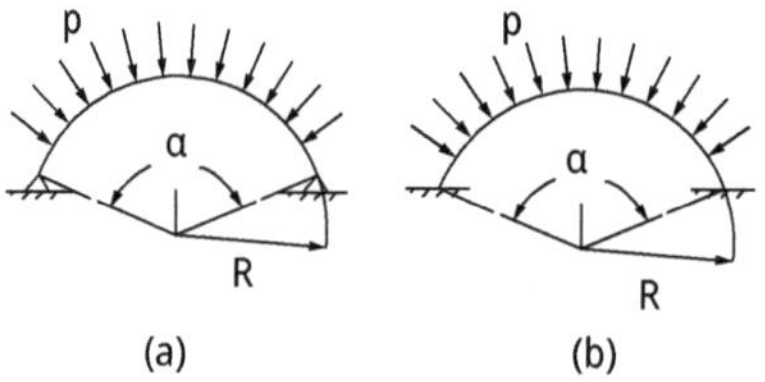

Fig. 16.7: A circular arch under uniform pressure, p, (a) under hinged–hinged ends, (b) under clamped–clamped ends.

Using eq. (16.2), we get

$$\frac{d^2w}{d\theta^2} + w = -\frac{MR^2}{EI} = -\frac{(pwR)\,R^2}{EI}$$
$$\Rightarrow \frac{d^2w}{d\theta^2} + \kappa^2 w = 0 \text{ where } \kappa^2 = 1 + \frac{pR^3}{EI} \tag{16.15}$$

The general solution for eq. (16.15) has the following form (see previous derivation for a ring):

$$w(\theta) = A\sin\tau\theta + B\cos\tau\theta \tag{16.16}$$

Constants A and B will be found based on the boundary conditions of the case. For the hinged–hinged case (Fig. 16.7a), we have

$$w(\theta = 0) = 0 \;\Rightarrow B = 0$$
$$w(\theta = \alpha) = 0 \;\Rightarrow A\sin\tau\alpha = 0 \tag{16.17}$$
$$\text{or } \sin\tau\alpha = 0 \rightarrow \tau\alpha = 2\pi \text{ and } \tau = \frac{2\pi}{\alpha}$$

Substituting the value of τ in eq. (16.15) provides the expression for the buckling pressure of the arch on hinged–hinged boundary conditions, namely

$$p_{cr} = \frac{EI}{R^3}\left[\left(\frac{2\pi}{\alpha}\right)^2 - 1\right] \tag{16.18}$$

Noting that $N = pR$, one can obtain the expression of a hinged–hinged arch under concentrated load applied at its center

$$N_{cr} = \frac{EI}{R^2}\left[\left(\frac{2\pi}{\alpha}\right)^2 - 1\right] \tag{16.19}$$

The buckling mode shape is presented in Fig. 16.8a. Note that for a parabolic flat arch under uniformly distributed pressure, p, one can neglect the variation of the compressive forces along the arch length leading to the following simple expression (see [1]):

$$p_{cr} = \left[\frac{4\pi^2 EI}{R^3\alpha}\right] \tag{16.20}$$

For the problem of a circular arch under uniform pressure, p, on two clamped ends, as presented in Fig. 16.7b and noting that at the top section of the arch two forces will act after buckling, a horizontal force $S = pR$ and a vertical shear force Q, the bending moment at any cross section of the arch is given by

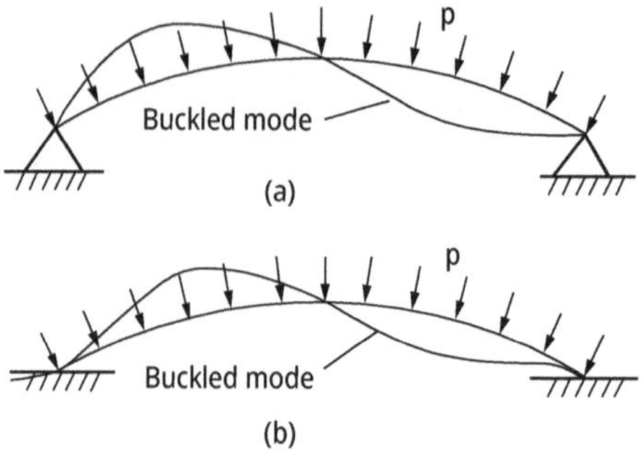

Fig. 16.8: Buckling modes for a circular arch under uniform pressure (not to scale), *p*: (a) hinged–hinged ends and (b) clamped–clamped ends.

$$M = Sw - QR\sin\theta = pRw - QR\sin\theta \tag{16.21}$$

Again using eq. (16.2) and substituting the expression for the bending moment (eq. (16.21)) leads to the following differential equation:

$$\begin{aligned}&\frac{d^2w}{d\theta^2} + w = -\frac{MR^2}{EI} = -\frac{R^2}{EI}(pRw - QR\sin\theta)\\&\Rightarrow \frac{d^2w}{d\theta^2} + \tau^2 w = \frac{QR^3\sin\theta}{EI} \text{ where } \tau^2 = 1 + \frac{pR^3}{EI}\end{aligned} \tag{16.22}$$

The solution for eq. (16.22) is found to be

$$w(\theta) = A\sin\tau\theta + B\cos\tau\theta + \frac{QR^3\sin\theta}{EI(\tau^2 - 1)} \tag{16.23}$$

To find the constants A and B and the shear force Q, we must enforce the boundary conditions of the problem, namely

$$\begin{aligned}&\text{at } \theta = 0 \text{ (the top of the arch) } w = \frac{d^2w}{d\theta} = 0\\&\text{at } \theta = \frac{\alpha}{2}\; w = \frac{dw}{d\theta} = 0\end{aligned} \tag{16.24}$$

The first condition shown in eq. (16.24) is satisfied by assuming the constant $B = 0$. Applying the second condition yields two algebraic equations, namely

$$\begin{aligned}&A\sin\tau\frac{\alpha}{2} + Q\frac{R^3\sin\frac{\alpha}{2}}{EI(\tau^2 - 1)} = 0\\&A\tau\cos\tau\frac{\alpha}{2} + Q\frac{R^3\cos\frac{\alpha}{2}}{EI(\tau^2 - 1)} = 0\end{aligned} \tag{16.25}$$

To find a unique solution, the determinant of the coefficients must be zero, namely

$$\det\begin{pmatrix} \sin\tau\frac{\alpha}{2} & \frac{R^3 \sin\frac{\alpha}{2}}{EI(\tau^2-1)} \\ \tau\cos\tau\frac{\alpha}{2} & \frac{R^3\cos\frac{\alpha}{2}}{EI(\tau^2-1)} \end{pmatrix} = 0 \tag{16.26}$$

Performing the calculations yields the following transcendental equation from which the critical pressure value can be found numerically,

$$\begin{aligned} &\sin\tau\frac{\alpha}{2}\cos\frac{\alpha}{2} - \tau\sin\frac{\alpha}{2}\cos\tau\frac{\alpha}{2} = 0 \\ &\Rightarrow \tau\tan\frac{\alpha}{2}\cot\tau\frac{\alpha}{2} = 1 \end{aligned} \tag{16.27}$$

Table 16.1 presents the solutions for the transcendental equation (16.27) for some values of the angle $\alpha/2$ (from [1])

Tab. 16.1: Solutions for the transcendental equation (16.27).

$\alpha/2$	30	60	90	120	150	180
τ	8.621	4/375	3	2.364	2.066	3.000

Substituting the values of τ from Tab. 16.1 into its definition $\tau^2 = 1 + \frac{pR^3}{EI}$ leads to the expression for the critical pressure

$$p_{\text{cr}} = \frac{EI}{R^3}\left(\tau^2 - 1\right) \tag{16.28}$$

Note that the value of eq. (16.28) will be always greater than the value of eq. (16.18). The buckling mode for a clamped–clamped arch under uniform pressure is depicted in Fig. 16.8b.

More results for other arch shapes can be found in the book authored by Karnovsky [26].

16.2.4 Vibrations of arches

The issue of vibrations of curved beams and arches has been heavily investigated and published in the literature (see typical examples in Refs. [26–38]). Already in 1887, Lamb [27] presented a study on curved bars vibrations and presented the equation of motion for such a structure, which later was used by other scholars to investigate the topic.

Consider a symmetrical circular arch with constant cross section having a central angle of 2α and radius R. I would be its moment of inertia, while m will stand for the mass per unit length of the arch. The planar case will have bending as well as radial vibrations of the arch (see [26–28]). W and V are the radial and the tangential displacements, respectively. Note that the following relationship exists between the two displacements (see Fig. 16.9):

$$W = \frac{dV}{d\varphi} \tag{16.29}$$

The equation of motion of the present problem can be written for the inextensional case as (see also [26–28]):

$$\frac{\partial^6 V}{\partial \varphi^6} + 2\frac{\partial^4 V}{\partial \varphi^4} + \frac{\partial^2 V}{\partial \varphi^2} = \frac{mR^4}{EI}\frac{\partial^2}{\partial t^2}\left(V - \frac{\partial^2 V}{\partial \varphi^2}\right) = \frac{mR^4}{EI}\left(\frac{\partial^2 V}{\partial t^2} - \frac{\partial^4 V}{\partial \varphi^2 \partial t^2}\right) \tag{16.30}$$

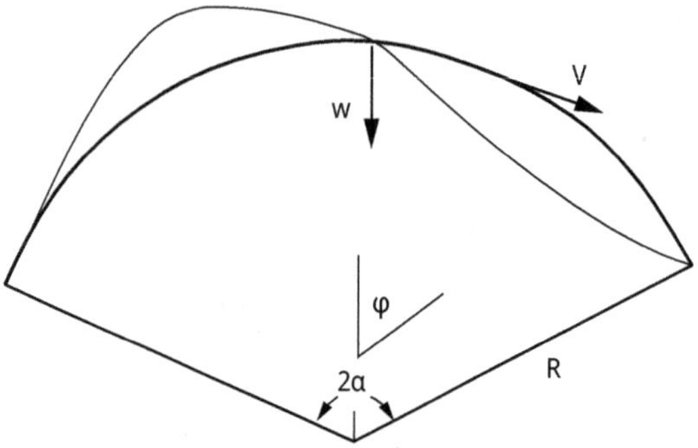

Fig. 16.9: A schematic circular arch with its notations.

Equation (16.30) can be applied to all circular shapes, including rings. For rings, one has to assume a solution in which the displacements are periodic with the angle φ leading to the following expression:

$$V(\varphi, t) = A\cos n\varphi \sin(\omega t + \alpha) \tag{16.31}$$

where A and α are constants to be determined from initial conditions of the problem, n is an integer and ω is the free vibration angular frequency, the unknown of the present case.

Inserting eq. (16.31) into eq. (16.30) leads to the natural frequencies for a ring (see also [26–28])

$$\tau^2 = \frac{n^2(n^2-1)^2}{n^2+1}, \text{ where } \tau^2 = \omega^2 R^4 \frac{mS}{EI}, \; S = \text{cross section of the arch}$$
$$\Rightarrow \omega = \frac{1}{R^2}\sqrt{\frac{EI}{mS}\frac{\sqrt{n^2+1}}{n(n^2-1)}} \quad n \neq 1 \tag{16.32}$$

For curved beams that are not rings, the following procedure should be followed:

The general solution of eq. (16.30) for the inextensional case (the tangential strain $= dV/d\varphi + W/R = 0$) applied on a circular arch with an aperture of $2a$ can be assumed to be of the following form:

$$V(\varphi, t) = \bar{V}(\varphi)\,\sin(\omega t + a) \tag{16.33}$$

The solution is then inserted into eq. (16.30), applying the relevant boundary conditions at both ends of the arch ($\varphi = \pm\, a$) to obtain the frequency determinant. For the extensional theory (which is the correct case for a circular arch), the following two equations, as written in [28], must be solved (instead of eq. (16.30)):

$$\frac{1}{R}\frac{\partial M}{\partial \varphi} - \frac{1}{R^2}\frac{\partial M}{\partial \varphi} = mS\frac{\partial^2 V}{\partial t^2} == \gg \frac{\partial}{\partial \varphi}\left[\frac{ES}{R}\left(\frac{\partial V}{R\partial \varphi} + \frac{W}{R}\right)\right] - \frac{1}{R^2}\frac{\partial M}{\partial \varphi} = mS\frac{\partial^2 V}{\partial t^2} \tag{16.34a}$$

$$-\frac{\partial^2 M}{R^2 \partial \varphi^2} - \frac{N}{R} = mS\frac{\partial^2 W}{\partial t^2} == \gg -\frac{\partial^2 M}{R^2 \partial \varphi^2} - \frac{ES}{R}\left(\frac{\partial V}{\partial \varphi} + \frac{W}{R}\right) mS\frac{\partial^2 W}{\partial t^2} \tag{16.34b}$$

where

$$M = EI\frac{\partial}{R\partial \varphi}\left[\frac{\partial W}{R\partial \varphi} - \frac{V}{R}\right] \tag{16.34c}$$

with N and M being the tangential force and the bending moment, respectively.

Performing the various mathematical operations in eqs. (16.34a)–(16.34c), their simplified form would look as

$$\begin{gathered}\left[1 + \left(\frac{I}{SR^2}\right)^2\right]\frac{\partial^2 V}{\partial \varphi^2} - \frac{I}{SR^2}\frac{\partial^3 W}{\partial \varphi^3} + \frac{\partial W}{\partial \varphi} = \frac{mR^2}{E}\frac{\partial^2 V}{\partial t^2} \\ -\frac{I}{SR^2}\frac{\partial^3 V}{\partial \varphi^3} + \frac{\partial V}{\partial \varphi} + \frac{I}{SR^2}\frac{\partial^4 W}{\partial \varphi^4} + W = -\frac{mR^2}{E}\frac{\partial^2 W}{\partial t^2}\end{gathered} \tag{16.35}$$

Assuming solutions for V and W having the general form as in eq. (16.33) and applying the boundary conditions of the relevant case and substituting in eqs. (16.35) will provide the frequency determinant from which the natural frequencies can be determined.

The results of the above approaches, inextensional and extensional, are presented in Fig. 16.10a, b and 16.11 for two boundary conditions pinned–pinned ($V = W = M = 0$ at $\varphi = \pm\, a$) and clamped–clamped ($V = W = \frac{\partial W}{\partial \varphi} = 0$ at $\varphi = \pm\, a$) as applied to circular arches. The thickness to mean radius h/R is presented for two values, 0.1 and 0.01 for the pinned–pinned case. The first four natural frequencies were calculated for the inextensional case, while for the extensional one, two symmetric and two antisymmetric (a total of four frequencies) are also presented in Fig. 16.10a, b and 16.11. The differences between the two approaches are clearly visible, and their usage is given to the discretion of the user.

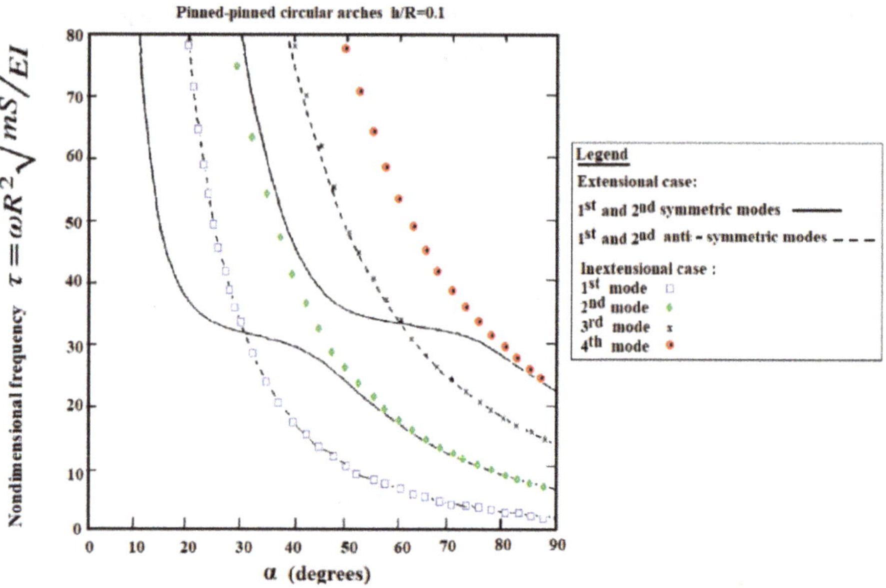

(a)

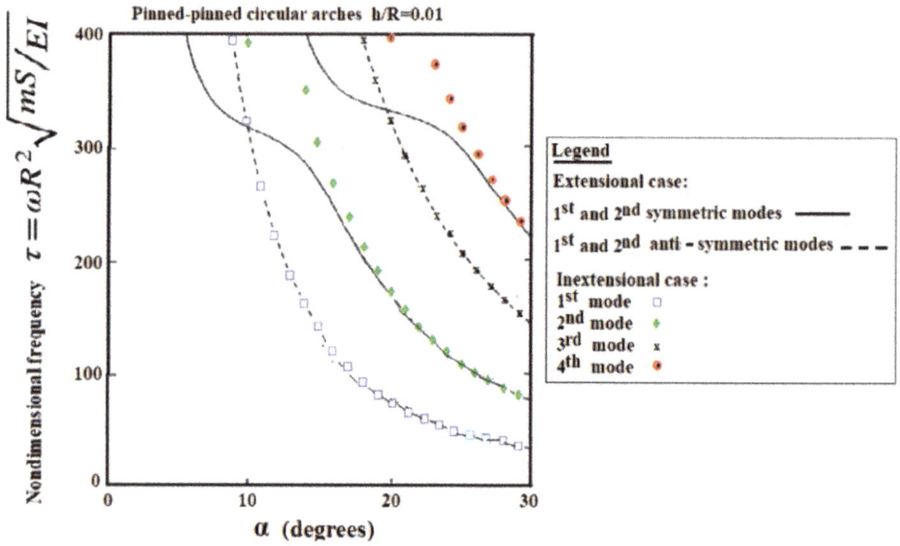

(b)

Fig. 16.10: Nondimensional frequency versus arc segment half angle α for pinned–pinned circular arches (a) with thickness to radius $h/R = 0.1$, (b) with thickness to radius $h/R = 0.01$ (adapted from [27]).

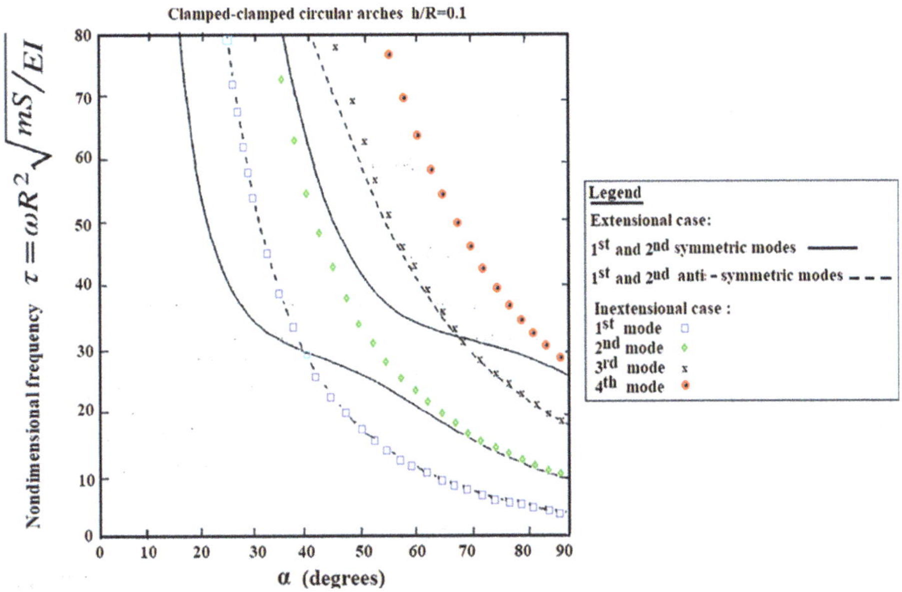

Fig. 16.11: Nondimensional frequency versus arc segment half angle α for clamped–clamped circular arches with thickness to radius $h/R = 0.1$ (adapted from [27]).

To enable calculation of natural frequencies for circular arches using the inextensional approach, Tab. 16.2 (adapted from [26]) is attached.

Tab. 16.2: Frequencies and mode shapes of uniform circular arches.

Boundary conditions	Mode type	Mode shape (schematic)	Parameter k
Pinned–pinned	First antisymmetrical		$\dfrac{4\pi^2-\alpha_0^2}{\sqrt{1+0.75\beta^2}}$
Pinned–pinned	First symmetrical		$\dfrac{4\pi^2-\alpha_0^2}{\sqrt{1+0.1652\beta^2}}$
Pinned–pinned	Second antisymmetrical		$\dfrac{4\pi^2-\alpha_0^2}{\sqrt{1+0.1875\beta^2}}$
Clamped–clamped	First antisymmetrical		$\sqrt{\dfrac{3{,}803.2-92.101\alpha_0^2+\alpha_0^4}{1+0.06054\alpha_0^2}}$

Tab. 16.2 (continued)

Boundary conditions	Mode type	Mode shape (schematic)	Parameter k
Clamped–clamped	First symmetrical		$\sqrt{\dfrac{14{,}620-197.84\alpha_0^2+\alpha_0^4}{1+0.01227\alpha_0^2}}$
Clamped–clamped	Second antisymmetrical		$\sqrt{\dfrac{39{,}942-343.16\alpha_0^2+\alpha_0^4}{1+0.02148\alpha_0^2}}$

$\omega_i=\dfrac{k}{R^2\alpha_0^2}\sqrt{\dfrac{Ei}{m}}$, α_0 is the central angle of the arch, measured in degrees, $\beta=\alpha_0/\pi$, EI is constant and m is the mass per unit length.

For further information regarding arches, the reader is referred to Appendix A containing the vibrations of complete rings as suggested by Timoshenko in 1937.

16.3 Static behavior, buckling and vibration of trusses

16.3.1 Static behavior of trusses

As already described in the introduction of the present chapter, Chapter 16, a truss is a structure composed of straight members connected through joints to yield a rigid structure which will not collapse due to the application of external forces. The joints are assumed to be pinned–pinned; therefore, each truss member will be loaded by compression or tension only. No moments or couples are allowed. The external loads are applied only at joints, and the weight of the truss is assumed to be distributed at the joints. Note that a truss can be either planar or a 3D structure. The present chapter will deal only with planar trusses, and the reader is referred to [39] for problems of 3D trusses. Note that a truss can be solved, namely to find all the forces in all its members, if the following equation is fulfilled: $m=2k-3$, where m stands for the total number of rods members and k is the number of joints. In this case the problem is *statically determined.* If the equation is not fulfilled, the case is *statically undetermined.* For the latter case, additional equations must be supplied, using continuity or energy methods to solve it. Some common trusses used in civil engineering are presented in Fig. 16.12.

To sum up, in this chapter two examples will be presented of how to solve the static behavior of trusses.

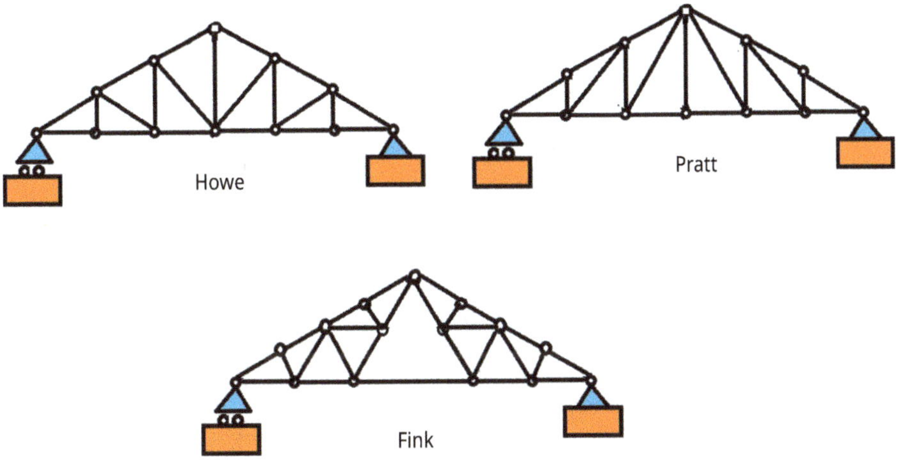

Fig. 16.12: Common trusses used in civil engineering.

Example 1: the method of joints

A simple truss having the shape of an isosceles triangle is formed by three members, AC, AB and AD, as displayed in Fig. 16.13. The truss has pinned–pinned boundary conditions at points A and C; therefore, a vertical reaction is at point A, R_1 and vertical and horizontal reactions at point C, R_2 and R_3, respectively would appear[1] (see Fig. 16.13). Before starting the solution procedure, one must check if the problem is *statically determined.* As we have three members $m = 3$ and three joints $k = 3$, the following equation $m = 2k-3$ is fulfilled and the problem can be solved.

The angles of the isosceles triangles can be easily found to be 33.7° at points A and C and 112.4° at point B.

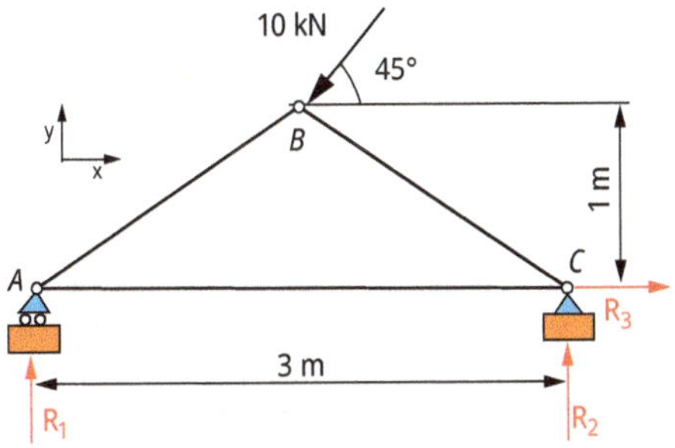

Fig. 16.13: A simple three members' truss.

1 Note that the direction of the reactions is arbitrary chosen, and their real directions will be finalized from the solution of the equilibrium of forces of the whole structure.

The equilibrium equations for the whole truss can be now written, and the values of the three reactions can be determined as follows:

$$\sum F_x = 0 \Rightarrow R_3 - 10\cos 45 = 0 \text{ then } R_3 = 7.07\text{kN}$$

$$\sum M_A = 0 \Rightarrow R_2 * 3 + 10\cos 45 * 1 - 10\sin 45 * 1.5 = 0 \text{ then } R_2 = 1.18\text{kN}$$

$$\sum F_Y = 0 \Rightarrow R_1 + R_2 - 10\sin 45 = 0 \text{ then } R_1 = 5.89\text{kN}$$

Now we can reach each joint and apply equilibrium equations in both the x and the y directions. Note that forces pointing outside the joint will cause compressive forces in the rod members, while forces pointing inside the joint will cause tensile forces in the relevant members. The direction of the forces at each joint is arbitrary chosen, with the correct direction to be obtained from the solution of the equilibrium equation. As for each joint, only two equilibrium equations can be written, the first joint to start with will be the one which has only two unknown forces, in our case joint A.

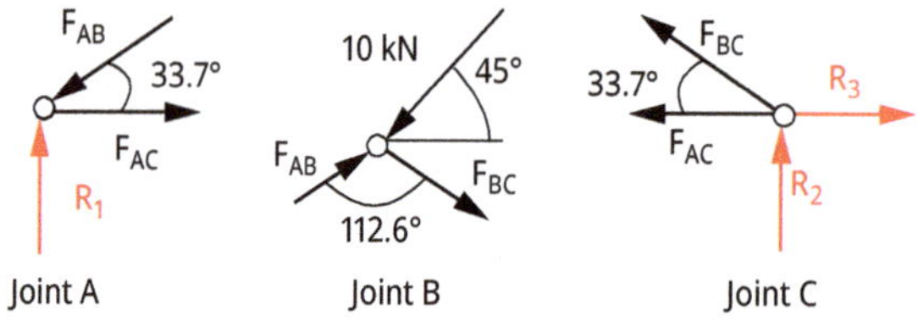

Fig. 16.14: The three joints' force equilibrium.

Joint A:

$$\sum F_Y = 0 \Rightarrow R_1 - F_{AB}\sin 33.7 = 5.89 - F_{AB}\sin 33.7 = 0 \text{ then } F_{AB} = 10.62\text{kN}[C]$$

$$\sum F_X = 0 \Rightarrow F_{AC} - F_{AB}\cos 33.7 = F_{AC} - 10.62\cos 33.7 = 0 \text{ then } F_{AC} = 8.84\text{kN}[T]$$

Joint C:

$$\sum F_Y = 0 \Rightarrow R_{A2} + F_{BC}\sin 33.7 = 1.18 + F_{BC}\sin 33.7 = 0$$

$$\text{then } F_{BC} = -2.13\text{kN}[C]$$

Note that the direction of F_{BC} is opposite to what is shown in Fig. 16.14. Also, the equilibrium in the x direction with the unknown F_{BC} force already found from equilibrium in the y direction will be identically satisfied and therefore is not used.

Example 2: the method of sections

The previous method, the method of joints, enables only two equilibrium equations which constraints the solution procedure. The second method, the method of sections, adds a third equilibrium equation, a moment equation, thus enabling solving for three unknowns at a joint. Note that the two methods can be used simultaneously, depending on the nature of the posed truss problem.

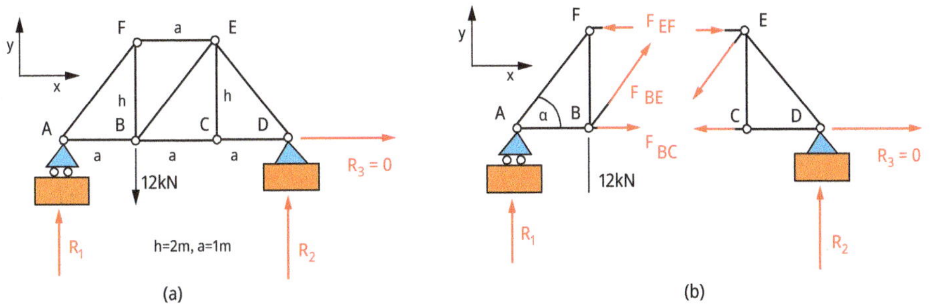

Fig. 16.15: (a) The planar truss, (b) application of the section method.

A planar truss is depicted in Fig. 16.15a. It is loaded by a force of 12kN at point B. To find the forces acting on members EF, BE and BC, we apply the section method: an imaginary cut is performed as shown in Fig. 16.15b; dividing the truss into two parts, each should be in equilibrium. The reactions R_1, R_2 and R_3 are first calculated

$$\sum M_A = 0 \Rightarrow R_2 * 3 + 12 * 1 = 0 \text{ then } R_2 = 4\text{kN}$$

$$\sum F_Y = 0 \Rightarrow R_1 + R_2 - 12 = R_1 + 4 - 12 = 0 \text{ then } R_1 = 8\text{kN}$$

Note that the reaction R_3 equals to zero, as no horizontal forces are applied on the truss.

Choosing the left side of the truss the equilibrium equations would be:

$$\sum F_x = 0 \Rightarrow F_{BC} - F_{EF} + F_{BE}\cos\alpha = 0$$

$$\sum F_y = 0 \Rightarrow R_1 - 12 + F_{BE}\sin\alpha = 8 - 12 + F_{BE}\sin\alpha = 0$$

$$\sum M_B = 0 \Rightarrow R_1 * 1 - F_{EF} * 2 = 8 * 1 - F_{EF} * 2 = 0 \text{ then } F_{EF} = 4\text{kN}[T]$$

Evaluating the angle α and its corresponding trigonometric functions yields

$$\tan\alpha = 2 \Rightarrow \alpha = 63.2350^{\circ} \text{ then } \sin\alpha = 0.8922, \cos\alpha = 0.4472$$

Inserting the values of the trigonometric functions into the first two equilibrium equations leads to:

$$F_{BE} = 4.4723\text{kN}[C] \text{ and } F_{BC} = 2\text{kN}[C]$$

where T and C represent tension and compression forces in the truss members, respectively.

16.3.2 Buckling and vibration of trusses

The stability and the dynamics of trusses have been extensively dealt with in the literature, and the reader is referred to typical references like [40–47] for the buckling issues and references [48–51] for vibrations of trusses.

For both topics, namely for buckling and vibrations of trusses, there are two evident behaviors, a local and a global one. Talking about local buckling of truss members, one must check all the compressed members against eventual loss of stability using the following simple formula

$$P_{cr} = \frac{\pi^2 EI}{L^2} \tag{16.36}$$

where EI is the bending rigidity of the rod or column which is a member of the truss, L is its length, and P_{cr} is the critical buckling load. The formula written in eq. (16.36) is also known as the Euler buckling load for columns. As can be seen, the slender the truss member is, the lower its Euler buckling load.

Slender trusses might buckle as a beam leading to global loss of stability of the whole structure.

To calculate the global buckling load of the truss, it is advised to use the homogenization approach as is described in [52] (see also Chapter 17).

The main idea of this approach is that the behavior of a slender truss can be treated as a solid isotropic column, with elastic constants representing it. The static and dynamic behavior of the equivalent isotropic column will be the same as that of the slender truss. However, local behavior, such as members' buckling, will not be reflected by this representation. This process is called in the literature *Homogenization* or *Smearing* and was used by many researchers to find equivalent properties of various structures with discontinuities, or other irregularities in the cross section of the column.

To perform the homogenization approach, tests have to be performed on the slender truss or numerical (finite element analysis) calculations yielding the maximal lateral deflection from which the equivalent bending rigidity of the truss $(EI)_{eqv.}$ can be evaluated. Remembering the maximal deflection of pinned–pinned beam loaded at its midspan $y_{max} = PL^3/(48EI)$, the equivalent bending rigidity of the slender truss can be evaluated using the calculated or measured y_{max} to yield $(EI)_{eqv.} = PL^3/(48y_{max})$. Then the global buckling load of the slender can be obtained from eq. (16.36). Note that for other boundary conditions, like a cantilever truss the equations would be $(EI)_{eqv.} = PL^3/(3y_{max})$ and the buckling load $P_{cr} = \pi^2(EI)_{eqv.}/(4L^2)$ (for more information, see Chapter 6).

Addressing the natural vibrations of a truss, a local and a global behavior is encountered as for the buckling issue discussed above. As all members of the truss are on pinned–pinned boundary conditions, the natural frequencies (see Chapters 7 and 11) can be calculated using the following equation:

$$\omega_i^2 = \frac{\pi^4 EI}{\rho A L^4} \tag{16.37}$$

where EI is the bending rigidity, ρA is the mass per unit length, and ω stands for circular frequency of the column. Equation (16.37) is valid for non-loaded members of the truss. For a loaded member, the following formulas should be used:

$$\left[\frac{\omega}{\omega_i}\right]^2 + \frac{P}{P_{cr}} = 1 \text{ for compressive loads} \tag{16.38}$$

$$\left[\frac{\omega}{\omega_i}\right]^2 - \frac{P}{P_{cr}} = 1 \text{ for tensile loads} \tag{16.39}$$

where $\omega_i^2 = \frac{\pi^4 EI}{\rho A L^4}$ and $P_{cr} = \frac{\pi^2 EI}{L^2}$.

For slender trusses, the application of the homogenization approach, discussed above for the buckling issue, can be applied also for the vibration case with the $(EI)_{eqv.}$ determined from finite elements calculations or tests and then inserted in eqs. (16.37) or (16.38) and (16.39) according to the solved case.

16.4 Static behavior, buckling and vibration of frames

16.4.1 Static behavior of frames

A truss with at least one of its individual members is a multi-force member which is defined as a frame. A multi-force member is defined as one with multiple forces and/or moments action on it. In contrast to the pinned–pinned members of a truss, the forces acting on the frame members will in general not be in the directions of the members. The way such frame structures are solved will be presented in the next example.

Example 3: solving a planar frame

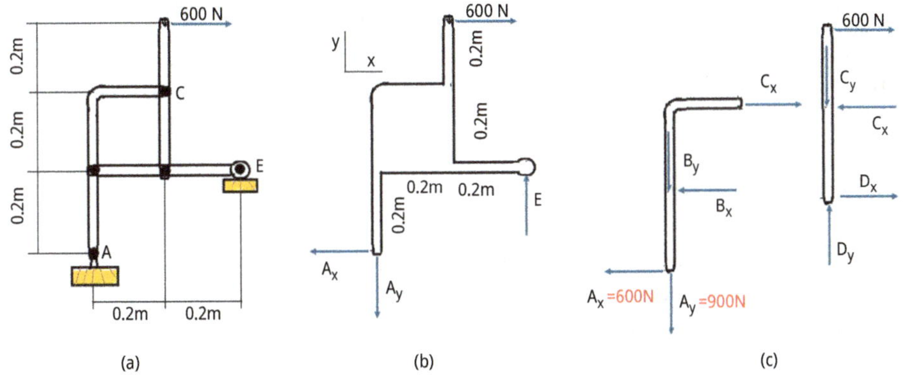

Fig. 16.16: (a) The planar frame, (b) entire frame and (c) solving for CD and ABC members.

It is required to find the magnitude of the load acting on the pin at point C, as shown in Fig. 16.16a. First, we solve for the reactions at points A and E for the entire frame, as presented in Fig. 16.16b:

$$\sum M_A = 0 \Rightarrow 0.4E - 0.6 * 600 = 0 \text{ then } E = 900N$$

$$\sum F_x = 0 \Rightarrow 600 - A_x = 0 \text{ then } A_x = 600N$$

$$\sum F_y = 0 \Rightarrow E - A_y = 900 - A_y = 0 \text{ then } A_y = 900N$$

With all the three reactions being solved for points A and E, we can now "dismantle" the frame into its members to solve the problem, as schematically shown in Fig. 16.16c. For member CD it can be written that

$$\sum M_D = 0 \Rightarrow 0.2C_x - 0.4 * 600 = 0 \text{ then } C_x = 1{,}200N$$

Solving for member ABC (see also Fig. 16.16a), we get

$$\sum M_B = 0 \Rightarrow 0.2C_y - 0.2 * 600 - 0.2 * 1{,}200 = 0 \text{ then } C_y = 1{,}800N$$

The magnitude of the force at point C will then be

$$C = \sqrt{1{,}200^2 + 1{,}800^2} = \sqrt{4{,}860{,}000} = 2{,}163.33N$$

To find the other unknowns of the problem, although not asked for, one has to apply equilibrium equations for each member of the frame. For instance, for member CD, we will have that

$$\sum F_x = 0 \Rightarrow 600 - C_x + D_x = 600 - 1{,}200 + D_x = 0 \text{ then } D_x = 600N$$

$$\sum F_y = 0 \Rightarrow D_y - C_y = D_y - 1{,}800 = 0 \text{ then } D_y = 1{,}200N$$

and for member ABC

$$\sum F_x = 0 \Rightarrow C_x - B_x - 600 = 1{,}200 - B_x - 600 = 0 \text{ then } B_x = 600N$$

$$\sum F_y = 0 \Rightarrow C_y - B_y - 900 = 1{,}800 - B_y - 900 = 0 \text{ then } B_y = 900N$$

Note that in case a member of the planar frame is in compression it must be calculated against buckling to prevent its loss of stability and cause the collapse of the frame. For more cases of planar frames, see Ref. [39].

16.4.2 Buckling of frames

The topic of buckling of various types of frames has been widely discussed in the literature, with typical examples to be found in Refs. [53–66]. The present chapter will present the basic concepts to understand the topic.

The three most methods used to calculate the buckling load of rigid frames are:

1. The energy method
2. The moment-distribution method.
3. Direct analytical solution based on slope deflection procedure.

Let us concentrate on the third method, as presented in [54].[2] Suppose a column ab is compressed axially by force P, then the end b would displace itself relative to end a as schematically shown in Fig. 16.17.

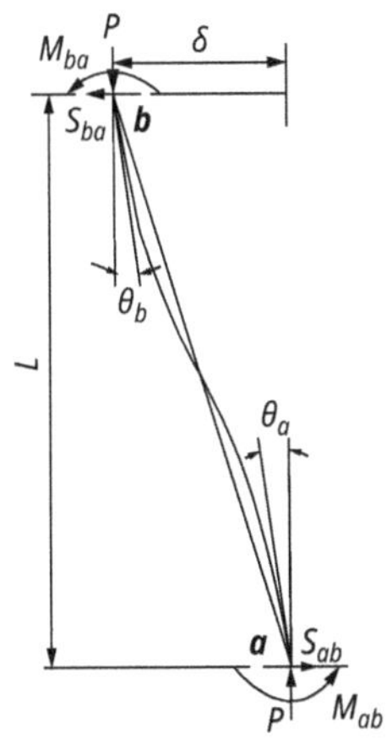

Fig. 16.17: A column under compression load P – reactions and displacements at its ends.

Applying the slope deflection approach (see [54]), the following relations can be written for the moments at each end of the column

$$M_{ab} = K\left[\alpha\theta_a + \beta\theta_b - (\alpha+\beta)\frac{\delta}{L}\right] \text{ and}$$

$$M_{ba} = K\left[\alpha\theta_{ab} + \beta\theta_a - (\alpha+\beta)\frac{\delta}{L}\right] \text{ with } K = \frac{EI}{L} \tag{16.40}$$

where E is the Young's modulus, I denotes the cross section moment of inertia, L is the length of the column, θ is the angle at the ends, δ is the relative displacement at the ends and α, β are constants depending on the sign and magnitude of the applied load and are given by the following relationships

2 See also Goldberg, J.E., Buckling of one-story frames and buildings, Journal of the structural Division, ASCE, Vol.86, Issue 10October 1960, pp. 53–85.

For axially compressive force

$$\alpha = \frac{\sin\lambda L - \lambda L\cos\lambda l}{\frac{2}{\lambda L}(1-\cos\lambda L) - \sin\lambda L} \quad \beta = \frac{\lambda L - \sin\lambda L}{\frac{2}{\lambda L}(1-\cos\lambda L) - \sin\lambda L} \tag{16.41}$$

For tensile applied load

$$\alpha = \frac{\lambda L\cosh\lambda L - \sinh\lambda L}{\frac{2}{\lambda L}(1-\cosh\lambda L)/\sin\lambda L} \quad \beta = \frac{\sinh\lambda L - \lambda L}{\frac{2}{\lambda L}(1-\cosh\lambda L)/\sinh\lambda L} \tag{16.42}$$

where $\lambda L = \pi\sqrt{\kappa}$ and $\kappa = \frac{P}{P_E}$ and $P_E = \pi^2\frac{EI}{L^2}$.

Note that a positive P would be compression force. Table 16.3 (from [54]) presents some of the values for the constants α, β as a function of κ.

Tab. 16.3: Slope deflection coefficients α, β for various values of the load parameter κ (adapted from [54]).

κ	α	β	κ	α	β
0.0	4.000	2.000	0.0	4.000	2.000
0.1	3.865	2.033	−0.1	4.131	1.968
0.2	.3.730	2.070	−0.2	4.255	1.938
0.3	3.589	2.109	−0.3	4.384	1.910
0.4	3.444	2.150	−0.4	4.502	1.883
0.5	3.295	2.194	−0.5	4.619	1.857
0.6	3.140	2.241	−0.6	4.736	1.834
0.7	2.981	2.291	−0.7	4.849	1.811
0.8	2.816	2.346	−0.8	4.959	1.789
0.9	2.645	2.404	−0.9	5.069	1.769
1.0	2.468	2.468	−1.0	5.175	1.749
1.2	2.090	2.610	−1.2	5.383	1.713
1.4	1.673	2.778	−1.4	5.583	1.681
1.5	1.457	2.873	−1.6	5.777	1.651
1.6	1.224	2.980	−1.8	5.964	1.623
1.8	0.717	3.224	−2.0	6.147	1.598
2.0	0.143	3.521	−2.5	6.580	1.544
2.2	−0.519	3.901	−3.0	6.990	1.499
2.4	−1.300	4.383	−4.0	7.750	1.430
2.5	−1.749	4.678	−5.0	8.420	1.380
2.6	−2.252	5.019	−7.0	9.620	1.300
2.8	−3.449	5.884	−9.0	10.690	1.260
3.0	−5.030	7.120			
3.1	−6.050	7.960			
3.2	−7.300	9.020			
3.3	−8.860	10.400			
3.4	-10.910	12.240			

Tab. 16.3 (continued)

K	*α*	*β*	*K*	*α*	*β*
3.5	−13.730	14.860			
3.6	−17.870	18.790			
3.7	−24.690	25.390			
3.8	−39.050	39.540			
3.9	−78.340	78.560			

The shear force can be obtained by equating moments of either end of the column and then substituting the expressions for M_{ab} and M_{ba} to yield

$$S_{ab} = -\frac{K}{L}(\alpha+\beta)\left(\theta_a + \theta_b - 2\frac{\delta}{L}\right) - P\frac{\delta}{L} \tag{16.43}$$

Let us assume that the base of the column, at point a, is elastically restraint by a rotational spring with a given constant $\bar{K}$. Then the moment at point (a) will be the multiplication of the spring constant by the angle θ_a which can be equated with the expression for M_{ab} from eqs. (16.40) to yield

$$M_{ab} = -\bar{K}\theta_a = K\left[\alpha\theta_a + \beta\theta_b - (\alpha+\beta)\frac{\delta}{L}\right], \theta_a = \frac{1}{\alpha + \frac{\bar{K}}{K}}\left[-\beta\theta_b - (\alpha+\beta)\frac{\delta}{L}\right] \tag{16.44}$$

This equation is substituted in the second equation of (16.40) and eq. (16.43) leading to

$$M_{ba} = K\left[\alpha^1_{ab}\theta_b - \gamma_{ab}\mu_{ab}\frac{\delta}{L}\right] \tag{16.45a}$$

$$S_{ab} = -\frac{K}{L}\gamma_{ab}\left(\mu_{ab}\theta_b - \mu^1_{ab}\frac{\delta}{L}\right) - P\frac{\delta}{L} \tag{16.45b}$$

where

$$\gamma_{ab} = \alpha+\beta, \alpha^1_{ab} = \alpha - \frac{\beta^2}{\alpha + \frac{\bar{K}}{K}},\ \mu_{ab} = 1 - \frac{\beta}{\alpha + \frac{\bar{K}}{K}},\ \mu^1_{ab} = 2 - \frac{\alpha+\beta}{\alpha + \frac{\bar{K}}{K}} \tag{16.46}$$

Note that the value of torsional stiffness can vary from $\bar{K} = 0$ for pinned (simply supported) boundary conditions to $\bar{K} \to \infty$ for a clamped case.

For pinned boundary conditions, the constants in eq. (16.46) will be

$$\alpha^1_{ab} = \alpha - \frac{\beta^2}{\alpha} = \frac{\gamma_{ab}}{\alpha}(\alpha-\beta),\ \mu_{ab} = \frac{\alpha-\beta}{\alpha},\ \mu^1_{ab} = 2 - \frac{\alpha+\beta}{\alpha} \tag{16.47}$$

and

$$M_{ba} = K\frac{\gamma_{ab}}{\alpha}(\alpha - \beta)\left[\theta_b - \frac{\delta}{L}\right] \tag{16.48a}$$

$$S_{ab} = -\frac{K}{L}\frac{\gamma_{ab}}{\alpha}(\alpha - \beta)\left[\theta_b - \frac{\delta}{L}\right] - P\frac{\delta}{L} \tag{16.48b}$$

For clamped boundary conditions, the constants of eq. (16.46) will simplify to

$$\alpha^1_{ab} = \alpha, \mu_{ab} = 1, \mu^1_{ab} = 2 \tag{16.49}$$

and

$$M_{ba} = K\left[\alpha\theta_b - \gamma_{ab}\frac{\delta}{L}\right] \tag{16.50a}$$

$$S_{ab} = -\frac{K}{L}\gamma_{ab}\left[\theta_b - 2\frac{\delta}{L}\right] - P\frac{\delta}{L} \tag{16.50b}$$

Based on the above equations, the buckling loads for frames having various shapes can be derived.

1. Symmetric buckling of frames: The frame shown in Fig. 16.18 would buckle in a symmetric mode, as displayed in the figure, if any lateral displacement is not allowed. Using the previous developed equations, the buckling load of this case can be evaluated. For this case, we have that $\delta = 0$ and $\theta_a = -\theta_c$.

At various joints, we can write using eqs. (16.50a) and (16.50b)

$$\sum M_B = 0 \text{ or } M_{BA} + M_{BC} = 0 ==> K\theta + 2K_{BC}\theta = 0; \alpha = -\frac{2K_{BC}}{K} \tag{16.51}$$

where $K_{BC} = EI_2/L$ and $K = EI_1/L$. The constant α can be evaluated from eq. (16.51), and then using its value, one can obtain the load parameter κ using Tab. 16.3. The frame critical buckling load will then be calculated using the following relationship:

$$P_{\text{cr}} = \kappa P_E = \kappa\frac{\pi^2 EI}{L^2} \tag{16.52}$$

2. Antisymmetric buckling of frames: A frame would buckle with an antisymmetric mode if the lateral displacement is allowed to happen (a schematic drawing displayed in Fig. 16.19).

Note that for this case we have θ (at B) = θ (at C), δ (at B) = δ (at C) and no lateral displacement for the horizontal column BC. Using the symmetry of the problem, we need to write only two equilibrium equations

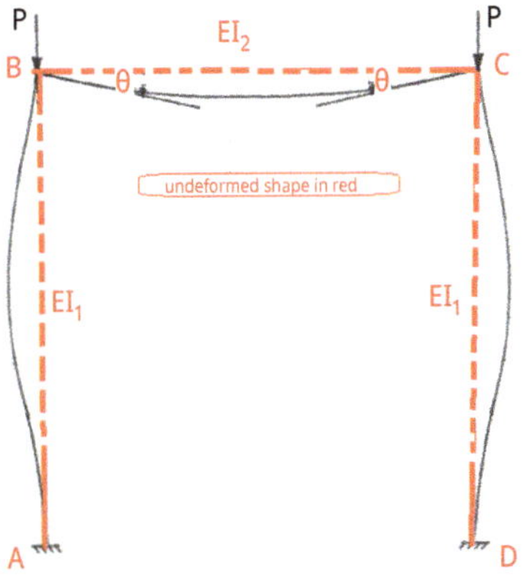

Fig. 16.18: A frame with clamped boundary conditions – a symmetric mode.

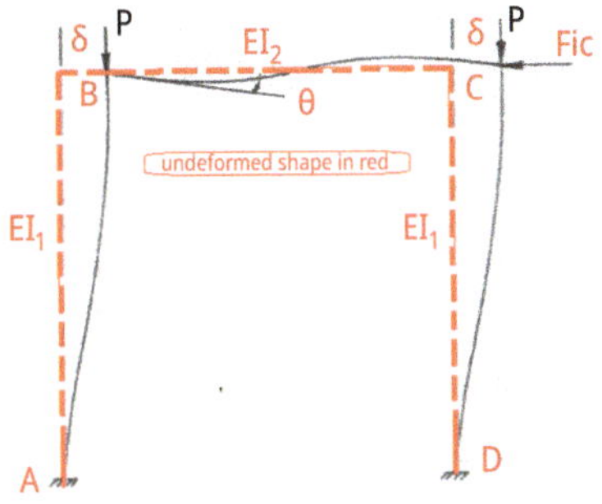

Fig. 16.19: A frame with clamped boundary conditions – an antisymmetric mode.

and the lateral force F_{ic} should vanish. This leads to the following equations:

$$\sum M_B = 0 \quad \text{or} \quad \sum M_c = 0 \tag{16.51}$$

$$\sum M_B = 0 \quad \text{or} \quad M_{BA} + M_{BC} = K\left[\alpha\theta_B - \gamma_{AB}\frac{\delta}{L}\right] + 6K_{BC}\theta_B \tag{16.52}$$

leading to

$$\theta_B = \frac{K\gamma_{AB}\delta}{L(K\alpha + 6K_{BC})} \tag{16.53}$$

Applying the second equilibrium equation (the vanishing of F_{ic}) provides the second relationship

$$-S_{AB} - S_{DC} - F_{ic} = 0 \text{ but } S_{AB} = S_{DC} \Rightarrow -S_{AB} = -S_{DC} = \frac{F_{ic}}{2} = 0 \tag{16.54}$$

Substituting eq. (16.50b) into eq. (16.54) yields

$$\frac{K}{L}\gamma_{AB}\left[\theta_B - 2\frac{\delta}{L}\right] + P\frac{\delta}{L} = \frac{F_{ic}}{2} = 0 \tag{16.55}$$

To solve eqs. (16.52) and (16.55) and obtain the critical load, a *trial-and-error procedure* should be applied. As pointed out in [54], for the frame of the present problem the critical load would be less than the Euler load where a given Young's modulus is given. So, an initial value is selected for the compression load P and the coefficients are calculated. Assuming δ to have a unit magnitude, the value of θ_B can be found from eq. (16.53). Substituting the values of θ_B and δ into eq. (16.55) provides the value of the force F_{ic} to be applied to deflect laterally the frame. The sign of the computed value of F_{ic} indicates whether the compression loads are greater or less than the critical loads. In case F_{ic} is positive, the assumed column loads are greater than critical load, as this force is now supporting the frame against further lateral deflection. If F_{ic} has a negative value, the assumed column loads are less than the critical load as its direction implies that the frame has a "reserve stiffness" ([54]). For $F_{ic} = 0$, the assumed column loads equal to the critical loads. Note that for the suggested *trial-and-error procedure* for a nonzero value for the F_{ic}, its sign would indicate the next trial value to be larger or smaller than the previous step.

3. Antisymmetric buckling of frames with one end clamped and one end pinned (simply supported): The frame dealt before for an antisymmetric buckling is again presented with a change of its boundary condition at point D (see Fig. 16.20), making the case non-symmetric (at point D the B.C. is pinned, while at point A it is clamped). For this case, we assume the lateral deflections at points B and C to have a unit magnitude and to be equal. Applying moments equilibrium at point B (the same can be done also for point C) yields

$$\sum M_B = 0 \text{ or } M_{BA} + M_{BC} = K\left[\alpha\theta_B - \gamma_{AB}\frac{\delta}{L}\right] + 6K_{BC}\theta_B \tag{16.56}$$

leading to

$$\theta_B = \frac{K\gamma_{AB}\delta}{L(K\alpha + 6K_{BC})} \tag{16.57}$$

The equilibrium shear equation in the horizontal direction can be written (as before) as

$$-S_{AB} - S_{DC} - F_{ic} = 0 \Longrightarrow -S_{AB} - S_{DC} = F_{ic} = 0 \tag{16.58}$$

Substituting eqs. (16.48b) and (16.50b) into eq. (16.58) yields

$$\frac{K}{L}\gamma_{ab}\left[\theta_b - 2\frac{\delta}{L}\right] + P\frac{\delta}{L} + \frac{K}{L}\frac{\gamma_{ab}}{\alpha}(\alpha - \beta)\left[\theta_b - \frac{\delta}{L}\right] + P\frac{\delta}{L} = F_{ic} = 0 \tag{16.59}$$

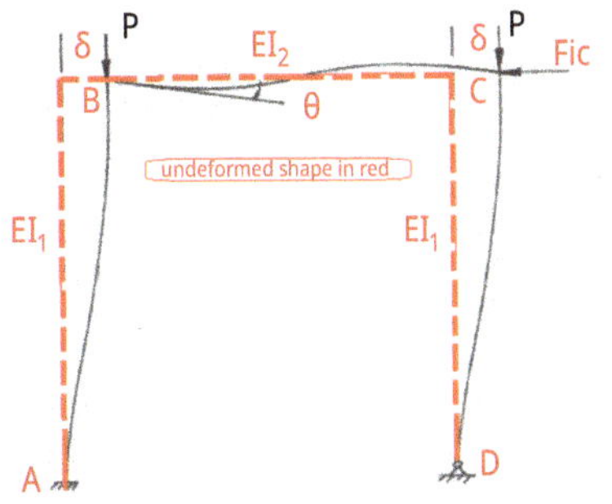

Fig. 16.20: A frame with clamped–pinned boundary conditions – an antisymmetric mode.

To determine the critical load of the present case the procedure described previously can be applied.

4. Antisymmetric buckling of a three-story single bay frame with pinned–pinned boundary conditions (simply supported): The frame depicted in Fig. 16.21 shows a three-story single bay frame resting on a pinned–pinned boundary condition. It is also known that $\theta_A = \theta_E$, $\theta_B = \theta_F$, $\theta_C = \theta_G$ and $\theta_D = \theta_H$. The lateral displacements at the same level are equal, namely $\delta_{CD} = \delta_{GH} = \delta$, $\delta_{BC} = \delta_{FG}$ and $\delta_{AB} = \delta_{EF}$. Also the moment equilibrium at points B, C, D, H, G or F is zero.

It can be shown that the rotations at the joints have the following forms (see [54]):

$$\theta_D = \frac{K_{CD}\gamma_{CD}\delta}{L_{CD}(K_{CD}\alpha_{CD} + 6K_{DH})} \tag{16.60a}$$

$$\theta_C = \frac{\frac{K_{CD}\gamma_{CD}\delta}{L_{CD}} + \frac{K_{CB}\gamma_{CB}}{L_{CB}}\delta_{CB}}{(K_{CB}\alpha_{CB} + K_{CD}\alpha_{CD} + 6K_{CG})} \tag{16.60b}$$

$$\theta_B = \frac{\frac{K_{BC}\gamma_{BC}\delta_{BC}}{L_{BC}} + \frac{K_{AB}\gamma_{AB}\delta_{AB}}{\alpha_{AB}L_{AB}}(\alpha_{AB} - \beta_{AB})}{K_{BC}\alpha_{BC} + 6K_{BF} + K_{BA}(\alpha_{AB} - \beta_{AB})\gamma_{AB}/\delta_{AB}} \tag{16.60c}$$

Accordingly, the shear equilibrium equations will be

$$\frac{K_{CD}}{L_{CD}}\gamma_{CD}\left(\theta_D - 2\frac{\delta_{CD}}{L_{CD}}\right) + P\frac{\delta_{CD}}{L_{CD}} = \frac{F_{ic}}{2} = 0 \tag{16.61a}$$

$$\frac{K_{BC}}{L_{BC}}\gamma_{BC}\left(\theta_C - 2\frac{\delta_{BC}}{L_{BC}}\right) + P\frac{\delta_{BC}}{L_{BC}} = F_{ic} = 0 \tag{16.61b}$$

$$\frac{K_{AB}}{L_{AB}}\gamma_{AB}\left(\theta_B - 2\frac{\delta_{AB}}{L_{AB}}\right) + P\frac{\delta_{AB}}{L_{AB}} = \frac{3F_{ic}}{2} = 0 \tag{16.61c}$$

To determine the critical load of the present case, the procedure described previously can be applied. Note that additional cases are described in [54].

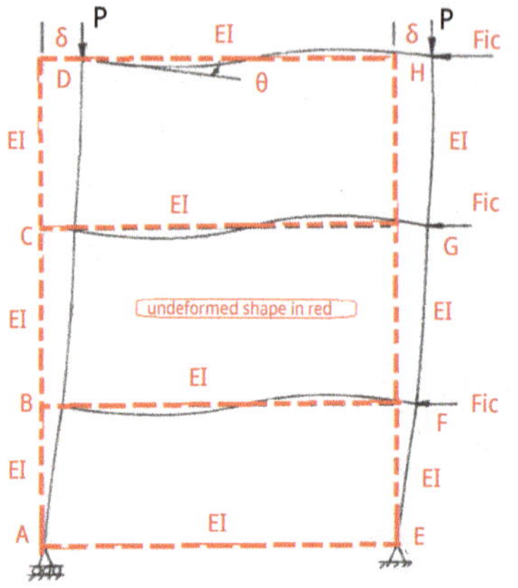

Fig. 16.21: A three-story single bay frame with pinned–pinned boundary conditions – an antisymmetric mode.

5. Rectangular frames – Timoshenko approach (see [2]): Based on the relationships developed in Section 6.6 of the present work, which reflects the early works of Timoshenko, the buckling of a symmetrical rectangular frame (see Fig. 16.22) can be determined.

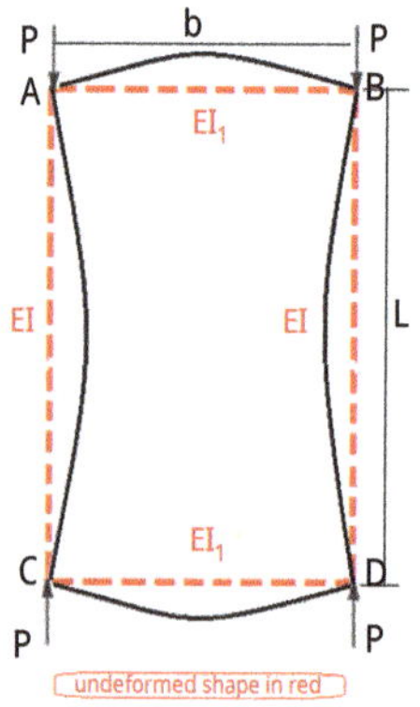

Fig. 16.22: A symmetrical rectangular frame axially compressed by forces P.

The frame is prevented to move laterally; thus, no sway is allowed. When the compression force P reaches the critical load, the vertical struts will begin to buckle (see the continuous line in Fig. 16.22). The buckling is accompanied by bending of the lateral columns, which apply reactive moments to the vertical struts, and thus resisting the buckling process. Therefore, the two vertical struts will behave like a beam on elastic supports at both ends.

To determine the critical load, the coefficient of the flexural rigidity at the ends of the struts can be found to be:

$$\mu = \frac{2EI_1}{b} \tag{16.62}$$

Then using the following equation (see [2])

$$\frac{\tan\kappa}{\kappa} = -\frac{2EI}{\mu L} = -\frac{Ib}{I_1 L} \quad \text{where } \kappa = \frac{\lambda L}{2} = \frac{L}{2}\sqrt{\frac{P}{EI}} \tag{16.63}$$

the critical compression load can be found for various cases. For instance, if the rectangular frame is composed of four identical struts, having the same length the bending rigidity, eq. (16.63) will simplify to the following:

$$\frac{\tan\kappa}{\kappa} = -1 \tag{16.64}$$

The smallest root for eq. (16.64) is $\kappa = \frac{\lambda L}{2} = \frac{L}{2}\sqrt{\frac{P}{EI}} = 2.029 \Longrightarrow P_{cr} = \frac{16.4674EI}{L^2}$

For rigid horizontal struts, we get $\tan\kappa = 0 \Longrightarrow \kappa = \pi$ and $P_{cr} = \frac{4\pi^2 EI}{L^2}$, namely the same critical value as the case of a column on clamped–clamped boundary conditions. For case $I_1 = 0$, we have $\kappa = \frac{\pi}{2}$ and $P_{cr} = \frac{\pi^2 EI}{L^2}$, as a pinned–pinned column.

For the case shown in Fig. 16.23, Timoshenko suggests the following procedure: First, the end restraint coefficient must be evaluated using eq. (16.65)

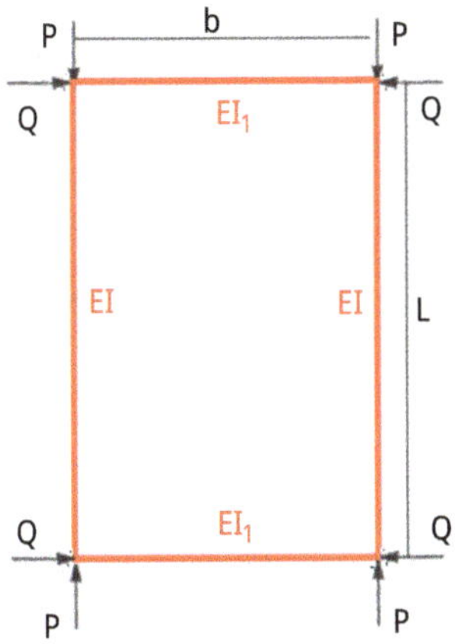

Fig. 16.23: A symmetrical rectangular frame, axially and laterally compressed by forces *P* and *Q*, respectively.

$$\mu = \frac{2EI_1}{b}\frac{\kappa_1}{\tan\kappa_1} \quad \text{where } \kappa_1 = \frac{\lambda b}{2} = \frac{b}{2}\sqrt{\frac{Q}{EI}} \tag{16.65}$$

Using eq. (16.63) with the new definition for μ (see eq. (16.65)) yields

$$\frac{\tan\kappa}{\kappa} = -\frac{2EI}{\mu L} = -\frac{Ib}{I_1 L}\frac{\tan\kappa_1}{\kappa_1} \quad \text{where } \kappa = \frac{\lambda L}{2} = \frac{L}{2}\sqrt{\frac{P}{EI}} \tag{16.66}$$

Knowing the right-hand side of eq. (16.66), the critical compressive load can be evaluated, provided the horizontal struts do not buckle first. If the critical value of Q is desired, then the following equation should be solved

$$\frac{\tan\kappa_1}{\kappa_1} = -\frac{I_1 L}{Ib}\frac{\tan\kappa}{\kappa} \tag{16.67}$$

Equation (16.67) is the same as eq. (16.66), and it defines the limiting critical values for the axial compression forces P and Q. For a square frame with all four struts having the same rigidity, we get a simple expression

$$\frac{\tan\kappa}{\kappa} = -\frac{\tan\kappa_1}{\kappa_1} \tag{16.68}$$

Equation (16.68) can be drawn on a graph (Fig. 16.24), showing stable and critical regions for the P versus Q values. As can be observed from Fig. 16.26, an increase in the critical value of P would cause a decrease in the critical value of Q and vice versa.

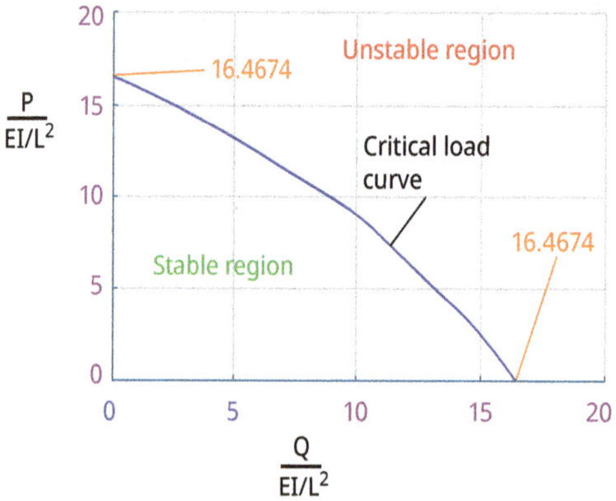

Fig. 16.24: Nondimensional force *P* versus nondimensional force *Q*.

The most common buckling of frames is summarized in Tab. 16.4.

16.4.3 Vibration of frames

The vibration of various types of frames has been widely discussed in the literature, with typical examples to be found in Refs. [67–79]. The present chapter will present the basic concepts to understand the topic.

A frame is assembled from bars which are interconnected to yield a stable structure. The frame is then attached to external boundary conditions, according to the functionality of the given structure. Addressing the natural vibrations of a frame,

Tab. 16.4: Common buckling of frames.

Case	Critical load	Remarks
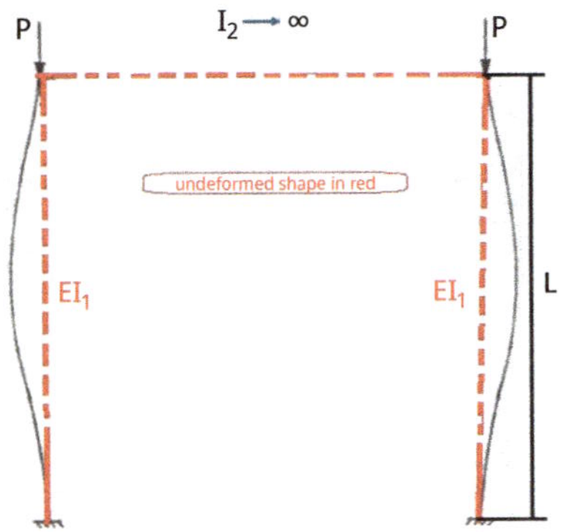	$P_{cr} = \dfrac{4\pi^2 EI_1}{L^2}$	Portal frame on clamped B.C. No sway.

(continued)

Tab. 16.4 (continued)

Case	Critical load	Remarks
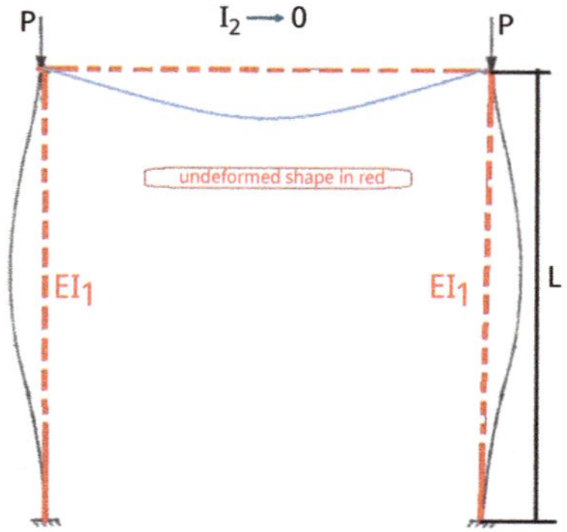	$P_{cr} = \frac{20.2EI_1}{L^2}$	Portal frame on clamped B.C. No sway.
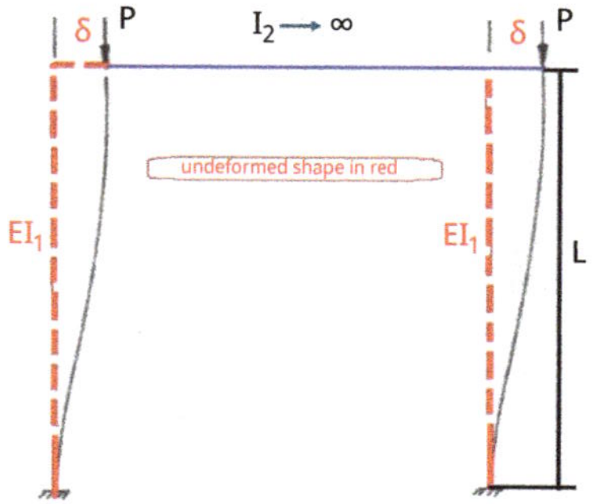	$P_{cr} = \frac{\pi^2 EI_1}{L^2}$	Portal frame on clamped B.C. With sway.

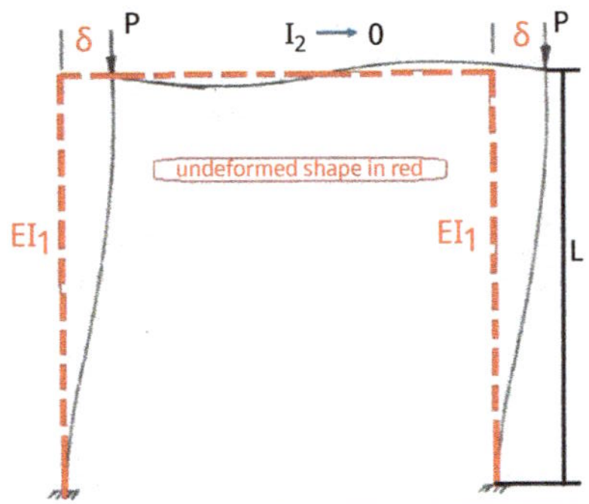	$P_{cr} = \frac{\pi^2 EI_1}{4L^2}$	Portal frame on clamped B.C. With sway.
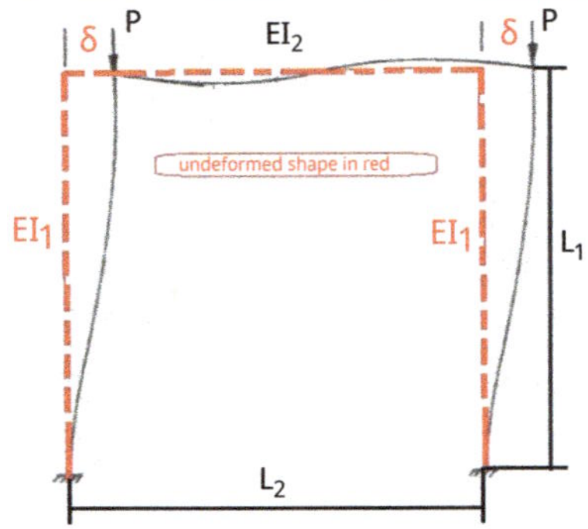	The characteristic equation: $-\frac{\tan \kappa_1 L_1}{\kappa_1 L_1} = \frac{I_1 L_2}{6 I_2 L_1}, \kappa_1^2 = \frac{P}{EI_1}$ If $I_1 = I_2 = I; L_1 = L_2 = L$ Then $\frac{\tan \kappa L}{\kappa L} = -\frac{1}{6}$ and $\kappa L = 2.71$ $P_{cr} = \frac{7.34EI}{L^2}$	Sideways buckling

(continued)

Tab. 16.4 (continued)

Case	Critical load	Remarks
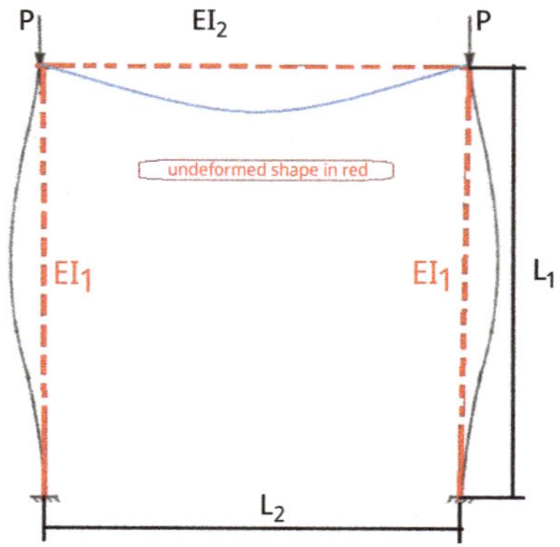	The characteristic equation: $2-2\cos\kappa_1L_1-\kappa_1L_1\sin\kappa_1L_1+$ $+\dfrac{I_1L_2\kappa_1}{2I_2}(\sin\kappa_1L_1-\kappa_1L_1\cos\kappa_1L_1)=0$ If $I_1=I_2=I;\,L_1=L_2=L$ Then $\kappa L\sin\kappa L+4\cos\kappa L+(\kappa L)^2\cos\kappa L=4$ and $\kappa L=5.02$ leading to $P_{cr}=\dfrac{25.2EI}{L^2}$	Symmetric buckling

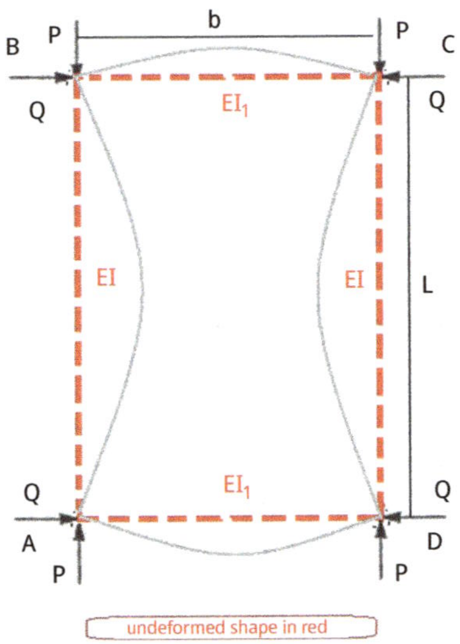	The characteristic equation: $\frac{\tan\kappa}{\kappa} = -\frac{2EI}{\mu L} = -\frac{Ib}{I_1 L}\frac{\tan\kappa_1}{\kappa_1}$ where $\kappa = \frac{\lambda L}{2} = \frac{L}{2}\sqrt{\frac{P}{EI}}$; $\kappa_1 = \frac{\lambda b}{2} = \frac{b}{2}\sqrt{\frac{Q}{EI}}$ If $I_1 = I_2 = I; L_1 = L_2 = L \frac{\tan\kappa}{\kappa} = -\frac{\tan\kappa_1}{\kappa_1}$ If $Q = 0$ then $\lim_{\kappa_1 \to \infty} \frac{\tan\kappa_1}{\kappa_1} = 1$ and the characteristic equation will be: $\frac{\tan\kappa}{\kappa} = -\frac{EI}{EI_1}\frac{b}{L}$ (1) If $I_1 = I; b = L$ then $\tan\kappa = -\kappa$ and $\kappa = 2.029 \Rightarrow P_{cr} = \frac{16.4674EI}{L^2}$	Symmetric buckling of rectangular rigid frames Timoshenko approach as described above this table

(continued)

Tab. 16.4 (continued)

Case	Critical load	Remarks
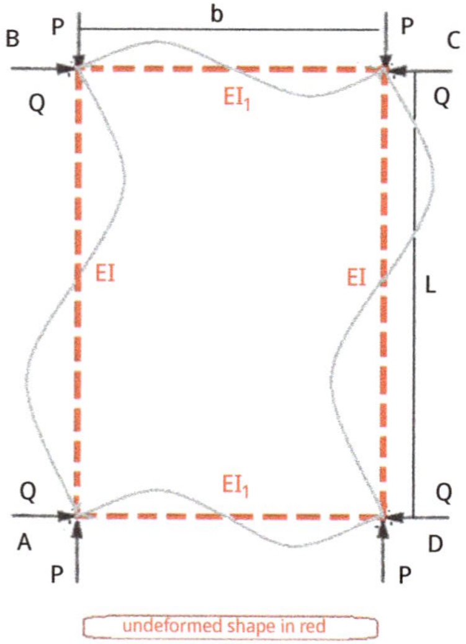	The characteristic equation: $\frac{b}{EI_1}\frac{1}{\kappa_1}\left(\frac{1}{\kappa_1}-\cot\kappa_1\right)=-\frac{L}{EI}\frac{1}{\kappa}\left(\frac{1}{\kappa}-\cot\kappa\right)$ If $Q=0$, the characteristic equation will be: $\frac{1}{\kappa}\left(\frac{1}{\kappa}-\cot\kappa\right)=-\frac{EIb}{3EI_1L}$ and for equal properties $b=L$ and $I=I_1$ The characteristic equation will simplify to $\frac{1}{\kappa}\left(\frac{1}{\kappa}-\cot\kappa\right)=-\frac{1}{3}$ or $\left(\frac{1}{\kappa}-\cot\kappa\right)=-\frac{\kappa}{3}$	Antisymmetric buckling of a rigid rectangular frame

local and global behavior is encountered. In contrast to a truss, which all its members are on pinned–pinned boundary conditions, the boundary conditions of each strut of the frame are elastic boundary conditions, with the deflection and the slope depending on the stiffness of the neighboring member. Therefore, to find the natural frequencies of the all the members of the frame, a finite element code should be applied to yield the eigenvalues and the eigenmodes of the frame.

For slender frames, the application of the homogenization approach, discussed in Section 16.3.2, can be applied also for the vibration case with the equivalent stiffness $(EI)_{eqv.}$ determined from finite elements calculations or tests. Then the natural frequencies of the equivalent beam can be evaluated depending on the boundaries of the beam (see Chapter 7 for closed-form formulas).

References

[1] Timoshenko, S.P. Buckling of flat curved bars and slightly curved plates, Journal of Applied Mechanics, 2(1), 1935, 17–20.

[2] Timoshenko, S.P. and Gere, J.M. Elastic stability, 2nd, McGraw-Hill, New York, 1963, 541.

[3] Byfield, M. Structural design from first principles, Chapter 4, CRC Press, Taylor & Francis Group, Boca Raton, 2018, 335.

[4] Boussinesq. Sur la résistance d'un anneau à la flexion (On the resistance to bending of a ring), in French, Comptes rendus hebdomadaires des seances de l'Academie de Science (France), 97, 1883, 843–844.

[5] Nair, R.S. Buckling and vibrations of arches and tied arches, Journal of Structural Engineering, 112(6), 1986, Paper No. 20700.

[6] Papangelis, J.P. and Trahair, N.S. Flexural-torsional buckling of arches, Journal of Structural Engineering, 113(4), 1987, Paper No. 21053.

[7] Wicks, P.J. General equations for buckling of thin, shallow arches of any shape, Journal of Structural Engineering, 117(2), 1991, Paper No. 25495.

[8] Pi, Y.-L. and Trahair, N.S. In-plane buckling and design of steel arches, Journal of Structural Engineering, 125, 1999, 1291–1298.

[9] Pi, Y.-L., Bradford, M.A. and Tin-Loi, F. Nonlinear analysis and buckling of elastically supported circular shallow arches, International Journal of Solids and Structures, 44, 2007, 2401–2425.

[10] Iles, D.C. Determining the buckling resistance of steel and composite bridge structures, Technical Report SCI Document ED008, 2012, 44.

[11] Abdelgawad, A., Anwar, A. and Nassar, M. Snap-through buckling of a shallow arch resting on a two-parameter elastic foundation, Applied Mathematical Modelling, Vil, 32, 2013, 7953–7963.

[12] Gua, Y.L., Zhao, S.-Y., Dou, C. and Pi, Y.-L. Out-of-plane elastic buckling of circular arches with elastic end restraints, Journal of Structural Engineering, 140, 2014, Article Id. 04014071, 9.

[13] Andersson, B. and Larsson, G. Verification of buckling analysis for Glulam arches, Master's Dissertation at Lund University, Sweden, 2014, 154.

[14] Sani, M.S.H.M., Tan, C.S., Muftah, F. and Tahir, M.Md. Experimental study on flexural behavior of cold-formed steel channels with curved section, ARPN Journal of Engineering and Applied Sciences, 11(6), 2015, 3655–3662.

[15] Sani, M.S.H.M., Tan, C.S., Muftah, F. and Tahir, M.M. Experimental study on flexural behavior of cold-formed steel channels with curved section, ARPN Journal of Engineering and Applied Sciences, 11(6), 2016, 3655–3662.
[16] Pi, Y.-L., Bradford, M.A. and Liu, A. Prebuckling analyses and knuckling of steel arches, in 8th international conference on steel and aluminium structures, Young, B. and Cai, Y. Eds., Hong Kong, China, 7–9, 2016, 12.
[17] Zilenaite, S. Comparative analysis of the buckling factor of the steel arch bridges, Engineering Structures and Technologies, 11(1), 2017, 11–16.
[18] Qurratulain, M.N. and Danish, M. Buckling of arch bridge, International Journal of Advance Research, Ideas and Innovations in Technology, 4(5), 2018, 427–430.
[19] Nazar, F.B. and Rajesh, K.N. Out-of-plane buckling I-section circular arches, B, in Proceedings of the International Conference on Systems, Energy & Environment (ICSEE) 2019, GCE Kannur, Kerala, July 2019, 6.
[20] Lonetti, P. and Pascuzzo, A. A practical method for the elastic buckling design of network arch bridge, International Journal of Steel Structures, 20(1), 2020, 311–329.
[21] Xi, K. and Zhang, Y. Design method of compression-bending arches with web openings considering local buckling of web, Journal of Physics: Conference Series, 1676, 2020, Paper Id. 012116, 7.
[22] Kiss, L.P., Jalalova, P. and Mehdiyev, Z. The planar buckling of pinned-fixed shallow arches, Journal of Theoretical and Applied Mechanics, 51, 2021, 437–451.
[23] Eroglu, U., Ruta, G., Paolone, A. and Tufekci, E. Buckling and post-buckling of parabolic arches with local damage, Ch. 6, in Modern trends in structural and solid mechanics 1-statics and stability, Challamel, N., Kaplunov, J. and Takewaki, I. Eds., ISTE Ltd and John Wiley & Sons, Inc., London, UK and Hoboken, USA, 2021, 121–144.
[24] Machacek, J. Buckling lengths of steel circular arches respecting non-uniform arch axial forces, Thin-Walled Structures, 180, 2022, Article No. 109916, 14.
[25] Rao, R., Ye, Z., Lv, J., Huang, Y. and Liu, A. Nonlinear instability behavior and buckling of shallow arches under gradient thermos-mechanical loads, Frontiers in Materials, 9, 2022, Article Id. 894260, 9.
[26] Karnovsky, I.A. Theory of arched structures- Strength, stability, vibration, Springer Nature, 2012, 460.
[27] Lamb, H. On the flexure and the vibrations of a curved bar, Proceedings of the London Mathematical Society, 19, 1887, 365–376.
[28] Chidamparam, P. and Leissa, A.W. Vibrations of planar curved beams, rings, and arches, Applied Mechanics Reviews, 46(9), 1993, 467–483.
[29] Auciello, N.M. and De Rosa, M.A. Free vibrations of circular arches- A review, Journal of Sound and Vibration, 176(4), 1994, 433–458.
[30] Plaut, R.H. and Johnson, E.R. The effects of initial thrust and elastic foundation on the vibration frequencies of a shallow arch, Journal of Sound and Vibration, 178(4), 1981, 565–571.
[31] Nair, R.S. Buckling and vibration of arches and tied arches, Journal of Structural Engineering, 112(6), 1986, 1429–1440.
[32] Franciosi, C. Free vibrations of arches in presence of axial forces, Journal of Sound and Vibration, 120(3), 1988, 609–616.
[33] Lee, B.K., Lee, J.Y., Choi, K.M. and Lee, T.E. Free vibrations with general boundary condition, KSCE Journal of Civil Engineering, 6(4), 2002, 469–474.
[34] Lee, B.K., Lee, T.E. and Ahn, D.S. Free vibrations with inclusion of axial extension, shear deformation and rotatory inertia in cartesian coordinates, KSCE Journal of Civil Engineering, 8(1), 2004, 43–48.
[35] Bozyigit, B., Yesilce, Y. and Acikgoz, S. Free vibrations analysis of arch-frames using dynamic stiffness approach, Vibro-Engineering Procedia, 30, 2020, 72–78.
[36] Outassafte, O., Adri, A., El Khouddar, Y., Rifai, S. and Benamar, R. Geometrically non-linear free and forced vibration of a shallow arch, Journal of Vibro-engineering, 23(7), 2021, 1508–1523.

[37] Nie, Z., Ren, X., Yang, Y., Fu, C. and Zhao, J. Free vibration analysis of arches with interval-uncertain parameters, MDPI Applied Sciences, 13, 2023, Paper Id. 12391, 16.
[38] Lee, J.K., Yoon, H.M., Oh, S.J. and Lee, B.K. Free vibration of axially functionally graded Timoshenko circular arch, Periodica Polytecnica Civil Engineering, 60(2), 2024, 445–458, 6(4), 2002, 469–474.
[39] Meriam, J.L., Kraige, L.G. and Bolton, J.N. Engineering mechanics, Vol. 1, Statics, SI version, Chapter 4/5, 8th edition, John Wiley & Sons Singapore Pte. Ltd., 2016, 531.
[40] Horne, M.R. The elastic lateral stability of trusses, The Structural Engineer, May 1960, 147–155.
[41] Renton, J.D. Buckling of long, regular trusses, International Journal of Solids and Structures, 9, 1973, 1489–1500, Pai, S.S. and Chamis, C.C. Probabilistic progressive buckling of trusses, NASA TM-105162, 1991, 16.
[42] Fedoroff, A. and Kouhia, R. Out-of-[lane elastic buckling of truss beams, Structural Engineering and Mechanics, 45(3), 2013, 613–619.
[43] Brown, H.J., Green, P.S., Ryan, J.L. and Reigles, D.G. Analytical investigation of the stability and post-buckling behavior of large-scale truss assemblies, Proceedings of the Annual Stability Conference Structural Stability Research Council, Toronto, Canada, March 25–28 2014.
[44] Wattanamankong, N., Petchsasithon, A. and Dhirasedh, S. Analysis of lateral buckling of bar with axial force accumulation in truss, 2nd International Conference on Civil Engineering and Materials Science, IOP Conf. Series: Materials Science and Engineering, 216, 2017, Id Article 012037, 6.
[45] Dirbas, W. Eigenvalue buckling and static behavior of truss-beam structure, Global Scientific Journals, 10(1), 2022, 866–894.
[46] Korakas, N. Buckling of trusses with eccentric joints, MSc. Thesis in Civil Engineering, Delft University of Technology, The Netherlands, 2022, 142.
[47] Manguri, A., Saeed, N., Szczepanski, M. and Jankowski, R. Buckling and shape control of prestressable trusses using optimum number of actuators, 13(3838), 2023, 13. www.nature.com/scientificreports/.
[48] Ghent, E.D., Jr. Modal characteristics of a vibrating truss, MSc. thesis in Engineering Mechanics, University of Missouri at Rola, Missouri, 1966, 48.
[49] Adams, E. and Plum, M. Free and forced vibrations of trusses by Fourier decomposition, and homotopy methods for nonlinear matrix eigenvalue problems (1) methods, Journal of Mathematical Analysis and Applications, 272, 2002, 333–353.
[50] Lahe, A., Braunbrück, A. and Klauson, A. An exact solution of truss vibration problems, Proceedings of the Estonian Academy of Sciences, 68(3), 2019, 244–263.
[51] Liu, X., Zhao, Y., Zhou, W. and Banerjee, J.R. Dynamic stiffness method for exact longitudinal free vibration of rods and trusses using simple and advanced theories, Applied Mathematical Modelling, 104, 2022, 401–420.
[52] Hakim, G. and Abramovich, H. Homogenization of multiwall plates- an analytical, numerical and experimental study, Thin-Walled Structures, 185, 2023, ID article 110583, 17.
[53] Vaswani, H.P. Model analysis method for determining buckling load of rectangular frames, Experimental Mechanics, 1(8), 1961, 55–64.
[54] Look, J.-O. Buckling of rigid frames, M.Sc. in Civil Engineering, Department of Civil Engineering, Kansas State University, Manhattan, Kansas, USA, 1963, 46.
[55] Douglas, J.M. Buckling strength of frames under primary bending, M.Sc. in Civil Engineering, Department of Civil Engineering University of Windsor, Ontario, Canada, 1964, 58.
[56] Schilling, C.G. Buckling of one-story frames, Engineering Journal, American Institute of Steel Construction, 20, 1983, 49–57.
[57] Christodoulou, A.A. and Kounadis, A.N. Elastica buckling analysis of a simple frame, Acta Mechanica, 61, 1986, 153–163.
[58] Özmen, G. and Girgin, K. Buckling lengths of unbraced multi-story frame columns, Structural Engineering and Mechanics, 19(1), 2005, 55–71.

[59] Lechner, A. Flexural buckling frames according to the new EC3 rules- a comparative parameter study, in Proceedings of stability and ductility of structures, Camotim, D., Silvestre, N. and Dinis, P.B. Eds., Lisbon, Portugal, 6–8 September 2006, 8.
[60] Burkholder, M.C. Performance based analysis of a steel braced frame building with buckling restrained braces, M.Sc. in Architecture with a specialization in Architectural Engineering, Faculty of California Polytechnic State University, San Luis Obispo, California, USA, 78.
[61] Tiainen, T. and Heinisuo, M. Buckling length of frame length, Journal of Structural Mechanics, 51(2), 2018, 49–61.
[62] Galishnikova, V.V. and Pahl, P.J. Analysis of frame buckling without sideways classification, Structural Mechanics of Engineering Constructions and Buildings, 14(4), 2018, 299–312.
[63] Krystosik, P. On the columns buckling length of unbraced steel frames with semi-rigid joints, Archives of Civil Engineering, LXVII(1), 2021, 539–556.
[64] Hu, Y., Khezri, M. and Rasmussen, K.J.R. Analytical solutions for buckling of space frames subjected to torsional loadings, Thin-Walled Structures, 173, 2022, Id article 108965, 25.
[65] Urruzola, J. and Garmendia, I. Calculation of linear buckling load for frames modeled with one-finite -element beams and columns, Computation, 109, 2023, Id article 1060109, 17.
[66] Abbas, M.K., Naser, Z.A., Abdulabbas, A.I. and Alwash, N.A. Simplified approach for determining buckling load in portal frames: Elastic stability analysis, IOP Conference Series: Earth and Environmental Science, 1374, 2024, Id article 012081, 8.
[67] Gladwell, G.M.L. The vibration of frames, Journal of Sound and Vibration, 1(4), 1964, 402–425.
[68] Howson, W.P. and Williams, F.W. Natural frequencies of frames with axially loaded Timoshenko members, Journal of Sound and Vibration, 26(4), 1973, 503–515.
[69] Przybylski, J., Tomski, L. and Rozanow, M.G. Free vibration of an axially prestressed planar plane, Journal of Sound and Vibration, 189(5), 1996, 609–624.
[70] Goicoechea, H.E., Filipich, C.P. and Rosales, M.B. Natural vibrations of plane frames under compression through a power series solution, in Asociacion Argentina de Mecánica Computacional, Vol. XXXII, Garino, C.G.G., Mirasso, A.E., Storti, M.A. and Tornello, M.E. Eds., Mendoza, Argentina, 19–22 Nov. 2013, 3433–3448.
[71] Ratazzi, A.R., Bambill, D.V. and Rossit, C.A. Dynamic behavior of framed structures with an elastic internal hinge, Blucher Mechanical Engineering Proceedings of the 10th World Congress on Computational Mechanics, 9–13 July 2012, Sao Paulo, Brazil, 1(1), May 2014, 16.
[72] Shirokov, V.S., Kholopov, I.S. and Solovejv, A.V. Determination of the frequency of natural vibrations of a modular building, Procedia Engineering, 154, 2016, 655–661.
[73] Moore, A.J. Dynamic characteristics and wind-induced response of a tall building, Ph.D. thesis, School of Civil Engineering, The University of Sidney, June 2016, 238.
[74] Khandelwal, D., Chandrakar, V.K.S., Porey, P.D. and Tomar, P.S. Vibrational analysis of framed structures, International Journal for Scientific Research and Development, 5(8), 2017, Id article 2321-0613, 5.
[75] Das, O., Ozturk, H. and Gonenli, C. Free vibration and buckling analysis of parabolic frame structures, International Journal of Scientific Research and Management, 7(12), 2019, 295–306.
[76] Martin, H., Maggi, C., Piovan, M., De Rosa, A. and Gutbrod, N.M. Natural vibrations and instability of frames: Exact analytical solutions using power series, Engineering Structures, 252, 2022, Id article 113663, 11.
[77] Genel, E.Ö., Tüfekci, M. and Tüfekci, E. Free vibrations of special frame structures: Analytical modelling and solution, Journal of Vibration and Control, 29(19–20), 2023, 4492–4502.
[78] Wu, Z., Ma, Q., Mei, Y., Sun, Z., Wang, Y. and Li, B. A new method for analysing dynamic characteristics of generalized planar frame structures under complex conditions, International Journal of Structural Stability and Dynamics, 24(3), 2024, Id article 2450024, 27.
[79] Urruzola, J. and Garmendia, I. Improved FEM natural frequency calculation for structural frames by local correction procedure, Buildings, 14, 2024, Id article 1195, 19.

Appendix A: vibration of rings

The solutions for vibration of the rings follow the work presented by Timoshenko.[3] It is assumed a ring with a radius R (see Fig. A1) with the dimensions of the cross sections being small in comparison with its radius.

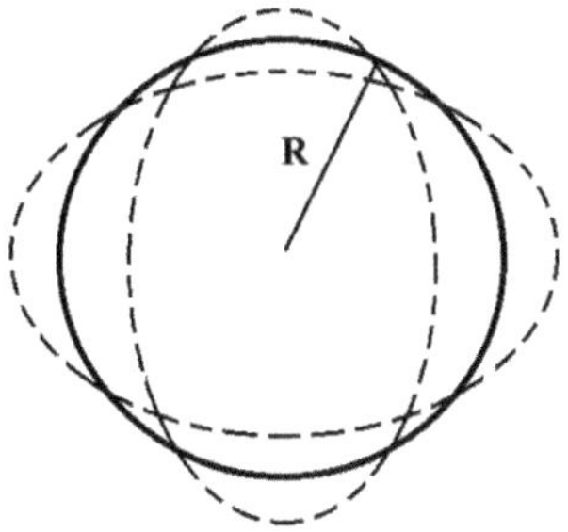

Fig. A1: Radial vibrations of a ring with radius R – the dashed lines are the horizontal and the vertical modes (not to scale).

First, the pure radial vibration will be presented. Assuming the cross section of the ring to be S and letting u to be the radial displacement, one can find the circumferential strain of the ring to be u/R. Applying the Lagrange's equation, the potential and the kinetic energies must be first determined.

The potential energy is then

$$V = \frac{1}{2}\int \sigma \cdot \epsilon d(\text{vol}) = \frac{1}{2}\int E\varepsilon \cdot \epsilon d(\text{vol}) = \frac{1}{2}E\frac{u^2}{R^2}S \cdot 2\pi R \tag{A.1}$$

while the kinetic energy is

$$T = m\frac{\dot{u}^2}{2} = \rho S \cdot 2\pi R\frac{\dot{u}^2}{2} \tag{A.2}$$

with ρ standing for the density volume of the ring and E its Young's modulus. Applying the Lagrange's equations leads to the equation of motion for the present case

$$\frac{d}{dt}\left(\frac{\partial T}{\partial \dot{u}}\right) - \frac{\partial T}{\partial u} + \frac{\partial V}{\partial u} = \rho S \cdot 2\pi R\ddot{u} + S \cdot 2\pi RE\frac{u}{R^2} = 0 \tag{A.3}$$

or

$$\ddot{u} + \frac{E}{\rho R^2}u = 0 \tag{A.4}$$

3 Timoshenko, S., Vibration problems in Engineering, Ch. 6, 2nd Ed., 5th printing, New York, D. van Nostrand Comp., Inc. 250 4th Avenue, USA, 1937, 470p.

The solution for eq. (A.4) is

$$u(t) = A\cos\tau t + B\cos\tau t \tag{A.5}$$

where $\tau = \sqrt{\frac{E}{\rho R^2}} ==> f = \frac{1}{2\pi}\sqrt{\frac{E}{\rho R^2}}$ is the natural frequency of the ring.

Note that a circular ring might also have other vibrations like longitudinal vibrations of rods. Assuming the number of wavelengths to the ring circumference to be j, the extensional-type frequencies can be calculated using the following formula

$$f = \frac{1}{2\pi}\sqrt{\frac{E}{\rho R^2}\sqrt{1+j^2}} \tag{A.6}$$

The torsional frequency of a ring, in which its cross section rotates out of plane by a small angle φ, can be written as

$$f = \frac{1}{2\pi}\sqrt{\frac{E}{\rho R^2}\frac{I_x}{I_p}} \tag{A.7}$$

where I_x and I_p are the moment of inertia of the ring's cross section relative to x axis (in the radial direction) and the polar moment of inertia, respectively. Note that for a circular cross section, the higher modes can be obtained from the following equation

$$f = \frac{1}{2\pi}\sqrt{\frac{E}{2\rho R^2}\sqrt{1+j^2}} \tag{A.8}$$

The flexural vibrations of a ring were also dealt with by Timoshenko in his book from 1937.[2]

He considered in-plane flexural vibrations without extension deriving the following formula for the natural frequencies of the ring

$$f_n = \frac{1}{2\pi}\sqrt{\frac{EI}{\rho S R^4}\frac{n^2(1-n^2)^2}{1+n^2}} \tag{A.9}$$

where I is a moment of inertia of the ring's cross section with respect to a principal axis at right angles to the ring's plane. Note that for $n = 1$ we get $f_n = 0$; namely, the ring vibrates as a rigid body. The fundamental mode of flexural vibration will be obtained for $n = 2$. For the case of bending and twisting of the cross section, eq. (A.9) is modified to be

$$f_n = \frac{1}{2\pi}\sqrt{\frac{EI}{\rho S R^4}\frac{n^2(1-n^2)^2}{1+n^2+\nu}} \tag{A.10}$$

where υ is the Poisson's ratio of the ring's material.

17 Additional aspects for the buckling and natural frequency relationship of thin-walled structures

17.1 Rayleigh approach

The equation of motion for an isotropic rod under compressive loading applied at its ends is:

$$EIv_{,xxxx} + N_x v_{,xx} + PSv_{,tt} = 0 \tag{17.1}$$

with v being the rod out-of-plane displacement, N_x the axial compressive load, S its cross section and P the mass per unit volume. Nondimensionalizing eq. (17.1) by EI gives:

$$v_{,xxxx} + k^2 v_{,xx} + c^2 v_{,tt} = 0 \tag{17.2}$$

where

$$k^2 = \frac{N_x}{\mathrm{EI}};\ c^2 = \frac{\mathrm{PS}}{\mathrm{EI}} \tag{17.3}$$

Instead of directly solving the differential equation presented by eq. (17.2), harmonic vibrations of the rod are assumed yielding the following expression:

$$v(x,t) = V(x)\sin\omega t \tag{17.4}$$

Substituting eq. (17.4) into eq. (17.2) yields

$$V_{,xxxx} + k^2 V_{,xx} - c^2\omega^2 V = 0 \tag{17.5}$$

Taking the case of uniform rod (EI = constant) on pinned–pinned ends, namely

$$V(0) = V(0)_{,xx} = V(L) = V(L)_{,xx} \tag{17.6}$$

The general solution for eq. (17.5) is:

$$V = A\sin\frac{\pi x}{L} \tag{17.7}$$

Note that eq. (17.7) satisfies also eq. (17.6).

Substituting eq. (17.7) into eq. (17.5) yields:

$$\omega^2 = \frac{\pi^2}{\rho A L^2}\left[N_{xE} - N_x\right] \text{ where } N_{xE} = \frac{\pi^2 EI}{L^2} \tag{17.8}$$

Nondimensionalizing both sides of eq. (17.8) by P_E gives:

https://doi.org/10.1515/9783111621104-017

$$\left(\frac{\omega}{\omega_1}\right)^2 = 1 - \frac{N_x}{N_{xE}} \tag{17.9}$$

where

$$\omega_1^2 = \frac{\pi^4 EI}{CSL^4} \tag{17.10}$$

For a clamped–clamped rod, the supposed shape of its buckling would be:

$$V(x) = \bar{A}\left(1 - \cos\frac{2\pi x}{L}\right) \tag{17.11}$$

Equation (17.11) satisfies the rod clamped–clamped ends, namely

$$V(0) = V_{,x}(0) = V(L) = V_{,x}(L) = 0 \tag{17.12}$$

Rayleigh's approach is next applied to yield an upper bound for the fundamental frequency of the compressed rod, using the mode expression presented by eq. (17.7).

Rayleigh's principle demands that the total potential energy would be equal to the kinetic energy of the vibrating rod (an isotropic material is also assumed) which yields the following expression, namely

$$\frac{1}{2}\int_0^L EI(\Phi_{,xx})^2 dx - \frac{N_x}{2}\int_0^L (\Phi_{,x})^2 dx = \frac{\omega^2}{2}\int_0^L \mathrm{PS}\Phi^2 dx \tag{17.13}$$

where Φ is the exact mode for a pinned–pinned loaded rod having a length L.

Note that eq. (17.13) is correct only if Φ is the compressed rod's vibration mode. If this does not hold, the equality presented by eq. (17.13) would transform into the following inequality

$$\int_0^L (V_{,xx})^2 dx - \frac{N_x}{EI}\int_0^L (V_{,x})^2 dx \geq \frac{\omega^2 \mathrm{PS}}{EI}$$

$$or \tag{7.14}$$

$$1 \geq \frac{\omega^2 c^2}{\int_0^L (V_{,xx})^2 dx} + \lambda^2 \frac{\int_0^L (V_{,x})^2 dx}{\int_0^L (V_{,xx})^2 dx}$$

where

$$\lambda^2 = \frac{N_x}{EI}, \quad c^2 = \frac{\mathrm{PS}}{EI} \tag{17.15}$$

As stated above, eq. (17.14) would provide an upper-bound approximation for the relation between the critical buckling load and the natural frequency of the compressed rod.

The normalized amplitude $\bar{A}$ of the assumed mode $V(x)$ for the clamped–clamped rod has the following form:

$$\int_0^L \bar{A}^2\left(1-\cos\frac{2\pi x}{L}\right)^2 dx = 1 \text{ and } \bar{A} = \sqrt{\frac{2}{3L}} \tag{17.16}$$

Inserting eq. (17.16) into eq. (17.14), the following inequality can be written as

$$1 \geq \frac{\omega^2 c^2}{\frac{16\pi^4}{3L^4}} + \lambda^2\frac{\frac{4\pi^2}{3L^2}}{\frac{16\pi^4}{3L^4}} \Rightarrow 1 \geq \frac{3\text{PS}\omega^2 L^4}{16\pi^4 EI} + \frac{N_X L^2}{4\pi^2 EI} \tag{17.17}$$

However, the buckling load and the first natural frequency for a clamped–clamped rod are

$$N_{xcr} = \frac{4\pi^2 EI}{L^2} \quad \omega_{n=1}^2 = (22.4)^2\frac{EI}{N_x L^4} \tag{17.18}$$

Using eq. (17.18), the inequality in eq. (17.17) can be written as

$$1 \geq 0.96582687\frac{\omega^2}{\omega_{n=1}^2} + \frac{N_x}{N_{xcr}} \quad \text{or} \quad 1.03538225\left(1-\frac{N_x}{N_{xcr}}\right) \geq \frac{\omega^2}{\omega_{n=1}^2} \tag{17.19}$$

The three graphs presented in Fig. 17.1 represent the nondimensional frequency squared versus the nondimensional axial compression for the exact mode of pinned–pinned rod (S-S exact on the graph), the exact mode clamped–clamped rod (C-C-exact) and the approximated mode clamped–clamped rod (C-C approx.). Note that for the C-C rod, we have an upper-level curve reaching above $N_x/N_{xcr} = 1$. For the pinned–pinned rod case, a linear relationship is drawn in Fig. 17.1, due to the fact that it uses the exact mode of a compressed rod.

17.2 Southwell's plot

Figure 17.2 presents schematically the behavior of a column under axial compression. As can be seen, there is a great difference between an ideal column and a tested one. The question is what valid comparison can be made between theory and experiment.

Ayrton and Perry [1] were the first to introduce imperfection in the loading and in the form of the column, yielding the following relation [4]:

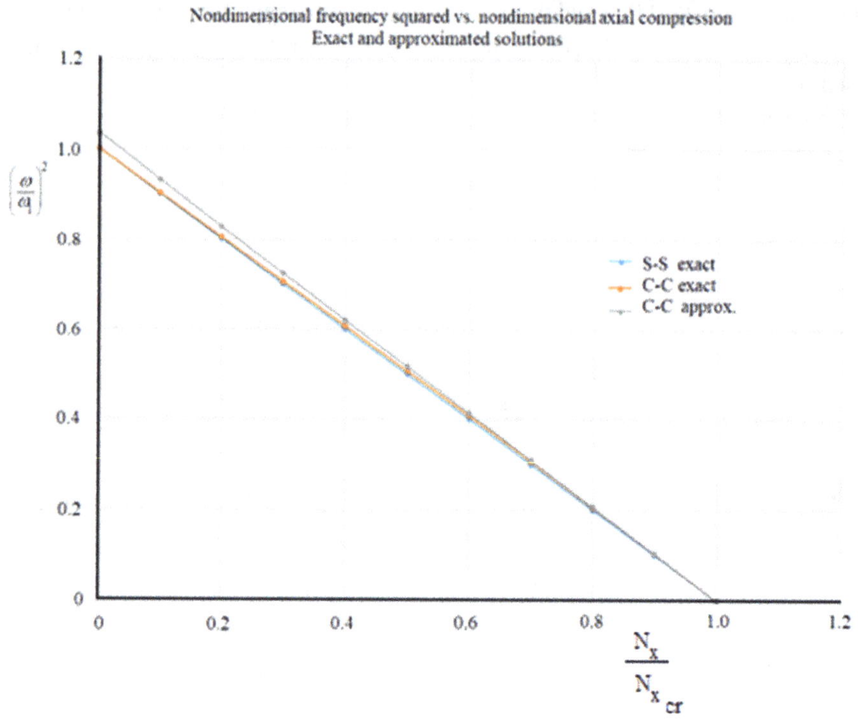

Fig. 17.1: Nondimensional circular frequency squared versus nondimensional axial compression for the assumed mode approach – a thin-walled rod.

$$y_1 = \frac{C}{1 - \frac{P}{P_{\text{Euler}}^{\text{cantilever}}}} \quad \text{where} \quad P_{\text{Euler}}^{\text{cantilever}} = \frac{\pi^2 EI}{4L^2} \tag{17.20}$$

where C is the initial geometric imperfection of a loaded cantilever strut; y_1 is the ordinate of the strut center line in the deformed configuration; E is the Young's modulus; I is the second moment of cross-sectional area; L is the length of the strut; and P is the axial compressive load.

Based on eq. (17.20), the central deflection of any strut on any boundary conditions is

$$\delta = \frac{C \cdot P}{P_{\text{Euler}} - P} \Rightarrow \frac{1}{\delta} = \frac{P_{\text{Euler}}}{C \cdot P} - \frac{1}{C}$$

$$\text{if} \quad x = \frac{1}{\delta}, y = \frac{1}{P} \quad \text{then} \tag{17.21}$$

$$x = \frac{P_{\text{Euler}}}{C} y - \frac{1}{C} \quad or \quad y = \frac{C}{P_{\text{Euler}}} x + \frac{1}{P_{\text{Euler}}}$$

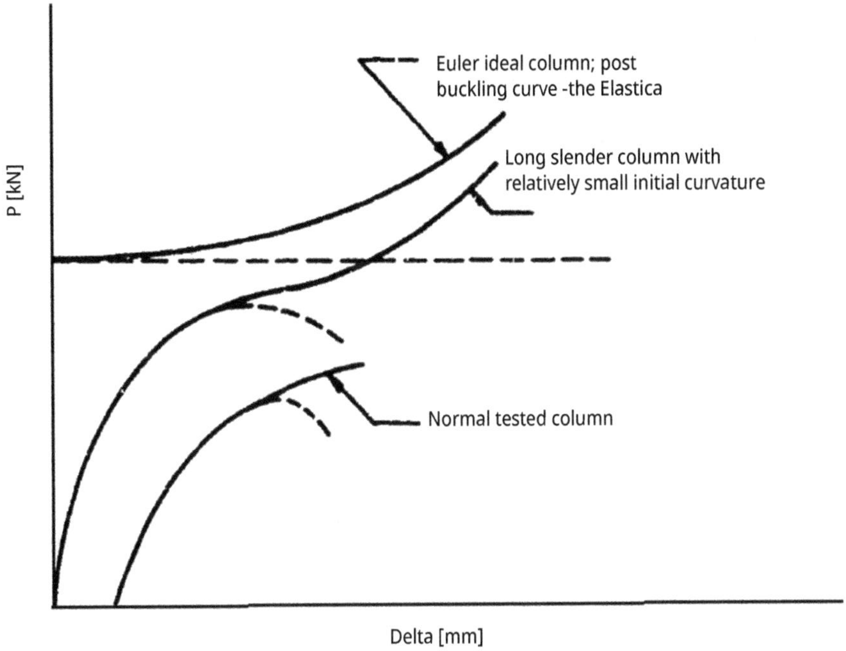

Fig. 17.2: Schematic behavior of compressed columns.

Then the inverse of the central deflection $1/\delta$ is linearly connected to $1/P$ with a slope C/P_{Euler}.

Another formulation stemming from eq. (17.20) is

$$\begin{gathered}\delta=\frac{C\cdot P}{P_{\text{Euler}}-P}\Rightarrow\delta\frac{(P_{\text{Euler}}-P)}{P}=C\\ \text{if}\quad \bar{x}=\frac{\delta}{P},\bar{y}=\delta\quad\text{then}\\ P_{\text{Euler}}\cdot\bar{x}-\bar{y}=C\quad\text{or}\quad\bar{y}=P_{\text{Euler}}\cdot\bar{x}-C\end{gathered}\tag{17.22}$$

It is clear from eq. (17.22) that the slope of the linear equations is the Euler critical load, while the intersection with the $\bar{y}$ axis (namely δ axis) is the effective initial geometric imperfection.

Southwell [2] also treated the question posed at the beginning of the chapter. He assumed a simply supported compressed strut, as presented in Fig. 17.3.

The governing equation, for the strut in the presence of initial geometric imperfection, y_0, is

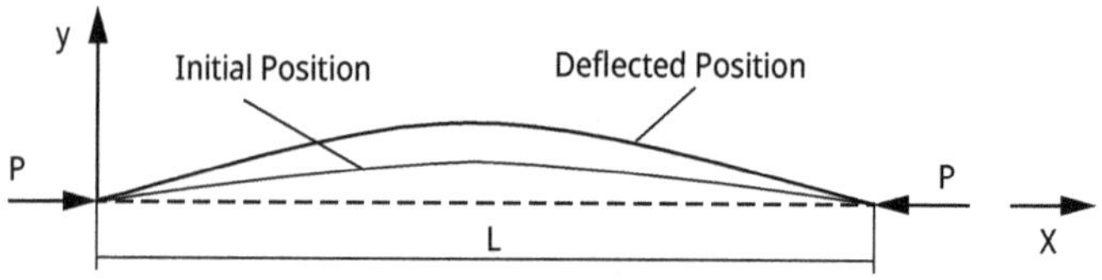

Fig. 17.3: Schematic strut assumed by Southwell [2].

$$EI\left(\frac{d^2y}{dx^2}-\frac{d^2y_0}{dx^2}\right)+Py=0\Rightarrow\frac{d^2y}{dx^2}+\lambda^2y=\frac{d^2y_0}{dx^2}$$
$$\text{where}\quad \lambda^2=\frac{P}{EI} \tag{17.23}$$

Assume that the initial imperfection and the lateral deflection have the following forms:

$$y=\sum_{m=1}^{\infty}Y_m\sin\frac{m\pi x}{L}\quad\text{and}\quad y_0=\sum_{m=1}^{\infty}\bar{Y}_m\sin\frac{m\pi x}{L} \tag{17.24}$$

Substitution of eq. (17.24) into eq. (17.23) leads to

$$Y_m=\frac{\bar{Y}_m}{1-\frac{P}{EI(m^2\pi^2/L^2)}}\rightarrow\frac{Y_m}{\bar{Y}_m}=\frac{1}{1-\frac{P}{m^2P_{\text{Euler}}}} \tag{17.25}$$

The deflection of the strut would be

$$y=\sum_{m=1}^{\infty}\bar{Y}_m\left(\frac{1}{1-\frac{P}{m^2P_{\text{Euler}}}}\right)\sin\frac{m\pi x}{L} \tag{17.26}$$

The maximal deflection would appear at the middle of the strut; therefore, its expression is given by substituting $x=L/2$ in eq. (17.26), yielding

$$\delta\equiv Y_1-\bar{Y}_1-Y_3+\bar{Y}_3+\cdots \tag{17.27}$$

As Southwell wrote in his manuscript [2] ". . . if P is a fairly considerable fraction of P_{Euler}," eq. (17.27) can be rewritten as

$$\delta=\frac{\bar{Y}_1}{1-\frac{P}{P_{\text{Euler}}}}-\bar{Y}_1 \tag{17.28}$$

Equation (17.28) can be modified to the following expression

$$\delta = \frac{\bar{Y}_1}{1 - \frac{P}{P_{\text{Euler}}}} - \bar{Y}_1 = \bar{Y}_1\left(\frac{1}{1 - \frac{P}{P_{\text{Euler}}}} - 1\right) = \bar{Y}_1 \frac{\frac{P}{P_{\text{Euler}}}}{1 - \frac{P}{P_{\text{Euler}}}}$$
$$\Rightarrow \delta\frac{P_{\text{Euler}}}{P} - \delta = \bar{Y}_1 \tag{17.29}$$

Southwell used eq. (17.29) to plot the ratio of the displacement to the applied load, δ/P, versus the displacement, δ. Donnell [3] used a third derivative of eq. (17.28) and plotted the load versus the ratio of the load to displacement. The differences between Southwell, the Donnell and the Ayrton and Perry representations are clearly presented in Fig. 17.4.

Note that, although developed for a strut under axial compression, scientists are using the Southwell method also for other types of loading, leading to buckling as described in [5, 6].

17.3 Novel excitation methods

A good excitation method would enable us to acquire the natural frequencies of a vibrating structure and lead to consistent results when applying VCT (vibration correlation technique) (see also Chapter 11).

Normally, an electromagnetic shaker (Fig. 17.5) would provide the necessary driving force to excite the specimen, as a function of frequency generator [7]. The response of the specimen can be recorded using a microphone and the resonance can be detected using Lissajous curves, with the excitation voltage and the specimen response voltage being supplied to the two axes of an oscilloscope (see Fig. 17.6).

Exciting the specimen using a loudspeaker would yield an ellipse, while using a shaker the Lissajous curve would be an eight (Fig. 17.6b), due to the "push–pull" movement of the shaker which is twice the response frequency.

Another important topic for the successful application of VCT is the mode shapes of the tested specimen. When using a microphone, its voltage can be used to draw the mode shape of the vibrating specimen. Using a single accelerometer to detect the response of the vibrating structure will not enable the recording of the modes, only the vibrating frequency.

A novel excitation method is the use of modal hammer as described in [8]. The method uses a special dedicated hammer to slightly impact the specimen, and its response is recorded using an accelerometer bonded on the tested specimen and connected to NI DAQ system (see Fig. 17.7). The excitation system is composed of an impact hammer (from National Instruments – NI), a single accelerometer and an NI DAQ system with a Me Scope Modal Software, to detect the excited natural frequencies and their respective mode shapes of the tested panel. The output is in the form of

The basic generic equation: $\qquad \delta = \dfrac{C \cdot P}{P_{Euler} - P}$

where

δ = mid span elastic deflection; $\qquad P$ = applied compressive load;

C = constant; $\qquad\qquad\qquad P_{Eur} = \dfrac{\pi^2 EI}{L^2}$ = Euler buckling load.

Ayrton&Perry (1886)

Formula: $\dfrac{1}{\delta} = \left[\dfrac{P_{Euler}}{P} - 1\right]\dfrac{1}{C}$ $\quad$ Variables: $\dfrac{1}{\delta}$ and $\dfrac{1}{C}$

Southwell (1932)

Formula: $\delta\left[\dfrac{P_{Euler}}{P} - 1\right] = C$ $\quad$ Variables: $\dfrac{\delta}{P}$ and δ

Donnell (1938)

Formula: $P_{Euler} - P = C\dfrac{P}{\delta}$ $\quad$ Variables: P and $\dfrac{r}{\delta}$

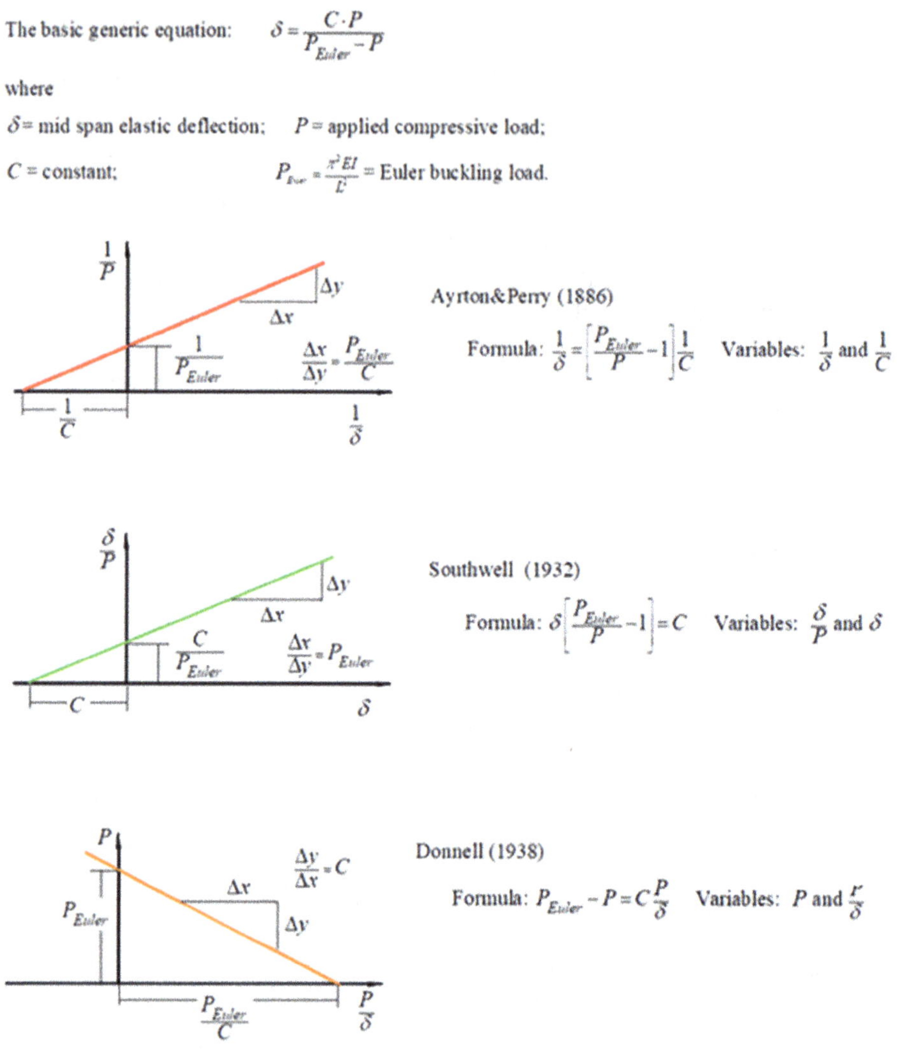

Fig. 17.4: Schematic linear representation strut of Ayrton and Perry [1], Southwell [2] and Donnell [3] models.

frequency response function (FRF) as shown in Fig. 17.8. Searching peaks on the imaginary FRF chart, see Fig. 17.8, would define the natural frequencies of the specimen.

The most sophisticated way to excite a given structure is the use of a laser as described in [9]. The test series were performed on various laminated composite shells at the laboratory of Aerospace Structures and Materials of the Delft University of Technology in the Netherlands. The test setup consists of an MTS 3,500 kN servo-hydraulic loading rig and the POLYTEC PSV 500 scanning laser vibrometer. The POLY-

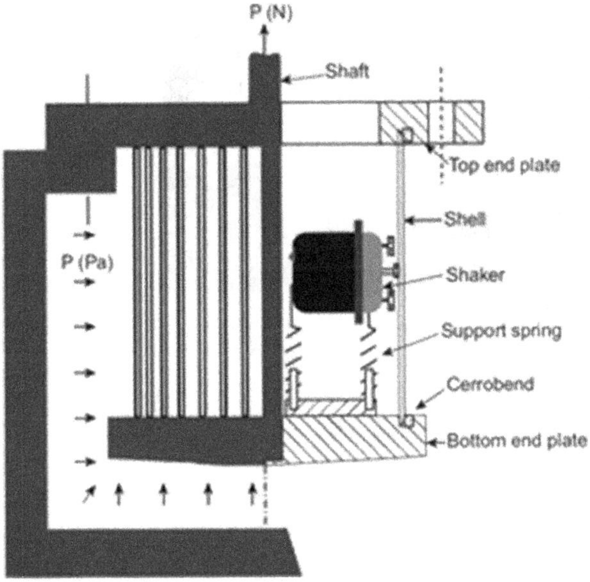

Fig. 17.5: An electromagnetic shaker.

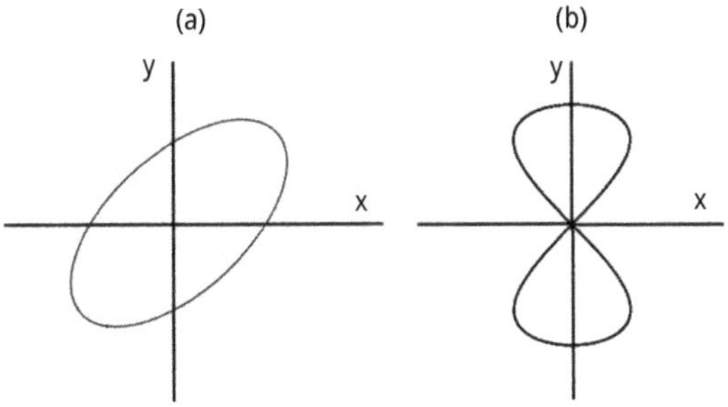

Fig. 17.6: Lissajous curves: (a) an ellipse and (b) an eight curve.

TEC setup is capable to exactly measure the natural frequencies and the vibration mode shapes of the tested cylinders.

The POLYTEC system has a laser scanning head, a data acquisition unit and a control unit. The excitation was done using a loudspeaker positioned at the rear part of the tested specimen and controlled via an amplifier by the POLYTEC system within a predefined frequency spectrum. The measurements were conducted using a 100e400 Hz frequency sweep, the range being fixed to include most of the low natural frequencies. The laser vibrometer can cover approximately 160° of the shell surfaces with high fidelity grid containing 450 points. After calibration of the scanning process, a typical measurement would include the following steps: (a) the axial compressive load level is applied, (b) the loudspeaker excites the laminated composite shell for a

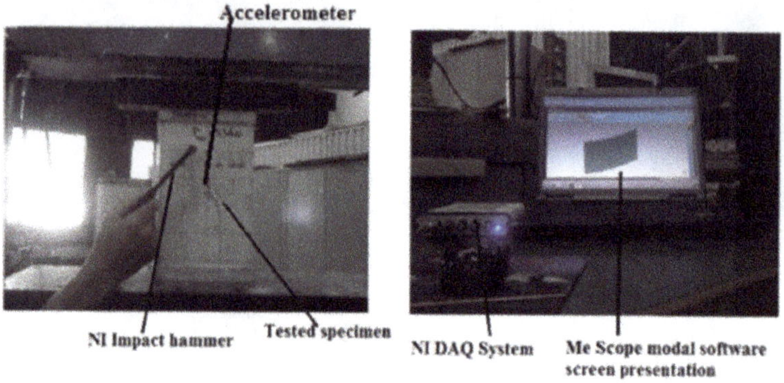

Fig. 17.7: The modal hammer excitation test setup (from [8]).

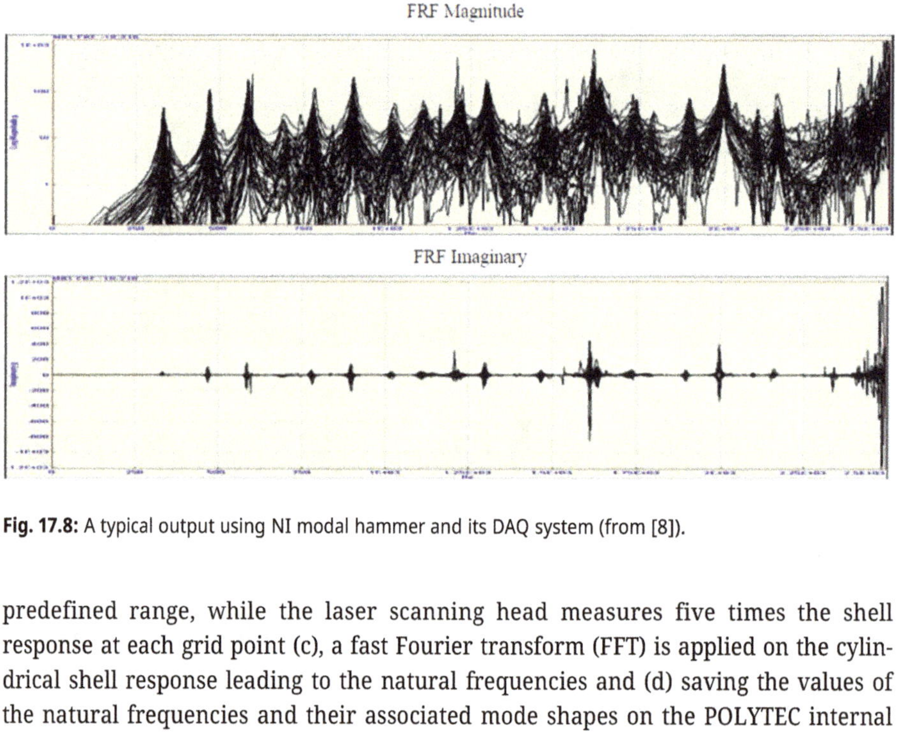

Fig. 17.8: A typical output using NI modal hammer and its DAQ system (from [8]).

predefined range, while the laser scanning head measures five times the shell response at each grid point (c), a fast Fourier transform (FFT) is applied on the cylindrical shell response leading to the natural frequencies and (d) saving the values of the natural frequencies and their associated mode shapes on the POLYTEC internal storage. Further details can be found in Ref. [9].

References

[1] Ayrton, W.E. and Perry, J. On struts, The Engineer, 62, 10 December, pp. 464–465 and 14 December pp. 513–514, 1886.

[2] Southwell, R.V. On the analysis of experimental observations in problems of elastic stability, Proceedings of the Royal Society, A, 135, 1932, 601–616.

[3] Donnell, L.H. On the application of Southwell's method for the analysis of buckling tests, in Stephen Timoshenko 60th Anniversary Volume, MacMillan Book Company, 1938, 27–38.

[4] Horton, W.H., Cundari, F.L. and Johnson, R.W., Applicability of the Southwell plot to the interpretation of test data obtained from stability studies of elastic column and plate structures, USAAVLABS Technical Report 69–32, 1971, 11.

[5] Singhal, M.K., Studies on the elastic stability of bodies, Ph.D. thesis, presented to the Faculty of the Division of Graduate Studies and Research, School of Aerospace Engineering, Georgia Institute of Technology, August 1973, 159.

[6] Kalkan, İ. Application of Southwell method on the analysis of lateral torsional buckling tests on reinforced concrete beams, International Journal of Engineering, Research and Development, 2(1), 2010, 58–66.

[7] Abramovich, H., Singer, J. and Weller, T. Repeated buckling and its influence on the geometrical imperfections of stiffened cylindrical shells under combined loading, International Journal of Non-Linear Mechanics, 37, 2002, 577–588.

[8] Abramovich, H., Govich, D. and Grunwald, A. Buckling prediction of panels using the vibration correlation technique, Progress in Aerospace Sciences, 78, 2015, 62–73.

[9] Labans, E., Abramovich, H. and Bisagni, C. An experimental vibration-buckling investigation on classical and variable angle tow composite shells under axial compression, Journal of Sound and Vibration, 449, 2019, 315–329.

18 Nonlinear numerical and experimental investigation of the behavior of rectangular multiwall plates under pressure

18.1 Introduction

The problem of large deflection of plates has attracted much attention since the end of the nineteenth century, due to its technical importance. Unfortunately, this problem has been found to be difficult to solve. The difficulties raised at early stage with Föppl equations for membrane large deflections [1], where no closed-form solution was found for the rectangular membrane case. In 1910 this equations set was enhanced by Theodor von Kármán [2], to include the bending resistance of plates.

The Föppl–von-Kármán equations set has challenged along the years many researchers. Nevertheless, only approximate solutions were developed, most of which are rather difficult to implement.

It was August Föppl himself to suggest an approximate approach. This approach is mentioned in Timoshenko [3], which mentions Föppl suggestion appearing in [4]. The approach is that the transverse distributed load q on the plate can be separated into two parts $q = q_1 + q_2$. The first part q_1 is balanced by the plate's bending and shearing resistance, calculated through the plate small deflection linear theory. The second part q_2 is balanced by the large deflections in-plane membrane forces only. Using the midpoint deflection w, this approximation is written as:

$$q = q_1 + q_2 = A \cdot w + B \cdot w^3 \tag{18.1}$$

The plate's small deflection coefficient A has been calculated in many previous studies, most of them using summation of Fourier series. The large deflection coefficient B, however, has no exact solution as it is ascribed to the difficult Föppl's membrane problem [1].

This expression (18.1) for the load–deflection behavior is quoted by many sources, such as Timoshenko [3] for square isotropic plates, Ugural [5], Wang and El-Sheikh [6] and many others.

One of them is Riber [7] who had suggested a "combined analytical solution" to find the constants in eq. (18.1). He used an energy method to obtain rather complex expressions for the coefficients A and B (see eq. (18.1)). He also presented simplified expressions for the constant B, but with some internal inconsistencies. Riber [7] as-

Note: The chapter is based on the manuscript: Hakim, G. And Abramovich, H., Multiwall rectangular plates under transverse pressure – A non-linear experimental and numerical study, Materials 2023, 16, Issue 5, ID article 2041, 27p.under a MDPI CC-BY license.

https://doi.org/10.1515/9783111621104-018

sumed in-plane immovable edges, which is not the case to be presented in the present study – see the BCs (boundary conditions) discussion later in the paper. Nevertheless, his B equation has inspired the presentation of a better expression for the coefficient B, later in the present chapter.

Other authors [8–15] present similar results. A detailed review for these references can be found in [16].

During the literature survey of the present chapter, several sources were found, referring to tests and calculation methods of plate's large deflection. To compare the results of these papers, it was necessary to normalize the various data to a common comparable structure. The structure was a thin square isotropic plate with movable edges and evenly distributed transverse load. The coefficients of eq. (18.1), a and B were calculated, considering the plate dimensions and the material properties. The result of this comparison has shown a considerable variability of the coefficient B. This variability was unexpected since most of the data were based on real laboratory tests that should respond in a similar way. This may demonstrate the fact that it is not easy to correctly measure this particular property. A full description of the comparison with a suggested explanation is presented in Appendix B.

18.2 Multiwall plates

18.2.1 Basic concepts

The structure of multiwall plates is two thin face sheets separated by an internal structure of ribs and walls. The plate is usually produced by extrusion, in which a melted material is pressed through a die with the required shape. The materials used are aluminum and various plastics. The result is a thick endless plate with a fixed cross section shape along the extrusion direction and width according to the equipment size. The plate is then cut to the desired length.

The plate is made of polycarbonate (PC), a tough transparent plastic. A typical 16 mm PC plate can be seen in Fig. 18.1.

The main application of PC multiwall plates is the glazing of architectural spaces, where both natural light and weather protection are required. As a result, the plates can be exposed to wind and snow loads which they safely resist.

Currently, no publication describes the general performance of these plates, except manufacturer's data sheets, which are very limited to specific products and specific applications. The available publications about latticed structures relate to specific shapes such as triangles and trapezoids and not a general approach as presented here, which can be considered novel.

The available approximated solutions for large deflections of plates are generally rather complicated and, in many cases, involve computational process which is not straightforward for engineers. Also, the nonlinear nature of load–deflection curves is

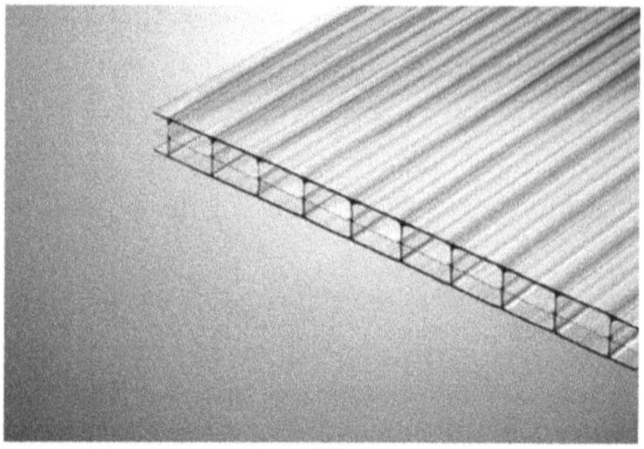

Fig. 18.1: A typical 16 mm multiwall PC plate.

not easily represented in these solutions. Most of the research works already done do not cover the full complexity of the multiwall plates, which are shear deformable and orthotropic. Therefore, an engineer who needs to design a system with multiwall plates will probably face serious difficulties which, it is hope, can be supplied by the present advanced study.

To calculate the multiwall plate response to distributed load, it is necessary to know the plate's equivalent elastic properties, its dimensions (length–width) and the boundary conditions. Looking at the multiwall structure, it is obvious that the plate is orthotropic for both bending and tension, and that its cross section is transverse shear deformable. In the present study, it is assumed that all necessary equivalent elastic properties are already known. One should note that a procedure to obtain these equivalent properties of the plate is presented in Hakim and Abramovich [17].

18.2.2 The coordinate system

The coordinates system is defined as presented in Fig. 18.2, where axis x is the extrusion direction and z axis is normal to the plate surface.

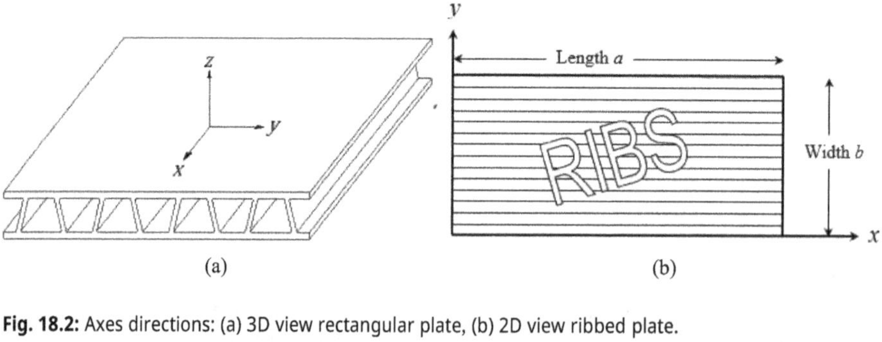

Fig. 18.2: Axes directions: (a) 3D view rectangular plate, (b) 2D view ribbed plate.

The original location of the axes may be set to any convenient place, as displayed in Fig. 18.2.

18.2.3 Boundary conditions (BCs)

For small deflection analysis, the assumption is that all in-plane stress, strains and deflections are negligible. Therefore, the BCs here ignore the in-plane conditions. The most used BCs are: free (F) – no restrictions, simply supported (S) – no z deflection but free rotation (no bending moments) and clamped (C) – no z deflection and no rotations (zero slope). S and C are the two extremes of the more complicated BC – flexible rotation support which is rarely used.

For large deflection analysis, the in-plane BCs must be considered. The two most common BCs are: immovable (I) – the plate edge is fixed to the support, and movable (M) – the plate edges are allowed to move. It is necessary to specify both in-plane movement directions – normal to the edge and parallel to the edge. The BC used later here is SSSS-M, in which the four S stands for the four plate sides simply supported and the M for the movable edges in both normal and parallel directions.

The movable (M) condition requires additional attention. When the Föppl's approximation is used, the plate in a large deflection regime is a membrane. Its deflection on movable boundary conditions should be then calculated. However, a well-known property of a membrane is that it cannot sustain in-plane compression forces, as it immediately wrinkles. However, a real plate does resist compression, as it has a bending rigidity and therefore a membrane with movable edges, which may have compression stress, has to be analyzed. Mathematically, it is possible (with the known difficulties of Föppl equations), but other practical problem would appear. Since this transversely loaded movable membrane is not a common case in the literature and perhaps even physically not possible, no previous scientific papers that would suggest a possible solution were found. Nevertheless, an expression is suggested later in the present chapter.

18.2.4 Methods used to define the coefficients *A* and *B* in eq. (18.1)

18.2.4.1 Analytical prediction for the coefficients *A* and *B*

The expressions that would predict the values for the coefficients A and B are next displayed. The variables used are:

- a [m] Plate length
  b [m] Plate width
  h [m] Plate thickness
  q [Pa] Distributed load
  w [m] Midpoint deflection
  m, n Summation indices

The plate equivalent properties are assumed to be known a priori:
- D_x [Nm] Plate x-direction bending rigidity
  D_y [Nm] Plate y-direction bending rigidity
  D_{xy} [Nm] Plate twist rigidity
  ν_x^b, ν_y^b Bending Poisson's ratios
  S_x, S_y [N/m] Transverse shear rigidity in x and y directions
  E_x^t, E_y^t [Pa] Equivalent plate tension E moduli in x and y direction
  ν_x^t, ν_y^t Tension Poisson's ratios

The x, y axes origin is set as shown in Fig. 18.2(b) – at the plate left corner.

18.2.4.2 Expression for coefficient *A* (small deflection cases)

The expression in eq. (18.2) was derived using the Libove and Batdorf NACA Report No. 899 [18].

$$w = \frac{q}{A} = \frac{16q}{\pi^6} \sum_{n=1,3,5\ldots}^{\infty} \sum_{m=1,3,5\ldots}^{\infty} \frac{\begin{bmatrix} \pi^4 K_8 \left(\frac{m}{a}\right)^4 + \pi^4 K_9 \left(\frac{m}{a}\right)^2 \left(\frac{n}{b}\right)^2 + \pi^4 K_{10} \left(\frac{n}{b}\right)^4 \\ -\pi^2 K_{11} \left(\frac{m}{a}\right)^2 - \pi^2 K_{12} \left(\frac{n}{b}\right)^2 + K_{13} \end{bmatrix} (-1)^{\frac{m+n}{2}-1}}{mn \begin{bmatrix} -\pi^2 K_1 \left(\frac{m}{a}\right)^6 - \pi^2 K_2 \left(\frac{m}{a}\right)^4 \left(\frac{n}{b}\right)^2 - \pi^2 K_3 \left(\frac{m}{a}\right)^2 \left(\frac{n}{b}\right)^4 \\ -\pi^2 K_4 \left(\frac{n}{b}\right)^6 + K_5 \left(\frac{m}{a}\right)^4 + K_6 \left(\frac{m}{a}\right)^2 \left(\frac{n}{b}\right)^2 + K_7 \left(\frac{n}{b}\right)^4 \end{bmatrix}} \tag{18.2}$$

with

$$K_1 = \frac{D_{xy}D_x}{2S_y};\quad K_2 = \frac{D_{xy}D_x}{2S_x} + \frac{D_xD_y - D_{xy}D_x\nu_y^b}{S_y};\quad K_3 = \frac{D_{xy}D_y}{2S_y} + \frac{D_xD_y - D_{xy}D_x\nu_y^b}{S_x}$$

$$K_4 = \frac{D_{xy}D_y}{2S_x};\quad K_5 = -D_x;\quad K_6 = -2\left[D_{xy}\left(1-\nu_x^b\nu_y^b\right) + D_x\nu_y^b\right];\quad K_7 = -D_y$$

$$K_8 = -\frac{D_{xy}D_x}{2S_xS_y};\quad K_9 = -\frac{D_xD_y - D_{xy}D_x\nu_y^b}{S_xS_y};\quad K_{10} = -\frac{D_{xy}D_y}{2S_xS_y} \tag{18.3}$$

$$K_{11} = \frac{D_{xy}\left(1-\nu_x^b\nu_y^b\right)}{2S_y} + \frac{D_x}{S_x};\quad K_{12} = \frac{D_{xy}\left(1-\nu_x^b\nu_y^b\right)}{2S_x} + \frac{D_y}{S_y};\quad K_{13} = -\left(1-\nu_x^b\nu_y^b\right)$$

The complete numerical analysis description is presented in appendix A.

18.2.4.3 Expression for coefficient *B* (large deflection cases)

As presented in eq. (18.1), coefficient B describes the plate's large deflection response, with the plate being considered as a thin membrane. Since normal membranes cannot have a movable (M) BC, it is complicated to find previous publications displaying an expression for the membrane deflection. Yet, Wang and El-Sheikh [6] have analyzed an isotropic rectangular plate and have presented an approximated expression for $q = Aw + Bw^3$ for M BCs, namely

$$q = \frac{\pi^6}{64}\left[4Dw\left(\frac{1}{a^2} + \frac{1}{b^2}\right)^2 + \frac{Ehw^3}{4}\left(\frac{1}{a^4} + \frac{1}{b^4}\right)\right] \tag{18.4}$$

Modifying this expression to an orthotropic plate and using only the B term suggests the following expression

$$B = k \cdot h \cdot \left(\frac{E_x{}^t}{a^4} + \frac{E_y{}^t}{b^4}\right) \tag{18.5}$$

where the coefficient k includes all the numerical factors.

Since eq. (18.5) is based only on the first term of multiple terms series, the result is not accurate enough to correctly represent the plate's response. Therefore, finite element analyses (FEAs) of 80 plates with various lengths and widths were performed. The FEA software was Femap with NX-Nastran version 2021.1 with its build-in nonlinear static analysis. The FEA results are given in Appendix C, while the FEA information is given in Appendix D. Typical plate arrangements and midpoint load–deflection graphs are shown in Fig. 18.3a–b.

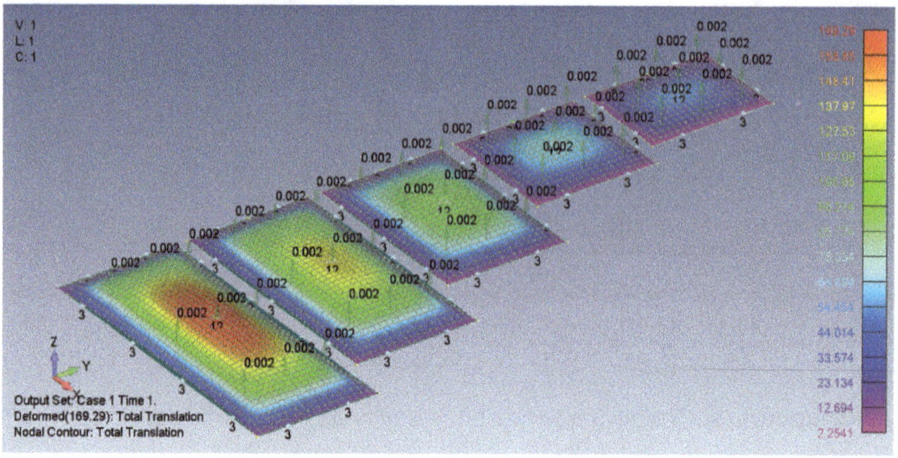

(a)

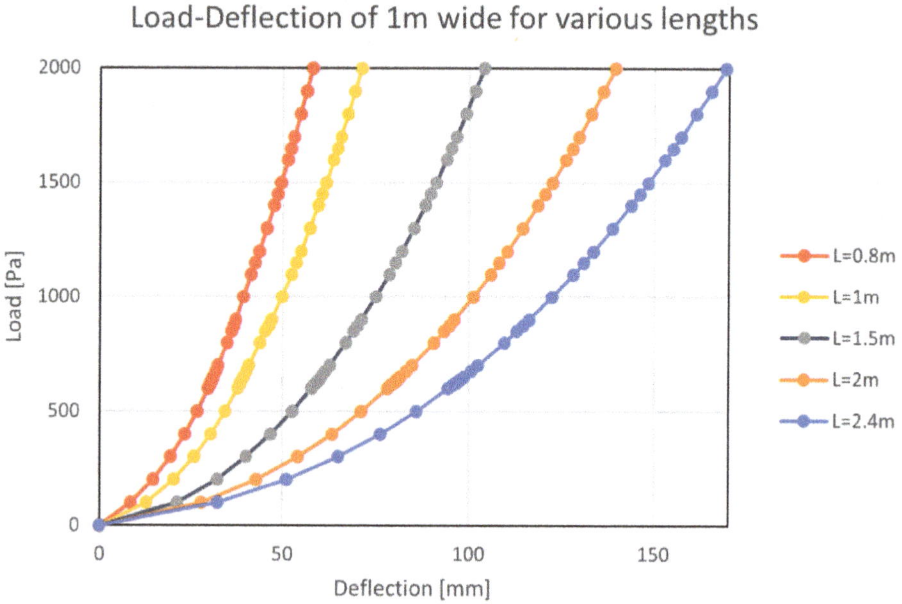

(b)

Fig. 18.3: (a) FEA for which the load is in [MPa] and the lateral deflection in [mm], (b) load–deflection curves.

The values of the coefficient B in eq. (18.1) were calculated with a least-squares regression from the FEA results, and the following expression is suggested for B:

$$B = h\sqrt{\frac{h}{a+b}\left(\frac{E_x^t + E_y^t}{(a+b)^4}\right)K\left(\frac{a}{b}\right)^p} \tag{18.6}$$

Note that the sum of the moduli and the sum of the length–width represent its averaged values, as the 2 division is included in K. The (a/b) is the aspect ratio of the plate. The K and p unitless values are $K = 201.44$, $p = -0.17165$, while the units of the variables are: B: [Pa/m^3], h: [m], a, b: [m], E: [Pa]. Note that eq. (18.6) describes well multiwall plates, but it is not necessarily suitable for other types of orthotropic plates.

18.3 Multiwall plates tests using a vacuum chamber

To check the multiwall plate response to transverse distributed load, a vacuum chamber test setup was designed and deployed, as presented in Fig. 18.4.

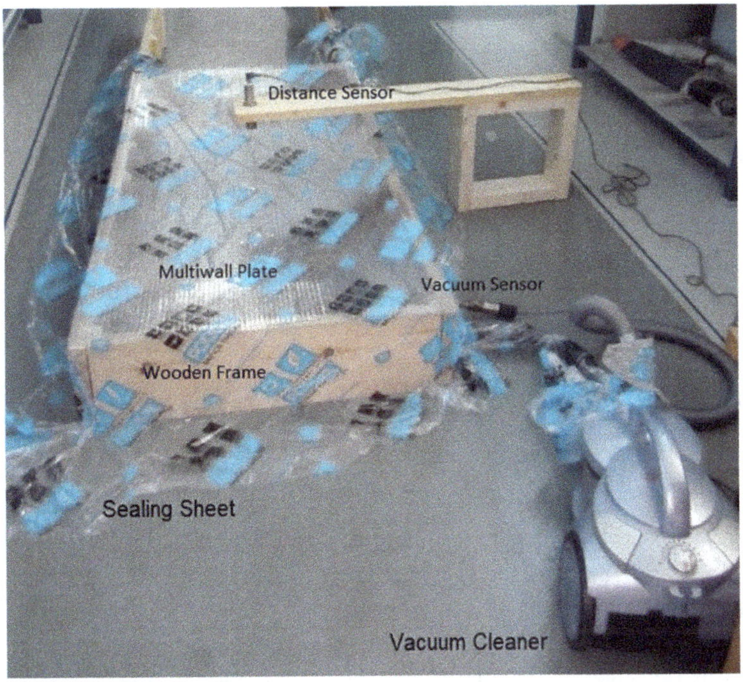

Fig. 18.4: The vacuum chamber test set-up.

A 35 mm-thick wooden frame encloses a rectangular space with the required dimensions. The multiwall plate is freely placed on the frame's edges. A thin plastic sheet covers the entire device and the floor nearby. A variable-speed vacuum cleaner is connected to the internal space through a drilled hole. The vacuum created causes the plastic sheet to seal all air leakages, allowing the vacuum level to gradually increase.

An electronic vacuum sensor measures the vacuum level through another hole in the wooden frame. The sensor is connected to a controller, which displays the data. Additionally, an ultrasonic distance sensor is placed 20 cm above the plate midpoint, measures the plate deflection and transmits it to the controller. The vacuum units are [Pa], and the distance units are [mm].

The controller has zero button that zeroes the vacuum and the distance values, as the test starts. During the test, the vacuum level gradually increased, while both load and deflection being recorded. Various local buckling phenomena are also closely monitored and recorded.

The plate edges are free to move on the wood frame, creating the M (movable) BC. Yet, as the vacuum level increases during the test, the thin plastic sealing sheet gets tension and presses the plate edge to the wood frame. This may change the BC from SSSS to be somewhat closer to CCCC.

Several tests were performed at the Krumbein Structures Laboratory, Faculty of Aerospace Engineering – Technion, with a typical test being next presented.

18.4 Experimental results

Two plates were tested: length x width 1.5 × 0.8 m and length x width 0.8 × 1.5 m.

The opening dimensions were 0.07 m less, i.e., 1.43 × 0.73 m.

Plate details:

- Type: 10 mm – PC double wall
  - Nominal area weight 1700 g/m^2

The measured results were least-squares fitted to the expression $q = Aw + Bw^3$ and were compared to the theoretical curves.

Part of the tested plates data are shown in Fig. 18.5, Tabs. 18.1 and 18.2.

Coefficient A was calculated using eq. (19.2) and the second coefficient, B, was found using eq. (18.6) and was compared to the measured values. The comparison is shown in Tab. 18.2.

As displayed in Tab. 18.2, the calculated coefficients comply with the measured one, with some differences.

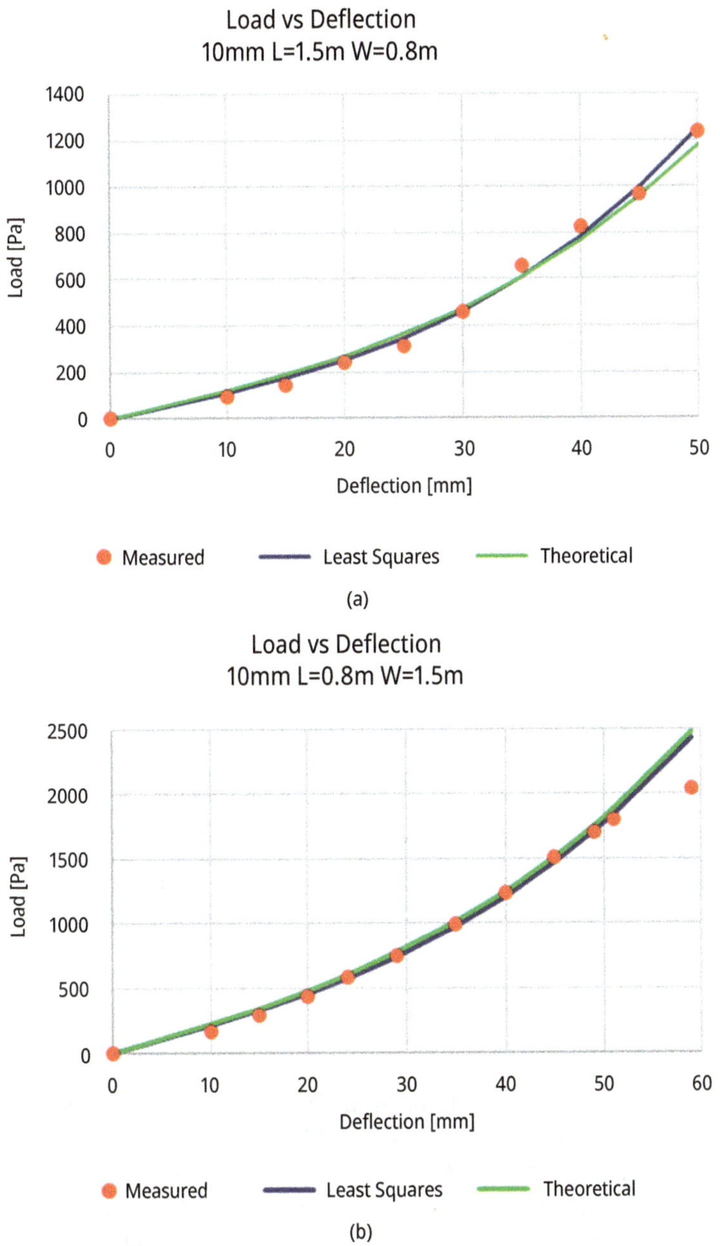

Fig. 18.5: Load–deflection graphs: (a) length 1.5 m, width 0.8 m, (b) length 0.8 m, width 1.5 m.

Tab. 18.1: The elastic constants of the multiwall plate.

Variable	Notation	Dimension	Value
Thickness	h	[mm]	10
Area Weight	W	[g/m^2]	1713
Walls Thickness	t_w	[mm]	1.154
Equivalent G	G_{eq}	[MPa]	100.348
For small deflection coefficient A:			
Bending	D_x	[Nm]	70.12106
Bending	D_y	[Nm]	54.10356
Bending	D_{xy}	[Nm]	8.362
Shear	S_x	[N/m]	59,890.03
Shear	S_y	[N/m]	1662.076
X Poisson ratio	v_x^b	--	0.380
Y Poisson's ratio	$v_y^b = D_y/D_x{}^*v_x^b$	--	0.293
For large deflection coefficient B:			
Tension	E_x^t	[MPa]	342.60
Tension	E_y^t	[MPa]	276.96
X Poisson ratio	v_x^t	--	0.38
Y Poisson's ratio	$v_y^t = E_y^t/E_x^{t*}v_x^t$	--	0.307

Tab. 18.2: Comparison of the A, B coefficient values.

Coefficient	Length 1.5 m, Width 0.8 m			Length 0.8 m, Width 1.5 m		
	Measured	Theoretical	% dif.	Measured	Theoretical	% dif.
A [Pa/m]	9,602	11,487	16.4%	19,061	21,805	12.6%
B [Pa/m^3]	5,370,163	3,475,816	35.3%	5,858,384	4,378,294	25.3%

18.5 FEA results

To find the response of the plates to transversal uniform load, 80 FEAs were performed: four plate types with four various widths and five different lengths (see in Fig. 18.3 one of the tests). The coefficients A and B for each plate were both theoretically calculated using eqs. 18.2 and 18.6, respectively, and found from the graphs drawn using eq. (18.1). A comparison of the analytical (theoretically) calculated A and B coefficients and the FEA-measured coefficients is presented in Fig. 18.6. The legend presented in Fig. 19.6b applies to all other graphs in Fig. 18.6.

The 45° lines represent a perfect agreement between the theory and measurements. As is shown in Fig. 18.6, a very good agreement between theory and analysis is found.

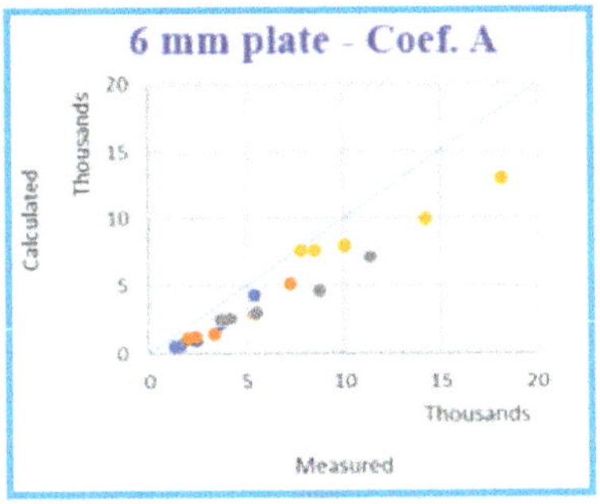

(a)

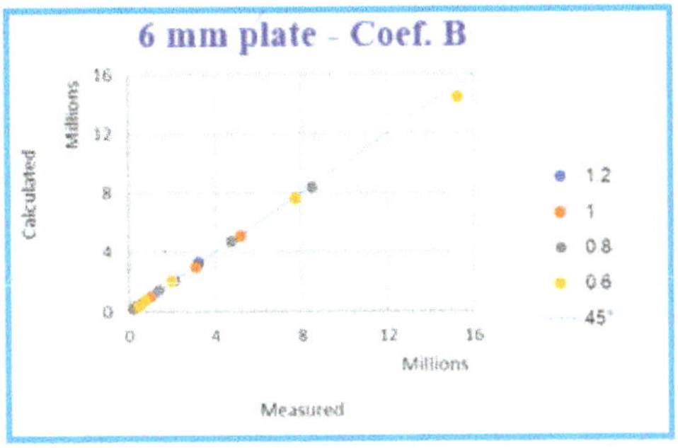

(b)

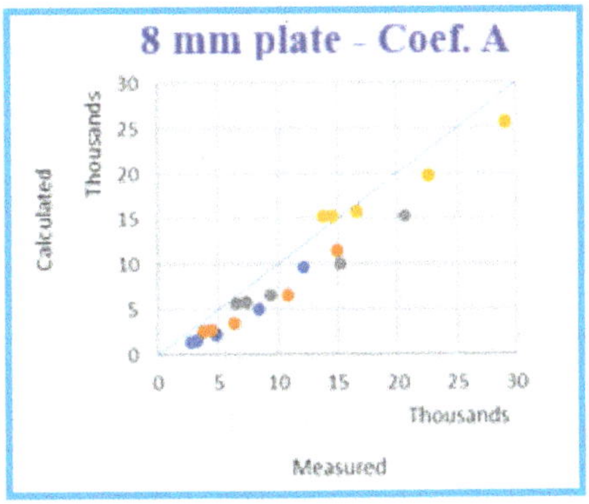

(c)

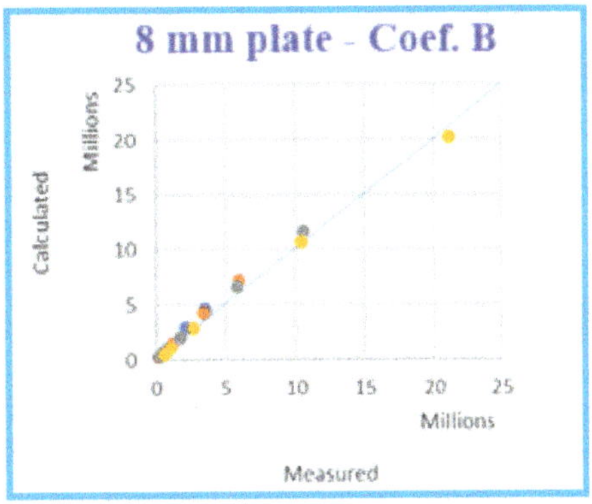

(d)

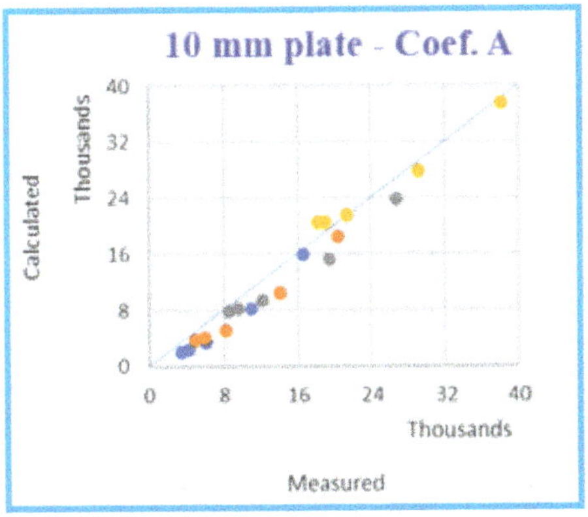

(e)

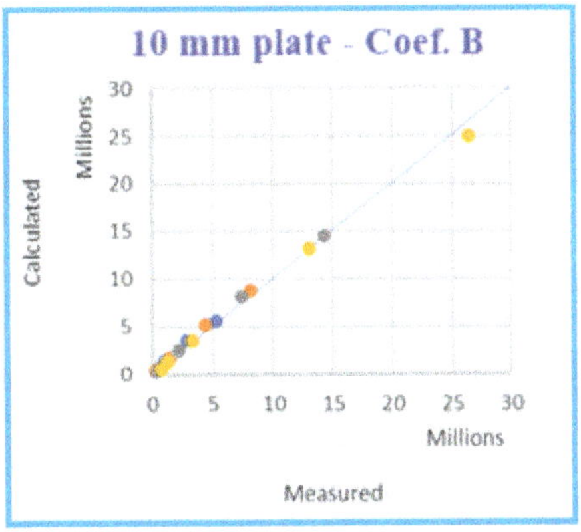

(f)

Fig. 18.6: Calculated analytical and FEA-measured *A*, *B* coefficients – a comparison of various tested plates: (a) and (b) 6 mm plates, (c) and (d) 8 mm plates, (e) and (f) 10 mm plates, (g) and (h) 16 mm plates.

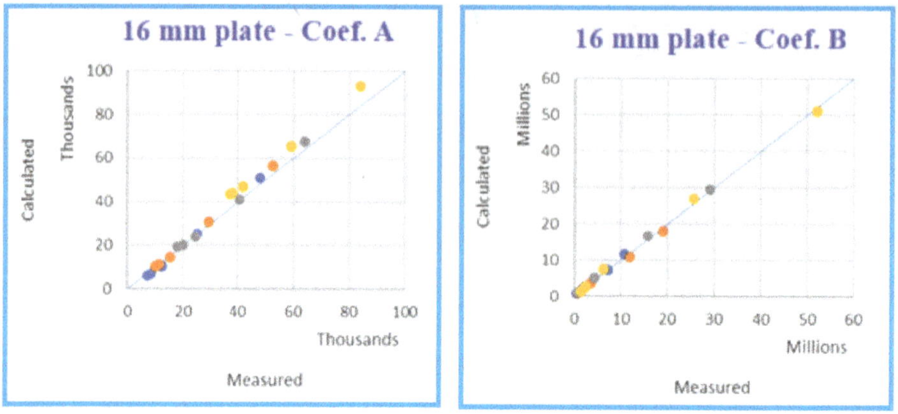

(g) (h)

Fig. 18.6 (continued)

18.6 Plazit-Polygal numerous multiwall plates' tests

One of the manufacturers of PC multiwall plates is Plazit-Polygal (Plaskolite). During the years 2001–2002, they performed a large number of vacuum loading tests like the one described in Chapter 19.4. Plazit-Polygal has allowed the authors to use the test data for research and publication purposes. This permission is very much appreciated.

From about 250 tests, 120 tests of plates of 6–16 mm thick were chosen. Each test has a set of measured load–deflection values for various widths/lengths. The analysis of every test was a linear least-squares regression that calculated the coefficients A and B in the expression $q = Aw + Bw^3$. The results are listed in Appendix C. The coefficient of determination R^2 (goodness of fit) in all tested cases was above 0.99. These very good correlations support the validity of the suggested A, B large deflection approximation eq. (18.1).

The equivalent elastic constants and moduli of the plates were measured and given with the test data in Appendix B. This information allows us to calculate the A and B coefficients according to the theory presented above. The measured and calculated A, B coefficients are compared in Fig. 18.7 and Fig. 18.8 (additional tests performed at Krumbein Structures Laboratory, Faculty of Aerospace Engineering – Technion) where the unit of A is [Pa/m] and the unit of B is [Pa/m^3].

Note that the legend presented in Fig. 18.7b applies to all other graphs in Fig. 18.7.

In Figs. 18.7 and 18.8 the horizontal axes are the A and B measured coefficients, while the vertical axes are the theoretically calculated coefficients. The 45° lines represent the location of the perfect agreement between the theory and measurements.

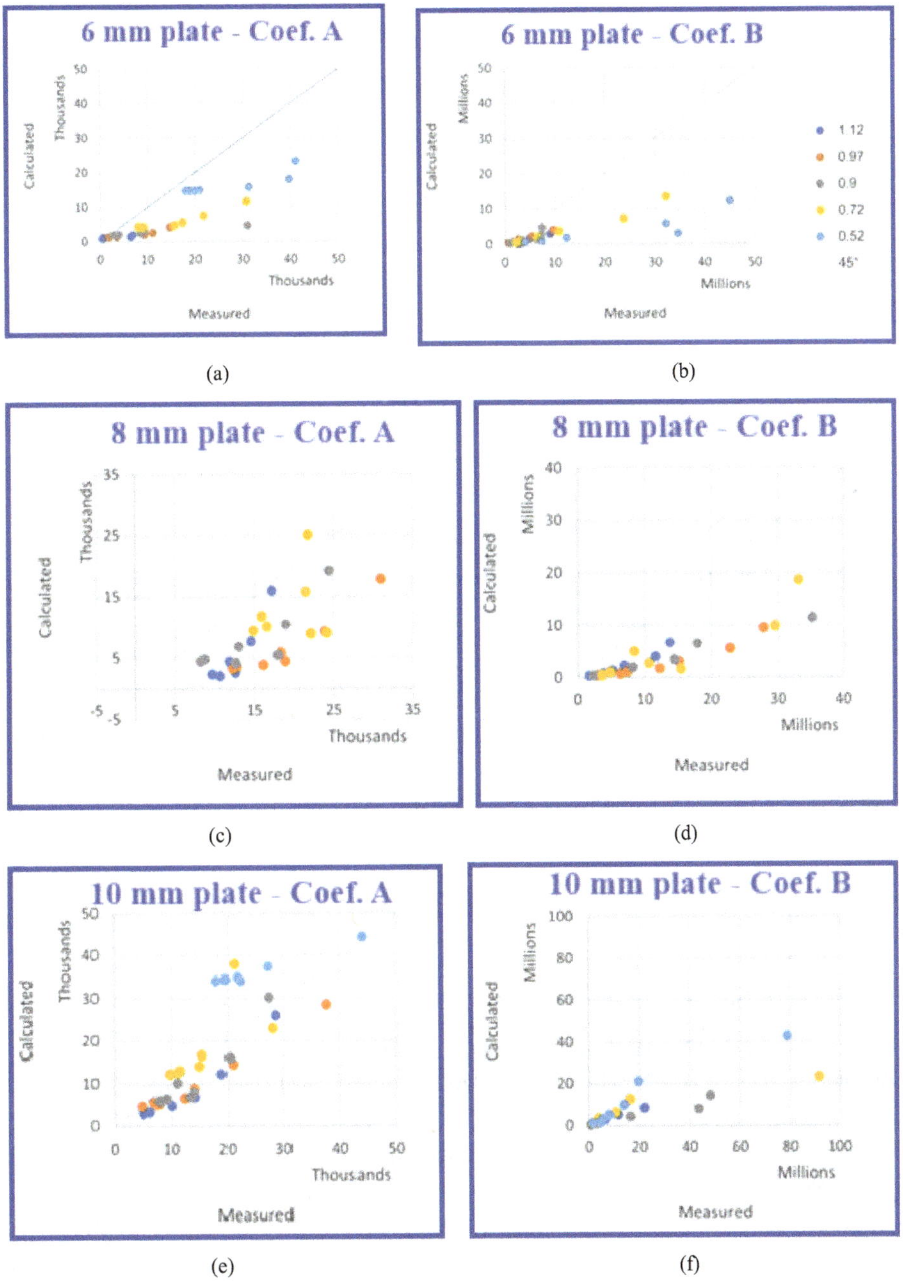

Fig. 18.7: Calculated analytical and test-measured (Plazit-Polygal specimens) *A*, *B* coefficients – a comparison of various tested plates: (a) and (b) 6 mm plates, (c) and (d) 8 mm plates, (e) and (f) 10 mm plates, (g) and (h) 16 mm plates.

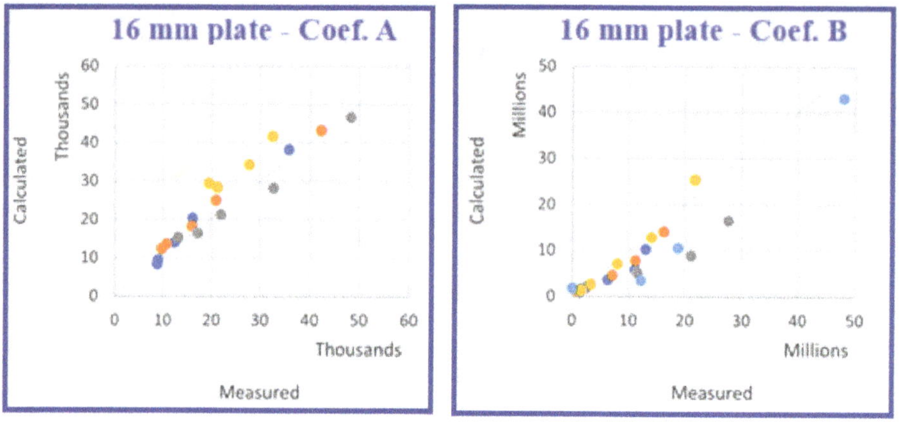

(g) (h)

Fig. 18.7 (continued)

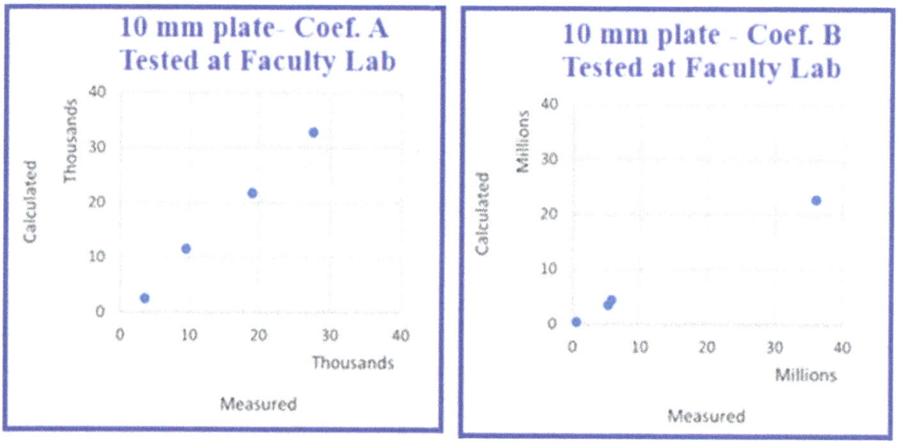

Fig. 18.8: Calculated analytical and test-measured *A*, *B* coefficients – a comparison of various plates performed at Krumbein Structures Laboratory, Faculty of Aerospace Engineering – Technion.

Note that the various colors are for various plate widths [m], shown in Fig. 18.6(b) graph legend.

Generally, the theory-measurement agreements are better for the coefficient *A* over the coefficient *B* and better for higher thickness over lower thickness.

Since Polygal's tests were performed more than 20 years ago, the tested samples are not available for verification anymore. Since many dimensions of the plates were missing in the records, the standard values were taken from the plate's data sheets. Nevertheless, the actual real plate dimension values almost always deviate from the standard values, as may occur in a real manufacturing. These deviations can be rather significant and, for sure, influence on the plate rigidity. This is probably the main source for the inconsistencies in the reported data.

Also, as the vacuum was measured at that time with a water manometer which is not accurate enough, it is very possible that errors do exist in the data. These errors can be seen in Appendix C, Tabs. C3, C4, C5, C6, where the measured coefficients A should monotonically increase while the length values decrease, but in several cases they unexpectedly decrease.

Yet, being very comprehensive with large number of length/width combinations, it still has some value for the understanding of the plate response to transverse pressure, besides proving the $q = Aw + Bw^3$ response.

Note that the last plate test (10 mm Plate Faculty Lab) was performed more recently under better-controlled environment and therefore presents a more accurate agreement.

18.7 Conclusions

Large deflection of multiwall plates under distributed transverse pressure and SSSS-M boundary conditions has been found to comply with the $q = Aw + Bw^3$ approximation rule, with a very good R^2 (goodness-of-fit) values.

The suggested expressions for coefficients A and B have good agreement with FEA results, while in actual tests appear to be more applicable for thicker plates.

The deviations in the presented experimental loading data may relate to the measurement technic. It is expected that better laboratory practice would lead to more accurate results.

References

[1] Föppl, A. Vorlesungen über technische, Mechanik, 5, 1907, 132–144.
[2] von Kármán, T. Art 27. Festigkeitsprobleme im maschinenbau, Encyklopedie der Mathematischen Wissenschaften, 4, 1910, 348–351.
[3] Timoshenko, S. and S, W.-K. Chapter 13. Large deflections of plates, in Theory of plates and shells, 2nd ed., McGraw-Hill Book Company, 1959 , 594. Reissued 1987.
[4] Föppl, A. and Föppl, L. Drang und Zwang – Eine höhere Festigkeitslehre für Ingenieure, 2nd ed., Vol. 2, R. Oldenbourg, München und Berlin, 1928.
[5] Ugural, C.A. Stresses in beams, plates, and shells, 4th ed., CRC Press, Taylor and Francis Group, Boca Raton FL, 2018, 618.
[6] Wang, D. and El-Sheikh, A.I. Large-deflection mathematical analysis of rectangular plates, Journal of Engineering Mechanics, 131(8), 2005, 809–821.
[7] Riber, H.J. Non-linear analytical solutions for laterally loaded sandwich plates, Composite Structures, 39(1–2), 1997, 63–83.

[8] Awrejcewicz, J., Krysko, V.A., jr, Kalutsky, L.A. and Krysko, V.A. Computing static behavior of flexible rectangular von Kármán plates in fast and reliable way, International Journal of Non-Linear Mechanics, 146, 2022, 32. ID article 104162.
[9] Battaglia, G., Di Matteo, A., Micale, G. and Pirrotta, A. Analysis of rectangular orthotropic membranes for mechanical properties identification through load- displacement data, Journal of Engineering Mechanics, 147(6), 2021, 18. ID article 04021028.
[10] Maier-Schneider, D., Maibach, J. and Obermeier, E. A new analytical solution for the load-deflection of square membranes, Journal of Microelectromechanical Systems, 4(4), 1995, 238–241.
[11] Niyogi, A.K. Nonlinear bending of rectangular orthotropic plates, International Journal of Solids Structures, 9, 1973, 1133–1139.
[12] Wang, M., Huang, X., Wang, X. and Qiu, X. An approximated solution to the finite deformation of an elastic rectangular plate under static and dynamic transverse loadings, International Journal of Impact Engineering, 155, 2021, 13. ID article 103916.
[13] Beatty, M.F. Ch. 2- Introduction to nonlinear elasticity, in nonlinear effects in fluids and solids, in Mathematical Concepts and methods in science and engineering, Plenum Press, New York, 1996, 13–112, 392p.
[14] Civalek, Ö. Geometrically nonlinear dynamic and static analysis of shallow spherical shell resting on two-parameters elastic foundations, International Journal of Pressure Vessels and Piping, 113, 2014, 1–9.
[15] Ogden, R.W. Non-linear elastic deformations, Dover Publications Inc, Mineola, New York, 1984, 532.
[16] Hakim, G. and Abramovich, H. Multiwall rectangular plates under transverse pressure – A non-linear experimental and numerical study, Materials, 16(5), 2023, 27. ID article 2041.
[17] Hakim, G. and Abramovich, H. Homogenization of multiwall plates- an analytical, numerical and experimental study, Thin-Walled Structures, 185, 2023, 17. ID article 110583.
[18] Libode, C. and Batdorf, S.B., A General Small-Deflection Theory for Flat Sandwich Plates, NACA Report No. 899, 1948, 139–156.
[19] Reddy, J.N. Theory and analysis of elastic plates and shells, 2nd Ed., CRC Press, Taylor & Francis Group, Boca Raton FL, 2007, 561.
[20] Yankelevsky, D., Feldgun, V. and Karinsky, Y. The mechanical behavior of glass plates, National Building Research Institute – Technion (Hebrew Language), 2017, 73.
[21] Pilkey, W.D. Formulas for Stress, Strain, and Structural Matrices, John Wiley & Sons, Inc, 2004, 1511.
[22] Levy, S., Bending of rectangular plates with large deflections, NACA Report 737, 1941, 19
[23] Hatsuo, I. On the large deflections of rectangular glass panes under uniform pressure, Bulletin of the Disaster Prevention Research Institute, Kurenai Kyoto University, 22(1), 1972, 1–7.
[24] ASTM E 1300-16, Standard practice for determining load resistance of glass in buildings, 2009.
[25] Scholes, A. and Bernstein, E.L. Bending of normally loaded simply supported rectangular plates in the large-deflection range, The Journal of Strain Analysis for Engineering Design, 4(3), 1969, 190–198.
[26] Keiser, R. Rechnerische und experimentelle Ermittlung der Durchbiegungenund Spannungen von quadratischen Platten bei freier Auflagerung an den Rändern, gleichmäßig verteilter Last und großen Ausbiegungen, ZAMM Journal of Applied Mathematics and Mechanics/Zeitschrift Für Angewandte Mathematik Und Mechanik (In German), 16(2), 1936, 73–98.
[27] Chia, C.-Y. Nonlinear analysis of plates, McGraw-Hill, Inc, 1980, 422.
[28] Brown, J.C. and Harvey, J.M. Large deflections of rectangular plates subjected to uniform lateral pressure and compressive edge loading, Journal Mechanical Engineering Science, 11(3), 1969, 305–317.

Appendix A: small deflection coefficient *A* – Libove's and Reddy's solutions

A multiwall plate has a considerable transverse shear flexibility in the width direction y, while it is relatively rigid to shear in the length direction x. Nevertheless, the analysis here considers both directions, omitting the shear in the length direction only at the end of the derivation.

Libove's Solution

In the following, Poisson's ratios are designated μ_x, μ_y as it appears in the source, while in the main paper they are designated v_x^b, v_y^b.

Libove's NACA Report No. 899 [18] presents the PDEs describing this problem.

The following equations are taken from [18]:

$$\left(N_x\frac{\partial^2}{\partial x^2}+N_y\frac{\partial^2}{\partial y^2}+2N_{xy}\frac{\partial^2}{\partial x\partial y}\right)w+\left(\frac{\partial}{\partial x}\right)Q_x+\left(\frac{\partial}{\partial y}\right)Q_y=-q \tag{A.1}$$

$$\left[-D_{xy}\frac{\partial^3}{\partial x\partial y^2}-\frac{D_x}{1-\mu_x\mu_y}\left(\mu_y\frac{\partial^3}{\partial x\partial y^2}+\frac{\partial^3}{\partial x^3}\right)\right]w+\left[\frac{1}{2}\frac{D_{xy}}{D_{Qx}}\frac{\partial^2}{\partial y^2}+\frac{D_x}{\left(1-\mu_x\mu_y\right)D_{Qx}}\frac{\partial^2}{\partial x^2}-1\right]Q_x+$$

$$\left[\frac{1}{2}\frac{D_{xy}}{D_{Qy}}\frac{\partial^2}{\partial x\partial y}+\frac{D_x\mu_y}{\left(1-\mu_x\mu_y\right)D_{Qy}}\frac{\partial^2}{\partial x\partial y}\right]Q_y=0 \tag{A.2}$$

$$\left[-D_{xy}\frac{\partial^3}{\partial x^2\partial y}-\frac{D_y}{1-\mu_x\mu_y}\left(\mu_x\frac{\partial^3}{\partial x^2\partial y}+\frac{\partial^3}{\partial y^3}\right)\right]w+\left[\frac{1}{2}\frac{D_{xy}}{D_{Qx}}\frac{\partial^2}{\partial x\partial y}+\frac{D_y\mu_x}{\left(1-\mu_x\mu_y\right)D_{Qx}}\frac{\partial^2}{\partial x\partial y}\right]Q_x+$$

$$\left[\frac{1}{2}\frac{D_{xy}}{D_{Qy}}\frac{\partial^2}{\partial x^2}+\frac{D_y}{\left(1-\mu_x\mu_y\right)D_{Qy}}\frac{\partial^2}{\partial y^2}-1\right]Q_y=0 \tag{A.3}$$

In order to have the equations written in a more readable form, we change the shear rigidity D_{Qx} to be S_x and the D_{Qy} to be S_y. After opening the parenthesis, we get:

$$N_x\frac{\partial^2 w}{\partial x^2}+N_y\frac{\partial^2 w}{\partial y^2}+2N_{xy}\frac{\partial^2 w}{\partial x\partial y}+\frac{\partial Q_x}{\partial x}+\frac{\partial Q_y}{\partial y}=-q \tag{A.4}$$

$$-\left[D_{xy}+\frac{\mu_y D_x}{\left(1-\mu_x\mu_y\right)}\right]\frac{\partial^3 w}{\partial x\partial y^2}-\frac{D_x}{\left(1-\mu_x\mu_y\right)}\frac{\partial^3 w}{\partial x^3}+\frac{D_{xy}}{2S_x}\frac{\partial^2 Q_x}{\partial y^2}+\frac{D_x}{\left(1-\mu_x\mu_y\right)S_x}\frac{\partial^2 Q_x}{\partial x^2}-Q_x+\left[\frac{D_{xy}}{2S_y}+\frac{\mu_y D_x}{\left(1-\mu_x\mu_y\right)S_y}\right]\frac{\partial^2 Q_y}{\partial x\partial y}=0 \tag{A.5}$$

$$-\left[D_{xy}+\frac{\mu_x D_y}{\left(1-\mu_x\mu_y\right)}\right]\frac{\partial^3 w}{\partial x^2\partial y}-\frac{D_y}{\left(1-\mu_x\mu_y\right)}\frac{\partial^3 w}{\partial y^3}+\left[\frac{D_{xy}}{2S_x}+\frac{\mu_x D_y}{\left(1-\mu_x\mu_y\right)S_x}\right]\frac{\partial^2 Q_x}{\partial x\partial y}+\frac{D_{xy}}{2S_y}\frac{\partial^2 Q_y}{\partial x^2}+\frac{D_y}{\left(1-\mu_x\mu_y\right)S_y}\frac{\partial^2 Q_y}{\partial y^2}-Q_y=0 \tag{A.6}$$

Some simplification can be done by taking N_x, N_y and N_{xy} to be zero, because no membrane forces exist in our problem.

Then one obtains:

$$\frac{\partial Q_x}{\partial x}+\frac{\partial Q_y}{\partial y}=-q \tag{A.7}$$

$$-\left[D_{xy}+\frac{\mu_y D_x}{\left(1-\mu_x\mu_y\right)}\right]\frac{\partial^3 w}{\partial x\partial y^2}-\frac{D_x}{\left(1-\mu_x\mu_y\right)}\frac{\partial^3 w}{\partial x^3}+\frac{D_{xy}}{2S_x}\frac{\partial^2 Q_x}{\partial y^2}+\frac{D_x}{\left(1-\mu_x\mu_y\right)S_x}\frac{\partial^2 Q_x}{\partial x^2}+\left[\frac{D_{xy}}{2S_y}+\frac{\mu_y D_x}{\left(1-\mu_x\mu_y\right)S_y}\right]\frac{\partial^2 Q_y}{\partial x\partial y}=Q_x \tag{A.8}$$

$$-\left[D_{xy}+\frac{\mu_x D_y}{\left(1-+\mu_x\mu_y\right)}\right]\frac{\partial^3 w}{\partial x^2\partial y}-\frac{D_y}{\left(1-\mu_x\mu_y\right)}\frac{\partial^3 w}{\partial y^3}+\left[\frac{D_{xy}}{2S_x}+\frac{\mu_x D_y}{\left(1-\mu_x\mu_y\right)S_x}\right]\frac{\partial^2 Q_x}{\partial x\partial y}+\frac{D_{xy}}{2S_y}\frac{\partial^2 Q_y}{\partial x^2}\;\frac{D_y}{\left(1-\mu_x\mu_y\right)S_y}\frac{\partial^2 Q_y}{\partial y^2}=Q_y \tag{A.9}$$

Equations (A.7), (A.8) and (A.9) are a linear set of three PDEs. It should be solved simultaneously for the unknowns Q_x, Q_y and w, for the independent load q. All the other coefficients are already known before.

According to Libode and Batdorf [18], this set can be separated to be three dual-variable PDEs for every unknown, in terms of the load q. One of them is for w, which is a sixth-order equation as shown in [18] (A.13a):

$$[D]w = -[M]q \tag{A.10}$$

where the operators $[D]$ and $[M]$, after the omission of N terms, are:

$$[D] = \frac{D_{xy}D_x}{2S_y}\frac{\partial^6}{\partial x^6} + \left(\frac{D_{xy}D_x}{2S_x} + \frac{D_xD_y - \frac{1}{2}D_{xy}D_x\mu_y - \frac{1}{2}D_{xy}D_y\mu_x}{S_y}\right)\frac{\partial^6}{\partial x^4 \partial y^2} +$$

$$\left(\frac{D_{xy}D_y}{2S_y} + \frac{D_xD_y - \frac{1}{2}D_{xy}D_x\mu_y - \frac{1}{2}D_{xy}D_y\mu_x}{S_x}\right)\frac{\partial^6}{\partial x^2 \partial y^4} + \frac{D_{xy}D_y}{2S_x}\frac{\partial^6}{\partial y^6} - D_x\frac{\partial^4}{\partial x^4} -$$

$$\left[2D_{xy}\left(1-\mu_x\mu_y\right) + D_x\mu_y + D_y\mu_x\right]\frac{\partial^4}{\partial x^2 \partial y^2} - D_y\frac{\partial^4}{\partial y^4} \tag{A.11}$$

$$[M] = \frac{D_{xy}D_x}{2S_xS_y}\frac{\partial^4}{\partial x^4} + \frac{D_xD_y - \frac{1}{2}D_{xy}D_x\mu_y - \frac{1}{2}D_{xy}D_y\mu_x}{S_xS_y}\frac{\partial^4}{\partial x^2 \partial y^2} + \frac{D_{xy}D_y}{2S_xS_y}\frac{\partial^4}{\partial y^4} -$$

$$\left(\frac{D_{xy}\left(1-\mu_x\mu_y\right)}{2S_x} + \frac{D_y}{S_y}\right)\frac{\partial^2}{\partial y^2} - \left(\frac{D_{xy}\left(1-\mu_x\mu_y\right)}{2S_y} + \frac{D_x}{S_x}\right)\frac{\partial^2}{\partial x^2} + \left(1-\mu_x\mu_y\right) \tag{A.12}$$

After replacing $D_y\mu_x$ with $D_x\mu_y$ in the above, according to Libode and Batdorf [18] (A.8), eq. (A.10) becomes:

$$\frac{D_{xy}D_x}{2S_y}\frac{\partial^6 w}{\partial x^6} + \left(\frac{D_{xy}D_x}{2S_x} + \frac{D_xD_y - D_{xy}D_x\mu_y}{S_y}\right)\frac{\partial^6 w}{\partial x^4 \partial y^2} + \left(\frac{D_{xy}D_y}{2S_y} + \frac{D_xD_y - D_{xy}D_x\mu_y}{S_x}\right)\frac{\partial^6 w}{\partial x^2 \partial y^4} +$$

$$\frac{D_{xy}D_y}{2S_x}\frac{\partial^6 w}{\partial y^6} - D_x\frac{\partial^4 w}{\partial x^4} - 2\left[D_{xy}\left(1-\mu_x\mu_y\right) + D_x\mu_y\right]\frac{\partial^4 w}{\partial x^2 \partial y^2} - D_y\frac{\partial^4 w}{\partial y^4} =$$

$$-\frac{D_{xy}D_x}{2S_xS_y}\frac{\partial^4 q}{\partial x^4} - \frac{D_xD_y - D_{xy}D_x\mu_y}{S_xS_y}\frac{\partial^4 q}{\partial x^2 \partial y^2} - \frac{D_{xy}D_y}{2S_xS_y}\frac{\partial^4 q}{\partial y^4} + \left(\frac{D_{xy}\left(1-\mu_x\mu_y\right)}{2S_x} + \frac{D_y}{S_y}\right)\frac{\partial^2 q}{\partial y^2} +$$

$$+\left(\frac{D_{xy}\left(1-\mu_x\mu_y\right)}{2S_y} + \frac{D_x}{S_x}\right)\frac{\partial^2 q}{\partial x^2} - \left(1-\mu_x\mu_y\right)q \tag{A.13}$$

or in a shorter form:

$$K_1\frac{\partial^6 w}{\partial x^6} + K_2\frac{\partial^6 w}{\partial x^4 \partial y^2} + K_3\frac{\partial^6 w}{\partial x^2 \partial y^4} + K_4\frac{\partial^6 w}{\partial y^6} + K_5\frac{\partial^4 w}{\partial x^4} + K_6\frac{\partial^4 w}{\partial x^2 \partial y^2} + K_7\frac{\partial^4 w}{\partial y^4} =$$

$$K_8\frac{\partial^4 q}{\partial x^4} + K_9\frac{\partial^4 q}{\partial x^2 \partial y^2} + K_{10}\frac{\partial^4 q}{\partial y^4} + K_{11}\frac{\partial^2 q}{\partial x^2} + K_{12}\frac{\partial^2 q}{\partial y^2} + K_{13}q \tag{A.14}$$

where the coefficients K_i are:

$$K_1 = \frac{D_{xy}D_x}{2S_y};\quad K_2 = \frac{D_{xy}D_x}{2S_x} + \frac{D_xD_y - D_{xy}D_x\mu_y}{S_y};\quad K_3 = \frac{D_{xy}D_y}{2S_y} + \frac{D_xD_y - D_{xy}D_x\mu_y}{S_x}$$

$$K_4 = \frac{D_{xy}D_y}{2S_x};\quad K_5 = -D_x;\quad K_6 = -2\left[D_{xy}\left(1-\mu_x\mu_y\right) + D_x\mu_y\right];\quad K_7 = -D_y$$

$$K_8 = -\frac{D_{xy}D_x}{2S_xS_y};\quad K_9 = -\frac{D_xD_y - D_{xy}D_x\mu_y}{S_xS_y};\quad K_{10} = -\frac{D_{xy}D_y}{2S_xS_y}$$

$$K_{11} = \frac{D_{xy}\left(1-\mu_x\mu_y\right)}{2S_y} + \frac{D_x}{S_x};\quad K_{12} = \frac{D_{xy}\left(1-\mu_x\mu_y\right)}{2S_x} + \frac{D_y}{S_y};\quad K_{13} = -\left(1-\mu_x\mu_y\right) \tag{A.15}$$

The following is a standard Fourier series solution of a linear PDE.

Since the deflection boundary conditions are all $w = 0$, we can assume a double sine series as the solution (Navier solution):

$$w(x,y) = \sum_{n=1}^{\infty}\sum_{m=1}^{\infty} W_{mn} \sin\frac{m\pi x}{a} \sin\frac{n\pi y}{b} \tag{A.16}$$

where W_{mn} are coefficients to be determined. Substituting eq. (A.16) into eq. (A.14) yields

$$\sum_{n=1}^{\infty}\sum_{m=1}^{\infty}\begin{bmatrix} -K_1\left(\frac{m\pi}{a}\right)^6 - K_2\left(\frac{m\pi}{a}\right)^4\left(\frac{n\pi}{b}\right)^2 - K_3\left(\frac{m\pi}{a}\right)^2\left(\frac{n\pi}{b}\right)^4 \\ -K_4\left(\frac{n\pi}{b}\right)^6 + K_5\left(\frac{m\pi}{a}\right)^4 + K_6\left(\frac{m\pi}{a}\right)^2\left(\frac{n\pi}{b}\right)^2 + K_7\left(\frac{n\pi}{b}\right)^4 \end{bmatrix} W_{mn}\sin\frac{m\pi x}{a}\sin\frac{n\pi y}{b} =$$
$$K_8\frac{\partial^4 q}{\partial x^4} + K_9\frac{\partial^4 q}{\partial x^2\partial y^2} + K_{10}\frac{\partial^4 q}{\partial y^4} + K_{11}\frac{\partial^2 q}{\partial x^2} + K_{12}\frac{\partial^2 q}{\partial y^2} + K_{13}q \tag{A.17}$$

This suggests that the equation's right-hand side should also be expanded into a double sine series:

$$q(x,y) = \sum_{n=1}^{\infty}\sum_{m=1}^{\infty} q_{mn} \sin\frac{m\pi x}{a} \sin\frac{n\pi y}{b} \tag{A.18}$$

So, eq. (A.17) becomes:

$$\sum_{n=1}^{\infty}\sum_{m=1}^{\infty}\left\{\begin{bmatrix}-K_1\left(\frac{m\pi}{a}\right)^6-K_2\left(\frac{m\pi}{a}\right)^4\left(\frac{n\pi}{b}\right)^2-K_3\left(\frac{m\pi}{a}\right)^2\left(\frac{n\pi}{b}\right)^4\\ -K_4\left(\frac{n\pi}{b}\right)^6+K_5\left(\frac{m\pi}{a}\right)^4+K_6\left(\frac{m\pi}{a}\right)^2\left(\frac{n\pi}{b}\right)^2+K_7\left(\frac{n\pi}{b}\right)^4\end{bmatrix}W_{mn} - \begin{bmatrix}K_8\left(\frac{m\pi}{a}\right)^4+K_9\left(\frac{m\pi}{a}\right)^2\left(\frac{n\pi}{b}\right)^2+K_{10}\left(\frac{n\pi}{b}\right)^4\\ -K_{11}\left(\frac{m\pi}{a}\right)^2-K_{12}\left(\frac{n\pi}{b}\right)^2+K_{13}\end{bmatrix}q_{mn}\right\}\sin\frac{m\pi x}{a}\sin\frac{n\pi y}{b}=0 \tag{A.19}$$

Since eq. (A.19) must exist at all points x,y in the domain, the coefficients of $\sin\frac{m\pi x}{a}\sin\frac{n\pi y}{b}$ must be zero for every m and n. This yields:

$$W_{mn}=\frac{\begin{bmatrix}K_8\left(\frac{m\pi}{a}\right)^4+K_9\left(\frac{m\pi}{a}\right)^2\left(\frac{n\pi}{b}\right)^2+K_{10}\left(\frac{n\pi}{b}\right)^4\\ -K_{11}\left(\frac{m\pi}{a}\right)^2-K_{12}\left(\frac{n\pi}{b}\right)^2+K_{13}\end{bmatrix}q_{mn}}{\begin{bmatrix}-K_1\left(\frac{m\pi}{a}\right)^6-K_2\left(\frac{m\pi}{a}\right)^4\left(\frac{n\pi}{b}\right)^2-K_3\left(\frac{m\pi}{a}\right)^2\left(\frac{n\pi}{b}\right)^4\\ -K_4\left(\frac{n\pi}{b}\right)^6+K_5\left(\frac{m\pi}{a}\right)^4+K_6\left(\frac{m\pi}{a}\right)^2\left(\frac{n\pi}{b}\right)^2+K_7\left(\frac{n\pi}{b}\right)^4\end{bmatrix}} \tag{A.20}$$

For an evenly distributed load $q(x,y)=q_o$, the coefficients q_{mn} are:

$$q_{mn}=\frac{16q_0}{\pi^2 mn}\quad m,n=1,3,5,7,\ \ldots \tag{A.21}$$

We are interested in the deflection at the middle point of the plate $w_{\max}$, where $x=\frac{a}{2};\ y=\frac{b}{2}$, and $\sin\frac{m\pi x}{a}\sin\frac{n\pi y}{b}=(-1)^{\frac{m+n}{2}-1}$.

The final solution is therefore:

$$w_{\max}=\frac{16q_0}{\pi^6}\sum_{n=1,3,5,\ldots}^{\infty}\sum_{m=1,3,5,\ldots}^{\infty}\frac{\begin{bmatrix}\pi^4K_8\left(\frac{m}{a}\right)^4+\pi^4K_9\left(\frac{m}{a}\right)^2\left(\frac{n}{b}\right)^2+\pi^4K_{10}\left(\frac{n}{b}\right)^4\\ -\pi^2K_{11}\left(\frac{m}{a}\right)^2-\pi^2K_{12}\left(\frac{n}{b}\right)^2+K_{13}\end{bmatrix}(-1)^{\frac{m+n}{2}-1}}{mn\begin{bmatrix}-\pi^2K_1\left(\frac{m}{a}\right)^6-\pi^2K_2\left(\frac{m}{a}\right)^4\left(\frac{n}{b}\right)^2-\pi^2K_3\left(\frac{m}{a}\right)^2\left(\frac{n}{b}\right)^4\\ -\pi^2K_4\left(\frac{n}{b}\right)^6+K_5\left(\frac{m}{a}\right)^4+K_6\left(\frac{m}{a}\right)^2\left(\frac{n}{b}\right)^2+K_7\left(\frac{n}{b}\right)^4\end{bmatrix}} \tag{A.22}$$

where the coefficients K_i are given in eq. (A.15).

Equation (A.22) presents the linear relationship between the load and the deflection for the small defection case, namely

$$w_{\max} = A \cdot q_0 \tag{A.23}$$

Reddy's solution

J.N. Reddy [19] has also solved the present problem. The numerical results of Reddy [19] are identical to those of Libove's NACA 899 [18].

Nevertheless, to complete this comparison, the details of Reddy's solution are repeated here with its original notations.

The elastic constant definitions used by Reddy are in Libove's terms:

$$D_{11} = \frac{D_x}{1-\mu_x\mu_y};\ \mathrm{D}_{12} = \frac{D_x\mu_y}{1-\mu_x\mu_y};\ \mathrm{D}_{22} = \frac{D_y}{1-\mu_x\mu_y};\ \mathrm{D}_{66} = \frac{1}{2}D_{xy};\ KA_{44} = S_y;\ KA_{55} = S_x;$$

$$\nu_{21} = \mu_y;\ \nu_{12} = \mu_x \tag{A.24}$$

There are several easing assumptions that simplify the original Reddy's expressions:

1. Deflections, strains, and rotations are small
2. No initial in-plane forces
3. Static state – no changes in time
4. No thermal loads
5. No elastic foundation

Under these assumptions, the expressions presented in [18] are:

(The original index $_0$ which represents middle plate value was omitted).

$$A_{11}\left(\frac{\partial^2 u}{\partial x^2} + \frac{\partial w}{\partial x}\frac{\partial^2 w}{\partial x^2}\right) + A_{12}\left(\frac{\partial^2 v}{\partial y \partial x} + \frac{\partial w}{\partial y}\frac{\partial^2 w}{\partial y \partial x}\right) + A_{66}\left(\frac{\partial^2 u}{\partial y^2} + \frac{\partial^2 v}{\partial x \partial y} + \frac{\partial^2 w}{\partial x \partial y}\frac{\partial w}{\partial y} + \frac{\partial w}{\partial x}\frac{\partial^2 w}{\partial y^2}\right) = 0$$

$$A_{66}\left(\frac{\partial^2 u}{\partial y \partial x} + \frac{\partial^2 v}{\partial x^2} + \frac{\partial^2 w}{\partial x^2}\frac{\partial w}{\partial y} + \frac{\partial w}{\partial x}\frac{\partial^2 w}{\partial y \partial x}\right) + A_{12}\left(\frac{\partial^2 u}{\partial x \partial y} + \frac{\partial w}{\partial x}\frac{\partial^2 w}{\partial x \partial y}\right) + A_{22}\left(\frac{\partial^2 v}{\partial y^2} + \frac{\partial w}{\partial y}\frac{\partial^2 w}{\partial y^2}\right) = 0$$

$$K_s A_{55}\left(\frac{\partial^2 w}{\partial x^2} + \frac{\partial \phi_x}{\partial x}\right) + K_s A_{44}\left(\frac{\partial^2 w}{\partial y^2} + \frac{\partial \phi_y}{\partial y}\right) + q(x,y) = 0$$

$$D_{11}\left(\frac{\partial^2 \phi_x}{\partial x^2}\right) + D_{12}\left(\frac{\partial^2 \phi_y}{\partial y \partial x}\right) + D_{66}\left(\frac{\partial^2 \phi_x}{\partial y^2} + \frac{\partial^2 \phi_y}{\partial y \partial x}\right) - K_s A_{55}\left(\frac{\partial w}{\partial x} + \phi_x\right) = 0$$

$$D_{66}\left(\frac{\partial^2 \phi_x}{\partial x \partial y} + \frac{\partial^2 \phi_y}{\partial x^2}\right) + D_{12}\left(\frac{\partial^2 \phi_x}{\partial x \partial y}\right) + D_{22}\left(\frac{\partial^2 \phi_y}{\partial y^2}\right) - K_s A_{44}\left(\frac{\partial w}{\partial y} + \phi_y\right) = 0 \tag{A.25}$$

The Navier solution with double Fourier series is:

$$w(x,y)=\sum_{n=1}^{\infty}\sum_{m=1}^{\infty}W_{mn}\sin\frac{m\pi x}{a}\sin\frac{n\pi y}{b}$$
$$\phi_x(x,y)=\sum_{n=1}^{\infty}\sum_{m=1}^{\infty}X_{mn}\cos\frac{m\pi x}{a}\sin\frac{n\pi y}{b} \tag{A.26}$$
$$\phi_y(x,y)=\sum_{n=1}^{\infty}\sum_{m=1}^{\infty}Y_{mn}\sin\frac{m\pi x}{a}\cos\frac{n\pi y}{b}$$

The load is expanded to:

$$q(x,y)=\sum_{n=1}^{\infty}\sum_{m=1}^{\infty}Q_{mn}\sin\frac{m\pi x}{a}\sin\frac{n\pi y}{b} \tag{A.26a}$$

Substituting the solution and the load into the equations above yields [9]

$$\hat{s}_{11}W_{mn}+\hat{s}_{12}X_{mn}+\hat{s}_{13}Y_{mn}=Q_{mn}$$
$$\hat{s}_{12}W_{mn}+\hat{s}_{22}X_{mn}+\hat{s}_{23}Y_{mn}=0 \tag{A.27}$$
$$\hat{s}_{13}W_{mn}+\hat{s}_{23}X_{mn}+\hat{s}_{33}Y_{mn}=0$$

where

$$\hat{s}_{11}=K_s\left(A_{55}\alpha_m^{\ 2}+A_{44}\beta_n^{\ 2}\right),\ \hat{s}_{12}=K_sA_{55}\alpha_m,\ \hat{s}_{13}=K_sA_{44}\beta_n,\ \hat{s}_{22}=D_{11}\alpha_m^{\ 2}+D_{66}\beta_n^{\ 2}+K_sA_{55},$$
$$\hat{s}_{23}=(D_{12}+D_{66})\alpha_m\beta_n,\ \hat{s}_{33}=D_{66}\alpha_m^{\ 2}+D_{22}\beta_n^{\ 2}+K_sA_{44} \tag{A.28}$$

and $\alpha_m=m\pi/a$, $\beta_n=n\pi/b$

Coefficients b are now defined as:

$$b_0=\hat{s}_{22}\hat{s}_{33}-\hat{s}_{23}\hat{s}_{23},\ b_1=\hat{s}_{23}\hat{s}_{13}-\hat{s}_{12}\hat{s}_{33},\ b_2=\hat{s}_{12}\hat{s}_{23}-\hat{s}_{22}\hat{s}_{13},\ b_{mn}=\hat{s}_{11}b_0+\hat{s}_{12}b_1+\hat{s}_{13}b_2 \tag{A.29}$$

The Fourier coefficients are [19]

$$W_{mn}=\frac{b_0}{b_{mn}}Q_{mn},\ X_{mn}=\frac{b_1}{b_{mn}}Q_{mn},\ Y_{mn}=\frac{b_2}{b_{mn}}Q_{mn} \tag{A.30}$$

This being the end of Reddy's text, the rest is a consequential result.

For an evenly distributed load $(x,y)=q_0$, the coefficients Q_{mn} are:

$$Q_{mn}=\frac{16q_0}{\pi^2mn}\quad m,n=1,3,5,7\ldots \tag{A.31}$$

At the plate's midpoint at location $x=\frac{a}{2};y=\frac{b}{2}$, there exists

$$\sin\frac{m\pi x}{a}\sin\frac{n\pi y}{b}=(-1)^{\frac{m+n}{2}-1} \tag{A.32}$$

The deflection at that point is the maximal deflection $w_{\max}$ which is:

$$w_{\max} = \sum_{n=1}^{\infty}\sum_{m=1}^{\infty} W_{mn}(-1)^{\frac{m+n}{2}-1} = \sum_{n=1}^{\infty}\sum_{m=1}^{\infty} \frac{b_0}{b_{mn}} Q_{mn}(-1)^{\frac{m+n}{2}-1}$$

$$= \frac{16q_0}{\pi^2} \sum_{n=1,3,5,\ldots}^{\infty} \sum_{m=1,3,5,\ldots}^{\infty} \frac{b_0}{b_{mn}} \cdot \frac{(-1)^{\frac{m+n}{2}-1}}{mn}$$

$$w_{\max} = \frac{16q_0}{\pi^6} \sum_{n=1,3,5,\ldots}^{\infty} \sum_{m=1,3,5,\ldots}^{\infty} \pi^4 \frac{b_0}{b_{mn}} \cdot \frac{(-1)^{\frac{m+n}{2}-1}}{mn} \tag{A.33}$$

where

$$\frac{b_0}{b_{mn}} = \frac{\hat{s}_{22}\hat{s}_{33} - \hat{s}_{23}\hat{s}_{23}}{\hat{s}_{11}b_0 + \hat{s}_{12}b_1 + \hat{s}_{13}b_2}$$

$$= \frac{\hat{s}_{22}\hat{s}_{33} - \hat{s}_{23}\hat{s}_{23}}{\hat{s}_{11}(\hat{s}_{22}\hat{s}_{33} - \hat{s}_{23}\hat{s}_{23}) + \hat{s}_{12}(\hat{s}_{23}\hat{s}_{13} - \hat{s}_{12}\hat{s}_{33}) + \hat{s}_{13}(\hat{s}_{12}\hat{s}_{23} - \hat{s}_{22}\hat{s}_{13})} \tag{A.34}$$

As in Libove's solution above, Reddy's solution (A.33) presents the linear relationship between the load and the deflection in a small deflection case: $w_{\max} = A \cdot q_0$.

Simplified Libove's solution

In multiwall plates, the x-direction shear rigidity is often very high: $S_x \to \infty$.

This causes the coefficients K_i to change:

$$K_1 = \frac{D_{xy}D_x}{2S_y}; \quad K_2 = \frac{D_xD_y - D_{xy}D_x\mu_y}{S_y}; \quad K_3 = \frac{D_{xy}D_y}{2S_y}$$

$$K_4 = 0; \quad K_5 = -D_x; \quad K_6 = -2\left[D_{xy}\left(1-\mu_x\mu_y\right) + D_x\mu_y\right]; \quad K_7 = -D_y$$

$$K_8 = 0; \quad K_9 = 0; \quad K_{10} = 0$$

$$K_{11} = \frac{D_{xy}\left(1-\mu_x\mu_y\right)}{2S_y}; \quad K_{12} = \frac{D_y}{S_y}; \quad K_{13} = -\left(1-\mu_x\mu_y\right) \tag{A.35}$$

This change simplifies the calculation a little bit:

$$w_{\max} = \frac{16q_0}{\pi^6} \sum_{n=1,3,5,\ldots}^{\infty} \sum_{m=1,3,5,\ldots}^{\infty} \frac{\left[-\pi^2K_{11}\left(\frac{m}{a}\right)^2 - \pi^2K_{12}\left(\frac{n}{b}\right)^2 + K_{13}\right](-1)^{\frac{m+n}{2}-1}}{mn\left[\begin{array}{l}-\pi^2K_1\left(\frac{m}{a}\right)^6 - \pi^2K_2\left(\frac{m}{a}\right)^4\left(\frac{n}{b}\right)^2 - \pi^2K_3\left(\frac{m}{a}\right)^2\left(\frac{n}{b}\right)^4 \\ +K_5\left(\frac{m}{a}\right)^4 + K_6\left(\frac{m}{a}\right)^2\left(\frac{n}{b}\right)^2 + K_7\left(\frac{n}{b}\right)^4\end{array}\right]} \tag{A.36}$$

For orthotropic plates that are rigid for shear deformation in both directions, eq. (A.36) reduces to

$$w_{\max} = \frac{16\left(\mu_x\mu_y - 1\right)q_0}{\pi^6} \sum_{n=1,3,5,\ldots}^{\infty} \sum_{m=1,3,5,\ldots}^{\infty} \frac{(-1)^{\frac{m+n}{2}-1}}{mn\left[-D_x\left(\frac{m}{a}\right)^4 - 2\left[D_{xy}\left(1-\mu_x\mu_y\right) + D_x\mu_y\right]\left(\frac{m}{a}\right)^2\left(\frac{n}{b}\right)^2 - D_y\left(\frac{n}{b}\right)^4\right]} \tag{A.37}$$

while for isotropic plates it further reduces to:

$$w_{\max} = \frac{16q_0b^4}{D\pi^6} \sum_{n=1,3,5,\ldots}^{\infty} \sum_{m=1,3,5,\ldots}^{\infty} \frac{(-1)^{\frac{m+n}{2}-1}}{mn\left(m^2 + \left(\frac{b}{a}\right)^2 n^2\right)^2} \tag{A.38}$$

and for a square plate ($a = b$):

$$w_{\max} = \frac{16q_0a^4}{D\pi^6} \sum_{n=1,3,5,\ldots}^{\infty} \sum_{m=1,3,5,\ldots}^{\infty} \frac{(-1)^{\frac{m+n}{2}-1}}{mn(m^2+n^2)^2} = 0.0040624\frac{q_0a^4}{D} \tag{A.39}$$

Appendix B: large deflection of thin square isotropic plate with distributed load and movable edges

Several references present information describing the large deflection of these plates. To compare the findings, it is necessary to normalize the results to a common format. Many articles present the following normalization:

$$q = Aw + Bw^3 = N_1\frac{D}{a^4}w + N_3\frac{Eh}{a^4}w^3 \tag{B.1}$$

which leads to:

$$\frac{qa^4}{Eh^4} = N_1\frac{1}{12(1-\nu^2)}\left(\frac{w}{h}\right) + N_3\left(\frac{w}{h}\right)^3 \tag{B.2}$$

where the notations are described in the chapter above.

This normalization allows to compare the results of any plate's material–dimensions combination.

CPT: Note that the small deflection linear Classical Plate Theory (CPT) of Navier's solution for this case, with large number of terms in the summation, has:

$$q = N_1\frac{D}{a^4}w = 246.16\frac{D}{a^4}w \xrightarrow{\text{yields}} \frac{qa^4}{Eh^4} = \frac{246.16}{12(1-\nu^2)}\left(\frac{w}{h}\right) \tag{B.3}$$

Timoshenko [3] states the first summation term coefficient only: $N_1 = 240.38$.

Yankelevsky D et al. [20] (2017, Hebrew language) offer simple approximated solutions for three BCs, including the one stated here. For a single degree of freedom model (SDOF), (p.7), the plate deflection is:

$$q = 389.64\frac{D}{a^4}w + 6.3238\frac{Eh}{a^4}w^3 \tag{B.4}$$

Walter D. Pilkey [21] (2005) offers an approximated solution for rectangular plate:

$$\frac{16a^4q}{\pi^6 D} = \left(1+\beta^2\right)^2 w_0 + \frac{3.88\beta^2\left(1-\nu^2\right)}{\left(\beta^2+0.6+1/\beta^2\right)h^2}w_0{}^3, \beta = a/b \tag{B.5}$$

Then for a square plate:

$$q = 240.35\frac{D}{a^4}w + 7.4723\frac{Eh}{a^4}w^3 \tag{B.6}$$

Levy Samuel [22] in NACA Report 737 solved the problem with Fourier series, creating an infinite system of nonlinear algebraic equations to be solved. The truncation makes the solution to be an approximated one. He did that (manually!) for the first several terms, for deflections up to 3.6 times the thickness. He also states that higher deflection requires more terms to converge accurately enough. The process is rather complex and is not easy to implement.

Using the data presented in Tab. C6 and at Fig. 18.7 belonging to the report [22], we have extracted the following load–deflection expression:

$$q = 240.475\frac{D}{a^4}w + 8.95617\frac{Eh}{a^4}w^3 \tag{B.7}$$

Ishizaki Hatsuo [23] (1972) did many actual loading tests of flat glass, deflections up to 10 times the thickness or breakage. The results were transferred to Excel sheet for analysis, and the resulting expression is:

$$q = 220\frac{D}{a^4}w + 1.7787\frac{Eh}{a^4}w^3 \tag{B.8}$$

The first number N_1 (small deflection coefficient) is reasonable, but the membrane coefficient N_3 is rather low, indicating more flexible plate than in other research.

ASTM E 1300 [24] (2009) “Standard Practice for Determining Load Resistance of Glass in Buildings” supplies data for flat glass under load. Using the same thicknesses and glass properties as in Ishizaki above and deflections up to 5 times the thickness, we get the following load expression:

$$q = 243.2\frac{D}{a^4}w + 2.29\frac{Eh}{a^4}w^3 \tag{B.9}$$

Scholes A. [25] (1969) did several actual tests that said to agree well with other real tests done by Kaiser Rudolf (1936) and Stippes M. (1959). He tested 3.25 mm-thick aluminum plate and deflections up to 3.3 times the thickness.

The expression found in his work is:

$$q = 260.3\frac{D}{a^4}w + 3.547\frac{Eh}{a^4}w^3 \tag{B.10}$$

Kaiser Rudolf [26] (1936) did actual test on 600 × 600 × 3.15 mm steel plate up to 2.57 times the thickness. Although only one load–deflection point is declared, some intermediate points are implicitly given. After correcting the wrong reported load data with the supplied water manometer information, the following expression was calculated:

$$q = 251.98\frac{D}{a^4}w + 3.366\frac{Eh}{a^4}w^3 \tag{B.11}$$

Chia Chuen-Yuan [27] (1980) presents similar results like Levy [22], using the first term only (a) and eight terms (b), showing expressions for this case with deflections up to 2 times the thickness:

$$\text{(a)} \quad q = 240.35\frac{D}{a^4}w + 3.7972\frac{Eh}{a^4}w^3 \tag{B.12}$$

$$\text{(b)} \quad q = 240.35\frac{D}{a^4}w_0 + 3.9008\frac{Eh}{a^4}{w_0}^3 \tag{B.13}$$

Brown J. C. [28] (1969) did several actual experiments with rectangular aluminum plates of 0.81, 1.02, 1.29, 1.63 mm thick with deflections up to 4.6 times the thickness. The reported results were converted to the standard form:

$$q = 307.9\frac{D}{a^4}w + 2.799\frac{Eh}{a^4}w^3 \tag{B.14}$$

Present study: To complete this comparison, the expression found in the present study is:

$$q = 246.16\frac{D}{a^4}w_0 + 18.6\sqrt{\frac{h}{a}}\cdot\frac{Eh}{a^4}{w_0}^3 \tag{B.15}$$

Table B1 summarizes the coefficients found by the various authors and presented in the literature.

It is obvious that although most of the data are based on actual experiments, a considerable variability exists here, especially in the large deflection coefficient N_3. The N_3 variability may be explained with the suggested h/a ratio.

This should be further checked for possible other explanations.

Tab. B1: Summary of the N_1 and N_3 coefficients found in the literature.

No.	References	Source	N_1	N_3
1	[3]	Navier's linear CPT	246.16	–
2	[20]	Yankelevsky D. et al.	389.64	6.3238
3	[21]	Walter D. Pilkey	240.35	7.4723
4	[22]	Levy Samuel	240.475	8.9562
5	[23]	Ishizaki Hatsuo	220	1.7787
6	[24]	ASTM E 1300	243.2	2.29
7	[25]	Scholes A.	260.3	3.547
8	[26]	Kaiser Rudolf	251.98	3.366
9	[27]	Chia Chuen-Yuan	240.35	3.7972
10	[28]	Brown J. C.	307.9	2.799
11	–	Present study – for $h/a = 0.025$	246.16	2.94

Appendix C: properties of plates, loading test and FEA data

Tab. C1: Properties of plates used in vacuum chamber tests.

			Polygal	Polygal	Polygal	Polygal	Faculty lab
Thickness	h	(mm)	6	8	10	16	10
Area weight	W	(g/m^2)	1,305	1,492	1,683	2,712	1,713
Wall thickness	t_w	(mm)	0.76	0.981	1.055	1.745	1.154
Equivalent Gt	G_t^{eq}	(MPa)	110.14	106.63	91.739	94.837	100.35
Equivalent Gb	G_b^{eq}	(MPa)	157.61	135.14	121.96	122.83	124.13
	For small deflection coefficient A:						
Bending:	D_x	(Nm)	18.438	40.896	70.421	234.91	70.121
	D_y	(Nm)	14.374	38.000	58.298	205.04	54.104
	D_{xy}	(Nm)	2.8370	5.7662	10.163	41.925	10.344
	S_x	(N/m)	10,272	51,480	70,445	244,549	59,890
	S_y	(N/m)	3,956.9	2,201.8	2,365.4	3,335.3	1,662.1
	v_x^b		0.38	0.38	0.38	0.38	0.38
	$v_y^b = D_y/D_x{}^*v_x^b$		0.296	0.353	0.315	0.332	0.293
	For large deflection coefficient B:						
Tension:			6/1,300	8/1,500	10/1,700	16/2,700	10/1,700
	E_x	(MPa)	435.00	373.00	336.60	339.00	342.60
	E_y	(MPa)	304.00	294.30	253.20	261.75	276.96
	v_x^t		0.38	0.38	0.38	0.38	0.38
	$v_y^t = E_y/E_x{}^*v_x^t$		0.266	0.300	0.286	0.293	0.307

Tab. C2: Properties of plates used in the FEA.

Thickness	h	(mm)	6	8	10	16
Area weight	W	(g/m^2)	1,300	1,500	1,700	2,700
Wall thickness	t_w	(mm)	0.76	0.981	1.055	1.745
Equivalent Gt	G_t^{eq}	(MPa)	110.14	106.63	91.739	94.837
Equivalent Gb	G_b^{eq}	(MPa)	157.00	135.87	123.19	122.28
	For small deflection coefficient A:					
Bending:	D_x	(Nm)	16.3625	35.4086	62.0025	205.3052
	D_y	(Nm)	12.7562	32.9011	51.3292	179.2000
	D_{xy}	(Nm)	2.8261	5.7971	10.2657	41.7391
	S_x	(N/m)	2,88,060	2,30,560	2,94,700	4,07,840
	S_y	(N/m)	3,840	2,240	2,400	3,200
	v_x^b		0.38	0.38	0.38	0.38
	$v_y^b = D_y/D_x{}^*v_x^b$		0.296		0.353	0.315
	For large deflection coefficient B:					
Tension:			6/1,300	8/1,500	10/1,700	16/2,700
	E_x	(MPa)	433.33	375.00	340.00	337.50
	E_y	(MPa)	304.00	294.30	253.20	261.75
	v_x^t		0.38	0.38	0.38	0.38
	$v_y^t = E_y/E_x{}^*v_x^t$		0.266	0.267	0.298	0.283

In the following tables, the unit of width/length is (m), the unit of A is (Pa/m) and the unit of B is (Pa/m^3).

Tab. C3: The 6 mm plate polygal vacuum tests A, B and theory calculated A, B.

Width (m)	Length (m)	*A*	*B*	Cal *A*	Cal *B*
1.12	2.32	452.9958	26,75,924.5	838.0078	2,35,076.9
1.12	1.92	600.2268	36,24,304.7	947.2736	423,544.7
1.12	1.42	6,495.11	3,478,986.3	1,340.64	1,001,306
1.12	1.17	6,866.385	5,022,146.1	1,892.093	1,650,043
1.12	0.92	8,989.318	9,007,183.6	3,341.16	2,892,938
0.97	2.32	1,805.044	689,141.94	1,394.157	280,299.7
0.97	1.92	3,589.357	1,571,819.9	1,510.433	518,885.1
0.97	1.42	9,345.895	2,682,268.9	1,947.39	1,284,722
0.97	1.17	11,092.35	5,344,922.6	2,553.413	2,183,605
0.97	0.92	14,659.1	9,617,604.9	4,106.67	3,979,952
0.9	2.32	2,846.633	1,042,440.4	1,833.756	3,04,838
0.9	1.92	3,992.159	2,319,479.3	1,950.029	572,014.2
0.9	1.42	8,516.378	6,275,772.6	2,409.381	1,449,841
0.9	1.17	9,318.191	7,315,412.6	3,048.365	2,503,720
0.9	0.92	30,982.47	7,447,190.6	4,668.252	4,656,408

Tab. C3 (continued)

Width (m)	Length (m)	*A*	*B*	Cal *A*	Cal *B*
0.72	2.32	7,953.529	2,283,356.5	4,250.167	380,062.2
0.72	1.92	9,081.767	3,334,754.5	4,334.273	740,756.6
0.72	1.42	15,608.04	6,558,901	4,830.433	2,006,872
0.72	1.17	17,281.05	10,784,310	5,576.681	3,628,605
0.72	0.92	21,693.31	23,693,730	7,445.51	7,160,451
0.72	0.72	30,733.39	32,097,457	11,636.92	13,408,479
0.52	2.32	17,991.24	4,017,196	14,732.46	488,196
0.52	1.92	19,992.22	7,342,052.5	14,648.92	998,573.1
0.52	1.67	18,881.28	12,322,531	14,669.48	1,663,546
0.52	1.42	20,924.37	3,4,600,895	14,902.81	2,951,309
0.52	1.17	31,266.07	32,159,471	15,684.94	5,676,341
0.52	0.92	39,697.58	44,985,911	17,947.36	12,157,585
0.52	0.72	41,093.97	110,628,017	23,031.12	24,851,580

Tab. C4: The 8 mm plate polygal vacuum tests A, B and theory calculated A, B.

Width (m)	Length (m)	*A*	*B*	Cal *A*	Cal *B*
1.12	2.32	10,615.63	1,714,038	1,998.142	326,809.6
1.12	1.92	9,658.971	3,702,746	2,240.471	588,822
1.12	1.67	12,600.17	3,940,898	2,546.095	887,351.3
1.12	1.42	12,449.55	5,276,544	3,126.229	1,392,040
1.12	1.17	11,817.76	6,953,640	4,378.486	2,293,929
1.12	0.92	14,593.63	11,593,336	7,702.517	4,021,832
1.12	0.72	17,232.15	13,744,086	15,946.62	6,673,482
0.97	2.32	12,245.53	2,858,078	3,233.376	389,679.3
0.97	1.92	12,727.31	6,404,959	3,482.44	721,366.5
0.97	1.67	16,119.69	7,413,828	3,812.679	1,110,141
0.97	1.42	18,888.9	12,222,746	4,445.654	1,786,052
0.97	1.17	18,311.34	15,101,302	5,799.834	3,035,700
0.97	0.92	23,887.15	22,780,661	9,315.615	5,533,025
0.97	0.72	30,987.88	27,779,780	17,806.58	9,546,048
0.9	2.32	12,629.23	2,398,253	4,174.936	423,793
0.9	1.92	8,274.929	3,639,945	4,417.666	795,227.9
0.9	1.67	8,759.203	4,771,837	4,756.228	1,236,838
0.9	1.42	17,964.97	82,29,842	5,415.468	2,015,603
0.9	1.17	12,983.75	14,426,740	6,827.399	34,80,732
0.9	0.92	18,981.43	17,899,379	10,459.87	6,473,451
0.9	0.72	24,429.51	35,284,855	19,115.65	11,400,203
0.72	2.32	22,097.52	3,652,679	8,994.497	528,371.7
0.72	1.92	24,121.8	4,877,503	9,143.279	1,029,818
0.72	1.67	14,853.55	15,352,886	9,456.951	1,650,391
0.72	1.42	16,587.32	1058,9949	10,156.51	2,790,001
0.72	1.17	15,934.62	8,425,231	11,731.1	5,044,573
0.72	0.92	21,440.99	29,554,347	15,759.63	9,954,631
0.72	0.72	21,748.56	33,151,357	25,043.26	18,640,790

Tab. C5: The 10 mm plate polygal vacuum tests A, B and theory calculated A, B.

Width (m)	Length (m)	*A*	*B*	Cal *A*	Cal *B*
1.12	2.32	5,279.765	892,109.2	2,897.421	403,685.8
1.12	1.92	6,410.03	1,607,430	3,286.038	727,332
1.12	1.67	5,226.576	4,155,473	3,775.575	1,096,085
1.12	1.42	10,148.55	4,496,982	4,709.153	1,719,493
1.12	1.17	14,394.48	6,498,713	6,742.018	2,833,536
1.12	0.92	19,032.32	11,718,453	12,205.65	4,967,897
1.12	0.72	28,726.07	21,991,590	25,938.38	8,243,301
0.97	2.32	5,008.443	1,513,391	4,596.355	481,344.5
0.97	1.92	7,734.193	1955,618	4,997.688	891,055.2
0.97	1.67	7,097.357	464,3788	5,523.879	1,371,281
0.97	1.42	12,476.48	5,000,107	6,532.75	2,206,189
0.97	1.17	14,198.68	10,234,183	8,706.231	3,749,795
0.97	0.92	20,930.16	15,246,556	14,421.05	6,834,571
0.97	0.72	37,527.73	19,321,951	28,430.69	11,791,587
0.9	2.32	8,175.153	1,308,174	5,870.023	523,482.9
0.9	1.92	9,144.266	2,979,196	6,264.635	982,291.3
0.9	1.67	13,540.09	2,623,564	6,804.076	1,527,783
0.9	1.42	14,152.08	4,766,794	7,850.599	248,9738
0.9	1.17	11,179.3	16,477,742	10,103.28	4,299,513
0.9	0.92	20,422.31	43,633,063	15,970.46	7,996,216
0.9	0.72	27,507.29	48,216,872	30,171.04	14,081,899
0.72	2.32	9,764.093	1,210,446	12,210.72	652,661.8
0.72	1.92	11,121.45	2,057,851	12,480.63	1,272,064
0.72	1.67	11,484.15	3,560,990	12,994.57	2,038,616
0.72	1.42	15,098.11	3,883,557	14,105.48	3,446,300
0.72	1.17	15,530.53	10,511,290	16,582.48	6,231,220
0.72	0.92	28,161.24	16,177,794	22,962.68	122,96,282
0.72	0.72	21,112.22	91,644,642	37,902.26	23,025,708
0.52	2.32	19,650.88	1,438,809	34,293.65	838,354.4
0.52	1.92	18,181.98	4,628,025	33,972.7	1,714,799
0.52	1.67	22,336.73	5,604,267	34,048.95	2,856,725
0.52	1.42	21,800.6	8,182,871	34,814.92	5,068,134
0.52	1.17	27,381.58	14,058,050	37,255.82	9,747,694
0.52	0.92	43,811.35	19,654,856	44,249.42	20,877,610
0.52	0.72	40,750.83	78,821,611	60,519.49	42,676,370

Tab. C6: The 16 mm plate polygal vacuum tests A, B and theory calculated A, B.

Width (m)	Length (m)	*A*	*B*	Cal *A*	Cal *B*
1.12	2.32	8,933.279	1,421,073	8,242.064	832,170.8
1.12	1.92	9,052.37	1,657,316	9,467.222	14,99,345
1.12	1.42	12,481.09	6,361,606	13,963.99	3,544,618
1.12	1.17	16,163.43	11,237,602	20,463.26	5,841,141
1.12	0.92	35,908.31	13,121,229	38,217.02	102,40,980

Tab. C6 (continued)

Width (m)	Length (m)	*A*	*B*	Cal *A*	Cal *B*
0.97	2.32	9890.783	1,140,765	12,333.83	992,258.9
0.97	1.92	10,953.37	2,569,300	13,586.46	1,836,850
0.97	1.42	16,100.85	7,337,956	18,328.21	4,547,908
0.97	1.17	21,039.46	11,380,187	25,112.24	7,729,946
0.97	0.92	42,337.42	16,439,904	43,280.84	14,089,003
0.9	2.32	13,151.49	1,481,442	15,239.14	1,079,124
0.9	1.92	17,183.46	2,547,259	16,475.18	2,024,927
0.9	1.42	21,875.81	11,601,028	21,323.35	5,132,426
0.9	1.17	32,690.09	21,102,044	28,261.72	8,863,153
0.9	0.92	48,325.19	27,733,684	46,679.89	16,483,654
0.72	2.32	21,365.54	1,614,211	28,516.63	1,345,418
0.72	1.92	19,605.54	3,484,898	29,481.13	2,622,273
0.72	1.42	27,959.72	8,239,599	34,419.8	7,104,312
0.72	1.17	32,710.82	14,309,190	41,783.57	12,845,236
0.72	0.92	48,831.18	21,922,037	61,033.98	25,347,948
0.52	2.32	40,168.46	122,204.1	68,615.16	1,728,210
0.52	1.92	39,558.79	12,311,136	68,271.49	3,534,942
0.52	1.42	34,807.32	18,890,827	71,902.62	10,447,613
0.52	0.92	56,908.98	48,134,886	99,778.62	43,037,770

Tab. C7: The 10 mm plate faculty lab vacuum tests A, B and theory calculated A, B.

Width (m)	Length (m)	*A*	*B*	Cal *A*	Cal B
1.13	2.33	3,658.214	648,934.7	2,479.251	413,461.1
0.73	1.43	9,602.007	5,370,164	11,487.23	3,475,817
1.43	0.73	19,061.98	5,858,385	21,805.59	4,378,295
0.73	0.73	27,613.32	36,089,724	32,835.42	22,731,859

Tab. C8: The 6 mm plate FEA A, B and theory calculated A, B.

Width (m)	Length (m)	*A*	*B*	Cal *A*	Cal *B*
1.2	2.4	1,352.5	195,200.78	583.00495	192,742.6
1.2	2	1,690.6	342,476.002	664.74488	337,874.6
1.2	1.5	2,435.8	771,141.611	943.8827	762,498.9
1.2	1	3,780.8	2,096,942.96	2,237.456	2,054,472
1.2	0.8	5,459.8	3,230,529.03	4,304.5028	3,277,946
1	2.4	2,020.9	237,038.006	1,106.511	241,597.5
1	2	2,427.3	432,414.367	1,196.5063	437,817.7
1	1.5	3,415.9	1,046,885.37	1,517.8621	1,044,852
1	1	5,436.3	3,136,751.28	2,956.0945	3,057,565
1	0.8	7,312.3	5,186,080.43	5,156.837	5,104,099

Tab. C8 (continued)

Width (m)	Length (m)	*A*	*B*	Cal *A*	Cal *B*
0.8	2.4	3,697.7	286,277.87	2,538.5252	305,448.5
0.8	2	4,208	537,577.951	2,619.1872	574,768.8
0.8	1.5	5,508.9	1,396,552.89	2,984.2191	1,463,440
0.8	1	8,723.5	4,727,916.95	4,665.1049	4,727,700
0.8	0.8	11,347	8,469,456.9	7,125.0737	8,345,855
0.6	2.4	7,826.7	440,304.645	7,656.8973	388,706.2
0.6	2	8,547.9	767,570.33	7,650.9892	763,622.2
0.6	1.5	10,104	2,014,716.54	7,949.0107	2,097,575
0.6	1	14,283	7,744,919.53	9,969.4304	7,645,229
0.6	0.8	18,211	1,53,10,356.5	12,938.884	14,487,347

Tab. C9: The 8 mm plate FEA A, B and theory calculated A, B.

Width (m)	Length (m)	*A*	*B*	Cal *A*	Cal *B*
1.2	2.4	2,831.98	209,943	1,391.957	268,758.5
1.2	2	3,318.587	382,338.8	1,571.233	471,129.2
1.2	1.5	4,829.209	836,190.2	2,188.248	1,063,221
1.2	1	8,415.004	2,126,931	5,050.915	2,864,736
1.2	0.8	12,239.55	3,542,515	9,634.085	4,570,738
1	2.4	3,826.083	282,870.4	2,576.69	336,881.3
1	2	4,546.522	509,018.9	2,766.492	610,488.9
1	1.5	6,349.96	1,214,195	3,460.85	1,456,932
1	1	10,850.79	3,487,875	6,598.992	4,263,441
1	0.8	15,045.61	6,025,413	11,425.3	7,117,108
0.8	2.4	6,466.841	378,227.5	5,618.453	425,914.5
0.8	2	7,303.275	683,105.1	5,771.521	801,452.3
0.8	1.5	9,412.111	1,737,764	6,521.063	2,040,607
0.8	1	15,196.44	5,805,936	10,074.37	6,592,261
0.8	0.8	20,606.35	10,550,171	15,335.99	11,637,383
0.6	2.4	13,854.28	637,013.4	15,284.07	542,008.4
0.6	2	14,604.9	1,072,049	15,231.35	1,064,788
0.6	1.5	16,662.53	2,699,289	15,760.38	2,924,839
0.6	1	22,717.44	10,523,348	19,741.63	10,660,436
0.6	0.8	29,149.22	21,212,249	25,735.51	20,201,022

Tab. C10: The 10 mm plate FEA A, B and theory calculated A, B.

Width (m)	Length (m)	*A*	*B*	Cal *A*	Cal *B*
1.2	2.4	3,471.776	266,698.2	2,074.733	332,895.2
1.2	2	4,277.692	464,471	2,367.607	583,559.8
1.2	1.5	6,167.916	10,34,814	3,378.928	1,316,949

Tab. C10 (continued)

Width (m)	Length (m)	*A*	*B*	Cal *A*	Cal *B*
1.2	1	11,004.35	2,840,307	8,155.409	3,548,379
1.2	0.8	16,640	5,287,136	15,925.41	5,661,502
1	2.4	5,041.094	342,169.1	3,752.39	417,274.8
1	2	5,969.138	613,404.1	4,063.639	756,176.4
1	1.5	8,251.718	1,477,673	5,190.444	1,804,615
1	1	14,203.62	4,449,369	10,353.93	5,280,871
1	0.8	20,480.89	8,169,863	18,434.41	88,15,540
0.8	2.4	8,484.767	470,300.6	7,924.851	527,555.1
0.8	2	9,522.598	841,360.8	8,188.214	992,711.5
0.8	1.5	12,163.73	2,145,383	9,403.726	2,527,578
0.8	1	19,432.33	7,450,490	15,143.81	8,165,442
0.8	0.8	26,721.05	14,272,233	23,786.57	14,414,536
0.6	2.4	18,260.27	779,447.6	20,585.35	671,353.6
0.6	2	19,010.76	1,319,853	20,543.94	1,318,889
0.6	1.5	21,417.6	3,330,184	21,477.9	3,622,825
0.6	1	29,186.41	13,096,619	27,827.71	13,204,450
0.6	0.8	38,296.02	26,541,955	37,407.44	25,021,808

Tab. C11: The 16 mm plate FEA A, B and theory calculated A, B.

Width (m)	Length (m)	*A*	*B*	Cal *A*	Cal *B*
1.2	2.4	7,300.703	539,625.82	6,043.828	680,603.8
1.2	2	8,756.195	962,335.5	6,968.5	1,193,087
1.2	1.5	12,494.57	2,256,154.1	10,164.26	2,692,501
1.2	1	25,073.1	7,181,867.8	25,486.15	7,254,656
1.2	0.8	48,174.93	107,14,430	50,808.47	11,574,934
1	2.4	10,234.71	754,218.1	10,192.25	853,117.9
1	2	11,604.14	1,394,012.3	11,158.47	1,546,002
1	1.5	15,309.64	3,595,719.6	14,616.61	3,689,533
1	1	29,275.91	11,815,887	30,707.89	10,796,736
1	0.8	52,687.53	18,995,236	56,414.56	18,023,361
0.8	2.4	18,117.44	910,655.82	19,449.97	1,078,586
0.8	2	19,822.02	1,640,282.7	20,303.23	2,029,597
0.8	1.5	24,555.24	4,306,897.9	23,945.99	5,167,631
0.8	1	40,674.54	15,820,108	41,107.7	16,694,237
0.8	0.8	64,001.02	29,188,219	67,574.55	29,470,502
0.6	2.4	37,083.77	1,376,602.7	43,775.8	1,372,582
0.6	2	37,921.72	2,428,218.6	43,952.9	2,696,468
0.6	1.5	41,970.62	6,360,732.5	47,090.34	7,406,861
0.6	1	59,121.26	25,612,031	65,386.7	26,996,483
0.6	0.8	83,851.03	52,164,977	93,201.39	51,157,057

Appendix D: FEA description

The analysis

The FEA software is the Siemens Simcenter Femap with Nastran Ver. 2021.1. The FEA process has three basic steps: model preparation, running the solver and post-processing.

The model preparation

To get correct results, we use a consistent system of units:
- Length – mm
- Force – N
- Stress, elastic modulus – MPa [N/(mm)^2]
- Shear rigidity – N/mm

Materials and property

In Femap, the 2D plate element property allows to define a different material for every one of the four elastic response modes:
- In-plane tension/compression and shear
- Out-of-plane bending and twist
- Cross section transverse shear deformation
- Coupling of membrane bending due to asymmetric structure

This ability is suitable to correctly represent the multiwall plate equivalent elastic constants. In our case, however, since the multiwall plate is orthotropic symmetric, the last coupling material is not necessary and therefore left ignored.

We define three 2D orthotropic materials: TensionMat, BendingMat and ShearMat, and fill their elastic moduli fields with the values found before (see Appendix C).

Geometry, boundary conditions (BC) and load

The geometry in use is rectangular surfaces, length in x direction and width in y direction. The edges BCs (constraints) are simply supported, in-plane movable on all edges, designated SSSS-M. In Femap, these constraints are designated "3" for z translation only.

The plate is loaded with an evenly distributed load, realized as pressure on the surface. This pressure load is operating perpendicular to the surface, meaning that it slightly changes its direction, while the plate deflects: it is a follower force. This type of load is best to simulate wind load or vacuum chamber test.

Meshing

Next, we mesh the model with the plate property. A relatively small number of elements are enough here, as the equivalent element considers only the major response, neglecting all plate's small features. A 50 × 50 mm element was used, resulting in various number of square Quad 4-noded elements. The default direction of the orthotropic element x direction is naturally x, so no explicit definition of the orientation is necessary.

More BCs after meshing

The analysis program requires that all rigid body degrees of freedom (DOFs) will be eliminated. Therefore, virtual BCs (not participating in the analysis) should be added to cancel these DOFs. The plate midpoint node gets 12 (no translation in x,y). Also, one plate edge midpoint gets a constraint that eliminates the free rigid body z rotation around the plate's midpoint and also eliminates z deflection (13 or 23).

Running the solver

Since large deflection with a nonlinear response is expected, it is necessary to use the nonlinear static solution method. The solver uses NX-Nastran SOL-106 routine.

Post-processing

Femap allows to get the load–deflection graphs at the requested nodes. The graph data are then transferred to Excel that produces graphs as shown in Fig. 18.3. The Excel sheet also performs theoretical least-squares regression analyses to find the coefficients A and B for each plate.

Figure D1 shows a typical FEA screen where the mesh, BCs, loads and deflections are presented.

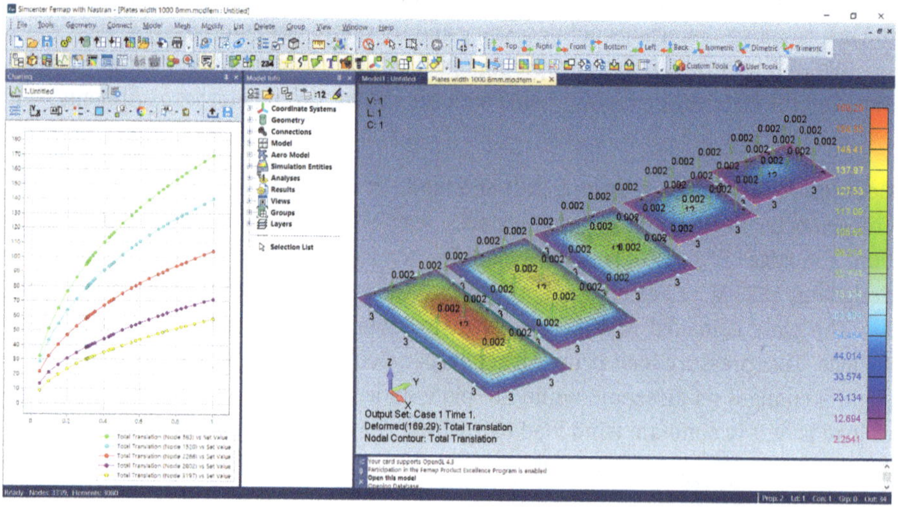

Fig. D1: Femap typical screen.

19 Investigation of the behavior of large aspect ratio thin rectangular plates under lateral pressure

19.1 Introduction

Thin and thick plates are known to be the basic blocks in various engineering sectors, like aerospace, civil and mechanical engineering. Moreover, these plates may undergo large deflections, and for some applications, their length-to-width ratio or aspect ratio (*AR*) tends to be large.

As a result, many research articles and books have been published on thin rectangular plates with transverse loads, for both small and large deflections. Many of them include (in various forms) a general statement that "long plates with a high *AR* can be treated as an infinitely long plate"; i.e., the remote ends do not affect the plate's mechanical performance, and the plate would experience its highest out-of-plane deflection (see also [1]). For instance, Timoshenko [2] states that "*the deflections of finite plates with b/a* < ⅔ (where *b* and *a* are the width and length of the plate, respectively) *are very close to those obtained for an infinitely long plate.*" Note that this statement would relate to *AR* > 1.5. Also, the stresses for *AR* = 2 are said to be 10% lower than those for an infinitely long plate. Reddy [3] provides information for the midpoint deflection and bending moments for plates with *AR* = 2 and less. Bakker et al. [4] refer to an *AR* of up to 2. Longer plates are not considered. Further information regarding the behavior of plates under loading can be found in [5–10]. Wang and El-sheikh [11] present deflection information up to *AR* = 10, restricting themselves to immovable boundary conditions (namely, the plate edges cannot move relative to their support frame). Shao [12], in his thesis, investigated high *AR* plates up to *AR* = 5, with movable edges, but the edges were forced to remain straight. The reason for this *AR* = 5 limit is that "*beyond this value, the behavior of the plate is nearly as a strip.*" Razdolsky [13] also investigated thin rectangular plates up to *AR* = 4, with simply supported immovable edges. He presents deflections and stresses in a graphical form for specific points on the plate. For the general case, a rather complex algorithm is suggested using multiple summations of trigonometric functions. Ostiguy and Evan-lwanowski [14] also checked the influence of the aspect ratio on the dynamic response and stability of rectangular plates under the large-deflection regime. They found that "*AR plays a crucial role in determining the stability of rectangular plates.*"

Note: The chapter is based on the manuscript: Hakim, G. and Abramovich, H., The behavior of long thin rectangular plates under normal pressure- A thorough investigation, Materials 2024, 17, ID article 2,902, 15p. under a MDPI CC-BY license.

https://doi.org/10.1515/9783111621104-019

The present chapter presents a thorough investigation into the behavior of long plates under normal pressure, challenging the above-quoted statement by checking the performance of high *AR* plates, for both deflections and stresses. Innovative, interesting results suggest another perspective for these long plates, showing higher deflections at specific mid-width points.

19.2 Materials, methods and results

To investigate the behavior of long plates with small deflections, the classical Navier solution for the midpoint small deflection of a thin rectangular plate was analyzed for large *AR*s. The solution presented in Reddy's work [3] is brought here again for clarification and is presented as follows:

$$w_{\max} = \frac{16qb^4}{D\pi^6} \sum_{n=1,3,5,\ldots}^{\infty} \sum_{m=1,3,5,\ldots}^{\infty} \frac{(-1)^{\frac{m+n}{2}-1}}{mn\left(m^2\left(\frac{b}{a}\right)^2+n^2\right)^2} \tag{19.1}$$

Then, the midpoint deflection expression was calculated for several aspect ratios (*AR*'s) $\left({}^{a}\!/_{b}\right)$ using the polycarbonate data presented in Tab. 19.1. The E and ν values of this isotropic linear elastic material were taken from manufacturer's commercial data sheets [15, 16].

Tab. 19.1: Structural and material data used to evaluate eq. (19.1) for various AR values.

Variable	Notation	Dimension	Value
Young's modulus	E	GPa	2.4
Poisson's ratio	ν	–	0.38
Thickness	*h*	m	0.005
Length	*a*	*Varies*	Various
Width	*b*	m	1
Distributed load	*q*	Pa	100
The calculated bending rigidity	$D = Eh^3/12(1-\nu^2)$	Nm	29.219
Summation indexes	*m* and *n*	–	1, 3, 5, . . ., 31

The calculated midpoint deflection for a square plate was found to be $w_{max} = 13.9$ mm.

The relative mid-deflection, defined as the ratio between the actual lateral displacement and the maximal deflection of a square plate, as a function of *AR* is depicted in Fig. 19.1.

From the graphs presented in Fig. 19.1, it is obvious that plates with a high finite *AR* (in the region of *AR* = 10) would deflect more than an infinitely long plate. The difference is just 1% and can be considered negligible, but it still contradicts the well-

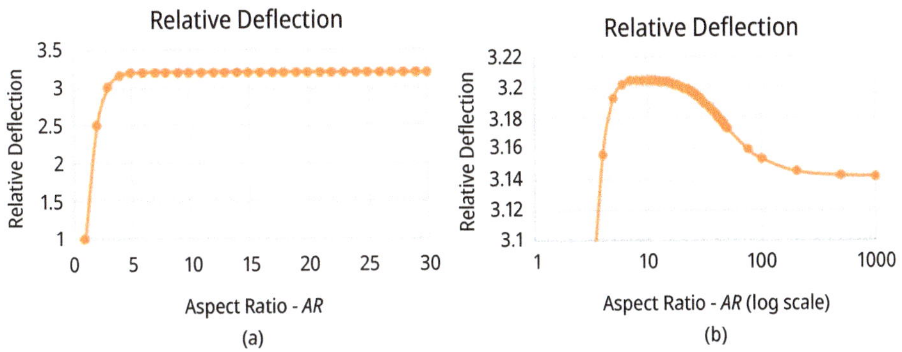

Fig. 19.1: Navier small-deflection theory – relative deflection versus *AR*: (**a**) full scale and (**b**) peak vicinity.

known general statement that "long plates with a high aspect ratio *AR* can be treated as an infinitely long plate."

To check the importance of the boundary conditions, an all-around clamped plate was investigated with increasing aspect ratios, in a similar way to that shown above for all-around simply supported boundary conditions. Since a simple solution for the case of all-around clamped boundary conditions is not available, a finite element analysis (FEA) method was used. The analysis code was Simcenter Femap with Nastran ver. 2021.1 from Siemens [17].

The dimensions and the mechanical properties, presented in Tab. 19.1, were used to yield the two graphs presented in Fig. 19.2. A pressure load of 25 Pa was used to have a small midpoint deflection of about half-plate thickness, while the midpoint deflection for a square plate was found to be $w_{max} = 1.1035$ mm.

As for the all-around simply supported case, as well as for the all-around clamped case, there is a small peak at $AR = 3$, which is about 0.5% larger than the infinite-length deflection case. So again, although the difference is just 0.5% and can be considered negligible, it still contradicts the well-known general statement that "long plates with a high aspect ratio, *AR*, can be treated as an infinitely long plate."

The present chapter will now proceed to investigate the large-deflection regime. While, for a small-deflection regime, the in-plane effects are generally negligible, for a large-deflection regime that exceeds the plate thickness, the in-plane effects play a major role and must be considered. The additional elastic energy required to strain the plate in its plane causes the plate to deflect less than expected when using the linear small-deflection theory.

To investigate the behavior of a thin plate with a high *AR*, in the large-deflection regime, rectangular plates were addressed. The geometry and the material data are presented in Tab. 19.1. Note that the length is defined in the x direction, while the width is in the y direction, as shown in Fig. 19.3. To enable large deflections, the plate was loaded with a uniform pressure, $q = 1{,}000$ Pa.

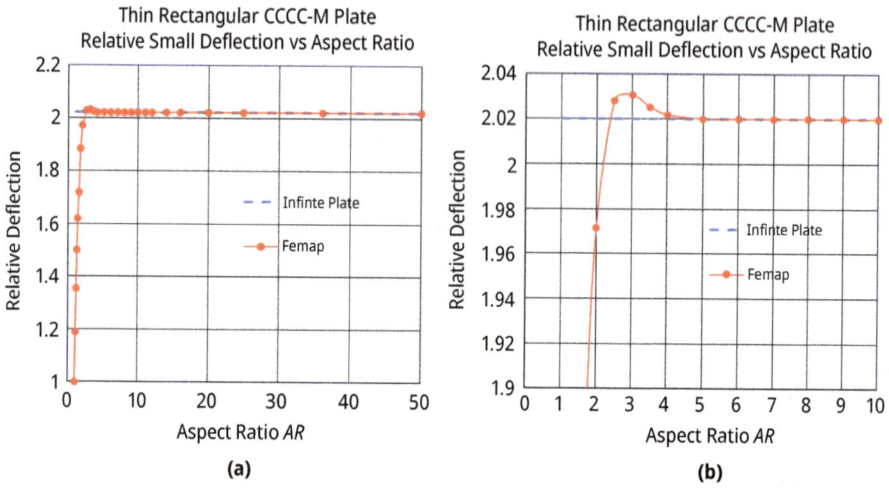

Fig. 19.2: Thin rectangular all-around clamped conditions in the small-deflection regime (FE results) relative deflection versus *AR*: (**a**) full scale and (**b**) peak vicinity.

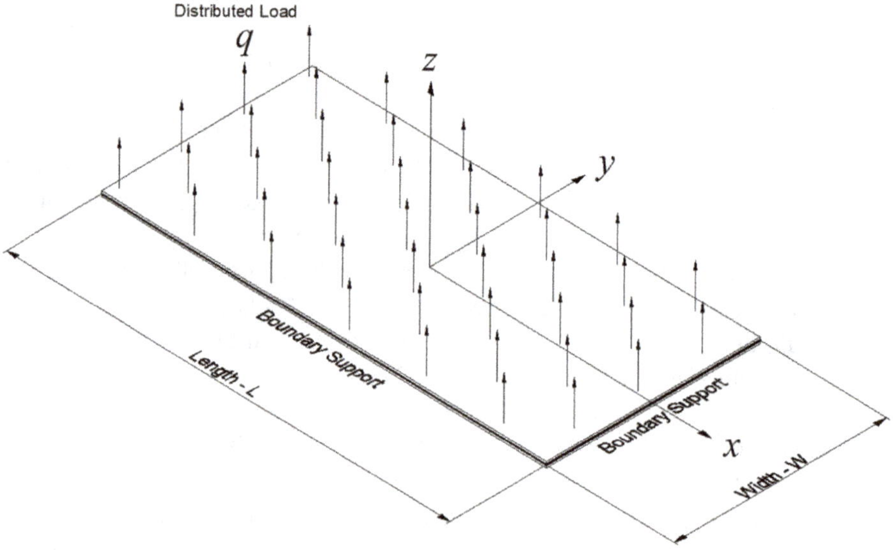

Fig. 19.3: The plate schematic model for the large-deflection regime.

In the plate's large-deflection regime, the in-plane boundary conditions (BCs) have a significant influence on both the deflection and the generated membrane stresses. Since there are many possible BC combinations, in the present study, we dealt with only two representative BCs: a four-side simply supported designated SSSS, where there are no bending moments at the various edges, and a four-side clamped designated CCCC, where no rotations are allowed at the various edges.

For the in-plane BC, we used two types, which are described hereafter. The movable four sides, where the plate edges can freely move relative to the support frame in both the parallel and perpendicular directions, are designated as **M**. The four immovable sides, where the plate edges cannot move relative to the support frame, are written as **I**.

Finite element analyses (FEAs) for several types of rectangular plates were performed. As for the case of small deflections presented above, the analysis code was Femap 2021.1 from Siemens, with Simcenter Nastran [17] as the code processor. A nonlinear static analysis was performed for quad-type plate elements with a 50 mm element size. The nonlinear code increased the load by 20 steps, while in each step, and the deflections and stresses were recalculated and used as a starting point for the next step. Each of these steps had internal iterations to verify its convergence. Upon completion of the run, all final deflections and membrane forces of the entire plate were transferred to an Excel sheet. Also, graphic pictures of the FEA results were saved for further processing. Then, the data and figures were further evaluated for significant findings. For further details, see Hakim and Abramovich [18].

One should note that during nonlinear analysis, due to the large-deflection status, the plate elements would change their special direction. Therefore, one must consider the load direction acting on the plate elements. Two types of distributed loads are commonly used within the FE code. The first one is a vector-oriented load direction, in which the load direction is fixed in space and does not follow the element direction's special changes. The second one is a pressure load type that operates perpendicularly to the element surface and follows the special change in the element direction, namely a follower force. These two load types represent clear physical situations. The vector-oriented distributed load represents snow load, where the load direction is the weight of the snow acting downward. The distributed pressure load represents a wind load that operates perpendicularly to the plate element and follows its special directions while loaded. In the present study, we used the vector-oriented distributed load, applying it to obtain the plate's response to distributed equal load.

Typical results are presented in Figs. 19.4–19.7, presenting the calculated out-of-plane deflections, the membrane x forces, the y forces and the shear xy forces for the SSSS-M case. The numbers near the plates are their respective *ARs*, while SFSF means that the two remote ends (left and right) are free and the two sides (up and down) are simply supported, representing an infinitely long side-supported plate.

The relatively high Y membrane force near the ends of the SFSF is probably related to the sudden discontinuity of the plate combined with the Poisson effect. Nevertheless, as the SFSF represents an infinitely long plate, this end effect is ignored here.

Based on the results presented in Fig. 19.4, the influence of the aspect ratio (*AR*) on the out-of-plane deflections was investigated.

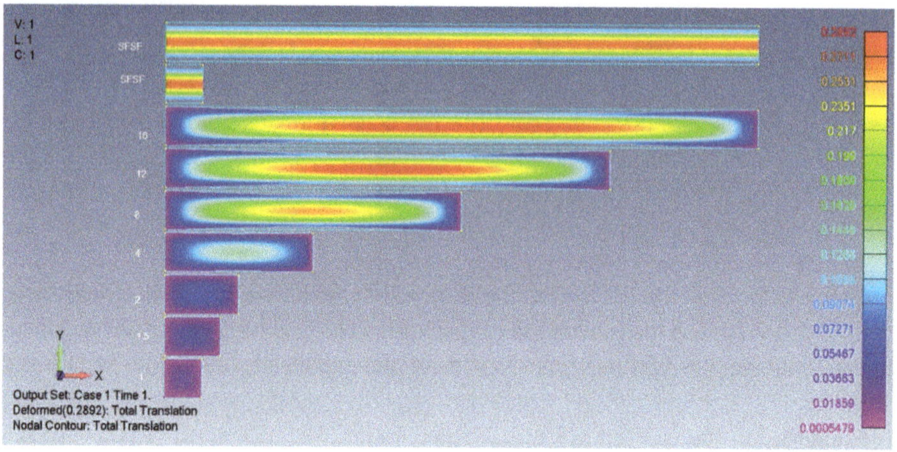

(a)

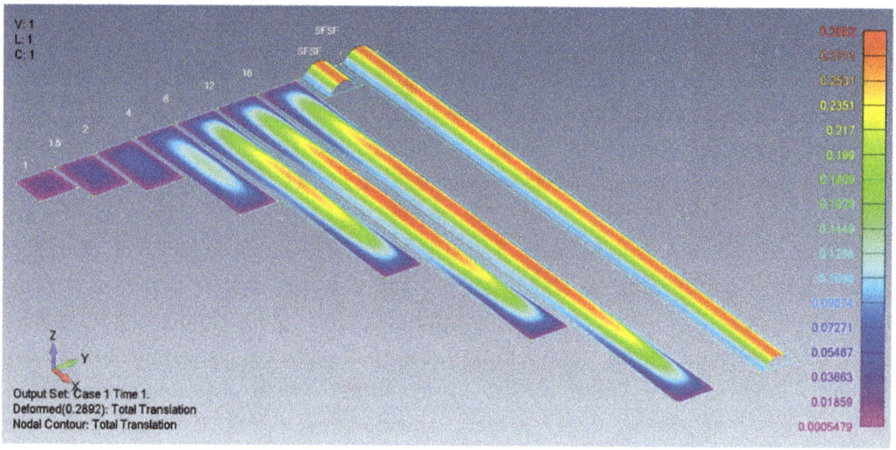

(b)

Fig. 19.4: Deflections of rectangular isotropic plates: (a) 2D view and (b) 3D view.

One may expect that the out-of-plane deflection for plates with a high aspect ratio would asymptotically approach the deflection of an infinitely long plate, with the long plate having the highest deflection.

However, checking the deflection against increasing values of *AR* reveals an interesting phenomenon for the movable BC (M) cases. To highlight it, the midpoint deflections for several plates were normalized by the deflection of a square plate ($AR = 1$). The relative deflection is defined as the ratio of the midpoint deflection to the midpoint deflection of a square plate. The resulting graphs are presented in Figs. 19.8 and 19.9.

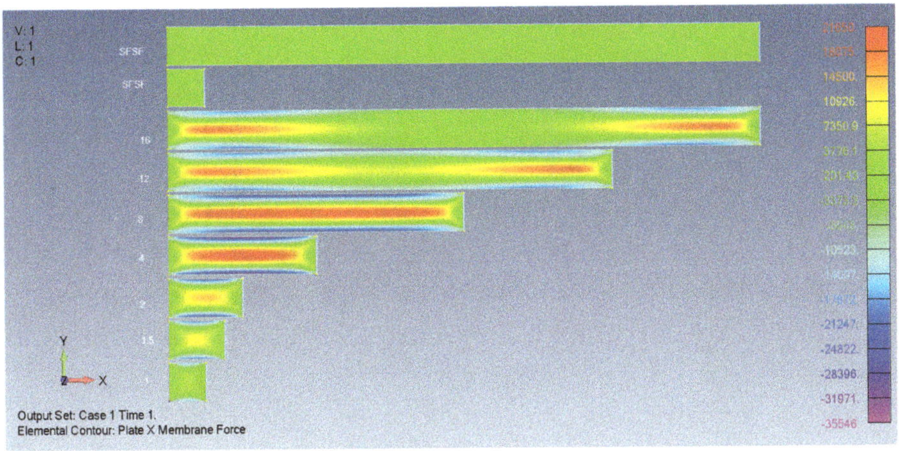

(a)

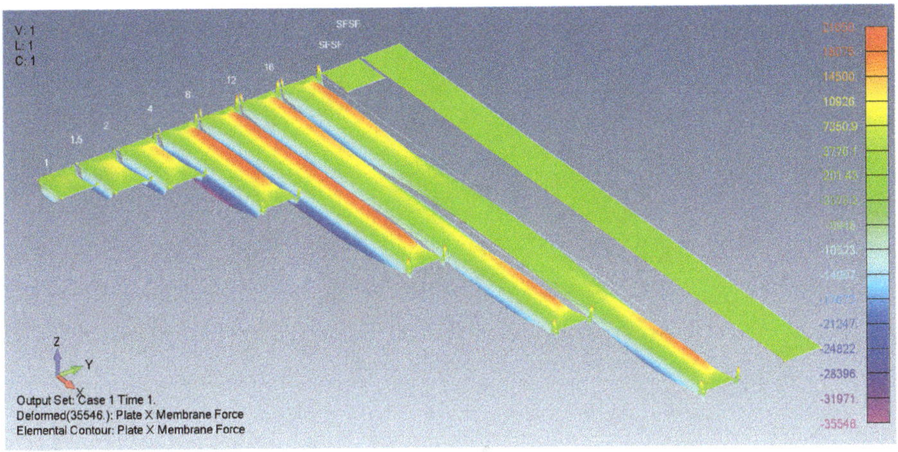

(b)

Fig. 19.5: *X* membrane forces of rectangular isotropic plates: (a) 2D view and (b) 3D view.

As shown in Fig. 19.8, the plate's deflections for $8 < AR < 20$ are higher than that of an infinitely long plate with a maximum $AR = 12$. This unexpected behavior persists with other Young's moduli and other Poisson's ratios ν, including $\nu = 0$.

Identical results were obtained using another piece of FEA software, Ansys 2023/R2, as shown in Fig. 19.8.

Ansys is a large piece of scientific analysis code described partially in [19, 20]. The section used in the present chapter was Ansys Workbench with its static structural finite element section. In the analysis, the large-deflection option was activated to correctly account for the in-plane strains.

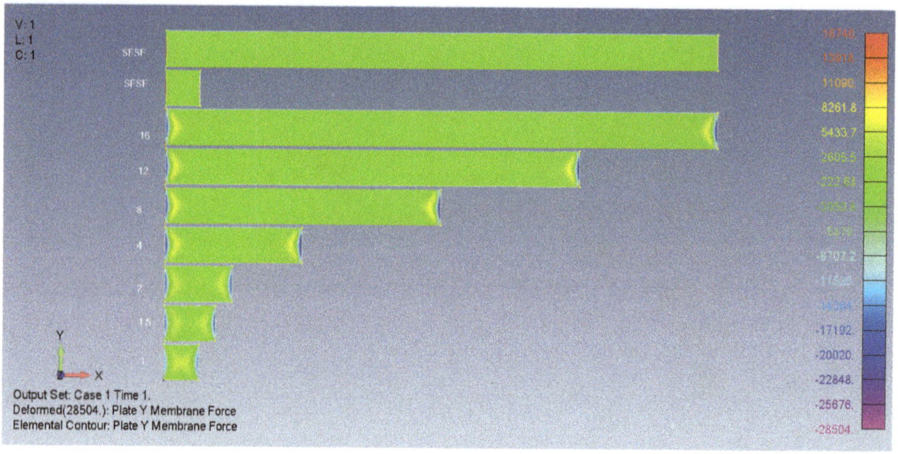

(a)

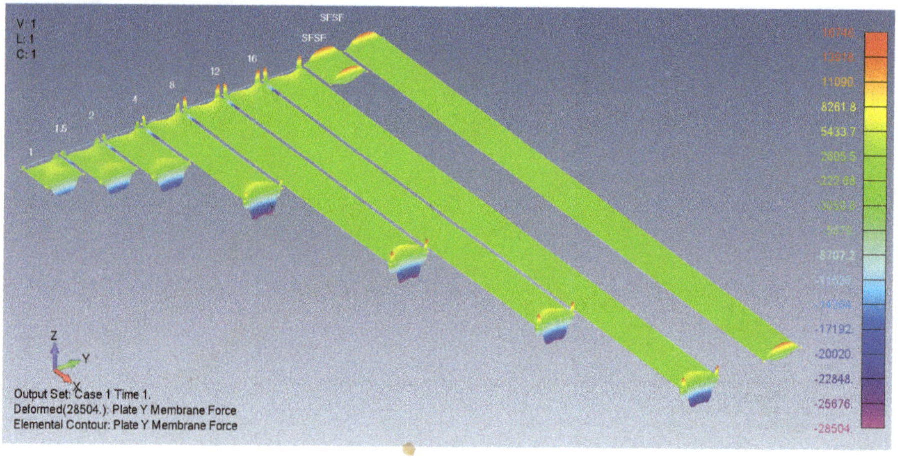

(b)

Fig. 19.6: *Y* membrane forces of rectangular isotropic plates: (a) 2D view and (b) 3D view.

To present the behavior shown in Fig. 19.8 more conveniently, an empirical formula describing the relative midpoint deflection, *RD*, as a function of *AR* using a minimum squares regression was performed. Excel's Solver add-in was used to minimize the squares sum by optimizing the formula coefficients. This empirical formula for the SSSS-M BC case is presented as follows:

$$RD = B + C \cdot \exp(D \cdot AR) \cdot \sin(E \cdot AR + F) \tag{19.2}$$

The optimal coefficients were found to be $B = 9.3078$, $C = -11.167$, $D = -0.20872$, $E = 0.24266$, and $F = 0.90972$, with an excellent correlation coefficient of $R = 0.99996$. The valid range of

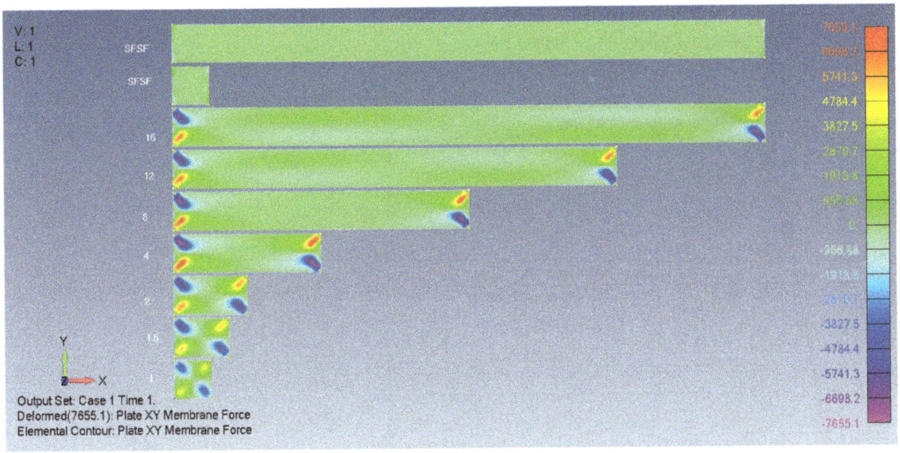

(a)

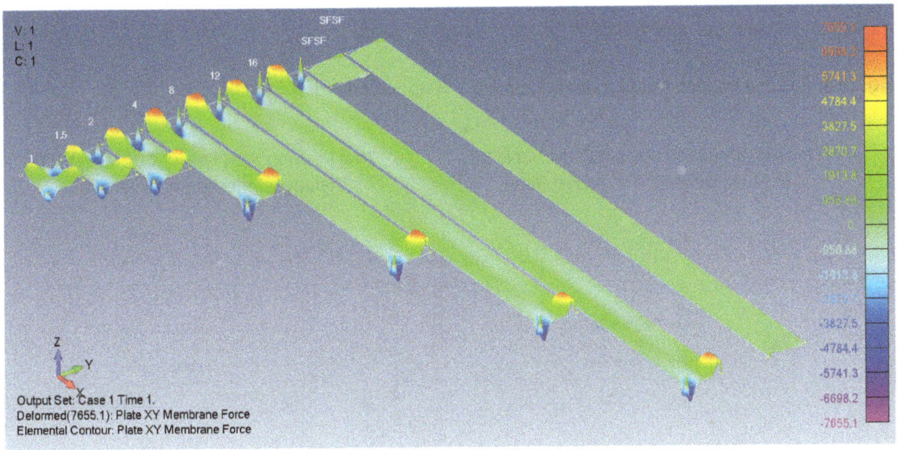

(b)

Fig. 19.7: *XY* membrane forces of rectangular isotropic plates: (a) 2D view and (b) 3D view.

AR in eq. (19.2) is $AR \geq 1$, while the *RD* value approaches *B* for high *AR*s. The influence of other material properties on this behavior is still to be checked.

The clamped movable CCCC-M case in Fig. 19.9 shows a similar behavior, but with another high range, $6 < AR < 15$, and a lower maximum point at $AR = 8$.

To demonstrate that the above phenomenon also occurs in other types of plates made of non-isotropic materials, 16 mm multiwall plates were analyzed. Multiwall plates consist of parallel thin-walled sheets connected by vertical ribs, extruded from various plastics to yield a light and yet rigid structure, as shown in Fig. 19.10.

A detailed description of these plates and their mechanical behavior under uniform pressure is presented in Hakim and Abramovich's work [18].

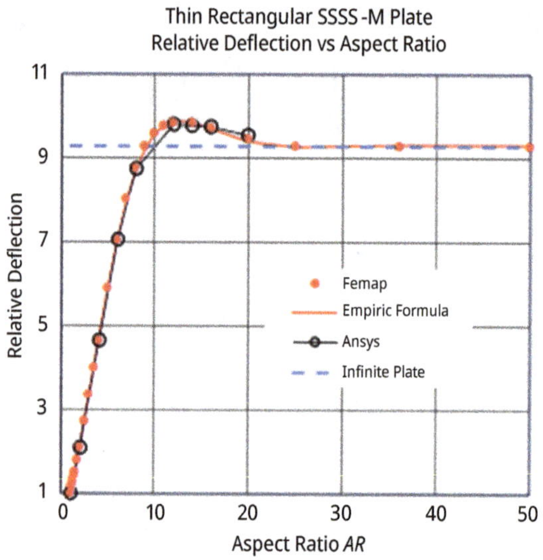

Fig. 19.8: Relative midpoint deflection for high *ARs* – FEAs and empirical formula.

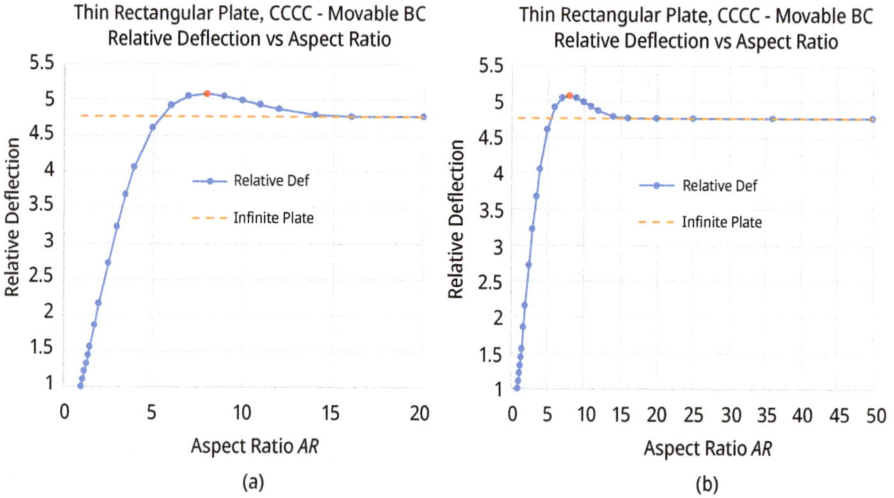

Fig. 19.9: CCCC-M's relative midpoint deflection for high *AR*: (a) up to *AR* = 20 and (b) up to *AR* = 50.

To correctly account for the complex internal structure, a homogenization procedure was applied, in which 10 independent elastic constants represented the global plate's elastic response as a solid plate, ignoring the small internal structure features. In Tab. 19.2, these constants are listed (see [21] for the homogenization process of the plate).

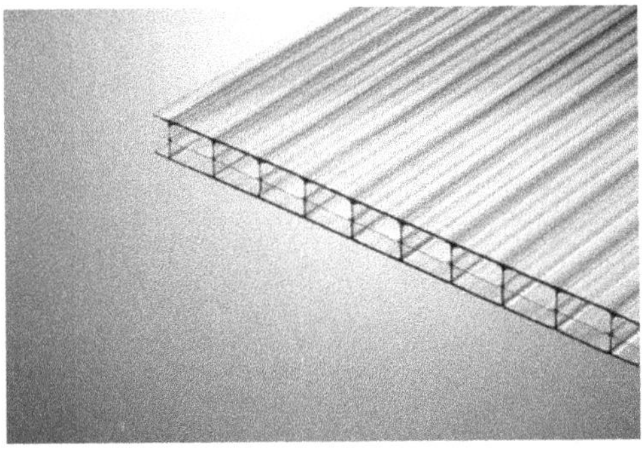

Fig. 19.10: A 16 mm triple-wall plate.

Tab. 19.2: The 16 mm multiwall elastic equivalent constants.

Variable	Notation	Units	Value
Y equivalent modulus	E_y^t	MPa	228.56
X equivalent modulus	E_x^t	MPa	318.77
Shear equivalent modulus	G_{xy}^t	MPa	83.417
Poisson's ratio (tension)	v_{xy}^t	–	0.37
X equivalent modulus	E_x^b	MPa	647.69
Y equivalent modulus	E_y^b	MPa	370.51
Shear equivalent modulus	G_{xy}^b	MPa	13.55
Poisson's ratio (bending)	v_{xy}^b	–	0.37
X shear rigidity	S_x	N/mm	136.27
Y shear rigidity	S_y	N/mm	3.8697

The superscript *t* stands for "tension" and *b* stands for "bending."

The above extruded plate, supported with the SSSS-M BC, was analyzed using the Femap code. The chart in Fig. 19.11 shows the same effect as that discussed above.

One should note that this behavior occurs only in plates with a movable BC (M), while plates with an immovable BC (I) do not show this phenomenon, as is presented in Fig. 19.12 for isotropic plates with properties and loading identical to the ones used for movable SSSS boundary conditions.

While trying to identify the origin of this behavior, the width midpoint deflections were checked along several long plates. The plates were identical but with different lengths. Figure 19.13 shows the mid-width deflection along the plates.

From these graphs, it is obvious that the high-deflection red points are located about six widths away from the plate ends. This might explain the maximum shown in Fig. 19.8 at $AR = 12$, where the two maxima meet each other.

To further investigate the topic in the large-deflection regime and the influence of the type of loading, namely the vector and pressure types described above, a comparison

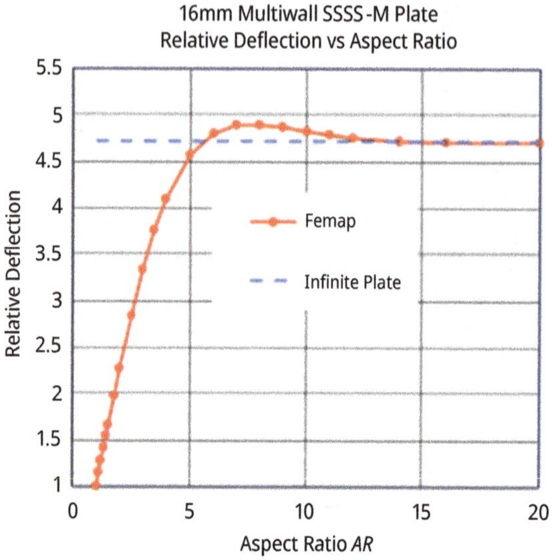

Fig. 19.11: A multiwall plate's higher deflection.

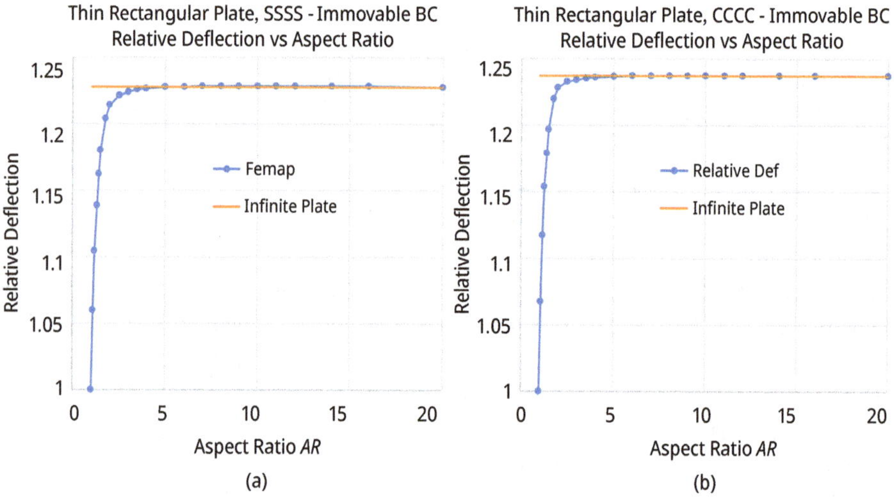

Fig. 19.12: Deflections of rectangular plates on immovable edges: (a) simply supported: SSSS-I; and (b) clamped: CCCC-I.

was made between the two cases. Figure 19.14 demonstrates that both load types present the same behavior, with the pressure load type presenting a smaller deflection.

To understand the difference in the lateral deflections when applying the two types of loading, as presented in Fig. 19.14, the sum of the vertical (in the direction of the z coordinate – see Fig. 19.3) nodal reaction forces at the plate edges was calculated.

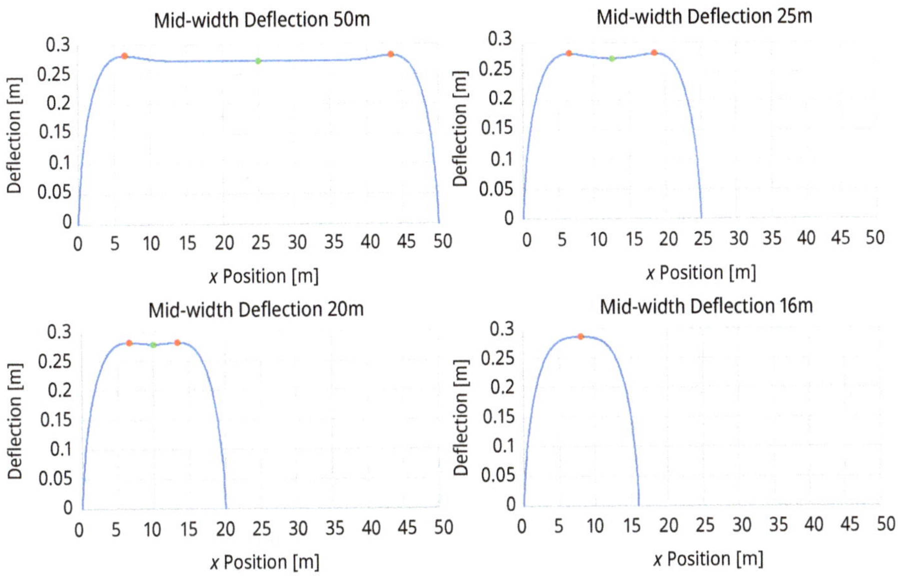

Fig. 19.13: Mid-width deflections of SSSS-M plate with various lengths.

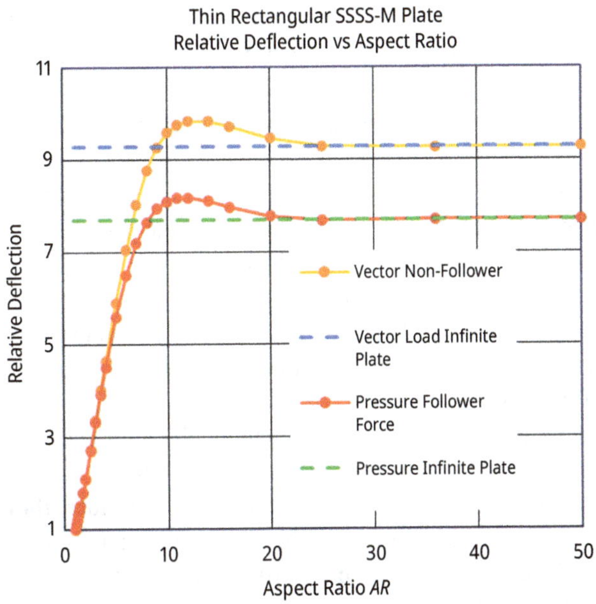

Fig. 19.14: Relative deflection versus *AR* for a fixed vector load and a follower pressure load.

The plate dimensions were 1 m × 12 m, and the load value was 1,000 Pa (one time with a fixed-direction vector load and one time with a pressure follower load). The reaction with the fixed vector load was exactly 12 kN (load multiplied by the area), while the reaction with the pressure follower load was only 10.7 kN. The reduced reaction in the second case was caused by the change in the element direction with the pressure force that followed the change. This reduction induced the smaller deflections shown in Fig. 19.14, as the vertical force was reduced.

19.3 Discussion

The classical question regarding thin rectangular plates, namely "Above which aspect ratio (AR) would the plate behave like an infinitely long plate?", cannot be answered intuitively. The present study clearly demonstrates that the influence of the plate's ends on the deflection and on the in-plane stresses penetrates from the ends into the plate's midpoint differently. Moreover, when the plate becomes longer, the end effects do not disappear but, rather, remain stuck to the moving-away ends.

In addition, for out-of-plane deflections, the end influence penetrates toward the plate's midpoint from each end by 5–6 times the width.

For tensile membrane *X* stresses (or forces), the influence also penetrates toward the midpoint by five times the width.

For tensile membrane *Y* stresses (or forces), the influence penetrates toward the midpoint by one width.

For shear membrane *XY* stresses (or force), the influence also penetrates toward the midpoint by one width.

Another significant finding stemming from the present study is that the midpoint deflection of a specific *AR* range is higher than that of an infinitely long plate. Higher deflections occur in SSSS-M's case six widths away from the ends toward the plate's midpoint. This is evident for the small-deflection linear Navier solution, all-around clamped plate with small-deflection and large-deflection regimes with movable edges. The authors believe that this is a new finding that has not been reached before in the literature and will encourage future research on it.

This innovative finding has some implications when designing plates for specified deflections. The common assumption is that infinitely long plates have the highest deflection, more than any rectangular plate with the same width, and therefore, they are considered the most conservative design. However, the common assumption described above might be wrong since the present findings prove that a long plate ($8 < AR < 20$ for a simply supported BC and $6 < AR < 15$ for a clamped BC) in the large-deflection regime would deflect more than an infinite plate, for both simply supported and clamped movable edges. Therefore, caution is suggested when making this common assumption while dealing with long plates under normal lateral pressure.

19.4 Conclusions

Based on the results presented in the present chapter, the following conclusions can be drawn:

- High *AR* rectangular plates (above a certain value) with movable edges cannot be considered infinitely long plates, but rather, the end influences are preserved near the ends.
- The midpoint deflections of the SSSS-M case with $10 < AR < 20$ and of the CCCC-M case with $6 < AR < 15$ are higher than that of an infinitely long plate. A plausible explanation for this phenomenon might be the unusual shape of the deflections for these unique boundary conditions.
- This behavior has also been detected for small-deflection theory based on the classical Navier solution and all-around clamped plates in the small-deflection regime.
- The influence of the plate's ends on the deflection and on the in-plane stresses penetrates from the far ends into the plate's midpoint in different ways.

References

[1] Hakim, G. and Abramovich, H. The behavior of long thin rectangular plates under normal pressure-A thorough investigation, Materials, 17, 2024, 15. ID article 2902.

[2] Timoshenko, S. and Woinowsky-Krieger, S. Theory of plates and shells, 2nd ed., 1959, Re-issued1987; McGraw-Hill Book Company: New York, NY, USA, 1987, 580.

[3] Reddy, J.N. Theory and analysis of elastic plates and shells, 2nd ed., CRC Press, Boca Raton, FL, USA; Taylor & Francis Group: Abingdon, UK, 2007, 561.

[4] Bakker, M.C.M., Rosmanit, M. and Hofmeyer, H. Approximate large-deflection analysis of simply supported rectangular plates under transverse loading using plate post-buckling solutions, Thin-Walled Struct, 46, 2008, 1224–1235.

[5] Bhaskar, K. and Varadan, T.K. Plates – theories and applications, Springer: Berlin/Heidelberg, Germany; ANE Books Pvt, Ltd., New Delhi, India, 2021, 282.

[6] McFarland, D.E., Smith, B.I. and Bernhart, W.D. Analysis of plates, 1st ed., Spartan Books, New York, NY, USA, 1972, 275.

[7] Onyeka, F.C., Okeke, E.T., Nwa-David, C.D. and Mama, B.O. Exact analytical solution for static bending of 3-D plate under transverse loading, Journal of Computational Applied Mechanics, 53, 2022, 309–331.

[8] Paik, J.K. Chapter 4, in Ultimate limit state analysis and design of plated structures, 2nd ed., John Wiley@Sons Ltd., Hoboken, NJ, USA, 2018, 179–270, 647.

[9] Bauer, F., Bauer, L., Becker, W. and Reiss, E.L. Bending of rectangular plates of finite deflections, Journal of Applied Mechanics, 32(4), 1965, 821–825.

[10] Brown, C.J., Goodey, R.J. and Rotter, J.M. Bending of rectangular plates subject to non-uniform pressure distributions relevant to containment structures, Proceedings of the Eurosteel 2017, Copenhagen, Denmark, 1(2–3), 13–15 September 2017, 1000–1009.

[11] Wang, D. and El-Sheikh, A.I. Large-deflection mathematical analysis of rectangular plates, Journal of Engineering Mechanics, 131, 2005, 809–821.

[12] Shao, C.J., Large Deflection of rectangular plate with high aspect ratio, Master's Thesis, University of California, Berkeley, California, 1964, 121.
[13] Razdolsky, A.G. Large deflections of elastic rectangular plates, International Journal for Computational Methods in Engineering Science and Mechanics, 16, 2015, 354–361.
[14] Ostiguy, G.L. and Evan-Iwanowski, R.M. Influence of the aspect ratio on the dynamic stability and nonlinear response of rectangular plates, Journal of Mechanical Design, 104(2), 1982, 417–425.
[15] Leverkusen, C.A.G., Makrolon 2407 MAS056 Datasheet. Available online:https://solutions.covestro.com/-/media/covestro/solution-center/products/datasheets/imported/makrolon/makrolon-2407_en_56977361-00009645-19213223.pdf
[16] Sabic Lexan 940A Datasheet. Available online: https://www.matweb.com/search/datasheet.aspx?matguid=cda0ab793fed4ed6a7e5df4902937a4b&ckck=1
[17] Siemens Simcenter Nastran basic nonlinear analysis user's guide, Siemens Industry Software: Plano, TX, USA, 2020. Available online: https://docs.plm.automation.siemens.com/data_services/resources/scnastran/2020_1/help/tdoc/en_US/pdf/basic_nonlinear.pdf
[18] Hakim, G. and Abramovich, H. Large deflections of thin-walled plates under transverse loading – investigation of the generated in-plane stresses, Materials, 15, 2022, 26. ID article 1577.
[19] Lee, -H.-H., Finite element simulations with ANSYS Workbench 2023: Theory, applications, case studies, SDC Publications: Mission, KS, USA, 2023; ISBN 1630576158, 9781630576158. Available online: https://www.amazon.com/Finite-Element-Simulations-ANSYS-Workbench/dp/1630576158.
[20] Ansys Internet Site. Available online: https://www.ansys.com/products/ansys-workbench (accessed on).
[21] Hakim, G. and Abramovich, H. Homogenization of multiwall plates – an analytical, numerical and experimental study, Thin-Walled Structures, 185, 2022, 17. ID article 110583.

Index

https://doi.org/10.1515/9783111621104-020

www.ingramcontent.com/pod-product-compliance
Lightning Source LLC
Chambersburg PA
CBHW060941210326
41598CB00031B/4691